T0399385

Hybrid Power

Generation, Storage, and Grids

CRC PRESS Series on
Sustainable Energy Strategies
by Yatish T. Shah

Hybrid Power: Generation, Storage, and Grids

Modular Systems for Energy Usage Management

Modular Systems for Energy and Fuel Recovery and Conversion

Thermal Energy: Sources, Recovery and Applications

Chemical Energy from Natural and Synthetic Gas

Other related books by Yatish T. Shah:

Energy and Fuel Systems Integration

Water for Energy and Fuel Production

Biofuels and Bioenergy: Processes and Technologies

Hybrid Power

Generation, Storage, and Grids

Yatish T. Shah

CRC Press is an imprint of the
Taylor & Francis Group, an **informa** business

First edition published 2021
by CRC Press
6000 Broken Sound Parkway NW, Suite 300, Boca Raton, FL 33487-2742

and by CRC Press
2 Park Square, Milton Park, Abingdon, Oxon, OX14 4RN

Library of Congress Cataloging-in-Publication Data

Names: Shah, Yatish T., author.
Title: Hybrid power : generation, storage, and grids / Yatish T. Shah.
Description: Boca Raton, FL : CRC Press, 2021. | Series: Sustainable energy strategies | Includes bibliographical references and index. | Summary: "Hybrid energy systems integrate multiple sources of power generation, storage, and transport mechanisms and can facilitate increased usage of cleaner, renewable, and more efficient energy sources. This book discusses hybrid energy systems from fundamentals through applications and discusses generation, storage, and grids. The book will be of interest to advanced students and researchers in academia, government, and industry seeking a comprehensive overview of the basics, technologies, and applications of hybrid energy systems"-- Provided by publisher.
Identifiers: LCCN 2020043335 (print) | LCCN 2020043336 (ebook) | ISBN 9780367678401 (hbk) | ISBN 9781003133094 (ebk)
Subjects: LCSH: Hybrid power systems.
Classification: LCC TK1041 .S45 2021 (print) | LCC TK1041 (ebook) | DDC 621.31/2--dc23
LC record available at https://lccn.loc.gov/2020043335
LC ebook record available at https://lccn.loc.gov/2020043336

ISBN: 978-0-367-67840-1 (hbk)
ISBN: 978-1-003-13309-4 (ebk)

Typeset in Times
by Deanta Global Publishing Services, Chennai, India

This book is dedicated to my wife Mary

Contents

Series Preface

While fossil fuels (coal, oil and gas) were the dominant sources of energy during the last century, since the beginning of the twenty-first century an exclusive dependence on fossil fuels is believed to be a non-sustainable strategy due to a) their environmental impacts, b) their nonrenewable nature, and c) their dependence on the local politics of the major providers. The world has also recognized that there are in fact ten sources of energy: coal, oil, gas, biomass, waste, nuclear, solar, geothermal, wind and water. These can generate our required chemical/biological, mechanical, electrical and thermal energy needs. A new paradigm has been to explore greater roles of renewable and nuclear energy in the energy mix to make energy supply more sustainable and environmentally friendly. The adopted strategy has been to replace fossil energy by renewable and nuclear energy as rapidly as possible. While fossil energy still remains dominant in the energy mix, by itself, it cannot be a sustainable source of energy for the long future.

Along with exploring all ten sources of energy, sustainable energy strategies must consider five parameters: a) availability of raw materials and accessibility of product market, b) safety and environmental protection associated with the energy system, c) technical viability of the energy system on the commercial scale, d) affordable economics, and e) market potential of a given energy option in the changing global environment. There are numerous examples substantiating the importance of each of these parameters for energy sustainability. For example, biomass or waste may not be easily available for a large-scale power system making a very large-scale biomass/waste power system (like a coal or natural gas power plant) unsustainable. Similarly, an electrical grid to transfer power to a remote area or onshore needs from a remote offshore operation may not be possible. Concerns of safety and environmental protection (due to emissions of carbon dioxide) limit the use of nuclear and coal-driven power plants. Many energy systems can be successful at laboratory or pilot scales, but may not be workable at commercial scales. Hydrogen production using a thermochemical cycle is one example. Many energy systems are as yet economically prohibitive. The devices to generate electricity from heat such as thermoelectric and thermophotovoltaic systems are still very expensive for commercial use. Large-scale solar and wind energy systems require huge upfront capital investments which may not be possible in some parts of the world. Finally, energy systems cannot be viable without market potential for the product. Gasoline production systems were not viable until the internal combustion engine for the automobile was invented. Power generation from wind or solar energy requires guaranteed markets for electricity. Thus, these five parameters collectively form a framework for sustainable energy strategies.

It should also be noted that the sustainability of a given energy system can change with time. For example, coal-fueled power plants became unsustainable due to their impact on the environment. These power plants are now being replaced by gas driven power plants. New technology and new market forces can also change sustainability of the energy system. For example, successful commercial developments of fuel cells and electric cars can make the use of internal combustion engines redundant in the vehicle industry. While an energy system can become unsustainable due to changes in parameters, outlined above, over time, it can regain sustainability by adopting strategies to address the changes in these five parameters. New energy systems must consider long-term sustainability with changing world dynamics and possibilities of new energy options.

Sustainable energy strategies must also consider the location of the energy system. On one hand, fossil and nuclear energy are high density energies and they are best suited for centralized operations in an urban area, while on the other hand, renewable energies are of low density and they are well-suited for distributed operations in rural and remote areas. Solar energy may be less affordable in locations far away from the equator. Offshore wind energy may not be sustainable if the distance from shore is too great for energy transport. Sustainable strategies for one country may be quite different from another depending on their resource (raw material) availability and local market potential. The current transformation from fossil energy to green energy is often prohibited by required infrastructure and the total cost of transformation. Local politics and social acceptance also play an important role. Nuclear energy is more acceptable in France than in any other country.

Sustainable energy strategies can also be size dependent. Biomass and waste can serve local communities well at a smaller scale. As mentioned before, the large-scale plants can be unsustainable because of limitations on raw materials. New energy devices that operate well at micro- and nanoscales may not be possible on a large scale. In recent years, nanotechnology has significantly affected the energy industry. New developments in nanotechnology should also be a part of sustainable energy strategies. While larger nuclear plants are considered to be the most cost effective for power generation in an urban environment, smaller modular nuclear reactors can be the more sustainable choice for distributed cogeneration processes. Recent advances in thermoelectric generators due to advances in nanomaterials are an example of a size-dependent sustainable energy strategy. A modular approach for energy systems is more sustainable at smaller scale than for a very large scale. Generally, a modular approach is not considered as a sustainable strategy for a very large, centralized energy system.

Finally, choosing a sustainable energy system is a game of options. New options are created by either improving the existing system or creating an innovative option through new ideas and their commercial development. For example, a coal-driven power plant can be made more sustainable by using very cost-effective carbon capture technologies. Since sustainability is time, location and size dependent, sustainable strategies should follow local needs and markets. In short, sustainable energy strategies must consider all ten sources and a framework of five stated parameters under which they can be made workable for local conditions. A revolution in technology (like nuclear fusion) can, however, have global and local impacts on sustainable energy strategies.

The CRC Press Series on Sustainable Energy Strategies will focus on novel ideas that will promote different energy sources sustainable for long term within the framework of the five parameters outlined above. Strategies can include both improvement in existing technologies and the development of new technologies.

Series Editor,
Yatish T. Shah

Preface

Energy is a basic need for society, and the energy industry has constantly evolved since the beginning of human civilization. During the last century, fossil fuel dominated the industry and efforts were made for centralized and large-scale power production. This led to the development of highly structured utility grid with reliable transport of electricity for the urban environment. The economy of scale dominated the industry, and energy storage was only relegated to large-scale hydropower storage.

Over the last 2 decades, the paradigms used during the last century came under serious scrutiny. There was a realization that fossil fuels are non-renewable and the supply of coal and oil is not sustainable over a long period. The use of internal combustion engine to generate both static and mobile power from fossil fuels led to unacceptable level of CO_2 emissions resulting into the ozone destruction and global warming concerns. Centralized power generation became problematic because it could not serve the power need of the rural and remote communities. All these factors led to the change in emphasis from fossil fuels and nuclear to more environmentally acceptable renewable sources like solar, wind, biomass, geothermal, hydro, and waste. Unlike fossil fuels, these sources are of low density and more distributed. This meant that there was also a need for the development of more distributed energy infrastructure. The distributed operation also became more important in order to serve the electricity need of rural and remote communities which were not served by the centralized grid.

The required changes mentioned above led to the realization that there are in fact ten sources of energy: coal, oil, gas, nuclear, biomass, waste, solar, wind, geothermal, and hydro, and going forward, we need to be less dependent on fossil fuels and more dependent on renewable fuels. There was also a realization that no single source of fuel satisfies all the requirements of power generation for the future. Furthermore, as the need for energy by developing country increases, efficiency of energy systems becomes more important. Thus the new energy world requires more use of renewable fuels, more distributed energy, more improvement in the efficiency of energy systems, less harm to the environment, and more care for the electricity need of rural, remote, and poor and developing communities. In order to harness renewable sources like solar and wind, which are unpredictable and intermittent, energy storage also became more important for sustainable power supply. As the energy storage was used for harnessing intermittent sources of energy like solar and wind, there was also realization that no single energy storage device satisfies all the requirements for power quality and stability in the grid.

The incorporation of renewable sources and the need to make energy systems more efficient and less harmful to the environment also required significant changes in the entire energy generation, storage, and transport infrastructure. First, the sources for power generation became hybrid where more than one source of power generation is used in either stationary or mobile energy system. As the use of energy storage became more prominent, its application in electric car and hybrid grids required hybrid energy storage devices. Lastly, in order to harness distributed resources (both in urban and rural environments), medium- to low-voltage microgrids are developed. Along with macrogrid and microgrids, the rural electrification required off-grid energy infrastructure which included bottom-up approaches of mini-grids, nanogrids and stand-alone systems. Thus along with multiple sources of power generation and hybrid storage devices, three levels of electricity transport—utility grid, microgrids, and off-grid—resulted in a hybrid grid electricity transport structure. The purpose of this book is to purposefully delineate these elements of hybrid energy (more specifically hybrid power) in its totality, which includes hybrid power generation, hybrid storage, and hybrid grid transport mechanisms. The book defines hybrid energy system (or hybrid power) in its broadest context which includes multiple sources of power generation, hybrid storage devices, and hybrid grid transport mechanisms. Hybrid energy system as defined here is the future of the energy industry because it can provide:

1. More options for local optimization of cost. Hybrid energy systems can be designed to achieve desired attributes at the lowest acceptable cost, which is the key to market acceptance.
2. Greater penetration of renewable sources in the overall energy mix. Hybrid energy systems can be designed to maximize the use of renewable resources, resulting in a system with lower emissions.
3. More efficient and less use of fossil fuels in the future. More efforts to reduce waste heat. More efforts to treat carbon emission by novel technologies.
4. More options for large centralized and small distributed operations. This will well serve both urban and rural and isolated communities. This will also allow better harvesting of distributed energy sources.
5. Less reliance on large-scale energy grids. Micro, mini, and nanogrids can operate independently or together with utility grid. This along with standalone systems will help more than a billion people who are currently without electricity.
6. More options for optimum uses of local energy resources, which can vary in different country. They provide flexibility in terms of the effective utilization of the renewable sources.
7. Greater role of energy storage to provide stable power and heating/cooling for all energy usages such as buildings, vehicles, district heating, and industrial heating.
8. Offset of negative aspects of many sources without their elimination from the overall mix.
9. Allowance of gradual transition to carbon-free energy world.
10. Less social and political tensions between available sources and the need for energy for all parts of the world.
11. Better management of sustainable and high-quality power.
12. A hybrid energy system can make use of the complementary nature of various sources, which increases the overall efficiency of the system and

improves its performance (power quality and reliability). For instance, combined heat and power operation, e.g., MT and FC, increases their overall efficiency or the response of an energy source with slower dynamic response (e.g., wind or FC) can be enhanced by the addition of a storage device with faster dynamics to meet different types of load requirements.

13. Reduced use of fossil fuel for buildings and transportation. Vehicle industry is also becoming more hybrid. The use of renewable sources for the buildings and use of distributed infrastructure will make consumers more prosumers, and society will move toward more zero-energy buildings.

The book is aimed at outlining basic as well as practical aspects of hybrid power. The content of the book is divided into seven chapters. Chapter 1 defines hybrid energy system in its broadest term and lays out reasons why it is the future of energy industry (in particular, power industry). Chapter 2 examines grid connected hybrid energy systems. The chapter also points out the usefulness of hybrid energy systems for home energy needs. Chapter 3 discusses the directions of mobile industry toward hybrid and electric cars. The chapter points out that the use of internal combustion engine in transportation industry may be on way out, and in the near future, it will be replaced by hybrid or electric vehicles in their many forms. Chapter 4 discusses various aspects of hybrid energy storage including its needs and benefits and its usages for hybrid and electric vehicles and in grids that are harnessing renewable energy sources. Chapter 5 examines various aspects of hybrid microgrids and points out the usefulness of microgrids to harness distributed renewable energy sources. Chapter 6 discusses various forms of off-grid hybrid power systems such as mini-grids, nanogrids, and stand-alone systems. The chapter also gives examples of various successful uses of hybrid renewable energy systems around the world, particularly for rural electrification. The chapter also discusses strategies for the use of hybrid renewable energy systems for rural electrification and gives a brief global overview on the use of off-grid energy for rural electrification. Finally, Chapter 7 examines the available literature for the simulation and optimization of hybrid renewable energy systems. Software tools and various theoretical methods for simulation and optimization, along with objectives for optimization and types of renewable systems that are examined, are outlined. It should be noted that the focus of this book is on hybrid power.

This book is thus aimed at the energy industry of the future. The seven chapters in the book cover various aspects of generation, storage, and grid transport of hybrid energy. The book should be good for a graduate course in hybrid energy or as a reference text for those who are pursuing research in academia, government, and industry on any aspects of hybrid energy.

Author Biography

Yatish T. Shah received his BSc in chemical engineering from the University of Michigan, Ann Arbor, USA, and MS and ScD in chemical engineering from the Massachusetts Institute of Technology, Cambridge, USA. He has more than 40 years of academic and industrial experience in energy-related areas. He was chairman of the Department of Chemical and Petroleum Engineering at the University of Pittsburgh, Pennsylvania, USA; dean of the College of Engineering at the University of Tulsa, Oklahoma, USA, and Drexel University, Philadelphia, Pennsylvania, USA; chief research officer at Clemson University, South Carolina, USA; and provost at Missouri University of Science and Technology, Rolla, USA, the University of Central Missouri, Warrensburg, USA, and Norfolk State University, Virginia, USA. He was also a visiting scholar at University of Cambridge, UK, and a visiting professor at the University of California, Berkley, USA, and Institut für Technische Chemie I der Universität Erlangen, Nürnberg, Germany. Dr. Shah has previously written ten books related to energy, seven of which are under "Sustainable Energy Strategies" book series by Taylor and Francis of which he is the editor. He has also published more than 250 refereed reviews, book chapters, and research technical publications in the areas of energy, environment, and reaction engineering. He is an active consultant to numerous industries and government organizations in the energy areas.

1 Basics of Hybrid Power

1.1 INTRODUCTION—CHRONOLOGICAL EVOLUTION OF ENERGY INDUSTRY

Energy is a basic need of human society. The energy industry is as old as human civilization. The industry has, however, evolved and changed in significant ways with the industrialization of society and due to the human craving for constantly improving the quality of life. In old civilization, the basic need for energy was heating and cooling, and the sources for energy were natural sources like sun, wind, water, and biomass. While there was no effort to harness energy from the sun, wind, and water, they provided heating and cooling. Wood was a major source of heating. Coal, oil, gas, and nuclear energy (the process of fission) were not discovered then. Hence, as shown in Table 1.1, up until 1850, the major sources of energy were the renewables mentioned above. The discovery of coal, oil, and gas and the inventions of electricity and the automobile (with an internal combustion (IC) engine), over subsequent years, significantly changed the landscape of energy usage (see Table 1.1) [1–6].

The shifts in sources of energy listed in Table 1.1 are due to many reasons. In 1800, the technologies to recover coal, oil, and gas from underground sources were not developed and their usages were not identified. Energy was largely used for heating purposes because electricity was not discovered. Significant changes in the source of energy occurred in the 19th and more in the 20th century; since the recovery of coal, oil, and gas; the discovery of electricity and nuclear energy; and the development of automobiles and other vehicles operated by internal combustion engines. All of this led to the explosion of the use of fossil fuels, to a lesser degree, the use of nuclear energy for power, and the buildup of a related infrastructure for fossil fuel recovery and grid for electricity transport. As shown in Table 1.1, in the 20th century, fossil fuels became the dominant source of energy. Fossil fuels and related infrastructures provided a stable delivery of power and heating and cooling needs. Coal-driven power plants and oil-driven automobiles became dominant forces in society.

While the acceptance of nuclear energy varied significantly across the globe during the 20th century due to its possible harmful effect to the environment (by nuclear waste and possible radiation effects in case of failure of nuclear power plants), at the end of the 20th century, there was also a realization that fossil fuel is non-renewable and it has a very harmful effect on the environment by the emission of carbon in the form of CO_2 and CH_4, which are inherent in the recovery, conversion, and usages of fossil fuels in a variety of ways. Global warming and other climate changes were strongly linked to the use of fossil fuels and man-made fossil fuel–driven energy technologies. More emphasis began to take place on the use of renewable and carbon-free energy sources such as biomass, waste, geothermal, solar, wind, and water.

TABLE 1.1
Energy Landscape 1800–2040 Calculated from EMR

Year	Fossil (%)	Nuclear (%)	Renewables (%)
1800	2	0	98
1850	10	0	90
1900	50	0	50
1950	76	0	24
2000	82	7	11
2040	75	9	16

Source: ExxonMobil, "The outlook for energy: A view to 2040," US Edition, ExxonMobil Report, ExxonMobil, Irving, TX, 2012. With permission.

Note: The numbers in the table are approximate calculations from the graphical data reported in the EMR [1–6].

Since the beginning of the 21st century, renewable sources have started replacing fossil fuels. The interest in nuclear energy (particularly for power) at large scale still did not gain significant momentum. With the use of renewable energy, carbon-free energy consumption in building, vehicle, and industrial applications began to rise. New innovations in renewable energy technologies, energy storage, and transport technologies and more emphasis on distributed energy and increased efficiencies in the automobile industry through the use of hybrid, electrical, and fuel cell-driven cars accelerated the use of renewable sources of energy. The penetration of renewable sources still faces many obstacles, and therefore, as pointed out in Table 1.1, the switch from fossil energy to renewable energy will occur in a slower but deliberate manner. The use of nuclear energy is also more likely to go through the change from large scale to more small, modular, and distributed scale. In short, however, societies are now looking at the energy industry that exploits energy from ten sources: coal, oil, gas, biomass, waste, nuclear, geothermal, solar, wind, and water, with prime considerations of economics, sustainable technologies, and environmental impact (particularly global warming and carbon emission) and the need to serve the entire global population [1–7].

1.1.1 Use of Energy and Its Global Consumption

Irrespective of the source (fossil, nuclear, or renewable), energy is needed for (a) stationary power, (b) heating and cooling for industrial and manufacturing processes, (c) heating and cooling of residential and commercial space, and (d) mobile systems. Energy is needed to improve industrial growth and quality of life. Just like sources, energy consumption in the world has also gone through significant changes (see Table 1.2). This growth in energy consumption has led to a significant increase in

TABLE 1.2
World Energy Consumption (Quadrillion BTU) [1–6]

Year	OECD countries	Non-OECD countries
1990	199	155
2000	234	171
2008	244	260
2015	250	323
2020	261	359
2025	270	402
2030	279	443
2035	288	482

CO_2 emission in the environment as shown in Table 1.3. The largest contributors of CO_2 to the environment are the developing nations.

The increase in CO_2 emission is due to carbon in energy sources. The distribution of carbon in various parts of the earth is illustrated in Table 1.4.

CO_2 emission is directly related to energy sources. The distribution of CO_2 emission for various energy sources for power generation is illustrated in Table 1.5. The life cycle of CO_2 emission of both renewable and non-renewable sources is illustrated in Figure 1.1. It is clear from this table and the figure that renewable and nuclear power generation is much less harmful to the environment than fossil fuels [1–6].

TABLE 1.3
Growth in End-use Demand and CO_2 Emissions in Sectors and Regions (1990–2040; 50-Year Span)

	Growth (1990–2040) (%)						
	Region						
Sector	NA	Europe	Russia/Caspian	Africa	Asia/Pacific	LA	Middle East
Residential/ Commercial	33.3	17.6	−33	214	90	100	800
Transport	24	36	−17	350	36	225	266
Industrial	6.6	−12	−25	183	225	200	300
Electricity							
Demand	18.2	56	20	800	614	600	500
CO_2 emission	−7	−27	−41	343	226	186	271

Source: ExxonMobil, "The outlook for energy: A view to 2040," US Edition, ExxonMobil Report, ExxonMobil, Irving, TX, 2012. With permission [1–6].

TABLE 1.4
Distribution of Organic Carbon in the Earth

Source of Carbon	Amount (10^5 g of Carbon)	Total Carbon (%)
Gas hydrates (onshore and offshore)	10,000	53.26
Recoverable and nonrecoverable fossil fuels (coal, oil, natural gas)	5,000	26.63
Soil	1,400	7.46
Dissolved organic matter in water	980	5.22
Land biota	830	4.42
Peat	500	2.68
Detrital organic matter	60	0.33
Atmosphere	3.6	0.0
Marine biota	3	0.0

Source: Englezos, P., Industrial & Engineering Chemistry Research, 32, 1251–1274, 1993; Kvenvolden, K.A., Chemical Geology, 71, 41–51, 1988. With permission; Tohidi, 2013, pers. comm.; Collet's work at USGS.

Note: This excludes dispersed organic carbon such as kerogen and bitumen, which equals nearly 1,000 times the total amount shown in the table [1–6].

TABLE 1.5
Relative Ranges of Greenhouse Gas Emissions from Different Electricity Generation Technologies [1–6]

Substance	Mean (% Relative to Lignite)	(Lower/Upper) Range (Tons of CO_2 eq/GWh)
Lignite	100	790/1,372
Coal	84.2	756/1,310
Oil	69.5	547/935
Natural gas	47.3	362/891
Geothermal	>5	40/1,822
Solar PV	8.1	13/731
Biomass	4.3	10/101
Nuclear	2.75	2/130
Hydroelectric	2.5	2/237
Wind	2.5	6/124

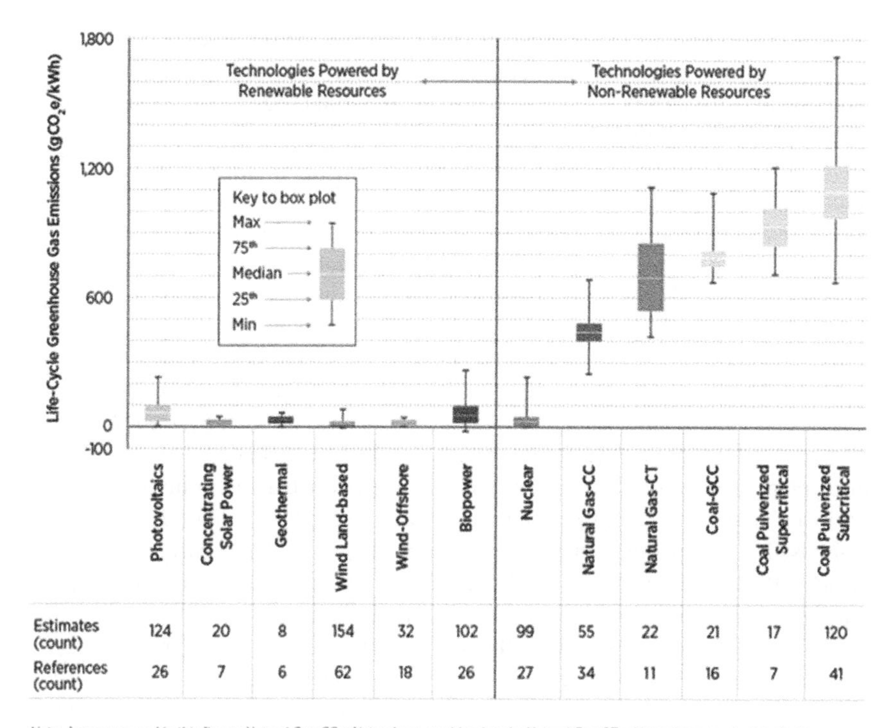

FIGURE 1.1 Summary of life cycle GHG emissions from electricity generation technologies *(Source: DOE 2015)* [8] Note: "References" refers to the number of literature citations that underlie each "box and whisker"; "estimates" exceed "references" because many literature citations include multiple, independent estimates.

1.2 SOURCES OF ENERGY AND THEIR LIMITATIONS

There are ten sources of energy: coal, oil, gas, biomass, waste, nuclear, geothermal, solar, wind, and water. Hydrogen is a derived source of energy because it only exists as compounds in nature, and it needs to be separated or generated from these compounds. For example, hydrogen can be generated from water by thermal dissociation. These sources are used to generate static and mobile power or heating and cooling needs for residential and commercial space and heating and cooling needs for industrial and manufacturing processes. We briefly assess here the pluses and minuses of these resources [1–6].

A. Coal

In the 20th century, coal was widely used to generate power by the process of coal combustion. The thermal energy generated from the combustion process is converted to power using the thermodynamic cycle. Coal is also extensively used to generate synthetic gases which can be either used to generate power (like in IGCC (integrated gasification combined cycle) process) or produce synthetic fuels or chemicals. Besides being an energy source, coal is also an important source to produce a variety of chemicals. The United States, along with Poland, India, China, and Australia, among others, possesses a significant amount of coal. The coal can be of different quality for power generation depending on its rank, ash and moisture contents, and other impurities. Most power plants use subbituminous and bituminous coals.

In terms of its use for energy, coal-based power plant is a very mature commercial technology. Similarly, thermal coal gasification is also a very mature technology. The construction and electricity costs from coal-based power plants are quite competitive, but here the economy of scale applies. Most coal-based power plants are large, and they are connected to large utility scale electrical grid. This set up generally does not require an energy storage device. They are thus centralized operations and serve well the urban environment where the electrical grid is well designed. Coal power plants generally do not serve well remote and isolated communities where distributed and small-scale operations away from the utility scale grid are required. Coal-based power plants require a reasonable amount of land, particularly if one considers the land use in mining and coal-preparation operations. The water requirement for coal-based power plants is also very excessive.

There are five major issues with the use of coal as a source of energy. First, coal is a non-renewable source of energy. Its very long-term reliability is questionable. Although the United States has a large reserve of coal, its long-term sustainability is questionable because some of the sources are also not easily recoverable by conventional mining operations. Many countries do not have enough coal to make this a reliable source of energy.

Second, while coal-based power plants are capable of generating electricity when needed, they are not well suited to provide electricity at small-scale and remote areas, where a suitable grid is not available. In short, coal-based power plants are not suitable for distributed power generation. Furthermore, coal-based power plants do not always have the ability to quickly respond to change in demand for electricity.

Third, coal-based power plants are inherently inefficient. Since they are based on the conversion of thermal energy to electrical energy, their efficiency is about 30%–40%. Even combined cycle operation has an efficiency of around 50%. Most sophisticated HELE (high efficiency low emission) supercritical coal-based power plants are about 60% efficient. This means that at least 40%–50% of energy is wasted as a waste heat. For large plants, cogeneration is difficult because waste heat has to be used locally since, unlike electricity, heat cannot be transported over long distance.

The fourth and most important disadvantage of coal-based power plants is the level of CO_2 emission generated by these plants. As shown in Figure 1.1, coal-based power plants are the worst in CO_2 emission. With the present day concern about carbon emission and its effect on the global warming, conventional coal-based power plants are rapidly becoming environmentally unacceptable. Several efforts are being made to capture carbon and sequester underground. These carbon capturing technologies are expensive, and they are making electricity cost noncompetitive. In recent years, ExxonMobil, in partnership with FuelCell Energy, has been trying to commercialize the technology of capturing CO_2 by carbon dioxide-based fuel cell. If successful, this hybrid process can not only reduce CO_2 emission but also generate additional power from the plant. This work is still in progress. Another approach to reducing CO_2 emission is to generate power by co-combustion of coal and biomass. The literature indicates that the right composition of coal–biomass (about 70% coal and 30% biomass) can be CO_2 neutral to the environment. Since biomass has a significantly different composition, reactivity, and low energy density than coal, co-combustion does require significant changes in the design of power plants. This can also lead to additional cost to electricity generation. Significant literature has been devoted to this subject, and this has been described in detail in my previous books [1–6].

The fifth reason against coal-based power plant is the other impurities in coal (such as sulfur, heavy metals, and ash) that have to be removed during coal-preparation or post-combustion operations. These operations can create enormous negative side effects on the power generation. High ash coal is common in some parts of the world, and this ash has to be discharged in some environmentally harmless manner. Besides their effects on the environment, the handling of these impurities adds cost to the power generation. To some extent, due to the presence of these impurities, each coal-based power plant has to be designed for specific types of coal, which makes plants less flexible.

All the arguments made for and against coal also apply to some degree to the use of oil shale for power and synthetic fuel.

B. **Oil**

Just like coal, oil recovered from different parts of the world varies significantly in composition. Crude oil, as it is commonly known, mainly comes from the Middle East, Russia, and some parts of the United States, among others. Like coal, oil is a non-renewable source and conventional oil is rapidly depleting. Secondary and tertiary oil and unconventional oils such as heavy oil, bitumen, and shale oil are harder to recover and refine. Like coal, oil is a good source for chemicals along with its use for energy. Unlike coal, oil can also be generated from gas (via FT synthesis) or coal and biomass. The oils produced from these different sources are very different in composition, and their usage will require different sets of technologies.

Unlike coal and gas, oil is largely used for heating and powering vehicles. The use of oil in large power plants is limited and it is replaced by liquid natural gas in some parts of the world like Indonesia and other East Asian countries. The largest use of oil is as gasoline, diesel fuel, and jet fuel for a variety of vehicle operations. Oil is a high-density energy fluid and very convenient to use for internal combustion engines, which are currently used in automobiles and jet engines. Oil is also extensively used for heating (No. 6 fuel oil) and backup diesel generator for power plants.

While the use of oil in current fossil fuel–dominated society is essential, it is unsustainable over a long period for a number of reasons. The first one is of course its non-renewable nature. Heavy oil (found largely in Venezuela and Canada) is hard to recover and refine and is not the most cost-effective. Over a long period, the use of oil for chemicals and other purposes will dominate.

Second, just like coal, the use of oil emits a significant amount of greenhouse gases as shown in Table 1.5 and Figure 1.1. Unlike in coal-based power plants, these greenhouse gases emitted by automobiles and airplanes are difficult to capture and remove. Transportation sector is one of the largest sources of CO_2 and other carbon emissions. The concerns about environment restrict the future use of oil in transportation and for other power and heating needs. In many parts of the world, oil is replaced by liquid natural and propane gases, which have less harmful effects on the environment.

In recent years, the use of an internal combustion engine for the vehicle industry has come under serious question. First, the internal combustion engine is highly inefficient for all the reasons mentioned earlier for coal. Only 30% of the energy consumed in an automobile is used for actual power to propel an automobile. The rest is waste heat coming out of the exhaust. Serious efforts are being made to recover energy in the waste heat (using thermoelectricity, recycling, etc.). However, these efforts require significant and costly changes in the design of automobiles.

The use of an internal combustion engine in the vehicle industry that uses oil faces stiff competition from hybrid cars, electric cars, and fuel cell powered (by hydrogen and natural gas) cars. More development efforts on hybrid cars, electric cars, and fuel cell powered cars are successfully pursued, and these cars are more efficient and emit less carbon in the environment. If these cars become cost-effective and durable, with the help of appropriate infrastructure, they can replace existing cars with internal combustion engines. Hybrid cars (based on electricity and fuel) are already successful; fuel cell is already used in forklifts and other heavy machinery cars, and electric cars are gaining more popularity. All this indicates that it is a matter of time until the vehicle industry will stop using environmentally harmful internal combustion engine in most vehicles. This can create a huge blow to the oil market and the use of oil in the energy industry. As indicated earlier, oil use is also facing stiff competition from cleaner and less environmentally harmful liquid natural gas and propane gas in transportation, power, and heating industries.

Finally, like coal, unconventional oil can also have other harmful side effects on the environment because of other impurities in heavy oil. The removal of these impurities makes the use of oil economically less competitive.

C. Gas

Natural gas is considered to be transition fuel for the energy industry. Gas, in general, can be transition as well as long-term fuel if natural gas (both from conventional and unconventional sources), synthetic gas, and hydrogen are included in the gas. Hydrogen (along with electricity) is considered to be the long-term fuel of the future. Unlike natural gas, hydrogen has to be generated from other fossil fuels or water.

Natural gas is a good substitute for both coal and oil because it is cleaner and more definable for chemical compositions. Properly processed natural gas creates little waste and emits about half of the CO_2 emitted by coal combustion. Natural gas power plants are easy to construct and very competitive for electricity cost and use much less land and water than coal-based power plant. Natural gas is easily available, and its conversion to power, synthetic fuel, or chemicals can be done both at large and small modular scale. Thus natural gas conversion can be more centralized or distributed.

In recent years, unconventional natural gas (particularly shale gas) has dominated the gas industry. While conventional gas is more and more recovered by secondary and tertiary processes, the success of "fracking" and horizontal wells has made recovery of unconventional gas like shale gas, coal bed methane, tight gas, deep gas, and gas from geopressurized zones more easier. In the United States, there is an explosion of shale gas and shale oil recovery which has made the United States the largest oil and gas producing country in the world. There is also a significant interest in recovering methane from gas hydrates in Alaska and from the bottom of the ocean. As shown in Table 1.4, there is more carbon in gas hydrates than in all other fossil fuels combined. If recovery technologies for all unconventional gases are successfully developed, these sources make natural gas an unlimited source of energy.

While the use of natural gas for power, synthetic fuels, and chemicals is very attractive and a commercially proven entity, there are several disadvantages in its use. First, natural gas-based power plants are inefficient in the same manner as coal-based power plant. On a large scale, the cogeneration of these plants would also be difficult because all waste heat has to be used locally. On a smaller scale, cogeneration can be effective in improving its efficiency. This, however, requires the transport of natural gas to isolated, remote, and rural areas. On a smaller scale, waste heat can be converted to electricity by the process of thermoelectricity. This will require additional investment.

Unconventional natural gas, like shale gas, raises several other issues. Shale gas is not pure (more than 90%) methane. It contains significant amounts of ethane, propane, carbon dioxide, and sometimes even higher hydrocarbons. These components must be removed so natural gas can be transported in the conventional gas pipelines. In recent years, the explosion of shale gas has also caused the explosion of ethane and propane worldwide. They need to be used as chemicals, reformed or combusted in combined cycle power plants. Unconventional gas like coal bed methane also generates a large amount of water that needs to be treated.

Natural gas is non-renewable. Although synthetic gas from coal and oil is an additional source of natural gas, by definition, natural gas recovered from underground is limited in supply. Natural gas is also not as easily accessible everywhere in the world as it is in the United States, Middle East, Russia, etc. Unlike coal and oil, natural gas is a low density energy source. This creates several issues for its storage and transport for its use in static and mobile applications. One approach taken is to compress or liquefy natural gas, but this adds significant cost to its use as a source of energy. Furthermore, in its use in automobiles and airplanes, where volume and weight are a premium, it is not as attractive and cost-effective as high-density oil energy.

While there are significant positive features of natural gas in the energy industry, it is still a source of carbon emission to the environment. As shown in Table 1.5 and Figure 1.1, while it emits half as much CO_2 compared to coal, it is still very significant compared to other renewable and nuclear sources. While this can be reduced by the use of carbon capture and sequestering technologies, they add to the cost of electricity and fuels. Unlike coal and oil, natural gas also emits methane into the atmosphere by leakages during recovery and conversion processes, gas flaring, and leakages from landfill gas and other natural agricultural and manure wastes. Methane leakage is more harmful to the environment (28–120 times more than CO_2 depending on how you estimate) than CO_2 [1–6]. There is a serious concern about methane leakage and its effect on global warming. Significant methane leakage also occurs from gas hydrates in the ocean, particularly when efforts are made to recover them. Lastly, the present method of "fracking" used for unconventional gas recovery has been linked to a series of minor earthquakes (particularly in the state of Oklahoma). This harmful effect on the environment has to be watched in the future. Thus, in the present political climate where carbon emission to the environment needs to be limited, natural gas is not the perfect source of energy.

D. **Biomass**

In the recent years, the interest in biomass has significantly increased because it is a renewable source of energy, and as shown in Table 1.5 and Figure 1.1, CO_2 emission from biomass-driven power or synthetic fuel plants is minimal. In fact, biomass conversion is CO_2 neutral during its life cycle because plants inhale CO_2 during the night. Biomass is also easily accessible everywhere.

Biomass is a low density energy source. It contains a large amount of water, and it is a very diverse source of energy with a wide variety of composition. Unlike coal, oil, and gas, it is difficult to store and transport. Biomass conversion plants are, therefore, most suitable at smaller, modular, and distributed scale. Large-scale biomass-based power plants are not possible unless biomass is combined with coal as mentioned before. Biomass power plants are thus less flexible and are generally located where biomass is easily accessible. Unlike coal and oil, biomass is difficult to store because, in the presence of water and oxygen, it biologically decays over time. Biomass is also difficult to transport because of its low density. The conversion of biomass to power is also less energy intensive than coal or oil because of its low density and large water content.

Biomass conversion also produces other gas and solid waste products that need to be handled. Biochar can have agricultural use. The land and water requirements for biomass conversion plants are rather large and not always available. The construction cost and electricity cost for biomass-based power plants are significant and may not be cost competitive to other sources of energy. The best use of biomass is heating and power generation at smaller scales. Biomass cannot replace coal, gas, or nuclear sources for large centralized power plants in the urban environment.

E. **Waste**

While there are different kinds of waste, the largest source of waste is the man-made municipal solid waste (MSW), which is largely cellulosic in nature. This, along with agricultural waste and human and animal manure, comprise of convertible cellulosic waste. While a significant portion of the waste is human made, it is renewable. In the United States, we generate more than 200 million tons of MSW per year. Worldwide, we generate about 1 billion tons of waste every year. Per capita, the United States generates eight times more waste than the rest of the world. This waste must be either converted to energy or buried in a landfill. The landfill waste also gets converted to methane over a long period by biological reactions, and efforts are being made to recover this methane and convert to power at local levels. Like biomass, MSW is also a low density energy source and is generally converted to gas or power at local levels. MSW is difficult to store and transport, so it must be converted in short spans of time. Very large-scale MSW power plants are not possible or desirable. Thus MSW-based power plant is a more distributed source of energy. There are numerous power plants of this nature all over the world. These power plants are expensive to construct, and the cost of electricity is higher than that for conventional fossil fuel sources. MSW conversion, however, handles unwanted waste and releases pressures for the additional need for landfill.

Although MSW largely contains biomass-based (cellulosic) materials, it also has numerous other impurities that lead to undesired air and solid byproducts that must be treated. Some of these products can be harmful to local residents if not properly treated. For this reason, the local community generally objects the installation of MSW conversion (gas or power) plants in its backyards. The handling of gaseous and solid byproducts can be expensive. Often MSW conversion plants are subsidized by the local government.

Besides cellulosic solid waste, waste heat is also an important source of energy. Waste heat is often generated because of the inefficiency of various thermodynamic processes. We actually use significantly less than 50% of the thermal energy we generate. Waste heat is converted to power or other needs by processes such as cogeneration and thermoelectricity. These processes make energy systems more hybrid.

In recent years, significant attention has also been given to gaseous waste, particularly CO_2, which causes significant harm to the environment. As will be shown in Chapter 2, significant efforts are also made to convert CO_2 to power or fuel. These conversions also make many energy systems hybrid. The conversions of all types of waste are done locally and on a small scale. Waste is difficult to transport over a long distance.

F. **Nuclear**

In the 20th century, large-scale nuclear power plants became successful in several countries like France, Russia, Japan, and to some extent in the United States because they produce electricity at a very competitive cost and they did not emit CO_2 or other harmful products in the environment. Worldwide, however, nuclear power plants did not get overwhelming acceptance due to the fear of accidents like the Chernobyl Disaster in Russia, the Three Mile Island accident in the United States, and Fukushima Daiichi nuclear disaster in Japan, as well as concerns about the disposal of nuclear waste.

The large-scale production of nuclear energy has several drawbacks. The construction cost for a large-scale nuclear power plant is very high and non-affordable in many developing nations. The electricity cost generated from large-scale power plant is, however, very favorable. While the land requirement for large-scale nuclear power plant is not high, the water requirement is very excessive. As shown in Table 1.5 and Figure 1.1, nuclear-based power plant emits very little CO_2 as well as other harmful gaseous products. A nuclear power plant can provide electricity when needed. However, the nuclear power plant cannot quickly respond to changes in demand. Just like coal-based power plants, nuclear-based power plants are also very inefficient because they are based on the same thermodynamic principle of converting thermal energy to electrical energy. It is also difficult to handle large-scale waste heat locally. Just like coal-based power plants, nuclear power plants work well with urban large-scale utility grid. The economy of scale applies to the nuclear power plant.

In order to address the issues of capital cost and efficiency, in recent years, significant efforts have been made to develop small modular-scale nuclear power plants which require lower capital costs (that can be handled by developing nations) and that can handle the needs of rural, isolated communities. The smaller reactors can also be more energy efficient because they can be easily adapted for cogeneration using heat for local needs such as district heating and industrial heating needs. Small reactors can be easily transportable and can be built away from the power site. As shown later, a small nuclear reactor can be combined with other renewable sources to provide carbon-free energy in a distributed manner. The licensing of such reactors could also be faster. The social and political acceptance of such reactors is, however, necessary.

G. **Geothermal**

This is a renewable source of energy. While there are certain parts of the world like the West Coast of the United States, Iceland, and Kenya that are easily amenable to this source of energy, with the present development of enhanced geothermal systems, this type of energy can be obtained in many parts of the world. The initial construction cost of the geothermal-based power plant is relatively high, particularly if an enhanced geothermal system several miles deep is used. Once built the electricity cost is reasonable (not very competitive) because the intensity of heat is generally lower than other sources, and one needs to use binary Organic Rankine cycle (ORC) to generate power which is not the most efficient. ORC also suffers from low thermal efficiency, and cogeneration is desirable to improve its efficiency. Because of comparatively low temperatures (250°C or below), it is equivalent to low density energy in terms of power production. The level of heat in geothermal energy is directly related to the depth of geothermal well; however, the deeper wells are more expensive. Geothermal heat is easily available when needed, and geothermal power plants are reasonably flexible and quickly respond to the change in demand. While it does not require significant land, its water demand can be significant. As shown in Table 1.5 and Figure 1.1, generally, geothermal energy emits 5% of CO_2 emitted by coal and 1% of sulfur dioxide and nitrous oxide. Turkish geothermal plant, however, emits 1,800 gm/kWh of CO_2, which is twice that emitted by the coal power plant. In general, geothermal energy is less harmful to the environment than fossil fuels.

Geothermal heat is extensively used for local heating and cooling needs through geothermal heat pumps. As shown later, geothermal energy can play a significant role in carbon-free energy buildings and community and district heating. Because of its low temperature, its industrial use is limited to low and medium temperature processes.

H. **Hydro**

Hydro energy is renewable, and as shown in Table 1.5 and Figure 1.1, it produces very little CO_2 and other air emissions as well as other types of waste products. This makes this type of energy very environment friendly in its operation. However, it requires significant land and water usages, and it is available only where the hydroelectric facility is available and has limited availability to generate electricity when needed. It is a flexible source of energy and can be operated on a large scale or small modular scale over a small river stream. Hydropower currently holds about 95% of stored electrical energy, but it can also be used to supply power at various scales. The construction cost for a large-scale hydropower plant is significant, and electricity generated from a large hydroelectric plant can be expensive.

In recent years, significant efforts have been made to generate power from ocean and sea waves using different types of wave converters as well as capturing energy from tidal waves. Small modular hydropower dams are also used to supply distributed power to smaller communities. These small hydropower plants are less expensive and create less environmental effects on land and aquatic life. The small hydropower plants as well as small-scale wave energy converters and tidal wave converters are, however, generally inefficient. The required infrastructure to capture power from hydro energy can be expensive. Hydro energy may not be available at all places. Commercial technologies for wave and tidal energies are not yet fully developed.

I. **Solar**

Solar energy is renewable but intermittent. Solar energy near the equator is available in abundance, but in general it is season dependent. As shown in Table 1.5 and Figure 1.1, solar energy is very environment friendly and produces very little CO_2 or other types of emissions in the air. Solar energy does not require water, but like all renewable sources of energy it is low density energy and requires large land use if it is used to generate large-scale power plant. The construction of a new solar power plant can be expensive, and the cost of electricity generated from the solar power plant can be high depending on the cost and efficiency of solar panels. Generally solar power plants require energy storage devices or other sources of energy to balance its intermittent nature.

A solar power plant cannot generate electricity when needed (like during night) and is not capable of quickly responding to changes in the demand. Solar power can be generated using thermal energy and converting that into electricity like other sources of thermal energy or by direct conversion of solar energy into electrical energy using photovoltaic cells (PV), thermophotovoltaic (TPV) cells, or thermoelectricity. All of these methods are inefficient, and significant works are ongoing to improve their efficiency and reduce their cost. Solar energy can also be used for heating and cooling, particularly for buildings and district heating. Research is also ongoing on the use of solar energy for vehicles. Solar energy, along with energy storage and geothermal energy, is a source for carbon-free zero energy buildings and communities. In recent years, the cost of electricity from solar energy has decreased considerably.

J. **Wind**

Like solar energy, wind energy is renewable but intermittent and depends on location, time, and season. Wind energy does not require water, and as shown in Table 1.5 and Figure 1.1, it emits very little CO_2 or other gases in the air. It also does not produce any other waste products. Wind energy is also low density energy and requires large areas of land to be generated in significant quantities. The construction cost is manageable, but the required power infrastructure can be challenging depending on the location of the wind farm (such as on the top of the mountains). In recent years, modular wind towers, blades, and turbines are built to cut the construction costs. Electricity cost from wind energy can be low. Wind energy is not always available to generate electricity when needed and is not capable of quickly responding to changes in demand. Like solar energy, wind energy generally requires other sources of energy or energy storage device to supply uninterrupted power. In recent years, the cost of electricity from wind energy has decreased considerably.

It is clear from the above description of ten sources of energy that no source is completely ideal for all situations. They all do not serve equally well both urban and isolated or rural environments. They are not equally suitable for large-scale operations in an urban environment and a small-scale distributed environment. They all have different types of issues with efficiency. Renewable sources generally require large capital investment and land compared to fossil and nuclear energy. The operating cost for renewable energy is always lower because it is free. Grid support is very important for large-scale operations. Small distributed energy can be used without grid or with small microgrids (MGs). Renewable energy sources require energy storage, particularly for solar and wind energy because they are intermittent. As shown later, just like the source of power generation, no single energy storage device completely serves the needs of all situations. This has resulted in the need for hybrid energy storage devices. Also, the traditional utility grid does not serve well the needs of rural and isolated environments where grid connections are not available. This has resulted in the upgrading of the traditional grid into hybrid grid transport involving utility grid, microgrid, and off-grid operations like mini-grid (for the rural environment), nanogrid (NG) for individual building or equipment, and stand-alone systems. Unlike off-grid, a microgrid can be connected to a mega grid or can operate in island mode. Both hybrid microgrid and hybrid off-grid operations are developed for harnessing distributed energy resources and, in particular, renewable energy resources.

Hybrid power and heat generations, hybrid storage, and hybrid grid transport are thus the elements of the future hybrid energy (power) systems.

There is one caveat to the above discussion. To some extent the economics and durability of various components of hybrid energy sources are competing with each other, and so the future prediction on the effectiveness of individual sources is difficult, particularly because nanotechnology is accelerating innovations in each source of energy recovery and conversion as well as in storage and transport capabilities. These advances can dramatically change the picture. If we examine the future of energy supply from the standpoint of (a) reliability and flexibility, (b) capital and operating costs, (c) energy efficiency, (d) environmental impact, and (e) long-term sustainability, it would be difficult to identify any single source of energy generation, storage, and transport that will have a permanent lasting effect. The sustainable future energy system requires a flexible hybrid energy system approach.

1.3 CRITERIA FOR SUSTAINABLE SOURCE OF ENERGY

Single or hybrid source of energy should be evaluated on the following five criteria that are generally accepted by the energy industry. The energy source should be evaluated by the collective measure of these five criteria. It should be realized that the customer need for energy varies from (a) large-scale urban environment where electrical energy and heating needs are satisfied by utility grids and fuel infrastructure, (b) to small-scale rural or isolated communities where large utility grid is not available and the demand for power and heating and cooling is at smaller and distributed levels, (c) to mobile energy where the demand is generally at a distributed level. What we will point out is that while some criteria are well satisfied by one or more sources, others are better satisfied by a different set of sources.

A. **Reliability and flexibility**

All customers for energy want reliability and flexibility in their energy supply. Reliability means that energy is available when they need in a stable manner so that there are no interruptions in their energy supply. Power and heating/cooling needs in an urban environment provided by large-scale coal or gas-based power plants or nuclear power plants with utility grid and gas or electricity infrastructure for heating and cooling needs were considered to be examples of reliable energy supply. However, the harms created by these power plants to the environment and the resulting political pressure have called into question their future reliability. The technologies used to generate and distribute power and gas are, however, commercially proven and reliable. These single source power and gas distribution methods do not work for remote and isolated communities because of a lack of electrical grid and gas infrastructure. For these communities different approaches have to be taken to receive stable supply of power and heating/cooling needs. Large-scale single renewable source for power and heating/cooling needs is not very reliable, particularly for solar and wind energy that are intermittent in their recovery. Due to the reasons articulated earlier, single source renewable energy is less reliable at large scale [9, 10].

Flexibility of energy source has many components. Flexible source means that suitable power or heating/cooling needs are provided if suddenly the demand is increased. Flexibility can also refer to the scale at which energy is provide – large or small scale. Flexibility also means whether or not energy can be provided at all locations and times, i.e., urban environment, isolate and rural environment, off seashore, mobile environment, day/night, different seasons, etc.

When one examines the sources outlined earlier, single source nuclear energy at large scale is not flexible to the sudden change in demand, coal-based power plant is marginally flexible, and gas-based power plants are very flexible to change in demand. They all can operate on a small scale, particularly nuclear- and gas-based power plants, but the cost of electricity can go up. They are all dependent on location due to their need for the grid. All renewable energy sources individually are not very flexible with respect to their size, demand need, and location. Except for large hydroelectric plant, all renewable energy sources individually are less flexible with respect to size and location.

B. **Affordability**

The construction of large-scale nuclear power plant is high and cannot be afforded by many developing nations. The construction costs for coal-based and particularly gas-based power plant are manageable. The electricity cost of nuclear-, coal- and gas-based power plants is competitive; however, if carbon capture and storage technologies are used for coal power plant, the electricity cost can rise by as much as 4¢–5¢ per kWh, making it less competitive. The construction of small-scale nuclear and renewable source power plants is affordable. However, large-scale solar, wind, geothermal, or hydropower plants are capital intensive. Both solar and wind power plants on a large scale have large land requirements. The operating costs for solar, wind, geothermal, and hydropower plants are minimal.

C. **Efficiency**

For large-scale coal-, gas-, nuclear-, or solar thermal-based power plants, efficiency is low due to thermodynamic restrictions. On a large scale, the implementations of cogeneration is also challenging because all waste heat has to be used locally. Small-scale thermal power plants, while costly, are more manageable for cogeneration. Low efficiency means larger fuel usage is required for the same level of energy production. Similar problems exist for all single source renewable or nuclear-based power plants at a smaller scale. Once again cogeneration is required in these cases which is somewhat easy to implement.

Efficiencies of other methods of power productions such as photovoltaic, thermophotovoltaic, and thermoelectric are also problematic. Significant research is being carried out to improve these efficiencies. Efficiency for conversion of wind energy to electricity is also low and needs improvement. While the efficiency of the hydroelectric process is somewhat higher, the similar efficiencies for conversion of wave and tidal energies to electricity are low. These require significant research before becoming commercial. While the United States and China use significant energy consumption to improve their GDP (gross domestic product), in future, focus on the improvement of efficiency can bring energy consumption down without affecting the GDP. The efficiency of heat requirements in various industrial processes such as glass and ceramic manufacturing, coal and oil-based chemical industry, etc., is also very low. Waste heat generated from various industrial processes needs to be minimized or used for other purposes. One option is to generate electricity from waste heat using thermoelectricity. The same thing can be done for the waste heat generated from automobiles. Lower fuel consumption by improving efficiency is also good for environment.

D. **Environment protection**

While there are many drivers for changes in energy industry like energy fractionation (better preference for distributed power), innovations in energy technologies and people's desire to be more mobile and dependence on personalized technologies like cell phone, the most important driver has been the need to protect environment. More specifically, there is a serious concern about carbon emission (in the form of carbon dioxide and methane) in the environment by man-made energy technologies. This has led to the desire to replace the use of fossil fuels by renewable sources of energy. In power generation industry, coal-based power plants are being replaced by gas-based power plants. Various carbon capture technologies are being examined. Unfortunately they all add significant cost to the produced electricity. While gas-based power plants are better alternatives to the coal-based power plants, they are still not the best long-term solution because gas-based power plants still emit half of CO_2 emitted by the coal-based power plants. Furthermore the recovery of shale gas (which has propelled the use of gas in the power plant) is also responsible for significant leakage of methane into the atmosphere which is even more harmful than CO_2. The use of oil in the transportation industries has also raised serious concerns due to emission of CO_2 by automobile and aircraft industries. Thus fossil fuels in general are considered not to be desirable sources for the protection of environment.

The generation of power by nuclear industry is a viable option because it emits very little carbon into the atmosphere. There is a significant movement to replace large nuclear power plants by smaller modular plants which (a) require less capital cost, (b) are easily movable, (c) can be built off-site, (d) can be used for cogeneration with the use of waste nuclear heat for local purposes, and (d) will require lower time to get licenses and build. Nuclear energy, however, needs to get more social and political acceptance.

The use of renewable sources of energy is the best option to protect the environment. Unfortunately, in general, they cannot be implemented at the scale of coal-, gas-, or nuclear-based power plants. Solar and wind energy require large land, and individually they are intermittent and cannot supply uninterrupted power. Renewable energy sources are best suited for distributed power and operate best with the support of energy storage. Thus, they are best used in a hybrid manner than as an individual source. The world requires deeper penetration of renewable energy in total energy mix, and this can only be achieved in a hybrid manner. In the car industry, hybrid cars, electric cars, and fuel cell-driven cars are replacing conventional internal combustion engine cars to reduce carbon emission. This change will continue to persist as technologies for hybrid, electric, and fuel cell cars advance.

E. **Long-term durability and sustainability**

The long-term durability and sustainability of energy source depends on a number of factors. First, the renewable nature of the source is an important factor. All fossil fuel sources are non-renewable so the availability of those sources for long term can be questionable. Second, the technologies used to generate power or provide heating and cooling needs can be very time dependent, particularly because the rapid innovation is caused by nanotechnology and other manufacturing factors. These changes can affect the economics of individual sources. For example, recent advances in solar and wind energy technologies have made the cost to generate electricity much cheaper and more competitive than the power generated by fossil fuels. This means that significant advances in cutting cost and improving durability of any technology can rapidly change the energy supply picture. Recent advances in fuel cell technology to cut cost and improve its durability have made it a very attractive option for the near future. To some extent, all ten sources of energy are competing for energy market, and dependence on any individual source over a long term can be risky. The same argument applies to energy storage technologies like lithium battery whose cost keeps rapidly declining. The long-term sustainability of any energy source is important for capturing the market and attracting new investments. Many times, long-term sustainability depends on the changing nature of social and political climates. Thus, it could depend on location and time. For example, nuclear energy continues to be an important source of energy in France due to its social and political support. The use of sugarcane to generate biofuel continues to be important in Brazil due to its economic and social and political support.

The long-term sustainability of energy sources also depends on the availability of infrastructure needed to capture the market. For example, fuel cell and hydrogen-based technologies will depend on the hydrogen infrastructure and its proliferation. The development of electric car will depend on the infrastructure for battery recharging. The development and sustainability of renewable energy technologies will depend on the availability of necessary infrastructure needed to capture the market. Unlike for fossil fuel-based energy, for renewable energy, energy storage and its infrastructure will play an important role in its expansion. One reason why the infusion of renewable energy in the fossil fuel-dominated energy world will be gradual is that the infrastructure needed for fossil fuel sources is well established but a similar infrastructure needed for renewable sources will take time and new investments. Long-term sustainability of any energy source depends on many factors such as durability of technology, scale of technology, social and political acceptance of technology that can change with time, required infrastructure, and most importantly the market. For all these reasons, relying on a single source is always risky. The classic example is the coal industry and the use of coal for energy.

1.4 FUTURE TRENDS IN ENERGY INDUSTRY

Energy industry is constantly evolving. While new technology (like fusion) can dramatically change the future energy industry, in general, energy industry has changed in an evolutionary manner rather than a revolutionary manner. Apart from the considerations of five criteria for sustainability mentioned above, strategies for sustainability should consider the future need of the society and the trend of energy industry. Here are some factors that can also help laying out strategies for the sustainable energy [1–6].

1. There is a movement of green energy in the world and this is not going to stop. Carbon emission and other environment impacts by human-made energy technologies will be both politically and socially questioned and objected. There is a global concern about carbon emission and its impact on global warming. This is significantly generated by the combustion technologies used in power plants and automobile engines. While coal and oil have some benefits (i.e., they are high-density energy sources and are still available in many parts of the world), they generate too much carbon emission to the likings of the society. In the long term, these sources would probably be used more for producing chemicals than energy. Gas is a viable source in the near future, but it also liberates significant amounts of CO_2 and CH_4. Gas will, however, be a transition fuel for some time due to its abundance availability (like shale gas) and ability to produce gas both from non-renewable and renewable sources.
2. The viability of nuclear energy on a large scale is only limited to some parts of the world. Nuclear waste and risk of nuclear plant failure are still very much of political and social concerns. Large-scale nuclear reactors require excessive capital cost to be affordable by developing countries. Large-scale reactors are also inefficient, not well suited for cogeneration, create a significant amount of nuclear waste, and require a significant amount of water. In recent years, small modular reactors have gained more acceptance because (a) they require less capital investment, (b) they are more suitable for cogeneration and hence have better thermal efficiency, (c) they are more flexible for change and upgrade, (d) they can be built off-site and can be moved to another site if needed, and (e) they fit better with the concept of distributed energy generation. Small modular nuclear plants thus have more sustainable possibilities.
3. Many predict that the future of energy industry lies in hydrogen and electricity. These are certainly more sustainable end games. Fuel cell which uses hydrogen is certainly a more sustainable power generation and energy storage option. Electricity will penetrate more and more both in static and mobile worlds.

4. Building industry is one of the largest polluters of carbon in the environment (anywhere between 25% and 40%). There is a worldwide trend to make buildings zero energy buildings with the help of renewable energy sources. This means more use of renewable sources like solar, wind, hydro, geothermal and biomass for power and heating and cooling needs of the buildings. This trend is likely to continue. This will also make energy consumers into energy prosumers, and it will have a pronounced effect on the transmission of electricity and electrical infrastructure as a whole.
5. Transportation industry is also becoming more electrical and shying more and more away from the internal combustion engine, which generates significant carbon emission to the environment. Use of fuel cell for power generation will continue to increase. Also, technology will radically change the way the transportation industry will operate. Within the next several decades, hybrid and electric vehicles (EVs) will dominate the transportation market.
6. Both material waste and waste heat generated by power plants, numerous manufacturing industries, residential communities, and transportation industry are becoming more of an issue for society. In future, more efforts will be made to convert these wastes into energy. This trend will continue indefinitely as part of the green energy effort.
7. The infusion of renewable and distributed energy sources for power generation will become more and more important. The electrical infrastructure will have to modify to accommodate this. This green revolution will be most supported globally by all countries. The use of microgrids to harvest distributed energy sources will become more prominent. The conventional grid will also have to be upgraded to make energy consumers into energy prosumers. The grid infrastructure will be more digitally controlled.
8. The electrical infrastructure will also be modified (from bottom up) to satisfy the power needs of rural, remote, and isolated communities. Off-grid energy will be aggressively developed, particularly in poor and developing countries like Africa, Asia, and South America. Mini-grids, nanogrids, and stand-alone systems will get some preference in the future. More energy will be consumed by developing nations in the world.
9. The role of energy storage in energy infrastructure will become more and more important. This will be particularly important when the contribution of renewable sources in power generation is significantly increased. More use of hybrid energy storage systems (HESS) will become evident as energy systems (in particular renewable energy systems (RES)) will be required to become more efficient and productive. This will happen both for static and mobile systems.
10. Distributed energy will gain more prominence in relation to centralized controlled energy. Energy systems will have to be more modular in nature so that it can adapt to changes most cost-effectively. Energy industry will become as modular as building, computer, and finance industries. The modular nature will allow more flexibility to adopt new changes, particularly by nanotechnology.

Any strategy for sustainable energy needs to be mindful of these trends in energy industry. It is clear that based on the ten factors described above, the major changes that will occur in the energy industry would be to have a larger share of renewable sources of energy in the overall energy mix and more efforts on the technologies that generate, store, manage, control, and transmit power in both static and mobile environments. The nature of centralized power distribution will change and more efforts will be made on distributed power generation. Power industry will have to be more prepared to handle low-density and distributed sources compared to high-density and centralized sources.

The above discussion also suggests that future energy industry will have less use of coal and oil and more dependence on renewable sources with gas acting as a transition fuel. While this transition will occur gradually, all strategies for sustainable energy should point toward more use of distributed renewable sources with electric power as a major game. In the long term, hydrogen will also be the desired fuel. Its market and infrastructure, however, will require some time. Accommodating more use of renewable fuel sources in existing energy infrastructure will require a different game plan than what has been used in centralized power generation and heating and cooling systems. Not only will power generation have to come from mixed sources; the role of energy storage and mechanisms of electricity transport will have to be different from the ones used thus far.

Renewable resources and clean alternative energy power generation technologies have attracted much attention because they have several advantages such as they allow less dependence on fossil fuel, have unrestrained availability of the resources which are free of cost, and produce lower harmful emissions to the atmosphere (i.e., environment friendly). Renewable energy sources, such as wind, solar, microhydro (MH), biomass, geothermal, ocean wave and tides, and clean alternative energy sources, such as fuel cells (FCs) and microturbines (MTs), will become better alternatives to conventional energy sources. Renewable energy sources, however, face some head winds due to their uncertainty, intermittency, and high initial cost. Management and control of renewable sources are also more complex. Recently, extensive research on renewable energy technology has been conducted worldwide which resulted in significant development in the renewable energy materials, decline in the cost of renewable energy technology, and increase in their efficiency.

To overcome the intermittency and uncertainty of renewable sources and to provide an economic, reliable, and sustained supply of electricity, a modified configuration that integrates these renewable energy sources and uses them in a hybrid system mode is proposed by many researchers. While the energy from renewable resources is available in abundance, due to its intermittency, hybrid combination and integration of two or more renewable sources make the best strategy. Hybrid combination takes the best advantage of their availabilities, operating characteristics, and thereby improves the system performance and efficiency. Hybrid renewable energy systems (HRES) are composed of one renewable and one conventional energy source or more than one renewable with or without conventional energy sources that works in stand-alone or grid-connected mode. Hybridization of different alternative energy sources can complement each other to some extent and achieve higher total energy efficiency than that could be obtained from a single renewable source. Multisource hybrid renewable energy systems, with proper control, have great potential to provide higher quality and more reliable power to customers than a system based on a single source. Due to this feature, hybrid energy systems have caught worldwide research attention [7–55].

The applications of hybrid energy systems in remote and isolated areas are as important as in grid-connected systems. In addition, the application of hybrid systems is becoming popular in distributed generation or microgrids, which are close to customers and distribute energy at medium to low voltage. Due to the advances in renewable energy technology which have improved their efficiency and reduced the cost and the advances in power electronic converters and automatic controllers which improve the operation of hybrid energy systems and reduce maintenance requirements, hybrid systems have become more practical and economical. Hybrid storage systems have added one more element to the success of hybrid energy systems, both in static and mobile environments. Hybrid energy systems are now becoming an integral part of the energy planning process to supply previously unelectrified remote areas [7–55]. Various hybrid energy systems have been installed in many countries over the last decade, resulting in the development of systems that can compete with conventional, fuel-based remote area power supplies for many applications.

The design process of hybrid energy systems requires the selection and sizing of the most suitable combination of energy sources, power conditioning devices, and energy storage system together with the implementation of an efficient energy dispatch strategy. The selection of a suitable combination from renewable technology to form a hybrid energy system depends on the availability of the renewable resources in the site where the hybrid system is intended to be installed. In addition to the availability of renewable sources, other factors may be taken into account for proper hybrid system design depending on the load requirements such as reliability, greenhouse gas emissions during the expected life cycle of the system, efficiency of energy conversion, land requirements, economic aspects, and social impacts. The unit sizing and optimization of a hybrid power system play an important role in deciding the reliability and economy of the system. With hydrogen and power as end games for the energy industry, hybrid energy approach for the use of renewable resources appears to be a very sustainable strategy. This book makes this case with broadly defined and articulated components of hybrid energy (power) which include mechanisms for sources, generation, storage, and transport of energy.

1.5 WHAT IS HYBRID ENERGY SYSTEM

While hybrid energy appears to be an important part of future energy industry, it is important to first define "hybrid energy (power) system." Hybrid energy system can be defined in a number of different ways. GE (General Electric) defines hybrid power as:

> Hybrid power plants usually combine multiple sources of power generation and/or energy storage and a control system to accentuate the positive aspects and overcome the shortcomings of a specific generation type, in order to provide power that is more

> affordable, reliable, and sustainable. Each application is unique, and the hybrid solution that works best for a specific situation will depend on numerous factors including existing generation assets, transmission and distribution infrastructure, market structure, storage availability, and fuel prices and availability.

This definition of hybrid energy system (or hybrid power) is somewhat restrictive. In this book, we define hybrid energy system in more general terms. First, hybrid energy system has two basic components: power and heating and cooling. GE's definition is more centered toward stationary power generation. Here we define hybrid energy system as one where single or multiple sources for power or heating and cooling result in single or multiple power or heating and cooling outputs. The system also includes energy storage and transport devices, both of which can also be hybrid like hybrid storage systems or hybrid grid systems which include utility grid as well as micro and/or off-grids. Holistically, a hybrid energy system must consider (a) both power and heat (cold), (b) sources, (c) generation technologies, (d) storage mechanisms, and (e) methods for energy (power) transmission and distribution.

Let's examine specific examples to further illustrate this definition of hybrid energy system. GE's definition of hybrid power includes multiple sources power generation (like wind and solar), with energy storage with and without grid. District heating with multiple renewable sources of heat with and without storage is another example of hybrid energy system where power may not be involved. A zero energy building with solar energy to generate power and heat and geothermal energy for heating, ventilation, and air conditioning (HVAC) system, with or without storage, is also another example of hybrid energy system. Here both power and heating and cooling are parts of hybrid energy system. Cogeneration (combined heat and power (CHP)) is another example of hybrid energy system to improve energy efficiency. ExxonMobil and FuelCell Energy's partnership to generate power by burning coal or natural gas and use fuel cell to generate more power from waste CO_2 is another form of hybrid energy system where power is generated in two different ways. Using multiple sources (including renewable and nuclear) to generate heat with and without thermal storage for industrial processes (such as glass making operation) with conversion of waste heat to generate electricity by the process of thermoelectricity is another form of hybrid energy system where industrial heating results in the waste heat to generate power. This is a reverse cogeneration process once again used to improve energy efficiency. Hybrid vehicles can also generate power by IC engine, fuel cell, or battery along with power generated via thermoelectricity of waste exhaust heat. An EV with energy storage (one or more) is also a hybrid operation.

Hybrid energy systems also include hybrid storage systems such as battery capacitors, battery flywheel etc. to match required power and energy density, charge-discharge time, cycle time etc. for the system. Hybrid energy systems also include hybrid grid systems which include hybrid energy and storage sources at three levels of grid: utility level mega grid, hybrid microgrid that can be either connected to mega grid or operated in islanded mode, and off-grid systems that include mini- and nanogrids and stand-alone systems. These systems are generally not connected to mega grid. Interconnection of nanogrids or mini-grids can be linked to microgrids or conventional utility grids. Holistic hybrid energy systems thus have three components: energy generation, storage, and transport.

The book demonstrates that all these elements, source for generation, storage, and grid transport, along with processes to generate additional energy (heat or power) to improve efficiency or to treat waste materials all together define the nature of hybrid energy systems. Such hybrid energy systems are (a) more reliable and flexible, (b) more efficient, (c) more affordable, (d) more environment friendly, and (e) ultimately more durable and sustainable over long term compared to the single component energy sources or systems. Hybrid energy system is essential for the deeper penetration of renewable and non-dispatchable energy (particularly solar and wind which are intermittent by their nature) sources in the overall energy mix in order to reduce carbon emission to the environment. Hybrid energy system is also important for the better use of nuclear heat and suitable power generation by a combination of nuclear and renewable sources. Hybrid storage system is the best way to improve quality of power over a long period. Three levels of hybrid grid transport is the best way to not only increase the level of renewable sources in the overall energy mix but to harness distributed energy resources and serve both urban and rural communities. The choice of hybrid energy system components, their integration and control, and their intelligent management and optimization are also part of assessment and discussion of this book.

1.6 ENERGY SUSTAINABILITY AND HYBRID ENERGY SYSTEMS

There is a lot of discussion in both academic and industrial communities on sustainability of energy sources over a long period. There are strong reasons for this; some of which can be articulated as follows:

1. Energy is a commodity that everybody needs to improve their quality of life. There is only limited supply and all nations (developed and developing) want to consume more energy. This cannot happen over a long period of time without making some fundamental changes in the way we recover sources and use them. Also as mentioned earlier, no single source is ideal for sustained energy consumption.
2. In order to preserve sources of energy over a long period, attention to efficiency of recovery and conversion is very important. This can vary significantly from source to source. High efficiency reduces the energy and fuel consumption for the same benefits which are always desirable. Besides inefficiency associated with various conversion equipment such as PV cell, thermoelectric materials etc., thermodynamic limitations for power generation and waste heat generated in both static and mobile power generation as well as heat used in various manufacturing processes must be recovered to improve efficiency of fuel usage. All of these inevitably lead to the use of hybrid energy systems in various processes. Efficiency of energy consumption is going to become more and more important as the energy consumption in developing countries rapidly grow. While some aspects of energy efficiency are material (like PV and TPV cells, thermoelectric, etc.),

equipment (used in various industrial processes), and process (like thermodynamic cycles used to convert thermal energy to electrical energy, fuel cell, etc.) related, in many cases, low efficacy results in significant amount of waste heat and materials. The improvement of efficiency will require extracting more power from waste heat or the use of waste heat by cogeneration. The conversion of waste material to more energy also means hybrid energy systems. Thus hybrid energy system by definition implies more efficient use of energy.

3. The cost of energy (both capital and operating) is very important because that partly determines the market it will capture. For new ventures, return on investment is also very important. Properly chosen, hybrid energy system allows better flexibility for the cost optimization. Careful optimization of hybrid sources allows each source to sustain longer.
4. Energy and environment protection must operate in harmony. There is a serious concern worldwide regarding the effects of carbon release to the environment by some sources and some technologies. This can seriously affect the sustainability of a particular source or technology over a long period. Carbon emission must be either prevented or treated. The treatment can lead to generation of hybrid power such as use of fuel cell to treat CO_2 by ExxonMobil gas-based power plants. Hybrid renewable energy system is well suited to reduce CO_2 emission. Environment protection will become increasingly more important for future energy usages. This means less use of fossil fuel and deeper penetration of renewable energies in the overall energy mix. Unlike fossil energy, renewable energy is in general of low density and time and location dependent (particularly solar and wind). This will force the use of multiple sources, including energy storage, to have steady and uninterrupted supply of energy which is particularly important for power generation.
5. To some extent, all ten sources of energy are competing with each other. Because of the rapid development of various energy technologies, it would be imprudent to rely on a single source over a sustained period of time. For example, the rapid development of solar and wind energy technologies have made them competitive in power production compared to gas-based power plants. Rapid development of fuel cell can further change this picture. Nanotechnology has further fueled this competition. In this fast changing environment, the use of hybrid energy system will ensure more sustainability.
6. Single source energy system generally has limited flexibility of scale and region coverage. Biomass power plants by itself cannot be on a very large scale. Solar and wind energy cannot be very large without access of large land. Since future dictates better balance between centralized and distributed energy systems, hybrid energy system can best facilitate achievement of this balance.
7. Hybrid storage systems and grid transport systems are necessities to counteract limitations of individual storage and large-scale grid transport systems.

8. Higher penetration of renewable sources in the overall energy mix is inevitable due to strong social and political interests in greening energy sources. This cannot occur in a flexible manner without the need for energy storage which results in hybrid energy system. Hybrid renewable energy sources are poised to best serve off-grid locations such as rural and remote communities.

The argument presented above indicates that sustainable future of energy industry requires hybrid sources, generation, storage, and transport to satisfy the energy need of both urban and rural or isolated environments. Hybrid energy sources can provide:

1. More options for local optimization of cost. Hybrid energy systems can be designed to achieve desired attributes at the lowest acceptable cost, which is the key to market acceptance.
2. Greater penetration of renewable sources in the overall energy mix. Hybrid energy systems can be designed to maximize the use of renewable resources, resulting in a system with lower emissions.
3. More efficient and less use of fossil fuels in the future.
4. More options for large centralized and small distributed operations. This will well serve both urban and rural and isolated communities. This will also allow better harvesting of distributed energy sources.
5. Less reliance on large-scale energy grids. Micro, mini, and nanogrids can operate independently or together with a utility grid. This along with stand-alone systems will help more than billion people who are currently without electricity.
6. More options for optimum uses of local energy resources, which can vary in different country. They provide flexibility in terms of the effective utilization of the renewable sources.
7. Greater role of energy storage to provide stable power and heating/cooling for all energy usages such as buildings, vehicles, district heating, industrial heating, etc.
8. Offsets negative aspects of many sources without their elimination from the overall mix.
9. Allow gradual transition to carbon-free energy world.
10. Less social and political tensions between available sources and need for energy for all parts of the world.
11. Better management of sustainable and high-quality power.
12. A hybrid energy system can make use of the complementary nature of various sources, which increases the overall efficiency of the system and improve its performance (power quality and reliability). For instance, combined heat and power operation, e.g., MT and FC, increases their overall efficiency or the response of an energy source with slower dynamic response (e.g., wind or FC) can be enhanced by the addition of a storage device with faster dynamics to meet different types of load requirements [,9, 12–22, 31–34, 40].

1.6.1 Challenges with Hybrid Renewable Energy Systems

Though a hybrid energy system has a bundle of advantages, there are some challenges and problems related to hybrid energy systems that have to be addressed:

1. Most of the hybrid systems require storage devices for which batteries are mostly used. These batteries require continuous monitoring and increase in cost, as the batteries' life is limited to a few years. It is reported that batteries' lifetime should increase to several years before they can be effectively used in hybrid systems.
2. Due to the dependence of renewable sources involved in the hybrid system on weather results in the load sharing between the different sources employed for power generation, the optimum power dispatch and the determination of cost per unit generation are not easy. However, optimization of hybrid energy system is essential. This should be obtained with several end objectives.
3. The reliability of power can be ensured by incorporating weather independent sources like diesel generator or fuel cell.
4. As the power generation from different sources of a hybrid system is comparable, a sudden change in the output power from any of the sources or a sudden change in the load can affect the system stability significantly.
5. Individual sources of the hybrid energy systems have to be operated at a point that gives the most efficient generation. In fact, this may not occur, due to the fact that the load sharing is often not linked to the capacity or ratings of the sources. Several factors decide load sharing like reliability of the source, economy of use, switching require between the sources, availability of fuel, etc. Therefore, it is desired to evaluate the schemes to increase the efficiency to a possible higher level.

Going forward, all of these challenges will need to be addressed.

1.6.2 Elements of Hybrid Energy (Power) System

The discussion of hybrid energy system is complex because there are several facets to the concept of hybrid energy system. I introduced this concept in my previous book [2] on "Energy and fuel systems integration." In this book various forms of hybrid nature of both fuels and energy systems are discussed. The book first considers the advantages of combining various fuel resources like coal, biomass, waste, and various forms of oils for downstream processes of combustion, gasification, liquefaction, and pyrolysis. The book illustrates that hybrid fuels (also designated as co-fuels in the book) offer many advantages to the downstream processes such as more favorable product slate, less impact on environment, more flexibility on scale, etc. The book also shows that co-digestion for anaerobic treatment of waste also improves methane and hydrogen production which are desirable for subsequent production of energy. For automobiles, hybrid fuels improve fuel flexibility and less environment impact. The book thus clearly articulates many advantages of hybrid fuels in energy industry.

Besides hybrid fuels (or as the book articulates as the integration of various fuels), the book also considers the role of combined heat and power (another form of hybrid energy system) for the nuclear industry. Cogeneration is another form of hybrid energy system that can be applied to both coal and gas power industries same as that for nuclear industry. The book also evaluates hybrid power generated from renewable sources. Finally, the book also considers the hybrid power used in the automobile industry.

Since the publication of the previous book [2]), the development and use of hybrid energy system has gained significant momentum and new advances. Besides the ones mentioned above, significant efforts are now made to develop hybrid power systems involving more than one energy source. The need for hybrid storage systems and hybrid grid transport systems has also been recognized, and these concepts are significantly advanced. Cogeneration has taken an alternate approach where waste heat is used to generate more power (like waste heat conversion to power by thermoelectricity) or even waste CO_2 to generate more power using fuel cell technology. Both hybrid and electric cars have made significant power electronics advances. New developments of utility grid, microgrid, nanogrid, and off-grid technologies and their interactions have allowed new possibilities for interconnections between centralized and distributed energy sources. These are also important for deeper penetration of renewable sources in the overall mix of the energy industry [9, 12–22, 31–34, 40]. They also make a hybrid energy system (or integrated energy system) a more sustainable strategy for the future.

The present book contains several elements of hybrid energy system. In literature, hybrid energy system (or more specifically hybrid power) is defined as the system in which more than one source of energy is used to generate energy (or power). As described earlier, while this definition makes most sense, it is not totally inclusive of all elements of hybrid energy (power) system. As discussed in this book, there are five basic elements of hybrid power system:

1. Hybrid power generated from more than one input source of energy like coal, oil, gas, biomass, solar, wind, etc. Energy storage can also be considered as one source of power. Going forward, the overall objective of centralized grid-connected energy is to (a) increase efficiency through cogeneration or conversion of waste heat to power by thermoelectricity, (b) reduce CO_2 emission by converting CO_2 to power via CO_2-based fuel cell, or (c) increase renewable energy contribution in grid power. All of these strategies will make centralized power more hybrid.
2. Hybrid power is generated in the mobile industry to improve the efficiency of vehicles and reduce its impact on the environment. Unlike in stationary power generation, the end objective here is to generate more mechanical energy for motion.
3. As we increase the contribution of renewable energy in the overall energy mix, the use of energy storage becomes more important. Just like the source

for power generation, no energy storage device completely satisfies the need for sustainable and high-quality power supply. The future for uninterruptable power supply with renewable sources for power will require hybrid energy storage (multiple storage) systems. Hybrid energy (power) storage systems will become an integral part of hybrid energy (power) system.

4. In order to capture distributed renewable and low-density energy sources, the grid transport will be modernized and become more hybrid. Hybrid microgrid will capture distributed energy sources at medium to low voltage in a more efficient manner. This hybrid microgrid either will be connected to macrogrid or can operate in islanded mode. When hybrid microgrid is connected to macrogrid, it can act as a backup in case of macrogrid failure.
5. In order to serve customers in remote and isolated areas, the off-grid hybrid energy systems are being developed. These systems include mini-, nano-, or picogrids and stand-alone systems. Unlike microgrids, these systems will not be connected to macrogrid. These systems are very useful for rural and remote area electrification.

The hybrid power system must consider all five elements of power generation, storage, and transport (grids) for both static and mobile applications.

1.6.3 Development of Renewables in Hybrid Energy Grid System

Off-grid renewable hybrid energy powered systems have been in use for many years, but most have been constructed on an ad hoc basis using components intended for other purposes. Recent technological developments in the micro-generation sector and electricity business restructuring have resulted in the growing interest in the use of micro-generation and microgrids. Purpose designed components and controllers are now available for off-grid systems, making possible the construction of efficient, controlled networks in a variety of configurations and sizes. The terms mini, micro, and nanogrids generally refer to the size and configuration of small to medium off-grid or grid-connected systems, which make use of renewable energy sources and can operate independently of grid-supplied power [9, 12–22, 31–34, 40].

Hybrid renewable energy (HRES) mini-grids have emerged as a game changer for rapid, cost-effective, and high-quality electrification in rural Africa. HRES mini-grids are easily installed, flexible, and can, in time, be connected to main power grids if and when such networks expand. The International Renewable Energy Agency (IRENA) estimates that mini-grids could supply 50% of the rural energy to the poor by 2030 [52]. Mini-grids involve small-scale electricity generation which serves a limited number of consumers via a distribution grid that can operate in isolation from the national transmission networks. Mini-grid systems are a viable solution for rural electrification, whenever the costs of mini-grid deployment are lower than for the extension of the national grid or stand-alone systems.

In addition to rural electrification, renewable energy (RE) mini- and microgrids are being considered for high- and medium-density urban applications, allowing community generation and storage of electricity. A recent proposal for SolarCity in the Western Cape is based on the microgrid concept. In total, using housing blocks as a basis for the microgrids, SolarCity's overall electricity network will be composed of 530 individual microgrids, which are combined to create a 10.78 MWp network. Each block's microgrid will include rooftop PV modules, inverters, smart electricity meters, and vanadium redox flow batteries for energy storage.

The increased complexity of microgrids in non-rural operations demands higher monitoring and control of various components, both in "off grid" operation and during synchronization with the main grid. By providing components with digital sensors and sophisticated controls, operations can be monitored and optimized in order to improve performance and enhance the quality of power supply. The microgrid concept has also been extended to industrial applications, although these are mainly grid-connected and make use of conventional generation as well as renewables.

Generally, mini-grids, nanogrids, and stand-alone systems are bottom-up approaches, independent of utility grids and largely used for rural electrification and for smaller operations like home or community energy systems. Hybrid microgrids are relegated to harness distributed energy sources both in urban and rural environments. What distinguishes the modern microgrid from older off-grid RE applications is the use of control systems or microgrid controllers. A common feature of RE or micro-sources-based distributed generation systems is the use of electronic interfaces which convert the energy from these sources to the grid voltage and frequency. These interfaces also provide enhanced flexibility for operation and energy management. In order to better organize these systems, the concept of a microgrid has been developed, which has a higher capacity and more control flexibility compared to single systems. A microgrid can operate in both grid-connected and stand-alone operation modes and benefit both the utility and customers with better reliability and power quality (see Chapter 5). The fundamentals of hybrid microgrids and the role of hybrid energy storage on microgrids are illustrated in detail in Chapters 4 and 5.

Regardless of what name these grid types go by, each has an important place in our energy future. And when used jointly as part of a broad, interconnected energy system, we all reap the benefits. For example, to optimize its distributed energy resources (DER) and improve power reliability for more than 10 million customers, Oncor, the largest public utility in Texas, invested in one of the most advanced microgrid solutions in the country. Today, the utility provides greater stability to an overtaxed grid, reduces outages, and delivers lower costs for its customers.

In the case of Montgomery County, Maryland, the ability to island from the grid during emergencies was a driving factor in its microgrid implementation. Following a devastating storm that left a quarter of the population and 71 county facilities without power for several days, county leaders sought the resilience of microgrids, but they got so much more. With the installation of two advanced microgrids, Montgomery County and its residents benefit from:

1. Increased resilience, upgraded aging electrical infrastructure, and enhanced sustainability
2. Creation of enough electricity to power 1,000 homes

3. Significant reduction in greenhouse gas emissions by 6,800 metric tons each year
4. Reliable power during unexpected outages, plus the ability to quickly "bounce back" to ensure personal safety, avoid costly damage, and minimize financial loss

Not only can mini-grid solutions enable better education, stronger commerce, and healthier living in rural communities; they play a critical role in preparing dispersed communities for a future energy ecosystem. Using decentralized energy technologies that will electrify rural populations who do not have access to electricity, we can improve agricultural activities through irrigation or crop conservation systems and change their overall economic conditions.

As the Navigant Research deployment tracker shows, microgrid deployment continues to rise in markets around the world contributing to a more decentralized energy distribution model. While mature energy economies look to modernize their infrastructure and provide more resilient energy, emerging economies are looking for access to reliable energy. In both cases, the use of DER and microgrid or mini-grid, nanogrid or stand-alone system technologies can increase the availability of abundant, reliable energy to unlock the full potential of our businesses and our communities. While hybrid energy sources, storage, and grid transport will make the future more complex, but on the whole, they are more sustainable and accessible to the entire world population.

1.7 DESIGN AND IMPLEMENTATION ISSUES FOR HYBRID RENEWABLE ENERGY SYSTEM (HRES)

There are two types of renewable energy sources; dispatchable like biomass, geothermal, and hydro and non-dispatchable like solar and wind. Unlike dispatchable renewable energy sources, non-dispatchable sources are available in most parts of the world and they are often preferred. To overcome the intermittency and uncertainty of non-dispatchable renewable sources and to provide an economic, reliable, and sustained supply of electricity, a modified configuration that integrates these renewable energy sources and uses them in a hybrid system mode is proposed by many researchers. Hybridization of different alternative energy sources can complement each other to some extent and achieve higher total energy efficiency than that could be obtained from a single renewable source. Multisource hybrid renewable energy systems, with proper control, have great potential to provide higher quality and more reliable power to customers than a system based on a single source. Hybrid Renewable Energy Systems (HRES) work in stand-alone or grid-connected mode [41]. Due to this feature, hybrid energy systems have caught worldwide research attention.

The applications of hybrid energy systems in remote and isolated areas are more relevant than grid-connected systems. Hybrid energy systems are now becoming an integral part of the energy planning process to supply previously unelectrified remote areas (see Chapter 6). In addition, the application of hybrid systems is becoming popular in distributed generation or microgrids, which recently have great concern. Due to advances in renewable energy technology which have improved their efficiency and reduced the cost and the advances in power electronic converters and automatic controllers which improve the operation of hybrid energy systems and reduce maintenance requirements, these advances have made hybrid systems practical and economical. Various hybrid energy systems have been installed [9, 12–22, 31–34, 40] in many countries over the last decade, resulting in the development of systems that can compete with conventional, fuel-based remote area power supplies in many applications.

Buildings, both residential and commercial, are a major source of environment pollution (about 30%–40% of CO_2 pollution is connected to the buildings). In recent years, a significant push is made toward zero energy buildings and other home energy systems using renewable sources. This push has also made residents of buildings not only energy consumers but also prosumers. This will create, as shown in Figures 1.2 and 1.3, enormous growth in both home and building energy management systems. The use of PV-battery and other hybrid renewable energy system is on meteoric rise.

The design process of hybrid energy systems requires the selection and sizing of the most suitable combination of energy sources, power conditioning devices, and energy storage system, together with the implementation of an efficient energy dispatch strategy [6]. The selection of the suitable combination from renewable technology to form a hybrid energy system depends on the availability of the renewable resources in the site where the hybrid system is intended to be installed. In addition to the availability of renewable sources, other factors may be taken into account for proper hybrid system design, depending on the load requirements such as, reliability, greenhouse gas emissions during the expected life cycle of the system, efficiency of energy conversion, land requirements, economic aspects, and

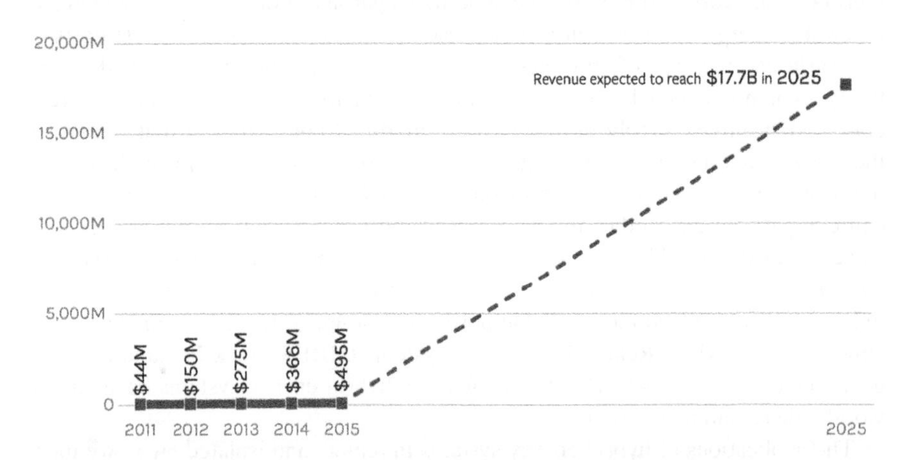

FIGURE 1.2 Growth in home energy management system revenue. Source: Advanced Energy Now, *2017 Market Report*, prepared by Navigant Research, 2017 [41].

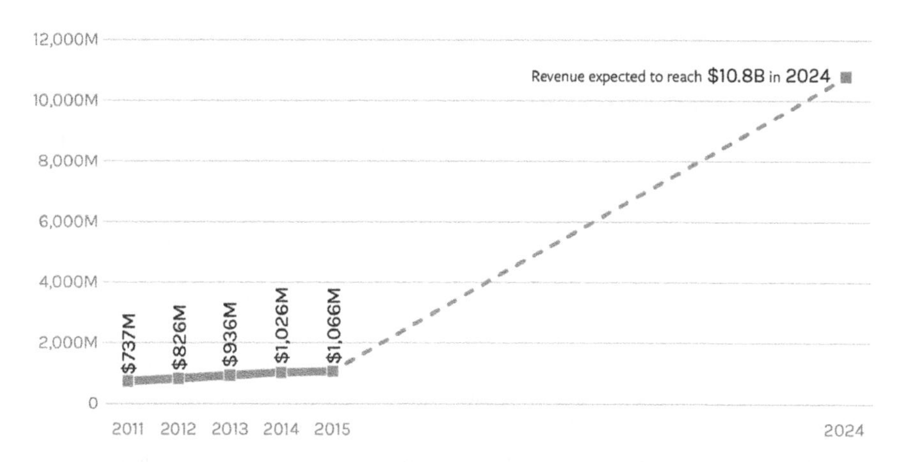

FIGURE 1.3 Growth in building energy management system revenue. Source: Advanced Energy Now, *2017 Market Report*, prepared by Navigant Research, 2017 [41].

social impacts [9, 12–22, 31–34, 40]. The unit sizing and optimization of a hybrid power system play an important role in deciding the reliability and economy of the system.

As mentioned above, integrating renewables into the electricity grid has obvious environmental benefits. But upgrading a vast infrastructure made up of thousands of individual utility companies and a web of high-voltage transmission lines is complicated. The grid was built under a one-way model, in which power is generated at centralized sites and sent through distribution lines to end users. Renewable energy generation facilities—particularly solar and wind—are more widely distributed. Solar and wind farms can feed into the main energy grid, and small-scale installations can be plugged directly into homes. This means that the power coming into the grid from such sources on very sunny or very windy days represents a supply of excess energy that can be redirected through the grid to where it's needed the most. In addition to demand-side management (DMS) tools like variable pricing for electricity use, new forms of smart grid information technology are being implemented by utility companies to balance out this two-way flow of energy.

Improving the grid to accommodate more renewable energy also means coordinating among utilities in neighboring areas to balance demand and supply and creating better forecasts of solar and wind output to anticipate generation levels over the course of days. Increasing the amount of renewable energy in the grid will also require a lot of batteries. Renewable energy sources sometimes generate more electricity than is needed in real time. To maintain a utility's capacity to provide precisely the right amount of electricity to its customers, large-scale lithium-ion batteries are now being added onto renewable power sites to store that power and distribute it as needed. These large batteries can hold multiple megawatt hours (MWh) of electricity; the average American home uses about 10 MWh per year. Getting more renewable energy into the mix of sources powering the US grid will hinge on equipping more utilities and energy producers with such batteries. A key priority moving forward will be to better incentivize the development of energy storage (homogeneous or hybrid).

The National Renewable Energy Laboratory (NREL) is exploring how to better integrate the three main grid regions that make up the US energy system. These three systems operate almost entirely independently of one another, with little ability for one to send excess electricity to help meet the demands of another. As part of its Interconnections Seam Study, NREL is working with national labs, universities, and industries to develop new ways of sharing a diverse pool of energy sources across these three systems. If successful, this effort could make it easier for abundant wind energy from Texas to power systems in states like Tennessee and Maine, or solar energy from the sunny Southwest to power homes in the cloudy Northwest.

Many believe that the need to transition the grid to better accommodate renewable energy sources will continue for the foreseeable future. Any change to the resource mix requires careful planning but we know that a high-penetration renewable grid is feasible through improved transmission and expanded deployment of energy storage and other advanced technologies. Currently RES accounts for at least 19.5% of the global electricity. We must integrate these increasing shares of renewables into our T&D (transmission and distribution) systems as soon as possible. Renewables are already saving time and cost for utilities. For example, in 2013, Idaho Power Company (IPC) received a $94 million grant from the US Department of Energy to modernize the grid, including the development of renewable energy integration tools. As part of the grant, IPC was able to improve its forecasting, using 15% of natural gas-fired reserves instead of 100%. This alone resulted in savings of approximately $50,000 for the utility and its customers, and IPC is seeing similar results on a consistent basis [41].

Another challenge is the injection of reactive power into the electric grid from outside power sources, such as solar energy. The utility controls the voltage levels of its system and injects energy into the grid when necessary to smooth out the natural swings in usage and keep the voltage at an acceptable level. But by renewable energies being injected into the grid, it can throw off a utility's synchronous generator, making it so that they can't track where the power is coming from. In order to incorporate renewable energies into their existing infrastructures, companies must address several primary issues that can make the system unstable:

Voltage management: It is especially important to control the voltage, and companies are looking to options such as Secondary and Tertiary Voltage Regulators; reactive power compensators such as the STATCOM (Static Synchronous Compensator); battery systems for storing energy reserves; and pole transformers for remote tap control.

Frequency control: A power system often has inconsistent frequencies when adding renewable energy. Companies are searching for solutions that can adjust frequency variations within a specified range by using a faster regulated response.

Controlling output fluctuations: Absorbing excess energy in cases of excessive output fluctuations will help companies maintain a smooth power curve.

Managing electric vehicle charging: This is an increasingly complex problem as electric vehicles become more widely adopted and rapid charging can cause a sudden increase in load to the system. Batteries can be used to store excess power, minimizing the negative effects of rapid charging on the grid.

Demand response *(DR)*: This is one of the more commonly known solutions for managing energy and is increasingly being applied to renewables as well. DR allows for the system to automatically request that users suppress power consumption during peak hours and can automatically shift any surplus power load to a later time.

All of these issues are currently being addressed as the need for HRES increases.

1.8 HYBRID ENERGY STORAGE

Energy storage encompasses a wide range of technologies and resource capabilities, and these differ in terms of cycle life, system life, efficiency, size, and other characteristics. Although battery technology has attracted a great deal of industry attention in recent years, pumped hydro technology still supplies the vast majority of grid-connected energy storage (>95%, see Table 1.6). The remaining categories combined comprise only 5% of installed capacity. Energy storage technologies can be broken down in five parts: electrochemical (like batteries, capacitors, and super capacitors), mechanical (like compressed air energy storage (CAES), adiabatic CAES, flywheels, and superconducting magnetic energy storage (SMES)), pumped hydro (like hydropower both large and small scale, gravel drain, and bulk gravitational), thermal storage (like molten salt, ice brick, heat battery, thermochemical storage, and phase-change material (PCM) storage), and hydrogen or synthetic natural gas storage. As shown in Table 1.7, several of these storage technologies are in demonstration or development stages [4, 6, 48–51].

More details on various energy storage technologies are outlined in Chapter 4 and in my previous book [6]. While there has been significant progress in various energy storage technologies, they are not ready for large-scale and widespread deployment.

TABLE 1.6
Energy Storage Distribution among Different Methods of Storage [4, 6]

Pumped hydro	95%	23.4 GW
Others (Remaining 5% is distributed as)		1.2 GW
Thermal storage	36%	431 MW
Compressed air	35%	423 MW
Battery	26%	304 MW
Flywheel	3%	40 MW

TABLE 1.7
Maturity of Electricity Storage Technologies [4, 6]

Deployed	Demonstration	Some Early Stage Technologies
Pumped hydro	Advanced Pb-acid and flow batteries	Adiabatic CAES
Compressed air energy storage (CAES)	Superconducting magnetic energy storage (SMES)	Hydrogen
Batteries (NaS, Li-ion, Pb-acid)	Electrochemical capacitors	Synthetic natural gas
Flywheels		
Thermal		

The main reason is that in spite of the large variety of energy storage system (ESS) technologies, no technology offers sufficient performance in respect of key figures of merit needed of an electrical energy storage medium. For instance, a high-performance ESS should exhibit high cycle efficiency, high power and energy storage capacity, low cost, high volumetric and/or gravimetric density, and long cycle life. The ESS technology of choice for many applications (especially those requiring high volumetric and/or gravimetric density) is battery storage. As shown in Figure 1.4, different energy storage technologies have different levels of energy stored with different power densities, and they all cannot be used in different applications. Even for various batteries, as shown in Table 1.8, different types of batteries have different characteristics. Again no single battery can simultaneously achieve all the desired characteristics of a high-performance ESS. Furthermore, no battery technology is in sight that can achieve these characteristics. So the focus is on finding ways to build ESS that comprise of different battery types so as to hide the weaknesses of each battery type, yet presenting the strongest features of each battery type.

Despite numerous research efforts in order to improve ESS capabilities over the past decade [48], a perfect ESS technology that copes the drawbacks in terms of all aspects is not to be expected to be developed in the near future. On the other hand, particular ESS applications require a combination of energy and power rating, charge and discharge time, life cycle, and other specifications that cannot be met by a single ESS technology. In order to increase the range of advantages that a single ESS technology can offer and at the same time enhance its capabilities without fundamental development of the storage mechanism and only via complementary use of the existing ESS technologies, more than one ESS technologies can be hybridized [4, 6, 48–51]. A hybrid energy storage system (HESS) is composed of two or more heterogeneous ESS technologies with matching characteristics and combines the power outputs of them in order to take advantage of each individual technology and at the same time hide their drawbacks. This is analogous to hybrid power generation outlined above where no single source is completely suitable in all situations.

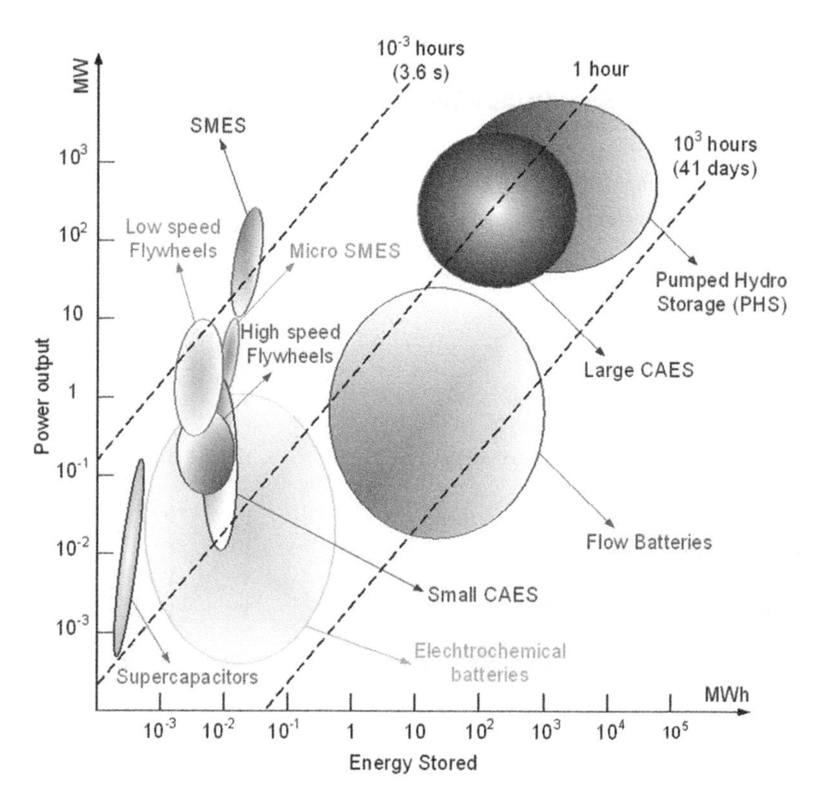

FIGURE 1.4 Fields of application of the different storage techniques according to stored energy and power output [51].

1.8.1 Need for Hybrid Energy Storage

A number of elements dictate the effectiveness of energy storage device. They range from power and energy density to output power rating and from self-leakage rate to cost per unit of stored energy, and from the life cycle of the storage element to the efficiency of the charge/discharge cycle. A brief comparison of power vs. energy density of few storage devices are illustrated in Figure 1.5. As can be seen, different devices have different power and energy density characteristics. While individual storage system satisfies some of these elements, just like sources for power generation, each storage device possess some drawbacks [27, 42, 48–51].

This motivates the need for a hybrid energy storage system comprised of heterogeneous types of energy storage elements organized in a hierarchical manner so as to hide the weaknesses of each storage element while eliciting their strengths. The hybrid storage system also offers solutions to challenges that one faces when dealing with the optimal design and runtime management of a energy storage system targeting some specific application scenario, for example, grid-scale energy management, household peak power shaving, mobile platform power saving, and more.

TABLE 1.8
Characteristics of Various Battery Technologies [1–6]

Application	Lead-Acid	NaS	Li-ion	Flow Batteries
Off-to-on peak intermittent shifting and firming	Y	Y	Y	Y
On-peak intermittent energy smoothing and shaping	Y	Y	Y	Y
Ancillary service provision	P	P	P	P
Black start provision	Y	Y	Y	Y
Transmission infrastructure	Y	Y	Y	Y
Distribution infrastructure	Y	Y	Y	Y
Transportable distribution-level outage mitigation	P	Y	Y	Y
Peak load shifting downstream of distribution system	Y	Y	Y	Y
Intermittent distributed generation integration	Y	P	P	P
End-user time-of-use rate optimization	P	P	P	P
Uninterruptible power supply	Y	Y	Y	Y
Microgrid formation	Y	Y	Y	Y

Note: Y: Definite suitability for application; P: Possible use for application; N: Unsuitable for application.

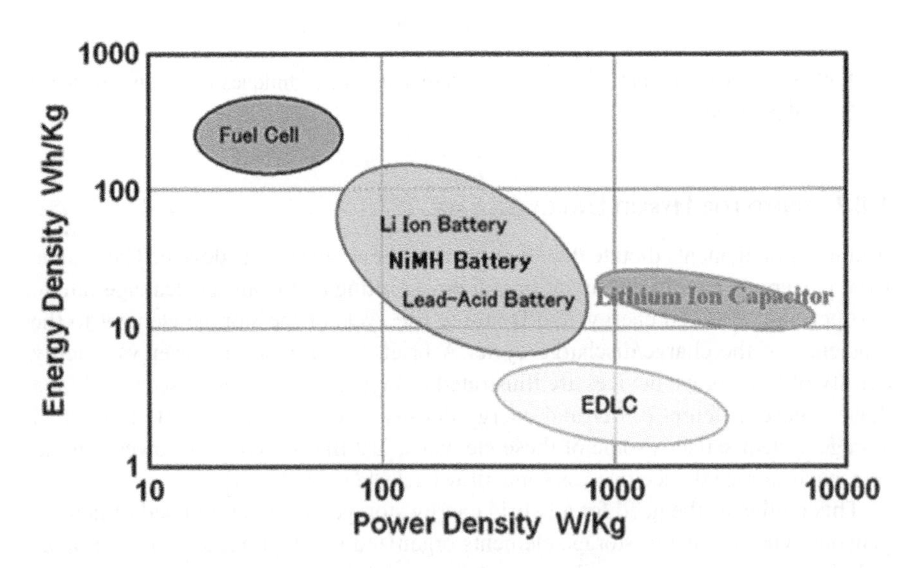

FIGURE 1.5 Energy and power density (Ragone plot) of different battery and other storage technologies [42].

Matching renewable generation intermittency to demand in an electricity supply system was the reintroduction of the ESS technologies in the power systems. Besides storing and smoothing renewable power, there are numerous advantages related to the advent of ESSs in the power systems. ESSs can increase power system operation and planning resiliency and efficiency by means of many applications including [26, 42, 48, 51]: energy time shift, supply capacity, load following, area regulation, fast regulation, supply spinning reserve, voltage support, transmission congestion relief, transmission and distribution upgrade deferral, power quality, renewable energy time shift, and others as described in Chapter 4.

In addition, rapid growth of the electric and hybrid vehicle technology opens a new era on ESS [49505127, 42, 48–51]. However, ESS solutions are not quite commercially or technologically mature in many features, thereby causing obstacles to their extensive utilization. The ESS technologies are different in terms of cost and technical properties such as [27, 42, 48–51] energy and power rating, volumetric and gravimetric energy density, volumetric and gravimetric power density discharge time, response time, operating temperature, self-discharge rate, round-trip efficiency, life time (years and cycles), investment power and energy cost, spatial requirement, environmental impact, and many others as described in Chapter 4 and in my previous book [6].

HESS (hybrid energy storage system) can refer to several different types of set up; the point in common is that two or more types of energy storage are combined to form a single system. As shown by Table 1.6, there is no single *energy storage* solution that is ideal for every grid-scale application. As explained by Greentech Media, they

> are typically designed for high-power applications (i.e., "sprinter" mode that provides lots of power in short bursts) or energy-dense applications (i.e., "marathon" mode that provides consistent lower power over long durations), and there are lifetime, performance, and cost penalties for using them in unintended ways.

HESS typically combines both "sprinter" and "marathon" storage solutions to fulfill applications that have diametrically opposed requirements, e.g., fast response vs. peak shaving. The potential for value stacking immediately jumps to mind. Hybrid storage offers other avenues for cost reductions; two or more systems can share much of the same power electronics and grid-connection hardware, reducing both upfront and maintenance costs.

Many combinations of storage technologies in hybrid storage systems are possible. These are described in detail in Chapter 4. Hybrid storage system is thus an important element of hybrid energy system.

1.9 HYBRID GRID TRANSPORT

It is important to realize that increased digitalization, the use of information technology, and power electronics are key enablers for the transformational change the electric power system encounters. Due to technological and societal developments, we can observe an acceleration of the changes in the interconnected electricity systems that are the largest man-made technical constructs on earth. Experts unanimously agree that the importance of electricity as "the fuel of choice" will increase by 60% over the next 35 years and it will be greener, more affordable, and more accessible than ever.

Figure 1.6 illustrates the transformation from traditional grid to the modern grid. In both cases, generated power is transported by high voltage transmission line to the medium voltage distribution system and finally to homes, offices and factories by low voltage delivery system where electricity is used. Unlike in the traditional grid, however, in the modern grid power is generated by multiple sources which include renewable energy sources and electricity flows bidirectionally.

The four key aspects or technology ingredients that shape the electricity system of the future and are commonly associated with hybrid grids are:

1. ***Mix of generation and storage***: By what technological means is the grid accommodating for the ways electricity is produced? This generation can be conventional (will probably be around for some time) but increasingly comes from variable renewables like wind and solar in all sizes, ranging from big offshore wind farms, to large solar fields, down to rooftop solar panels. Utility grid can also have hybrid power generation by converting waste heat into power by thermoelectricity or converting CO_2 into power by fuel cell technology. For renewable sources, energy storage would be very important for uninterruptable and high-level power quality. As mentioned above, mix energy storage devices will, in general, be required. So called utility grids will definitely be hybrid going forward.
2. ***Mix of transmission and distribution***: By what technological means is the electricity transported to the point where it is used? Here we can think of overhead lines and underground cables but also local generation with microturbines with or without storage, mobile sources like batteries from

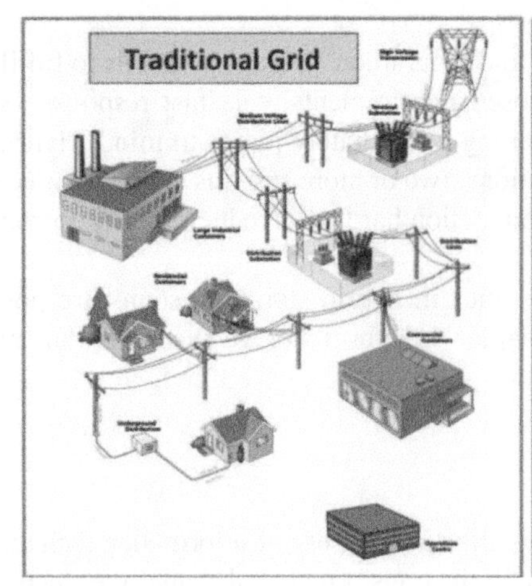

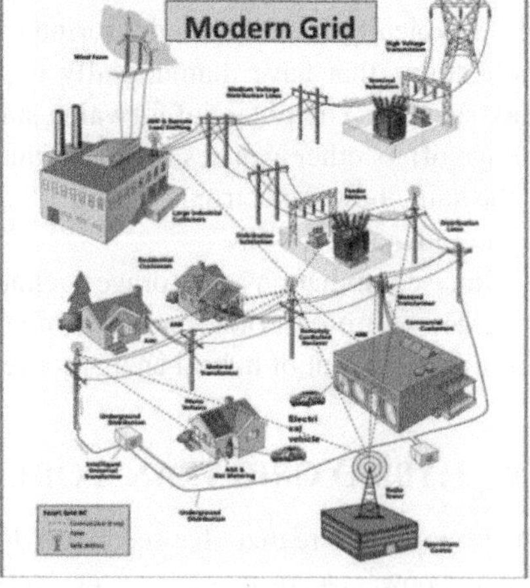

FIGURE 1.6 Grid transformation introduces complex resources [41].

electric vehicles, and even wireless for low energy and/or short distances. The transmission will occur at medium- and low-voltage levels for distributed energy resources through hybrid microgrids which can be connected to utility grid or can operate in islanded mode. The energy transport will also be managed in off-grid mode where small independent mini-grids, nanogrids, or stand-alone systems will provide electricity to rural and remote communities. Thus, there will be three levels of power transmission and distributions.

3. ***Mix of delivery***: In what technological "form" or characterization is the electricity to be delivered during intermediate stages or ultimately to the point of use? Is it direct current, alternating current (frequencies 16.7, 50, 60, and 400 cycles/second are presently widely used), high frequency, or even micro waves.
4. ***Mix of control***: How is a safe, stable, and reliable electricity system operation at all levels guaranteed? The protection and control is governed by the increasing digitalization, combining prediction of electricity generation and load, and sensoring/monitoring at global/regional and local level together with an optimal combination of slow (mechanical) and fast (power electronic) actuators.

By definition a *hybrid grid* is characterized with those four key aspects with three levels of grid operations. *A hybrid grid accommodates changing mixes of generation and storage, transmission and distribution, delivery form, and control technologies.* Due to technological and societal developments, the electric power system as we know it today changes into a flexible (future proof), reliable, sustainable, and affordable hybrid grid that encompasses the four key aspects, and it is our challenge to manage this transformation. As shown in Figure 1.6, new grid will not look like old grid and this transformation will introduce complex resources. The research, development, and demonstration needs required for this modernization, as outlined by US Department of Energy, is illustrated in Table 1.9.

1.9.1 Components of Hybrid Grid System

There is a lot of confusion regarding the nomenclature of macrogrid, microgrid, mini-grid, nanogrid, and stand-alone electrical systems. In its Q4 2018 Microgrid Deployment Tracker, Navigant Research reported 2,258 microgrid projects, representing nearly 20 GW of capacity across seven geographies. Interestingly, Navigant includes both grid-interactive microgrids and remote microgrids or mini-grids in its tracker. However, these two grid types are quite distinct and are deployed to meet very different energy needs. To increase the development of reliable, resilient energy, we must understand the grid varieties available to address local energy needs. When thinking about the decentralization of energy, many define grid types based on their size, but that can be misleading. The definition of mini-, micro, and nanogrids vary from source to source and often overlaps. The general definitions are:

TABLE 1.9
Moving from Traditional to Modern Electric Power Systems—RD&D Needs [41]

Electric Systems	Characteristics: Traditional	Characteristics: Modern	RD&D Needs
Generation	• Centralized • Dispatchable • Large thermal plants • Mechanically coupled	• Centralized and distributed • More stochastic • Efficient and flexible units • Electronically coupled	• Planning tools • Energy storage • Control coordination • Flexible thermal generators
Transmission	• supervisory control and data acquisition (SCADA) for status visibility (sampling, not high definition) • Operator-based controls (primarily load following and balancing) • Destabilizing effects • Congestion, despite underutilized capacity (limited flow control) • Threats/vulnerabilities not well defined	• High fidelity, time synchronized measurements • Breadth and depth in visibility • Automatic control • Switchable network relieves capacity constraints • Threats are considered and risks are appropriately managed	• Multi-terminal high-voltage direct current (HVDC) • Low-cost power flow controller technologies • Next-generation energy management systems (EMS) • Integrated planning tools • Security • Low-cost bulk storage
Distribution	• Limited visibility • Limited controllability • Radial design (one-way flow), floating on transmission • Increasing fault currents and voltage issues stressing system • Aging assets (unknown effects)	• Enhanced observability • Local, autonomous coordination • Network design and two-way flow • Backbone of delivery system • Self-healing • Active monitoring of asset conditions	• Security • Microgrids • Advanced distribution management systems (DMS) • Distribution and asset sensors • Solid-state transformer • Smart voltage regulation equipment • Community storage
Customers	• Uniformly high reliability, but insensitive to upstream issues • Energy consumers (kWh) • Predictable behavior based on historical needs and weather • Interconnection without integration • Growing intolerance to sustained outages	• Customer-determined reliability/PQ • Prosumers (integrated) • Variable behavior and technology adoption patterns • Plug/play functionality • Kept informed during outages (and before) • Hybrid AC/DC distribution • Data access (outage/usage)	• Single-customer microgrids • Building EMS • DER integration • Security • Transactive controls • Behind-the-meter storage • Low-cost sensors

A. **Regional or central macrogrid**

In a modern energy economy such as North America, Europe, or China, the central or regional grid acts as a manager of energy for a large population. It manages electricity supply and voltage to ensure reliable energy generation is provided to all tenants of the grid infrastructure. The power generation generally occurs centrally and it is transmitted by high-voltage lines and distributed at mid- and low-voltage distribution lines to the customers. Central macrogrid is prevalent in all major urban environment. It is often called utility grid or smart grid. In recent years, due to infusion of renewable sources, macrogrid is becoming more and more hybrid. There are also other reasons for the development of hybrid character of macrogrids. These are all described in detail in Chapter 2.

B. **Microgrid**

Unlike a completely off-grid model, a microgrid provides an interactive and functional relationship between the central grid and its users. This is an important distinction. Much like how microeconomics is a scale and has a behavioral relationship with macroeconomics, this interactive relationship allows a microgrid to be connected to and use the services of the central grid and can support services to the grid when it's beneficial to do so. A microgrid can also island from the grid and operate as a mini-grid would, maximizing the benefits to both the central grid and end users. Microgrids can be deployed in a variety of sizes and locations from a single building to an entire municipality. Microgrids are similar to mini-grids but operate at a smaller size and generation capacity (1–50 kW). Microgrids serve a concentration of consumers and also use distribution lines. In addition to rural, urban, and industrial sector, microgrids are appearing on campuses, such as universities, hospitals, military establishments, and business parks in urban environment. The major purpose of microgrid is to harness distributed and renewable energy sources at medium- and low-voltage environment which can be directly used by the customers. Because they carry largely renewable energy sources, they generally tend to be hybrid. Fundamentals of hybrid microgrids and their management and control systems are described in detail in Chapter 5.

B-1 Rural systems

The rural microgrid (RMG) usually has a single generation unit or multiple co-located units, where alternative sources of supply such as prime movers are included. The MG usually relies on renewable energy generation for its prime source of energy, and more recent systems include storage. The RMG is traditionally used to supply small villages with electricity, and the grid is bounded by the village or settlement boundaries. Solar PV is usually used as the primary source of energy, but the development of micro and pico-hydro systems has led to the increasing use of hydropower as a primary source. Generation units are usually located separately from consumers.

The RMG concept has also been applied to small remote towns, such as in Namibia, where solar PV has been installed to reduce the usage of the prime generator, diesel power. Although this case qualifies as a remote urban mini- or microgrid, the grid was not originally designed as such. The RMG concept is being increasingly applied to rural agriculture and small industrial establishments, where the availability of electricity has made it possible to mechanize previous manual operations.

B-2 Urban/industrial/commercial microgrids (UMG)

The urban MG differs from the rural MG in that multiple sources of energy at different locations are incorporated into the grid. Bulk storage of electricity at a single location as well as prime mover diesel-powered generation is incorporated, and the grid may include cogeneration or conventional sources such as biogas or biomass. The energy sources may also be co-located at a consumer's premises or at a separate location such as a community solar, wind farm, or biogas plant.

C. **Off-grid Systems**

Access to macrogrid is not available in all parts of the world, particularly in rural and remote environments. There are about 1.16 billion people in this world without any excess to electricity, mainly in rural and remote communities of developing nations. In order for them to access electricity, off-grid systems which include mini-grids, nanogrids, picogrids, or stand-alone systems of electricity are developed. These are bottom-up approaches to the development of electricity distribution system. While these systems can be developed in urban environment and be connected to macro or microgrids, their major purpose is to provide electricity to communities with no access to conventional grids. This is the third level of electricity distribution system, and it is described in detail in Chapter 6.

C-1 Mini-grid

By contrast to microgrid, a mini-grid is often characterized by its use in remote locations where there is no central grid available. It is also often described as off-grid since it does not require a connection to the larger macrogrid. In emerging energy economies such as Africa and India, rural communities have found success using mini-grids that can operate autonomously or when connected to a localized distribution network. Using DER such as solar or wind, or more likely diesel generators, these mini-grids function exactly like a macrogrid, simply on a smaller scale. Not only can mini-grid solutions enable better education, stronger commerce, and healthier living in rural communities, they play a critical role in preparing dispersed communities for a future energy ecosystem. One definition of a mini-grid put forward by the World Bank: *Isolated, small-scale distribution networks typically operating below 11 kilovolts (kV) that provide power to a localized group of customers and produce electricity from small generators, potentially coupled with energy storage system* [52, 55].

There are three main approaches to conduct rural electrification in a competitive and effective way: mini-grids, nanogrids, and stand-alone systems [11, 52–55]. All of these systems operate independently of the national electricity grid and are thus known as "off-grid systems." A mini-grid, also sometimes referred to as an isolated grid, is an off-grid system that involves small-scale electricity generation and which serves a limited number of consumers via a distribution grid that can operate in isolation from national electricity transmission networks (Mini-Grid Policy Toolkit, 2014 [52, 55]). Mini-grids can supply electricity to concentrated settlements, including domestic, business, and institutional customers, with power at or above grid quality level.

Clean energy mini-grids (CEMGs) utilize one or several renewable energies (solar, hydro, wind, biomass) to produce electricity. Backup power can be supplied by electricity stored in, for example, batteries or otherwise by diesel. Storage provides or absorbs power to balance supply and demand and to counteract the moment-to-moment fluctuations in customer loads and unpredictable fluctuations in generation. All mini-grids are hybrid in nature.

A mini-grid used for HRES refers to a low-voltage network involving one or more small-scale electricity generation unit(s) or micro-sources (MS), such as photovoltaic (PV), fuel cells, microturbines, small wind turbines (WT), and storage devices (flywheels, super capacitors, batteries), and the distribution of electricity to a limited number of customers via a distribution grid that can operate in isolation from national electricity transmission networks and supply groups of consumers with electricity at grid quality level. The consumer group may be a rural settlement, a residential estate, a commercial or industrial park, a university campus, a medical center, or any other consumers grouped together for a common purpose. The generating capacity of the mini-grid varies with definitions but a range of between 50 kW and 1 MW seems to be reasonable. Some definitions go as high as 10 MW. The main distinguishing characteristic is that the grid derives power from its own power sources and serves a concentrated number of consumers. The mini-grid uses distribution lines.

C-2 Nanogrids

The nanogrid (NG) is a relatively new concept and applies to a grid serving a single customer or building. Like mini-grids, they are not connected to macrogrids, and like mini-grids, they are considered as off-grids. The NG generally has a single generation unit and does not use transmission or distribution lines and may make use of DC reticulation. As with microgrids, there are two focuses in the development of nanogrids: the urban/commercial sector and rural communities. Navigant Research came up with a definition of a nanogrid, a term originated by Lawrence Berkeley National Laboratory.

> *A small electrical domain connected to the grid of no greater than 100 kW and limited to a single building structure or primary load or a network of off-grid loads not exceeding 5 kW, both categories representing devices (such as DG, batteries, EVs [electric vehicles], and smart loads) capable of islanding and/or energy self-sufficiency through some level of intelligent DER management or controls.*

Either nanogrid or mini-grid can be connected to microgrid or operated in island mode. Off-grid operations generally do not follow any definition of grid operation. Thus hybrid grid transport involves utility grid, hybrid microgrid, hybrid mini-grid, nanogrid, or off-grid. These grids can be connected or operated in island modes.

A nanogrid is different from a microgrid in its control and delivery mechanisms. Although some microgrids can be developed for single buildings, they mostly interface with the utility. Some aren't even fully islandable. A nanogrid, however, would be "indifferent to whether a utility grid is present." Rather, it would be a mostly autonomous DC-based system that would digitally connect individual devices to one another, as well as for power generation and storage within the building. A nanogrid is a single domain of power—for voltage, capacity, reliability, administration, and price. Nanogrids include storage internally; local generation operates as a special type of nanogrid. A building-scale microgrid can be as simple as a network of nanogrids, without any central entity [43–47].

The nanogrid is conceptually similar to an automobile or aircraft, which both house their own isolated grid networks powered by batteries that can support electronics, lighting, and internet communications. Uninterruptible power supplies also perform a similar function in buildings during grid disturbances. Essentially, it would allow most devices to plug into power sockets and connect to the nanogrid, which could balance supply with demand from those individual loads. It presumes digital communication among entities, embraces DC power, and is only intended for use within (or between) buildings. Building-scale microgrids are built on a foundation of nanogrids and pervasive communication. The system capacity could be anywhere from a few kilowatts to hundreds of kilowatts [43–47].

According to Nordman [44], there are lots of potential benefits to structuring local DC power distribution in this way. Conversion losses would be cut, investments in inverters and breakers would be reduced, and device-level controls would enable a much more nimble way to match generation or storage capabilities with demand. The building would also theoretically be immune to problems that are more likely to be encountered with a local microgrid or the broader centralized grid. However, obstacles currently outweigh those benefits by a significant degree. Nordman thinks nanogrids should be universal, meaning they would operate on the exact same communications and voltage standards. But because nanogrids are still mostly conceptual, no organization has attempted to create those standards. Nanogrids can't really scale without them. There's also another structural issue to deal with on the grid. Theoretically, these localized, autonomous systems could be scaled without much interference with the utility. But Nordman [44] also envisions nanogrids being tied to larger microgrids, which are ultimately tied to the central grid. That inevitably brings utilities into the picture, and they likely wouldn't have much incentive to support a system designed to drastically cut their electricity sales. That's why growth in nanogrids will likely occur in developing countries with weak grid services where it makes sense to power buildings in isolation. The concept is compelling. It's also so completely different from the status quo that it currently has little chance of scaling in this grid-centric world.

A building is often only as intelligent as the electrical distribution network it connects with. That's why smart buildings are often seen as an extension of the smart grid. Meters, building controls, intelligent lighting and HVAC systems, distributed energy systems, and the software layered on top are indeed valuable for controlling localized energy use within a building. But in many cases, the building relies on the utility or regional electricity grid to value those services. Some analysts consider these technologies as the "enterprise smart grid" because of their interaction with the electricity network. Nanogrid can help when there is no supporting centralized grid [43–47]. Arrival of the concept of smart grids and electricity market introduces the demand-side management (DMS) which leads to controllable loads. Furthermore, by the rapid penetration of RESs, such as PV and wind generation systems, operating as distributed generation systems (DGSs), many issues about utilization of these units have emerged. Smart homes, with the presence of these units and issues about DMS, have gathered so much attention to managing the energy from multiple sources, loads, and the utility grid. The paper by Hagh and Aghdam [47] introduces a multi-port converter to be used in smart homes and hybrid AC/DC nanogrids. Besides energy management among generation units and energy storage devices, this converter is capable of power management of appliances of the home using various communication networks. Furthermore a comparison between optimally tuned proportional integral derivative (PID) and fractional order proportional integral derivative (FOPID) controllers has been carried out. Genetic algorithm (GA) is used for optimizing the parameters of the controllers.

The business case for nanogrids echoes many of the same arguments used on behalf of microgrids. These smaller, modular, and flexible distribution networks are the antithesis of the bigger is better, economies of scale thinking that has guided energy resource planning over much of the past century. Nanogrids take the notion of a bottom-up energy paradigm to extreme heights. In some cases, nanogrids help articulate a business case that is even more radical than a microgrid; in other cases, nanogrids can peacefully coexist with the status quo. Many believe that the linking of batteries to distributed solar PV systems is a game changer. When combined with solar or another form of on-site generation, nanogrids can even help commercial buildings to become completely self-sustaining with regard to electricity [43–47]. *Nanogrids offer improved flexibility over the power grid's traditional limitations.* In addition to helping companies to avoid suits and fines, the reduced complications of integrating nanogrids into a facility's energy profile means that acquiring permissions and completing installation will take place over a far shorter time frame than that of a traditional microgrid [43–47]. There are two types of nanogrids:

1. Rural nanogrids are an extension of the rooftop solar concept to provide community as well as individual benefits. The technology is applicable to closely located dwelling units, which may have been provided with individual solar PV in the past. Typical rural examples would be rural schools or clinics, which have for many years now been provided with solar PV systems and storage. Most to date use AC inverters to drive AC loads. Nanogrid developments for rural networks use community storage and energy control centers (ECC) to provide power to community centers as well as individuals.
2. Commercial nanogrids are being developed to manage energy in single or multiple buildings, in an effort to reduce electrical consumption. Less extensive than a microgrid, a nanogrid can be as small as the energy management

system for an entire building. A nanogrid includes the generating source, in-house distribution, and energy storage functions and can be extended to multiple buildings.

Development is focused on AC nanogrids for today's technology and DC nanogrids for the sustainable homes and buildings of tomorrow. AC nanogrids incorporate smart appliances, lighting and HVAC with on-site power generation, and an ECC. The ECC acts as a data acquisition unit, collecting and recording the power flow data from the grid connection and the smart appliances in the building. If the architecture is changed from a single AC system and connected directly to DC renewable energy sources with storage elements to DC loads, power losses and cost can be significantly reduced. A DC nanogrid would start with fewer power converters, has a higher overall system efficiency, and would be easier to interface to renewable energy sources. There are no frequency stability and reactive power issues, and less conduction loss. Consumer electronics, electronic ballasts, LED lighting, and variable-speed motor drives can be more conveniently powered by DC.

Characteristic of the modern nanogrid is the use of DC distribution and DC appliances. This was limited in the past by the fact that most small PV systems operated on a nominal voltage of 12 V. The availability of panels operating at higher voltages (36 V and above) makes DC distribution more feasible. The other major development is LED lighting, which works off DC. In addition to lighting, radios, computers, and television sets, all work on DC. The higher voltage also makes DC refrigeration more feasible. A DC nanogrid with two DC voltage levels is envisaged: a high-voltage (380 V) DC bus powering HVAC, kitchen loads, and other major home appliances, and a number of low-voltage (48 V, 24 V, etc.) DC buses powering small tabletop appliances, computers and entertainment systems, and LED lighting. Nanogrids at individual level are often termed pico grid.

C-3 Stand-alone Systems

Stand-alone systems are small electricity systems, which are not connected to a central electricity distribution system and provide electricity to individual appliances, homes, or small productive uses such as a small business. They thus serve the needs of individual customers, while utilizing locally available renewable resources. Due to price drop, stand-alone off-grid systems powered by biomass, small wind, small hydro, and small solar power are becoming more and more common [13, 23–26, 35, 37, 39, 45].

To extend the time of use, energy storage systems have become more popular. Storage is typically implemented as a battery bank. Power drawn directly from the battery is often extra low voltage (DC), and this is used especially for lighting as well as for DC appliances. An inverter is used to generate AC low voltage which powers standard appliances [54]. Stand-alone systems can be differentiated into pico, home, and productive systems. Pico systems are used to power individual appliances like lights, TV, radio, etc. Home systems are used to power individual households. Productive systems are used to power a small business, clinic, hotel, factory, etc.

Mini-grids, nanogrids, and stand-alone systems are in the vast majority of cases more cost competitive than extension of the national grid network. As rural areas in developing and emerging countries are often located far away from the national grid in difficult terrain or on islands, extending the national grid to rural areas is normally extremely costly and technically difficult, whereas off-grid systems are flexible, easy to use, and adaptable to local needs and conditions. With appropriate training, they can also be operated by local technicians, which in turn leads to local employment. Mountainous and forest areas as well as small islands, for instance, with difficult access for machinery, require more time and resources to install transmission lines, whereas off-grid systems are easier and less costly to implement and can use local renewable energy sources to provide electricity. In general, mini-grids, nanogrids, and stand-alone systems are part of off-grid systems [13, 23–26, 35, 37, 39, 45].

1.10 ORGANIZATION OF THE BOOK

The present first volume of the two volumes book will further advance the concepts originally laid out in my previous book [2]. This book is focused on the hybrid power system and will lay the arguments that no single fuel or energy source is perfect for the sustainable power production or no single energy storage device or grid transport mechanism serves the needs of all the possible end games for the power industry. Hybrid (or integrated) power systems are the future of the power industry. The second volume will analyze the role of hybrid energy systems for the decarbonization of ten different industries.

The present book covers the five elements of hybrid power system in five chapters. Chapter 2 discusses issues related to hybrid power systems that are connected to the utility grid. Mobile hybrid energy systems are discussed in Chapter 3. The fundamentals of hybrid storage systems, their design, management and control methods, and several examples where hybrid energy storage systems are currently used to make hybrid power systems more productive and efficient and actual applications of hybrid energy storage for static and mobile systems are discussed in Chapter 4. Chapter 5 discusses in details fundamentals, management, and control aspects of hybrid microgrid operations that can be either grid-connected or operated in islanded mode. The chapter also evaluates the market, challenges, and value added propositions of hybrid microgrids. The last subject is evaluated with sample case studies. Chapter 6 discusses the fundamentals of off-grid (which include mini-grids and nanogrids and stand-alone hybrid energy) systems that are predominantly used for rural electrification or for isolated buildings and equipment. These are not grid-connected systems, and they are the third leg of hybrid grid systems developed from bottom up. Chapter 6 also gives a brief global landscape of the use of off-grid energy systems. Finally, simulation and optimization methods for hybrid power systems are discussed in Chapter 7. The optimization can occur in a number of different ways and include a number of different objective functions. A brief literature review on some of these options is presented.

REFERENCES

1. Shah YT. *Water for Energy and Fuel Production*. CRC Press, Taylor and Francis group, New York; 2014.
2. Shah YT. *Energy and Fuel Systems Integration*. CRC Press, Taylor and Francis group, New York; 2015.
3. Shah YT. *Chemical Energy from Natural and Synthetic Gas*. CRC Press, Taylor and Francis group, New York; 2017.
4. Shah YT. *Thermal Energy: Sources, Recovery and Applications*. CRC Press, Taylor and Francis Group, New York; 2018.
5. Shah YT. *Modular Systems for Energy and Fuel Recovery and Conversion*. CRC Press, Taylor and Francis Group, New York; 2019.
6. Shah YT. *Modular Systems for Energy Usage Management*. CRC Press, Taylor and Francis Group, New York; in Press.
7. Ibrahim M, Khair A, Ansari S. A review of hybrid renewable energy systems for electric power generation. *International Journal of Engineering Research and Applications*. 2015;5(8):42–8. www.ijera.com.
8. Wiser R, Barbose G, Heeter J, Mai T, Bird L, Bolinger M, et al. A retrospective analysis of the benefits and impacts of U.S. renewable portfolio standards. Golden, CO: Lawrence Berkeley National laboratory; 2016. A report from National Renewable Energy Laboratory (NREL), Technical Report TP-6A20-65005 January 2016, Contract Nos. DE-AC36-08GO28308 (NREL) and DE-AC02-05CH11231. Available from: www.nrel.gov/publications.
9. Kamjoo A, Maheri A, Putrus GA. Reliability criteria in optimal sizing of stand-alone hybrid wind-PV-battery bank system. In: 2nd International Symposium on Environment Friendly Energies and Applications, EFEA, 25 Jun 2012–27 Jun 2012, Newcastle upon Tyne, UK; 2012. pp. 184–9. IEEE.
10. Reliability Test System Task Force of the Application of Probability Methods Subcommittee. IEEE reliability test system. IEEE Transactions on Power Apparatus and Systems. 1979;PAS-98(6):273–82.
11. Rycroft, M. The development of renewable energy based mini-, micro- and nano-grids. EE publishers, articles: Energize; September 13, 2016. Website report.
12. Lazarov VD, Notton G, Zarkov Z, Bochev I. Hybrid power systems with renewable energy sources types, structures, trends for research and development. In: Proceedings of International Conference ELMA, Sofia, Bulgaria; 2005. pp. 515–20.
13. Bajpai P, Dash V. Hybrid renewable energy systems for power generation in stand-alone applications: A review. *Renewable and Sustainable Energy Reviews*. 2012; 16(5):2926–39.
14. Nehrir MH, Wang C, Strunz K, Aki H, Rama Kumar R. A review of hybrid renewable/alternative energy systems for electric power generation: Configurations, control, and applications. *IEEE Transactions on Sustainable Energy*. 2011;2(4):392–403.
15. Colson CM, Nehrir MH. Evaluating the benefits of a hybrid solid oxide fuel cell combined heat and power plant for energy sustainability and emissions avoidance. *IEEE Transactions on Energy Conversion* 2011; 26(1):140–8.
16. Nema P, Nema RK, Rangnekar S. A current and future state of art development of hybrid energy system using wind and PV-solar: A review. *Renewable and Sustainable Energy Reviews*. 2009;13(8):2096–2103.
17. Wichert B. PV-Diesel hybrid energy systems for remote area power generation: A review of current practice and future developments. *Renewable and Sustainable Energy Reviews*. 1997;1(3): 209–28.
18. Rahman S, Kwa-sur T. A feasibility study of photovoltaic-fuel cell hybrid energy system. *IEEE Transactions on Energy Conversion*. 1988; 3(1):50–5.

19. Asadi E, Sadjadi S. Optimization methods applied to renewable and sustainable energy: A review. *Uncertain Supply Chain Management*. 2017;5:1–26. DOI: 10.5267/j.uscm.2016.6.002.
20. Buckeridge JS, Ding JJ. Design considerations for a sustainable hybrid energy system. *Transactions of the Institution of Professional Engineers New Zealand: Civil Engineering Section*. 2000;27(1):1–5.
21. Hong YY, Lian RC. Optimal Sizing of hybrid wind/PV/diesel generation in a stand-alone power system using Markov-based genetic algorithm. *IEEE Transactions On Power Delivery*. 2012;27(2):640–7.
22. Reddy YJ, Kumar YP, Raju KP. Real time and high fidelity simulation of hybrid power system dynamics. In: Proceedings of IEEE International Conference On Recent Advances In Intelligent Computational Systems; September 22, 2011. IEEE.
23. Zhang L, Barakat G, Yassine A. Design and optimal sizing of hybrid PV/wind/diesel system with battery storage by using DIRECT search algorithm. In: 15th International Power Electronics and Motion Control Conference, EPE-PEMC 2012 ECCE Europe, Novi Sad, Serbia; 2012.
24. Kabouris J, Contaxis GC. Autonomous system expansion planning considering renewable energy sources—A computer package. *IEEE Transactions on Energy Conversion*. 1992;7(3):374–81.
25. Kabouris J, Contaxis GC. Optimum expansion planning of an unconventional generation system operating in parallel with a large scale network. *IEEE Transactions on Energy Conversion*. 1991; 6(3):394–400.
26. Kusakana K, Vermaak HJ, Numbi BP. Optimal sizing of a hybrid renewable energy plant using linear programming. In: IEEE Power and Energy Society Conference and Exposition in Africa: Intelligent Grid Integration of Renewable Energy Resources (PowerAfrica), 09–13 July 2012, Johannesburg, South Africa; 2012. IEEE Vehicle Power and Propulsion Conference, Oct. 9–12, 2012, Seoul, Korea.
27. Etxeberria A, Vechiu I, Camblong H, Vinassa JM. Hybrid energy storage systems for renewable energy sources integration in microgrids: A review. In: Conference Proceedings IPEC, Singapore; 2010. pp. 532–7. IEEE.
28. Goyal M, Gupta R. Operation and control of a distributed microgrid with hybrid system. 2012 IEEE 5th India International Conference on Power Electronics (IICPE); 2012. IEEE.
29. Bashir M, Sadeh J. Size optimization of new hybrid stand-alone renewable energy system considering a reliability index. In: 2012 11th International Conference on Environment and Electrical Engineering, May 18–25, 2012, Venice, Italy; 2012. IEEE. DOI: 10.1109/EEEIC.2012.6221521.
30. Reddy YJ, Kumar YP, Raju KP, Ramsesh A. Retrofitted hybrid power system design with renewable energy sources for buildings. *IEEE Transactions on Smart Grid*. 2012;3(4):2174–87.
31. Solanki C. *Solar Photovoltaics: Fundamentals, Technologies and Applications*. New Delhi: PHI Learning; 2011.
32. Connoly D, Lund H, Mathiesen BV, Leahy M. A review of computer tools for analyzing the integration of renewable energy into various energy systems. *Applied Energy*. 2010; 87:1059–82.
33. NREL (National Renewable Energy Laboratory): HOMER. The Micropower Optimization Model [Online]. Available from: www.homerenergy.com.
34. Afzal A. Performance analysis of integrated wind, photovoltaic and biomass energy systems. In: World Renewable Energy Congress, May 8–13, 2011, Linköping, Sweden; 2011. Linköping University.
35. Ruberti T. Off-grid hybrids: Fuel cell solar-PV hybrids. *Refocus*. 2003;4(5): 54–57.

36. Abdull Razak NA, bin Othman MM, Musirin I. Optimal sizing and operational strategy of hybrid renewable energy system using HOMER. The 4th International Power Engineering and Optimization Conference, Shah Alam, Malaysia; 2010.
37. UNCTAD (United Nations Conference on Trade and Development). *Renewable Energy Technologies for Rural Development*. UNCTAD current studies on science, technology and innovation. New York and Geneva: United Nations; 2010.
38. Roy A, Kedare SB, Bandyopadhyay S. Optimum sizing of wind-battery systems incorporating resource uncertainty. *Applied Energy*. 2010;87(8):2712–27.
39. Canol A, Jurado F, Sánchez H. Sizing and energy management of a stand-alone PV/hydrogen/battery-based hybrid system. In: 2012 International Symposium on Power Electronics, Electrical Drives, Automation and Motion; 2012.
40. Huang R, Low SH, Topcu U, Chandy KM. Optimal design of hybrid energy system with PV/ wind turbine/ storage: A case study. Available from: www.ijera.com.
41. Smart grid system report. Washington, DC: U.S. Department of Energy; November 2018. A 2018 report to Congress.
42. Ragone plot. Wikipedia, The Free Encyclopedia; 2020 [last visited May 11, 2020].
43. Asmus P. Nanogrids vs. microgrids: Energy storage a winner in both cases [October 21, 2015]. Available from: www.forbes.com/sites/pikeresearch/2015/10/26/nanogrids-vs-microgrids/. A website report.
44. Nordman B. Networked local power distribution with nanogrids. Golden, CO: Lawrence Berkeley National Laboratory [April 29, 2013]. Available from: https://eta-intranet.lbl.gov/sites/default/files/ microgridMAY2013.pdf. A website report.
45. Burmester D, Rayudu R, Seah W, Akinyele D. A review of nanogrid topologies and technologies. *Renewable and Sustainable Energy Reviews*. 2016;67:760–75. DOI: 10.1016/j.rser.2016.09.073.
46. Burmester D. Nanogrid topology, control and interactions in a microgrid structure [Ph.D. thesis in Computer Science]. New Zealand: Victoria University of Wellington. https://researcharchive.vuw.ac.nz/xmlui/bitstream/handle/10063/.../thesis_access.pdf?
47. Hagh MT, Aghdam FH. Smart hybrid nanogrids using modular multiport power electronic interface. In: 2016 IEEE Innovative Smart Grid Technologies: Asia (ISGT-Asia), 28 November–1 December, Melbourne, VIC, Australia; 2016. IEEE. DOI: 10.1109/ISGT-Asia.2016.7796456.
48. Kim Y, Wang Y, Chang N, Pedram M. Computer-aided design and optimization of hybrid energy storage systems. Foundations and Trends in Electronic Design Automation. 2013;7(4):247–338. DOI: 10.1561/1000000035.
49. Ericson SJ, Rose E, Jayaswal H, Cole WJ, Engel-Cox J, Logan J, et al. Hybrid Storage Market Assessment: A JISEA White Paper. Golden, CO: National Renewable Energy Laboratory; October 2017. NREL/MP-6A50-70237, Contract No. DE-AC36-08GO28308.
50. Serpi A, Porru M, Damiano A. A novel highly integrated hybrid energy storage system for electric propulsion and smart grid applications, Chapter 4. Intech Publication; 2018. dx.doi.org/10.5772/intechopen.73671, www.intechopen.com.
51. Ibrahim H, Ilinca A. Techno-economic analysis of different energy storage technologies. InTech paper; 2013. www. Intech.com. DOI: 10.5772/52220.
52. Renewable mini-grids innovation landscape brief. Abu Dhabi: IRENA; 2019.
53. ARE (Alliance for Rural Electrification). Hybrid mini-grids for rural electrification. 2011. Available from: www.ruralelec.org/fileadmin/DATA/Documents/06_Publications/Position_papers/ARE_Mini-grids_-_Full_version.pdf.
54. ARE. Best practices of the alliance for rural electrification. Alliance for Rural Electrification; 2013. Available from: www.ruralelec.org/fileadmin/DATA/Documents/06_Publications/ARE_Best_Practises_2013_FINAL.pdf.
55. Rethinking Energy-2017-accelerating global energy transformation. Abu Dhabi: IRENA; 2017.

2 Utility Grid with Hybrid Energy System

2.1 INTRODUCTION

The most basic form of power generation and transport during the last century was centralized power generation by coal, gas, or nuclear power plants and transmission of power by centralized utility or smart grid. Although this system served well urban communities, the operations were thermally inefficient, harmful to environment, and did not serve the communities who did not have access to smart grid. As pointed out in Chapter 1, efforts to integrate renewable and distributed low-density energy sources (like solar and wind) in smart grid have forced grid to be multifaceted and multi-functional in a number of different ways. First, the inclusions of renewable energy sources and homogeneous or hybrid energy storage devices made the utility grid hybrid. Second, in order to capture distributed low-density energy sources, localized microgrids are created which can either be connected to a main grid or be operated independently. Third, to provide power supply to very remote areas, off-grid hybrid renewable energy operations are created. It is clear that energy generation, storage, and transport structures became more multifaceted and hybrid in nature. Efforts to make power generation more efficient (i.e., cogeneration) or to better handle the emission of harmful CO_2 emission also made grid more multi-functional and hybrid. In this chapter, we address the development of hybrid utility or smart grid. Subsequent chapters will address the issues of hybrid storage, microgrid, and off-grid operations.

As pointed out above, smart grid can become a hybrid energy system in a number of different ways. In cogeneration, hybrid energy is produced from one source of power generation by utilizing waste heat for heating/cooling purposes or for generating an additional source of power. In recent years, significant efforts are being made to generate additional power from waste heat from numerous micro and macroscale operations using thermoelectricity. Large-scale power generation from coal, gas, or nuclear energy produces significant amount of waste heat, which can also be used to generate additional power (such as in combined cycle power generation) or used as excess heat for industrial or domestic heating and cooling needs. Hybrid power can also be produced, as shown by ExxonMobil, by using CO_2 produced from the power plant to generate more power using fuel cell technology. Thus, in these cases, hybrid energy system either improves process efficiency or reduces harmful environment impact.

The future energy industry will require less use of fossil fuel and deeper penetration of renewable sources of energy in the overall energy mix. The present book shows that the use of renewable sources (particular solar and wind energy which are intermittent) will require a hybrid mode to supply power or heat in a stable manner. Thus, a hybrid energy system will be the preferred mode for deeper penetration of renewable sources. My previous book [1] showed that hybrid energy can also be provided by using co-fuel (coal and biomass) or co-energy (nuclear and renewable sources) for power generation.

As mentioned in Chapter 1, this book divides grid transport into three levels: large macrogrids, microgrids, and off-grids which include mini-grids, nanogrids, and stand-alone systems for rural areas, islands, and remote locations. Large grids may be further divided into developed and developing grids. Thus grid system for power transport can be sub-divided as:

- ***Large developed grid***: The US electricity grid is an example of a large developed grid, providing reliable energy at a low cost. Primary storage markets for large grids include ancillary services, transmission deferral, and customer demand charge reductions. Increasing reliability and reducing diesel fuel use are not primary concerns. The use of hybrid energy in a large developed grid is discussed in this chapter.
- ***Large developing grid***: India is an example of a large developing grid. A large developing grid provides low-cost energy, relative to the cost of a diesel generator, but blackouts are common. Increasing reliability is an important market for developing grids. The use of hybrid energy for a large developing grid is discussed in this chapter [2].
- ***Microgrid***: Examples may include universities, hospitals, and military bases. A microgrid is connected to a larger grid but has the ability to produce its own electricity for demand charge and resilience purposes. Microgrids may be further subdivided by the size of the load. Microgrids are created to harness distributed energy resources at medium- to low-voltage levels. Chapter 5 describes in detail the workings of hybrid microgrids.
- ***Off-grid (mini-grids, nanogrids, and stand-alone systems) systems in rural, islands and remote locations***: Examples include mines, off-grid communities, and, of course, islands. Islands and remote locations are not connected to a larger grid and generally face higher energy costs because most energy production comes from diesel generators. Although, as shown in Chapter 6, in recent years, HRES is penetrating more and more in off-grid operations, reducing diesel fuel use and increasing level of HRES with the help of energy storage are the primary objectives of the off-grid market.

Besides three levels of grid transport mentioned above, a hybrid energy system also contains homogeneous or hybrid storage devices which are particularly important to harness hybrid renewable sources. This is illustrated in details in Chapter 4. In this chapter we focus on the changing nature of the large developed or developing macrogrid.

2.2 HYBRID SMART GRID

A *smart grid* is an electrical grid which includes a variety of operation and energy measures including smart meters, smart appliances, renewable energy resources, and energy efficient resources. Electronic power conditioning and control of the production and distribution of electricity are important aspects of the smart grid. Rollout of smart grid technology also implies a fundamental reengineering of the electricity services industry, although typical usage of the term is focused on the technical infrastructure. The first alternating current power grid system was installed in 1886 in Great Barrington, Massachusetts. At that time, the grid was a centralized unidirectional system of electric power transmission, electricity distribution, and demand-driven control [3].

In the 20th century, local grids grew over time and were eventually interconnected for economic and reliability reasons. By the 1960s, the electric grids of developed countries had become very large, mature, and highly interconnected, with thousands of "central" generation power stations delivering power to major load centers via high capacity power lines which were then branched and divided to provide power to smaller industrial and domestic users over the entire supply area. The topology of the 1960s grid was a result of the strong economies of scale: large coal-, gas-, and oil-fired power stations in the 1 GW (1,000 MW) to 3 GW scale were still found to be cost-effective, due to their efficiency-boosting features that can be cost-effective only when the stations became very large.

Power stations were located strategically to be close to fossil fuel reserves (either the mines or the wells themselves, or else close to rail, road, or port supply lines). Siting of hydroelectric dams in mountain areas also strongly influenced the structure of the emerging grid. Nuclear power plants were sited for the availability of cooling water. Fossil fuel-fired power stations were initially very polluting and were sited as far as economically possible from population centers, once electricity distribution networks permitted it. By the late 1960s, the electricity grid reached the overwhelming majority of the population of developed countries, with only outlying regional areas remaining "off-grid."

Metering of electricity consumption was necessary on a per-user basis in order to allow appropriate billing according to the (highly variable) level of consumption of different users. Because of limited data collection and processing capability during the period of growth of the grid, fixed-tariff arrangements were commonly put in place, as well as dual-tariff arrangements where nighttime power was charged at a lower rate than daytime power. The motivation for dual-tariff arrangements was the lower nighttime demand. Dual tariffs made possible the use of low-cost nighttime electrical power in applications such as the maintaining of "heat banks" which served to "smooth out" the daily demand and reduce the number of turbines that needed to be turned off overnight, thereby improving the utilization and profitability of the generation and transmission facilities. The metering capabilities of the 1960s grid meant technological limitations on a degree to which price signals could be propagated through the system. From the 1970s to the 1990s, growing demand led to increasing numbers of power stations. In some areas, supply of electricity, especially at peak times, could not keep up with this demand, resulting in poor power quality including blackouts, power cuts, and brownouts. Increasingly, electricity was depended on for industry, heating, communication, lighting, and entertainment, and consumers demanded ever higher levels of reliability.

Toward the end of the 20th century, electricity demand patterns were established: domestic heating and air conditioning led to daily peaks in demand that were met by an array of "peaking power generators" that would only be turned on for short periods each day. The relatively low utilization of these peaking generators (commonly, gas turbines were used due to their relatively lower capital cost and faster start-up times), together with the necessary redundancy in the electricity grid, resulted in higher costs to the electricity companies, which were passed on in the form of increased tariffs.

Since the early 21st century, opportunities to take advantage of improvements in electronic communication technology to resolve the limitations and costs of the electrical grid have become apparent. Concerns about energy efficiency of large-scale power plants and emission of CO_2 took the front seat, and this resulted in the search for methods for improving efficiency and reduction of CO_2 emission. Technological limitations on metering no longer forced peak power prices to be averaged out and passed on to all consumers equally. In parallel, growing concerns over environmental damage from fossil fuel-fired power stations led to a desire to use large amounts of renewable energy. The desire to make buildings zero-energy buildings took the front seat. Also transportation industry became more hybrid and electrical to reduce harmful emission caused by internal combustion engines. All of these factors made power generation and storage in smart grid more hybrid and multi-functional. Dominant forms such as wind power and solar power are highly variable, and so the need for more sophisticated control systems became apparent, to facilitate the connection of sources to the otherwise highly controllable grid. Power from photovoltaic cells (and to a lesser extent, wind turbines) also, significantly, called into question the imperative for large, centralized power stations. The rapidly falling costs point to a major change from the centralized grid topology to one that is highly distributed, with power being both generated *and* consumed right at the limits of the grid. Finally, growing concern over terrorist attacks in some countries led to calls for a more robust energy grid that is less dependent on centralized power stations that were perceived to be potential attack targets.

The first official definition of smart grid was provided by the Energy Independence and Security Act of 2007 (EISA-2007), which was approved by the US Congress in January 2007 and signed to law by President George W. Bush in December 2007. Title XIII of this bill provides a description, with ten characteristics, that can be considered a definition for smart grid, as follows:

> "It is the policy of the United States to support the modernization of the Nation's electricity transmission and distribution system to maintain a reliable and secure electricity infrastructure that can meet future demand growth and to achieve each of the following, which together characterize a Smart Grid: (1) Increased use of digital information and controls technology to improve reliability, security, and efficiency of the electric grid. (2) Dynamic optimization of grid operations and resources, with full cyber-security. (3) Deployment and integration of distributed resources and generation, including renewable resources. (4) Development and incorporation of demand response, demand-side resources, and energy-efficiency resources. (5) Deployment of 'smart' technologies (real-time, automated, interactive technologies that optimize the

physical operation of appliances and consumer devices) for metering, communications concerning grid operations and status, and distribution automation. (6) Integration of 'smart' appliances and consumer devices. (7) Deployment and integration of advanced electricity storage and peak-shaving technologies, including plug-in electric and hybrid electric vehicles, and thermal storage air conditioning. (8) Provision to consumers of timely information and control options. (9) Development of standards for communication and interoperability of appliances and equipment connected to the electric grid, including the infrastructure serving the grid. (10) Identification and lowering of unreasonable or unnecessary barriers to adoption of smart grid technologies, practices, and services."

The European Union Commission Task Force for Smart Grids also provided their smart grid definition.

A common element to most definitions is the application of digital processing and communications to the power grid, making data flow and information management central to the smart grid. Various capabilities result from the deeply integrated use of digital technology with power grids. Integration of the new grid information is one of the key issues in the design of smart grids. Electric utilities now find themselves making three classes of transformations: improvement of infrastructure, called the *strong grid* in China; addition of the digital layer, which is the essence of the *smart grid*; and business process transformation, necessary to capitalize on the investments in smart technology. Much of the work that has been going on in electric grid modernization, especially substation and distribution automation, is now included in the general concept of the smart grid [1–5].

Smart grid technologies emerged from earlier attempts at using electronic control, metering, and monitoring. In the 1980s, automatic meter reading was used for monitoring loads from large customers and evolved into the Advanced Metering Infrastructure of the 1990s, whose meters could store how electricity was used at different times of the day. Smart meters added continuous communications so that monitoring can be done in real time and can be used as a gateway to demand response-aware devices and "smart sockets" in the home. Early forms of such demand-side management technologies were dynamic-demand aware devices that passively sensed the load on the grid by monitoring changes in the power supply frequency. Devices such as industrial and domestic air conditioners, refrigerators, and heaters adjusted their duty cycle to avoid activation during times the grid was suffering a peak condition. Beginning in 2000, Italy's Telegestore Project was the first to network large numbers (27 million) of homes using smart meters connected via low bandwidth power line communication. Some experiments used the term broadband over power lines (BPL), while others used wireless technologies such as mesh networking promoted for more reliable connections to disparate devices in the home as well as supporting metering of other utilities such as gas and water.

Monitoring and synchronization of wide area networks were revolutionized in the early 1990s when the Bonneville Power Administration expanded its smart grid research with prototype sensors that are capable of very rapid analysis of anomalies in electricity quality over very large geographic areas. The culmination of this work was the first operational Wide Area Measurement System (WAMS) in 2000. Other countries are rapidly integrating this technology—China started having a comprehensive national WAMS when the past 5-year economic plan completed in 2012.

The earliest deployments of smart grids include the Italian system Telegestore (2005), the mesh network of Austin, Texas (since 2003), and the smart grid in Boulder, Colorado (2008) [3–5].

2.2.1 Features of the Smart Grid

The smart grid represents the full suite of current and proposed responses to the challenges of electricity supply. Because of the diverse range of factors, there are numerous competing taxonomies and no agreement on a universal definition. Nevertheless, one possible categorization can be given as [2–5]:

A **Reliability**

The smart grid makes use of technologies such as state estimation that improve fault detection and allow self-healing of the network without the intervention of technicians. This will ensure a more reliable supply of electricity and reduced vulnerability to natural disasters or attacks. Although multiple routes are touted as a feature of the smart grid, the old grid also featured multiple routes. Initial power lines in the grid were built using a radial model, later connectivity was guaranteed via multiple routes, referred to as a network structure. However, this created a new problem: if the current flow or related effects across the network exceed the limits of any particular network element, it could fail, and the current would be shunted to other network elements, which eventually may fail also, causing a domino effect. A technique to prevent this is load shedding by rolling blackout or voltage reduction (brownout).

B **Flexibility in network topology**

Next-generation transmission and distribution infrastructure will be better able to handle possible *bidirectional energy flows*, allowing for distributed generation (DG) such as from photovoltaic panels on building roofs, but also charging to/from the batteries of electric cars, wind turbines, pumped hydroelectric power, the use of fuel cells, and other sources. Classic grids were designed for a one-way flow of electricity, but if a local sub-network generates more power than it is consuming, the reverse flow can raise safety and reliability issues. A smart grid aims to manage these situations.

C **Efficiency**

Numerous contributions to overall improvement of the efficiency of energy infrastructure are anticipated from the deployment of smart grid technology, in particular including *demand-side management*, for example, turning off air conditioners during short-term spikes in electricity price, reducing the voltage when possible on distribution lines through Voltage/VAR Optimization (VVO), eliminating truck-rolls for meter reading, and reducing truck-rolls by improved outage management using data from Advanced Metering Infrastructure systems. The overall effect is less redundancy in transmission and distribution lines and greater utilization of generators, leading to lower power prices.

D **Load adjustment/load balancing**

The total load connected to the power grid can vary significantly over time. Although the total load is the sum of many individual choices of the clients, the overall load is not necessarily stable or slow varying. For example, if a popular television program starts, millions of televisions will start to draw current instantly. Traditionally, to respond to a rapid increase in power consumption, faster than the start-up time of a large generator, some spare generators are put on a dissipative standby mode. A smart grid may warn all individual television sets, or another larger customer, to reduce the load temporarily (to allow time to start up a larger generator) or continuously (in the case of limited resources). Using mathematical prediction algorithms, it is possible to predict how many standby generators need to be used, to reach a certain failure rate. In the traditional grid, the failure rate can only be reduced at the cost of more standby generators. In a smart grid, the load reduction by even a small portion of the clients may eliminate the problem.

While traditionally load balancing strategies have been designed to change consumers' consumption patterns to make demand more uniform, developments in energy storage and individual renewable energy generation have provided opportunities to devise balanced power grids without affecting consumers' behavior. Typically, storing energy during off-peak times eases high demand supply during peak hours. Dynamic game-theoretic frameworks have proved particularly efficient at storage scheduling by optimizing energy cost using their Nash equilibrium [3–5].

E **Peak curtailment/leveling and time-of-use pricing**

To reduce demand during the high-cost peak usage periods, communications and metering technologies inform smart devices in the home and business when energy demand is high and track how much electricity is used and when it is used. It also gives utility companies the ability to reduce consumption by communicating to devices directly in order to prevent system overloads. Examples would be a utility reducing the usage of a group of electric vehicle charging stations or shifting temperature set points of air conditioners in a city. To motivate them to cut back use and perform what is called *peak curtailment* or *peak leveling* (see Figure 2.1), prices of electricity are increased during high-demand periods and decreased during low-demand periods [2–5]. It is thought that consumers and businesses will tend to consume less during high-demand periods if it is possible for consumers and consumer devices to be aware of the high-price premium for using electricity at peak periods. This could mean making trade-offs such as cycling on/off air conditioners or running dishwashers at 9 pm instead of 5 pm. When businesses and consumers see a direct economic benefit of using energy at off-peak times, the theory is that

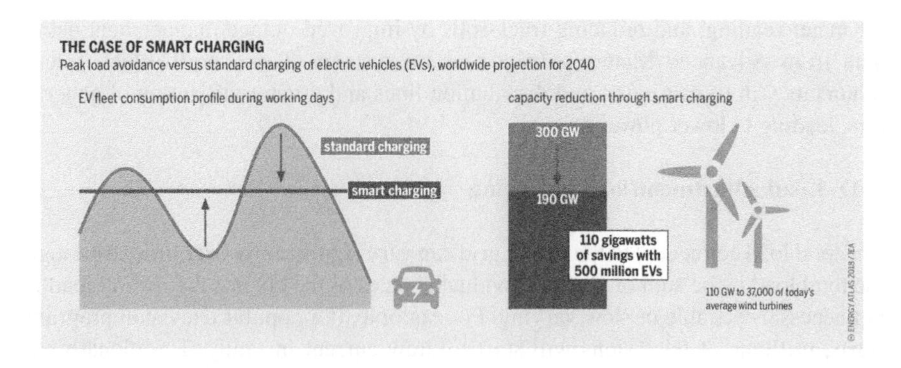

FIGURE 2.1 Peak load avoidance by smart charging of electric vehicles [3].

they will include energy cost of operation into their consumer device and building construction decisions and hence become more energy efficient.

F **Sustainability**

The improved flexibility of the smart grid permits greater penetration of highly variable renewable energy sources such as solar power and wind power, even without the addition of energy storage. For very large level of RES penetration, energy storage appears necessary. Current network infrastructure is not built to allow for many distributed feed-in points, and typically even if some feed-in is allowed at the local (distribution) level, the transmission-level infrastructure cannot accommodate it. Rapid fluctuations in distributed generation, such as due to cloudy or gusty weather, present significant challenges to power engineers who need to ensure stable power levels through varying the output of the more controllable generators such as gas turbines and hydroelectric generators. Smart grid technology is a necessary condition for very large amounts of renewable electricity on the grid for this reason.

G **Market enabling**

The smart grid allows for systematic communication between suppliers (their energy price) and consumers (their willingness-to-pay) and permits both the suppliers and the consumers to be more flexible and sophisticated in their operational strategies. Only the critical loads will need to pay the peak energy prices, and consumers will be able to be more strategic when they use energy. Generators with greater flexibility will be able to sell energy strategically for maximum profit, whereas inflexible generators such as base load steam turbines and wind turbines will receive a varying tariff based on the level of demand and the status of the other generators currently operating. The overall effect is a signal that awards energy efficiency and energy consumption that is sensitive to the time varying limitations of the supply. At the domestic level, appliances with a degree of energy storage or thermal mass (such as refrigerators, heat banks, and heat pumps) will be well placed to "play" the market and seek to minimize energy cost by adapting demand to the lower cost energy support periods. This is an extension of the dual-tariff energy pricing mentioned above.

H **Demand response support**

Demand response support allows generators and loads to interact in an automated fashion in real time, coordinating demand to flatten spikes. Eliminating the fraction of demand that occurs in these spikes eliminates the cost of adding reserve generators, cuts wear and tear and extends the life of equipment, and allows users to cut their energy bills by telling low-priority devices to use energy only when it is cheapest.

Currently, power grid systems have varying degrees of communication within control systems for their high-value assets, such as in generating plants, transmission lines, substations, and major energy users. In general, information flows one way, from the users and the loads they control back to the utilities. The utilities attempt to meet the demand and succeed or fail to varying degrees (brownouts, rolling blackout, and uncontrolled blackout). The total amount of power demand by the users can have a very wide probability distribution which requires spare generating plants in standby mode to respond to the rapidly changing power usage. This one-way flow of information is expensive; the last 10% of generating capacity may be required as little as 1% of the time, and brownouts and outages can be costly to consumers.

Demand response can be provided by commercial, residential loads, and industrial loads [1–5]. For example, Alcoa's Warrick Operation is participating in MISO as a qualified demand response resource, and the Trimet Aluminium uses its smelter as a short-term mega-battery. Latency of the data flow was a major concern, with some early smart meter architectures allowing actually as long as 24 hours delay in receiving the data, preventing any possible reaction by either supplying or demanding devices.

I **Platform for advanced services**

As with other industries, use of robust two-way communications, advanced sensors, and distributed computing technology will improve the efficiency, reliability, and safety of power delivery and use. It also opens up the potential for entirely new services or improvements on existing ones, such as fire monitoring and alarms that can shut off power, make phone calls to emergency services, etc.

J **Provision megabits, control power with kilobits, and sell the rest**

The amount of data required to perform monitoring and switching one's appliances off automatically is very small compared with that already reaching even remote homes to support voice, security, internet, and TV services. Many smart grid bandwidth upgrades are paid for by over-provisioning to also support consumer services and subsidizing the communications with energy-related services or subsidizing the energy-related services, such as higher rates during peak hours, with communications. This is particularly true where governments run both sets of services as a public monopoly. Because power and communications companies are generally separate commercial enterprises in North America and Europe, it has required considerable government and large-vendor effort to encourage various enterprises to cooperate. Some, like Cisco, see opportunity in providing devices to consumers very similar to those that they have long been providing to industry. Others, such as Silver Spring Networks or Google, are data integrators rather than vendors of equipment. While the AC power control standards suggest power line, networking would be the primary means of communication among smart grid and home devices, the bits may not reach the home via broadband over power lines (BPL) initially but by fixed wireless.

2.2.2 Technology

The bulk of smart grid technologies are already used in other applications such as manufacturing and telecommunications and are being adapted for use in grid operations [3–5].

- **Integrated communications:** Areas for improvement include substation automation, demand response, distribution automation, supervisory control and data acquisition (SCADA), energy management systems, wireless mesh networks and other technologies, power line carrier communications, and fiber optics. Integrated communications will allow for real-time control, information, and data exchange to optimize system reliability, asset utilization, and security.
- **Sensing and measurement:** Core duties are evaluating congestion and grid stability, monitoring equipment health, energy theft prevention, and control strategies support. Technologies include advanced microprocessor meters (smart meter) and meter reading equipment, wide-area monitoring systems (typically based on online readings by distributed temperature sensing combined with real-time thermal rating (RTTR) systems), electromagnetic signature measurement/analysis, time-of-use and real-time pricing tools, advanced switches and cables, backscatter radio technology, and digital protective relays. Many in the power systems engineering community believe that the Northeast blackout of 2003 could have been contained to a much smaller area if a wide-area phasor measurement network had been in place.
- **Distributed power flow control:** Power flow control devices clamp onto existing transmission lines to control the flow of power within. Transmission lines enabled with such devices support greater use of renewable energy by providing more consistent, real-time control over how that energy is routed within the grid. This technology enables the grid to more effectively store intermittent energy from renewables for later use.
- **Smart power generation using advanced components:** Smart power generation is a concept of matching electricity generation with demand, using

multiple identical generators which can start, stop, and operate efficiently at chosen load, independently of the others, making them suitable for base load and peaking power generation.

- **Matching supply and demand:** Called load balancing, it is essential for a stable and reliable supply of electricity. Short-term deviations in the balance lead to frequency variations and a prolonged mismatch results in blackouts. Operators of power transmission systems are charged with the balancing task, matching the power output of all the generators to the load of their electrical grid. The load balancing task has become much more challenging as increasingly intermittent and variable generators such as wind turbines and solar cells are added to the grid, forcing other producers to adapt their output much more frequently than has been required in the past. First two dynamic-grid stability power plants utilizing the concept were ordered by Elering and were built by Wärtsilä in Kiisa, Estonia (Kiisa Power Plant). The purpose of the plants was to "provide dynamic generation capacity to meet sudden and unexpected drops in the electricity supply." They were ready during 2013 and 2014, and their total output was 250 MW.
- **Power system automation:** This enables rapid diagnosis of and precise solutions to specific grid disruptions or outages. These technologies rely on and contribute to each of the other four key areas. Three technology categories for advanced control methods are distributed intelligent agents (control systems), analytical tools (software algorithms and high-speed computers), and operational applications (SCADA, substation automation, demand response, etc.). Using artificial intelligence programming techniques, the Fujian power grid in China created a wide-area protection system that is rapidly able to accurately calculate a control strategy and execute it. The voltage stability monitoring and control (VSMC) software uses a sensitivity-based successive linear programming method to reliably determine the optimal control solution.

Smart grid provides IT-based solutions which the traditional power grid is lacking. These new solutions pave the way for new entrants that were traditionally not related to the energy grid. Technology companies are disrupting the traditional energy market players in several ways. They develop complex distribution systems to meet the more decentralized power generation due to microgrids. In addition, an increase in data collection, bringing many new possibilities for technology companies as deploying transmission grid sensors at a user level and balancing system reserves [3–5]. The technology in microgrids makes energy consumption cheaper for households than buying from utilities. Additionally, residents can manage their energy consumption easier and more effectively with the connection to smart meters [3–5]. However, the performances and reliability of microgrids strongly depend on the continuous interaction between power generation, storage, and load requirements [3–5]. A hybrid system combining renewable energy sources (with or without gas) and energy storage can work well in a microgrid setup. As a consequence of the entrance of the technology companies in the energy market, utilities and DSO's need to create new business models to keep current customers and to create new customers. This resulted in (a) focus on a customer engagement strategy, (b) creation of alliances with new entered technology companies, and (c) more focus on renewable energy sources. This also resulted in major new programs such as IntelliGrid, Grid 2030, Modern Grid Initiative (MGI), GridWise, GridWise architectural council, GridWorks, Pacific Northwest Smart Grid Demonstration Project, SolarCities and Smart Grid Energy Research Center (SMERC). Significant new approaches to smart grid modeling efforts occur. In short, modern smart grid is becoming more and more hybrid, complex and looks forward to serving innovative needs where consumers become prosumers, buildings become "zero energy buildings," and electric cars and hybrid electric cars will need more charging from grid. Hybrid energy grid is the future [3–5].

2.2.3 Economics

In 2009, the US smart grid industry was valued at about $21.4 billion—by 2014, it exceeded at least $42.8 billion. Given the success of the smart grids in the United States, the world market grew at a faster rate, surging from $69.3 billion in 2009 to $171.4 billion by 2014. With the segments set to benefit the most will be smart metering hardware sellers and makers of software that is used to transmit and organize the massive amount of data collected by meters [3–5].

The size of smart grid market was valued at over $30 billion in 2017 and is set to expand over 11% CAGR to hit $70 billion by 2024. The growing need to digitalize the power sector driven by aging electrical grid infrastructure will stimulate the global market size. The industry is primarily driven by favorable government regulations and mandates, along with a rising share of renewables in the global energy mix. According to the International Energy Agency (IEA), global investments in digital electricity infrastructure were over $50 billion in 2017.

A 2011 study from the Electric Power Research Institute concluded that investment in a US smart grid will cost up to $476 billion over 20 years but will provide up to $2 trillion in customer benefits over that time. In 2015, the World Economic Forum reported a transformational investment of more than $7.6 trillion by members of the OECD is needed over the next 25 years (or $300 billion per year) to modernize, expand, and decentralize the electricity infrastructure with technical innovation as key to the transformation. A 2019 study from International Energy Agency estimated that the current (depreciated) value of the US electric grid is more than $1 trillion. The total cost of replacing it with a smart grid was estimated to be more than $4 trillion. If smart grids are deployed fully across the United States, the country expects to save $130 billion annually.

As customers can choose their electricity suppliers, depending on their different tariff methods, the focus of transportation costs will be increased. Reduction of maintenance and replacements costs will stimulate more advanced control. A smart grid precisely limits electrical power down to the residential level, networks small-scale distributed energy generation and storage devices, communicates information on operating status and needs, collects information on prices and grid conditions, and moves the grid beyond central control to a collaborative network. All of this will result in more hybrid energy system.

2.2.4 Concerns and Challenges

Most opposition and concerns have centered on smart meters and the items (such as remote control, remote disconnect, and variable rate pricing) enabled by them. Where opposition to smart meters is encountered, they are often marketed as "smart grid" which connects smart grid to smart meters in the eyes of opponents. Specific points of opposition or concern include:

- Consumer concerns over privacy, e.g., use of usage data by law enforcement
- Social concerns over "fair" availability of electricity
- Concern that complex rate systems (e.g., variable rates) remove clarity and accountability, allowing the supplier to take advantage of the customer
- Concern over remotely controllable "kill switch" incorporated into most smart meters
- Social concerns over Enron style abuses of information leverage
- Concerns over giving the government mechanisms to control the use of all power using activities
- Concerns over radio frequency (RF) emissions from smart meters

While modernization of electrical grids into smart grids allows for optimization of everyday processes, a smart grid, being online, can be vulnerable to cyberattacks. Transformers which increase the voltage of electricity created at power plants for long-distance travel, transmission lines themselves, and distribution lines which deliver the electricity to its consumers are particularly susceptible. These systems rely on sensors which gather information from the field and then deliver it to control centers, where algorithms automate analysis and decision-making processes. These decisions are sent back to the field, where existing equipment executes them [3–5]. Hackers have the potential to disrupt these automated control systems, severing the channels which allow generated electricity to be utilized [3–5]. This is called a denial of service or DoS attack. They can also launch integrity attacks, which corrupt information being transmitted along the system, as well as desynchronization attacks, which affect when such information is delivered to the appropriate location [3–5]. Additionally, intruders can again access via renewable energy generation systems and smart meters connected to the grid, taking advantage of more specialized weaknesses or ones whose security has not been prioritized. Because a smart grid has a large number of access points, like smart meters, defending all of its weak points can prove difficult. There is also concern about the security of the infrastructure, primarily that involving communications technology. Concerns chiefly center around the communications technology at the heart of the smart grid. Designed to allow real-time contact between utilities and meters in customers' homes and businesses, there is a risk that these capabilities could be exploited for criminal or even terrorist actions. One of the key capabilities of this connectivity is the ability to remotely switch off power supplies, enabling utilities to quickly and easily cease or modify supplies to customers who default on payment. This is undoubtedly a massive boon for energy providers, but also raises some significant security issues. Cybercriminals have infiltrated the US electric grid before on numerous occasions. Aside from computer infiltration, there are also concerns that computer malware like Stuxnet, which targeted SCADA systems which are widely used in industry, could be used to attack a smart grid network.

Electricity theft is a concern in the United States where the smart meters being deployed use RF technology to communicate with the electricity transmission network. People with knowledge of electronics can devise interference devices to cause the smart meter to report lower than actual usage. Similarly, the same technology can be employed to make it appear that the energy the consumer is using is being used by another customer, increasing their bill.

The damage from a well-executed, sizable cyberattack could be extensive and long lasting. One incapacitated substation could take from nine days to over a year to repair, depending on the nature of the attack. It can also cause an hours-long outage in a small radius. It could have an immediate effect on transportation infrastructure, as traffic lights and other routing mechanisms as well as ventilation equipment for underground roadways are reliant on electricity. Additionally, infrastructure which relies on the electric grid, including wastewater treatment facilities, the information technology sector, and communications systems could be impacted.

The December 2015 Ukraine power grid cyberattack, the first record of its kind, disrupted services to nearly a quarter of a million people by bringing substations offline. The Council on Foreign Relations has noted that states are most likely to be the perpetrators of such an attack as they have access to the resources to carry one out despite the high level of difficulty of doing so. Cyber intrusions can be used as portions of a larger offense, military or otherwise. Some security experts warn that this type of event is easily scalable to grids elsewhere. The insurance company Lloyd's of London has already modeled the outcome of a cyberattack on the Eastern Interconnection, which has the potential to impact 15 states, put 93 million people in the dark, and cost the country's economy anywhere from $243 billion to $1 trillion in various damages.

Some experts argue that the first step to increasing the cyber defenses of the smart electric grid is completing a comprehensive risk analysis of existing infrastructure, including research of software, hardware, and communication processes. Additionally, as intrusions themselves can provide valuable information, it could be useful to analyze system logs and other records of their nature and timing. Common weaknesses already identified using such methods by the Department of Homeland Security include poor code quality, improper authentication, and weak firewall rules. Once this step is completed, some suggest that it makes sense to then complete an analysis of the potential consequences of the aforementioned failures or shortcomings. This includes both immediate consequences as well as second- and third-order cascading impacts on parallel systems. Finally, risk mitigation solutions, which may include simple remediation of infrastructure inadequacies or novel strategies, can be deployed to address the situation. Some such measures include recoding of control system algorithms to make them more able to resist and recover from cyberattacks or preventative techniques that allow more efficient detection of unusual or unauthorized changes to data. Strategies to account for human error which can compromise systems include educating those who work in the field to be wary of strange USB drives, which can introduce malware if inserted, even if just to check their contents [3]. Other solutions include utilizing transmission substations, constrained SCADA networks, policy-based data sharing, and attestation for constrained smart meters [2–5].

Before a utility installs an advanced metering system, or any type of smart system, it must make a business case for the investment. Some components, like the power system stabilizers (PSS) installed on generators are very expensive, require complex integration in the grid's control system, are needed only during emergencies, and are only effective if other suppliers on the network have them. Without any incentive to install them, power suppliers don't. Most utilities find it difficult to justify installing a communications infrastructure for a single application (e.g., meter reading). Because of this, a utility must typically identify several applications that will use the same communications infrastructure—for example, reading a meter, monitoring power quality, remote connection and disconnection of customers, enabling demand response, etc. Ideally, the communications infrastructure will not only support near-term applications, but unanticipated applications that will arise in the future. Regulatory or legislative actions can also drive utilities to implement pieces of a smart grid puzzle. Each utility has a unique set of business and regulatory and legislative drivers that guide its investments. This means that each utility will take a different path to creating their smart grid and that different utilities will create smart grids at different adoption rates.

Some features of smart grids draw opposition from industries that currently are or hope to provide similar services. An example is competition with cable and DSL internet providers from broadband over power line internet access. Providers of SCADA control systems for grids have intentionally designed proprietary hardware, protocols, and software so that they cannot interoperate with other systems in order to tie its customers to the vendor.

The incorporation of digital communications and computer infrastructure with the grid's existing physical infrastructure poses challenges and inherent vulnerabilities. According to *IEEE Security and Privacy Magazine*, the smart grid will require that people develop and use large computer and communication infrastructure that supports a greater degree of situational awareness and that allows for more specific command and control operations. This process is necessary to support major systems such as demand response wide-area measurement and control, storage and transportation of electricity, and the automation of electric distribution.

Various "smart grid" systems have dual functions. This includes Advanced Metering Infrastructure systems which, when used with various software, can be used to detect power theft and, by process of elimination, can detect where equipment failures have taken place. These are in addition to their primary functions of eliminating the need for human meter reading and measuring the time of use of electricity. The worldwide power loss including theft is estimated at approximately $200 billion annually. Electricity theft also represents a major challenge when providing reliable electrical service in developing countries [2–5].

2.2.5 Why Hybrid Smart Grid?

The future operation of smart grid will be hybrid for a number of reasons. There is a significant pressure to reduce the use of fossil fuel and increase the use of renewable energy. This requires a number of strategies [3, 5]:

1. **Conservation of energy.** This, of course, depends on the human behavior which cannot be always controlled.
2. **Make power generation more efficient.** This requires use of waste heat for heating and cooling or for more power generation. This strategy will make grid and the energy system hybrid.
3. **Reduce CO2 emission.** One method is to generate power from CO_2 like CO_2 fuel cell. CO_2 can also be transformed into fuel. This will once again make grid and the energy system hybrid.
4. **Use smart grid to make more use of RES like solar and wind for power generation.** This will reduce fossil fuel use. The power generated from RES can be used to support future HEV and EV vehicles. The power can also be generated in a distributed manner from zero-energy buildings. The penetration of RES to generate power will require hybrid sources including energy storage. Battery storage appears to be the preferred mode for the future.
5. **Since RES are more distributed, one method for capturing these distributed energy sources is to create microgrid operated at medium- to low-voltage mode that can be connected to a smart grid.** Microgrid can also be operated in islanded mode in case the smart grid fails. The use of microgrid will also make the overall grid operation more hybrid.

Besides the need for efficiency improvement and cutting down CO_2 emission and increase in renewable energy sources in the overall energy mix, the future hybrid smart grids will also be affected by the changes that are occurring in the transportation industries. As shown in the next chapter, hybrid energy system is also going to take over future transportation industries. HEV and EV vehicles will replace current internal combustion vehicles (ICV) vehicles. The charging of these vehicles will have a significant impact on the operation of smart grids. The charging can occur through a direct connection to the grid or through an EV-home connection. Some of the efforts made in this direction are described in the next chapter. The last four reasons for the development of hybrid grid outlined above are further described below.

2.3 HYBRID ENERGY SYSTEMS FOR FOSSIL FUEL-BASED POWER GENERATION

Hybrid energy system from fossil fuel-based power generation processes (both static and mobile) can be generated by converting either (a) waste heat or (b) CO_2 to power or useful fuels. Here we describe both of these in some details.

2.3.1 Hybrid Energy System Created by Conversion of Waste Heat to Power by Thermoelectricity

As we pushed toward less fossil fuel consumption for power generation, the conversion efficiency becomes an important issue. For example, the total energy efficiency of a conventional thermal power plant using steam turbines is approximately 40%; the best modern combined cycle plant using a gas turbine and a steam turbine is between 50% and 60%. In vehicles using gasoline-powered combustion engines, the conversion efficiency is about 30%, and diesel-powered combustion engines achieve about 40% efficiency. Similarly the use of thermal energy in industrial processes is also highly inefficient and these processes, in general, generate a large amount of waste heat. While these different sources of waste heat can be used for local heating and cooling needs by cogeneration, thermoelectric generators (TEG) have the potential to recover them as power and to make a major contribution to reducing fossil fuel consumption [6–12]. As a consequence of lower energy consumption and higher total energy efficiency, TEG also can help reduce CO_2 and other greenhouse gas emissions.

A thermoelectric power generator is a solid-state device that provides direct energy conversion from thermal energy (heat) due to a temperature gradient into electrical energy based on "Seebeck effect" [6–12]. The thermoelectric power cycle, with charge carriers (electrons) serving as the working fluid, follows the fundamental laws of thermodynamics and intimately resembles the power cycle of a conventional heat engine. Thermoelectric power generators offer several distinct advantages over other technologies [6–12]:

- They are extremely reliable (typically exceed 100,000 hours of steady-state operation) and silent in operation since they have no mechanical moving parts and require considerably less maintenance.
- They are simple, compact, and safe.
- They are very small in size and virtually weightless.
- They are capable of operating at elevated temperatures.
- They are suited for small-scale and remote applications typical of rural power supply, where there is limited or no electricity.
- They are environmentally friendly.
- They are not position dependent.
- They are flexible power sources.
- They produce no chemical substance, operate in silence, and are reliable.

On the other hand, the most significant disadvantage is the low-energy efficiency of TEG (typically 5%) [7]. In addition, the energy efficiency and the released output power are temperature dependent. Modern automated combustion systems need to be connected to the electricity grid. Boilers with an integrated thermoelectric generator can utilize waste heat from the furnace. The CHP (combined heat and power) units that are created provide an independent source of electric energy and enable efficient use of fuel. Low-conversion efficiency has been a major cause in restricting their use in electrical power generation to specialized fields with extensive applications where reliability is a major consideration and cost is not.

In general, the cost of a thermoelectric power generator essentially consists of the device cost and operating cost. The operating cost is governed by the generator's conversion efficiency, while the device cost is determined by the cost of its construction to produce the desired electrical power output [6–12]. Since the conversion efficiency of a module is comparatively low, thermoelectric generation using waste heat energy is an ideal application. In this case, the operating cost is negligible compared to the module cost because the energy input (fuel) cost is cheap or free. Therefore, an important objective in thermoelectric power generation using waste heat energy is to reduce the cost-per-watt of the devices. Moreover, cost-per-watt can be reduced by optimizing the device geometry, improving the manufacture quality, and simply by operating the device at a larger temperature difference [7]. In addition, in designing high-performance thermoelectric power generators, the improvement of thermoelectric properties of materials and system optimization has attracted the attention of many research activities [6–12]. Their performance and economic competitiveness appear to depend on successful development of more advanced thermoelectric materials and thermoelectric power module designs.

Thermoelectric generators are semiconductor devices based on thermoelectric effects that can convert thermal energy directly into electricity. When a temperature gradient is established between junctions of materials, e.g., one junction is heated and the other cooled, a voltage (Seebeck voltage) is generated. The thermocouple that is created can be connected to a load to provide electric power. Thus, based on this Seebeck effect, thermoelectric devices can act as electrical power generators, as shown in the literature [6–12]. The equation that dictates the performance of TEG can be expressed as

$$\mathbf{ZT = S^2T/\mu k} \tag{2.1}$$

While the thermoelectric materials' dimensionless figure of merit ZT is a well-defined metric to evaluate thermoelectric materials, it can be a poor metric for maximum thermoelectric device efficiency because of the temperature dependence of the Seebeck coefficient "S," the electrical resistivity "k," and the thermal conductivity "μ," where "T" is the absolute temperature. Historically the field has used a thermoelectric device dimensionless figure of merit ZT to characterize a device operating between a hot side temperature "T_h" and cold side temperature "T_c." While there are many approximate methods to calculate ZT from temperature-dependent materials properties, an exact method uses a simple algorithm that can be performed on a spreadsheet calculator [6–12]. The conversion efficiency from heat to power is a function of operating temperature difference. An increase in the temperature difference provides an increase in heat available for conversion, so large temperature differences are desirable [7]. Only materials which possess $ZT > 0.5$ are regarded as thermoelectric materials [7]. Established thermoelectric materials can be divided into groups depending upon the temperature range of operation [6–12]:

- Low-temperature materials, up to around 450 K
- Medium-temperature materials, from 450 K up to around 850 K
- High-temperature materials, from 850 K up to around 1,300 K

Alloys based on bismuth in combinations with antimony, tellurium, and selenium are low-temperature materials. Medium-temperature materials are based on lead telluride and its alloys. High-temperature materials are fabricated from silicon germanium alloys [7].

The TEG device is composed of one or more thermoelectric couples. The simplest TEG consists of a thermocouple, comprising a pair of P-type and N-type thermoelements or legs connected electrically in series and thermally in parallel. The differentiation between N- and P-doped materials is important. The single and multistage configurations of TEG are described in Figures 2.2a, and 2.2b respectively. The details on these configurations are given in numerous references [6–12].

The sizes of conventional thermoelectric devices vary from 3 mm^2 by 4 mm thick to 75 mm^2 by 5 mm thick. Most of thermoelectric modules are not larger than 50 mm in length due to mechanical consideration. The height of single-stage thermoelectric modules ranges from 1 to 5 mm. The modules contain from 3 to 127 thermocouples [7]. There are multistage thermoelectric devices designed to meet requirements for large temperature differentials. Multistage thermoelectric modules can be up to 20 mm in height, depending on the number of stages. The power output for most of the commercially available thermoelectric power generators ranges from microwatts to multi-kilowatts [7]. For example, a standard thermoelectric device consists of 71 thermocouples with a size of 75 mm^2 can deliver electrical power of approximately 19 W [6–12]. The maximum output power from a thermoelectric power generator typically varies depending on temperature difference between hot and cold plates and module specifications, such as module geometry (i.e., cross-sectional area and thermoelement length), thermoelectric materials, and contact properties. The maximum power output increases parabolically with an increase in temperature difference. For a given temperature difference, there is a significant variation in maximum power output for different modules due to variation in thermoelectric materials, module geometry, and contact properties. The maximum power output follows a clear trend and increases with decrease in the thermoelement length for a given module cross-sectional area. As demonstrated in my previous book [11], the performance of TEG depends on the materials and TEG configurations. Significant literature [6–10, 12], along with my previous book [11], describes the progress made on materials and TEG configurations over the past several decades. Here we will mainly focus on the applications of TEG for various types of waste heat.

A. **Applications**

About 70% of energy in the world is wasted as heat and is released into the environment with a significant influence on global warming [7, 13]. The waste heat energy released into the environment is one of the most significant sources of clean, fuel-free, and cheap energy available. The unfavorable effects of global warming can be

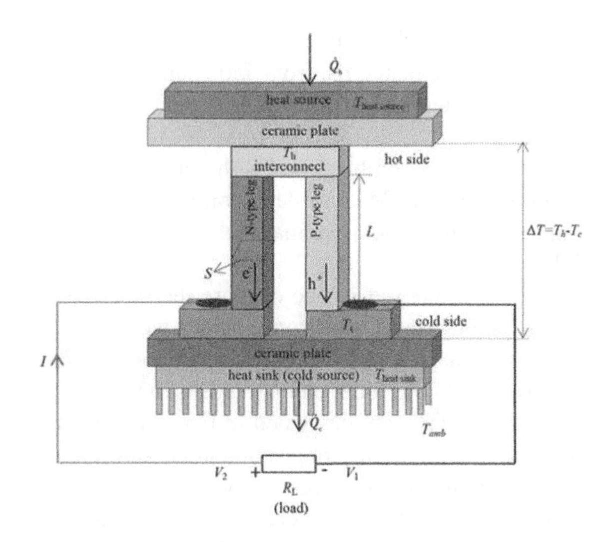

(a) single stage

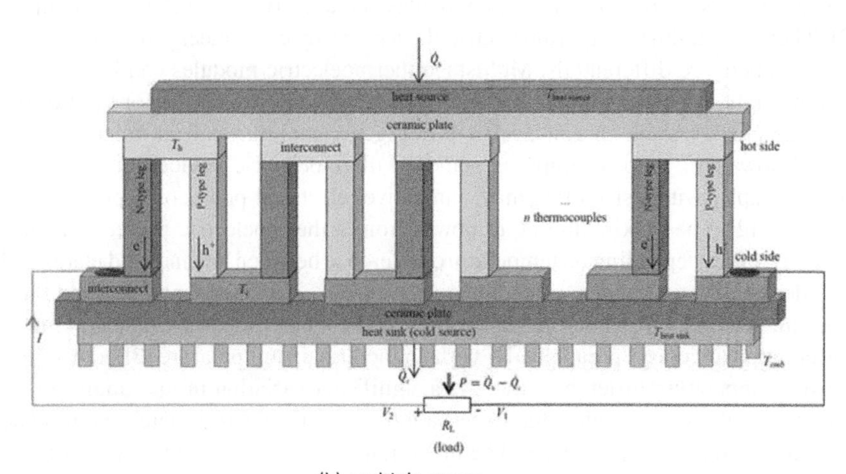

(b) multiple stages

FIGURE 2.2 Schematic diagram showing components and arrangement of a typical single-stage and multiple stages thermoelectric power generator [7].

diminished using the TEG system by harvesting waste heat from residential, industrial, and commercial fields [6–12].TEG is substantially used to recover waste heat in different applications ranging from μW to MW (see Figure 2.3). Different waste heat sources and temperature ranges for thermoelectric energy harvesting are shown in Table 2.1 [7]. What makes thermoelectricity most interesting with regard to hybrid energy is that here not only processes generating power generate additional power from waste heat, but processes using heat as a source of energy can also generate power from the waste heat. This is more like a reverse cogeneration form of hybrid energy system.

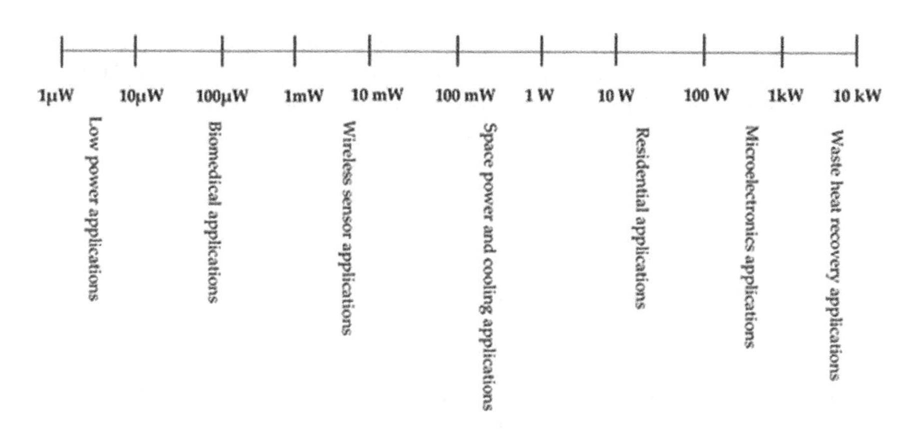

FIGURE 2.3 Energy conversion applications [7].

TABLE 2.1
Different Waste Heat Sources and Temperature Ranges for Thermoelectric Harvesting Technology [7]

Temperature Ranges (°C)	Temperature (°C)	Waste Heat Sources
High temperature (>650)	650–760	Aluminum refining furnaces
	760–815	Copper reverberatory furnace
	760–110	Copper refining furnace
	620–730	Cement kiln
		Hydrogen plants
Medium temperature (230–650)	315–600	Reciprocating engine exhausts
	425–650	Catalytic crackers
	425–650	Annealing furnace cooling systems
Low temperature (25–100)	32–55	Cooling water
	27–50	Air compressors
	27–88	Forming dies and pumps

Enormous quantities of waste heat generated from various sources are continuously discharged into the earth's environment, much of it at temperatures which are too low to recover using conventional electrical power generators. Thermoelectric power generation, which presents itself as a promising alternative green technology, has been successfully used to produce electrical power in a range of scales directly from various sources of waste heat energy. Some of them are described here. The applications can be divided into microscale and macroscale. At microscale, waste heat can be converted to electric power with a micro thermoelectric generator. Waste human body heat can be used to power a thermoelectric "watch battery". It is estimated that 2,875 thermoelements connected in series would be required to obtain the 2 V required to operate the watch [7, 14, 15]. The TEG devices are especially suitable for waste heat harvesting for low-power generation to supply electric energy for microelectronic applications [6–12, 16–31]. Glatz et al. [16] presented a novel polymer-based wafer level fabrication process for micro thermoelectric power generators for the application on nonplanar surfaces. Furthermore, it can be used for various microelectronic devices, like wireless sensor networks, mobile devices (e.g., MP3 player, smartphones, and iPod), and biomedical devices. The thermoelectric energy harvesters are microelectronic devices made of inorganic thermoelectric materials, at different dimensions, with a lifetime of about five years [7, 14, 15]. A TEG to be applied in a network of body sensors has been presented in several references [7, 14, 15].

At macroscale, TEG can be used for domestic waste heat applications and industrial waste heat applications. Rowe [6, 17] reported that a waste heat-based thermoelectric power generator is used in a domestic central heating system with the modules located between the heat source and the water jacket. It was concluded that two modules based on PbTe technology when operated at hot and cold side temperatures of 550°C and 50°C, respectively, would generate the 50 W required to power the circulating pump [6, 17]. Waste heat energy can also be utilized proportionally from 20 kW to 50 kW wood- or diesel-heated stoves [6–12, 18], especially during the winter months in rural regions where electric power supply is unreliable or intermittent to power thermoelectric generators.

The most amounts of heat are emitted and released into the atmosphere in the form of flue gases and radiant heat energy with a negative impact to the environmental pollution (emissions of CO_2) by industrial processes [6–12]. For this reason, thermoelectric harvesters are good candidates to recover waste heat from industries [6–12] and convert it into useful power (e.g., to supply small sensing electronic device in a plant).

Utilization of TEGs in the industrial field is beneficial from two points of view:

- In the industrial applications where recoverability of the waste heat by the conventional system (radiated heat energy) is very difficult to be done.
- In the industrial applications where the use of thermoelectric materials reduces the need for maintenance of the systems and the price of the electric power is low, even if the efficiency is low [6,6–12, 19].

The results of a test carried out on a TEG system attached at a carburizing furnace (made of 16 Bi_2Te_3 modules and a heat exchanger) are shown by Kaibe et al. [20]. The system harvested about 20% of the heat (P=4 kW). The maximum electrical output power generated by TEG was approximately 214 W, leading to thermoelectric conversion efficiency of 5%. Aranguren et al. [32] built a TEG prototype in which TEG was attached at the exhaust of a combustion chamber, with 48 modules connected in series and two different kinds of finned heat sinks, heat exchangers, and heat pipes. TEGs are also useful for recovery of waste heat from the cement rotary kiln to generate electricity, considering that the rotary kiln is the main equipment used for large-scale industrial cement production [21]. The electric output power evaluation of a TEG system attached to an industrial thermal oil heater is presented by

Barma et al. [22]. The impact of different design and flow parameters were assessed to maximize the electrical output power. The estimated annual electrical power generation from the proposed system was about 181,209 kWh. The thermal efficiency of the TEG based on recently developed thermoelectric materials (N-type hot forged Bi_2Te_3 and P-type $(Bi,Sb)_2Te_3$ used for the temperature range of 300 K–573 K) was enhanced up to 8.18%.

Most of the recent research activities on applications of thermoelectric power generation have been directed toward utilization of industrial waste heat [6–12]. Vast amounts of heat are rejected from industry, manufacturing plants, and power utilities as gases or liquids at temperatures which are too low to be used in conventional generating units (<450 K). In this large-scale application, thermoelectric power generators offer a potential alternative of electricity generation powered by waste heat energy that would contribute to solving the worldwide energy crisis, and at the same time, help reduce environmental global warming. In particular, the replacement of by-heat boiler and gas turbine by thermoelectric power generators makes it capable of largely reducing capital cost, increasing stability, saving energy source, and protecting environment [7].

Min and Rowe [23, 24] reported that New Energy and Industrial Technology Development Organization (NEDO) designed low-temperature waste heat power generator that consisted of an array of modules sandwiched between hot and cold water-carrying channels. When operated using hot water at a temperature of approximately 90°C and cold flow at ambient temperature, Watt-100 generates 100 W at a power density approaching 80 kW/m^3. In this application, the system was scalable, enabling 1.5 kW of electrical power to be generated [6–12]. Thermoelectric power generators have also been successfully applied in recovering waste heat from steel manufacturing plants. In this application, large amounts of cooling water are typically discharged at constant temperatures of around 90°C when used for cooling ingots in steel plants. When operating in its continuous steel casting mode, the furnace provides a steady-state source of convenient piped water which can be readily converted by thermoelectric power generators into electricity. It was reported [6–12] that total electrical power of around 8 MW would be produced employing currently available modules fabricated, using Bi_2Te_3 thermoelectric modules technology.

Another application where thermoelectric power generators using waste heat energy have potential use is in industrial cogeneration systems [19]. For example, Yodovard et al. [25] assessed the potential of waste heat thermoelectric power generation for diesel cycle and gas turbine cogeneration in the manufacturing industrial sector in Thailand. The data from more than 27,000 factories from different sectors, namely, chemical product, food processing, oil refining, palm oil mills petrochemical, pulp and paper rice mills, sugar mills, and textiles, were used. It is reported that gas turbine and diesel cycle cogeneration systems produced electricity estimated at 33% and 40% of fuel input, respectively. The useful waste heat from stack exhaust of cogeneration systems was estimated at approximately 20% for a gas turbine and 10% for the diesel cycle. The corresponding net power generation was about 100 MW. The potential for thermoelectric power from various industrial processes is depicted in Table 2.2 [11].

TABLE 2.2
Potential of Thermoelectric Power from Various Manufacturing and Process Operations [11]

Manufacturing Process Industry	Process Heating Energy Use (TBtu/year)	Process Heating Energy Losses (TBtu/year)	Estimated Recoverable Heat Range (TBtu/year)	Estimated Thermoelectric Potential (TBtu/year)	Estimated Thermoelectric Potential (GWh/year)
Petroleum refining	2,250	397	40–99	1–2	291–727
Chemicals	1,460	328	33–82	1–2	240–601
Forest products	980	701	70–175	2–4	513–1,280
Iron and steel	729	334	33–84	1–2	245–612
Food and beverage	518	293	29–73	1–2	215–537
Glass	161	88	9–22	0–1	64–161
Other manufacturing	1,110	426	43–107	1–3	312–780
All manufacturing	**7,208**	**2,567**	**257–642**	**7–16**	**1,880–4,626**

Source: © Waste Heat from Incineration of Solid Waste Applications

Recently, the possibility of utilizing the heat from incinerated municipal solid waste has also been considered. For example, in Japan, the solid waste per capita is around 1 kg per day and the amount of energy in equivalent oil is estimated at 18 million kJ by the end of the 21st century. It was reported by [6–12, 19] that an on-site experiment using a 60 W thermoelectric module, installed near the boiler section of an incinerator plant, achieved an estimated conversion efficiency of approximately 4.4%. The incinerator waste gas temperature varied between 823 K and 973 K, and with forced air cooling on the cold side, an estimated conversion efficiency of approximately 4.5% was achieved. An analysis of a conceptual large-scale system burning 100 ton of solid waste during a 16-hour day indicated that around 426 kW could be delivered [6–12, 19]. In the management of waste heat in incineration applications, the thermoelectric modules are typically placed on walls of the furnace's funnels. This construction can eliminate the by-heat furnace, gas turbine, and other appending parts of steam recycle [19].

B. ***TEGs attached to the solar systems***

Solar TEG (STEG) systems, PV systems, and concentrating solar power plants can generate electricity by using the solar heat. A STEG is composed of a TEG system sandwiched between a solar absorber and a heat sink as shown in Figure 2.4. The

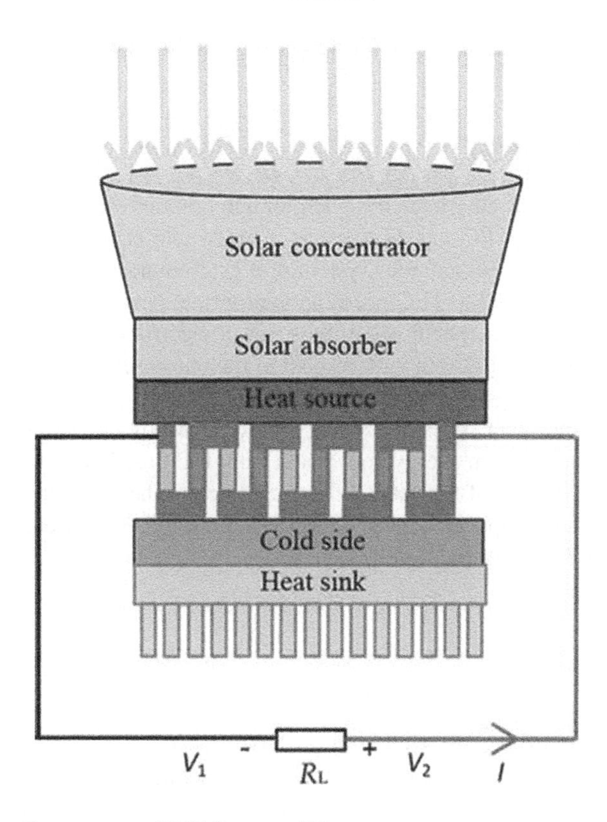

FIGURE 2.4 Components of STEG system [7].

solar flux is absorbed by the solar absorber and is concentrated into one point. Then, the heat is transferred through TEG by using a pipeline and is partially converted into electrical power by the TEG. A heat sink rejects the excess heat at the cold junction of the TEG to keep a proper ΔT across the TEG [26, 27].

Due to the development of the thermoelectric materials, a solar TEG with an incident flux of 100 kW/m^2 and a hot side temperature of 1,000°C could obtain 15.9% conversion efficiency. The solar TEG is very attractive for stand-alone power conversion. The efficiency of a solar TEG depends on both the efficiency with which sunlight is absorbed and converted into heat and the basic TEG efficiency. Furthermore, the total efficiency of a solar TEG is also influenced by the heat lost from the surface. The efficiency of solar TEG systems is relatively small due to the low Carnot efficiency provoked by the reduced temperature difference across the TEG and the reduced ZT [28]. Its improvement needs to rise temperature differences and to develop new materials with high ZT like nanostructured and complex bulk materials (e.g., a device with $ZT=2$ and a temperature of 1,500°C would lead to obtaining a conversion efficiency about 30.6%) [29]. According to the literature survey, both residential and commercial applications gain much more interest in the regions of incident solar radiation of solar TEGs. This can be explained by the fact that most of the heat released at the cold side of the TEG can be used for domestic hot water and space heating [27].

C. ***Grid integration of TEG***

Most TEG applications have been designed for autonomous operation within a local system. In general, a TEG can be seen as a renewable energy power generation source that supplies an autonomous system or a grid-connected system. To be suitable for grid connection, the TEG needs an appropriate power conditioning system. This power conditioning system has to be a power electronic system, with specific regulation capabilities, different with respect to the ones used for solar photovoltaic and wind power systems [30], because the TEG operating conditions are different with respect to the other renewable energy sources. Molina et al. [31] proposed a control strategy to perform energy conversion from DC to AC output voltage, which maintains the operation of the thermoelectric device at the maximum power point (MPP). In the same proposal, active and reactive power controls were addressed by using a dedicated power conditioning system.

2.3.2 Hybrid Energy System Created by Conversion of CO_2 to Power or Fuel

2.3.2.1 Exxon-Mobil- Fuel Cell Energy Partnership

The greatest opportunity for future large-scale deployment of carbon capture and sequestration (CCS) may be in the natural gas power generation sector, since capturing CO_2 from coal-fired generation is roughly twice as expensive. In 2016, ExxonMobil announced a partnership with FuelCell Energy, Inc. to advance new technology that may substantially improve CCS efficiency, effectiveness, and affordability for large natural gas-fired power plants. The research indicates that by applying this new technology, more than 90% of a natural gas power plant's carbon dioxide emissions could be captured [33–36].

Scientists at ExxonMobil and FuelCell Energy, Inc. are jointly pursuing new technology that could reduce the costs associated with current CCS processes, by increasing the amount of electricity a power plant produces while simultaneously delivering significant reductions in carbon dioxide emissions. At the center of these efforts is a carbonate fuel cell. Laboratory tests have indicated that applying carbonate fuel cells to natural gas power generation could capture carbon dioxide more efficiently than current, conventional CCS technology.

During conventional carbon capture processes, a chemical reacts with the carbon dioxide to extract it from power plant exhaust. Steam is then required to release the carbon dioxide from the chemical—steam that would otherwise be used to move a turbine. The effect is to decrease the amount of electric power the turbine can generate. Using fuel cells to capture carbon dioxide from power plants can result in a more efficient separation of carbon dioxide from power plant exhaust with an increased output of electricity. Power plant exhaust is directed to the fuel cell, replacing air that is normally used in combination with natural gas during the fuel cell power generation process. As the fuel cell generates power, the carbon dioxide becomes more concentrated, allowing it to be more easily and affordably captured from the cell's exhaust and stored.

ExxonMobil's research indicates that a typical 500 MW power plant using a carbonate fuel cell may be able to generate an additional 120 MW of power, while current CCS technology actually consumes about 50 MW of power. ExxonMobil has been assessing a number of carbon capture technologies for many years and believes that carbonate fuel cell technology offers a great potential. The technology's capability has been tested in the laboratory, and data from those simulations is currently under analysis. Further development will involve a more detailed examination of each component of the system and optimization of the system as a whole [33–36].

In theory, carbon capture is simple. The carbon dioxide produced when fossil fuels are burned to produce electricity is captured and then stored deep underground instead of being released into the atmosphere, where they become heat-trapping greenhouse gases. Because large amounts of energy are required to concentrate carbon dioxide molecules together so that they can be caught, current carbon capture technology is expensive. It's just one of several technical and economic hurdles facing large-scale use of carbon capture. The fuel cell could be a fundamental shift in carbon capture because it can trap the gas while also generating electricity. This is important in power generation, where every percentage increase in efficiency matters. Fuel cell will also considerably reduce emission of CO_2 into the atmosphere.

When natural gas is burned in a gas turbine, the exhaust produced is only about 4% carbon dioxide. Carbonate fuel cells can grab that carbon dioxide and concentrate it into a stream that is around 70–80% carbon dioxide while creating more electricity at the same time. Further processing increases the carbon dioxide concentration to over 95%. While that's a promising start, ExxonMobil and FuelCell Energy are planning to test and improve the technology to further increase its efficiency and demonstrate it at larger scale. The goal is to minimize emissions while maximizing power output. Such a treatment of CO_2 will also make power plants hybrid energy systems.

The unique feature of these fuel cells is that they act as a CO_2 pump while using natural gas to generate power. As CO_2 stream is passing through the fuel cell membrane, it is getting concentrated so that it can then go into storage, while generating additional electricity. FuelCell Energy's scalable and affordable carbon capture solution captures carbon emissions from existing coal- or gas-fired power plants, while simultaneously producing power. The solution is scalable so the amount of carbon capture can be increased over time. The fuel cells also destroy approximately 70% of the plant's smog-producing pollutants. The capability to efficiently and affordably capture carbon while simultaneously producing power enables emission compliance in a manner that provides a revenue stream and returns on capital from the sale of power produced by the fuel cells. This process is called "Sure Source."

Benefits of Sure Source carbon capture include [33–36]:

- **Scalable and affordable:** Fuel cell plants can be added incrementally in a cost-effective manner; starting with as little as 5% capture with no appreciable change in the cost of power and with minimum capital outlay. To achieve 90% capture, the cost of power for a coal-fired plant is increased by only $0.02/kWh, while conventional carbon capture technologies for coal-fired power plants almost double the cost of power.
- **Produces, rather than consumes, additional power:** The process generates additional power—cleanly and efficiently—during the carbon capture process, contributing to the existing coal plant's total output. In comparison, conventional carbon capture technologies *consume* about 20% of the plant's overall power output.
- **Destruction of pollutants:** Captures and separates CO_2 from the flue gas of coal- and gas-fired power plants while simultaneously destroying approximately 70% of the smog-producing nitrogen oxide (NOx).
- **Return on investment:** Generates a return on capital rather than an increase in operating expense, extends the life of existing coal-fired power plants, and enables low-carbon utilization of domestic coal and gas resources.
- **Proven solution:** Millions of megawatt hours of ultra-clean power are generated by Sure Source power plants globally.

Since Sure Source power plants produce power efficiently and with virtually zero emissions, the net result is a compelling carbon capture solution for preventing the release of greenhouse gases by coal- or gas-fired power plants, while simultaneously increasing overall net efficiency and clean power output in an affordable manner for ratepayers.

CCS has long been an ultimate goal for many energy companies, for both money saving and environmental reasons. It is the process by which the carbon dioxide, which would otherwise be released as waste from power plants into the atmosphere, is captured, compressed, and injected underground for permanent storage. Fuel cells have been a rapidly expanding option for mini-grids and distributed generation because they produce electricity directly from a chemical reaction, and devices are generally fairly small and easy to install and transport. In this case, the fuel cells would run directly off the power plant emissions, removing them from the air and, in turn, producing additional electricity to feed it back into the system or indeed sell off.

Many facilities have been able to capture CO_2 emissions since the 1970s, but in very energetically exhaustive ways. Existing CCS technology actually consumes up to 25% of electricity at a power plant, equating to a large amount of money. With power plants already facing financial challenges from the growing interest in renewable energy, falling oil prices, and an increasing unpopularity of coal, a 25% jump in power generation isn't usually feasible. For example, Southern Company built a coal-fired power plant in Kemper, Mississippi, which was supposed to incorporate CCS. The project ended up costing $7billion in total, three times the original estimate. In 2015, ExxonMobil claimed that it captured 6.9 million metric tons of CO_2 using the CCS process, which is the equivalent amount of fumes from over a million cars. Researchers think that pursuing new CCS technology could actually help to reduce costs. While current CCS processes are associated with added expenditures, by combining with fuel cell, the new method could increase the amount of electricity a power plant produces. In addition to the cost saving and environmental benefits, carbonate fuel cells also produce a chemical feedstock called "syngas," which is primarily made of hydrogen and can be used as a fuel for internal combustion engines.

In September 2015, FuelCell Energy announced a $23.7 million cost-shared project with the Department of Energy to demonstrate that its technology could capture 90% of the CO_2 from a small stream of coal exhaust and concentrate it to 95% purity. In the first phase of the project, a modified version of its commercially available 2.8 MW SureSource 3000 fuel cell system will capture 54 metric tons of carbon dioxide per day at the Barry plant. That's just a small fraction of the CO_2 emitted by the facility. To put things in perspective, a typical 500-MW coal plant emits about 3.3 million metric tons of carbon dioxide per year—which works out to about 9,000 metric tons per day—and so capturing 90% of those emissions would require about 400 MW of fuel cell capacity [33–36].

This leads to one potential setback for the technology in that the amount of CO_2 that can be captured depends heavily on how many fuel cells are in operation. For example, a 500 MW combined cycle plant would require at least 120 MW of fuel cells to achieve a level of 90% carbon capture. An equivalent coal plant might need as much as 400 MW of fuel cells because coal plants are generally less efficient and emit higher levels of CO_2. That's a lot of fuel cells. In terms of capital investment, fuel cell power is three to four times as expensive as conventional coal power. With a carbonate fuel cell, the cogenerated heat and electricity could boost the host plant's power output by 80%. Advocates point out that when full life cycle costs are taken into account, along with the difficult-to-quantify environmental benefits, the balance begins to favor such a combined cycle. Either way, FuelCell Energy and Exxon scientists will focus on increasing the efficiency in separating the CO_2 from gas turbine exhausts and are likely to learn a lot in the process. They are working to better understand the chemical processes and working out how they respond to different compositions and concentrations of the flue gas. If successful, the next steps will be to launch a pilot project for more testing and then integrate it to a larger scale pilot facility after. Eventually, the goal is to grid-tied build a 2 MW–3 MW demonstration plant that would run alongside a coal-powered plant. Table 2.3 compares conventional CCS technology with CCS based on CO_2 fuel cell technology [33–36].

Table 2.3 Comparison of performance of coal power plant with various options for carbon capture [33–36].

500 MW coal power plant with no carbon capture: Emits 3.6 million tons of CO_2 per year. This is equivalent to more than 685,000 cars annually. Cost of electricity 6¢ /kWr.

TABLE 2.3
Comparison of the Three Scenarios Analyzed [37]

Scenario	Thermal Product	Storage
#1—Electric boiler	Variable	None
#2—Electric thermal storage	Constant	Thermal storage from electricity
#3—Electric boiler thermal storage	Constant	Thermal storage from electricity and thermal sources

500 MW coal power plant with conventional carbon capture: Carbon capture by conventional absorption technology consumes significant amount of power plant's output (20% or 100 MW) to capture 90% of CO_2. This causes cost of electricity to increase by 80% and increase in pollutants by 25% (lbs/MWh). The cost of electricity would be 11¢/kWh.

500 MW coal power plant with fuel cell CO_2 capture: In this process, flue gas from the coal plant is routed into the fuel cells, which then concentrate and capture CO_2 as a side reaction during power generation. The coal plant remains at full power while the fuel cells affordably capture CO_2 and destroy approximately 70% of the coal plant NOx emissions. This will increase total power generation by 80% to 900 MW, increase cost of electricity by 33%, and decrease pollutants by 78% (lbs/MWh). The cost of electricity would be 8¢/kWh.

2.3.2.2 MCFC Trigeneration Plant

Rinaldi et al. [38] examined another hybrid energy approach by looking at the possibility of separating and recovering CO_2 in a biogas plant that co-produces electricity, hydrogen, and heat. Exploiting the ability of a molten carbonate fuel cell (MCFC) to concentrate CO_2 in the anode exhaust stream reduces the energy consumption and complexity of CO_2 separation techniques that would otherwise be required to remove dilute CO_2 from combustion exhaust streams. Three potential CO_2 concentrating configurations are numerically simulated to evaluate potential CO_2 recovery rates: (1) anode oxidation and partial CO_2 recirculation, (2) integration with exhaust from an internal combustion engine, and (3) series connection of molten carbonate cathodes initially fed with internal combustion engine (ICE) exhaust. Physical models were calibrated with data acquired from an operating MCFC tri-generating plant. Results illustrated a high compatibility between hydrogen co-production and CO_2 recovery with series connection of molten carbonate systems offering the best results for efficient CO_2 recovery. In this case, the carbon capture ratio (CCR) exceeds 73% for two systems in series and 90% for three MCFCs in series. This remarkably high carbon recovery is possible with 1.4 MWe delivered by the ICE system and 0.9 MWe and about 350 kg/day of H_2 delivered by the three MCFCs.

The systems described could be readily deployed using existing commercial systems. Moreover, the first and the second configurations, while appearing more complex, could be realized in a single system with some valve design and controls for switching between ICE exhaust and CO_2 injection from anode recirculation. A series configuration of MCFC technology coupled with an ICE achieves outstanding carbon recovery (exceeding 90%), with minimal parasitic load: that of the production of O_2 in an ASU and the cooling fan in the second condenser.

Carbon separation and hydrogen co-production processes are compatible and benefit from the carbonate ion charge-carrying properties of a molten carbonate fuel cell. Hydrogen co-production reduces fuel-to-electric efficiency due to the pressure-swing adsorption (PSA) parasitic load and lower operating fuel utilization used to generate excess hydrogen. All three configurations achieved notable carbon removal with minimal parasitic load. Configuration 1 (stand-alone system) required a high CO_2 utilization factor (>75%) in order to separate a sufficient carbon quantity. Possible drawbacks to carbon recovery from a MCFC tri-generation system include complex thermal integration due to additional heat exchange steps as well as reduced fuel-to-electric efficiency due to the low fuel utilization and high parasitic electric load of the hydrogen separation unit (HSU)and any additional compression of CO_2 and H_2 products not considered in this analysis.

2.3.2.3 CO_2 to Fuel and Value-added Products

After more than 40 years of global competitive research on the conversion of CO_2 to fuel, the first pilot demonstration to successfully take CO_2 from thin air to the petrol pump turns out to be based on rather conventional materials and well established processing technology [39]. The stakes have been raised in the quest for a commercially viable CO_2-to-fuel technology with the recent announcement, by the Audi–Sunfire–Climeworks consortium of companies, operating in Germany and Switzerland, of a pilot project for producing diesel fuel from CO_2 and H_2O and renewable energy sources, such as wind or solar or hydropower. The CO_2 supply was obtained mainly from a biogas plant supplemented with some CO_2 captured directly from the air. A pilot plant in Dresden is destined in the months ahead to produce around 160 l/day of the synthetic diesel, which they have dubbed "blue crude" [39].

The process employed for making "blue crude" actually involves five main steps: the first requires capture of CO_2 from thin air; the second involves electricity generation from renewable sources; the third uses electrical energy to electrochemically split water into a hydrogen (H_2) and oxygen (O_2) mixture from which the H_2 must be separated; the fourth uses this H_2 to reduce the CO_2 captured from the atmosphere, to form a mixture of carbon monoxide (CO) and water (H_2O) in a high-temperature and high-pressure reverse water gas shift (RWGS) process, from which the H_2O and unreacted CO_2 is separated; and the fifth mixes the CO with the H_2 to form synthesis gas (syngas), which is subsequently converted in a high-temperature and high-pressure Fischer–Tropsch (F-T) catalytic procedure, to generate the long-chain hydrocarbons comprising the "blue crude." It appears from the literature publication that about 80% of the "blue crude" produced by the F-T process is suitable for synthetic diesel making, which is more analogous to crude oil where the other fraction is mainly lighter hydrocarbons, which cannot be used in diesel fuel. The literature indicates that while the electrochemical, adsorption, separation, and catalytic methods employed to make "blue crude" powered only by renewable energy sources are rather conventional in nature, the technology nevertheless works, can be scaled up, and has set today's benchmark standard for converting CO_2 to fuel. For any technology to compete commercially with "blue crude" technology, it will have to be more straightforward in materials and process engineering with less steps involved and more energy efficient and economically attractive [39].

One way to transcend the complexity and efficiency of the multistep Audi CO_2+H_2O to fuel process is to discover a "dream" catalyst dubbed the "techno-leaf," which is able to directly convert gaseous CO_2+H_2O under mild conditions to a fuel, using just the energy contained in sunlight as the renewable power source. The gas-phase "techno-leaf" CO_2+H_2O to fuel process, not to be confused with the aqueous phase "artificial leaf," in essence, combines steps 2 and 3 in the five-step process described above, with step 4 if we stop at CO and step 5 if we make fuels such as CH_3OH, $(CH_3)_2O$, HCO_2H, or CH_4. If the "techno-leaf" could also capture CO_2 and H_2O from thin air and operate with low concentrations of CO_2 in the presence of O_2, like the "real leaf," it is a very serious "dream" to combine all five steps of the above mentioned process in one operation. With revolutionary advances in materials, chemistry, and catalysis to enhance the CO_2-to-fuel efficiency, this vision of the "techno-leaf" process could, in principle, be reduced to practice.

In the Audi process, much of its CO_2 is captured from a concentrated stream from an industrial biogas plant, which is much more energy efficient than from the ambient air. In the short term, this makes practical and economic sense. Though for a long-term sustainable solution, ideally, the co-extraction of CO_2 and H_2O from ambient air by a champion adsorbent would provide attractive logistical benefits for the production of synthetic fuels, especially in regions with limited or no fresh H_2O resources. In this context, amine-functionalized cellulose fashioned in nanofiber form has recently been identified as a promising sorbent that is able to concurrently extract CO_2 and H_2O vapor from ambient air with demonstrated potential for industrial-scale applicability. The favorable adsorption and desorption capacities and energetics at low-partial pressures and tolerance to moisture in air bode well for the successful application of this CO_2 and H_2O co-extraction process. By combining this process with the conversion of CO_2 and H_2O to fuel, energy can be conserved and costs reduced, because the need to transport CO_2 and consume fresh water resources is eliminated.

Kumaravel et al. [40] examined the photochemical conversion of carbon dioxide (CO_2) into fuels and value-added products. The conversion of carbon dioxide (CO_2) into fuels and value-added products is one of the most significant inventions to address global warming and energy needs. Photoelectrochemical (PEC) CO_2 conversion can be considered as an artificial photosynthesis technique that produces formate, formaldehyde, formic acid, methane, methanol, ethanol, etc. Recent advances in electrode materials, mechanisms, kinetics, thermodynamics, and reactor designs of PEC CO_2 conversion were comprehensively reviewed in this study. The adsorption and activation of CO_2 intermediates at the electrode surface are the key steps for improving the kinetics of CO_2 conversion. PEC efficiency could be upgraded through the utilization of 2D/3D materials, plasmonic metals, carbon-based catalysts, porous nanostructures, metal–organic frameworks, molecular catalysts, and biological molecules. The defect engineered (by cation/anion vacancy, crystal distortion, pits, and creation of oxygen vacancies) 2D/3D materials, Z-scheme heterojunctions, bioelectrodes, and tandem photovoltaic–PEC reactors are suitable options to enhance the efficiency at low external bias.

2.4 COAL-BASED HYBRID POWER PLANTS

Besides generating hybrid power plants by the use of waste heat and waste gas stream as described above, coal-based hybrid power plants can also be accomplished by the combined use of coal and other non-renewable fuels like gas and oil shale or renewable fuels like biomass and solar. Cofiring of coal and biomass and coal and shale oil have been extensively described in my previous book [1], and they will not be repeated here. Here, we briefly examine cofiring of coal and natural gas and coal–solar hybrid for power generation reported in an excellent report by Mills from IEA [41].

2.4.1 Cofiring of Coal and Natural Gas

An option attracting the interest of some power utilities is that of cofiring natural gas in coal-fired boilers. This hybrid technique can be instrumental in improving operational flexibility and reducing emissions. Both coal- and gas-based generation are vital in powering many of the world's developed and emerging economies. Both are used to provide a secure and uninterrupted supply of electricity, which is needed to ensure that economies and societies can develop and prosper. In some countries, coal provides much of the power, whereas in others, gas dominates. However, there are many instances where the national energy mix includes combinations of the two. Each brings its own well-documented advantages and disadvantages, but recent years have seen a growing interest in means by which these two fuels might be combined in an environment-friendly and cost-effective way. In many countries, environmental legislation continues to be introduced to reduce levels of SO_2, NOx, particulates, mercury, and, more recently, CO_2, and increasingly operators face the dilemma of what to do with their plants. One alternative is to modify the plant so that natural gas can be added, helping to keep it in service. Cofiring with gas seems a promising option for at least some existing coal-fired plants. Importantly, some conversions allow the ratio of coal to gas to be varied, providing a useful degree of flexibility in terms of fuel supply and plant operation [41].

Potentially, gas can be added to an existing system in a number of ways. Some replace a portion of the main coal feed, whereas others use gas as a means for minimizing emissions of species such as NOx. The amount of gas used for some types of application will normally be less if the plant is converted to actually cofire gas in the true sense. This allows much greater volumes to be fed directly into a boiler and burned simultaneously with the coal feed. Depending on the individual plant and operational requirements, the existing unit can be configured in a number of ways so that cofiring becomes a practical proposition. It may be a case of replacing oil-fired igniters or warm-up guns with gas-fired equivalents, one of the simplest options. But should a plant operator wish to consistently put even more gas through his plant, it is likely that gas firing will need to be incorporated into the main burner system; there are dual- or multi-fuel burners suitable for this, available commercially from a number of suppliers [41, 42].

Various types of burner assemblies are offered commercially, some designed specifically for cofiring. For example, Breen Energy Solutions [41, 42] of the United States has developed a proprietary system known as dual orifice cofiring. This was developed as an effective means for handling variable gas input to a coal-fired boiler. Potentially, there are a number of possible inlets for the gas supply. Feeding it at the igniter has merit as the gas is needed in this location. However, feeding in more gas than the device is designed to handle can cause problems. For example, it can create significant competition between the gas and coal flames for available oxygen. Furthermore, doubling or tripling the gas throughput can also place the resultant flame near the center of the boiler, minimizing water wall steam creation and increasing superheat/reheat steam temperatures [43].

Other US manufacturers also produce dual-fuel burners. For example, Texas-based Forney and Storm Technologies have developed a proprietary system known as the Eagle Air burner. This is a wall-fired boiler dual-fuel burner that operates on coal or natural gas. Available in a range of capacities, it can operate on 100% coal or natural gas and incorporates multiple zones of secondary air that allow for combustion and NOx tuning. Advantages cited include enhanced fuel flexibility, the ability to stage the gas and coal for improved combustion and emissions, and improved off-peak low-load operation (meaning reduced cycling) [41, 43].

Replacing a percentage of coal feed with gas will clearly help reduce overall plant emissions of SO_2, NOx, particulates, mercury, and CO_2. Breen Energy Solutions [42] claims that replacement of 35% of coal feed using their cofiring system can reduce SO_2/SO_3 emissions by up to 35%, NOx emissions by 45%, particulates by 35%, mercury by 35%, and CO_2 by 20%. Stable low-load operation can also be achieved. In the United States, the National Fire Protection Association also requires that two pulverizers are kept in service. However, cofiring gas at low load can avoid this.

Natural gas can also be added for reburning. Reburning technology was originally developed for NOx combustion control, primarily on coal-fired furnaces. It is a staged fuel approach that uses the entire volume of a furnace, rather than the control of NOx production/destruction within the flame envelope. Reburn is a three-stage combustion process that takes place in primary, reburn, and burnout zones. In the primary zone, pulverized coal is fired through conventional or low-NOx burners operating at low excess air. A second fuel injection is made in a region of the boiler after the coal combustion, creating a fuel-rich reaction zone (the reburn zone). Here, reactive radical species are produced from the natural gas that react chemically with the NOx produced in the primary zone, reducing it to molecular nitrogen. The partial combustion of the natural gas in this reburn zone results in high levels of CO. A final addition of overfire air, creating the burnout zone, completes the overall combustion process.

In practice, the technique usually involves the splitting of the boiler's combustion zone by installing a second level of burners above the primary combustion zone. Typically, up to 25% of the total heat input is injected into this reburn zone, creating fuel-rich conditions in the region of primary combustion. Within the reburn zone, NOx formed in the combustion zone is partially reduced to elemental nitrogen. The formation of additional NOx is limited due to the lower oxygen concentrations and lower combustion temperature in the reburn zone. Most coal-fired power plants that have deployed reburn systems use natural gas as the reburn fuel. Although other fuels have been used for reburning, gas usually provides the greatest NOx reduction performance as it is easy to inject and control and does not contain any fuel nitrogen. Natural gas reburn can reduce NOx emissions by up to 70%. In recent years, some technological advances have increased the effectiveness of reburning, such as dual-fuel orifice cofiring technology. Alongside this, Breen Energy Solutions has also developed a system known as fuel lean gas reburning (FLGR) [41]. With these new developments, reburning systems can provide a means for incorporating a sizable amount of natural gas into an existing coal-fired power plant, thus providing a number of environmental and operational benefits.

There are various pluses and minuses involved in cofiring process. Some of the main attractions of cofiring include [41, 43]:

- **Possible adaptation/reuse of existing infrastructure and control systems**. Many coal-fired power plants already use natural gas as a start-up or backup fuel, so the necessary infrastructure and control systems for feeding gas to the boiler may already be in place.
- **Enhanced fuel flexibility**. Cofiring removes total reliance on a single source of fuel, creating fuel flexibility. If a problem arises with availability of one fuel, the plant has the ability to maintain operations by switching to the other. Similarly, increases in the price of either fuel can be countered by changing the cofiring ratio such that the cheaper fuel predominates.
- **Cost savings can be achieved by switching to the cheapest fuel at the time**. It may also be possible to switch to lower cost coals without the risk of lost capacity.
- **Some emission control upgrades may be reduced in scope, delayed, or avoided**, depending on the coal quality, level of gas cofiring intended, and the regulatory environment.
- **Improved operational flexibility**. Cofiring can reduce warm-up times, allowing the unit to be brought on line faster than an unmodified equivalent, as well as enabling faster ramp-up. A faster start-up can help minimize higher emissions sometimes experienced during this phase. This technique allows some US power plants to comply with the federal Mercury and Air Toxins Standards.
- **Cofiring can provide a significant reduction in the minimum unit load achievable,** an important factor for many coal-fired power plants. Also, by cofiring a significant amount of gas when in low-load conditions, the minimum operating temperature of a plant's selective catalytic reduction (SCR) unit (where fitted) can be maintained.

- **Less coal throughput will reduce wear and tear on pulverizers and coal handling systems**, as well as reduce associated operation and maintenance (O&M) costs.
- **Reduced coal throughput generates fewer solid and liquid plant wastes**. Bottom ash, fly ash, FGD scrubber sludge or gypsum, mill rejects, and various other waste products will be reduced as well as handling and disposal costs.
- **Emissions of CO_2, NOx, SO_2, and particulates to air will be proportionately lower**. O&M costs of environmental control systems such as FGD, SCR units, and electrostatic precipitators (ESPs) or bag filters are likely to be lower. In the case of SCR systems, ammonia use will be less and extended catalyst life is likely. Where mercury control systems are in place, less activated carbon will be required.
- **More public acceptance**.
- **Keeps coal supply sector alive with employment and saves money in competitive market through fuel flexibility**. Also cofiring can get credits for greenhouse gas abatement measures.

As with any technology, there are drawbacks. An obvious requirement is that the coal-fired plant has an adequate source of natural gas available at an acceptable price. If the plant already uses gas in some way, existing infrastructure may be adequate. If not, additional supply and control equipment may be required. Depending on the overall length and any local constraints, costs for a new gas pipeline can be considerable. A major attraction often cited for cofiring is the low price of natural gas. Although this is currently the case in the United States, gas is much more expensive and less readily available in some other economies. Even in the United States, there are concerns that prices could increase significantly in the future as political and environmental pressures on hydraulic fracturing and investments in gas export facilities could drive up the price of gas, closer to those seen in Europe. Higher gas prices could cancel out any advantages and cost savings provided by cofiring [41, 43].

In the United States, the Environmental Protection Agency (EPA) has suggested that, under some circumstances, cofiring could be an alternative to applying partial carbon capture and storage (CCS) to coal-fired power plants. The EPA has advised that new emissions standards could be met by cofiring ~40% natural gas in highly efficient supercritical pulverized coal power plants. However, some industry observers think that more than this level would be needed and that, for this to be achievable, the boiler would need to be specifically designed to operate in this manner.

Various technical issues will need to be considered when a switch to cofiring is contemplated. These include [41, 43]:

- Due to the high hydrogen content of natural gas (~25% mass), latent heat losses resulting from the production of water during the combustion process can be much higher than for all but the wettest coals.

- Differences in the flame temperature, gas mass flow, soot and ash reflectivity, and slag levels on boiler waterwalls can result in heat transfer imbalances throughout the steam generator.
- In some locations, supply restrictions may limit the maximum level of gas available. Furthermore, seasonal restrictions may apply, giving priority to other applications such as home heating.
- A major risk associated with cofiring is often poor natural gas burner placement—if not located appropriately, it can result in excessive temperatures or incomplete combustion in certain areas [41].
- There can be significant impacts on heat transfer in the boiler between coal and gas. In some cases, original heat transfer surfaces may be inadequate for full natural gas firing. The heat transfer characteristics for natural gas versus coal vary significantly—coal has more radiant heat transfer and gas has more convection. If modifications are not made to the existing heat transfer surfaces or alternative operating conditions identified, problems with metallurgy can arise and major plant components run the risk of becoming unreliable. It can be expensive to make significant changes to boiler heat transfer surfaces. Overcoming this requires evaluation across the complete spectrum of load dispatch and cycling scenarios [44].

The cost of generating electricity from coal or gas can be similar. Even slight changes in fuel price can result in significant swings in production costs, and this can create market opportunities for utilities that have both gas- and coal-fired assets. Cofiring can be a possible option, allowing pricing and market conditions to drive the fuel choice and mix. Substituting some coal input with gas is considered to be a low-risk option, allowing utilities to better meet changing market requirements.

A number of utilities have already adopted gas cofiring, and others are considering converting some of their coal plants such that they can operate on a mix of the two. In the United States, a number of plant operators are currently investigating the test firing of natural gas to determine the long-term feasibility of either full conversion or dual-fuel firing. Some are engaged in feasibility studies to evaluate the possible ramifications of cofiring. Others have already switched. In economies where electricity demand fluctuates, a power plant that can cycle quickly to meet peaks and troughs, and also ramp down during periods of low demand, is more likely to be profitable. However, most coal-fired units can only operate as low as 30%–35% load and still sustain good combustion, restricting the plant's ability to cycle. Furthermore, coal plants can be slow to cycle up to full load: it can take 12 hours or more to ramp up to load from a cold start. A plant capable of switching to gas at low loads and taking load down even farther, then switch back to coal at higher loads, could have a significant advantage over the competition [41].

The biggest near-term potential market for cofiring appears to be the United States, where a number of utilities are looking to extend the working lives of their plants while simultaneously reducing their environmental footprint. Cofiring can help reduce emissions, improve operational flexibility, and allow faster start-ups, bringing plants on line more quickly and cleanly. Apart from the United States, there are a limited number of plants in countries such as Indonesia and Malaysia that cofire coal and gas and also new projects in development. Other parts of the world where both coal and gas are easily accessible can also be potential future markets for cofiring [41].

2.4.2 Coal–Solar Hybrid Power Plant

Mills [41, 43] explored the concept of combining solar with conventional coal-fired power generation. This approach offers a route to combining renewable energy with inexpensive stable output from existing (or new build) thermal generation assets. In suitable locations, solar radiation can be harnessed and used to raise steam that can be fed into an existing conventional coal-fired power plant (a coal–solar hybrid). In such a system, solar thermal energy can be used to produce high-pressure and high-temperature steam that can be integrated into an existing power plant's steam cycle in several ways such that power output is boosted and/or coal consumption reduced.

Most existing solar thermal designs operate at ~300°C–400°C, lower than that of a typical modern coal-fired power plant (operating at 500°C or more). Thus, the temperature of the steam from the solar field is not high enough, and further heat must be provided before it can be fed to the plant's steam turbine(s). Feeding steam produced by the solar collection system directly into the main turbine can increase the overall efficiency of the plant by making the best use of the steam output from the solar field. However, the conditions of the steam generated by this must be matched to the coal-fired steam turbine cycle—this can be an engineering challenge [41, 43]. Alternatively, solar thermal energy can be used to heat the feedwater prior to entering the boiler. In a conventional steam power plant, as feedwater enters the feedwater heater, steam is extracted from the steam turbine to heat it. When solar heat is added to the feedwater, less steam is extracted from the turbine; this reduces coal input, increases the unit electrical output, or both.

Mills [43] points out that potentially, there are a number of points where steam generated from solar power could be fed into a conventional coal-fired power plant. This will be influenced partly by the type of solar collection system employed, as these tend to operate at different temperature ranges. Thus, candidate locations for steam from trough-based collectors include feedwater heating, low-pressure cold reheat, and high-pressure steam between the evaporator and superheater. In the case of power towers, possible locations are similar to those from troughs, but would also allow higher temperature admissions to hot reheat or main steam circuits. Possible locations from systems using Compact Linear Fresnel Reflectors include feedwater heating and low-pressure cold reheat [43]. Thus, the main applications are preheating of boiler feedwater, additional preheating of feedwater downstream from the top preheater, and the production of intermediate pressure (IP) steam or main steam [43]. According to Mills [43], the merit order of the various hybridization modes is: (a) solar preheating of high-pressure feedwater, (b) additional solar preheating of feedwater, (c) solar heating of low-pressure feedwater, and (d) solar production of high- and intermediate-pressure steam .

Some existing coal plants are particularly well suited to hybridization as they already allow operation in boost mode. As a turbo generator often has a capacity margin, this is achieved by closing the highest pressure steam extraction (but with an efficiency penalty). To date, feedwater heating has been the focus of several projects [41, 43, 45–47]. On coal-based power plants, cycle efficiency is improved by preheating feedwater before it enters the boiler. The preheating is carried out through the use of a train of preheaters that extract steam from the turbine at various pressure levels. By replacing the highest pressure steam extractions with solar steam (fully or partially), water preheating can be maintained while expanding more steam through the turbine, thereby boosting its power output [41, 43].

According to Mills [43], it is anticipated that some developing countries will build concentrated solar power (CSP) plants as well as new coal-fired units; they may already operate the latter. Thus, there are a significant number of potential sites, both existing and newly built, in countries that benefit from a good supply of solar energy [43]. The incorporation of solar energy into an existing coal-fired power station has the potential to increase overall plant efficiency, reduce coal demand and CO_2 emissions, plus minimize the problem of solar power's variability. At night, or when solar intensity is low, the output from the coal plant can be increased accordingly, allowing the combination to operate on an uninterrupted basis, 24 hours a day. When adequate solar intensity resumes, the coal plant can be ramped down once again. Alternatively, the increased steam flow produced by the solar boiler can be fed through the existing steam turbine, boosting output (so-called "solar boost"). This method of incorporating solar energy will cost less than an equivalent stand-alone CSP plant as many of the systems and infrastructure of the coal plant, such as steam turbine and grid connection, are already in place. A stand-alone plant requires all such systems in order to function. The levelized cost of energy (LCOE) from a coal–solar hybrid will be lower than that of a stand-alone CSP plant and be able to compete with that produced by PV systems [41, 43].

Solar collection systems normally operate using focusing mirrors or similar that track the sun's path and concentrate solar radiation onto a central point(s) where the heat is transferred to a heat transfer fluid. This is then used to raise steam that is fed into the coal plant and expanded through conventional steam turbines. As noted, the heat generated can be fed into the water/steam circuit of the coal plant to boost power output and/or reduce coal demand. Steam can be injected at several possible places in a conventional Rankine cycle. Rankine cycle is used in the vast majority of conventional steam-based thermal power plants where an operating fluid is continuously evaporated and condensed.

As pointed out by Mills [41, 43], steam turbines in most power plants have an excess capacity margin that allows increased output. By incorporating solar-derived steam, the amount of steam bled off the turbine for feedwater heating (which creates an efficiency penalty) can be reduced. This makes more steam available for expansion through the turbine, increasing its output (boost mode). The main modifications required to the coal-fired plant's steam cycle are often limited largely to the heat recovery steam generator (HRSG) which must be capable of handling the steam coming from the solar steam generation system.

In recent years, a number of coal–solar hybrid projects have been developed or proposed; some have focused on solar boost via feedwater heating, whereas others have adopted alternative hybridization configurations such as solar boost with superheated (SH) steam fed into a cold reheat pipe and coal saving with additional feedwater preheating after the power plant's top preheater [43, 48]. Clearly, any solar-based systems can only operate effectively in locations where the daily level of sunshine is adequate. A further requirement for coal–solar hybrids is the availability of suitable land, close to the existing power plant, needed for the solar collection system. Up to several thousand hectares may be required. Usually, CSP technology involves concentrating sunlight onto a receiver that contains a heat transfer medium, either oil-based fluid or a molten salt. Several types of solar collection devices are available commercially, and the choice will be influenced by factors such as land availability. The solar collector field can comprise 30%–50% of the cost of a CSP plant.

Around the world, in recent years the application of CSP has grown steadily. Many industry observers consider that the technology can provide a route for harnessing the huge solar resource with potentially better dispatchability than via PV cells. Also, required CSP can be built at lower costs, if it is integrated with conventional fossil fuel power plants, by "coal–solar hybridization" process. The hybridization will allow a path to implementing and maturing the technology, at lower cost than for new greenfield installations. An additional bonus of hybridizing is that by eliminating part of the coal feed, plant emissions can be reduced. Under some circumstances, the addition of a renewable energy source to a coal-fired plant ("greening" existing coal-fired assets) could allow access to feed-in tariffs or other forms of subsidy. Compared to conventional fossil fuel-fired power plants, the cost of electricity produced from solar power remains high. However, potentially, solar augmentation could provide the lowest cost option for adding solar power to an existing generation fleet [41, 43, 45–48].

Some existing coal plants are particularly well suited to hybridization as, assuming that the turbo generator has the corresponding capacity margin, they already allow a "boost mode" by closing the highest pressure steam extraction. Hybridizing in this manner could provide a power boost without extra coal consumption. If the solar potential exceeds the turbine's extra capacity, coal saving is possible. In current coal–solar hybrid plants, solar steam feeds only the highest pressure preheater, but other hybridization concepts could be adopted and combined to increase the solar share, especially on greenfield projects [43]. Such solar boosters increase capacity and energy generation without extra coal consumption, and with virtually no other extra cost than that of the solar field [43, 45–48].

A. **Advantages of coal–solar hybridization**

Depending on the particular circumstances, the main advantages cited by Mills [41, 43] for coal–solar hybridization are:

- The higher initial investment is balanced by reduced fuel consumption or increased power output.

- Combining the two technologies allows “greening” of existing coal-fired power assets.
- Hybridization can provide both dispatchable peaking and base load power to the grid at all times. CSP coupled with conventional thermal capacity (with or without thermal storage) can offer that capability.
- Hybrid technologies could help meet renewable portfolio standards and CO_2 emissions reduction goals at a lower capital cost than deployment of stand-alone solar plants. Capex (Captial expense) is less for the same capacity.
- Siting solar technology at an existing fossil fuel plant site can shorten project development timelines and reduce transmission and interconnection costs.
- Solar thermal augmentation can lower coal demand, reducing plant emissions and fuel costs per MWh generated.
- Solar augmentation can boost plant output during times of peak demand. According to US studies carried out by Electric Power Research Institute (EPRI), potentially, a solar trough system could provide 20% of the energy required for a steam cycle.
- Hybridization will reduce the level of coal and ash handling, reducing load on components such as fabric filters, pulverizing mills, and ash crushers. It could also avoid the requirement to upgrade fabric filters or electrostatic precipitators (ESPs).
- Solar input could provide some level of mitigation against difficult coal contracts, such as wet coal, fines, and variable coal quality.
- The majority of solar plant components could be sourced locally, helping boost local economies.
- Rapid deployment—depending on size and configuration, hybrid plants could be completed in less than two years from notice to proceed.
- Hybridization could be used to extend the lifespan of existing thermal facilities—for example, where regulatory changes require a coal-fired plant to reduce emissions or face closure.
- Hybridization could avoid certain limitations and restrictions applied to new greenfield site projects.
- Hybrid plants will benefit from the general cost reductions that CSP technology is achieving. Many of these will also be directly applicable to hybrid plants.

B. **Disadvantages of coal–solar hybridization**

Although coal–solar hybridization can provide some benefits, according to Mills [41, 43] there are obvious criteria that must be met:

- The location must receive good solar intensity for extended periods, both on a daily and yearly basis. This is not always the case.
- A suitable area of land close to the existing thermal power plant is required. It must meet certain criteria in terms of hectares available, topography, and issues such as shading.

- The land will no longer be available for other purposes such as agriculture.
- A solar add-on will require capital investment.
- There will be additional costs for operation and maintenance of the solar component, such as mirror washing.
- The scale of most coal–solar hybrid projects has so far been low, mainly because these have been retrofits at existing coal-fired plants. Practical issues have tended to limit the solar contribution to ~5%. A new power plant, designed and built based on the hybrid concept from the outset, could possibly accommodate up to 30%–40% solar share.

Each potential project brings its own combination of advantages and disadvantages. Although there are certain areas that will be common to all, there are various factors that will be specific to each individual site—projects need to be examined on a site-by-site basis. The addition of a thermal storage system, used to store excess solar heat harvested during daylight hours, can be important for solar thermal power generation. When demand dictates, heat can be reclaimed and used to raise steam. This is a major advantage over PV-based generation as it is easier to store large amounts of heat than electricity. The ability to reclaim heat at night means that electricity from a plant incorporating thermal storage can usually be considered dispatchable, whereas that from a PV plant is not.

According to Mills [43], not all coal–solar hybrids would necessarily opt to include thermal storage. The main reason for its addition is to increase the capacity factor, which increases the utilization of the power plant and thereby improving the overall economics. Usually, around half of the cost of a stand-alone CSP plant comes from the power island, with the balance from the solar equipment. In a hybrid plant, the power block is shared with the coal-fired portion and is therefore already available. Consequently, there may be less incentive to add thermal storage to a hybrid. However, adding storage can increase the utilization of the solar portion, making it easier to control and improve the plant's reliability. The viability of a coal–solar hybrid will be influenced by numerous site-specific factors, and the addition of thermal storage is likely to depend on the specific project. Various challenges remain for the large-scale deployment of coal–solar hybrid systems—these may be political, technical, or financial. From a purely practical point of view, any solar-based power generation system needs a consistent source of sunlight of adequate intensity, and this clearly limits possible locations. Although arguably less important for PV systems, it is crucial for those based on solar thermal technology. Coal–solar hybrids require land close to the power plant for the solar collection system, possibly up to several thousand hectares. However, land immediately around a power plant is sometimes unattractive for other purposes, so may be readily available. A possible complication is that, often, coal-fired plants are located near a source of water needed for cooling. This means that they are rarely located in arid high-desert sites that are well suited to solar thermal applications [41, 43].

It appears that any outstanding technical problems associated with hybridization can be overcome. However, depending on the individual circumstances, the biggest issue may simply be economic viability. In some locations, hybrids may have to compete directly with other forms of power generation such as natural gas-fired plants. At the moment, gas prices in some parts of the world remain low, and this will undoubtedly make investment in hybrids more uncertain [47]. However, there are situations where alternatives are much more limited, and here, coal–solar hybrids could find useful niche markets. More details of this concept are described in detail in an excellent reports by Mills [41, 43].

2.5 HYBRID ENERGY SYSTEM FOR GRID-CONNECTED NUCLEAR-BASED POWER PLANT

Just like coal-based power plant, nuclear-based power plants are also inefficient. Significant amount of nuclear heat is wasted and emitted in the atmosphere. This nuclear heat can be used for industrial heating purposes by cogeneration or it can also be used to generate additional power by thermoelectricity as described above. Use of nuclear waste heat for industrial purposes is described in detail in my previous book [1] and will not be repeated here. Here we, however, briefly summarize co-power generation by nuclear-wind power hybrid, studied by NREL and Idaho national lab, as reported by Ruth et al. [37].

NREL and Idaho national lab examined N-R HESs (nuclear-renewable hybrid energy systems) managed by a single entity that links a nuclear reactor that generates heat, a thermal power cycle for heat-to-electricity conversion, at least one renewable energy source, and an industrial process that uses thermal and/or electrical energy [37]. Texas Panhandle is considered as the case location. Here N-R HES does not directly include an industrial process but rather sells a thermal product (steam or a high-temperature heat transfer fluid) to one or more industrial customers such as those found in an industrial park. The study added the option to convert wind power to thermal energy so that energy from both the wind power plant and the nuclear reactor can be sold in both thermal and electrical forms. In addition, the study added an option to store thermal energy from any source: nuclear generated, electricity generated by the wind power plant, or grid electricity. Including thermal energy storage enables the N-R HES to provide a constant flow of thermal energy, enabling the customer or customers to operate at steady state.

The study considered three scenarios for the analysis. Each included a light water small modular nuclear reactor (LW-SMR), a thermal power cycle that converts nuclear-generated thermal energy to electricity, a wind power plant, and equipment to convert electricity to thermal energy (e.g., an electric boiler). The scenarios differ in whether they include thermal energy storage and whether the storage unit can store thermal energy from both the nuclear subsystem and electricity or only electricity-generated thermal energy. The scenarios also differ in whether the system provides the thermal product at a constant rate or if it varies over time, forcing the user of that thermal energy to either ramp or have a backup thermal generation system. Many industries such as chemical plants, metallurgical plants, and petroleum refineries run continuous flow systems and thus require a constant heat source. Other industrial users such as food processors may be able to operate when heat is available [37].

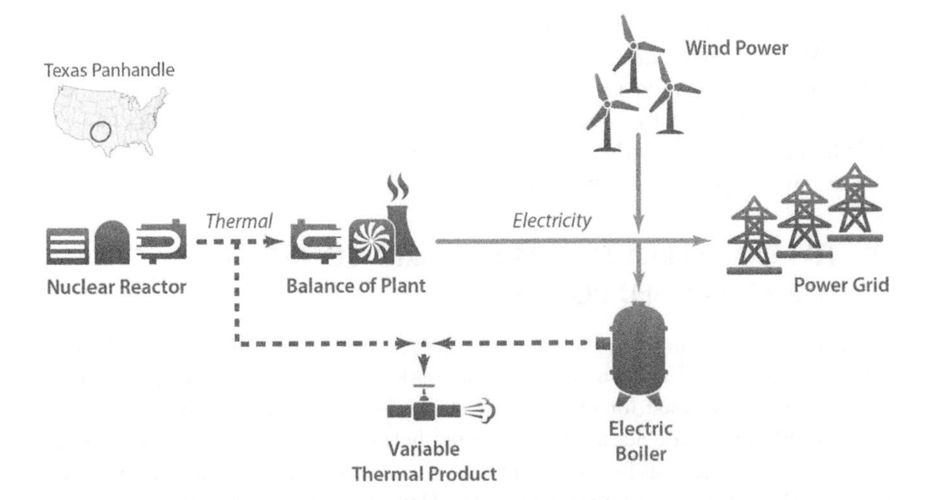

FIGURE 2.5 N-R HES Scenario #1: The electric boiler converts wind-generated electricity to thermal energy when the electricity price is very low [37]. Dotted line arrows indicate the flow of thermal energy and solid line arrows indicate electricity.

Figure 2.5 shows the first scenario's full configuration. This N-R HES includes an electrical boiler that converts the wind electricity to thermal energy. The study assumes that the thermal product's generation does not need to be constant. For this scenario, the study identified the optimal configuration from eight options under various electricity price vectors and thermal energy prices [37]:

- The nuclear reactor produces only thermal energy.
- The nuclear reactor and thermal power cycle produce electricity, possibly in conjunction with thermal energy.
- The wind power plant produces only electricity.
- The wind power plant and electric boiler produce thermal energy, possibly in conjunction with electricity sold to the grid.
- The nuclear reactor produces only thermal energy and the wind power plant produces only electricity.
- The nuclear reactor and thermal power cycle (balance of plant) produce electricity, possibly in conjunction with thermal energy, and the wind power plant produces electricity.
- The nuclear reactor produces only thermal energy and the wind power plant and electric boiler produce thermal energy, possibly in conjunction with electricity sold to the grid.
- The nuclear reactor and thermal power cycle (balance of plant) produce electricity, possibly in conjunction with thermal energy, and the wind power plant and electric boiler produce thermal energy, possibly in conjunction with electricity sold to the grid. This configuration is the full N-R HES.

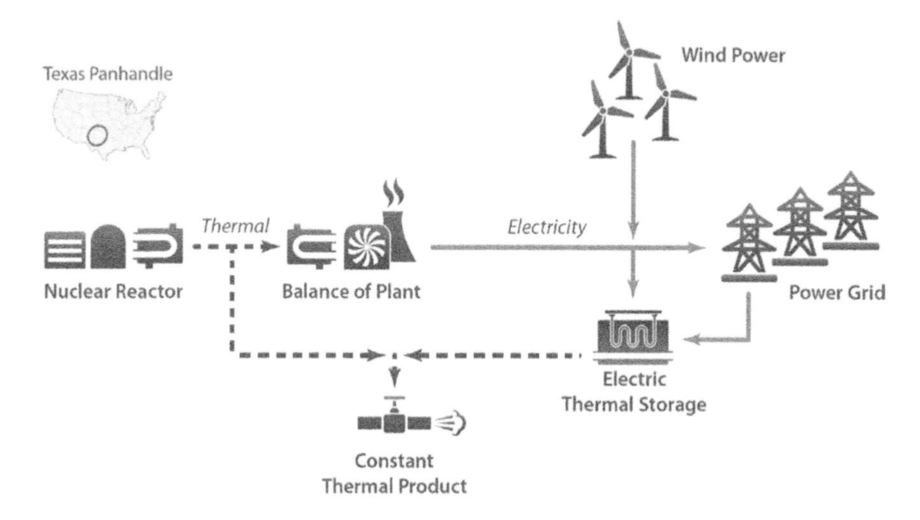

FIGURE 2.6 N-R HES Scenario #2: The electric thermal storage unit both converts wind-generated electricity to thermal energy when the electricity price is very low and stores that energy so that the thermal product's rate is constant [37]. Dotted line arrows indicate the flow of thermal energy and solid line arrows indicate electricity.

Figure 2.6 shows the second scenario's full configuration. In Scenario 2, the study replaced the electric boiler in Scenario 1 with an electric thermal storage unit—a unit operation that both converts electricity to thermal energy and stores that thermal energy. Because the electric thermal storage unit can store thermal energy, the study allowed it to use grid electricity in addition to wind-generated electricity so that the storage unit is not sized to provide a constant flow during the longest periods without wind. The thermal product is constant instead of ramping as in the first scenario because the storage can absorb the variability of generation.

The study based the electric thermal storage unit on Firebrick Resistance Heated Energy Storage (FIRES), as proposed by Stark et al. [49]. The FIRES concept consists of a ceramic firebrick storage medium of relatively high-heat capacity and density and a maximum operating temperature of approximately 1,800°C. Ruth et al. [37] chose ceramic firebrick because of its low estimated cost, high durability, and large sensible heat-storage capacities. The firebrick was "charged" using resistance heating when the price of electricity was low. Multilayer insulation surrounds the firebrick and allowed for thermal expansion. The system was expected to be similar to high-temperature firebrick industrial recuperators. Some parts of the world use a similar technology for home heating, although they are much smaller units. Alternatively, the firebrick could be heated directly by thermal energy from the nuclear reactor. Evaluation of that alternative was outside the scope of the analysis by Ruth et al. [37].

Figure 2.7 shows the third scenario's full configuration. The third scenario built upon both the first and second scenarios. In Scenario 3, the study used the boiler from the first scenario to convert both wind-generated and grid electricity to thermal

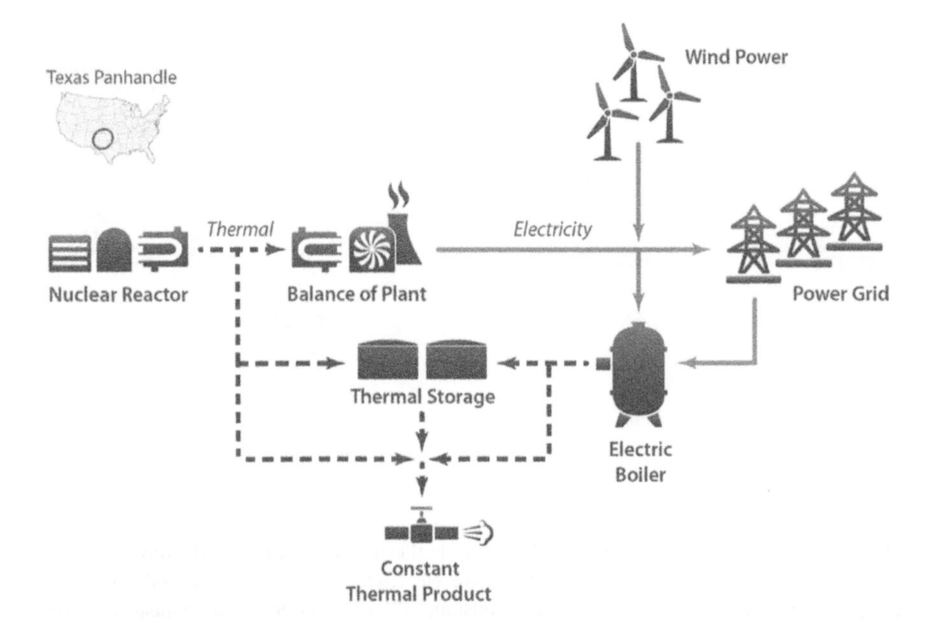

FIGURE 2.7 N-R HES Scenario #3: The electric boiler converts electricity to thermal energy when the electricity price is very low. The thermal storage unit stores thermal energy from both the nuclear reactor and the electric boiler. The thermal product's rate is constant [37]. Dotted line arrows indicate the flow of thermal energy and solid line arrows indicate electricity.

energy. A molten salt thermal storage unit was also included. Unlike the second scenario, the study used a thermal storage unit that can store thermal energy from either the electric boiler or the nuclear reactor. Like the second scenario, the study held the thermal product constant.

Table 2.3 compares the three scenarios that are analyzed and reported by Ruth et al. [37]. The first scenario was titled "Electric Boiler" and was not forced to provide the thermal product at a constant rate (the amount of heat provided can vary from hour to hour). The second scenario was titled "Electric Thermal Storage" and was required to provide the thermal product at a constant rate. In that scenario, stored thermal energy can be generated only by electricity. The third scenario was titled "Electric Boiler Thermal Storage" and also was required to provide the thermal product at a constant rate. Stored thermal energy can be generated by either electricity or the nuclear reactor.

The analysis carried out by NREL and Idaho national lab showed that N-R HES configurations could be profitable, primarily because the projected cost of nuclear generation of heat is less than natural gas generation cost projections under the analysis assumptions. Nuclear generation of heat does not emit carbon dioxide, so, with the cost of carbon, it would be more economical than natural gas generation. Even though nuclear generation of the thermal product exclusively was the optimal configuration at the base case electricity and thermal product prices, the benefits of the N-R HES's flexibility are apparent at higher electricity and thermal product prices.

That flexibility allows the conceptual N-R HES to support resource adequacy on the grid if capacity payments are sufficient. Flexibility also allows the N-R HES to maximize profitability by switching between products depending upon the value of each product [37].

2.6 DRIVERS, BENEFITS, ISSUES, AND STRATEGIES TO INCREASE CONTRIBUTION OF HRES IN THE GRID

As shown in Chapter 1, no single form of power generation is optimal in all situations. Wind and solar power generation are intermittent but consume no fuel and emit no greenhouse gases. Natural gas-fueled generation emits greenhouse gases but it is dispatchable (i.e., has output that can be readily controlled between maximum rated capacity or decreased to zero) to help balance supply and demand. It also has a relatively low CAPEX ($/kW). Hydropower often requires dedicating significant amounts of land area but is renewable and dispatchable. The chapter showed that the best way to harness sustainable power is to use more than one source with or without energy storage. Hybrid power plants usually combine multiple sources of power generation and/or energy storage and a control system to accentuate the positive aspects and overcome the shortcomings of a specific generation or storage type, in order to provide power that is more affordable, reliable, and sustainable. Each application is unique, and the hybrid solution that works best for a specific situation will depend on numerous factors including existing generation and storage assets, transmission and distribution infrastructure, market structure, and fuel prices and availability [50–52].

There is a strong desire to move away from fossil energy sources for power generation because they are non-renewable and they cause significant damage to the environment. There are two strategies that can be followed to reduce dependence on fossil fuel and its effect on environment. The first strategy is based on reducing energy consumption by applying energy savings and efficiency programs. As shown earlier, using waste heat from coal, gas, or nuclear-based power plants for cogeneration or additional power generation by thermoelectricity is one approach. The effect of fossil fuel on environment can also be reduced by converting CO_2 into more power or fuel as described earlier. Both of these approaches result in hybrid energy grid. A second strategy is to use more renewable energy sources with or without conventional sources and energy storage devices. This strategy also results in hybrid energy grid. The use of hybrid renewable energy sources has attracted much attention because it has several advantages such as, less dependence on fossil fuel, availability of the resources which are free of cost, and lower harmful emissions to the atmosphere (i.e., environmental friendly). Renewable energy sources, such as wind, solar, microhydro (MH), biomass, geothermal, ocean wave and tides, and clean alternative energy sources, such as fuel cells (FCs) and microturbines (MTs), have become better alternatives for conventional energy sources [50–52]. However, in comparison to conventional energy sources, renewable energy sources are individually less competitive and less reliable due to their uncertainty and intermittency due to dependence on weather and high initial cost. Unlike fossil fuels, they are also low-density and more distributed energy sources. Recently, extensive research on renewable energy technology has been conducted worldwide which resulted in significant development in the renewable energy materials, decline in the cost of renewable energy technology, and increase in their efficiency. Renewable sources like solar and wind generally require use of a storage device for stable power generation.

To overcome the intermittency and uncertainty of renewable sources and to provide an economic, reliable, and sustained supply of electricity, a modified configuration that integrates these renewable energy sources and uses them in a hybrid system mode with energy storage is proposed by many researchers. The energy from renewable resources is available in abundance but intermittent in nature; hybrid combination and integration of two or more renewable sources with energy storage make the best utilization of their operating characteristics and improve the system performance and efficiency. Unlike coal and nuclear energy, renewable energy resources are low-density energy and operate best in hybrid form.

Hybrid Energy Systems (HES) can be composed of one renewable and one conventional energy source or more than one renewable with or without conventional energy sources, or one renewable source with homogeneous or hybrid storage systems that works in grid-connected mode [50–52]. Hybridization of different alternative energy sources can complement each other to some extent and achieve higher total energy efficiency than that could be obtained from a single renewable source. Multi-source (which include energy storage) hybrid renewable energy systems, with proper control, have great potential to provide higher quality and more reliable power to customers than a system based on a single source. Due to this feature, hybrid energy systems have caught worldwide research attention. It is now believed that hybridization of renewable sources is the best way to increase contribution of renewable energy in the overall energy mix.

The design process of hybrid energy systems requires the selection and sizing of the most suitable combination of energy sources, power conditioning devices, and energy storage system, together with the implementation of an efficient energy dispatch strategy [50–52]. The selection of a suitable combination from renewable technology to form a hybrid energy system depends on the availability of the renewable resources on the site where the hybrid system is intended to be installed. In addition to availability of renewable sources, other factors may be taken into account for proper hybrid system design, depending on the load requirements such as reliability, greenhouse gas emissions during the expected life cycle of the system, efficiency of energy conversion, land requirements, economic aspects, and social impacts [50–52]. The unit sizing and optimization of a hybrid power system play an important role in deciding the reliability and economy of the system. In recent years, significant progress has been made to handle all of these issues. It is now clear that going forward grid will be hybrid due to enormous appetite to capture HRES to save environment.

In general, inclusion of hybrid energy, particularly those involving renewable resources in the grid, has many advantages which include:

1- **A hybrid energy system can make use of the complementary nature of various sources**, which increases the overall efficiency of the system and improves its performance (power quality and reliability). For instance, combined heat and power operation, e.g., MT and FC, increases their overall efficiency or the response of an energy source with slower dynamic response (e.g., wind or FC), which can be enhanced by the addition of a storage device with faster dynamics to meet different types of load requirements [50–52].

2- ***Lower emissions***: Hybrid energy systems can be designed to maximize the use of renewable resources, resulting in a system with lower emissions.

3- ***Acceptable cost***: Hybrid energy systems can be designed to achieve desired attributes at the lowest acceptable cost, which is the key to market acceptance.

4- ***Flexibility***: Hybrid energy systems provide flexibility in terms of the effective utilization of the renewable sources.

Although a hybrid system has a bundle of advantages, there are some issues and problems related to hybrid systems that have to be addressed:

1. Even when grid connected, most of hybrid systems require storage devices for which batteries are mostly used. These batteries require continuous monitoring, and there is increase in cost, as the batteries' life is limited to a few years. It is reported that the battery lifetime should increase to around a few years for its economic use in hybrid systems. In some cases, multiple energy storage devices may be required. The conventional grid operation is not very stable with a large level of infusion of renewable energy sources, particularly non-dispatchable sources like solar and wind.
2. Due to dependence of renewable sources (like solar and wind) involved in the hybrid system on weather results in the load sharing between the different sources employed for power generation, the optimum power dispatch and the determination of cost per unit generation are not easy. Long-term accurate model predictions are difficult.
3. The reliability of power can be ensured by incorporating weather independent sources like diesel generator or fuel cell or the addition of energy storage. This, however, adds another level of complexity in the operation of the grid.
4. The stability and power quality are an issue. As the power generation from different sources of a hybrid system is comparable, a sudden change in the output power from any of the sources or a sudden change in the load can affect the system stability and the level of power quality significantly.
5. Individual sources within a hybrid system have to be operated at a point that gives the most efficient generation. In fact, this may not occur due to the fact that the load sharing is often not linked to the capacity or ratings of the sources. Several factors determine the load sharing like the reliability of the source, economy of use, switching required between the sources, availability of fuel, etc. Therefore, it is desired to evaluate the schemes to increase the efficiency to as high level as possible.

In spite of various pluses and minuses of renewable hybrid energy, in recent years, due to overarching considerations of efficiency and carbon emission, hybrid renewable energy sources are increasingly used in grid-connected operations. Three basic strategies have been adopted to increase the contribution of HRES in the utility grid [50–52].

1. Use of HRES in residential and other buildings for power and heating/cooling purposes: The connection of this method to the grid can make the consumers more as prosumers where they can not only strive toward zero-energy building but also make net profit by sending excess energy back to the grid. The new plugs in electric cars are also used to connect houses with cars and grids. This approach reduces the use of fossil fuel for building and transportation needs (see Chapter 3).
2. Use of energy storage, largely battery, to expand the contribution of HRES (largely solar and wind) in the grid: The energy storage smoothens the instability caused by intermittent sources and allows stable power delivery by the grid with increasing levels of hybrid energy sources. This strategy has clearly resulted in a significant growth in the battery market. As shown in Chapter 4, energy storage can be homogeneous or hybrid. The use of hybrid energy storage has found increasing acceptance, because no single storage device satisfies all the requirements of the storage device.
3. Use of microgrid at medium voltage to harness distributed low-density renewable energy sources: This is a top-down approach where DERs are connected to medium-voltage grid distribution lines rather than transmission lines through the use of microgrid. The microgrid is connected to the utility grid. Lower voltage distribution microgrid offers several advantages to harness hybrid renewable energy sources compared to macrogrid, and it can operate independently if the main grid fails for some reason. As shown in Chapter 5, this method of harnessing distributed energy sources has many positive features both in urban and in rural environments. For rural electrification, off-grid hybrid energy (one not connected to macrogrid) can be used with the help of hybrid renewable resources. This subject is discussed in Chapter 6.

Besides adopting these three strategies, a number of issues need to be resolved, such as predicting solar and wind power input at different time scales, weather-dependent variability of RES, grid stability, role and location of storage on the grid, voltage and frequency control, proper power electronics between macrogrids and microgrids, etc. Furthermore, the insertion of HRES in the grid or microgrid is affected by several regulatory issues at federal and state level, federal and state incentives, and various financial and planning perspectives. In the following four sections, along with three strategies mentioned above, management and control strategies of HRES in the grid are discussed in detail. As mentioned above, the three strategies are also discussed in detail in the remainder of the book.

2.7 ROLE OF HRES IN GRID-CONNECTED BUILDINGS AND VEHICLES

Despite technological advances in small-scale hybrid renewable energy systems, there are very few studies that model the economic decision-making of a household which generates energy from multiple renewable sources. Tervo et al. [53] examine the costs and benefits of a residential photovoltaic system with a lithium-ion battery. A household that produces and uses energy from multiple sources, including renewables and storage, is known as a "hybrid-prosumer." HRES overcomes the inconsistent supply of a single renewable source. For instance, a photovoltaic–wind system is more likely to consistently produce power than a photovoltaic system alone because peak operating times for wind and solar systems occur at different times of the day and year. Thus, rather than wind and solar acting as substitute energy sources, as a hybrid system, they could create synergy in the production of electricity [54]. The potential for generating electricity when needed will be higher with hybrid than a single energy source. The hybrid nature can also be provided by the use of energy storage which provides a reliable backup when and if consumption exceeds production. Energy storage can help reduce the size of other components (e.g., photovoltaic panels or wind turbines) and cut down costs. Connecting HRES to the smart grid allows the homeowner to measure the electricity sent back to the grid [54–56]. Furthermore, an HRES can meet the demand for energy with a lower environmental footprint and contribute to a distributed and diversified energy infrastructure [54, 56].

Several projects all over the world have demonstrated the application of small-scale hybrid energy technologies. For example, Frostburg State University in Maryland, United States, showcases a grid-tied residential size solar–wind system [57–59]; Yuan Ze University in Taiwan owns a small-scale photovoltaic–wind–fuel cell system [59]; and Pamukkale University in Turkey demonstrates a hybrid photovoltaic–hydrogen fuel cell–battery system designed to meet demand from non-fossil fuels [60]. In recent years, commercial HRES developers have introduced products targeting the residential sector [56–60]. For example, WindStream Technologies, Inc. is a US-based developer of renewable energy generation products, and since May 2015, the company has commercialized a 1.2-kW system of solar panels and wind turbines that are suitable for grid-tied residential installations [61]. Another example is General Electric Company which has recently commercialized a solar–wind and a hydro-wind system in several countries including the United States [62]. According to an industry research report conducted by Global Market Insights, Inc. and authored by Gupta and Bais [58], the global hybrid solar–wind market was valued at $700 million in 2015, where the US market accounted for close to 28%. The report also finds that, from 2013 to 2015, the generation of energy from grid-connected hybrid solar–wind installations in the United States was increased by 24% [58]. Given that hybrid energy technologies have a solid emerging demand, it is timely to consider their impact on other energy-related decisions such as energy efficiency, energy services, and energy consumption, particularly in buildings and vehicles.

Despite the substantial research on the techno-economic analysis of residential HRES, economic approaches have been restricted to cost analysis. Deshmukh and Deshmukh [63] find that cost analysis is the most popular tool used to select among different types and sizes of HRES. Specifically, the life cycle cost of a system is calculated as the present value of lifetime costs. Lifetime costs include initial installation (e.g., component and system cost), replacement cost (e.g., batteries and/or inverters may need to be replaced), and operating and maintenance cost, less any salvage value. Typically, calculations are made on an annual basis, and the lifetimes assumed for the systems differ widely across studies [64]. Many studies rely on the calculation of the levelized cost of energy (LCOE) as an indicator for the financial performance of a small-scale HRES. The LCOE is usually computed as the ratio of the total annualized cost of a system to either the annual electricity generated by the system or the annual electricity consumed by the household. The present value of lifetime cost is divided by the capital recovery factor to find total annualized cost. A system with the lowest LCOE is considered to be cost-effective in relation to others [64].

Syed et al. [65] focused on identifying and monetizing the annual benefits of a photovoltaic–wind system for a representative house in Canada. The system generates a total of 7,720–8832 kWh of energy annually, but since it does not have storage, excess energy is sent back to the grid. They find that the photovoltaic–wind system generates CAD 381.7 annually in electricity bill savings and CAD 340.7 annually in credit for sending surplus energy back to the grid. The study also finds that a house with the photovoltaic–wind hybrid system generates 56% less greenhouse gases compared to a fully grid-dependent house. This is because greenhouse gas emissions from the generation of electricity at fossil fuel-based plants are reduced with the hybrid system, due to the reduction in the electricity import from the grid. They suggested that more benefit–cost studies are needed to comprehensively evaluate the economic costs and benefits of adopting HRES in the residential sector.

Ghenai and Bettayeb [66] examined grid-tied solar PV/fuel cell hybrid power system for university building. Based on the simulation study, the results showed that the grid-tied solar PV/fuel cell hybrid power system offers a good performance for the tested system architectures. With a solar PV capacity of 500 kW and fuel cell capacity of 100 kW integrated with the grid, the total energy generated from grid-tied renewable energy system to meet the desired load is 26% from the grid (purchase), 42% produced from the solar PV, and 32% from the fuel cell. From the total annual power produced, 95% is used to meet the AC primary load of the building and 5% is the sellback to the grid. The proposed grid-tied solar PV/fuel cell hybrid power system with the sale of electricity back to the grid has a high renewable fraction (40.4%), low levelized cost of energy ($71/MWh), and low carbon dioxide emissions (133 kg CO_2/MWh).

The LCOE is an approach that can be used to rapidly evaluate different types and sizes of HRES; it has practical applications for the hybrid energy industry. However, it has several drawbacks when used to evaluate a household's energy decision-making process. First of all, according to Bazilian et al. [67], the metrics used in the economics of renewable energy production are not standardized because they are defined in different ways based on the type of available data. Second, cost analyses simplify the hybrid-prosumer's decision-making to choosing the least cost option. Options are either framed as a hybrid versus a single renewable energy source (e.g., photovoltaic–wind battery versus photovoltaic battery) or different types of HRES (e.g., photovoltaic–wind battery versus hydro-wind battery). However, when a hybrid-prosumer considers the adoption of a HRES, he/she is actually making several other decisions simultaneously or close to each other. These other decisions may have a confounding effect on the choice of the type and size of a given system. In addition, the adoption of a HRES with a specific combination of renewable sources may affect the hybrid-prosumer's other energy-related decisions such as energy-efficiency improvements, energy service production and consumption, and the level of energy consumption.

Energy-efficiency improvements: For the residential sector, demand for energy efficiency and the market for renewable energy generation are not independent. Oftentimes, when households consider on-site generation of a renewable energy, they also consider improving the energy efficiency of their homes [68]. For example, in California, a majority of solar photovoltaic homeowners upgraded the energy efficiency of their homes and/or appliances before, or in conjunction with installing solar photovoltaic panels [65]. Thus, one can think of energy generation as a demand shifter in the market for energy efficiency. In the context of policy making, McAllister [69] argues that both energy efficiency and renewable energy programs are needed to achieve a net-zero-energy objective.

Energy services: Another important factor for hybrid-prosumers is the level of energy services they can produce from their investments in energy. Fell [70] defines energy services as "those functions performed using energy which are means to obtain or facilitate desired end services." The most common examples of energy services include space heating, cooling, lighting, water heating, and refrigeration. Hybrid-prosumers are producers of energy services in a sense that they transform energy (renewable and non-renewable sources) into energy services by using conversion technologies such as furnaces, space heaters, and pumps [71, 72]. Hybrid-prosumers also derive utility from the consumption of energy services and their demand for such services is directly affected by the amount and type of energy used as an input. The implicit value of energy services rarely enters into cost analyses such as the LCOE.

Energy consumption: The decision to adopt an on-site energy generating system may directly affect energy consumption in the residential sector. After households invest in an energy generating system, such as a HRES, they may exhibit a different load profile. On the one hand, households may increase energy consumption post-generation similar to the rebound effect of energy efficiency [61, 62]. According to the rebound effect, improvement in the technical efficiency of technologies reduces the shadow price of energy services, which in turn increases demand for energy input. With respect to renewable energy generation, this implies that renewable

> generation may reduce the marginal or average cost of energy input, and hence increase its use. McAllister [69] argues that this is entirely due to an income effect, where households "facing reduced total and/or marginal cost of electricity due to the installation of an energy generating system would, in theory, increase overall electricity consumption (presumably prioritizing those end uses with the highest marginal utility)." The income effect indicates redistribution of the savings from the electricity bill to the overall use of more energy.

Based on large dataset from San Diego, California, McAllister [69] finds that the overall electricity consumption trend among solar photovoltaic adopters is that long-term consumption (two and three years post-generation) may be higher than consumption before generation. McAllister [69] finds that this increase in energy consumption post-generation is not very large overall (less than 5%), and although it is hard to determine the causation, it is observed for households which install relatively large-sized generation systems. Such households are "more interested in covering most or all of their consumption and may not be interested in reducing consumption, or may also be involved in home expansion or other energy intensive activities" [69]. Fikru et al. [73] find that energy consumption of a prosumer could be higher than the energy consumption of a comparable grid-dependent household. This is because households that generate energy have a lower valuation for energy services because of their ability to generate energy on site. Lower shadow price for energy services implies higher demand for such services which requires higher energy input, keeping other factors constant. On the other hand, after adopting an on-site energy generating system, the household may decrease energy consumption. For example, the household may make efforts to coordinate the timing of generation with consumption. Energy storage may also contribute to the most efficient use of energy.

2.7.1 Home-Based Solar Electricity Connected to the Grid

As mentioned above, there are three modes of electricity transport: utility grid-tied, hybrid microgrid, and off-grid. The off-grid includes mini- or nanogrid as well as stand-alone systems. Hybrid microgrid can be connected to utility grid or it can operate in islanded mode. Off-grid systems are generally not connected to microgrid but this may be possible in future. While in principle, macro, micro, mini, and nanogrids can be interconnected, at present, connections between microgrids and macrogrids are most relevant. Off-grid systems which are largely used for rural electrification or for individual buildings or equipment are not connected to conventional utility grid. As shown in Chapter 1 and this chapter, in urban areas, the electricity is transported by high-voltage utility grid. In recent years, this grid is often considered to be smart grid because it operates bidirectionally and smartly manages on time supply and demand functions, taking into account emergency and unexpected needs. This mode of transport in urban environment includes high-voltage transmission, transmission at small and middle voltage, and ultimately, distribution to end users. Utility grid

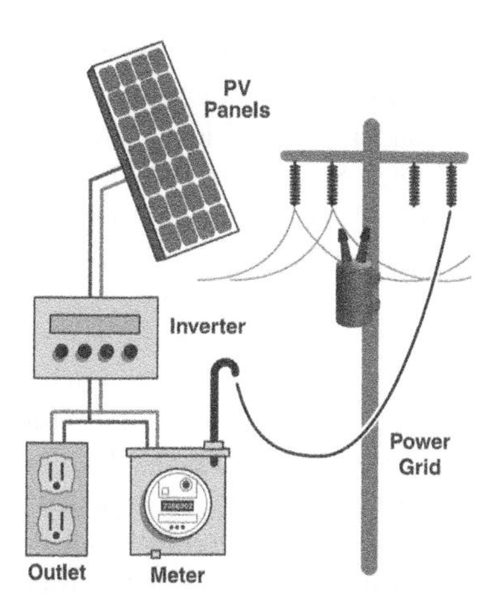

FIGURE 2.8 Utility grid-tied operation [74]. *Homeowners can be given credit by their local power companies for the electricity produced at their homes through "net metering" programs. Solar electric systems sometimes produce more electricity than your home needs. This extra electricity is either stored in batteries or fed into the utility grid.*

also provides energy storage by spinning reserve. As shown in Figure 2.8, grid-tied solar system can provide electricity to home owners using utility grid.

The utility grid offers several advantages:

1. **Save more money with net metering**

A grid connection allows one to save more money with solar panels through better efficiency rates, net metering, plus lower equipment and installation costs: batteries, and other stand-alone equipment, are required for a fully functional off-grid solar system and add to costs as well as maintenance. Grid-tied solar systems are therefore generally cheaper and simpler to install. Solar panels often generate more electricity than consumption. With net metering, homeowners can put this excess electricity onto the utility grid instead of storing it themselves with batteries. Net metering (or feed-in tariff schemes in some countries) plays an important role in how solar power is incentivized. Many utility companies are committed to buying electricity from homeowners at the same rate as they sell it themselves. In recent years utility grid-tied operation has become more hybrid in order to accommodate a larger share of renewable energy in the overall energy mix.

2. **The utility grid is a virtual battery**

Electricity has to be spent in real time. However, it can be temporarily stored as other forms of energy (e.g., chemical energy in batteries). Energy storage typically comes with significant losses. The electric power grid is in many ways also a battery, without the need for maintenance or replacements, and with much better efficiency rates. In other words, more electricity (and more money) goes to waste with conventional battery systems. According to Energy Information Administration (EIA) data [74], national, annual electricity transmission and distribution losses average about 7% of the electricity that is transmitted in the United States. Lead-acid batteries, which are commonly used with solar panels, are only 80%–90% efficient at storing energy, and their performance degrades with time. Additional perks of being grid-tied include access to backup power from the utility grid (in case the solar system stops generating electricity for one reason or another). At the same time, one can help to mitigate the utility company's peak load. As a result, the efficiency of our electrical system as a whole goes up.

There are a few key differences between the equipment needed for grid-tied, off-grid, and hybrid solar systems. Standard grid-tied solar systems rely on (a) grid-tie inverter (GTI) or micro-inverters: a solar inverter regulates the voltage and current received from solar panels. Direct current (DC) from solar panels is converted into alternating current (AC), which is the type of current that is utilized by the majority of electrical appliances. In addition to this, grid-tie inverters, also known as grid-interactive or synchronous inverters, synchronize the phase and frequency of the current to fit the utility grid (nominally 60 Hz). The output voltage is also adjusted slightly higher than the grid voltage in order for excess electricity to flow outward to the grid. Micro-inverters go on the back of each solar panel, as opposed to one central inverter that typically takes on the entire solar array. There has recently been a lot of debate on whether micro-inverters are better than central (string) inverters. Micro-inverters are certainly more expensive but in many cases yield higher efficiency rates. (b) Power meter: power meter, often called a net meter or a two-way meter, is capable of measuring power going in both directions, from the grid to the house and vice versa.

The process for connecting resident with HRES to grid is described in an excellent report by Department of Energy [75]. As shown in Figure 2.9, connecting home directly to grid will require coordination among all participants [52].

RE can be better absorbed if electric vehicle charging and discharging is done strategically. Conventional power supplies can be sent as needed to match demand and provide ancillary services for grid stability. Contribution to grid by RES is increasing although these sources are intermittent by nature. This is now an operational challenge to balance the intermittency of RES. Electric vehicles (EVs) offer a scope to manage demand and potentially mitigate the amount of curtailed energy by controlling when EVs are charged. Different types of charger such as AC/DC, slow/fast, are discussed in European standard [76]. Integration of ESSs in EVs charging station has grown with AC but the DC system has higher energy efficiency with improvement of up to 10% [76] with less number of conversion stages taking generation from RES. The important communication system makes it all possible in a coordinated manner. It communicates with the smart metering system present on the MG and on the EV charging station, through Modbus on TCP/IP connection, using the internal LAN, and with the ES converters, through the CAN protocol.

For large market penetration of plug-in electric vehicles (PHEV), load control by smart charging is encouraged. It can reduce the size of central storage devices. Clement-Nyns et al. [77] examined the impact of vehicle-to-grid (V2G) on the distribution grid. Pang et al. [78] investigated charging and discharging of PHEV in a

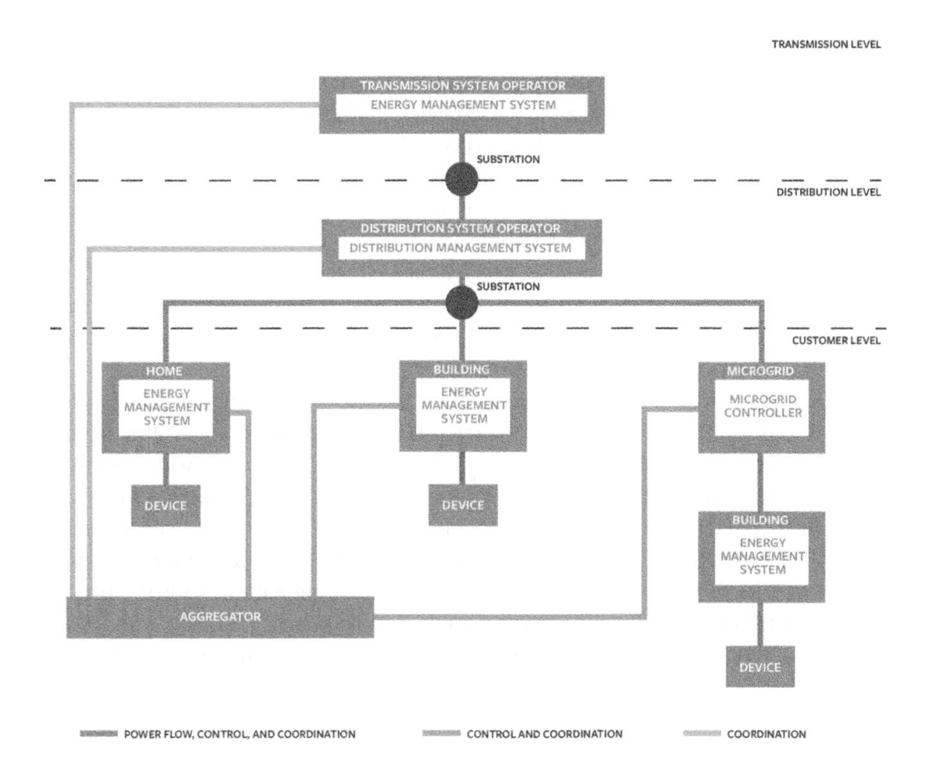

FIGURE 2.9 Coordination among participants [52].

cooperative manner that helps voltage control and reduces congestion. They evaluated battery electric vehicles (BEVs)/PHEVs as dispersed energy storage for vehicle-to-building (V2B) uses in the smart grid. The PHEVs as dynamically configurable dispersed storage can operate in V2B. Based on the distinctive attribute of the battery, the benefits of using PHEVs as energy storage for DSM and outage management are deliberated by researchers [76]. The faster charging is yet to come up. The parking time can be utilized for charge or discharge mode when required. *This distributed energy system, which bundles electric vehicle range extension with personal energy storage, has some intriguing elements to it.* Oak Ridge National Laboratory (ORNL) has explored additive manufacturing with integrated energy between home and car. A system called Verd2GO appears to be yet another battery swapping scheme on the surface, but upon closer look, is actually akin to a swappable EV "range extender" that doubles as a portable power solution. These systems are described in great detail in my previous book [76] and other literature [79–82].

2.8 ROLE OF BATTERY STORAGE FOR INCREASING LEVEL OF HRES IN THE GRID

In recent years, battery technology has been found to be the most favorable for grid usage. A battery ES with its own specific features can serve a particular usage when time, space, portability, and size are some of the factors. In this section, we evaluate role of battery storage for increasing levels of HRES in grid. The section reviews battery ES in view of the latest technologies, advantages, sizing, efficiency, price, and life cycle assessment. In particular, the discussion includes subjects such as battery costs, balance of system costs, project costs, hybrid system costs, and other value streams such as energy arbitrage, frequency regulation, spinning reserves, generation capacity, transmission deferral, demand charge reductions, resilience and reliability, and decreased diesel generation. Market considerations, regulatory framework, and federal and state incentives for battery storage are also assessed. Finally some successful case studies demonstrating role of battery storage are outlined and additional perspectives for storage for HRES in the grid based on research reported in the literature are briefly reviewed.

In recent years, battery storage installation has grown worldwide. This is illustrated in Figure 2.10. Luo et al. [84] carried out an in-depth study of the electrochemical properties of BES (battery energy storage) by examining a wide variety of storages. The study indicated that energy capacity and the self-discharge or capacity fade of BES systems (BESSs) affect the suitable storage duration. The study by Dunn et al. [85] has shown that BES can go forward for ancillary service if its cost reduces. Knap et al. [86] evaluated the technical viability of Li-ion batteries for the inertial response (IR) in grids with ample contribution of wind power. Leadbetter and Swan [87] indicated that one particular BESS cannot be suitable for all the short, medium, as well as long-term support services. Only Li-ion can serve for short duration support. For distributed storage and medium duration support, Pb-acid and Li-ion batteries are most suitable. A lithium-antimony-lead (Li-Sb-Pb) liquid metal battery is proposed by Wang et al. [88], which has higher current density, longer cycle life, and simpler manufacturing of large-scale stationary storage systems.

Battery storage is becoming an important component of the energy grid [89–95]. A correlation can be drawn between the growth in HRES and the expected growth in battery storage. Since 2004, world variable renewable energy installed capacity (largely solar and wind) has grown 25% annually [92]. In the United States, wind and solar have grown from producing 0.37% of total generation in 2004 to 6.5% in 2016. The market for battery storage today in many ways reflects the market for

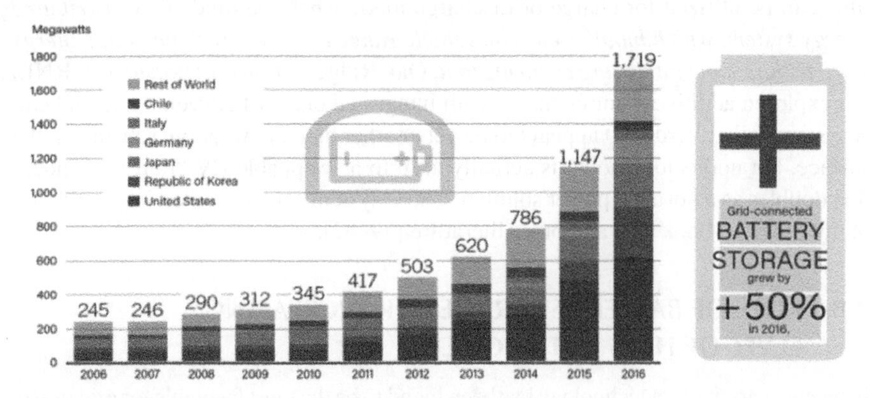

FIGURE 2.10 Battery storage installations [83].

renewable energy in 2004. The growth in battery storage, however, is expected by some to be even more dramatic [89–95]. Global battery storage capacity increased sevenfold in 10 years and by 50% in 2016 alone [83, 89–98]. In the next 6 years, US battery capacity is forecasted to grow 22-fold [89–95].

Reductions in battery costs are making the economics of battery storage more compelling [83, 96, 97]. Prices for lithium-ion batteries declined by 14% annually from 2007 to 2014 for a total price decline of more than 65% in 7 years, and similar future cost reductions are expected [89–95]. Batteries are dispatchable, have fast response times, have zero end use emissions, and face fewer siting restrictions than most conventional generation. Thus, battery storage can provide value to both utilities and customers through a number of generation, transmission, and distribution applications. However, current high costs and regulatory restrictions may be barriers to overcome for battery storage technologies if the projected growth is to be realized. A battery hybrid—a battery system paired operationally with a generation system—can help overcome the cost and regulatory limitations of stand-alone storage. Along with increasing the performance of services that batteries already provide, battery hybrids are able to capture value streams where stand-alone batteries cannot.

A. **Costs**

Falling battery costs are quickly improving the economics of battery storage. Driven largely by economies of scale from increasing electric vehicle sales, battery prices fell by 65% from 2010 to 2015 [89–95]. Total capital costs for an 8-h storage system are projected to decline by 34%–81%, with an expected decrease of 57% by 2050 [94]. In other analysis, a recent paper by Schmidt et al. [94] uses estimates of battery pack prices settling around $175 per kWh and total installed capital costs around $340 per kWh. Depending on deployment rates, these prices are expected to be reached between 2027 and 2040 [89–95]. The costs of battery and balance of systems (BOS) [95] over last several years are illustrated in Figure 2.11a,b.

B. **Balance of system costs**

While the battery pack is the most costly single component, balance-of-systems (BOS) costs make up a larger share of the total cost than the battery itself [83, 96, 97]. BOS costs consist of all non-battery related costs. BOS costs include hardware, labor, permitting, overhead, customer acquisition, and construction. Container and inverter expenditures represent the two largest portion of total expenditures; however, customer acquisition (CA) and engineering, procurement, and construction (EPC) expenses are also considerable [83, 96, 97]. Since 2015, total $ per KW cost has decreased from about 680 to about 400.

C. **Project costs**

Costs vary by size of project, type of battery, and duration of storage. A wide array of battery compositions exists, each with different price points and technical

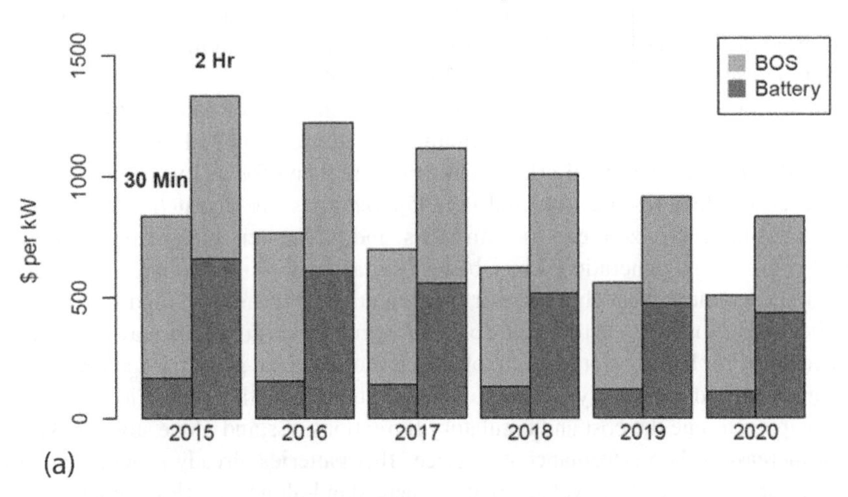

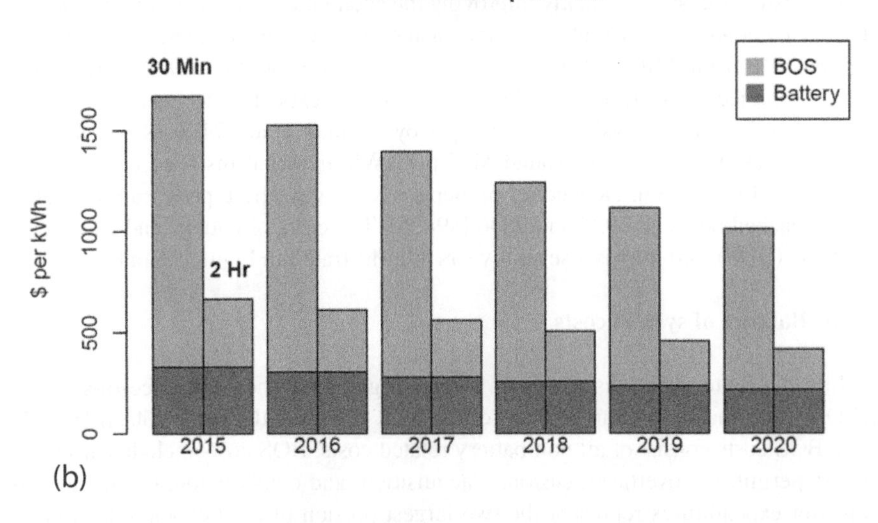

FIGURE 2.11 (a) Current and forecasted total systems cost per kW of storage for a 30-minute and a 2-hour utility-scale system. [83]; (b) Current and forecasted total systems cost per kWh of storage for a 30-minute and a 2-hour utility-scale system [83].

characteristics. Even among lithium-ion batteries, prices and technical parameters vary by battery composition and system configuration. In addition to battery type, two characteristics are important to consider when determining total costs—battery size and storage duration. In general, larger battery projects are cheaper on a per-kilowatt (kW) or per-kilowatt-hour (kWh) basis due to economies of scale. In 2016, on average all-in costs for commercial and industrial behind-the-meter uses were 25% more than larger utility-scale projects [83].

Short duration storage is best suited for high-power applications such as frequency regulation and spinning reserves, while long duration storage is best suited for applications with longer runtime requirements, such as providing resilience or capacity [83]. BOS costs generally scale with maximum output (kilowatts) while battery pack costs scale with output duration (kilowatt hours). The available results indicate that the 30-minute system has lower costs when compared on a per-kilowatt basis due to lower battery costs, while the 2-hour system has lower costs when compared on a per-kilowatt-hour basis due to lower BOS costs.

D. **Costs for hybrid systems**

Engineering synergies can make hybrid systems cheaper than building a separate storage and generator. Cost reductions from hybridization can be separated into soft cost savings and savings from sharing hardware. Soft costs such as permitting and customer acquisition scale by project; therefore, these costs are reduced when the storage and generation system are built at the same time. Labor costs and administrative overhead are also lower for hybrid systems than for separate systems [83, 95]. Hardware cost savings from hybridization come from shared hardware and reduced efficiency losses. As displayed in Figure 2.12, solar photovoltaic (PV)-storage hybrids

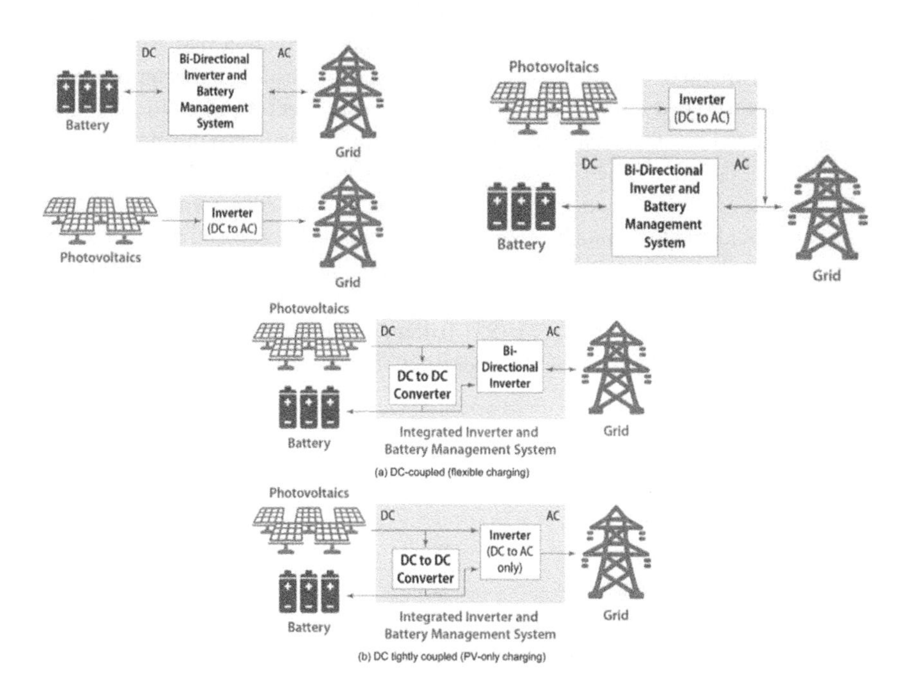

FIGURE 2.12 Coupling architectures of solar with storage systems (left) independent coupling, (center) AC coupling, (right) DC coupling. [96]

can share an inverter, leading to additional savings. Furthermore, peak production, which would otherwise be clipped due to generation exceeding inverter capacity, can now go toward recharging the battery. Charging efficiency also increases due to avoided DC-AC conversion losses.

Savings from hybridizing can be significant. A recent National Renewable Energy Laboratory (NREL) report estimates that installing a separate residential PV and battery system is 18% more expensive than simultaneous installation [83]. Although the percentages for utility-scale savings are lower, they are still positive and non-negligible. Coupling a battery with utility PV saves an estimated 36% of battery BOS costs, which relates to roughly 8% of total project installation costs [83]. Furthermore, rapidly declining battery costs mean the fraction of total costs from BOS is increasing [94, 95]. This increases the relative benefits of hybridization. Solar panels seldom produce at their maximum capacity. It is therefore optimal to have an inverter loading ratio (ILR), the ratio of solar capacity to inverter capacity, greater than one [83, 96, 97]. A larger ILR increases the inverter capacity factor but also increases the percentage of electricity that must be clipped due to insufficient inverter capacity.

E. **Other value streams**

Batteries can ramp quickly, have zero end use emissions, face fewer siting restrictions than traditional generators, and are dispatchable. Fast ramping makes storage well suited to provide ancillary services such as frequency regulation [83]. Storage may be sited downstream of transmission nodes to reduce congestion and defer transmission upgrades. Finally, storage can provide behind-the-meter power to customers when it is needed most. This allows batteries to provide resilience during blackouts and reduce demand and time-of-use (TOU) charges to customers [83].

Along with the cost reductions from co-location, hybridizing storage with a generator can provide added value by combining the rapid response and dispatchability of storage with the long potential runtime of the generator. Hybrids also benefit from policy incentives and may be better situated to navigate a changing regulatory framework. As pointed out by Ericson et al. [83], battery storage and battery hybrid storage has eight different applications on the operation of smart grid [83]:

1. Energy arbitrage
2. Frequency regulation
3. Spinning reserves
4. Generation capacity
5. Transmission deferral
6. Demand charge reductions
7. Resilience and reliability
8. Decreased diesel generation

In the following discussion, each section contains a brief summary of the application, approximate market size, how storage can provide the service, and how hybridization can add value [83, 96, 97]. The discussion follows the excellent report of Ericson et al. [83].

1. **Energy arbitrage**

A battery participating in energy arbitrage stores energy when prices are low and sells energy when prices are high. The possible market size for energy arbitrage is large, but revenues are not sufficient to fully support current battery costs [83, 96, 97]. Energy arbitrage is best suited as a secondary revenue stream paired with other services to increase profitability.

Energy arbitrage pairs well with value streams such as generation capacity, transmission deferral, demand charge reductions, and resilience and reliability, which only use the battery a portion of the time. Pairing storage with variable generation can increase revenues from energy arbitrage. Periods of high production from variable generation increase line congestion and may exceed line capacity, leading to low or even negative localized prices. Pairing storage with variable generation allows the battery to charge during these periods of low prices. As the penetrations of variable renewables on utility grids increase, it can be expected that this value stream will grow. Hybridization also benefits behind-the-meter applications in areas without full-net metering laws. The battery can be charged when energy would otherwise be curtailed or sold back to the grid at low rates [83].

2. **Frequency regulation**

Frequency regulation is the automatic response of power to a change in frequency due to an imbalance between generation and load [83]. The fast response characteristics of batteries make them well suited for this service. Storage is already playing an important role in some regulation markets. PJM, a US regional transmission organization (RTO), has seen an especially large influx of frequency regulation from batteries and flywheels due to market changes to better compensate resources providing fast response. Storage currently accounts for nearly 40% of PJM fast-response regulation capacity requirements [83, 96, 97].

Market rule changes enacted in 2016 have significantly decreased the profitability of storage in PJM, lowering the estimated yearly revenue of a 20 MW/5 MWh system from $623 in 2014 to $86 today [83]. The PJM rule changes demonstrate the importance of market structure and regulations on the value of storage, along with the risk that policy change poses for storage projects. While frequency regulation is currently profitable in some locations, the market remains small. The total US market for frequency regulation is around 2 GW [83, 99], so it may be saturated if storage is deployed on a large scale.

3. **Spinning reserves**

To maintain grid reliability in the case of an unexpected plant outage, utilities are required to keep generation capacity that is partially loaded but may come online quickly. Spinning reserves have the requirements that they must be synchronized to the grid and must be able to ramp to full capacity within 10 minutes. The average US market size for spinning reserves was 5.4 GW in 2014 [83, 100].

Spinning reserves provide significantly higher revenue per MWh of total capacity than non-spinning reserves. This is due to the added cost of keeping generators synchronized with the grid. Storage, especially hybrid storage, is well suited to the spinning reserves market. Traditional generators providing spinning reserves must run at part capacity to remain synchronized to the grid, which is less efficient than when they are running at full capacity. The fast response times of batteries mean they are always synchronized with the grid, even when not discharging. While spinning reserve prices alone do not currently support storage capacity costs, it may be combined with other value streams. For example, transmission deferral, demand charge reductions, and peaking capacity rarely require the battery to discharge. The rest of the time, the battery may sell its capacity into the spinning reserves markets.

The profitability of using storage for spinning reserves is hindered by the high costs of storage capacity and by the requirement that spinning reserve assets be able to discharge for an extended period. Requirements vary by region, ranging from 30-minute to 2-hour minimum runtimes [83, 100]. Pairing storage with a generator to provide spinning reserves is currently profitable in some markets. A hybrid battery and natural gas generator allow for participation in the spinning reserves market without requiring a long duration battery or a generator inefficiently operating at part load. The fast response of the battery complements the low cost of capacity of the natural gas generator. The battery discharges while the generator ramps and the generator produces for the rest of the period requirement [83, 96, 97].

4. **Capacity**

To ensure long-term grid reliability, some markets pay generators for generation capacity. Capacity payments are potentially a large source of revenue for battery storage and battery hybrids. For example, capacity payments in PJM for 2016 accounted for 22% of total wholesale electricity payments and can greatly increase the profitability of storage [83, 101]. Furthermore, because capacity is only needed for a few hours during days with especially high peak demand, capacity payments are well suited to be stacked with other services. However, capacity markets can be quite volatile.

The primary barrier for storage to sell into capacity markets is a regulatory structure that was originally designed without considering the characteristics of battery storage. There are different capacity market requirements across US RTOs/independent system operators (ISOs). California independent system operator (CAISO) and New York independent system operator (NYISO) both have 4-hour requirements to participate in capacity markets. While ISO-New England and PJM do not have minimum-duration requirements, they both have a "no-excuses policy," which requires capacity resources to provide their capacity obligation for the duration of performance events or face significant financial penalties. Because performance periods in these markets have no maximum duration, storage with limited discharge duration faces considerable risk [83, 101].

Declining battery costs are beginning to improve the economics of longer duration batteries. The year 2016 saw a shift from short duration storage providing frequency response to longer duration storage providing capacity. In response to the Aliso Canyon gas leak, more than 100 MW of 4-hour duration storage came online to provide capacity in California [83, 101]. Due to the much larger size of capacity markets as compared with frequency regulation markets, the trend toward longer duration storage is expected to continue.

Battery hybrids selling into the capacity market may be able to benefit from increased capacity payments. While 4 hours of storage are mandated to sell into many capacity markets, most of the time, the peak demand period lasts for less than 4 hours. A battery discharging for 1 hour could meet an estimated 46% of peak demand periods, while a battery with 2 hours of capacity could fully produce for 66% of peak demand periods [83, 101]. Pairing a battery that can discharge in less than 4 hours with variable generation, such as wind or solar, could increase the hybrid's capacity factor above what the generator and battery, de-rated to meet the 4-hour requirement, could provide. Pairing batteries with variable generation may also add value in markets with performance penalties. Hybridization can decrease the risk of performance penalties for both battery and generator. The battery faces the risk of having insufficient energy while the variable generation faces the risk of insufficient production during the performance period. The battery can firm total output while the generator increases output duration [83, 96, 97].

Increasing capacity payments will potentially be an important revenue stream for battery hybrids in the near future. However, some markets prohibit battery hybrids from selling into capacity as a single resource and, in markets where they are allowed, battery hybrids face significant burdens to prove increased reliability in order to receive increased capacity payments [83, 101].

5. **Transmission deferral**

The electricity grid is in constant need of repair and expansion. As demand changes, transmission lines can become congested and must be upgraded. However, maximum loads occur for only a few hours per year, and in many cases, the highest annual load occurs on a single day of the year [83, 96, 97]. Storage sited downstream of congested nodes can defer or eliminate the need for transmission upgrades. Transmission deferral can be very valuable, with savings exceeding $500/kW/year in some cases [83, 102]. Furthermore, because storage providing transmission upgrade deferral only needs to discharge for a few days per year, transmission deferral can be stacked with other services to increase profitability.

Hybridization with distributed PV can add value in areas where the duration of maximum load is uncertain or exceeds the battery maximum discharge time. The dispatchability of the battery complements the PV generation to provide dispatchable distributed generation. Storage can also be paired with wind generation to reduce transmission requirements for wind farms. High wind production often occurs during periods of low demand, and the highest quality wind resources are often located far from demand centers, requiring the construction of new transmission lines to access the resource [83]. The variable nature of wind production means transmission lines are underutilized during periods of low production, and energy is curtailed when high production exceeds line capacity. Co-locating storage with wind can reduce transmission requirements, which can lead to significant cost savings [83]. Importantly, storage charged with eligible renewables receives an added 30% tax credit, further improving the economics [83].

6. **Demand charge reduction**

Unlike energy charges, which are based on total energy used, demand charges are based on the highest instantaneous, daily, weekly, monthly, or yearly load. The market for demand charges is large and potentially profitable. Nearly every commercial and industrial facility faces demand charges, and demand charges can constitute more than 50% of a commercial customer's bill [83, 96, 97]. Demand charges lead to high marginal cost of electricity. The dispatchable nature of storage allows batteries to supply power during these periods. The yearly value of battery storage for demand charge reduction ranges from $50/kW to $250/kW, depending on the rate structure [83, 90–94, 96, 97]. Battery storage is an especially economical option for customers facing high demand charge rates or demand charge ratchets and for customers who are close to the cutoff between two rates.

Demand charge reduction is one of the most valuable uses of storage for many behind-the-meter customers. Cost savings for a representative California affordable housing apartment complex has been shown to be nearly three times for solar plus storage compared to solar only [83, 96, 97, 103]. To effectively reduce demand charges, the battery must have sufficient capacity to discharge during the peak period. Pairing a battery system with a PV system can allow for a smaller battery to achieve the same demand charge reductions [83, 96, 97]. The PV system reduces most of the demand peak while the battery discharges during periods of cloud cover or when solar resource is not available (i.e., evening).

Hybrid systems are not applicable in all cases. The profitability of behind-the-meter storage is dependent on the demand charge rate, and the profitability of a PV system is dependent on the size and structure of TOU charges. Solar panels are best suited for regions with a high TOU rate that corresponds to peak solar production. Solar-storage hybrids are best suited for areas where high demand charges intersect with TOU charges that are conducive to solar [83, 96, 97].

7. **Resilience and reliability**

Some businesses and government facilities require that a critical load be met at all times. The economic consequences of a blackout for data centers and many industrial processes can be devastating. For hospitals, other critical care facilities, and critical public services such as police and fire stations, continued power is essential. Resilience and reliability differ by grid characteristics and outage type. Resilience is the ability to continue power during a natural disaster or major power outage. Resilience commonly implies a stable grid in a developed country, which has infrequent but potentially lengthy power outages. Reliability refers to the ability to maintain consistent power on an unreliable grid. Reliability commonly implies an unstable grid in a developing country, which has frequent but often short power outages. While self-generation is required for both resilience and reliability, battery requirements and profitability vary between the two.

There is no formal market for resilience, so size and value of resilience are hard to quantify, but resilience is still valuable. Between 2003 and 2012, weather-related power outages are estimated to have cost the US economy between $18 billion and $33 billion annually [83, 104]. To provide resilience, the United States has more than 170 GW of distributed generators built for emergency generation during outages [83, 96, 97].

8. **Decreased diesel generation**

While diesel generators are traditionally used for backup power, hybridized storage solutions may be a more effective way to provide resilience and reliability. For example, in New York City, regulations prohibit backup generators from storing more than 250 gallons of fuel at any location, meaning they can run out of energy during extended blackouts [83, 96, 97]. Furthermore, for areas with unreliable grids, fuel and maintenance costs from frequent generator use can increase electricity costs. In some international markets, the lower reliability of utility power creates stronger drivers for the use of hybridized storage solutions instead of traditional fossil-fueled backup generators. Solar-storage hybrids have also been the solution of choice for energy access solutions and remote health care in developing countries and for enabling small and medium enterprises. Storage paired with solar and a diesel generator allows for critical loads to be met with a smaller generator and less fuel [83, 96, 97]. This increases resilience length and can reduce costs. Storage-solar diesel hybrids do have higher capital costs than stand-alone diesel generators, but reduced fuel use and reduced generator wear, along with added benefits from the solar-storage component, can make up for higher initial costs [83, 96, 97].

2.8.1 Market Considerations

There are several market opportunities for battery storage and battery hybrids. Each market is differentiated by primary application and project location. Not all generator pairings are applicable for all applications or locations. This section provides an overview of key market considerations for battery storage and storage hybrids. The discussion, once again, follows the excellent report by Ericson et al. [83].

A. ***Storage location on grid***

As pointed out earlier, storage location on grid can have significant effect on system performance. Along with the type of grid, storage applications change depending on where on the grid the battery is located. Battery storage can be located at the transmission level, distribution level, or behind the meter. Not all services can be provided at all locations. Large utility-scale batteries are located at the transmission level and used primarily for frequency regulation, capacity, and energy arbitrage. Smaller distributed batteries are primarily used for demand charge reduction, transmission deferral, and resilience or reliability. However, distributed storage may provide wholesale services as well.

In general, the closer to the customer load a battery is sited, the more services it can perform [83]. For example, storage providing transmission deferral must be sited downstream of congestion, and batteries used for demand charge reduction must be sited behind the meter. Although battery storage is more modular than most electricity resources, it is still affected by economies of scale. As such, large-scale projects have lower marginal costs than smaller projects. In 2016, batteries used for commercial demand charge reduction were, on average, 25% more expensive than utility-scale battery storage [83, 101]. Wholesale markets also have minimum size requirements to participate. While small storage can be aggregated, coordination and customer procurement adds to aggregation costs [83, 96, 97].

B. ***Markets for utility-scale storage***

The largest potential application for utility-scale battery storage in large developed and developing grids is as a replacement for peaking plants [83, 96, 97]. Along with replacing peaking plants, storage-gas turbine hybrids have widespread potential market applications for providing spinning reserves. While currently only used in Southern California, these hybrid systems can offer attractive economics in most markets where regulations allow for hybrid participation.

Large islands represent attractive markets for utility-scale storage because they have especially high fuel costs. Islands with large current or planned renewable penetration are especially attractive for battery hybrid systems because high levels of renewable penetration allow the battery to be cheaply charged during periods when energy would otherwise be curtailed.

California represents the biggest US market for battery storage due to its storage mandate and favorable market conditions. California currently accounts for 54% of total US storage deployments [83, 105]. Furthermore, nearly 80% of current projects under development or contracted in the United States are in California [83].

C. ***Markets for distributed storage***

Distributed storage can reduce demand charges, defer transmission upgrades, and provide resilience and reliability. In large developed grids, areas with high demand charges offer the best opportunities for distributed storage. New York and California present the largest US markets for distributed storage due to high demand charges, distributed storage mandates, and tax incentives for storage. However, there are many regions across the country where high demand charges present opportunities for distributed storage. Demand charges are highest in California and lowest in several parts in the middle of the country, particularly the regions with low population density [83].

For batteries providing customer services, large facilities that have economies of scale offer the best market opportunities. Batteries with sufficient capacity, either 100 kW or 1 MW depending on the region, can also participate in capacity, wholesale energy, and frequency regulation markets. Large potential energy storage customers include universities and college campuses, hospitals, laboratories, large office buildings, industrial sites, water treatment plants, municipalities, energy cooperatives, hotels and resorts, mining operations, and military bases. In large developing grids, providing reliability represents the largest market potential. Batteries paired with solar and diesel generators are becoming increasingly attractive options for providing reliable power. Small islands and remote locations present attractive markets for distributed storage. The battery pairs with diesel generation and renewable energy to reduce fuel costs.

Table 2.4 summarizes the major markets for battery storage and battery hybrids as reported by Ericson et al. [83]. Primary application refers to the value stream that provides the majority of the revenue. Secondary applications refer to value streams that can be stacked to increase profitability. Table 2.5 illustrates various estimated market sizes as reported in various NREL reports [83, 96, 97].

TABLE 2.4
Major Hybrid Markets [83]

Primary Application	Primary Market	Battery Location	Secondary Applications	Generator Pairings
Capacity	Large grids	Utility/large distributed	Energy arbitrage, frequency regulation	Wind, PV
Frequency regulation	Large grids	Utility/large distributed	–	–
Spinning reserves	Large grids	Utility	Frequency regulation, black start	Natural gas turbines
Transmission deferral	Large grids	Utility (distributed)	Energy arbitrage, frequency regulation, black start (demand charge reduction, resilience)	Wind, PV
Demand charge reductions	Large grids and islands	Distributed	Resilience, transmission deferral, energy arbitrage	PV, CHP
Resilience (reliability)	Developed grids (developing grids)	Distributed	PV self-consumption, demand charge reduction	PV, diesel-PV
Decreased diesel generation	Islands and remote locations	Utility/ distributed	Decreasing diesel ramping[a]	PV, diesel-PV, wind

[a] Decreased diesel generation refers to the battery reducing wear and tear on the generator.

TABLE 2.5
Estimated Market Size [83]

	US Market Potential (MW of capacity)[a]			World Market Potential (MW of capacity)		
Primary Application	Market Upper Bound[b]	Market Estimate[c]	Hybrid Estimate[d]	Market Upper Bound [e]	Market Estimate [e]	Hybrid Estimate
Capacity	18,000	9,000	4,500	112,000	40,000	20,000
Frequency Regulation	2,000	600	0	12,000	3,000	0
Spinning Reserves	6,000	400	400	37,000	2,500	2,500
Transmission Deferral	10,000	2,000	1,000	62,000	12,000	6,000
Demand-Charge Reductions	32,000	8,000	5,000	200,000	30,000	20,000
Resilience and Reliability	9,200	1,300	1,000	57,000	8,000	6,000
Decreased Diesel Generation[e]	–	–	–	50,000	15,000	15,000
Total	77,200	21,300	11,900	530,000	110,500	69,500

[a] Measured in cumulative capacity additions.
[b] US Market Upper Bound Estimates from Sandia National Laboratory [60].
[c] Market Estimates from literature [83] and communication with industry experts.
[d] Hybrid numbers for battery capacity in hybrid.
[e] Estimates from literature [83].

2.8.2 Regulatory Framework

Energy markets were originally designed with a separation between generation, transmission, and consumption. Battery storage does not fit readily into any of these categories, although it can provide generation, transmission, and customer services; therefore, in many jurisdictions, current market regulations do not fully compensate storage for its services [83, 91].

Succinctly, resource participation in electric markets is governed by participation models that consist of market rules for different types of resources. In November 2016, Federal Energy Regulatory Commission (FERC) filed a notice of proposed rulemaking to establish participation models for utility-scale and aggregated distributed storage [83, 91]. FERC proposed rules to address the fact that "current tariffs that do not recognize the operational characteristics of electric storage resources serve to limit the participation of electric storage resources in the organized wholesale electric markets and result in inefficient use of these resources" [83, 91].

Regulatory reform is continuously evolving, suggesting that battery storage projects must not only consider which services can be provided but also which services may be monetized today and in the future. Market rules addressing customer-sited storage are often less mature than wholesale market structures, which prohibits clear approaches for remuneration of possible multiple value streams to the distributed system owner and negatively impact the system economics. Combining battery storage with generation can allow the hybrid resource to meet regulatory requirements, such as minimum production times and minimum ramp rates, that the storage or generator could not meet on its own.

2.8.3 Federal and State Incentives

Policy incentives vary by region and are often different for utility-scale and behind-the-meter applications. Furthermore, some incentives are only applicable to batteries paired with renewables. At the federal level, storage charged with renewable energy is eligible for a 30% federal investment tax credit (ITC) and an accelerated depreciation schedule. The structure of the ITC and the modified accelerated cost recovery system (MACRS) indicate that battery storage charged with at least 50% renewable energy follows a 5-year depreciation schedule instead of a 7-year schedule. Storage that is charged from at least 75% renewable energy receives a tax credit equal to 30% of the portion charged by renewables. Because the storage and generation must be in close proximity and under common ownership to receive the tax benefits, these tax incentives are especially beneficial to renewable hybrid systems.

At the state level, different levels of policy support have led to very different levels of storage deployment. California has several energy storage incentives and mandates. Assembly Bills 2514 and 2868 mandate more than 1.3 GW of energy storage procurement by 2020, and California's Self-Generation Incentive Program (SGIP) provides incentives for behind-the-meter storage. Combined, these have led to California supporting the bulk of utility-scale and behind-the-meter storage in the United States. The majority of battery deployments are located in California, Hawaii, and the Northeast. California is the largest market, with the bulk of storage projects installed to provide capacity after the Aliso Canyon gas leak [83, 106]. Resilience programs in the Northeast, such as the New Jersey Resiliency Bank, are also beginning to provide grants for solar-storage projects to provide backup power during natural disasters [83].

2.8.4 Successful Case Studies

This section provides four case studies on storage applications outlined by Ericson et al. [83]. Each case study provides insight into uses for battery storage and battery hybrids and the types of markets battery storage and battery hybrids that are likely to serve in the coming years [107–109].

A. **Kaua'i solar-storage hybrid projects**

Both the AES Corporation and Tesla recently constructed solar-storage hybrid projects on the Hawaiian island of Kaua'i. In 2017, the Kaua'i Island Utility Cooperative (KIUC) signed a power purchase agreement with AES for a 28-MW PV system paired with a 20-MW/100-MWh battery [110]. Tesla also opened a 13-MW solar farm combined with a 13-MW/52-MWh battery installation in 2017 (see Figure 2.13). High costs of power from petroleum-fired generators led KIUC to aggressively adopt renewable generation. KIUC currently provides 36% of its generation from renewables and plans to reach 70% by 2030 [110]. During peak solar hours, 77% of electricity is already produced by solar energy. Without storage, continued solar additions would lead to increased solar curtailment. The high levels of renewable penetration also cause a significant amount of ramping on the system [83].

The battery systems smooth generation and allow for a higher level of total generation to be produced from solar. Pairing the battery with solar both reduces cost and allows the battery to receive the 30% federal ITC. The combination of high electricity prices from petroleum-fired generators, high penetration of renewables, and the federal ITC makes the Kaua'i project economically viable [83].

B. **Sterling Municipal Light Department**

In 2016, the Sterling Municipal Light Department (SMLD) in Sterling, Massachusetts, built a 2-MW/3.9-MWh battery system to pair with an existing 3.4-MW solar array. SMLD, a wholesale aggregator of power in the ISO-New England region, uses the storage system for power resilience and demand charge reduction.

FIGURE 2.13 Telsa solar-storage facility in Kaua'i. [83]

Because the solar-storage system provides backup power for critical response functions in the case of a power outage, SMLD received a $1.46 million grant from the Massachusetts Department of Energy Resources as part of the Community Clean Energy Resiliency Initiative. SMLD also received a $250,000 grant from the US Department of Energy. While SMLD did not benefit from reduced construction costs because the solar panels were built at a different date than the battery storage, hybridization still provides resilience benefits that the battery or PV alone could not deliver. Along with resilience, the battery provides value through a combination of arbitrage, frequency regulation, and demand charge reductions. ISO-New England has monthly regional network service payments and a yearly capacity payment [111]. It was estimated that the projected revenue was $288,076 ($/MW/year) [83, 111].

More than two-thirds of yearly revenues for the SMLD project come from the battery operating a few times each year to reduce demand charges. This fact demonstrates the importance of demand charge on battery storage profitability. However, the resilience grants also significantly improved the project economics. At an estimated total system cost of $1.7 million per MW of capacity, the project has a payback period of less than 7 years before the grants for resilience are considered. With the grants, the payback period is cut roughly in half to a 3–3.5-year payback period [83].

C. **Irvine Ranch Water District**

In September 2016, Advanced Microgrid Solutions (AMS) and Irvine Ranch Water District (IRWD) in Irvine, California, signed a private-public partnership for the installation of a 7-MW/34-MWh battery storage system. Based on the contract, AMS installed batteries at 11 water treatment facilities and pumping stations to provide capacity and reduce demand charges [83, 96, 97, 102]. The batteries were aggregated to sell into capacity and ancillary service markets as a single unit. AMS owned and operated the batteries to manage requests from Southern California Edison for load reduction as part of a 10-year power purchasing agreement. At the same time, AMS received payment from IRWD for demand charge reductions. The water district benefited from expected annual cost savings of $500,000 [83].

D. **Southern California Edison spinning reserves**

In April 2017, the first of GE's 50-MW LM6000 gas turbines paired with a 10-MW/4.3-MWh battery storage system came online. The hybrid system provides spinning reserves and primary frequency response while the turbine is offline. The storage provided power during the turbine's 5-minute ramp time, and the generator provides power during the rest of the required runtime. This allows the turbine to sell its entire 50 MW of capacity into the spinning reserves market while offline. The turbines were retrofitted to reduce start-up times from the usual 10 minutes to 5 minutes in order to reduce the required battery size.

The battery addition generated $1.4 million in additional yearly revenue from spinning reserves [112]. At around $5 million for the battery and related BOS, the payback period is a little over 3 years. In addition to attractive economics, the hybrid system reduced water use and air emissions. The demonstration that battery storage can be profitably paired with generators to provide spinning reserves opens up a large potential marketplace for storage hybrids [83].

2.8.5 Important Takeaways for Battery Storage and Battery Hybrids

Navigant Research estimates 94 GW of world installed battery capacity by 2025, with 21.6 GW installed in 2025, putting their estimates in line with our forecast of 110.5 GW of capacity by 2028 [102, 113]. Similarly, estimates of 27.4 GW for distributed solar-storage hybrids by 2026 coincide with our estimates of 26 GW of solar-storage hybrids for resilience and demand charge reductions [83, 91, 102]. Greentech Media Research (GTM) forecasts US annual battery storage deployments to be 7.2 GWh by 2022, equating to roughly 2.5 GW of annual capacity additions. Thus GTM estimates of cumulative US additions are within the same range of 21.3 GW by 2028. The majority of utility-scale battery deployments will be for energy capacity as a replacement for peaking plants. Demand charge reduction also has a large market potential. Finally, battery hybrids to reduce diesel fuel use are expected to have a large market potential in the near future. Some of the highlights on use of battery to increase level of HRES in grid can be summarized as follows [83, 96, 97]:

A. **Battery costs**

- Lithium-ion battery costs fell by more than 65% between 2010 and 2015, and costs are expected to continue to decline.
- Battery costs are most significant for applications such as transmission deferral and selling into capacity markets, which may require 4 or more hours of discharge duration. BOS costs are relatively higher for short duration applications, such as frequency regulation.
- Battery hybrids can reduce BOS costs by building one hybrid unit instead of two individual storage and generator units.
- Solar-storage hybrids can further reduce costs by sharing an inverter and other system equipment.

B. **Value streams**

- Battery storage can provide multiple value streams, including energy arbitrage, frequency regulation, spinning reserves, generation capacity, transmission deferral, demand charge reductions, resilience and reliability, and island and off-grid generation.
- Combining value streams can increase profitability.

- Hybrids work best when the strengths of battery storage—fast response and dispatchability—are paired with generator strengths—lower capacity costs and unlimited duration.
- A storage-gas turbine hybrid can sell into spinning reserves while the generator is offline, eliminating the need for the turbine to inefficiently run at part capacity.
- Depending on regulatory decisions, paired storage-variable generation hybrids may be able to increase the capacity factor of both.
- Pairing a solar, storage, and diesel generator can be economic in areas with high fuel costs, such as islands and remote locations.
- Solar-storage hybrids can more effectively reduce demand charges than either a battery or PV system alone.
- Storage hybrids can provide grid resilience during power outages.

C. **Regulatory framework and policy incentives**

- Currently in the United States, the 30% investment tax credit provides an incentive to hybridize battery storage with renewables.
- Further regulatory changes beneficial to storage, especially at the federal level, are still several years away.
- In the nearer term, state mandates and incentives can improve battery storage and battery hybrid economics.
- Grants and initiatives for power resilience can improve the economics of battery hybrids.

D. **Market considerations and market potential**

- The largest application for utility-scale storage is peaking capacity, while the largest application for distributed storage is for demand charge reductions and resiliency.
- The near-term market opportunities for utility-scale storage include large islands and markets with high capacity payments.
- Utility-scale battery hybrids can help meet specific market regulations, such as ramp requirements and minimum discharge times, and avoid nonperformance penalties. Tax credits and incentives also increase the profitability of hybrid projects.
- The customers for distributed storage and hybrid projects are those with large energy loads, such as demand aggregators, universities, mines, and municipalities with sufficient load to support economies of scale.
- Hybrids used for resilience and reliability could be especially beneficial for unreliable grids and customers for whom blackouts are especially costly or endangering.
- Customers in areas with both large demand charge reductions and high TOU charges that coincide with solar production could benefit most from solar-storage hybrids.

2.8.6 Additional Perspectives on the Role of Storage for HRES in the Grid

While battery has been the preferred storage device for HRES in the grid, there are some other research studies and perspectives reported in the literature on the role of energy storage for HRES in the grid. Some of these studies are briefly described here.

The paper by Luo et al. [84] provides an overview of the current development of various types of electrical energy storage technologies. While the study is largely focused on the electrical energy storage outside the grid through extensive assessment of various EES technologies, the study indicates that the Li-ion battery has relatively high power/energy densities and specific power/energy, which has resulted in the current broad range of development, particularly in small-scale EES applications and grid-level storage. The study also indicates that the cycle efficiencies of EES technologies have been continuously improved with time through development efforts leading to technology breakthroughs, and most commercialized techniques normally have medium-to-high cycle efficiencies. The energy capacity and the self-discharge of EES systems are the major factors in deciding the associated suitable storage duration. While the increasing level of hybrid renewable energy systems in the grid requires energy storage for power quality and stability, as pointed out in Chapter 1, no ES technology claims high in all aspects; each technology has its own limitation in performance when used for grid connection. System capacity, type of application, and the cost of peak time electricity decide the storage capacity. A wide variety of such technology may be required to address the issues arising during grid connection for increasing level of HRES. While, as shown earlier, batteries are preferred as storage device on the grid, significant literature is published to point out reasons for other options.

Fathima and Palanisamy [114] examined optimized sizing, selection, and economic analysis of battery energy storage for grid-connected wind-PV hybrid system. An optimized sizing methodology for battery ES is used to cater peak shaving and ramp rate limiting in the power dispatch using bat algorithm. This is used in a grid-connected solar–wind hybrid power system to combat loss of power. Different types of battery ES such as lead-acid (Pb-acid), Li-ion, flow batteries, and sodium sulfur (NaS) were considered and analyzed in a comparative fashion. The integrated system was then tested with an efficient battery management strategy which prevents overcharging/discharging of the battery. The study was successfully applied to a case study in India.

Hittinger et al. [115] examined grid storage technologies applied for four purposes such as frequency regulation, power smoothing for wind as base load plant, power smoothing for load following, and peak shaving and arrived at a conclusion that the power accumulation capacity is vital for frequency regulation, whereas the energy capacity influences energy intensive applications like peak shaving. The study indicated that decreased capital cost, increased power capability, and increased efficiency all would improve the value of an energy storage technology and each has cost implications that vary by application. Based on their analysis they suggested hybrid storage system for grid-connected HRES and implied that each hybrid storage combination is different and blanket statements on superiority of a given hybrid system are not always appropriate. The transient stability of a distributed generation – battery supercapacitor has been carried out by Srivastava et al. [116]. Korada et al. [117] have developed a three-level grid-adaptive power management strategy (GA-PMS) in MG with RES–battery supercapacitor to support grid.

A compressed air energy storage (CAES) and wind energy system is used by Jirutitijaroen [118]. He tried to time-shift wind energy to maximize the daily revenue by stochastic dynamic programming (SDP) for forecasting generation and price. With a similar objective, Salas and Powell [119] added an approximate dynamic programming (ADP) algorithm that showed the proficiency of designing near-optimal control policies for a large number of heterogeneous storage devices in a time-dependent environment with good accuracy at par with stochastic and dynamic models when demand variability is additionally taken. Madlener and Latz [120] examined economics of centralized and decentralized compressed air energy storage for enhanced grid integration of wind power and concluded that the economic feasibility of a centralized CAES is more viable than the distributed wind turbines CAES.

Koller et al. [121] showed the effect of the grid-connected 1 MW BESS on frequency reserves, peak clipping, and islanded operation of an MG. Grid-forming and grid-following inverters for the variable RES are detailed by Kroposki et al. [122]. They indicated that achieving 100% variable renewable energy grids will require (a) better ways of matching supply and demand over multiple timescales, (b) significant curtailment, and (c) proper operations with very high instantaneous penetrations of variable renewable energy sources. As AC power systems evolve from synchronous generator-dominated systems to inverter-dominated ones, one must ensure that these technologies operate in a compatible manner. This includes designing inverter-based systems to provide system stability and additional grid services necessary for proper AC power system operations. These requirements for grid stability should be incorporated at all sizes of inverters because a future grid might have extremely large numbers of small, highly distributed variable renewable energy systems. The study indicated that with proper control considerations, inverter-based systems can not only maintain or improve grid stability under a variety of contingencies but also dramatically improve the response characteristics of power systems and increase operational stability.

Malysz et al. [123] proposed an online optimal operation for BESS (battery energy storage system) based on a mixed-integer linear program (MILP) over a rolling horizon window. Whereas, after a detailed study of different batteries, Müller et al. [124] suggested Li-ion batteries of LFP-C type are economical in the long run for large capacities for stationary applications with RES on grids. Bose et al. [125] studied the optimal placement of large-scale energy storage on power grids with both conventional and wind generation. The solution technique for this infinite horizon problem assumed cyclic demand and generation profiles using a semi-definite relaxation of AC optimal power flow. Changes in storage allocation in the network are studied as a function of total storage budget and transmission line-flow constraints. These questions are investigated using an IEEE benchmark system with various generation portfolios. The study indicated that the line-flow limits have a significant effect. Bhandari et al. [126] showed that hybridizing PV-wind with microhydro power plants into a single mini-grid has been practically applied in Nepal and this has increased the reliability and meets the load in an environment-friendly way.

Hearn et al. [127] examined grid-level flywheel energy storage system (FESS) and showed that locating it at the transformer and higher levels in the grid will reduce its size by inherent power smoothing by the pool. Denholm and Hand [128] showed that the ability to exchange power with neighboring grids, load shifting, and storage can successfully deal with high penetration of renewable. Kerestes et al. [129] showed that peak shaving can be dealt with by gas-powered generation and load leveling (flat profile) by coal-fired or battery or pump storage. Williams et al. [130] put forth DSM (demand-side management) by adding heat pumps and thermal storage to PV that adds on energy independence of the house. Palizban et al. [131] pointed out that a hybrid of different energy storages can serve multiple purposes in a cost-effective way.

The literature described above indicates that, while batteries are preferred energy storage devices for grids with HRES, in general, basic characteristics of energy storage, its location, and its nature (homogeneous or hybrid) on grids are very important for various aspects of power quality and stability of grids with HRES.

2.9 MANAGEMENT AND CONTROL ISSUES OF HRES IN THE GRID

The insertion of HRES in the grid at significant level encounters several challenges. Here we address some of them.

2.9.1 Weather-Dependent Variability of Renewable Resources

The wind energy and PV are expected to have the lion's share of the HRES for utility grid. So, the future energy source is pivoted on the in-depth realization of their variability. Resource variability is a multifaceted notion expressed by a range of distinctive characteristics. Simultaneously, research to date tells that there is restricted knowledge about the variability of the future power system. The variable attribute of climatic fluctuations is the reason for the inconsistency of the RES and creates uncertainty in the energy production on the range of seconds, hours, and even days. It is estimated that clouds limit up to 70% of daylight-hour solar energy potential. Grid sometimes deals with aggregation over massive areas, and this mitigates the variability of every single RES.

Presently a large variation is tackled via switching in fast-acting conventional sources depending on the climate forecasts on a minute-by-minute and hourly basis. Such variability can additionally be taken care by setting up large-scale storage on the grid or by the long-distance transmission of RE linking to larger pools of such generations in order to equalize regional surplus or shortfall nearby in future. Graabak et al. [132] have addressed the variability characteristics such as: (i) distribution long term, (ii) distribution short term, (iii) step changes, (iv) autocorrelation, (v) spatial correlation, (vi) cross-correlation, and (vi) predictable pattern. Distribution can be short term (minutes, less than 1 hour) or long term (1 hour or more). These terminologies carry their own implications.

Many such related papers refer to "step changes" as a variability characteristic. These are the alteration in the available resource that takes place in small-time steps of minutes to some hours. Another variable characteristic is autocorrelation [133] which figures out the statistical relation among values of the same parameter in a series. The relationship of wind speed information between different locations and the corresponding relationship of solar irradiance for different locations are under study by several projects. This spatial correlation is perceived as one of the instrument to gauge variability characteristics. Wind and solar sources may also show one kind of diurnal and seasonal trends.

Power from sun, wind, and ocean additionally exhibit predictable seasonal patterns recognized as a distinguishing variability characteristic. Pattern forecast for this trend of wind and sun is complicated, and it is a subject matter taken up in many papers. In a precise study, Tande et al. [134] have viewed reanalysis data set for illustrating wind variability characteristics. With information of a temporal resolution of 6 hours and a spatial resolution of 2.5 hours in each latitude and longitude, a two-dimensional linear interpolation of neighboring locations is utilized to get wind speeds at the chosen sites. Both offshore and onshore information can be dealt with in this way for explaining the variability. It is apparent that entry of offshore wind generation and its variability will noticeably affect the grid.

In the study performed by Wiemken et al. [135], record from 1995 extracted from 100 monitored PV systems (rooftop plants 1–5 kW) with a 5-minute time resolution ensembled for 243 kW (grid connected) is used. A model is developed taking onshore wind and PV energy generation for the period 2001–2011 across 27 nations in Europe. The data were taken from NASA for hourly values of wind speed and solar irradiance documented at a spatial resolution of 0.5°E/W and 0.66°N/S. The generations from wind and PV translated from the climatic record were later on combined to structure regional or nation-specific datasets. The model first considered PV and wind sources to contribute half of the energy supply of the total requirement. Further PV share in the wind/PV proportions of 0, 20, 40, and 60% was investigated.

2.9.2 Power Generation Forecasts

Contribution of wind energy has been the largest share out of the renewable energies and is expected further growth. For responsible and sustainable growth of wind energy industry, reliability, robustness, and stability are important factors. As wind energy integration to the grid is in MW scale, in future it may function as base load plant. So, the decision of economic load dispatch will largely be affected by proper forecasting of wind power. The objective is to improve accuracy in forecasting wind speed and power 1 day ahead so that it becomes reliable, which will be a benefit to the load dispatch centers as well as installation of additional wind turbines onshore and offshore.

Wind forecasting has been taken up in literature by various researchers. The forecasting for power may be very short term (within 2.5 seconds), short term (10 minutes to 1 hour), long term (15 minutes to 3 hours), or a day ahead (24 hours). Forecasting wind speed is an important factor, based on which planning of new wind farm depends. Specifically for offshore wind farms, the safety requirement is less advanced. The error in forecasting wind power 1 day ahead is more compared to the short term because wind speed and power prediction requires more computation time and this needs to be improved. Research has shown a good result from the hybrid method. The researchers are oriented to make wind power predictable. When the wind is predictable, it becomes reliable, which will be a benefit to the load dispatch centers for economic load dispatch as well as the installation of additional wind turbines onshore and offshore.

Going through the available tools and the accuracy, the methods/prediction models are broadly divided into physical-, statistical-, and artificial intelligence-based methods [136]. Out of various statistical methods such as curve fitting, statistical approximation autoregressive integrated moving average (ARIMA), seasonal ARIMA, extrapolation with periodic function, and methods of finding probability density functions (PDF) have been evaluated by the coefficient of determination [137]. Different software models have been developed such as WPMS, WPPT, Prediktor, ARMINES, Previento, Zephyr, AWPPS, Ewind, and ANEMOS and adopted in different countries [138]. Some of them are hybrid methods. Prediction of offshore extreme wind is important for the protection of offshore wind system so that such sites can be avoided during planning. Method of independent storms (MIS) stands better as compared to the other three in the study by An et al. [139]. In the study by Wang et al. [140], the extreme wind has been estimated by the combination of swarm optimization with the traditional methods which added improvement. The available software has their limitations up to how many meteorological data required, precision in numerical weather prediction (NWP), different accuracy indices for short- and long-term prediction, etc. Intelligent techniques such as Artificial Neural Network (ANN) [141], Fuzzy, Support Vector Machine (SVM), Wavelet, Hilbert-Huang transform, data mining techniques [142], swarm optimization combining the statistical methods of time series prediction with improvement in nonlinear node functions, and training algorithms have given good results as compared to statistical/any method alone [142, 143]. Combination of Fuzzy and ANN takes less prediction time and thus gives faster result. It has been remarked that grouping wind farms for wind forecasting can give better result [142]. Instead of predicting the wind speed exactly, prediction into lower and upper bounds method (LUBE) [144] in prediction interval with defined confidence level gives a better result in performance indices. Wind speed has been estimated by radial basis function (RBF) neural network, and wind turbine has been appropriately controlled for maximization of wind power [145]. "Anti-phase correlation" of wind speed and solar radiation has been found after wavelet analysis, implying that wind and solar energy can complement each other in generating electricity [146].

Smart grid performs also with the penetration of PV and has to consolidate its performance figures in the presence of variability. Many researchers report on the novel hybrid intelligent algorithm for PV forecasting, taking its fluctuating behavior. In this regard, wavelet transforms (WT), stochastic learning, remote sensing method, and fuzzy adaptive resonance theory mapping (ARTMAP) (FA) network are often used. Forecasting accurately improves system efficiency also. As the numerical prediction depends on the weather data, which is provided by sensors, reduction of dependence on sensors for wind speed rather than the estimation method of the wind speed for sensor-less control is needed. Different capacities of battery, wind, and PV are considered to check which proportion of each component is economical for a specific location, Dhahran in Saudi Arabia, taking historical weather data during the demand of different months in a year for the wind-PV hybrid power system (WPVHPS) [147]. The addition of wind generation is more economical than PV. The addition of more battery can reduce the diesel generation and time of use.

Prediction is vital for energy management. The energy management functions in a wind-battery system are to (1) charge the battery from wind, (2) supply the load from wind power, (3) trade the wind/battery power to the grid, (4) buy power from grid and store in the battery or supply it to the load, and (5) supply the local load from battery. The day ahead electricity rate and wind energy are forecasted through Wavelet-ARMA of time series, breaking it into smooth subseries. The state of charge (SoC) of battery is predicted in a longer time horizon. In another case study of Turkey [148], based on the 15 years of data of global solar radiation distribution, no relationship between the distribution of annual time lapse and solar energy and solar radiation intensity are established.

The solar and wind energy potentials are surveyed for five sites in Corsica [149]. From this study, two sites with the desirable trait are chosen, and the sizing and the economics for an isolated hybrid PV/wind system are compared. The trend is dependent on site-specific resource analysis. The sites with more wind potential have less cost of energy and more feasibility. Energy management system (EMS) is an integration of all the algorithms, procedures, and devices to control and reduce the usage and the cost of energy used to deliver the load with its specifications. A critical review by Mahesh and Sandhu [150] pointed out that most of the EMS for RES is concerned with the flow and the control of power and efficient battery utilization for its durability. However, a full-fledged control approach is yet to be developed.

Wu et al. [103] proposed optimal scheduling of the PV system for saving the time-of-use (TOU) cost. Sichilalu et al. [151] focused on a net-zero-energy building by demand-side management. The energy management of a grid-connected WPVHPS has been introduced in hardware by Li et al. [152]. In this study, the hardware, communication, and how to meet its requests and functions are emphasized. The system could manage both grid-connected mode and stand-alone mode. EMS for both stand-alone and grid-connected hybrid RES are reviewed by Olatomiwa et al. [153]. EMS based on linear programming, intelligent techniques, and fuzzy-logic controllers is discussed for various combinations. In the study by Sehar et al. [154], an EMS for controlling end-user building loads, AC, light, and ice storage discharge, with adequate solar rooftop PV systems in groups to absorb plug-in electrical vehicle penetration using practical charging situations is developed without delaying EV charging. The EMS is developed by Merabet et al. [155] for a microgrid with RES that checks net excess generation, battery power, and SoC and takes the decision whether to charge/discharge the battery, reduce PV generation, shed load, or increase generation of PV by maximum power point tracking (MPPT) to control load end voltage. Boukettaya et al. [156] developed a supervisory control in an MG with WPVHPS, a flywheel energy storage system (FESS). Reihani et al. [157] studied the EMS for a MW-range battery energy storage system (BESS) with actual grid data serving for peak load shaving, power smoothing, and voltage regulation of a distribution transformer.

A distributed algorithm that extracts renewable energy sources on high priority through monitor and prediction of generation and loads online is proposed by Mohamed and Mohammed [158]. It works to reduce cost and improve system stability. Lucas and Chondrogiannis [159] reports a battery management system (BMS) based on physics-based models of lithium-ion (Li-ion) batteries and vanadium redox-flow (VRF) BESS. Lawder et al. [160] demonstrated a VRF storage device for frequency regulation and peak-shaving tasks. Multiple BMSs are required in order to reach the desired capacities at grid-level demand. A part of the (EMS) in order to achieve specific operational objectives was described by Nick et al. [161].

Gelazanskas et al. [162] reviewed demand-side management (DSM) and DR, including incentives, noncritical load scheduling, and peak shaving methods. Vasiljevska et al. [163] demonstrated an EMS in a medium-voltage (MV) network with several MGs by a hierarchical multilevel decentralized arrangement. A power management system (PMS) was proposed for a PV-battery-based hybrid DC/AC MGs for both grid-connected and island modes by Yi et al. [164]. It balances the power flows and regulates bus voltage automatically under different operating circumstances.

2.9.3 Nature of Control

A. **Centralized control**

A grid-connected hybrid system with battery is studied and tested for centralized control under three scenarios by Abbassi and Chebbi [165]. The control strategy developed could maximize the utilization of the hybrid system. Centralized control requires fast communication and supercomputing to handle a large amount of data in a short time. This is less reliable due to single point attack risk. A new topology of WPVHPS is proposed by Singaravel et al. [166]. In this topology, the sources are connected together to the grid via only a single boost converter–inverter setup.

B. **Distributed control**

It is very suitable for grid-integrated renewable sources. Alagoz et al. [167] describe that DERs are gradually increasing in count with each consumer turning into a prosumer. This can take the best out of it if there is a bidirectional interaction between DERs. A service-oriented infrastructure can be formed by a tree-like user-mode network (UMN). Li et al. [168] studied the coordination control of a WPVHPS and a proton exchange membrane fuel cell (PEMFC). A grid-connected WPVHPS was proposed by Hong and Chen [169]. The pitch control of the wind turbine uses radial basis function network-sliding mode (RBFNSM), and the MPPT of PV system uses general regression neural network (GRNN).

For control of the voltage and frequency at the point of common coupling firefly algorithm based proportional integral (PI), and PID controllers are used by Chaurasia et al. [170]. Bendary and Ismail [171] showed that the modified adaptive accelerated particle swarm optimization (MAAPSO) predicts better than particle swarm optimization (PSO) for proportional integral derivative (PID) and fractional-order PID battery-charge controller.

C. **Hybrid control**

This type is regarded as a combination of centralized and distributed control and is more versatile. Qi et al. [172] reviewed supervisory model predictive control (MPC) and developed it in distributed architecture taking two spatially distributed wind and PV subsystems, each with storage, in a DC power grid, with a local load connected. For a WPVHPS with fuel cell, Baghaee et al. [173] developed a direct control scheme in a hybrid AC/DC structure that deploys a harmonic virtual impedance loop and compensates voltage.

D. **Control communication**

Dynamic interaction between transmission and distribution systems caused due to transformations in power systems make control vulnerable. This is also happening in case of integration of renewable power plants to grid. Vision for perfect grid management can never undermine the importance of control communication. If the output of a renewable energy power plant is greater than 10% of the line capacity, temporary unavailability [174] can adversely affect power grid stability and so demands a communication. It is important to develop an intelligent, self-adaptive, dynamic, and open system. So, a multi-agent system (MAS) was proposed by Jun et al.[175] to handle the energy management of the hybrid PV-wind generation system in which each agent with a RES reacts intelligently to changes.

For the energy control in a distributed manner, energy routers can serve dynamically the energy distribution in the grid, where the whole structure can be termed as energy internet [176]. For peak load and outage, a building integrated PV (BIPV), mainly for self-feeding of buildings equipped with PV array and storage, is studied in a DC MG by Cao and Yang [177]. Hierarchical control is designed by Petri nets (PNs) interface for a four-layered EMS that regards the grid availability and user's commands. The layers are human-machine interface (HMI), prediction, cost management, and operation.

2.9.4 Power Quality Issues

The power quality is also an issue with the PV-wind hybrid power system integrated to the grid. A well-written review on the subject has been published by

Badwawi et al. [178] for problems and related solutions for this system in grid-connected condition. Voltage and frequency fluctuation and harmonics are major power quality issues with a severe effect on the weak grid. Appropriate design and advanced fast control can solve these issues. Filters, control of PWM inverter, and droop control can also be a solution.

2.9.5 Long-Distance Transmission

All countries emphasize on the use of clean and alternate energy. As discussed in previous subsections, with the rapid development of RES, fresh set of technological requirements pops up on the grid: the location of RE resources distant from load centers and the power variability. The characteristics and its control of the electricity grid need a modification to integrate RE [179]. At present, many countries lack affordable storage facilities for renewable power. But on a positive note, the excess power is transmitted through the national grid by internal transmission lines. However, connectivity to the national grid should be even or balanced. The large-scale intermittency demands to switch in fast-acting conventional reserves on the basis of climatic forecasts on short- to long-time frame, by setting up grid-scale storage or by long-distance transmission of RE generation connecting to larger reserves for resources in order to equalize regional and local surplus or shortfall.

Long-distance transmission capacity is necessary to dispatch a huge quantity of renewable power a thousand kilometer or more across the country. The construction of transmission tower is given low priority with historically low investment in transmission due to community concern over the required right of way in more dense urban areas. Further many long transmission lines are aged and of inadequate capacity. Both remote solar PV and wind energy generation require "Green power Superhighways." HVDC transmission [179] and use of superconductors [179] are costly alternatives as RE itself costs more to the user. HVDC lines offer transient as well as short-term voltage stability. Variability of the source can be well managed via an extensive and robust transmission line network. The transmission capacities based on power electronics devices start to change the grid characteristics and control requirements. The key power electronic technology has a high impact on the power quality because of its fast control and sensitivity to fault and other abnormal conditions of the grid.

Hence, research is still going on HV superconducting cable for long-distance transmission of RE [107]. In the present day of renewable energy, the grid has to serve national character. With more urbanization and industrialization, the reduction of carbon dioxide emission has been essential and requires long-distance delivery of renewable power [108]. Rooftop PV can reduce the need for long-distance transmission, but have a higher cost than wind or concentrating solar power, and with small but considerable aesthetic sense. The gradual entry of big wind and solar generation demand huge spending of money in improving the capacity and efficiency of long-distance electricity transmission. Many researchers feel that at present there is a growing gap between the grid system and control technologies and power electronics equipment design capability.

2.10 ADDITIONAL FACTORS INFLUENCING HRES IN THE GRID

Besides management and control issues for HRES in the grid that are outlined above, there are a number of other factors that play an important role for HRES in the grid. These include the role of distributed energy resources and use of microgrid to manage HRES in the grid; the role of planning, regulatory, project finance, and technical perspective; and the role of HRES in developing utility grid such as one in India.

2.10.1 Roles of Distributed Energy Resources and MG for HRES in the Grid

A renewable hybrid energy system comprises one or more renewable energy sources, a power conditioning device, a controller, and one or more energy storage systems. When such hybrid renewable energy sources (HRES) are integrated into the grid, variable output due to the stochastic nature of input may lead to instability and power quality issues. In this changing scenario, microgrids (MGs) have come up as a solution to maintain power supply in small scale as an autonomous entity in the event of grid failure. It has complementing resources or different DG sources in combination with storage with power electronic interface. Distributed energy resource (DER) can be either a distributed generator or distributed energy storage. Under its spectrum, it can be PV, wind, heat pumps, combined heat and power (CHP) generation, energy storage (ES), fuel cells (FCs), electric vehicles (EVs), energy efficiency (EE), and demand response (DR). The behavior of the resources, such as EE, DR, heat pumps, and EVs, is user dependent. Further, the PV source has no inertia. So ES and FCs can provide more reliability and flexibility to the grid if operated in a manner coincident with grid needs that respect storage limitations.

These DGs have made the grid more resilient, efficient, environment-friendly, flexible, less vulnerable, easier to control, immune to issues at some other location, slow gradual capital investment, and integratable to grid with minimal disturbance to existing loads during commencing. Participation of DERs in operation can be profitable because load shifting without grid upgradation curtails peak demand. EVs and MGs can provide ancillary services. Under normal operation of the grid, varying capabilities of the DERs support voltage and reactive power whereas under fault voltage and frequency ride through capability are expected. Under such fault, the inverter must respond as per requirement. With the coordination of inverter-based resources in a group, it is possible that the DERs counteract to grid contingencies such as voltage and frequency deviations and assist in fast recovery. So they are termed virtual inertia. But, at the same time, some issues are of concern and have drawn the attention of researchers. They are mainly due to stochastic nature such as load following, power vs. energy profile in storage, stability, reliability, cost, control architecture, autonomous control, power quality issues, and grid interconnection.

Considering these issues, Kusakana [109] carried out a focused feasibility study of the unit commitment for reliable power supply and modeling of energy systems of PV, wind, and diesel generator. In the past decade, more significant development has taken place with various combinations of sources and storage. Optimization in all respect of wind energy for grid integration has been thoroughly reviewed by Behera et al. [180] and observed to have good success. The control topology and the objectives have also changed in recent years. In addition to other reviews, control aspects and reliability issues with such sources are discussed by Khare et al. [181]. The application of evolutionary technique and game theory in hybrid renewable energy is also presented.

2.10.2 Regulatory, Project Finance, and Technical Perspective

The renewable electricity demand is predicted to add up to 20% more within the next 5 years. They can have the quickest development within the power sector, providing nearly one-third of the requirement in 2023. Further, there is a forecast to exceed 70% of world electricity produced, primarily by PV and followed by wind, hydropower, and bio-energy. Hydropower remains the biggest of such supply, meeting 16% of the world electricity demand by 2023, followed by wind (6%), PV (4%), and bio-energy (3%). Energy storage for grid applications lacks a sufficient regulatory history.

Whereas active regulation of voltage was not permitted and the DERs had to trip on abnormal voltage or frequency, participation in voltage and frequency control was desirable due to a gradual increase in the percentage of DER in power system. This was resolved in 2003. The first amendment to this came after a decade (11 years) but the second one came just after 4 years of the first [182]. This comes in line with the steeper increase of DER penetration than the previous decade. As the DERs are geographically dispersed, the communication interface between the DER and the main grid and in between the DERs has been an additional demand of the hour for smooth and reliable coordinated control.

Some of the distribution grid safety demands are (1) short trip times, (2) ride through with momentary cessation, (3) voltage rise concerns, (4) protection coordination, and (5) islanding concerns for the safety of workers. Bulk system reliability demands (1) long trip times, (2) ride through without momentary cessation, and (3) reactive power support. Increasing penetration of unconventional generation to grid is reducing system inertia which can degrade system frequency stability. So, active power output is modulated in response to frequency deviation (default droop 0.05 p.u. frequency for 1 p.u. active power change). Voltage benchmarks standard for voltage fluctuations is within ±5% at the customer end. As a DER exports active power, the voltage rises and the profile is disturbed and quality is compromised.

Current grid standards massively need that low-power KW range single-phase PV systems supply at unity power factor with maximum power point tracking (MPPT) and detect fault and island from the grid in such situation [183]. However, loss of these generations under grid faults gives rise to voltage flickers, power outages, and an unstable system. So grid code amendments for increased entry of PV systems in the distribution grid are expected. The standards have undergone a significant review for low-tension interconnection in many countries. Also, reactive power can be supported either by changing the tap setting of the transformer or by PV inverters with advanced control strategies to maintain the grid voltage.

Investments in RES for utility are normally assessed from regulatory, project finance, and technical perspectives. The regulatory requirement is satisfied by utility compliance as well as reduction of the associated cost. The budget estimate looks at the investment and benefits of the particular project. The technical assessment deeply goes through the safety concern of the specific technology involved and its operation. Besides these project specific assessments for RES, physical benefits of transmission and storage and the effect in the integrated picture of the grid is also important. It is therefore always recommended to go for an integrated approach for full exploitation of renewable generation and electricity storage with respect to transmission and distribution [184]. And this in line with the state utility cannot be undermined also.

The United States alone has more than 3,000 utilities, 8 electric reliability councils, and thousands of engineering, economic, environmental, and land use regulatory authorities [185]. Because the profitability of a storage project depends on national, state, and local policies, further analysis is required before embarking on any specific project.

2.10.3 Role of Planning

Planning wind-PV hybrid power system (WPVHPS) involves a cost-effective design on priority. The various aspects that are optimally adjusted before commencing are size, fluctuation of load, and generation. But, some design considerations such as tilt angle of PV panel and a hub height of wind turbine too have importance. Besides the priority objective, when the reliability of supply is seen, the optimum number of units plays an important role. The years of service life is also important in planning. Graphical construction and probabilistic approaches in combination with an optimization method are used for planning. Planning has become a multi-objective optimization with multidimension.

Yang et al. [186] optimally designed wind-solar-battery system for the minimal annualized cost satisfying the limit of loss of power supply probability (LPSP). The five factors such as number PV module, wind turbine, and battery units, module inclination, and height of wind turbine have been optimized by genetic algorithm (GA). The result is indicative that the minimum number of wind turbine with some batteries and PV panels with the location-dependent tilt angle is a good solution.

After going through various traditional approaches for their suitability for wind-PV hybrid systems, Sinha and Chandel. [187] review recent trends in optimization techniques for solar photovoltaic–wind-based hybrid energy systems. Abbassi and Chebbi [165] discuss energy management strategy for a grid-connected wind-solar hybrid system with battery storage and policy for optimizing conventional energy generation. The statistical probability density functions are considered for wind speed and irradiation. Discrete Fourier transform (DFT) of the output power to different fast and slow components is done. Monte Carlo simulation (MCS) for different scenarios is very useful for confirming a design for such stochastic variations of generation and load. One contribution of the storage in such system is toward the frequency management. In a similar line, Arabali et al. [188] suggest a new strategy to meet the controllable heating, ventilation, and air conditioning (HVAC) load with a hybrid RES and ES system. From recorded weather data and load stochastic model of the wind generation, PV generation, and load are developed by Fuzzy C-Means (FCM) clustering, dividing data into ten clusters to show seasonal variations. A multi-objective GA is employed to get the optimal size, cost, and availability DC microgrid systems with PV and wind [189]. When planned with high-temporal resolution data, there was increased control, improved export, availability of power, and decreased variability than for hourly data set. The diesel generator is initially thought as an alternate supply once power fails because it is well transferrable, standard, and has a high power-to-weight ratio. When various DERs are integrated into the system, these can affect the voltage profile of the system and demands frequent tap change, but if the voltage is set based on one fixed point, there may be an overvoltage at another. During planning in addition to overall operational cost, the capacity of the capacitor bank or the power factor correction equipment and the inverter control are also to be considered.

2.10.4 Role of HRES in Developing Indian Grid

According to the International Energy Agency, investment in renewables in India exceeded that of fossil fuel-based power generation in 2017. In India, the grid-interactive PV-wind generation of 688.42 MW is added in 2018–2019 with a cumulative of 64.5 GW till March 2019 [190]. Till the end of the financial year 2017–2018, the total RE installed was 70 GW whereas it is 79 GW at the end of the financial year 2018–2019. The latest RE update has major contributors, which are PV (ground mounted and rooftop) (36.2%) and wind (45.3%) and biomass (both bagasse and non-bagasse cogeneration) (12.5%). Small contributions were also made by small hydro (5.8%) and waste to power (0.2%). A 41 MW (25 MW PV + 16 MW wind) with storage is under construction in Andhra Pradesh, India. This pilot project will work on efficient grid management through real-time monitoring of ramps, peak shifting, and matching of load and generation profiles. India targets 175 GW of installed capacity from RES by the year 2022, which includes 100 GW of PV and 60 GW of wind. To this effect, India's Ministry of New and Renewable Energy (MNRE) released the National Wind-Solar Hybrid Policy in May 2016. It is framed to support large grid-connected WPVHPS for optimal and efficient utilization of transmission infrastructure and land, reducing the variability in renewable power generation and achieving better grid stability. Superimposition of wind and solar resource can complement variability of both. As per the policy, a wind-PV plant is defined as a hybrid plant if one satisfies at least one-fourth of the rated power capacity of the other. Different configurations and use of technology for AC, DC integration with storage are encouraged with incentives as specified therein.

The Central Electricity Authority is empowered to frame the standards for connectivity and sharing of transmission lines, etc. for such systems. So in India, case studies on hardware with grid interaction are limited to academics. A case study of Barwani [191] found that PV-wind-battery-DG hybrid system is the most optimal solution when cost and emission are the main targets. The work by Robinson et al.

[192] involves the development of the RE-based hybrid system for electricity that can supply desired power continuously throughout the year irrespective of fluctuation of energy available from stand-alone systems. The energy assessment has been done using Homer simulation tool for developing a small solar–wind hybrid system, at National Institute of Engineering Centre for Renewable Energy and Sustainable Technologies (NIECREST), Mysuru, India. The WPVHPS was fully charged during the daytime and thereafter the performance was checked by connecting to 596-W load through the 1,500-kVA inverter and energy meter. The WPVHPS was able to supply energy for 3 hours roughly in the evening.

2.10.5 Closing Thoughts for HRES in Grid

In summary, renewable energy is environmentally, socio-ethically, and economically sustainable compared with the dominant centralized and non-renewable energy generation systems. However, the techno-economic limitations for ever-growing renewables' share of power generation in the majority of the countries are alike. The RES is not currently cost-competitive with base load coal-fired power and is geographically dispersed. However, it leads over a conventional generation in low emissions of air pollutants, free fuel, and a low gestation period.

Traditionally, the electric power system is not intended to handle RE generation and storage. But with the rapid growth in the alternate energy sector, the integration of the distributed energy and RES into an electric power grid can be done in many ways along with power quality solution. The power electronic technology plays a significant part in the integration of RES into the electrical grid. They offer exclusive competence over conventional interconnection technologies. They further provide additional power quality and voltage/reactive power support.

It has been discussed that utility RE investments are typically evaluated from regulatory, project finance, and engineering viewpoints. The regulatory evaluation focuses on ensuring utility conformance to RES and that expenses are kept judiciously limited. From a finance perspective, the return on the investment within disjunctive limits of the funding and cash flows for a particular project are evaluated. The technical evaluation determines the engineering and operational safety of the project and the specific technologies deployed. While these approaches are essential for investors, utilities, regulators, and ratepayers, they do not scope out the goodness that a RES can convey beyond the boundaries of a given project, such as the usefulness of transmission and storage and the organizational plus point of bringing an integrated grid.

Variability of RES occurs due to the nature of the climate. Therefore the uncertainty in the generation affecting up to 70% of daytime solar capacity due to passing clouds, and 100% of wind capacity on calm days, is much greater than the somewhat expected variations of a few percent in demand that system operators handle. It necessitates a more complicated voltage and frequency regulation. The larger the RE entrant, the more complicated (sometimes unattainable) is the management of this challenge. Spatial aggregation of RES greatly lessens forecast errors, just as it lessens variability. This may be due to the spatial smoothening effect. The forecast error rises further as the time range of the forecast is expanded. Forecasting techniques are improving constantly. But this requires a better weather model and better data collection and processing. In contrast to the convention fossil fuel power sources, selecting a site to exploit certain RES has few or no degrees of freedom. In other words, RE such as wind and PV, are site constrained. Transmission needs to be extended to these sources, not the other way around. Future distribution systems will contain MGs, and hence it is necessary to understand the steady-state and transient operating conditions of such systems to appraise their effects on the present grid.

Control system is the key element for flexible operation, high efficiency, and superior power **quality in RE integration. In this regard, the control system fetches real-time states through local** measures and via the communication, takes actions to attain the control objectives (for instance, maximum power extraction, output voltage and frequency regulation, reactive power compensation, etc.), and at last sends commands to the actuators, usually power electronic converters. Challenges in control design and realization, energy management strategies, communication layout and protocols, and topologies for power electronics-based distributed RES are all addressed in brief here. The energy storage by batteries in grid-level applications guides both transmission and generation services to the grid. It mitigates the unpredictability of generation. It has been emphasized to conduct a review of the technological potential for a range of battery chemistries. More detailed discussions on MG, off-grid hybrid RES and hybrid storage systems are given in Chapters 4 to 6. Lastly, it should be emphasized again that, because of some demerits and irrevocable externalities in conventional energy production, it has become essential to go for and uphold technologies and insist on RES. Power generation using RES should be enhanced in order to reduce the operational cost of power generation and protect environment.

REFERENCES

1. Shah YT. *Energy and Fuel Systems Integration*. New York: CRC Press; 2016.
2. Nejad MF, Saberian A, Hizam H, Radzi MA, Ab Kadir MZ. Application of smart power grid in developing countries. In: 2013 IEEE 7th International Power Engineering and Optimization Conference (PEOCO); 2013. pp. 427–31. IEEE.
3. Smart grid. *Wikipedia, The Free Encyclopedia*. 2020 [last visited May 14, 2020].
4. National Energy Technology Laboratory. NETL modern grid initiative—Powering our 21st-century economy. United States Department of Energy, Office of Electricity Delivery and Energy Reliability: 17. Retrieved 2008-12-06.
5. Yu FR, Zhang P, Xiao W, Choudhury P. Communication systems for grid integration of renewable energy resources. *IEEE Network*. 2011;25(5):22–9.
6. Rowe DM. *CRC Handbook of Thermo-Electrics*. 1st ed. New York: CRC Press; 1995. 701 p.
7. Ensescu D. Thermoelectric energy harvesting: Basic principles and applications. InTech open access paper;2019. DOI: 10.5772/intechopen.83495. http://www.intech.com.
8. Riffat SB, Ma X. Thermoelectrics: A review of present and potential applications. *Applied Thermal Engineering*. 2003;23:913–935.
9. Patil D, Arakerimath RR, Dr. A review of thermoelectric generator for waste heat recovery from engine exhaust. *International Journal of Research in Aeronautical and Mechanical Engineering*. 2013;1:1–9. [36] Waste Heat Recovery: Bureau of Energy Efficiency, pp. 173.

10. Ismail BI, Ahmed WH. Thermoelectric power generation using waste-heat energy as an alternative green technology. *Recent Patents on Electrical Engineering (Continued as Recent Advances in Electrical & Electronic Engineering)*.2009; 2(1):27–39.
11. Shah YT. *Thermal Energy: Sources, Recovery and Applications.* New York: CRC Press, Taylor and Francis Group; 2018.
12. Zheng XF, Liu CX, Yan YY, Wang Q. A review of thermoelectrics research—Recent developments and potentials for sustainable and renewable energy applications. *Renewable and Sustainable Energy Reviews.* 2014;32:486–503.
13. Zevenhovena R, Beyeneb A. The relative contribution of waste heat from power plants to global warming. *Energy.* 2011;36:3754–62. DOI: 10.1016/j. energy.2010.10.010.
14. Gyselinckx B, Van Hoof C, Ryckaert J, Yazicioglu RF, Fiorini P, Leonov V. Human++: Autonomous wireless sensors for body area networks. In: Custom Integrated Circuits Conference, September 18–21, 2005, San José, CA; 2006. pp. 13–9. IEEE.
15. Wang Z, Leonov V, Fiorini P, Van Hoof C. Realization of a wearable miniaturized thermoelectric generator for human body applications. *Sensors and Actuators A: Physical.* 2009;156(1): 95–102. DOI: 10.1016/j.sna.2009.02.028.
16. Glatz W, Muntwyler S, Hierold C. Optimization and fabrication of thick flexible polymer based micro thermoelectric generator. *Sensors and Actuators A: Physical.* 2006;132:337–45.
17. Rowe DM. Thermoelectrics, an environmentally-friendly source of electrical power. *Renewable Energy.* 1999;16:1251–6.
18. Gao HB, Huang GH, Li HJ, Qu ZG, Zhang YZ. Development of stove-powered thermoelectric generators: A review. *Applied Thermal Engineering.* 2016;96:297–310. DOI: 10.1016/j. applthermaleng.2015.11.032.
19. Champier D. Thermoelectric generators: A review of applications. *Energy Conversion and Management.* 2017;140:167–81. DOI: 10.1016/j. enconman.2017.02.070.
20. Kaibe H, Makino K, Kajihara T, Fujimoto S, Hachiuma H. Thermoelectric generating system attached to a carburizing furnace at Komatsu Ltd., Awazu Plant. In: AIP Conference Proceedings; 2012. Vol. 1449, pp. 524–27. DOI: 10.1063/1.4731609.
21. Luo Q, Li P, Cai L, Zhou P, Tang D, Zhai P, et al. A thermoelectric waste-heat recovery system for Portland cement rotary kilns. *Journal of Electronic Materials.* 2015;44(6):1750–62. DOI: 10.1007/s11664-014- 3543-1.
22. Barma MC, Riaz M, Saidur R, Long BD. Estimation of thermoelectric power generation by recovering waste heat from biomass fired thermal oil heater. *Energy Conversion and Management.* 2015;98:303–13.
23. Min G, Rowe DM, Kontostavlakis K. Thermoelectric figure-of-merit under large temperature differences. *Journal of Physics D: Applied Physics.* 2004;37:1301–4.
24. Min G, Rowe DM. Ring-structured thermoelectric module. *Semiconductor Science and Technology.* 2007;22:880–3.
25. Yodovard P, Khedari J, Hirunlabh J. The potential of waste heat thermoelectric power generation from diesel cycle and gas turbine cogeneration plants. *Energy Sources.* 2001;23:213–24.
26. Omer SA, Infield DG. Design and thermal analysis of two stage solar concentrator for combined heat and thermoelectric power generation. *Energy Conversion & Management.* 2000;41:737–56.
27. Kraemer D, McEnaney K, Chiesa M, Chen G. Modeling and optimization of solar thermoelectric generators for terrestrial applications. *Solar Energy.* 2012;86(5):1338–50. DOI: 10.1016/j. solener.2012.01.025.
28. Olsen ML, Warren EL, Parilla PA, Toberer ES, Kennedy CE, Snyder GJ, et al. A high-temperature, high- efficiency solar thermoelectric generator prototype. *Energy Procedia.* 2014;49:1460–9. DOI: 10.1016/j. egypro.2014.03.155.

29. Baranowski LL, Snyder GJ, Toberer ES. Concentrated solar thermoelectric generators. *Energy & Environmental Science.* 2012;5:9055–67. DOI: 10.1039/C2EE22248E.
30. Date A, Date A, Dixon C, Akbarzadeh A. Progress of thermoelectric power generation systems: Prospect for small to medium scale power generation. *Renewable and Sustainable Energy Reviews.* 2014;33: 371–81. DOI: 10.1016/j.rser.2014.01.081.
31. Molina MG, Juanicó LE, Rinalde GF. Design of innovative power conditioning system for the grid integration of thermoelectric generators. *International Journal of Hydrogen Energy.* 2012;37(13): 10057–63. DOI: 10.1016/j. ijhydene.2012.01.177.
32. Aranguren P, Astrain D, Rodriguez A, Martinez A. Experimental investigation of the applicability of a thermoelectric generator to recover waste heat from a combustion chamber. *Applied Energy.* 2015;152:121–30. DOI: 10.1016/j.apenergy.2015.04.077.
33. Carbon capture-Fuel cell energy. Fuel Cells;2020. A website report.
34. Proctor D. Exxon mobil extends deal for fuel cell carbon capture project. Exxon Mobil, Power; 2019. A website report.
35. ExxonMobil, fuel cell energy expand agreement to optimize carbonate fuel cell technology for large-scale carbon capture. A website report by Green Car Congress; November 10, 2019.
36. Rexed I. Carbonate fuel cells [Ph.D. thesis]. Stockholm, Sweden: KTH Royal Institute of Technology School of Chemical Science and Engineering; 2014.
37. Ruth M, Cutler D, Flores-Espino F, Stark G, Jenkin T. The economic potential of three nuclear-renewable hybrid energy systems providing thermal energy to industry. Golden, Co: National Renewable Energy Laboratory; December 2016. Technical Report NREL/TP-6A50-66745, Contract No. DE-AC36-08GO28308.
38. Rinaldi G, McLarty D, Brouwer J, Lanzini A, Santarelli M. Study of CO2 recovery in a carbonate fuel cell tri-generation plant. *Journal of Power Sources.* 2015;284:16–26. DOI 10.1016/j.jpowsour.2015.02.147., https://escholarship.org/uc/item/0r97943n.
39. Ozin G. Race for a CO2 to fuel technology. In: *Advanced Engineering News (Energy/Environment).* Audi-Sunfire-Climeworks consortium of companies; April 30, 2015. A website report.
40. Kumaravel V, Bartlett J, Pillai SC. Photoelectrochemical conversion of carbon dioxide (CO2) into fuels and value-added products. ACS Energy Letters. 2020;5(2):486–519. https://doi.org/10.1021/acsenergylett.9b02585.
41. Mills S. Combining solar power with coal-fired power plants, or cofiring natural gas. London, UK: IEA Clean Coal Center; October 2017. Report No.: CCC/279.
42. Breen Energy Solutions. Natural gas cofire solutions. 2014 [Accessed June 6, 2017]. Available from: http://breenes.com/solutions-services/combustion-products/natural-gas-co-fire-technologies/.
43. Mills S. Combining solar power with coal-fired power plants, or cofiring natural gas. Clean Energy. 2018; 2(1):1–9. https://doi.org/10.1093/ce/zky004.
44. Gossard S. Coal-to-gas plant conversions in the U.S. *Power Engineering.* 2015;119(6);3.
45. Miller K. *Hybrid Solar Thermal Integration at Existing Fossil Generation Facilities.* Johannesburg, South Africa: Black & Veatch; Aug 2013. 31 pp. Available from: http://www.eskom.co.za/AboutElectricity/.../G3CSP_FossilHybridMillerFINAL.pptx.
46. Roos T. *Solar Thermal Augmentation of Coal-Fired Power Stations.* South Africa: Council for Scientific and Industrial Research (CSIR); 2015. 18p. Available from: http://www.fossilfuel.co.za/conferences/2015/Independent-Power-Generation-in-SA/Day-2/Session-3/01Thomas-Roos.pdf.
47. Miser T. Synergy through hybridity: Harnessing multiple energy sources for diversified benefits. *Power Engineering.* 2016;120(5);2.
48. Siros F. Hybridization of thermal plants is a great driver to increase the CSP share in the global energy mix. In: IEA Technology Roadmap, 1st Workshop, February 3–4, 2014, Paris, France; 2014. 4 pp. International Energy Agency.

49. Stark DC, Curtis D, Ibekwe R, Forsberg C. Conceptual Design and Market Assessment of Firebrick Resistance Heated Energy Storage (FIRES) – Avoiding Wind and Solar Electricity Price Collapse to Improve Nuclear, Wind, and Solar Economics International Congress on Advanced Nuclear Power Plants (ICAPP 2016) Paper 16622, San Francisco, California, April 17–20, 2016.
50. Department of Energy and National Energy Technology Laboratory. Environmental impacts of smart grid. Available from: http://www.netl.doe.gov/File%20LibraryResearch/Energy%20Analysis/Publications/Envimpact_SmartGrid.pdf.
51. U.S. Department of Energy. Smart grid savings and grid integration of renewables in Idaho. Available from: http://energy.gov/sites/prod/files/2013/06/f1/IdahoPowerCaseStudy.pdf.
52. Smart grid system report. Washington, DC: U.S. Department of Energy; November 2018. A 2018 report to Congress.
53. Tervo E, Agbim K, DeAngelis F, Hernandez J, Kim HK, Odukomaiyaa A. An economic analysis of residential photovoltaic systems with lithium ion battery storage in the United States. *Renewable and Sustainable Energy Reviews.* 2018;94:1057–66.
54. QVARTZ. Emergence of Hybrid Renewable Energy Systems. 2019 [Accessed July 9, 2019]. Available from: https://qvartz.com/media/2019/hybrid_renewableenergy.pdf.
55. Couture T, Barbose G, Jacobs D, Parkinson G, Chessin E, Belden A, Wilson H, Barrett H, Rickerson W. *Residential prosumers: Drivers and policy options (re-prosumers).* Berkeley, CA: Meister Consultants Group; Lawrence Berkeley National Laboratory; 2014.
56. Center for Sustainable Systems. U.S. Renewable Energy Factsheet 2018, Pub. No. CSS03-12. University of Michigan [Accessed July 9, 2019]. Available from: http://css.umich.edu/sites/default/files/U.S._Renewable_Energy_Factsheet_CSS03-12_e2018.pdf.
57. Soysal OA, Soysal HS. A residential example of hybrid wind-solar energy system: WISE. In: Proceedings of the Power and Energy Society General Meeting-Conversion and Delivery of Electrical Energy in the 21st Century, July 20–24, 2008, Pittsburgh, PA, USA; 2008. pp. 1–5.
58. Gupta A, Bais AS. Hybrid solar wind market size by product (stand-alone, grid connected), end use (residential, commercial, industrial), industry analysis report, regional outlook (U.S., Canada, U.K, Germany, China, India, Australia, Japan, South Africa, Nigeria, Tanzania, Chile, Brazil), application potential, price trends, competitive market share & forecast, 2016–2024 [Accessed July 9, 2019]. Available from: https://www.gminsights.com/industryanalysis/hybrid-solar-wind-market. Report No.: GMI830.
59. Cetin E, Yilanci A, Oner Y, Colak M, Kasikci I, Ozturk HK. Electrical analysis of a hybrid photovoltaic-hydrogen/fuel cell energy system in Denizli, Turkey. *Energy and Buildings.* 2009;41:975–81.
60. Markham D. Hybrid rooftop wind and solar generator now available in U.S. for early adopters. 2015 [Accessed July 9, 2019]. Available from: https://www.treehugger.com/wind-technology/hybrid-rooftop-wind-and-solar-generator-now-availableus-early-adopters.html.
61. General Electric Renewable Energy Website. More than just a trend [Accessed July 9, 2019]. Available from: https://www.gerenewableenergy.com/hybrid.now-available-us-early-adopters.html.
62. Grepperud S, Rasmussen I. A general equilibrium assessment of rebound effects. *Energy Economics.* 2004;26:261–82.
63. Deshmukh MK, Deshmukh SS. Modeling of hybrid renewable energy systems. *Renewable and Sustainable Energy Reviews.* 2008;12:235–49.
64. Gupta A, Saini RP, Sharma MP. Modelling of hybrid energy system—Part I: Problem formulation and model development. *Renewable Energy.* 2011;36:459–65.
65. Syed AM, Fung AS, Ugursal VI, Taherian H. Analysis of PV/wind potential in the Canadian residential sector through high-resolution building energy simulation. *International Journal of Energy Research.* 2009;33:342–57.

66. Ghenai C, Bettayeb M. Grid-tied solar PV/fuel cell hybrid power system for university building. *Energy Procedia*. 2019;159:96–103. https://doi.org/10.1016/j.egypro.2018.12.025.
67. Bazilian M, Onyeji I, Liebreich M, MacGill I, Chase J, Shah J, Gielen D, Arent D, Landfear D, Zheng S. Re-considering the economics of photovoltaic power. *Renewable Energy*. 2013;53:329–38.
68. Gupta A, Saini RP, Sharma MP. Modelling of hybrid energy system—Part III: Case study with simulation results. *Renewable Energy*. 2011;36:474–81.
69. McAllister JA. Solar adoption and energy consumption in the residential sector [UC Berkeley Electronic thesis and dissertation]. 2012 [Accessed July 9, 2019]. Available from: http://digitalassets.lib.berkeley.edu/etd/ucb/text/McAllister_berkeley_0028E_12779.pdf.
70. Fell MJ. Energy services: A conceptual review. *Energy Research and Social Science*. 2017;27:129–40.
71. Guertin C, Kumbhakar SC, Duraiappah AK. *Determining Demand for Energy Services: Investigating Income-Driven Behaviors*. Winnipeg, MB: International Institute for Sustainable Development; 2003.
72. Hunt LC, Ryan DL. *Economic Modelling of Energy Services: Rectifying Mis-Specified Energy Demand Functions*. Surrey, UK: Surrey Energy Economics Center; 2014.
73. Fikru MG, Gelles G, Ichim AM, Kimball JW, Smith JD, Zawodniok MJ. An economic model for residential energy consumption, generation, storage and reliance on cleaner energy. *Renewable Energy*. 2018;119:429–38.
74. Connecting your solar electricity system to the utility grid. Golden, CO: National Renewable Energy Laboratory; July 2002. A website NREL report, grid, DOE/GO-102002-1594.
75. Grid connected renewable energy sourced building. Washington, DC: Department of Energy; 2020. A website report.
76. Shah YT. *Modular Systems for Energy Usage Management*. New York: CRC Press; 2020.
77. Clement-Nyns K, Haesen E, Driesen J. The impact of vehicle-to-grid on the distribution grid. *Electric Power Systems Research*. 2011;(1):185–92. DOI: 10.1016/j.epsr.2010.08.007.
78. Pang C, Dutta P, Kezunovic M. BEVs/PHEVs as dispersed energy storage for V2B uses in the smart grid. *IEEE Transactions on Smart Grid*. 2012;(1):473–482. DOI: 10.1109/tsg.2011.2172228.
79. Irish S. V2G: The role for EVs in future energy supply and demand. A website report [March 17, 2017]. Available from: http://www.renewableenergyfocus.com.
80. Mortice Z. Integrated energy systems: This building and car create a symbiotic relationship to leave the electric grid behind. Oak Ridge, TN: Oak Ridge National Laboratory; January 21, 2016. A website report.
81. AMIE Demonstration Project—Oak Ridge National Laboratory. OakRidge, TN: Oak Ridge National Laboratory; 2016. A website report. Available from: https://web.ornl.gov/sci/eere/amie/.
82. AMIE. Oak Ridge, TN: Oak Ridge National Laboratory; 2016. A website report. Available from: https://web.ornl.gov/sci/eere/amie/media/AMIE-DemonstrationProject.pdf.
83. Ericson SJ, Rose E, Jayaswal H, Cole WJ, Engel-Cox J, Logan J, et al. Hybrid storage market assessment: A JISEA white paper. Golden, CO: National Renewable Energy Laboratory. NREL/MP-6A50-70237, Contract No. DE-AC3608GO28308.
84. Luo X, Wang J, Dooner M, Clarke J. Overview of current development in electrical energy storage technologies and the application potential in power system operation. *Applied Energy*. 2015:511–36. DOI: 10.1016/j.apenergy.2014.09.081.

85. Dunn B, Kamath H, Tarascon JM. Electrical energy storage for the grid: A battery of choices. *Science*. 2011;334(6058):928–35. DOI: 10.1126/science.1212741.
86. Knap V, Sinha R, Swierczynski M, Stroe DI, Chaudhary S. Grid inertial response with Lithium-ion battery energy storage systems. In: IEEE 23rd International Symposium on Industrial Electronics (ISIE), Istanbul, Turkey; 2014. pp. 1817–22. DOI: 10.1109/isie.2014.6864891.
87. Leadbetter J, Swan LG. Selection of battery technology to support grid-integrated renewable electricity. *Journal of Power Sources*. 2012;376–86. DOI: 10.1016/j.jpowsour.2012.05.081
88. Wang K, Jiang K, Chung B, Ouchi T, Burke PJ, Boysen DA, et al. Lithium-antimony-lead liquid metal battery for grid-level energy storage. *Nature*. 2014;514(7522):348–50. DOI: 10.1038/nature13700.
89. McKinsey & Company. *An Integrated Perspective on the Future of Mobility*. New York, NY: Bloomberg New Energy Finance; 2016.
90. Elgqvist E, Anderson K, Settle E. *Federal Tax Incentives for Battery Storage Systems*. Golden, CO: National Renewable Energy Laboratory; 2017.
91. Federal Energy Regulatory Commission. *Electric Storage Participation in Markets Operated by Regional Transmission Organizations and Independent Systems Operators*. Notice of Proposed Regulation. Section 12. Washington, DC: Federal Energy Regulatory Commission; November 2016.
92. *Advanced Batteries for Utility-Scale Energy Storage*. Chicago, IL: Navigant Research; 2016.
93. Nykvist B, Nilsson M. Rapidly falling costs of battery packs for electric vehicles. *Nature Climate Change*. 2015;5(4):329–32.
94. Schmidt O, Hawkes A, Gambhir A, Staffell I. The future cost of electrical energy storage based on experience rates. *Nature Energy*. 2017;2(8):1–8.
95. Feldman D, Margolis R, Denholm P, Stekli J. *Exploring the Potential Competitiveness of Utility-Scale Photovoltais Plus Batteries with Concentrating Solar Power, 2015–2030*. Golden, CO: National Renewable Energy Laboratory; 2016.
96. Dykes K, King J, DiOrio N, King R, Gevorgian V, Corbus D. Opportunities for research and development of hybrid power plants. Golden, CO: National Renewable Energy Laboratory; May 2020. Technical Report No.: NREL/TP-5000-75026, Contract No. DE-AC36-08GO28308.
97. Denholm PL, Eichman JD.and Margolis RM, Evaluating the technical and economic performance of PV plus storage power plants. Golden, CO: National Renewable Energy Laboratory; 2017. NREL/TP-6A20-68737. Available from: https://www.nrel.gov/docs/fy17osti/68737.pdf.
98. REN21. *Renewables 2017 Global Status Report*, 2017.
99. Monitoring Analytics, LLC. *State of the Market Report for PJM 2016*. 2017.
100. Zhou Z, Levin T, Conzelmann G. *Survey of U.S. Ancillary Services Markets*. Lemont, IL: Argonne National Laboratory; 2016.
101. Denholm P, Jorgenson J, Hummon M, Jenkin T, Palchak D, Kirby B,et al. *The Value of Energy Storage for Grid Applications*. Golden, CO: National Renewable Energy Laboratory; 2013.
102. Eller A, Dehamna A. *Energy Storage for the Grid and Ancillary Services*. Chicago. IL: Navigant Research; 2016.
103. Wu Z, Xia X. Optimal switching renewable energy system for demand side management. *Solar Energy*. 2015;278–88. DOI: 10.1016/j.solener.2015.02.001.
104. Department of Energy. *The Potential Benefits of Distributed Generation and Rate-Related Issues that may Impede their Expansion*. Washington, DC: U.S. Department of Energy; 2007.

105. Simpkins Travis, Anderson Kate, Cutler Dylan, Olis Dan. *Optimal Sizing of a Solar-PlusStorage System for Utility Bill Savings and Resiliency Benefits.* NREL, 2016.
106. Munsell Mike. *In Shift to Longer-Duration Applications, US Energy Storage Installations Grow 100% in 2016.* gtm Research, 2017.
107. Trevisani L, Fabbri M, Negrini F. Long distance renewable energy sources power transmission using hydrogen-cooled MgB2 superconducting line. *Cryogenics.* 2007(2):113–20. DOI: 10.1016/j.cryogenics.2006.10.002.
108. Oyedepo SO, Agbetuyi AF, Odunfa KM. Transmission network enhancement with renewable energy. *Journal of Fundamentals of Renewable Energy and Applications.* 2014;(5):1–11. DOI: 10.4172/20904541.1000145.
109. Kusakana K. Optimal scheduled power flow for distributed photovoltaic/wind/diesel generators with battery storage system. *IET Renewable Power Generation.* 2015;(8):916–24. DOI: 10.1049/iet-rpg.2015.0027.
110. Bloomberg. *Case Study: AES Delivers Firm Solar with $110/MWh PV-and-storage PPA in Hawaii.* Bloomberg New Energy Finance, 2017.
111. Byrne R, Hamilton S, Borneo D, Olinsky-Paul T, Gyuk I. *The Value Proposition for Energy Storage at the Sterling Municipal Light Department.* Livermore, CA: Sandia National Laboratories; 2017.
112. Martin Chris. *Slap a Battery on a Gas Turbine and Make an Extra $1.4 Million.* Bloomberg, 2017.
113. Alex Eller, Dehamna Anissa. *Market Data: Commercial & Industrial Energy Storage.* Navigant Research, 2016.
114. Fathima H, Palanisamy K. Optimized sizing, selection, and economic analysis of battery energy storage for grid-connected wind-PV hybrid system. *Modelling and Simulation in Engineering.* 2015;2015:1–16. DOI: 10.1155/2015/713530.
115. Hittinger E, Whitacre JF, Apt J. What properties of grid energy storage are most valuable? *Journal of Power Sources.* 2012; 206:436–49. DOI: 10.1016/j.jpowsour.2011.12.003.
116. Srivastava AK, Kumar AA, Schulz NN. Impact of distributed solar wind power system generations with energy storage devices on the electric grid. *IEEE Systems Journal.* 2012;6(1):110–7. DOI: 10.1109/jsyst.2011.2163013.
117. Korada N, Mishra MK. Grid adaptive power management strategy for an integrated microgrid with hybrid energy storage. *IEEE Transactions on Industrial Electronics.* 2017;64(4):2884–92. DOI: 10.1109/tie.2016.2631443.
118. Shu Z, Jirutitijaroen P. Optimal operation strategy of energy storage system for grid-connected wind power plants. *IEEE Transactions on Sustainable Energy.* 2016;5(1):190–9. DOI: 10.1109/tste.2013.2278406.
119. Salas DF, Powell WB. Benchmarking a scalable approximate dynamic programming algorithm for stochastic control of grid-level energy storage. *INFORMS Journal on Computing.* 2018;30(1):106–23. DOI: 10.1287/ijoc.2017.0768.
120. Madlener R, Latz J. Economics of centralized and decentralized compressed air energy storage for enhanced grid integration of wind power. *Applied Energy.* 2013;101:299–309. DOI: 10.1016/j.apenergy.2011.09.033.
121. Koller M, Borsche T, Ulbig A, Andersson G. Review of grid applications with the Zurich 1MW battery energy storage system. *Electric Power Systems Research.* 2015;120:128–35. DOI: 10.1016/j.epsr.2014.06.023.
122. Kroposki B, Johnson B, Zhang Y, Gevorgian V, Denholm P, Hodge BM, et al. Achieving a 100% renewable grid: Operating electric power systems with extremely high levels of variable renewable energy. *IEEE Power and Energy Magazine.* 2017; 15(2):61–73. DOI: 10.1109/mpe.2016.2637122.
123. Malysz P, Sirouspour S, Emadi A. An optimal energy storage control strategy for grid-connected microgrids. *IEEE Transactions on Smart Grid.* 2014;(4):1785–96. DOI: 10.1109/tsg.2014.2302396.

124. Müller M, Viernstein L, Truong CN, Eiting A, Hesse HC, Witzmann R, et al. Evaluation of grid-level adaptability for stationary battery energy storage system applications in Europe. Journal of Energy Storage. 2017;:1–11. DOI: 10.1016/j.est.2016.11.005.
125. Bose S, Gayme DF, Topcu U, Chandy KM. Optimal placement of energy storage in the grid. In: 51st IEEE Conference on Decision and Control (CDC), Maui, HI, USA; 2012. pp. 5605–12. IEEE. DOI: 10.1109/cdc.2012.6426113.
126. Bhandari B, Lee KT, Lee CS, Song CK, Maskey RK, Ahn SH. A novel off-grid hybrid power system comprised of solar photovoltaic, wind, and hydro energy sources. *Applied Energy*. 2014;133:236–42. DOI: 10.1016/j.apenergy.2014.07.033.
127. Hearn CS, Lewis MC, Pratap SB, Hebner RE, Uriarte FM, Chen D, et.al. Utilization of optimal control law to size grid-level flywheel energy storage. *IEEE Transactions on Sustainable Energy*. 2013;4(3):611–18. DOI: 10.1109/tste.2013.2238564.
128. Denholm P, Hand M. Grid flexibility and storage required to achieve very high penetration of variable renewable electricity. *Energy Policy*. 2011; 39(3):1817–30. DOI: 10.1016/j.enpol.2011.01.019.
129. Kerestes RJ, Reed GF, Sparacino AR. Economic analysis of grid level energy storage for the application of load leveling. In: 2012 IEEE PES Gen Meeting, San Diego, CA, USA; 2012. pp. 1–9. DOI: 10.1109/pesgm.2012.6345072.
130. Williams CJ, Binder JO, Kelm T. Demand side management through heat pumps, thermal storage and battery storage to increase local self-consumption and grid compatibility of PV systems. In: 3rd IEEE PES Innovative Smart Grid Technologies Europe (ISGT Europe), Berlin, Germany; 2012. pp. 1–6. DOI: 10.1109/isgteurope.2012.6465874.
131. Palizban O, Kauhaniemi K. Energy storage systems in modern grids—Matrix of technologies and applications. *Journal of Energy Storage*. 2016; 6:248–259. DOI: 10.1016/j.est.2016.02.001.
132. Graabak I, Korpås M. Variability characteristics of European wind and solar power resources—A review. *Energies*. 2016;449:1–31.
133. Coker P, Barlow J, Cockerill T, Shipworth D. Measuring significant variability characteristics: An assessment of the three UK renewables. *Renewable Energy*. 2013: 53:111–20. DOI: 10.1016/j.renene.2012.11.013.
134. Tande JO, Korpås M, Warland L, Uhlen K, Van Hulle F. Impact of trade wind offshore wind power capacity scenarios on power flows in the European HV network. In: Proceedings of the 7th International Workshop on Large-Scale Integration of Wind Power and on Transmission Networks for Offshore Wind Farms, May 2008, Madrid, Spain; 2008. pp. 26–7.
135. Wiemken E, Beyer HG, Heydenreich W, Kiefer K. Power characteristics of PV ensembles, experience from the combined power production of 100 grid connected PV systems distributed over the area of Germany. *Solar Energy*. 2001; 70(6):513–18. DOI: 10.1016/S0038-092X(00)00146-8.
136. Lei M, Shiyan L, Chuanwen J, Hongling L, Yan Z. A review on the forecasting of wind speed and generated power. *Renewable and Sustainable Energy Reviews*. 2009; 13(4):915–920. DOI: 10.1016/j.rser.2008.02.002.
137. Carta JA, Ramirez P, Velazquez S. A review of wind speed probability distributions used in wind energy analysis: Case studies in the Canary Islands. *Renewable and Sustainable Energy Reviews*. 2009; 13(5):933–955. DOI: 10.1016/j.rser.2008.05.005.
138. Wang X, Guo P, Huang X. A review of wind power forecasting models. *Energy Procedia*. 2011; 12:770–778. DOI: 10.1016/j.egypro.2011.10.103.
139. An Y, Pandey MD. A comparison of methods of extreme wind speed estimation. *Journal of Wind Engineering and Industrial Aerodynamics*. 2005; 93(7):535–545. DOI: 10.1016/j.jweia.2005.05.003.
140. Wang J, Qin S, Jin S, Wu J. Estimation methods review and analysis of offshore extreme wind speeds and wind energy resources. *Renewable and Sustainable Energy Reviews*. 2015; 42:26–42. DOI: 10.1016/j.rser.2014.09.042.

141. Ata R. Artificial neural networks applications in wind energy systems: A review. *Renewable and Sustainable Energy Reviews.* 2015; 49:534–62. DOI: 10.1016/j.rser.2015.04.166.
142. Jung J, Broadwater RP. Current status and future advances for wind speed and power forecasting. *Renewable and Sustainable Energy Reviews.* 2014; 31:762–77. DOI: 10.1016/j.rser.2013.12.054.
143. Tascikaraoglu A, Uzunoglu M. A review of combined approaches for prediction of short-term wind speed and power. *Renewable and Sustainable Energy Reviews.* 2014; 34:243–54. DOI: 10.1016/j.rser.2014.03.033.
144. Kavousi-Fard A, Khosravi A, Nahavandi S. A new fuzzy-based combined prediction interval for wind power forecasting. *IEEE Transactions on Power Systems.* 2016; (1):18–26. DOI: 10.1109/TPWRS.2015.2393880
145. Qiao W, Zhou W, Aller JM, Harley RG. Wind speed estimation based sensorless output maximization control for a wind turbine driving a DFIG.*IEEE Transactions on Power Electronics.* 2008;(3):1156–169. DOI: 10.1109/TPEL.2008.921185.
146. Chang TP, Liu FJ, Ko HH, Huang MC. Oscillation characteristic study of wind speed, global solar radiation and air temperature using wavelet analysis. 2017:650–7. DOI: 10.1016/j.apenergy.2016.12.149.
147. Elhadidy MA. Performance evaluation of hybrid (wind/solar/diesel) power systems. *Renewable Energy.* 2002;(3):401–13. DOI: 10.1016/S0960-1481(01)00139-2.
148. Coskun C, Oktay Z, Dincer I. Estimation of monthly solar radiation distribution for solar energy system analysis. *Energy.* 2011;(2):1319–23. DOI: 10.1016/j.energy.2010.11.009.
149. Notton G, Diaf S, Stoyanov L. Hybrid photovoltaic/wind energy systems for remote locations. *Energy Procedia.* 2011; 6:666–77. DOI: 10.1016/j.egypro.2011.05.076.
150. Mahesh A, Sandhu KS. Hybrid wind/photovoltaic energy system developments: Critical review and findings. *Renewable and Sustainable Energy Reviews.* 2015; 52(C):1135–47. DOI: 10.1016/j.rser.2015.08.008.
151. Sichilalu SM, Xia X. Optimal energy control of grid tied PV-diesel-battery hybrid system powering heat pump water heater. *Solar Energy.* 2015:243–54. DOI: 10.1016/j.solener.2015.02.028.
152. Li G, Chen Y, Li T. The realization of control subsystem in the energy management of wind/solar hybrid power system. In: 3rd International Conference on Power Electronics Systems and Applications (PESA), May 2009, Hong Kong, China; 2009. pp. 1–4.
153. Olatomiwa L, Mekhilef S, Ismail MS. Energy management strategies in hybrid renewable energy systems: A review. *Renewable and Sustainable Energy Reviews.* 2016; 62:821–35. DOI: 10.1016/j.rser.2016.05.040.
154. Sehar F, Pipattanasomporn M, Rahman S. Coordinated control of building loads, PVs and ice storage to absorb PEV penetrations. *International Journal of Electrical Power & Energy Systems.* 2018; 95:394–404. DOI: 10.1016/j.ijepes.2017.09.009.
155. Merabet A, Ahmed KT, Ibrahim H, Beguenane R, Ghias AM. Energy management and control system for laboratory scale microgrid based wind-PV-battery. *IEEE Transactions on Sustainable Energy.* 2017;8:145–54. DOI: 10.1109/TSTE.2016.2587828.
156. Boukettaya G, Krichen L. A dynamic power management strategy of a grid connected hybrid generation system using wind, photovoltaic and flywheel energy storage system in residential applications. *Energy.* 2014;71:148–59. DOI: 10.1016/j.energy.2014.04.039.
157. Reihani E, Sepasi S, Roose LR, Matsuura M. Energy management at the distribution grid using a battery energy storage system (BESS). *International Journal of Electrical Power & Energy Solar Wind Power System.* 2016;77:337–44. DOI: 10.1016/j.ijepes.2015.11.035.
158. Mohamed A, Mohammed O. Real-time energy management scheme for hybrid renewable energy systems in smart grid applications. *Electric Power Systems Research.* 2013;96:133–43. DOI: 10.1016/j.epsr.2012.10.015.

159. Lucas A, Chondrogiannis S. Smart grid energy storage controller for frequency regulation and peak shaving, using a vanadium redox flow battery. *International Journal of Electrical Power & Energy Systems*. 2016;80:26–36. DOI: 10.1016/j.ijepes.2016.01.025.
160. Lawder MT, Suthar B, Northrop PW, De S, Hoff CM, Leitermann O, et al. Battery energy storage system (BESS) and battery management system (BMS) for grid-scale applications. *Proceedings of the IEEE*. 2014;102:1014–30. DOI: 10.1109/jproc.2014.2317451.
161. Nick M, Cherkaoui R, Paolone M. Optimal allocation of dispersed energy storage systems in active distribution networks for energy balance and grid support. *IEEE Transactions on Power Systems*. 2014;29:2300–10. DOI: 10.1109/tpwrs.2014.2302020.
162. Gelazanskas L, Gamage KA. Demand side management in smart grid: A review and proposals for future direction. *Sustainable Cities and Society*. 2014; 11:22–30. DOI: 10.1016/j.scs.2013.11.001.
163. Vasiljevska J, Lopes JP, Matos MA. Integrated micro-generation, load and energy storage control functionality under the multi micro-grid concept. *Electric Power Systems Research*. 2013; 95:292–301. DOI: 10.1016/j.epsr.2012.09.014.
164. Yi Z, Dong W, Etemadi AH. A unified control and power management scheme for PV-battery-based hybrid microgrids for both grid-connected and islanded modes. *IEEE Transactions on Smart Grid*. 2017;9(6):5975–85. DOI: 10.1109/tsg.2017.2700332.
165. Abbassi R, Chebbi S. Energy management strategy for a grid-connected wind-solar hybrid system with battery storage: Policy for optimizing conventional energy generation. *International Review of Electrical Engineering*. 2012;2:3979–90.
166. Singaravel MR, Daniel SA. MPPT with single DC–DC converter and inverter for grid-connected hybrid wind-driven PMSG-PV system. *IEEE Transactions on Industrial Electronics*. 2015;8:4849–57. DOI: 10.1109/tie.2015.2399277.
167. Alagoz BB, Kaygusuz A, Karabiber A. A user-mode distributed energy management architecture for smart grid applications. *Energy*. 2012;1:167–77. DOI: 10.1016/j.energy.2012.06.051.
168. Li X, Jiao X, Wang L. Coordinated power control of wind-PV-fuel cell for hybrid distributed generation systems. In: The SICE Annual Conference, September 2013, Nagoya, Japan; 2013. pp. 150–5. IEEE.
169. Hong CM, Chen CH. Intelligent control of a grid-connected wind-photovoltaic hybrid power systems. *International Journal of Electrical Power & Energy Systems*. 2014;55:554–61. DOI: 10.1016/j.ijepes.2013.10.024.
170. Chaurasia GS, Singh AK, Agrawal S, Sharma NK. A meta-heuristic firefly algorithm based smart control strategy and analysis of a grid connected hybrid photovoltaic/wind distributed generation system. *Solar Energy*. 2017;150:265–74. DOI: 10.1016/j.solener.2017.03.079.
171. Bendary AF, Ismail MM. Battery charge management for hybrid PV/wind/fuel cell with storage battery. *Energy Procedia*. 2019:107–116. DOI: 10.1016/j.egypro.2019.04.012. Grid-Connected Distributed Wind-Photovoltaic Energy Management: A Review. DOI: http://dx.doi.org/10.5772/intechopen.88923.
172. Qi W, Liu J, Christofides PD. Distributed supervisory predictive control of distributed wind and solar energy systems. *IEEE Transactions on Control Systems Technology*. 2013;2:504–12. DOI: 10.1109/TCST.2011.2180907.
173. Baghaee HR, Mirsalim M, Gharehpetian GB, Talebi HA. A decentralized power management and sliding mode control strategy for hybrid AC/DC microgrids including renewable energy resources. *IEEE Transactions on Industrial Informatics*. 2017:1– . DOI: 10.1109/tii.2017.2677943.
174. Gregory DC, Alesi LH, Crain JA. Distributed hybrid renewable energy power plant and methods, systems, and computer readable media for controlling a distributed hybrid renewable energy power plant [USA patent]; 2012.

175. Jun Z, Junfeng L, Jie W, Ngan HW. A multi-agent solution to energy management in hybrid renewable energy generation system. *Renewable Energy*. 2011;5:1352–63. DOI: 10.1016/j.renene.2010.11.032.
176. Xu Y, Zhang J, Wang W, Juneja A, Bhattacharya S. Energy router: Architectures and functionalities toward energy internet. In: IEEE International Conference on Smart Grid Communications (SmartGridComm), Brussels, Belgium; 2011. pp. 31–6. DOI: 10.1109/smartgridcomm.2011.610234.
177. Cao J, Yang M. Energy internet—Towards smart grid 2.0. In: Fourth International Conference on Networking and Distributed Computing, Los Angeles, CA; 2013. pp. 105–10. DOI: 10.1109/icndc.2013.10.
178. Badwawi RA, Abusara M, Mallick T. A review of hybrid solar PV and wind energy system. *Smart Science*. 2015;(3):127–138. DOI: 10.1080/23080477.2015.11665647.
179. Sun J, Li M, Zhang Z, Xu T, He J, Wang H, et al. Renewable energy transmission by HVDC across the continent: System challenges and opportunities. *CSEE Journal of Power and Energy Systems*. 2017;4:353–64. DOI: 10.17775/CSEEJPES.2017.01200.
180. Behera S, Sahoo S, Pati BB. A review on optimization algorithms and application to wind energy integration to grid. *Renewable and Sustainable Energy Reviews*. 2015; ( ):214–27. DOI: 10.1016/j.rser.2015.03.066.
181. Khare V, Nema S, Baredar P. Solar-wind hybrid renewable energy system: A review. *Renewable and Sustainable Energy Reviews*. 2016:58,23–33. DOI: 10.1016/j.rser.2015.12.223.
182. Fuel Cells, Photovoltaics, Dispersed Generation, and Energy Storage. IEEE standard for interconnection and interoperability of distributed energy resources with associated electric power systems interfaces. IEEE Std. 2018:1547–2018. DOI: 10.1109/ieeestd.2018.8332112.
183. Yang Y, Enjeti P, Blaabjerg F, Wang H. Wide-scale adoption of photovoltaic energy: Grid code modifications are explored in the distribution grid. *IEEE Industry Applications Magazine*. 2005;5:21–31. DOI: 10.1109/mias.2014.2345837.
184. Zame KK, Brehm CA, Nitica AT, Richard CL, Schweitzer III GD. Smart grid and energy storage: Policy recommendations. *Renewable and Sustainable Energy Reviews*. 2018;82:1646–54. DOI: 10.1016/j.rser.2017.07.01.
185. Regulatory Assistance Project. *Electricity Regulation in the US: A Guide*. Montpelier, Vermont: Regulatory Assistance Project; 2011.
186. Yang H, Zhou W, Lu L, Fang Z. Optimal sizing method for stand-alone hybrid solar–wind system with LPSP technology by using genetic algorithm. *Solar Energy* 2003;82:354–467.
187. Sinha S, Chandel SS. Review of recent trends in optimization techniques for solar photovoltaic–wind based hybrid energy systems. *Renewable and Sustainable Energy Reviews*. 2015;50:755–69.
188. Arabali A, Ghofrani M, Etezadi-Amoli M, Fadali MS, Baghzouz Y. Genetic-algorithm based optimization approach for energy management. *IEEE Transactions on Power Delivery*. 2013;28(1):162–70.
189. Arabali A, Ghofrani M, Etezadi-Amoli M, Fadali MS. Stochastic performance assessment and sizing for a hybrid power system of solar/wind/energy storage. *IEEE Transactions on Sustainable Energy*. 2014;5(2):363–71.
190. Available from: https://mnre.gov.in/physical-progress-achievements.
191. Sawle Y, Gupta SC, Kumar Bohre A. PV-wind hybrid system: A review with case study. *Cogent Engineering*. 2016;3(1):1189305. DOI: 10.1080/23311916.2016.1189305.
192. Robinson P, Gowda AC, Sameer S, Patil S. Development of renewable energy based hybrid system for electricity generation—A case study. *International Journal of Latest Technology in Engineering, Management & Applied Science*. 2017;VI(VIIIS):46–52.

3 Hybrid Power for Mobile Systems

3.1 INTRODUCTION

The widespread application of hydrocarbon-based transportation has been raising global issues such as an increase in the demand for non-renewable petroleum production, high gasoline prices, inefficient power production, and CO2 emission leading to climate change. Hence, searching for highly efficient, safe, and clean alternative solutions to this method of transport has been among the most emphasized challenges attracting the attention of researchers in both the environment and transportation sectors. A well-knit and coordinated transportation provides mobility to people and goods. The transportation sector mainly consists of road, railway, ships and aviation, where road transportation consumes 75% of the total energy spent on transportation. The automobile industry plays a significant role in the economic growth of the world and hence affects the entire population. Since vehicles mostly run on internal combustion engine (ICE), the transportation industry is accountable for 25%–30% of the total greenhouse gas emission [1]. A comparison of the efficiency of ICE and electric motor is illustrated in Figure 3.1 [2]. ICE works on the process of fuel combustion resulting in the production of various gases like CO2, NO2, NO, and CO [3] which cause environmental degradation in the form of greenhouse effect and are responsible for their adverse effect on human health. To overcome this, the transportation industry is trying hard to manufacture vehicles that can be run on alternate power sources. Electric vehicles (EVs) were tried as a solution in 1881 where battery alone propelled the vehicle and therefore required a bulky battery pack. Absence of an ICE handicapped these vehicles with a short driving range [4]. Hybrid electric vehicles (HEVs) were conceptualized to bridge the power of ICE and the emission-free nature of EVs [5]. HEVs offer better fuel efficiency over ICE-based vehicles and generally work in charge-sustaining (CS) mode where the state of charge (SOC) of the battery is maintained throughout the trip.

The issue with CS mode is that its charging efficiency relies mainly on regenerative braking and gasoline, so plug-in HEVs (PHEVs) were conceptualized as another possible solution [6]. Accordingly, the development of innovative technologies for the utilization of (plug-in) hybrid electric vehicles ((P)HEVs) and full-electric vehicles (EVs) was undertaken. To achieve a seamless transition from traditional internal combustion engine (ICE)-based vehicles to fully electric vehicles, (P)HEV technologies have recently been employed not only in passenger cars but also in heavy-duty vehicles [3]. Unlike HEVs, PHEVs have the additional facility to be charged externally through power outlets. Most of the power in a PHEV is derived from an electric motor (EM) which acts as a primary source, while ICE

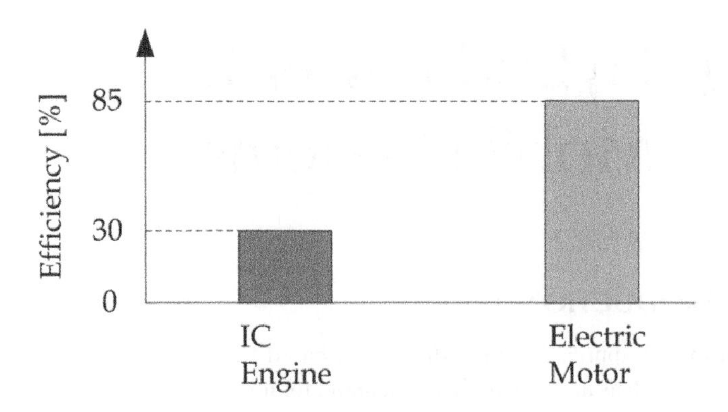

FIGURE 3.1 Efficiency of IC engine vs. electric motor [2]

acts as a backup. As the battery SOC reaches a particular threshold, the PHEV behaves like a regular HEV and the ICE kicks in and acts as a primary power source. The PHEVs mainly work in charge depletion (CD) mode where the SOC is depleted up to a threshold level. PHEVs extend the all-electric range, improve local air quality, and may also have grid connection capability. In recent years, fuel cell vehicles have also taken forefront. Since they run on hydrogen, power generation is once again pollution-free. Generally it is also accompanied by energy storage to increase its power intensity during the acceleration period. FCHEV (fuel cell-based hybrid electric vehicle) is another form of small- and large-scale hybrid vehicle that will be developed and commercialized in the near future. Unlike PHEVs, these kinds of cars will require hydrogen production or storage capabilities within the vehicle and external hydrogen infrastructure. Many of electrical vehicles are accompanied by energy storage devices to improve vehicle performance under all conditions [6].

EVs (full electric vehicles) have recently received significant interest owing to recent revolutions in charging infrastructures and the viability of controllable loads supporting the grid in vehicle-to-grid (V2G), vehicle-to-building (V2B), vehicle-to-home (V2H), and vehicle-to-infrastructure (V2I) applications (see Chapter 2). Another possible approach for extending the electric range of an HEV is to allow continuous charging of the battery while running. The emergence of solar-driven HEVs (PVHEVs) leads to continuous charging of batteries by means of solar energy, which minimizes the usage of gasoline and hence reduces environmental pollution. The reduction in pollution by various types of hybrid vehicles mentioned above is depicted in Section 3.10.

Hybrid and electric vehicles have been demonstrated as auspicious solutions for ensuring improvements in fuel-saving and emission reductions. Robust and affordable batteries are a primary challenge for hybrid vehicles. Various HEV battery compositions have been tried in the past with the best results from lithium-ion derivatives. Three levels of integration of battery packs are possible in vehicles: (1)

singular battery cells, (2) modules, comprised of individual battery cell, and (3) battery packs, comprised of modules. Battery should be able to supply high power over short periods and must be capable of enduring millions of transient shallow cycles over vehicle life [6]. To extend the range and life of a battery, it can be interfaced with an ultracapacitor (UC) which permits longer life cycle, higher rate of charge/discharge, and lower internal resistance which result in lesser heat loss and better reliability. UC improves the efficiency cycle to around 90% from 80% [7]. The combination of battery and UC forming a hybrid energy storage system (HESS) is more efficient as compared to their individual performances. The use of HESS in EV is described in Chapter 4. The fundamental requirement for traction motors used in HEVs is to generate propulsion torque over a wide speed range. Two most commonly used motors in HEV propulsion are the permanent magnet motor (PM) and the induction motor (IM). More details on various types of motor used in hybrid cars are given in Section 3.8. The market for hybrid and electric cars is rapidly expanding. Various types of hybrid cars are now manufactured and are available by automobile companies like Audi, BMW, Chevrolet, Ford, Honda, Mercedes, McLaren, Nissan, Mitsubishi, Hyundai, Porsche, Tesla, and Toyota.

This chapter expands the concepts on hybrid energy for vehicles outlined in my previous book [1]. In that book the concepts of automobiles operating with multiple and flex fuels, hybrid energy cars using internal combustion engine and electricity and grid-integrated vehicles were examined. These concepts are further expanded here to give a description of a broader spectrum of the use of hybrid energy for mobile systems. The topics related to hybrid energy for mobile systems that are considered here are:

1. Multi and flex hybrid cars fuel with or without conversion of waste heat from the exhaust to electricity (The subject of waste heat to power was discussed in detail in Chapter 2. This chapter expands the application of this concept to mobile systems.)
2. Different types of hybrid cars involving fuels (including hydrogen) and electricity
3. Different types of HEVs/EVs such as mild hybrid, full hybrid, plug-in hybrid, battery-driven EV, Fuelcell hybrid, and all electric
4. Role of power trains, power converters, controllers, and motors
5. Role of homogeneous and hybrid energy storage (Role of hybrid energy storage in EV is discussed in detail in Chapter 4.)
6. Role of solar energy on electric car
7. Hybrid vehicles other than passenger cars
8. Economics and market potential of hybrid cars
9. Fuel consumption and efficiency and environment impacts of hybrid cars

3.2 TYPES OF HYBRID VEHICLES

The concept of hybrid vehicles can be applied to all different types of vehicles besides automobiles [2, 6, 8–23]. Mopeds, electric bicycles, and even electric kick scooters are a simple form of a hybrid vehicle, powered by an internal combustion engine or an electric motor and the rider's muscles. Here we briefly summarize the development of the concept of hybrid energy as it is applied to different types of vehicles [2, 6, 8–23].

A. **Bicycle**

In a parallel hybrid bicycle, human and motor torques are mechanically coupled at the pedal or one of the wheels, e.g. using a hub motor, a roller pressing onto a tire, or a connection to a wheel using a transmission element. Most motorized bicycles, mopeds are of this type. In a series hybrid bicycle (SHB) (a kind of chainless bicycle) the user pedals a generator, charging a battery or feeding the motor, which delivers all of the torque required. They are commercially available, being simple in theory and manufacturing.

A *series hybrid electric-petroleum bicycle* (SHEPB) is powered by pedals, batteries, a petrol generator, or plug-in charger—providing flexibility and range enhancements over electric-only bicycles. A SHEPB prototype made by David Kitson in Australia in 2014 used a lightweight brushless DC electric motor from an aerial drone and a small hand-tool-sized internal combustion engine, and a 3D printed drive system and lightweight housing, altogether weighing less than 4.5 kg. Active cooling keeps plastic parts from softening. The prototype used a regular electric bicycle charge port.

B. **Motorcycles**

Peugeot HYmotion3 compressor, a hybrid scooter is a three-wheeler that uses two separate power sources to power the front and back wheels. The back wheel is powered by a single cylinder 125 cc, 20 bhp (15 kW) single cylinder motor while the front wheels are each driven by their own electric motor. When the bike is moving up to 10 km/h only electric motors are used on a stop–start basis reducing the amount of carbon emission. SEMA launched Yamaha in 2010, with Honda following a year later, fueling a competition to reign in new customers and set new standards for mobility. Each company provided the capability to reach 60 miles (97 km) per charge by adopting advanced lithium-ion batteries. These proposed hybrid motorcycles could incorporate components from the upcoming Honda Insight car and its hybrid powertrain.

C. **Automobiles, Taxis, and Light Trucks**

In 1899, Henri Pieper developed the world's first petro-electric hybrid automobile. In 1900, Ferdinand Porsche developed a series-hybrid using two motor-in-wheel-hub arrangements with an internal combustion generator set providing electric power. When the term *hybrid vehicle* is used, it most often refers to a Hybrid electric automobile. These encompass such vehicles as Saturn Vue, Toyota Prius, Toyota Yaris, Toyota Camry Hybrid, Ford Escape Hybrid, Toyota Highlander Hybrid, Honda

Insight, Honda Civic Hybrid, Lexus RX 400 h and 450 h, Hyundai Ioniq, and others. A petroleum-electric hybrid most commonly uses internal combustion engines (using a variety of fuels, generally gasoline or Diesel engines) and electric motors to power the vehicle. The energy is stored in the fuel of the internal combustion engine and an electric battery set. As shown later, there are many types of petroleum-electric hybrid drivetrains, from *Full hybrid* to Mild hybrid, which offer varying advantages and disadvantages. In 2000, North America's first hybrid electric taxi was put into service in Vancouver, British Columbia, operating a 2001 Toyota Prius which traveled over 332,000 km (206,000 mi) before being retired. In 2015, a taxi driver in Austria claimed to have covered 1,000,000 km (620,000 mi) in his Toyota Prius with the original battery pack.

D. **Buses**

Hybrid technology for buses has seen increased attention since recent battery developments decreased battery weight significantly. Drivetrains consist of conventional diesel engines and gas turbines. Some designs concentrate on using car engines to save on engineering and training costs, although, recent designs have focused on using conventional diesel engines already used in bus designs. As of 2007, several manufacturers were working on new hybrid designs, or hybrid drivetrains that fit into existing chassis offerings without major re-design. A challenge to hybrid buses may still come from cheaper lightweight imports from the former Eastern block countries or China, where national operators are looking at fuel consumption issues surrounding the weight of the bus, which has increased with recent bus technology innovations such as glazing, air conditioning and electrical systems. A hybrid bus can also deliver fuel economy through the hybrid drivetrain. Hybrid technology is also being promoted by environmentally concerned transit authorities.

E. **Trucks**

In 2003, GM introduced a hybrid diesel-electric military (light) truck, equipped with a diesel electric and a fuel cell auxiliary power unit. Hybrid electric light trucks were introduced in 2004 by Mercedes Benz (Sprinter) and Micro-Vett SPA (Daily Bimodale). In mid-2005 Isuzu introduced the Elf Diesel Hybrid Truck in the Japanese Market. Other produced heavy vehicles are:

1. Big mining machines like the Liebherr T 282B dump truck or Keaton Vandersteen LeTourneau L-2350 wheel loader are powered that way. Also there were several models of BelAZ (7530 and 7560 series) in USSR (now in Belarus) since the mid-1970s.
2. NASA's huge Crawler-Transporters are diesel-electric.
3. Mitsubishi Fuso Canter Eco Hybrid is a diesel-electric commercial truck.
4. Azure Dynamics Balance Hybrid Electric is a gasoline-hybrid electric-medium-duty truck based on the Ford E-450 chassis.

5. Hino Motors (a Toyota subsidiary) has the world's first production of hybrid electric truck in Australia (110 kW or 150 hp diesel engine plus a 23 kW or 31 hp electric motor).

Other hybrid petroleum-electric truck makers are DAF Trucks, MAN with MAN TGL Series, and Nissan Motors and Renault Trucks with Renault Puncher.

F. **Military Vehicles**

Some 70 years after Porsche's pioneering efforts in hybrid-drivetrain-armored fighting vehicles in World War II, the US Army's manned ground vehicles of the Future Combat System all use a hybrid electric drive consisting of a diesel engine to generate electrical power for mobility and all other vehicle subsystems. Other military hybrid prototypes include the Millenworks Light Utility Vehicle, the International FTTS, HEMTT model A3, and the Shadow RST-V.

G. **Locomotives**

William H. Patton filed a patent application for a gasoline-electric hybrid rail-car propulsion system in early 1889 and for a similar hybrid boat propulsion system in mid-1889. In May 2003, JR East started test runs with the so-called NE (new energy) train and validated the system's functionality (series hybrid with lithium-ion battery) in cold regions. In 2004, Railpower Technologies ran pilots in the United States with the so-called Green Goats. Railpower offers hybrid electric road switchers, as does GE. Diesel-electric locomotives may not always be considered HEVs, not having energy storage on board, unless they are fed with electricity via a collector for short distances (for example, in tunnels with emission limits), in which case they are better classified as dual-mode vehicles.

H. **Marine and Other Aquatic Vehicles**

For large boats that are already diesel-electric, the upgrade to hybrid can be as straightforward as adding a large battery bank and control equipment; this configuration can provide fuel saving for the operators and be more environmentally sensitive. Producers of marine hybrid propulsion include eCycle Inc and Solar Sailor Holdings.

I. **Aircraft**

A hybrid electric aircraft is an aircraft with an hybrid electric powertrain, needed for airliners as the energy density of lithium-ion batteries is much lower than aviation fuel. By May 2018, there were over 30 projects, and short-haul hybrid-electric airliners were envisioned from 2032. The most advanced are the Zunum Aero 10-seater, the Airbus E-Fan X demonstrator, the VoltAero Cassio, and modified Bombardier Dash 8.

3.3 TYPES OF HYBRID RENEWABLE SOLAR VEHICLES

In recent years the use of renewable energy, particularly solar energy, is extensively explored in vehicle industry. Some of the recent developments made for hybrid solar electric vehicles are briefly illustrated in this section [9, 10].

A. Solar Electric Car

A *solar vehicle* is an electric vehicle powered completely or significantly by direct solar energy. Usually, photovoltaic (PV) cells contained in solar panels convert the sun's energy directly into electric energy. The term "solar vehicle" usually implies that solar energy is used to power all or part of a vehicle's propulsion. Solar power may also be used to provide power for communications or controls or other auxiliary functions. Solar vehicles are not sold as practical day-to-day transportation devices at present, but are primarily demonstration vehicles and engineering exercises, often sponsored by government agencies. However, indirectly solar-charged vehicles are widespread and solar boats are available commercially. Solar cars depend on PV cells to convert sunlight into electricity to drive electric motors. Unlike solar thermal energy which converts solar energy to heat, PV cells directly convert sunlight into electricity [9, 10].

The design of a solar car is severely limited by the amount of energy input into the car. Solar cars are built for solar car races and also for public use. Even the best solar cells can only collect limited power and energy over the area of a car's surface. This limits solar cars to ultra-light composite bodies to save weight. Solar cars lack the safety and convenience features of conventional vehicles. The area of photovoltaic modules required to power a car with conventional design is too large to be carried on board. A prototype car and trailer "solar taxi" has been built. According to the website, it is capable of 100 km/day using $6m^2$ of standard crystalline silicon cells. Electricity is stored using a nickel/salt battery. A stationary system such as a rooftop solar panel, however, can be used to charge conventional electric vehicles. It is also possible to use solar panels to extend the range of a hybrid or electric car, as incorporated in the Fisker Karma, available as an option on the Chevy Volt, on the hood and roof of "Destiny 2000" modifications of Pontiac Fieros, Italdesign Quaranta, Free Drive EV Solar Bug, and numerous other electric vehicles, in both concept and production. In May 2007 a partnership of Canadian companies led by Hymotion added PV cells to a Toyota Prius to extend the range. SEV claims 20 miles per day from their combined 215 W module mounted on the car roof and an additional 3kWh battery [8–16].

It is also technically possible to use photovoltaic technology, (specifically thermophotovoltaic (TPV) technology) to provide motive power for a car. Fuel is used to heat an emitter. The infrared radiation generated is converted to electricity by a low band gap PV cell (e.g. GaSb). A prototype TPV hybrid car was even built. The "Viking 29" was the World's first thermophotovoltaic (TPV) powered automobile, designed and built by the Vehicle Research Institute (VRI) at Western Washington University. Efficiency would need to be increased and cost decreased to make TPV competitive with fuel cells or internal combustion engines [8–16].

B. **Hybrid Plug-In Solar Electric Car**

An interesting variant of the electric vehicle is the triple hybrid vehicle—the PHEV that has solar panels to assist. The 2010 Toyota Prius model had an option to mount solar panels on the roof. They power a ventilation system while parked to help provide cooling. There are many applications of *photovoltaics in transport* either for motive power or as auxiliary power units, particularly where fuel, maintenance, emissions, or noise requirements preclude internal combustion engines or fuel cells. Due to the limited area available on each vehicle either speed or range or both are limited when used for motive power. There are limits to using photovoltaic (PV) cells for cars. Power from a solar array is limited by the size of the vehicle and area that can be exposed to sunlight. This can also be overcome by adding a flatbed and connecting it to the car and this gives more area for panels for powering the car. While energy can be accumulated in batteries to lower peak demand on the array and provide operation in sunless conditions, the battery adds weight and cost to the vehicle. The power limit can be mitigated by the use of conventional electric cars supplied by solar (or other) power, recharging from the electrical grid [8–11, 13, 17–19].

While sunlight is free, the creation of PV cells to capture that sunlight is expensive. Costs for solar panels are steadily declining (22% cost reduction per doubling of production volume). Even though sunlight has no lifespan, PV cells do. The lifetime of a solar module is approximately 30 years. Standard photovoltaics often come with a warranty of 90% (from nominal power) after 10 years and 80% after 25 years. Mobile applications are unlikely to require lifetimes as long as building integrated PV and solar parks. Current PV panels are mostly designed for stationary installations. However, to be successful in mobile applications, PV panels need to be designed to withstand vibrations. Also, solar panels, especially those incorporating glass, have significant weight. In order for its addition to be of value, a solar panel must provide energy equivalent to or greater than the energy consumed to propel its weight [8–11, 13, 17–19].

C. **Electric Car with Solar Assist/Solar Taxi**

A Swiss project, called "Solartaxi", has circumnavigated the world. This is the first time in history an electric vehicle (not self-sufficient solar vehicle) has gone around the world, covering 50,000 km in 18 months and crossing 40 countries. It is a roadworthy electric vehicle hauling a trailer with solar panels, carrying a 6 m^2-sized solar array. The Solartaxi has Zebra batteries, which permit a range of 400 km without recharging. The car can also run for 200 km without the trailer. Its maximum speed is 90 km/h. The car weighs 500 kg and the trailer weighs 200 kg. In recent years, solar electrical vehicle is adding convex solar cells to the roof of hybrid electric vehicles [9–11, 13, 17–19].

D. **Hybrid Solar Buses**

Solar buses are propulsed by solar energy, all or part of which is collected from stationary solar panel installations. The Tindo bus is a 100% solar bus that operates as free public transport service in Adelaide City as an initiative of the City Council. Bus services which use electric buses that are partially powered by solar panels installed on the bus roof, intended to reduce energy consumption and to prolong the life cycle of the rechargable battery of the electric bus, have been put in place in China [9–11, 13, 17–19]. Solar buses are to be distinguished from conventional buses in which electric functions of the bus such as lighting, heating, or air-conditioning, but not the propulsion itself, are fed by solar energy. Such systems are more widespread as they allow bus companies to meet specific regulations, for example the anti-idling laws that are in force in several of the US states, and can be retrofitted to existing vehicle batteries without changing the conventional engine.

E. **Hybrid Solar-Powered Rapid Transit Vehicles**

Railway presents a low rolling resistance option that would be beneficial of planned journeys and stops. PV panels were tested as APUs (Auxilliary Power Unit) on Italian rolling stock under EU project PVTRAIN. Direct feed to a DC grid avoids losses through DC to AC conversion. DC grids are only to be found in electric powered transport: railways, trams, and trolleybuses. Conversion of DC from PV panels to grid alternating current (AC) was estimated to cause around 3% of the electricity being wasted; PVTrain concluded that the most interest for PV in rail transport was on freight cars where on board electrical power would allow new functionality:

1. GPS or other positioning devices, so as to improve its use in fleet management and efficiency,
2. Electric locks, a video monitor, and remote control system for cars with sliding doors, so as to reduce the risk of robbery for valuable goods,
3. ABS brakes, which would raise the maximum velocity of freight cars to 160 km/h, improving productivity.

The Kismaros—Királyrét narrow-gauge line near Budapest has built a solar-powered railcar called 'Vili'. With a maximum speed of 25 km/h, 'Vili' is driven by two 7 kW motors capable of regenerative braking and powered by 9.9m2 of PV panels. Electricity is stored in on-board batteries [20]. In addition to on-board solar panels, there is the possibility to use stationary (off-board) panels to generate electricity specifically for use in transport [9–11, 13, 17–19].

A few pilot projects have also been built in the framework of the "Heliotram" project, such as the tram depots in Hannover Leinhausen and Geneva (Bachet de Pesay). The 150 kW_p Geneva site injected 600 V DC directly into the tram/trolleybus electricity network which provided about 1% of the electricity used by the Geneva transport network at its opening in 1999. On December 16, 2017, a fully solar-powered train was launched in New South Wales, Australia. The train was powered using onboard solar panels and onboard rechargeable batteries. It held a capacity of 100 seated passengers for a 3 km journey. The invention by Dearborn [18] provides a means by which rail (railroad) transportation operators can generate megawatts of carbon-free electrical power from the space over rail tracks and right-of-way without buying or leasing significant amounts of sun accessible space or power transmission right of way.

F. **Hybrid Power Management and Distribution in Space Mission**

Space power systems are composed of power generation, energy storage, and power distribution and management. Power generation includes hybrid mix of solar arrays, fuel cells, and thermodynamic engines. Energy storage includes batteries but may include the evolving flywheel technology. Power distribution and management (PMAD) connects the power generation and energy storage to the user loads. It regulates the power and handles the delivery of power. It also provides the primary system fault detection, fault isolation, and rerouting of power. These are typically designed as separate subsystems. The general approach is to integrate certain hybrid PMAD regulator and health management functions into power generation and energy storage. Embedding these functions combined with plug-and-play features enabled them to act as independent self-contained modular subsystems. This makes them more portable so they can be moved to different parts of the vehicle or another vehicle entirely [24, 25].

There are a number of factors that affect the effectiveness of the modular hybrid PMAD design. For example, international space station (ISS) requires high degree of modularity driven by scalability and supportability needs. The ISS design incorporates modularity using assembly level of hardware which is the preferred level of replacement for ISS [24, 25]. Both low- and high-level modules are portable and have embedded functions and related software. Using many small blocks to satisfy the need of a spacecraft means many interconnections, which drives up the mass of harnessing while reducing overall reliability. Using large blocks means a tendency for overcapacity and excessive mass. An assortment of modular elements with varied capacity enables designers to mix and match modules to arrive at a system that meets mission needs [24, 25]. Hybrid nature of the PMAD design ensures flexibility and realibility during space mission.

Current practice in Power Management and Distribution design is to develop mission-specific solutions. Hybrid PMAD elements (e.g., load and bus regulators, energy storage subsystem interfaces, and protection) are designed to optimally meet these unique system and subsystem requirements for each mission and cannot be used in any other vehicles. PMAD system is composed of two types of electronic devices: power electronics and control electronics. The power electronics (includes electromechanical devices) conduct and direct the current and tend to scale in relation to the power loads. Control electronics manages the power electronics with low-power mixed signal (digital and analog) devices that scale somewhat independently of power loads. While different control techniques are required for different applications, regulation functions can all be addressed with common power converter designs. That is, solar array regulation, bus regulation, battery charge/discharge control, and load power regulation can all be done with common DC-DC converters (referred to as Flexible Power Modules). These modules are used in sufficient numbers to meet the operational requirements. An advantage of the hybrid modular approach is the ability to continue operation but at a reduced capacity after a failure. Oftering [25] points out that this hybrid modular approach for PMAD has worked very effectively for NASA. More details on hybrid modular power management for space vehicles is described in my previous book [24] and by Oftering [25].

3.4 BATTERY ELECTRIC VEHICLES

As shown in Chapter 4, both homogeneous and heterogeneous energy storage are very important for the future development of HEV/EV. One of the most important storage devices for HEV/EV is the battery. The prospect for widespread introduction of full-performance all-electric vehicles depends on significant advancements of the battery technologies, and the commercial viability of these vehicles depends on a battery cost breakthrough. Advances in electric motors, power electronics, and batteries for automotive applications, which have resulted from the development and production of hybrid vehicles, have renewed interest in the development of battery electric vehicles. However, the cost, low energy density, and required charging time of batteries will continue to constrain the introduction of BEVs. The high–low speed torque performance of electric motors gives the BEV a potential acceleration advantage over conventional internal combustion engine-powered vehicles, and this can be an attractive feature for some customers [2,14, 23]. Battery-operated hybrid car with single motor and two motors are illustrated in Figures 3.2a and 3.2b, respectively [2].

A review of zero-emission vehicle technology commissioned by the California Air Resources Board (CARB) concluded that commercialization (tens of thousands of vehicles) of full-performance battery electric vehicles would not occur before 2015 and that mass production (hundreds of thousands of vehicles) would not occur before 2030 [26, 27]. These projections were based on the continued development of lithium-ion (Li-ion) battery technology leading to reduced cost, higher energy densities, and reduced charging times, all of which allow greater range. They pointed to a possible role for a limited range, city electric vehicle (CEV), which could meet the requirements of a majority of household trips. However, recent BEV introductions suggest that progress in the technology and acceptance of Li-ion batteries may be more rapid than the CARB study concluded [2, 14, 23].

Early commercial application of Li-ion battery technology to vehicles includes the Tesla Roadster, a high-performance sports car. This vehicle, of which about 1,000 have been sold, has a fuel consumption of 0.74 gal/100 miles (energy equivalent basis, EPA combined city/highway). The manufacturer claims a range of 244 miles (also EPA combined city/highway) and a useful battery life of more than 100,000 miles. The base price of $128,000 indicates the continuing problem of battery cost when used in near full-performance vehicles. Tesla announced that it will produce and sell, at about half the price of the Roadster, a five-passenger BEV, the Tesla S, with a range of 160, 230, or 300 miles, depending on optional battery size. Nissan has also announced production of its Leaf EV, a five-passenger car with a range of 100 miles. This vehicle has a Li-ion battery with a total storage capacity of 24 kWh.

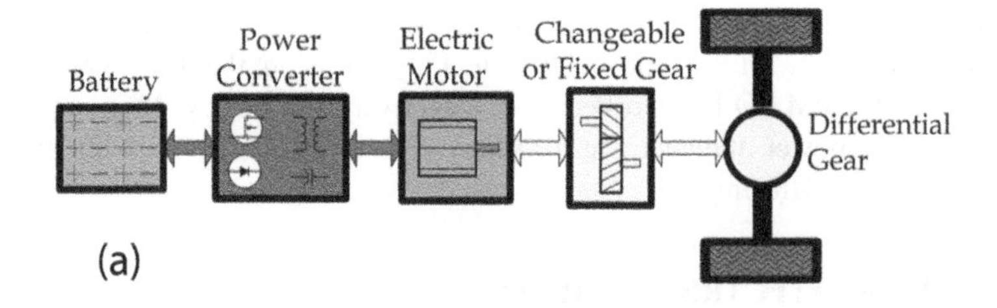

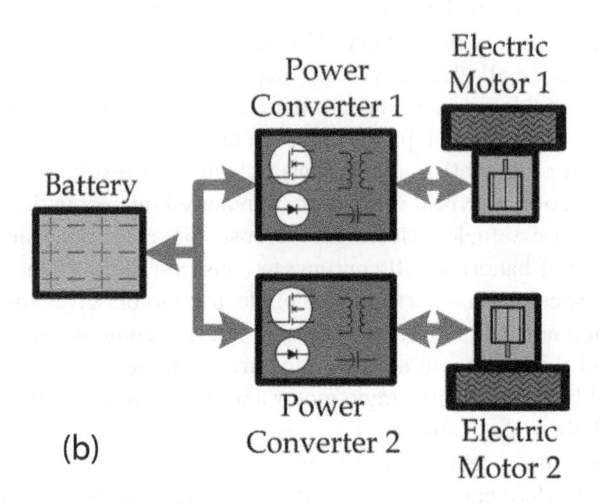

FIGURE 3.2 (a) Battery-operated hybrid car with single motor [2]. (b) Battery-operated hybrid car with two motors [2]

Within the horizon of this study, the most likely future for large numbers of battery electric vehicles in the United States is in the limited-range, small-vehicle market. Range extended electric vehicles (hybrids and PHEVs) are more likely to satisfy the electricity-fueled full-performance—market, from both cost and technological considerations, over the next 15 years [2, 14, 23, 26, 27].

3.4.1 Continuously Outboard Recharged Electric Vehicle (COREV)

Some battery electric vehicles (BEVs) can be recharged while the user drives. Such a vehicle establishes contact with an electrified rail, plate, or overhead wires on the highway via an attached conducting wheel or other similar mechanisms. The BEV's batteries are recharged by this process—on the highway—and can then be used normally on other roads until the battery is discharged. For example, some of the battery-electric locomotives used for maintenance trains on the London Underground are capable of this mode of operation.

Developing a BEV infrastructure would provide the advantage of virtually unrestricted highway range. Since many destinations are within 100 km of a major highway, the BEV technology could reduce the need for expensive battery systems. Unfortunately, private use of the existing electrical system is almost universally prohibited. Besides, the technology for such electrical infrastructure is largely outdated and, outside some cities, not widely distributed. Updating the required electrical and infrastructure costs could perhaps be funded by toll revenue or by dedicated transportation taxes.

3.4.2 Development of Battery Technology

In spite of the significant progress that battery technology has experienced in the last 20 years, the battery is still the most challenging technology in the design of hybrid vehicles. Table 3.1 illustrates the dramatic difference between the energy densities of today's commercial batteries and gasoline, diesel fuel, ethanol, compressed natural gas, and hydrogen. In the past, all production hybrid vehicles used batteries employing nickel-metal-hydride (NiMH) chemistry. It is anticipated that the NiMH battery will be replaced by Li-ion batteries in the future. The acceptability of today's hybrid vehicles has been shown to be strongly dependent on the price of gasoline, as evidenced by the rapid growth of hybrid sales in 2008, when gasoline prices were high, and the fact that hybrid sales dropped dramatically in early 2009 when prices returned to lower values. The key to improving the competitive position of hybrid vehicles of the HEV and PHEV types is the commercial development of batteries with parameters that are substantially better than those of today's batteries, leading to reduced cost and size. The required parametric improvements are as follows [2, 8, 14, 23, 26, 27]:

TABLE 3.1
Volumetric and Gravimetric Energy Densities of Different Energy Storage Mechanisms [2, 8, 14, 26–30]

Storage mechanism	Volumetric energy density (Wh/L)	Gravimetric energy density (Wh/kg)
Lead Battery	about 100	about 20–25
Nickel metal hydride battery	about 150	about 40–50
Li-ion battery	about 200	about 75–85
Hydrogen absorbing alloy *2 Wt%)	about 3400	about 700
CNG (20 MPa)	about 2300	about 10500
High-pressure hydrogen (35 Mpa)	about 1000	about 15000
Ethanol	about 5600	about 8000
Gasoline	about 9500	about 10500
Bio-diesel	about 9800	about 10000
Diesel	about 10200	about 10000

1. Higher cycle life at increased SOC variation,
2. Higher energy density,
3. Higher power density, and
4. Lower cost.

Table 3.2 shows the desirable characteristics of batteries suitable for the HEV, the PHEV, and the all-electric (EV or BEV) vehicles. The HEV uses electric propulsion primarily as an assistant to the IC engine, thus requiring a battery with a high power capability but relatively little energy capacity, i.e., a high power to energy (P/E) ratio. To preserve battery life and maintain the capacity to recover charge through regenerative braking, the battery is cycled over a relatively small state of charge. This mode of operation is known as charge sustaining (CS). The PHEV is expected to provide [2, 8, 14, 23, 26, 27] some degree of electric-only range. Its battery must therefore contain sufficient energy to provide this range. The battery may be allowed to expend all of its stored energy to achieve this range goal, in which case the battery is said to be operated in the charge-depleting (CD) mode. The power requirement of this battery is not much different from that of the HEV battery, but because of the higher energy requirement, the P/E ratio is smaller. The BEV requires an even higher energy capacity battery than the PHEV, the value depending on the desired driving range. Since the BEV has no IC engine, its battery cannot be charged during driving, and therefore it cannot operate in a CS mode. In all cases the SOC variation is limited to a specified range by the vehicle manufacturer to preserve battery cycle life. Table 3.2 shows unused energy for EV, HEV, and PHEV. Thus the usable energy is less than the battery-rated (or "nameplate") capacity. Despite substantial improvements in the packaging and performance of lead-acid batteries, their energy and power densities are still considerably inferior to those of NiMH. And while other chemistries, like Li-air, have theoretically better performance than Li-ion, they need more development [2, 8, 14, 23, 27].

TABLE 3.2
Energy Capacity, State-of-Charge Variation, and Relative Power Density to Energy Density Ratios for Batteries Applicable to Full-Hybrid (HEV), Plug-in Hybrid (PHEV), and All-Electric (EV) Vehicles [2, 6, 8–27]

Vehicle	Charge depleting	Charge sustaining	Unused energy	Battery size	P/E(kW/kWh)
EV	80%	0%	20%	>40 kWh	2.0
PHEV	65%	10%	25%	about 5–15kWh	5–15
HEV	15% uncharged capacity	10%	65%	about 1–2 kWh	15–20

A. **NiMH Batteries**

The highest-performance battery currently available in commercially significant quantities for HEVs and PHEVs uses NiMH chemistry. Despite significant improvements in lifetime and packaging, these batteries are still expensive and heavy, and in applications they are restricted to an SOC range of about 20 percent to preserve battery cycle life. Because of their relatively poor charge/discharge efficiency, special consideration must be given to their thermal management. The NiMH chemistry also exhibits a high rate of self-discharge. The most technically advanced NiMH battery used in the Toyota Prius has a weight of 45 kg and an energy capacity of 1.31 kWh. This results in a usable energy of approximately 0.262 kWh when applied with an SOC variation of 20 percent.

B. **Li-Ion Batteries**

A *lithium-ion battery* or *Li-ion battery* (abbreviated as *LIB*) is a type of rechargeable battery. Lithium-ion batteries are commonly used for portable electronics and electric vehicles and are growing in popularity for military and aerospace applications. They can however be a safety hazard since they contain a flammable electrolyte, and if damaged or incorrectly charged can lead to explosions and fires [27–30]. Chemistry, performance, cost, and safety characteristics vary across LIB types. Lithium iron phosphate ($LiFePO\backslash_4$), lithium ion manganese oxide battery ($LiMn_2O_4$, Li_2MnO_3, or LMO), Lithium Nickel Cobalt Aluminum Oxide (LiNiCoAlO2, "NCA") and lithium nickel manganese cobalt oxide ($LiNiMnCoO_2$ or NMC) offer lower energy density but longer lives and less likelihood of fire or explosion. NMC and its derivatives are widely used in electric vehicles.

The increasing demand for batteries has led vendors and academics to focus on improving the energy density, operating temperature, safety, durability, charging time, output power, and cost of lithium ion battery technology. Cathode materials are generally constructed from $LiCoO_2$ or $LiMn_2O_4$. The cobalt-based cathodes are ideal due to their high theoretical specific heat capacity, high volumetric capacity, low self-discharge, high discharge voltage, and good cycling performance. Limitations include the high cost of the material, and low thermal stability. Manganese cathodes are attractive because manganese is cheaper and because it could theoretically be used to make a more efficient, longer-lasting battery if its limitations could be overcome. Limitations include the tendency for manganese to dissolve into the electrolyte during cycling leading to poor cycling stability for the cathode. Cobalt-based cathodes are the most common; however, other materials are being researched with the goal of lowering costs and improving battery life. As of 2017, $LiFePO_4$ is a candidate for large-scale production of lithium-ion batteries such as electric vehicle applications due to its low cost, excellent safety, and high cycle durability.

The relative gravimetric energy densities of Li-ion, NiMH, and Pb-acid are approximately 4, 2, and 1, respectively. An additional advantage of the Li systems is their high cell potential, approximately 3 times that of NiMH. This means that 66 percent fewer Li-ion cells are required to achieve a given battery voltage. The ecologically benign materials in the Li-ion systems are also an advantage. A disadvantage of Li-ion cells is that the requirement for cleanliness in the manufacturing environment is considerably more severe than for NiMH cells. This increases manufacturing costs. Another critical issue is how the performance of Li-ion batteries is impacted by low and high temperatures [27–30].

NCA-Graphite, LFP-Graphite, and MS-TiO—represent the most promising Li-ion systems currently under development. NCA-Graphite has moderate cost, good life potential, and fair safety. The lithium-iron phosphate (LFP) system is currently receiving a great deal of attention because of its stability, potentially lower material costs, and its application in power tools. The manganese-spinel-lithium-titanate system (MS-TiO) is the safest of any being studied because of the mechanical stability of the spinel structure, but its cell voltage is considerably lower than those of the NCA and LFP systems. However, it has the highest charge/discharge efficiency, and it is predicted to be the lowest-cost system.

The needs of HEVs and PHEVs are quite distinct. HEVs need high power density and long cycle life over a very small excursion of the SOC. For example, the Prius battery has a nominal rating of 1.3 kWh but it uses only 260 Wh in +/−10 percent excursions around 50 percent SOCs. On the other hand, the larger energy requirement of the PHEV argues for a battery with a higher energy rating and the capability of deeper cycling. The Volt, the PHEV being developed by GM, uses a 16-kWh battery to meet its advertised all-electric range of 40 miles. This is a substantial challenge to achieve at acceptable weight, volume, and cost. The Li-ion chemistry comes closest to meeting it, given the present state of battery development [2, 8, 11, 23, 27–29].

3.4.3 Development of Battery Capacity and Charging Infrastructure

The growing fleet of electric vehicles and increasing battery capacity per vehicle is driving the development of another new market: charging infrastructure. Rapid deployment of DC fast charging stations at public places goes hand in hand with deployment of plug-in HEV and EV vehicles. Charging stations with charging power levels of 100–200 kW are expected to become a mainstream in the coming years. The charging stations with higher power, up to 350 kW and more, are already available and can significantly reduce the charging time as battery technologies and thermal management of the battery packs allow such fast charging. A modular design of such big power charging stations enables the charging of several cars simultaneously [2, 8, 11, 23, 27–29].

3.5 TYPES OF HYBRID SYSTEMS

In line with the broader definition of hybrid energy outlined in Chapterl, here we define hybrid energy for mobile systems in broad terms. Hybrid energy is defined as the multiple source of fuel or energy that includes cofuels, cogeneration, and use of energy storage (like battery or super capacitors) along with one or more sources of energy. A conventional definition of hybrid vehicle is one that uses two or more distinct types of power, such as (i) internal combustion engine to

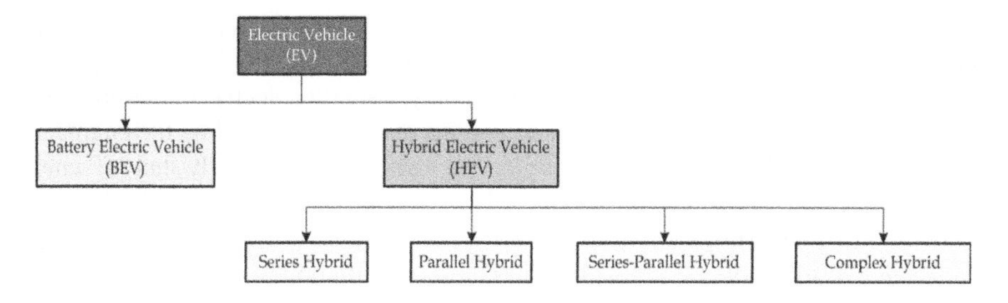

FIGURE 3.3 Types of hybrid electric vehicles [2].

drive an electric generator that powers an electric motor, e.g. in diesel-electric trains using diesel engines to drive an electric generator that powers an electric motor, and (ii) submarines that use diesels when surfaced and batteries when submerged. Other means to store energy include pressurized fluid in hydraulic hybrids. The types of hybrid electric vehicles are graphically illustrated in Figure 3.3 [2]

The basic principle with hybrid vehicles is that the different motors work better at different speeds; the electric motor is more efficient at producing torque, or turning power, and the combustion engine is better for maintaining high speed (better than typical electric motor). Switching from one to the other at the proper time while speeding up yields a win-win in terms of energy efficiency, as it translates into greater fuel efficiency, for example. Conversion of waste heat to power also improves energy efficiency. Similar principles apply for energy storage device where battery is good to provide energy density while supercapacitors are good to provide power density. In a given driving cycle, EV needs power density for acceleration and energy density for long distance. The varieties of hybrid electric designs can be differentiated by the structure of the hybrid vehicle drivetrain, the fuel type, energy storage devices, and the mode of operation.

In 2007, several automobile manufacturers announced that future vehicles will use aspects of hybrid electric technology to reduce fuel consumption without the use of the hybrid drivetrain. Regenerative braking can be used to recapture energy and store power electrical accessories, such as air conditioning. Shutting down the engine at idle can also be used to reduce fuel consumption and emissions without the addition of a hybrid drivetrain. In both cases, some of the advantages of hybrid electric technology are gained while additional cost and weight may be limited to the addition of larger batteries and starter motors. There is no standard terminology for such vehicles, although they may be termed mild hybrids.

Hybrids-Electric vehicles (HEVs) combine the advantage of gasoline *engines* and electric *motors*. The key areas for efficiency or performance gains are regenerative braking, dual power sources, and less idling. In *Regenerative Braking,* the drivetrain can be used to convert kinetic energy (from the moving car) into stored electrical energy (batteries). The same electric motor that powers the drivetrain is used to resist the motion of the drivetrain. This applied resistance from the electric motor causes the wheel to slow down and simultaneously recharge the batteries. In *Dual Power,*

power can come from either the engine, motor, or both depending on driving circumstances. Additional power to assist the engine in accelerating or climbing might be provided by the electric motor. Or more commonly, a smaller electric motor provides all of the power for low-speed driving conditions and is augmented by the engine at higher speeds. In *Automatic Start/Shutoff, the engine is* automatically shut off when the vehicle comes to a stop and restarts it when the accelerator is pressed down. This automation is much simpler with an electric motor. We briefly examine here different modes of hybrid energy (as defined earlier) used in automobiles [2, 8, 14].

A. **Hybrid Fuel (Dual Mode)**

In addition to vehicles that use two or more different devices for propulsion, some also consider vehicles that use distinct energy sources or input types ("fuels") using the same engine to be hybrids, although to avoid confusion with hybrids as described above and to use correctly the terms, these are perhaps more correctly described as dual mode vehicles. For example, Ford Escape Plug-in Hybrid with a flexible fuel capability to run on E85 (ethanol) [20].

1. Some electric trolleybuses can switch between an on-board diesel engine and overhead electrical power depending on conditions (see dual mode bus). In principle, this could be combined with a battery subsystem to create a true plug-in hybrid trolleybus, although as of 2006, no such design seems to have been announced.
2. Flexible-fuel vehicles can use a mixture of input fuels mixed in one tank typically gasoline and ethanol, methanol, or biobutanol.
3. Liquified petroleum gas and natural gas are very different from petroleum or diesel and cannot be used in the same tanks, so it would be impossible to build an (LPG or NG) flexible fuel system. Instead, vehicles are built with two, parallel, fuel systems feeding a single engine. These are bi- or dual-fuel vehicles. For example, some Chevrolet Silverado 2500 HDs can effortlessly switch between petroleum and natural gas, offering a range of over 1000 km (650 miles). While the duplicated tanks cost space in some applications, the increased range, decreased cost of fuel, and flexibility where LPG or CNG infrastructure is incomplete may be a significant incentive to purchase. With a growing fueling station infrastructure, a large-scale adoption of these bi-fuel vehicles could be seen in the near future.
4. Some vehicles have been modified to use another fuel source if it is available, such as cars modified to run on autogas (LPG) and diesels modified to run on waste vegetable oil that has not been processed into biodiesel.
5. Power-assist mechanisms for bicycles and other human-powered vehicles can also be operated by multiple fuels.

Hybrid vehicles might also use an internal combustion engine running on biofuels, such as a flexible-fuel engine running on ethanol or engines running on biodiesel. In 2007 Ford produced 20 demonstration Escape Hybrid E85s for real-world testing in fleets in the United States. Also as a demonstration project, Ford delivered in 2008 the first flexible-fuel plug-in hybrid SUV to the U.S. Department of Energy (DOE), a Ford Escape Plug-in Hybrid, capable of running on gasoline or E85. The Chevrolet Volt plug-in hybrid electric vehicle would be the first commercially available flex-fuel plug-in hybrid capable of adapting the propulsion to the biofuels used in several world markets such as the ethanol blend E85 in the U.S., or E100 in Brazil, or biodiesel in Sweden. Hyundai introduced in 2009 the Hyundai Elantra LPI Hybrid, which is the first mass production hybrid electric vehicle to run on liquefied petroleum gas (LPG).

B. **Fluid Power Hybrid**

Hydraulic hybrid and pneumatic hybrid vehicles use an engine to charge a pressure accumulator to drive the wheels via hydraulic (liquid) or pneumatic (compressed air) drive units. In most cases the engine is detached from the drivetrain, serving solely to charge the energy accumulator. The transmission is seamless. Regenerative braking can be used to recover some of the supplied drive energy back into the accumulator [20].

1. **Petro-Air Hybrid**
   A French company, MDI, has designed and has running models of a petro-air hybrid engine car. The system does not use air motors to drive the vehicle, being directly driven by a hybrid engine. The engine uses a mixture of compressed air and gasoline injected into the cylinders. A key aspect of the hybrid engine is the "active chamber", which is a compartment heating air via fuel doubling the energy output. Tata Motors of India assessed the design phase towards full production for the Indian market and moved into "completing detailed development of the compressed air engine into specific vehicle and stationary applications".
2. **Petro-Hydraulic Hybrid**
   Petro-hydraulic configurations have been common in trains and heavy vehicles for decades. The auto industry recently focused on this hybrid configuration as it now shows promise for introduction into smaller vehicles. In petro-hydraulic hybrids, the energy recovery rate is high and therefore the system is more efficient than electric battery-charged hybrids using the current electric battery technology, demonstrating a 60%–70% increase in energy economy in US Environmental Protection Agency (EPA) testing. The charging engine needs only to be sized for average usage with acceleration bursts using the stored energy in the hydraulic accumulator, which is charged when in low energy demanding vehicle operation. The charging engine runs at optimum speed and load for efficiency and longevity. Under tests undertaken by the US Environmental Protection Agency (EPA), a hydraulic hybrid Ford Expedition returned 32 miles per US gallon (7.4 L/100 km; 38 mpg-$_{imp}$) City and 22 miles per US gallon (11 L/100 km; 26 mpg-$_{imp}$) highway. UPS currently has two trucks in service using this technology.

Although petro-hydraulic hybrid technology has been known for decades, and used in trains and very large construction vehicles, high costs of the equipment precluded the systems from lighter trucks and cars. In the 1990s, a team of engineers working at EPA's National Vehicle and Fuel Emissions Laboratory succeeded in developing a revolutionary type of petro-hydraulic hybrid powertrain that would propel a typical American sedan car. The test car achieved over 80 mpg on combined EPA city/highway driving cycles. The petro-hydraulic hybrid system has faster and more efficient charge/discharge cycling than petro-electric hybrids and is also cheaper to build. The accumulator vessel size dictates total energy storage capacity and may require more space than an electric battery set. Any vehicle space consumed by a larger size of accumulator vessel may be offset by the need for a smaller sized charging engine, in HP and physical size.

C. **Plug-in Hybrid**

The principal difference between the previously described HEV variants and the PHEV is that the latter is fitted with a larger battery that can be charged from the electric utility grid ("plugged in") and that operates in a charge-depleting mode; that is, the state of charge of the battery is allowed to vary over a much larger range, 50 percent being typically proposed. The significant fuel consumption benefit is obtained during urban driving when the vehicle can be driven on electric power only. Once the all-electric range has been achieved and the battery discharged to its lowest allowable state of charge, the vehicle is operated in the charge-sustaining mode and differs little from the HEV. While the micro and ISG hybrids offer some improvement in fuel consumption for a relatively modest cost, it is the power-split HEV and PHEV architectures that promise a significant improvement. The PHEV also offers the long-term potential for displacing fossil fuels with other primary energy sources such as nuclear or renewable sources of electricity, depending on the fuel source of the electric grid from which the PHEV draws electricity [2, 8, 13, 20, 24]. The Energy Independence and Security Act of 2007 defines a plug-in hybrid as a light-, medium-, or heavy-duty vehicle that draws motive power from a battery with a capacity of at least 4 kilowatt-hours and can be recharged from an external source of electricity. The plug-in hybrid is usually a general fuel-electric (parallel or serial) hybrid with increased energy storage capacity, usually through a lithium-ion battery, which allows the vehicle to drive on an all-electric mode a distance that depends on the battery size and its mechanical layout (series or parallel).

D. **Hybrid Fuel Cell Vehicle**

Fuel cell vehicles have the potential to significantly reduce greenhouse gas emissions (depending on how hydrogen is produced and what raw materials are used to produced hydrogen) as well as U.S. dependence on imported oil over the long term. A recent report [22, 30] states that if issues of cost, durability and hydrogen infrastructure are resolved, fuel cell vehicles have enormous potential. It is already used in forklift and some larger trucks. Its use in conventional automobiles will take some time. However, during last few years significant progress has been made.

While hydrogen required for fuel cell can be generated within car, most present cars are designed to store hydrogen and replenish the storage tank from hydrogen filling stations. This will require significant investment in hydrogen filling infrastructure. Hydrogen can also be generated from methanol or natural gas within the car. While the development effort is moving at a rapid rate, it will be several years before significant numbers of fuel cell car will be on the road. While fuel cell car has high energy density, it will require a source of high power density. Thus, like all HEV and EV, fuel cell car will also be hybrid [22, 30].

The development of fuel cell vehicles at all sizes at commercial scale still requires more progress on the following issues:

1. Higher cost of fuel cells compared to other energy converters,
2. Lack of a hydrogen distribution infrastructure,
3. Need for a low carbon source of hydrogen (biomass or water electrolysis using electricity produced with low emissions),
4. Need to demonstrate acceptable durability and reliability, and weight and volume of an on-board hydrogen storage tank sized for a range of 300–400 miles.
5. High cost per kWh power production. The target is $2.0/kWh.

E. **Other forms of Hybrid Vehicles**

Hybrid vehicles are further differentiated by the relative sizes of the IC engine, battery, and motor. In all cases, however, an economically and functionally significant component of the system is the power electronic subsystem necessary to control the electrical part of the drive train. The hybridization of diesel (compression ignition; CI) vehicles is expected to have somewhat lower efficiency benefits than hybridization of gasoline vehicles, in part because conventional CI vehicles already exhibit lower fuel consumption than comparable gasoline vehicles. Further, CI vehicles also have very low fuel consumption at idle, making the benefits of idle-stop less attractive. Conventional CI power trains are more expensive than their gasoline counterparts, which, when added to the cost of hybridization, makes a CI hybrid power train very expensive for the additional fuel consumption reductions provided over and above just moving to a hybrid or CI power train alone. As a result, it is unlikely that original equipment manufacturers (OEMs) will offer a wide array of CI hybrids. The most likely levels of CI hybridization will be idle-stop and, perhaps, some mild hybrids [2, 6, 8].

F. **Hybrid Power from Waste Heat of Vehicle Exhaust Gases**

Just like hybrid power can be generated using waste heat from various static sources by thermoelectricity (as shown in the previous chapter), it can also be obtained from

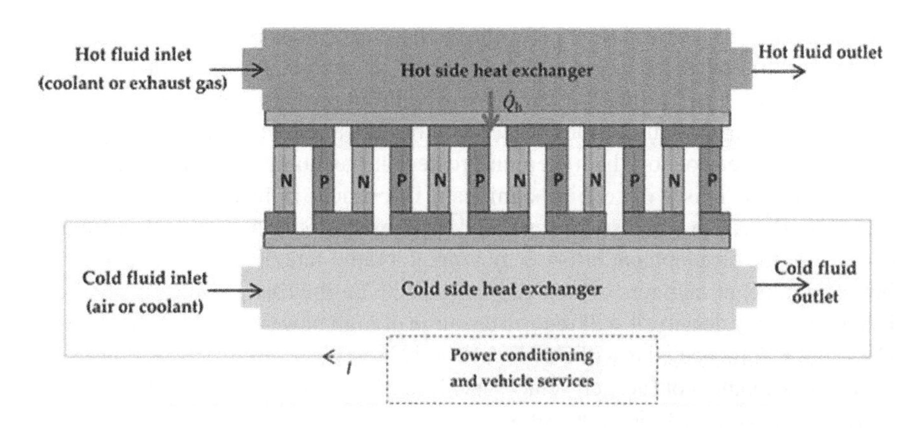

FIGURE 3.4 The main components of an automotive TEG system [32].

vehicle industry. Here we briefly examined power generation from various vehicle industries using thermoelectricity [31–46].

F-1 **Road Applications**

The automotive industry is considered as the most attractive sector in which TEGs (thermoelectric generators) are used to recover the lost heat (see Figure 3.4) [32]. Various leading automobile manufacturers develop TEGs for waste recovery to reduce the costs of the fuel for their vehicles [31–46]. It has been demonstrated that vehicles (the gasoline vehicle and hybrid electric vehicles) have inefficient internal combustion engines [32]. Fuel combustion is used in a proportion of 25% for vehicle operation; 30% is lost into the coolant and 40% is lost as waste heat with exhaust gases. In this case, the TEG technology could be an option to recuperate the waste heat energy for gasoline vehicles and hybrid electric vehicles. A significant power conversion could be achieved by combining cooling system losses with the heat recovery from automobile exhausts. The use of TEG systems with an energy conversion of 5% would raise the electrical energy in a vehicle by 6% (5% from exhaust gases and 1% from the cooling system) [31–37].

A TEG with ZT = 1.25 and efficiency of 10% can recover about 35–40% of the power from the exhaust gas where the power generated can help to increase the efficiency to up 16% [32]. The components where TEGs could be attached in a vehicle are the exhaust system and the radiators. In this case, the amount of waste heat is decreased and exhaust temperatures are reduced. These aspects require more efficiency from the TEG device. Furthermore, the design of such power conversion system takes into account various heat exchangers mounted on the TEG device. These systems have a lifecycle of 10–30 years and the materials accumulated on their surfaces from the exhaust gas, air, or coolant represent a major concern in order to not damage their proper operation [47]. Important testing is helpful to confirm the reliability of TEG systems in automotive applications. Furthermore, the design requires knowing the maximum electric output power and conversion efficiency from TEG systems [31–37].

The main components of the automotive TEG that considers waste heat like their energy source are one heat exchanger which takes heat from engine coolant and the exhaust gases and release it to the hot side of the TEG. The total TEG system involves heat exchanger which takes the heat from the TEG and releases it to the coolant or to the air; the electrical power conditioning and the interface unit to supply the electric output power of the TEG system to the automobile electric system. Supporting the TEG system are the secondary components (e.g., the electronic unit, the electric pump, sensors system, valves, fans, and so on) depending on the vehicle design and application type [31–37].

Since 1914 the possibility of using thermoelectric power generation to recover some of waste heat energy from reciprocating engines has been explored. In this exploration, the exhaust gases in the pipe provide the heat source to the thermoelectric power generator, whereas the heat sink (cold side) is suggested to be provided by circulation of cooling water. More recently, Taguchi [38] invented an exhaust gas-based thermoelectric power generator for an automobile application. In this patent, a pump supplies cooling water through each of the cooling water circulation paths. The cooling water circulation path includes a cooling water pipe arranged along the exhaust pipe to pass the cooling water. At stacks, a plurality of thermoelectric generation elements is attached to the exhaust pipe and the cooling water pipe successively in a direction from the upstream toward downstream of the exhaust gas. The cooling water pipe and the exhaust pipe pass the cooling water and the exhaust gas, respectively, in opposite directions so that the downstream stack has an increased difference in temperature between the exhaust pipe and the cooling water pipe, and the stacks provide power outputs having a reduced difference, and hence an increased total power output. This invention is proposed to provide increased thermoelectric conversion efficiency without complicated piping [38].

A comprehensive theoretical study concluded that a thermoelectric generator powered by exhaust heat could meet the electrical requirements of a medium sized automobile [31–37]. Among the established thermoelectric materials, those modules based on PbTe technology have been found the most suitable for converting waste heat energy from automobiles into electrical power [31–37]. Wide-scale applications of thermoelectrics in the automobile industry would lead to some reductions in fuel consumption, and thus environmental global warming, but this technology is not yet proven. Smith and Thornton [40] reported that TE device efficiencies are low (~5%); however, thin-film and quantum-well technologies offer the possibility of higher efficiency in the future (~10%–15%). In their study, four vehicle platforms were considered: a midsize car, a midsize sport utility vehicle, a Class 4 truck, and a Class 8 truck. A simple vehicle and engine waste heat model showed that the Class 8 truck presents the least challenging requirements for TE system efficiency, mass, and cost. This is because Class 8 trucks have a relatively large amount of exhaust waste heat, low mass sensitivity, and travel a high number of miles per year, all of which help to maximize fuel savings and economic benefits.

A driving and duty cycle analysis for the Class 8 truck elucidates trade-offs in system sizing and shows the strong sensitivity of waste heat, and thus TE system electrical output, to vehicle speed and driving cycle. It is not feasible for a TE system to replace the alternator, as too little waste heat is available during city driving and/or idling. Together with a typical alternator, a TE system could enable the electrification of 8%–15% of a Class 8 truck's accessories, providing 2%–3% fuel savings. Additional electrification would require a larger alternator and battery to augment the TE system so that adequate electrical power is available during low-speed driving and idling. Achieving an economic payback in three years dictates that the TE system cost less than roughly $450/kW, requiring an almost tenfold reduction from today's costs. Such a cost reduction might be enabled in the future by thin-film devices that use expensive TE junction materials more efficiently. NREL has also examined the use of carbon nanotubes for the TE generator (TEG) [39].

Thacher et al. [41] carried out the feasibility of the TEG system installed in the exhaust pipe in a light truck by connecting a series of 16 TEG modules. The experimental results showed good performance of the system at high speeds. Hsiao et al. [42] carried out an analytical and experimental assessment of the waste heat recovery system from an automobile engine. The results showed better performance by attaching TEGs to the exhaust pipe than to the radiators. Hsu et al. [43] introduced a heat exchanger with 8 TEGs and 8 air-cooled heat sink assemblies, obtaining a maximum power of 44 W. An application to recover waste heat has been developed by Hsu et al. [44], for a system consisting of 24 TEGs used to convert heat from the exhaust pipe of a vehicle to electrical energy. The results show a temperature increase at the hot side T_h from 323 to 403 K and a load resistance of 23–30 Ω to harvest the waste heat for the system. Tian et al. [45] theoretically analyzed the performance between a segmented TEG (Bi_2Te_3 used in low-temperature region and Skutterudite in high-temperature areas) used to recover exhaust waste heat from a diesel engine and traditional TEG. They found that a segmented TEG is suitable for large temperature difference and a high-temperature heat source, and has a higher potential for waste heat recovery compared to the traditional device. Meng et al. [46] addressed the automobile performance when applying TEG in exhaust waste heat recovery. The results showed that the effects of the different properties and the heat loss to the environmental gas on performance are considerable.

The conversion efficiency for the TEG system could be in the range of 5–10% [31–37]. The researchers' attention is focused on the development of new thermoelectric materials that offer improved energy conversion efficiency and a working temperature range more significant than for internal combustion engines. Even if the cost of the bismuth telluride is relatively high, the technical feasibility of TEGs for the automobile industry is widely demonstrated, making it very attractive. The goal of the manufacturers is to develop TEG systems with automated production and low-cost thermoelectric materials [31–37].

F-2. Air Applications

A considerable amount of heat is released into the atmosphere from air vehicles (turbine engines from helicopters and aircraft jet engines) [32]. To obtain a

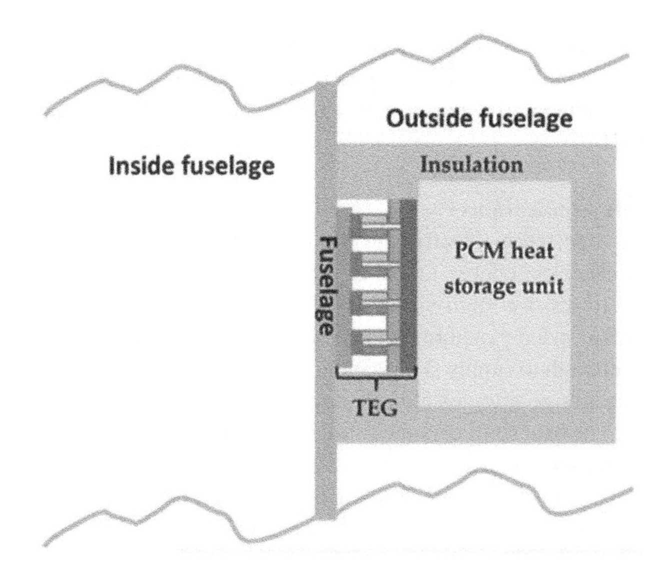

FIGURE 3.5 Schematic of the thermoelectric harvesting system in an aircraft [32].

significant reduction of the gas pollutant into the environment, it is necessary to achieve a significant reduction of electricity consumption and utilization of the available energy in these types of vehicles. Implicitly, this leads to the reduction of operating cost. To power these air vehicles, TEG systems are used on fixed-wing aircraft with a significant temperature difference of around 100°C across the TEG (see Figure 3.5).

TEG for energy harvesting uses the available temperature gradient and collects sufficient energy to power up an energy wireless sensor node (WSN) to be autonomous. This WSN is used for health monitoring systems (HMS) in an aircraft structure. The main components of a WSN are the energy source and the wireless sensor unit. An in-depth review of WSN mechanisms and applications is presented by Samson et al [48]. A TEG energy harvesting captures enough energy for a wireless sensor. One side of the TEG is fixed directly to the fuselage and the other side is attached to a phase-change material (PCM) heat storage unit to obtain a temperature difference during take-off and landing. PCM is considered an essential element for the heat storage unit because it can maximize the ΔT of the TEG system to solve the low TEG conversion efficiency. In this case, electrical energy is generated [32]. Water is an adequate PCM for heat storage. The temperature difference across the TEG is obtained from the slow changing temperature of the heat storage unit and the rapidly changing temperature of the aircraft fuselage. A lot of energy is produced during the PC, through latent heat. An application of Bi_2Te_3 modules on turbine nozzles has been addressed by Kousksou et al. [49]. Even though the electric power that can be harvested may be significant, the weight of the cold exchanger is still excessive for the specific application.

Future applications in aircraft may be envisioned in locations in which there are hot and cold heat flows, especially with the use of light thermoelectric materials. However, one of the main issues that remain is the weight of the heat exchangers [32].

F-3. Spacecraft Applications

The radioisotope thermoelectric generators (RTGs) are a solid and highly reliable source of electrical energy to power space vehicles being capable of operating in vacuum and to resist at high vibrations. RTGs are used to power space vehicles for distant NASA space expeditions (e.g., several years or several decades) where sunlight is not enough to supply solar panels. The natural radioactive decay of plutonium-238 releases huge amounts of heat, suitable for utilization in RTGs to convert it into electricity. The heat source temperature in this case is about 1000°C [32]. At this high temperature, semiconductor materials used in RTG can be silicon germanium (Si Ge), lead tin telluride (PbSnTe), tellurides of antimony, germanium, and silver (TAGS) and lead telluride (PbTe).

F-4. Marine Applications

Up to now, just a few surveys have been performed in the marine industry due to the lack of clear and stringent international rules at the global level. The marine transport has a significant influence on climate change because of a large amount of greenhouse gas emissions [32]. The naval transport generates a wide amount of waste heat, used to provide thermal energy onboard and seldom electrical energy. The heat sources on the marine vessels are the main engine, lubrication oil cooler, an electrical generating unit, generator and incinerators. The utilization of waste heat onboard is for heating heavy fuel oil and accommodation places, and for freshwater production. The main engine represents the principal source of waste heat. Board incinerators are used for burning the onboard waste instead of being thrown overboard to pollute sea water. The incinerators are the most favorable TEG systems due to the availability of their high-temperature differences. The specialists' attention is focused on the future design and optimization of high-power density TEGs for the marine environment, as well as on the development of hybrid thermoelectric ships considered as green platforms for assessing the efficiency of TEGs [32].

3.6 DEGREE OF HYBRIDIZATION

Hybrid vehicles can also be differentiated based on the degree of hybridization. Major breakdown of the degree of hybrid is illustrated in Figure 3.6 [2].

A. **Mild Hybrid**

Mild hybrid is a vehicle that cannot be driven solely on its electric motor, because the electric motor does not have enough power to propel the vehicle on its own. Mild hybrids include only some of the features found in hybrid technology, and

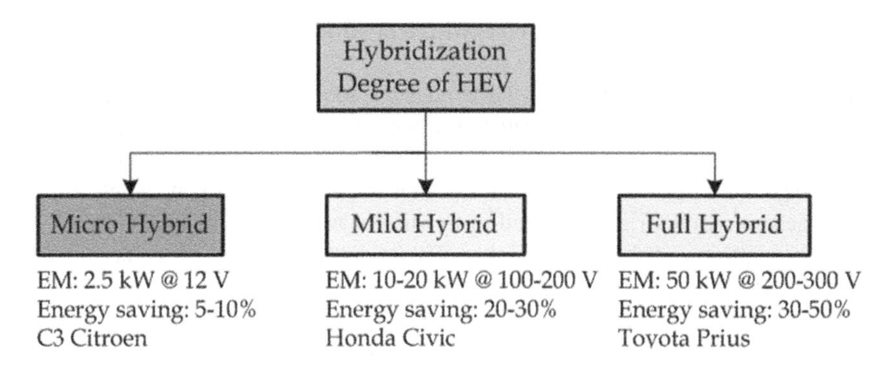

FIGURE 3.6 Degree of hybridization [2]

usually achieve limited fuel consumption savings, up to 15 percent in urban driving and 8–10 percent overall cycle. A mild hybrid is essentially a conventional vehicle with oversize starter motor, allowing the engine to be turned off whenever the car is coasting, braking, or stopped; also the car can restart quickly and efficiently. The motor is often mounted between the engine and transmission, taking the place of the torque converter, and is used to supply additional propulsion energy when accelerating. Accessories can continue to run on electrical power while the gasoline engine is off, and as in other hybrid designs, the motor is used for regenerative braking to recapture energy. As compared to full hybrids, mild hybrids have smaller batteries and a smaller, weaker motor/generator, which allows manufacturers to reduce cost and weight.

Honda's early hybrids, including the first generation Insight, used this design, leveraging their reputation for design of small, efficient gasoline engines; their system is dubbed Integrated Motor Assist (IMA). Starting with the 2006 Civic Hybrid, the IMA system now can propel the vehicle solely on electric power during medium speed cruising. Another example is the 2005–2007 Chevrolet Silverado Hybrid, a full-size pickup truck. Chevrolet was able to get a 10% improvement on the Silverado's fuel efficiency by shutting down and restarting the engine on demand and using regenerative braking. General Motors has also used its mild BAS Hybrid technology in other models such as the Saturn Vue Green Line, the Saturn Aura Greenline, and the Malibu Hybrid [2, 8].

B. **Mild Parallel Hybrid**

These types use a generally compact electric motor (usually <20 kW) to provide auto-stop/start features and to provide extra power assist during the acceleration, and to generate on the deceleration phase (aka regenerative braking). On-road examples include Honda Civic Hybrid, Honda Insight 2nd generation, Honda CR-Z, Honda Accord Hybrid, Mercedes Benz S400 BlueHYBRID, BMW 7 Series hybrids, General Motors BAS Hybrids, Suzuki S-Cross, Suzuki Wagon R and Smart fortwo with micro hybrid drive [2, 8].

C. **Belt-Driven Alternator/Starter (Micro or Mild Hybrid)**

In the belt-driven alternator/starter (BAS) design, sometimes known as a micro or mild hybrid, the starter and generator of a conventional vehicle are replaced by a single belt- or chain-driven larger machine, capable of both starting the engine and generating electric power. In some BAS designs, in addition to the new belt-driven starter generator, the original geared-to-flywheel starter is retained for cold starts. Fuel consumption is reduced by turning off and decoupling the engine at idle and during deceleration. In some designs, particularly those that have replaced the belt with a chain for increased torque transmission, both electric vehicle launch and some degree of braking energy regeneration are possible.

This mode of operation is known as idle-stop, and while not technically qualifying as a hybrid since the motor/generator provides no or little tractive power, it is included in this chapter for completeness. Idle-stop designs reduce fuel consumption by up to 6 percent in urban driving with SI engines [50]. For SI engines having variable valve timing to reduce inlet throttling loss the benefit may be less than 6 percent. For CI engines, the benefit of idle-stop drops to about 1 percent because CI engines are more efficient at idle due to their lack of inlet throttling.

The BAS design is not quite simple as it first appears. Maintaining hydraulic pressure in the automatic transmission is necessary for smooth and rapid restart, and safety issues related to unexpected restart must be considered. The company ZF has designed a transmission that provides a means of maintaining hydraulic pressure using a "hydraulic impulse storage device" that appears to address the transmission problem which is also addressed in existing designs by an electrically driven hydraulic pump [2, 8].

D. **Full Hybrid**

Full hybrid, sometimes also called a strong hybrid, is a vehicle that can run only on a combustion engine, an electric motor, or a combination of both. Ford's hybrid system, Toyota's Hybrid Synergy Drive, and General Motors/Chrysler's Two-Mode Hybrid technologies are full hybrid systems. The Toyota Prius, Ford Escape Hybrid, and Ford Fusion Hybrid are examples of full hybrids, as these cars can be moved forward on battery power alone. A large, high-capacity battery pack is needed for battery-only operation. These vehicles have a split power path allowing greater flexibility in the drivetrain by interconverting mechanical and electrical power, at some cost in complexity.

The full hybrid (HEV) has sufficient electrical energy storage and a powerful enough electric motor to provide significant electrical assist to the IC engine during acceleration and regeneration during braking. There are several architectural approaches to achieving a full hybrid, the three in current production being the integrated starter/generator (ISG) or integrated motor assist (IMA), the power split, and the two-mode. These are all parallel or power split designs. The HEV may also provide a limited electric-only range if the battery capacity and motor size are sufficient.

The ratio of electric to mechanical power provided for propulsion of an HEV varies with driving conditions and the state of charge of the battery. This operational feature is accomplished with sophisticated computer controls. Commercially available HEVs such as the Toyota Prius, Honda Civic, Nissan Altima, or Ford Escape can support a limited all-electric range at limited speeds. In these vehicles the battery is operated in a charge-sustaining (CS) mode; that is, the state of charge (SOC) of the battery is allowed to vary over a very narrow range, typically 15–20 percent, to ensure long battery life. The IC engine operates over a narrow speed/load range to improve efficiency, and regeneration is employed to recover braking energy. According to Toyota, the contributions of stop-start, regenerative braking, and engine modifications to fuel consumption improvements are approximately 5, 10, and 30 percent, respectively [2, 8].

E. **ISG/IMA Hybrid**

In the ISG/IMA design, the starter and generator are replaced by a larger electrical machine connecting the engine and transmission. These vehicles generally use a larger battery and a higher voltage (e.g., 140 V) than the BAS. Additionally, the motor/generator and battery are powerful enough to provide electrical launch from a stop and the ability to support some degree of electric-only travel. In its simplest form the ISG is mechanically fixed to the IC engine crankshaft, but in some designs a second clutch isolates the engine and the electrical machine to enable larger regeneration of braking energy. When incorporating an effective regenerative braking system, the ISG hybrid achieves a fuel consumption reduction of 34 percent in the combined driving cycle, as demonstrated by the Honda Civic. A part of the improved fuel consumption comes from vehicle modifications, including the use of a smaller, more efficient SI engine [2, 8]. A comparison of architecture, performance, and their applications for various hybrids is given in Tables 3.3 and 3.4.

3.7 POWER ELECTRONICS OF HYBRID VEHICLES

The term *power electronics* refers to the semi-conductor switches and their associated circuitry that are used to control the power supplied to the electrical machines or to charge the battery in an HEV or PHEV. For purposes of driving electric motors these circuits function as an inverter, changing the battery direct voltage into an alternating voltage of controlled amplitude and frequency. For charging the propulsion battery they function as a controlled rectifier, changing the ac voltage of the machine to the dc value required by the battery. The direction of power flow is either into or out of the battery, depending on vehicle mode of operation. Plug-in hybrids also require power electronic circuits to convert the ac main voltage to a precise dc voltage to charge the propulsion battery.

Power electronic circuits known as DC-DC converters change the propulsion battery dc voltage to the dc voltage appropriate to charging the accessory battery (i.e., the standard 12 V battery retained to power vehicle accessories). A DC-DC converter may also be used to increase system efficiency by stepping up the propulsion

TABLE 3.3
Summary of Architectures and their Application [2, 8]

Architecture	Loss	Efficiency	Complexity	Hybridization	Sizing of Computation time	Component math. complexities
Series	VH	L	L	Full HEV, PHEV	VH	L
Parallel	H	M	M	Micro, mild and Full HEV	H	M
Series-Parallel	M	H	H	Full HEV, PHEV	M	H
Complex or Complete Hybrid	L	VH	VH	Full HEV	M	VH

TABLE 3.4
Comparison of Various HEVs* [2, 8]

Type of HEV	Elec. Power for mild distance	Electric power for long distance(10-40miles)-Recharge on grid	Energy savings	Elec. Power(kW)	Car example
Micro and Micro-mild			5–10%, up to 25% in city traffic	1.5–10	PSA C2
Mild Hybrid			10–25%	5–20	Honda civic
Full Hybrid	X		25–40%	30–75	Toyota Prius
PHEV	X	X	50–100%	70–100	GM Volt
EV	X	X	100%	30–100	Nissan Leaf

* All HEVs have start/stop: Stop engine idle when a vehicle slows down and comes to a stop feature, regenerative braking, and additional electric power for few seconds.

battery voltage before it is supplied to the inverter. The latest Toyota Prius uses such a design. Both inverter and DC-DC converter technologies are well developed for industrial and other applications. The special problems for hybrid vehicles are cost, cooling, and packaging. Although the ambient environment for automotive electronics is much harsher than that in industrial or commercial applications, the cost in the automotive application is required to be lower. The significant improvement made in Toyota hybrid after 2005 is due in large measure to the increased switching frequency made possible by the higher-speed motor and higher voltage introduced in 2005. These changes reduce the physical size of magnetic components and improve the utilization of silicon devices. Both these consequences result in improved packaging density.

The inevitable global shift towards the EV & HEV development brought the number of challenges that were not applicable to a standard vehicle powered only by an internal combustion engine (ICE). Even in the HEV, where the ICE is still present, the entire system and its operation has gone through a dramatic change. Smooth torque transition among the ICE, E-motor/generator and the not-electrical part of the breaking system is essential. New failure modes, some of them potentially disastrous, have been introduced. Consequently a number of precautionary systems and strategies became inevitable [51].

The proper operation of the legacy low-voltage vehicle electrical system required the introduction of a DC-DC converter. This unit accepts a portion of the energy stored in the high voltage traction battery and converts it into low voltage required by the vehicle (typically 12 V). A low voltage battery for this system remains a necessity, thought its capacity may be reduced.

In most instances at least a small AC/DC converter/On-board charger is also required as a functional block of the EV or HEV. The environmental challenges typically dictate liquid cooling of the power electronics, E-motor(s) and if present also the ICE. In the case of HEV, sharing of the cooling system between the ICE and EV components of the vehicle is desirable and beneficial but challenging for the EV design. The heating and cooling environmental equipment has been replaced by a more efficient and smaller heat pump.

Additional power convertors are an essential part of the EV and HEV electrical system. The DC-DC converter, usually sized between 1.5 kW and 3.5 kW, performs the conversion from the HV energy storage to the 12 V legacy vehicle system. Resonant, zero-voltage switching (ZVS) converter is recommended for this application since it maintains high efficiency (typically 96–98%) for a wide range of loading. Because the DC-DC converter seldom operates at its nominal power, its operating regime is important for the economical use of the stored energy [52]. Cycling the converter is based on the condition of the 12 V battery and cycling two or more (smaller size) converters may serve better [51]. In order to enable recharging from a regular electrical outlet, the on-board charger (AC/DC) of typical size up to 3.5 kW is used. Due to the regulations, it must be a power factor-corrected (PFC) device. Even if the efficiency of the charger is not as critical as in the case the DC-DC, the ZVS topology is recommended for saving the energy. The charger efficiency is due to the PFC slightly lower, about 94%, if resonant conversion is used. Since the vehicle liquid cooling system is inactive during the charging process, it is recommended that the charger is air-cooled. Both converters are controlled from the central processor, usually using CAN buss.

The overall power electronics system for small to medium HEV contains a 12 V battery, 2–4 kW DC-DC connector, 3–6 kW charger, 40–120 kW inverter, a main controller, and an HMI. Also the 400 V battery is the energy storage of the entire system, it may be charged by the on board charger (AC/DC) using a regular outlet or larger fast charger, external to the vehicle. Battery disconnect is a safety device collocated with the battery. It connects or disconnects both, the positive as well as the negative poles of the battery from the rest of the system. Most vehicles activate ON device only during emergencies to prevent fire or explosion. Normally, this device is OFF and is controlled by the vehicle driver. The HV DC bus distributes the power to the inverter, DC-DC converter, and accepts the power from the AC/DC during the charging period. The low voltage battery and the entire 12 V legacy system are supplied by the DC-DC converter. This system, a low voltage system, also provides power to all the other components of the EV/HEV control system. The main controller along with HMI provides high-level control and supervision to all components of the system. Each individual Blok has its own controller that performs the local control and reports to the main controller. The HV DC bus uses the HV conductor, shielded to contain the radiated EMI. This bus is floating with respect to the chassis. A safety device similar to ground fault interrupt (GFI) is used to detect possible leakage current. If the fault is detected, the battery disconnect separates the battery from the rest of the electrical system. More details on overall power electronic system are given by Drobnik and Jain [51].

3.7.1 Hybrid Power Train Systems

Just like hybrid stationary power, hybrid vehicles use multiple sources of energy for power. Hybrid vehicles achieve reduced fuel consumption by incorporating in the drive train, in addition to an internal combustion (IC) engine, both an energy storage device and a means of converting the stored energy into mechanical motion. Some hybrids are also able to convert mechanical motion into stored energy. In its most general sense, the storage device can be a battery, flywheel, compressible fluid, elastomer, or ultra capacitor. The means of converting energy between storage and mechanical motion is through the use of one or more motors/generators (e.g., electric, pneumatic, hydraulic). In motor mode, these devices convert stored energy into mechanical motion to propel the vehicle, and in generator mode, these devices convert vehicle motion into stored energy by providing part of the vehicle braking function (regeneration). Similarly, a fuel cell vehicle is also a hybrid in which the internal combustion engine is replaced by the fuel cell, but this system will likely need supplemental energy storage to meet peak power demands and allow the fuel cell to be sized for the average power requirement. In this chapter, hybrid vehicle designs employing an internal combustion engine and battery-energy storage are considered. Battery electric and fuel cell vehicles (BEVs and FCVs) are also other alternative power trains.

Hybrid electric vehicles incorporate a battery, an electric motor, and an internal combustion engine in the drive train. In its most effective implementation this configuration permits the IC engine to shut down when the vehicle is decelerating and is stopped, permits braking energy to be recovered, and permits the IC engine to be downsized and operated at more efficient operating points. It should be emphasized that the benefits of hybrids are highly dependent on the drive cycle used to measure fuel consumption. For example, a design featuring only idle-stop operation, which shuts off the internal combustion engine when the vehicle is stopped, will demonstrate a large improvement on the city cycle portion of the Federal Test Procedure (FTP), where stop-start behaviors are simulated, but virtually no improvement on the highway cycle.

In addition to the introduction of an electric motor, hybrid designs may include the functions of idle-stop and regenerative braking, and the IC engine is frequently downsized from that in its equivalent conventional vehicle. For a hybrid vehicle, these operational and physical changes can result in an increase in fuel economy (mpg) of between 11 and 100 percent or a decrease in fuel consumption (gallons per 100 miles driven) of between 10 and 50 percent, depending on the vehicle class. Hybrid vehicles are the fastest-growing segment of the light-duty vehicle market, although they still make up less than 3 percent of the new car market in the United States.

Depending on the architectural configuration of the motors, generators, and engines, hybrid designs fall into four classes—series, parallel, mixed series/parallel, and full hybrid. The third design is commonly known as power split architecture. Schematics of these architectures are shown in Figures 3.7, 3.8 3.9, and 3.10, respectively. Within each class there are variations of implementation. Broadly defined, the series hybrid uses the internal combustion engine for the sole purpose of driving a generator to charge the battery and/or powering an electric drive motor. The electric motor provides all the tractive force. Energy flows from the IC engine through the generator and battery to the motor. In the parallel and mixed series/parallel designs, the IC engine not only charges the battery but is also mechanically connected to the wheels and, along with the electric motor, provides tractive power. Full hybrids combine all of these features.

A. Series Hybrid

A series- or serial-hybrid vehicle is driven by an electric motor, functioning as an electric vehicle while the battery pack energy supply is sufficient, with an engine tuned for running as a generator when the battery pack is insufficient (see Figure 3.7). There is typically no mechanical connection between the engine and the wheels, and the primary purpose of the range extender is to charge the battery. Because there is no mechanical connection between the IC engine and the wheels, the motor and the battery must be sized for the vehicle's full torque and power requirements. The

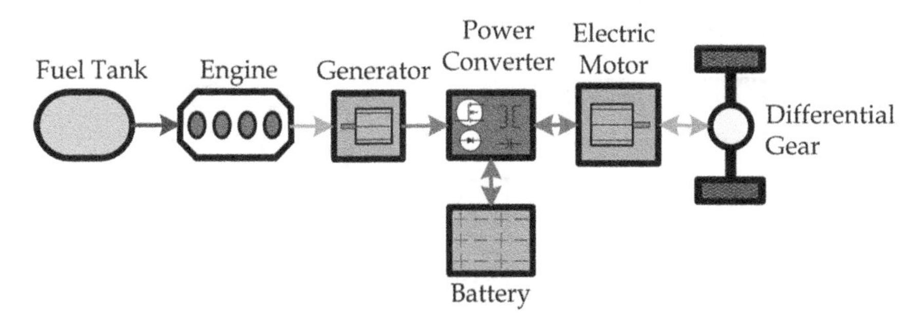

FIGURE 3.7 Powertrain for series hybrid [2]

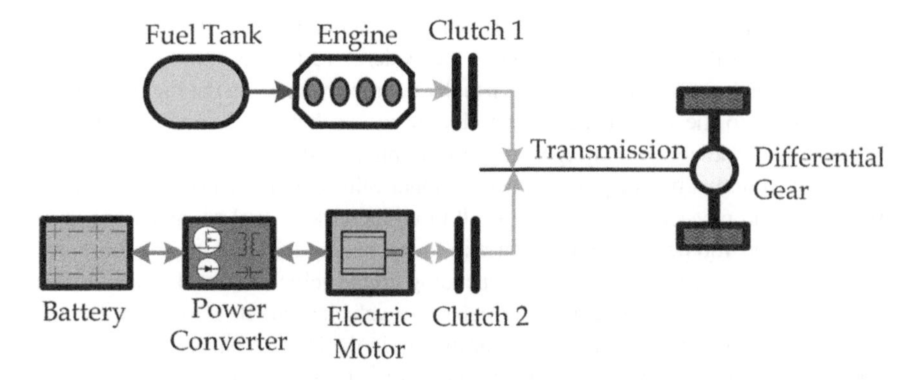

FIGURE 3.8 Powertrain for parallel hybrid [2]

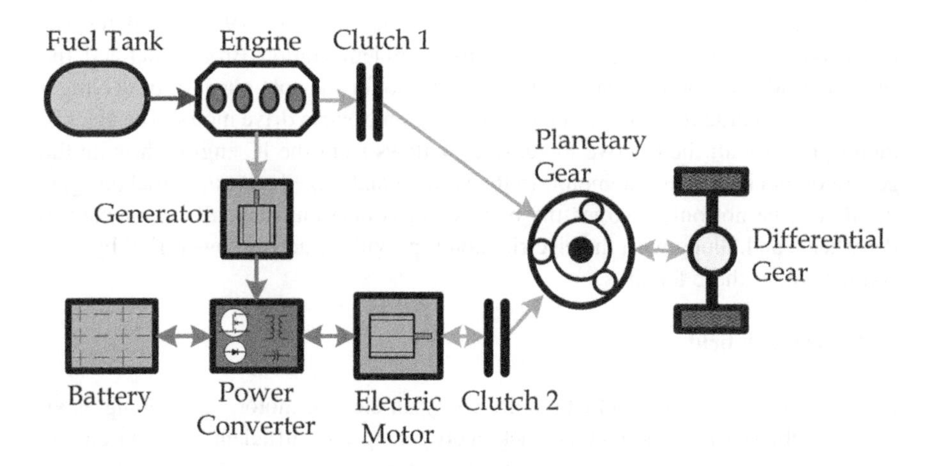

FIGURE 3.9 Powertrain for series-parallel-mixed hybrid [2]

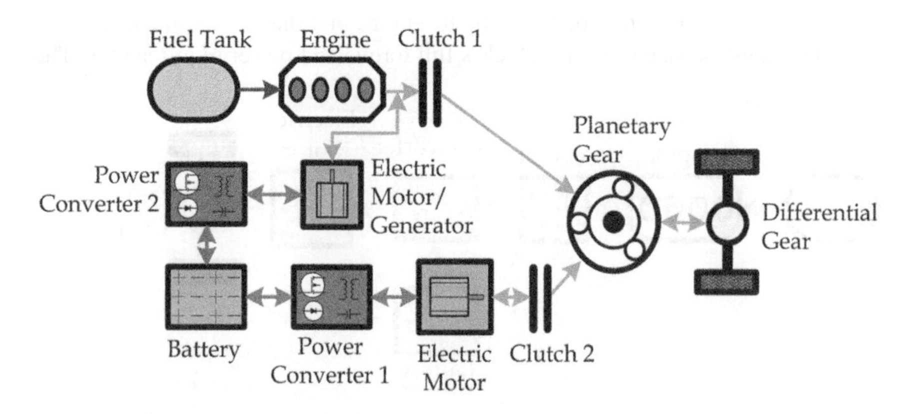

FIGURE 3.10 Powertrain for complete (full) hybrid [2]

advantages of this configuration are that a smaller engine can be used since it is not required to provide the power needed for acceleration, and the engine can be optimized with respect to fuel consumption. Series-hybrids have also been referred to as extended range electric vehicle, range-extended electric vehicle, or electric vehicle-extended range (EREV/REEV/EVER). In series hybrids, only the electric motor drives the drivetrain, and a smaller ICE (also called range extender) works as a generator to power the electric motor or to recharge the batteries. They also usually have a larger battery pack than parallel hybrids, making them more expensive. Once the batteries are low, the small combustion engine can generate power at its optimum settings at all times, making them more efficient in extensive city driving

The BMW i3 with Range Extender is a production series-hybrid vehicle. It operates as an electric vehicle until the battery charge is low, activates an engine-powered generator to maintain power, and is also available without the range extender. The Fisker Karma was the first series-hybrid production vehicle. When describing cars, the battery of a series-hybrid is usually charged by being plugged in—but a series-hybrid may also allow for a battery to only act as a buffer (and for regeneration purposes), and for the electric motor's power to be supplied constantly by a supporting engine. Series arrangements have been common in diesel-electric locomotives and ships. Ferdinand Porsche effectively invented this arrangement in speed-record-setting racing cars in the early 20th century, such as the Lohner-Porsche Mixte Hybrid. Porsche named his arrangement "System Mixt" and it was a wheel hub motor design, where each of the two front wheels was powered by a separate motor. This arrangement was sometimes referred to as an *electric transmission*, as the electric generator and driving motor replaced a mechanical transmission. The vehicle could not move unless the internal combustion engine was running.

In 1997 Toyota released the first series-hybrid bus sold in Japan. GM introduced the Chevy Volt series plug-in hybrid in 2010, aiming for an all-electric range of 40 mi (64 km), though this car also has a mechanical connection between the engine and drivetrain. Supercapacitors combined with a lithium ion battery bank have been used by AFS Trinity in a converted Saturn Vue SUV vehicle. Using supercapacitors they claim up to 150 mpg in a series-hybrid arrangement.

B. **Parallel Hybrid**

In a parallel hybrid vehicle an electric motor and an internal combustion engine are coupled such that they can power the vehicle either individually or together. Most commonly the internal combustion engine, the electric motor, and gear box are coupled by automatically controlled clutches (see Figure 3.8). For electric driving the clutch in between the internal combustion engine and motor is open while the clutch to the gear box is engaged. While in combustion mode the engine and motor run at the same speed.

Honda's Integrated Motor Assist (IMA) system as found in the Insight, Civic, Accord, as well as the GM Belted Alternator/Starter (BAS Hybrid) system, found in the Chevrolet Malibu hybrids, are examples of production parallel hybrids. The internal combustion engine of many parallel hybrids can also act as a generator for supplemental recharging. As of 2013, commercialized parallel hybrids use a full size combustion engine with a single, small (<20 kW) electric motor, and a small battery pack as the electric motor is designed to supplement the main engine, not to be the sole source of motive power from launch. But after 2015 parallel hybrids with over 50 kW are available, enabling electric driving at moderate acceleration. Parallel hybrids are more efficient than comparable non-hybrid vehicles especially during urban stop-and-go conditions where the electric motor is permitted to contribute and during highway operations. The first mass production parallel hybrid sold outside Japan was the 1st generation Honda Insight.

C. Power-Split or Series-Parallel Hybrid

In a power-split hybrid electric drive train there are two motors: a traction electric motor and an internal combustion engine. The power from these two motors can be shared to drive the wheels via a power split device, which is a simple planetary gear set (see Figure 3.9). The ratio can be from 100% for the combustion engine to 100% for the traction electric motor, or anything in between, such as 40% for the electric motor and 60% for the combustion engine. The combustion engine can act as a generator charging the batteries. Power-split hybrids have the benefits of a combination of series and parallel characteristics. As a result, they are more efficient overall, because series hybrids tend to be more efficient at lower speeds and parallel at high speeds; however, the cost of power-split hybrid is higher than a pure parallel. Examples of power-split (referred to by some as "series-parallel") hybrid powertrains include 2007 models of Ford, General Motors, Lexus, Nissan, and Toyota. In each of the hybrids above it is common to use regenerative braking to recharge the batteries [2, 8].

Modern versions such as the Toyota Hybrid Synergy Drive have a second electric motor/generator connected to the planetary gear. In cooperation with the traction motor/generator and the power-split device this provides a continuously variable transmission. On the open road, the primary power source is the internal combustion engine. When maximum power is required, for example to overtake, the traction electric motor is used to assist. This increases the available power for a short period, giving the effect of having a larger engine than actually installed. In most applications, the combustion engine is switched off when the car is slow or stationary thereby reducing curbside emissions. Passenger car installations include Toyota Prius, Ford Escape, and Fusion, as well as Lexus RX400h, RX450h, GS450h, LS600h, and CT200h.

The power-split hybrid design, typified by the Toyota Prius, the Ford Escape, and the Nissan Altima, incorporates a differential gear set that connects together the IC engine, an electrical generator, and the drive shaft. The drive shaft is also connected to an electric motor. This mechanical configuration incorporating the addition of a generator provides the flexibility of several operational modes. In particular the wheels can be driven by both the IC engine and the electric motor, with the motor's power coming from the generator, not the battery. The car is thus driven in both series and parallel modes simultaneously, which is not a possible mode for the

ISG design. This operational mode allows the IC engine operation to be optimized for maximum reduction in fuel consumption. The vehicles that use this power split design show a range of fuel consumption reduction from 10 to 50 percent. The low end of this range is the Toyota Lexus, the design of which is optimized for performance, not low fuel consumption. In Chapter 9, where the committee estimates fuel consumption benefits for vehicle classes, the Lexus is not used in the range of benefits for the power split design. This gives the fuel consumption benefits from the power split design a range of 24–50 percent.

General Motors (GM) is working with BMW and Chrysler on a different split hybrid architecture that uses the so-called two-mode system [2, 8]. This also splits the power flow from the engine but uses more clutches and gears to match the load to the drive and minimize electrical losses. The claim is that by using multiple gears the drive is more efficient in real-world driving situations and reduces fuel consumption when towing a trailer or driving at high speed. Toyota is using a similar approach with one or two gears in its latest hybrid systems. The fuel consumption reduction for the two-mode power split design, characterized by the Chevrolet Tahoe and Saturn Vue, ranges from 25 to 29 percent. However, the committee thinks that other implementations of the two-mode system could provide a maximum fuel consumption benefit of about 45 percent.

A complete hybrid combines all the features mentioned above (see Figure 3.10).

3.7.2 Architecture of HEV

The key components in an HEV consist of an electric motor (EM), battery, convertor, ICE, fuel tank, and control board. These components can be categorized into three groups [6]: (a) Drivetrains, which physically integrate the ICE power source and electric drive; (b) Battery/energy storage system (ESS), which emphasizes large or modest energy storage and power capabilities; and (c) Control system, which instructs electric systems/ICE and manages the HESS. These components can be integrated in different ways and sizes which results in variation in vehicle design. As described in detail earlier, based on the component integration, drivetrains mainly include series, parallel, and power split designs. Jennings et al. [53] classified the HEV's architecture into six different categories, which are mild/microparallel, parallel, series, power split, combined, and through-the-road (TTR) hybrids.

In series HEV, the power sources provide electrical energy at DC bus, which is then converted to traction power. In parallel HEVs, traction power can be supplied by ICE or EM alone, or together by both the sources. The EM is used to charge the HESS by means of regenerative braking. The parallel mild HEV is an ideal option as they provide a prime trade-off between the cost of vehicle and its performance. Complex HEVs incorporate features of both parallel as well as series architecture. They are almost like the series-parallel hybrid except for the variance in power flow of the motor, which is bidirectional in complex hybrid and unidirectional in series-parallel HEVs. The disadvantage of complex hybrid is its complexity in design [6].

Architecturally, PHEV is similar to HEVs except for a large-sized onboard battery, having high energy density and efficiency. The combination of CS and CD modes requires a more complex control strategy than in an HEV. PHEVs begin operation in CD mode, and as soon as the battery reaches a threshold value of SOC, the battery shifts to CS mode until the vehicle is parked and recharged. The architecture of a solar-driven HEV (PVHEV) is similar to the PHEV except for an additional photovoltaic (PV) panel, which charges the battery during a sunny day. To extract the maximum power from PV panels, the maximum power point tracker (MPPT) algorithms are applied.

A. **Patterns of Power Flow**:

HEV contains two powertrains. In Powertrain 1, powerflow is unidirectional, while in Powertrain 2, powerflow is bidirectional. There are nine patterns of power flows in HEVs to encounter the load demand [6]: (a) powertrain 1 alone delivers power to load; (b) powertrain 2 alone delivers power to the load; (c) both powertrains (1 and 2) deliver power to load at the same time;(d) powertrain 2 obtains power from load (regenerative braking); (e) powertrain 2 obtains power from powertrain 1;(f) powertrain 2 obtains power from powertrain 1 and load at the same time; (g) powertrain 1 delivers power to load and to powertrain 2 at the same time; (h) powertrain 1 delivers power to powertrain 2, and powertrain 2 delivers power to load; and (i) powertrain 1 delivers power to load, and load delivers power to powertrain 2.

Various papers have been published by researchers on the architecture of hybrid vehicles, and they are well reviewed by Singh et al. [6]. The published research includes the addition of a small ICE/generator to the battery-powered EV to develop a series-hybrid drivetrain. The series-hybrid drivetrain has some prominent benefits: (1) the ICE and the driven wheels are not coupled mechanically, which compels ICE to operate at its narrow optimal region; (2) single-torque source operation simplifies the speed control; (3) the torque-speed characteristic of EM obviates multigear transmission; and (4) easy drivetrain control, simple structure, and easy packaging. However, it suffers from some drawbacks. They are as follows: (1) the conversion of energy takes place in two steps, i.e., mechanical to electrical through generator and vice versa through motor, and hence results in more energy losses; (2) two electric machines are required, i.e., generator and motor separately; and (3) a big-size traction motor is required. The series hybrid drivetrain is mostly used in heavy vehicles such as buses, trucks, and military vehicles. This configuration was also considered for the analysis of hybrid lithium high-energy battery. The literature also includes simulation of the PHEV for series and parallel architectures and concludes that during powering mode, the operating points of the motor for parallel PHEV are more concentrated in the extended high-speed, high-efficient region. In case of parallel PHEV, regenerative braking takes place in a high efficiency region which is not the case in the series counterpart. While analyzing the effect of different functions on energy management strategies apropos of both architectures, the parallel configuration was found to be superior [6].

The literature also considers power split for the analysis of the through-the-road (TTR) architecture, a subcategory of parallel architecture. The split-parallel architecture for TTR and its control challenges are detailed in the work by Zulkifli et al. [54]. Miller [55] employed power split architecture as it provides better liberty of power control. He also showed that electronic continuous variable transmission (e-CVT) was more efficient than mechanical continuous variable transmission (CVT). Cheng and Dong [56] simulated a model of the power split PHEV powertrain and a TTR hybrid electric powertrain and their prototypes were investigated based on various parameters. A quasi-static model was used to investigate and evaluate vehicle performance, fuel economy, emissions, and supervisory control of the passenger car. A low-frequency vehicle powertrain dynamics model was used to evaluate the vehicle dynamics, acceleration, and braking performance of a racecar. Several other aspects of patterns of power flow are reviewed in detail by Singh et al. [6].

A summary of various architectures and their application is given in Table 3.3. Table 3.4 provides a brief summary of comparison of various HEVs/EVs. There are various architectures available for HEVs, but since the complex hybrid involves bidirectional power flow, it is more suitable and beneficial compared to the rest. Various types of power conversions for EV/HEV are illustrated in Figure 3.11 [2].

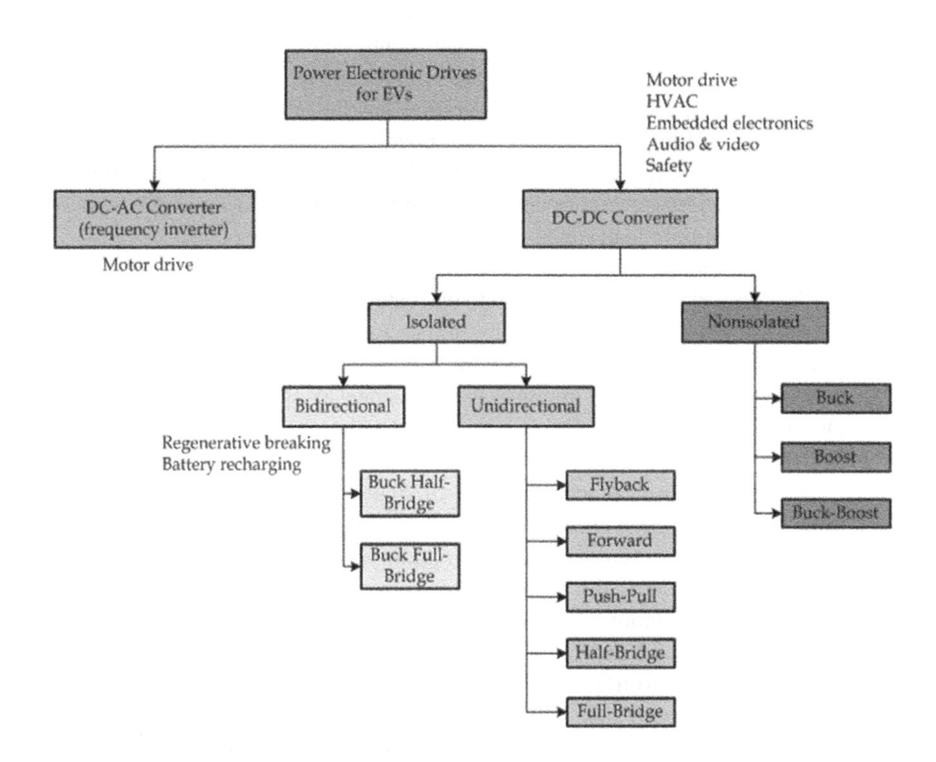

FIGURE 3.11 Power conversions for EV [2]

B. **Bidirectional DC-AC Converter**

Power converters are proliferated in all kinds of applications to increase controllability and efficiency in automotive applications. As pointed out by Singh et al. [6], the bidirectional converter is essential in hybrid vehicles to convert DC from the battery/UC/fuel cell (FC) or their combination into AC that is given to the motor drive. An extensive research has been carried out on DC-AC converters including single-stage single-phase, single-stage three-phase or zero voltage switching inverters [6]. Various motor drives used in EVs and HEVs have also been proposed [6]. There are various topologies of traction inverters such as voltage source inverter (VSI), current source inverter (CSI), impedance source converter (ZSI), and soft switching [6]. These are all reviewed in detail by Singh et al. [6]. Here we present a brief summary of their review.

1. **Current Source Inverter**

   The CSI can be used for the speed control of AC motors, especially induction motors with varying load torque. There are three types of CSIs: (a) single-phase CSI, (b) auto-sequential commutated mode single-phase inverter (ASCI), and (c) three-phase CSI. CSIs offer several advantages. The circuit for CSI is simple. It uses only converter-grade thyristors having reverse-blocking capability and is able to withstand high-voltage spikes during commutation. An output short-circuit or simultaneous conduction in an inverter arm is controlled by the 'controlled current source' i.e., a current-limited voltage source in series with a large inductance. Finally, the converter/inverter combined configuration has an inherent four-quadrant operation capability without any extra power component. CSIs also have some drawbacks. It suffers from the drawback of having limited operating frequency and hence cannot be used for uninterruptible power supply systems. Also, at light loads and high frequency, these inverters have sluggish performance and stability problems [6].

   These inverters can be divided into two categories: one is to reduce switch count, and another is to reduce capacitance for HEVs. The reduced switch scheme faces the challenge of cost and efficiency, whereas reduced capacitance reduces the cost and improves the power density of the traction inverter. Different types and use of CSI are articulated in the review of Singh et al. [6].
2. **Voltage Source Inverter**

   VSIs can be used practically in both single- and three-phase applications. VSIs have good speed range, multiple motor controls from a single unit, and a simple regulator design. In addition, there are some disadvantages of the VSI; e.g., the power factor decreases as the speed decreases, and it induces harmonics, cogging, and jerky start and stop motions. There are three types of VSIs: (a) single-phase half-bridge inverter, (b) single-phase full-bridge inverter and (c) three-phase VSI.

   VSIs offer several advantages. It has low-power consumption and high-energy efficiency up to 90%. It has high-power handling capability and no temperature variation- and aging-caused drifting or degradation in linearity.

It is easy to implement and control and it is compatible with today's digital controller. There are several drawbacks of VSIs. It has attenuation of the fundamental component of the waveform. It drastically increases switching frequencies and hence creates stresses on switching devices and it generates harmonic components. Several other aspects of the use of VSI in EV are discussed in detail by Singh et al. [6].

3. **Impedance Source Inverter**

Impendence source inverter (also called Z-source inverter, ZSI for short) has been considered an efficient candidate in vehicle applications such as drive system reliability, which leads to an increase in the range of inverter output. The ZSI is one of the most promising power electronics converter topologies suitable for motor drive applications [6]. It has the properties of buck and boost in single-stage conversion. A special Z network composed of two capacitors and two inductors connected to the well-known three-phase inverter bridge allows working in buck or boost mode using the shoot-through state. The ZSI improves the stability and safety of a brushless DC (BLDC) motor drive system under complex conditions. As pointed out by Singh et al. [6], ZSI offers several advantages. It provides the desired AC voltage output regardless of the input voltage and yields high-voltage utility factor. It overcomes voltage sags without any additional circuits and minimizes the motor ratings to deliver the required power. Finally, it improves the power factor and reduces harmonic current and common mode voltage of the line.

There are three topologies of ZSIs, namely basic, bidirectional, and high-performance ZSIs. Replacement of the input diode by a bidirectional switch in the basic version results in bidirectional ZSI topology. The bidirectional ZSI is able to exchange energy between AC and DC energy storage. Being a basic topology, the variable frequency (VF) ZSI cannot work in regenerative mode and hence cannot charge the battery due to which the output voltage is low. However, the continuous input current by VF-ZSI is suitable for photovoltaic (PV) applications. To perform rectangular wave modulation for motor drive control, an improved circuit topology of ZSI depending on the drive condition of the HEVs/EVs has been discussed in the literature [6]. Cao et al. [57] have developed current-fed quasi-ZSI (CF-qZSI) with high efficiency by using reverse-blocking IGBT for HEV applications. CF-qZSI is able to achieve bidirectional power flow and voltage buck operation as it has a diode and an LC network in its design. Another bidirectional DC-AC converter has been proposed for HEV applications, which can be used as a single input single/multiple output. Other aspects of ZSI and its applicability to EV/HEV are reviewed in detail by Singh et al. [6].

4. **Three-Phase Bridge and Controller**

The most common topology of the EV and HEV inverter is the three-phase bridge [47]. It uses IGBTs with anti-parallel diodes as a switch. There are topologies specific to improving quality of waveform (THD), efficiency, and reliability of the traction system. These include multi-phase (more than 3-phase), multi-level (more than two voltage levels), or resonant (using resonance to limit switching losses) topologies. These enhanced topologies are more complex and require many more components [58]. It is a cost adder, hard to justify in the extremely cost-sensitive automotive arena. Nevertheless it may be an excellent option for racing or military vehicle.

The semiconductor switch is typically implemented by a silicon-based IGBT and diode. When new materials such as SiC or GaN become more rugged and affordable the silicon will be undoubtfuly replaced. The principal of operation of the standard six-switch topology with two-level modulation is simple. The semiconductor switches are turned on and a sine wave of the desired frequency is approximated by using pulse-width modulation (PWM). The saw-tooth wave form is compared with a reference sine wave, resulting in the PWM that is applied to the drivers and the IGBTs. In order to generate sine waves with acceptable distortion, the switching frequency of the individual switches must be substantially higher than the desired output frequency [59]; typical multiplier is 4 xs to 12 xs. Today more sophisticated digital principals are used in generating the PWM signal. Often it is beneficial to vary the switching frequency with load to optimize the switching loss.

Since the IGBT is a unidirectional device an anti-parallel diode is used to enable current conduction in the opposite direction. The state of each IGBT is controlled by an IGBT driver. There is a benefit to accompany the standard inverter with a DC-DC converter, typically in boost topology. The converter conditions the voltage from energy source(s) before it is applied to the inverter. The advantages include: enabling of high-resolution PWM, reduction of the torque ripple, elimination of acoustic noise, better utilization of the inverter semiconductors, and finally higher energy density and added safety (should the voltage at the inverter DC terminals be higher than the DC bus voltage, the DC-DC converter prevents the current flow).

The historical purpose of the driver is to modify sufficiently the incoming PWM signal and to apply it to the IGBT gate. Since the emitter of the high side (HS) IGBTs are referenced to each phase, which is constantly moving between the negative and the positive rail, HS drivers must not only have functional isolation but also safety isolation. The same is true about the power supply passing on to the housekeeping power to the secondary site of those drivers. Even if not functionally necessary, often the lover side (LS) switches are isolated. Isolation of the LS devices improves the signal integrity, safety, and electromagnetic emission (EMI). The PWM signal generated by the controller is brought to the primary side of the isolator. The secondary side is referenced to the emitter of each IGBT and connected to the gate. Typically a voltage of 15 is used to turn ON the IGBT and a voltage of -8 is used to turn OFF the IGBT. Some manufacturers reduced the negative voltage to -5 V. Besides the ON/OFF control of the IGBT the driver may accommodate additional protection, safety, and reporting functions. The switching time can be manipulated by controlling the gate voltage and current obtained.

The controller is the brain of the inverter, typically implemented by a microcontroller unit (MCU). It accepts control commands from the driver in the form of throttle and breaks along with other inputs and turns them into organized, conditioned PWM signals for each IGBT driver. There are several control strategies that perform the vehicle motor control. Field Oriented Control (FOC) is widely used. FOC provides independent control of the torque and magnetic flux [60]. FOC may use sensors such as encoder or resolver to determine the exact position and speed of the rotor or be sensorless. Functional safety is extremely important for the road vehicle and must be built into the control system architecture. The housekeeping power supply for the controller, sensors, and gate driver is typically provided by the 12 V legacy vehicle system. This is one of the weak spots of the system [51].

3.7.3 Economics and Market for Power Electronics

Power electronics are a key technology for hybrids and represent 20% of the material costs. It is even bigger for EV cars. As mentioned above, HEV/EV power devices are used in DC-DC converters and DC-AC inverters. There are various configurations depending on the hybrid version and car makers' choices. Inverters are roughly the same for full hybrid, plug in hybrid, and EV cars with an average power of 50 kW. This application alone represents 74% of the total power module market for HEV and EV cars in 2009. IGBT is the device of choice for such high-power applications and represents 80% of the total HEV/EV power module market. Standard voltage of IGBT devices is 650 V but there is a trend to increase it. It is still unknown if it will be 700/800 V or directly 1.2kV which is already a standard.

The HEV/EV power module market stood at $300M in 2009 and is expected to grow strongly until 2020 at a growth rate close to 30% to reach $5B in 2020. Today, the power module market is mainly dominated by Toyota, which manufactures the module internally. With the near-universal involvement of other car makers, semiconductor companies (Infineon, Fuji, Mitsubishi, STM) will enter the market and will take a big market share in the power device pie [61].

The cost of power module depends on the type of hybrid car. For micro HEV, start/stop module plus DC-DC booster option runs less than $50 which is the total cost for power module. For mild HEV, the DC-AC inverter plus DC-DC booster option runs greater than about $100 which is its total cost. The same device for full HEV, PHEV, and EV runs greater than $400. Battery charger for PHEV and EV cost greater than $200. For all hybrid models DC-DC converter at 14 V cost less than $100. The total power module average price content/car for full HEV can be greater than $450 and for PHEV and EV, greater than $600. Standard ICE power device applications are not considered (oil pump, steering, braking, and HVAC). Auxiliaries' inverters have not been considered due to low amount of power devices. As HEV and EV remain expensive, car makers and tier-one suppliers want to cut the cost. Power modules represent about 50% of the inverter and converter cost so power module cost reduction is the main goal of all the market players. It is expected that the power module average cost will be reduced by more than 25% in the coming years.

Up to now, Toyota was dominating the HEV market and power module value chain. With the market growth and arrival of many players at the different levels (car makers, tier-one suppliers, and semi-conductor companies), the landscape will change drastically. Automotive tier-one suppliers invest heavily in HEV/EV powertrain and will play an important role in the HEV/EV power device value chain: Bosch, Continental, Valeo, Delphi, Denso, and Hitachi. They have the knowledge of specific automotive requirements that are very stringent for power devices. Some of them design the power modules themselves to cut the cost. At the same time, semiconductor companies try to climb the value chain by developing new power modules. Hence, it will be a hard time in the next years for power module manufacturers to find a significant place on the HEV/EV market.

Several companies (Mitsubishi Rohm, Toyota) have developed inverter prototype based on SiC diodes and switches that show significant size reduction with silicon devices. The SiC has clear advantages for HEV/EV applications (better power density, less losses, and higher operating temperature) but cost pressure for automotive industry is a big challenge. To succeed, the availability of SiC switches is paramount because it would allow the reduction of the cooling systems cost. At the same time, the cost of SiC devices would need to be significantly reduced and the passive components and packaging need to be adapted to support high operating temperatures. If the cost of SiC devices can be reduced, then SiCs may be an option for HEVs and EVs. It may be introduced first in EV applications that are more sensitive to losses to gain distance range. GaN is another possible option because of its better performance/cost ratio compared to SiC. Toyota and many other companies evaluate this solution and consider that if SiC cost can't be reduce, it would be an affordable substrate especially for inverter application that is very cost sensitive [61].

Together with the evolution of the electrified car, the underlying power electronics is evolving. Different trends have been observed, impacting the power electronic technologies as well as the power electronic supply chain. The growing demand is causing supply issues, and the huge volume of power electronic devices required for EV/HEV can impact the technology choice and also the adoption of innovative technologies (SiC transistors and innovative substrates). Electric cars are becoming bigger and more powerful, bringing specific requirements to the traction inverters. High-power inverters are typically based on high-end semiconductor dies, power module designs, and packaging solutions. These inverters differ on power ranges and sizes, built specially for Mild Hybrid to Battery Electric vehicles. In this large possible market, there is a place for *IGBT* and MOSFET modules, as well as for other components in some cases. Each player will choose the optimal solution for its inverter. SiC technology offers higher device efficiency compared to silicon devices. SiC transistors have already been used in EV/HEV onboard chargers and DC-DC converters and, since the introduction of Tesla Model 3 to the market, also in the traction inverter [62].

Specific EV/HEV challenges include frequent thermal cycling and call for using advanced materials such as Si3N4 AMB substrate in the high-power modules. Silver sintering is increasingly being accepted as a suitable substitute for the more conventional soldering die attach method. Double-side cooling and integrated substrates are also proposed solutions to increase the thermal dissipation of high-power-density modules. In 2018, the power module market for EV/HEV already accounted for 23.7 percent of the total power module market and it is expected to increase with a 14.4 percent CAGR 2017–2023 [61, 62].

Bosch, Valéo, Schaeffler, Continental, and other automotive tier-one companies develop and commercially offer an integrated solution for EV/HEV traction, the so-called electric axle. The electric axle (e-axle) is a compact, integrated product which includes electric motors, gears, and power electronics. An e-axle eases the electrification of conventional vehicles and associates the optimized performance and new functionalities with a compact design. The e-axle represents also an excellent opportunity for Tierl companies to increase their margins and market shares by offering a more complete solution for car electrification to automotive OEMs. This high level of integration brings new challenges concerning especially the form factor of different sub-systems and their thermal management. Specific products and technology solutions are required to fully optimize the e-axle on the system level in terms of efficiency, power density, cost, and reliability [61, 62].

The general trend toward higher power in cars would naturally lead to a preference for high-end materials. However, another trend observed in the automotive industry can partially change this picture—the use of more motors per vehicle. Instead of one high-power motor (and associated high-power inverter), some car manufacturers have developed electrified vehicles with two motors: one motor can be used to power the front axle and the second one the rear axle. This enables four-wheel driving operation and the motor and inverter can be used in the optimal operation mode with the highest efficiency. For example, instead of one 100 kW inverter, one can use one 80 kW and one 20 kW inverter. In the 80 kW inverter one can use high-end die (high-end silicon or SiC) and packaging (power module with Si3N4, sintering die attach, double side cooling, etc.). The focus will be on high performance and reliability. In the 20 kW inverter, which is of low power and has low heat generation with low heat dissipation challenges, one can use more conventional packaging and die solutions. For lower power, discrete components are often used instead of power modules. The use of a combination of a high- and a low-power inverter leads to the opportunity for both high-end and low-end power electronic solutions.

All in all, there is a lot of development ranging from device, module packaging to system integration to reach the best performance with an optimal form factor and cost. Moreover, there are other axes of development, such as current sensors, passives, and batteries that are also being closely looked at when designing EV vehicles, to properly define the full system requirements and optimize the complete set of components. HEV and EV power electronics market is booming and may use SiC or GaN technologies in the near future [61, 62].

3.8 TYPES OF ELECTRICAL MACHINES (MOTORS)

In *split path* vehicles (Toyota, Ford, GM, and Chrysler) there are two electrical machines, one of which functions as a motor primarily, and the other functions as a generator primarily. One of the primary requirements of these machines is that they are very efficient, as the electrical portion of the energy must be converted from the engine to the generator, through two inverters, through the motor again, and then to the wheels.

There is an extensive need of advanced motors and generators to meet the aggressive targets in terms of efficiency, power density, and cost of the drivetrain in HEVs. The specification of the motor/generator depends on its usage, like in light-/medium-/heavy-duty vehicles, off/on highway vehicles, and locomotives. The performance of the machine depends mainly on vehicle duty cycle, thermal characteristics, and the cooling mechanism implemented. Invention of power converter topologies for drive control has advanced the traction systems for EVs over recent years. Various types of motors for EV are illustrated in Figure 3.12 [2].

These days, the switched reluctance motor (SRM) is receiving much attention in EV/HEV applications. These motors have various advantages such as easier control, rugged construction, better fault tolerance capability, and outstanding torque-speed characteristics. Very low production cost of BLSR motors (even lower than that for induction motors), together with some other important characteristics (e.g., wide speed range), makes them a serious candidate for driving EVs. An SRM is well suited for applications where constant power is required over a wide operating region. There are, however, several drawbacks of SRM such as electromagnetic interference, torque ripple, high noise, and requirement of a special convertor topology. High noise and fluctuation in torque might be compensated for with a more complex and expensive controller. In order to overcome the drawbacks of SRM such as low torque/power density, and high torque ripple, a novel modular (M)-SRM with hybrid magnetic paths for simple structure, low cost, and improved dynamic performance has been presented for EV applications. There is a significant body of research for SRM reported in the literature. This literature is well reviewed by Singh et al. [6].

Brushless DC motors (BLDC) are widely used in the hybrid car market. BLDC motors are theoretically the result of reversing the position of the stator and rotor of PM DC motors. Specifically, they are of a type called an interior permanent magnet (IPM) machine (or motor). These machines are wound similarly to the induction

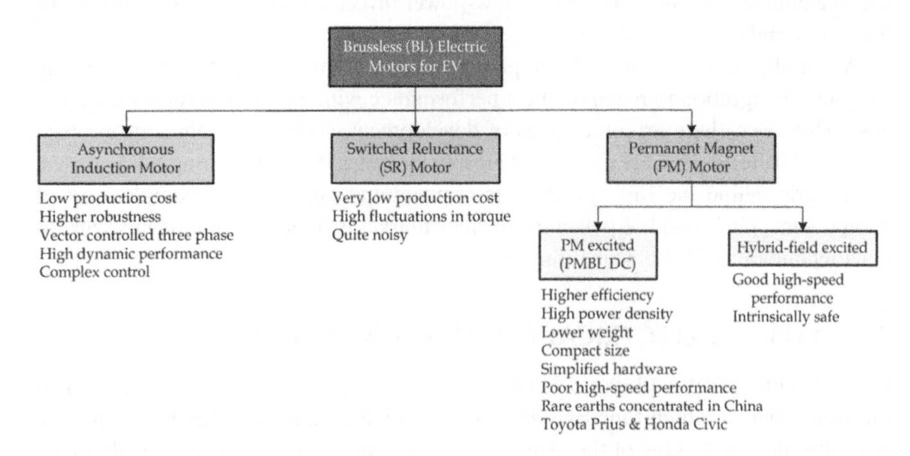

FIGURE 3.12 Types of motors for EV [2]

motors found in a typical home, but (for high efficiency) use very strong rare-earth magnets in the rotor. These magnets contain neodymium, iron, and boron, and are therefore called Neodymium magnets. Cutting edge U.K. motors that are now being produced are using Neodymium Permanent Magnet technology. As security of supply returns, it is certain that there will accordingly be a return to superior motor designs that NdFeB Permanent Magnets enable [2, 8]. Their main advantages are high efficiency, compactness, and high energy density. The literature [6] has shown that a 5-phase brushless fault-tolerant hybrid-excitation motor can be used for EV applications. BLDC can also be used for EV by means of a fuzzy-tuned PI speed control. A brushless dual-rotor flux switching permanent magnet motor can be used for PHEV application. A double-stator permanent magnet BLDC (PMBLDC) can also be engaged for HEV application. These and other modifications, research, and development reported in the literature are described in great detail by Singh et al. [6].

Permanent magnet synchronous motors (PMSMs) prove to be strong contenders to IM in HEV applications. Their benefits include lesser heating, higher power density, and higher efficiency. However, PMSMs suffer from a major drawback of demagnetization due to armature reaction. A comparison of the PM motor and IM for HEV application with the help of advanced vehicle simulator (ADVISOR) tool has been carried out in the literature and the results indicate superior performance of PM motors, in terms of standard performance indices such as traction capabilities and fuel efficiency. A PMSM also shows better result for HEV when clubbed with HESS [6]. A novel PMSM drive system with a bidirectional ZSI is proposed in the literature [6] and tested by keeping in view the feasibility and effectiveness for an EV system. The application of a bidirectional ZSI to PMSM drive system improves the reliability since it is a single-stage structure and the shoot-through states can no longer destroy the inverter.

Permanent magnet brushless dc motor (PMBL) is a very promising technology that has been in wide use with EVs. It appears that this drive type will be a major market leader, though automakers outside China should be cautious and seek drive alternatives, as long as world reserves of rare earths used in the permanent magnets are practically totally situated in China, whose government could apply export restrictions. Hybrid-field-excited PMBL offers superior performance, as field can be strengthened and weakened. The penalty for this choice is higher production cost and increased control complexity.

Finally IMs are widely accepted for the propulsion of EVs, due to their reliability, ruggedness, low maintenance, and high efficiency. In addition, from a safety point of view, these motors get de-excited during fault in inverter. IMs offer a higher power density and better efficiency when compared to the DC machine. IM provides a wide range of speed with good efficiency. The new design of IM for EVs has been proposed with improved speed torque curve. The field-oriented control (FOC) algorithm needs accurate estimation of motor state variables in order to ensure full-torque performances and good efficiency of IM. For this purpose, a two-rotor resistance estimation method has also been presented in the literature [6].

Squirrel cage rotor and the three-phased, asynchronous induction motors currently dominate the industrial applications. Their relatively low-cost, high robustness, and good dynamic performance make them a good candidate for driving EVs as well. As a matter of fact, they are utilized in a number of commercial EVs. However, the dynamic performance needed by EVs is met by induction motor at a relatively high price, for the necessary vector control is a highly complex technique. Furthermore, there are drive alternative, that better satisfy specific EVs' demands such as high torque and power density, high efficiency over a wide torque and speed range, and wide-constant-power operating capacity.

A brief study has been carried out on various methods of controlling the traction motors used in hybrid/electric vehicles and this is also summarized by Singh et al. [6]. With the growing demand for switching over to renewable energy resources, PV technology is charging ahead of the other alternatives. The various salient features of it being noiselessness, pollution-free, immune to direct contamination, and its simplicity in operation makes it the preferred choice. The various structures of PV systems and their suitability in hybrid vehicles have been discussed in the literature. The I-V characteristics of PV cells are nonlinear, and there exists only one maximum power point (MPP). By interfacing the power, electronic devices with PV system, the efficiency can be increased along with MPP controller. Many algorithms have been put forward to track this MPP, but among all, constant voltage tracking method is the most traditional method but has limitations during varying temperatures. To overcome them, perturb and observe (P&O) and incremental conductance (IC) methods are most widely used. In case of P&O, the load impedance varies periodically and senses the change in the direction of power, whereas in IC, the impedance is monitored to detect whether MPP has reached. Once again more details on MPPT algorithms used in HEVs are described by Singh et al. [6].

In some cases, manufacturers are producing HEVs that use the added energy provided by the hybrid systems to give vehicles a power boost, rather than significantly improved fuel efficiency compared to their traditional counterparts. The trade-off between added performance and improved fuel efficiency is partly controlled by the software within the hybrid system and partly the result of the engine, battery, and motor size. In the future, manufacturers may provide HEV owners with the ability to partially control this balance (fuel efficiency vs. added performance) as they wish, through a user-controlled setting.

With the possible exception of microhybrids, all vehicles use permanent magnet alternating current motors. Since the battery capacity is the key limitation for hybrid vehicles, electrical machine efficiency is of paramount importance. Most systems employ "buried magnet" rotating machine configurations with expensive rare-earth high-strength magnets. GM and Honda are using flat wire for the armature winding to increase efficiency. Although rectangular conductors are common for large machines, their use in relatively small machines shows the extent to which manufacturers are going to get better efficiency. Rotating machine technologies and designs are well developed, and the automotive application challenge is to lower their manufacturing cost. Because rotating machines are such a mature component, the cost of their manufacture in high volumes is driven principally by the cost of materials. Thus their cost is relatively unresponsive to technology developments. Major improvements in volumetric power density can be achieved by increasing the speed of the motor. This volumetric improvement results in material reduction but generally also in increased losses. High-speed motors also require a gear set to match the mechanical speed required of the drive train. While the design of the motor/inverter system is an optimization problem, no technology breakthroughs that would radically improve the state of the art are foreseen [2, 8].

Computers have been used to control emissions and optimize efficiency of conventional power trains. In addition to engine control, controllers in hybrid vehicles monitor the state of charge of the battery and determine power flows to and from the battery and engine. The control task is more complex for the PHEV where there is a greater opportunity to optimize the tradeoff between electric and IC engine use with respect to fuel consumption. One suggested approach is to have the controller predetermine the propulsion profile from expected route data provided by the driver or an off-board wirelessly connected server. Vehicle computers are powerful enough to handle these tasks, and no technical problems are expected.

3.9 ECONOMICS AND MARKET FOR HYBRID CARS

The economics and market for EV/HEV cars are extensively examined in the literature [63–72]. According to Frost and Sullivan [63], global sales of BEV and PHEVs climbed from 1.2 million in 2017, a 52% increase compared to 2016, to 1.6 million in 2018 and it was predicted to rise to 2 million in 2019, 7 million in 2020, 30 million in 2030, and 100 million in 2050. The share of these vehicles globally had increased from 0.5% in 2014 to 1.7% in 2017. In 2017, China had 48% of market share and Europe had 26%. In terms of EV sales by country, China was once again the leader of the pack with over 600,000-unit sales, far ahead of the United States which racked up 200,000.

Full and mild HEV sales accounted for 2.8 million units in 2018, a 22% year-to-year increase. Several European car manufacturers also launched their 48 V mild hybrid models in 2017. This cost-effective solution, which electrifies vehicle auxiliary systems and at the same time reduces CO2 emissions, proliferated in 2018–2019 among all European carmakers, followed by the Chinese ones. *The industry forecast a 50% CAGR for the 2017–2023 period, for mild hybrids, because these low-cost electrified vehicle models are attractive. Their approach can be easily implemented in any car, from city cars to higher-end luxury models."*

It is expected that the penetration of EVs/PHEVs will be around 35–47% of the new cars by 2040. It is observed that agencies have provided different statistics based on the growth rates, and thus, there are no unique data available on the long-term market share of these vehicles. The data on BEVs and XHEVs (including full HEVs and PHEVs) are taken from Frost and Sullivan [63] and Morgan Stanley & Co. [64, 65]. It can easily be inferred that the share of green vehicles will increase over the coming years. The six European countries, i.e., Germany, France, Norway, Netherlands, the UK, and Sweden, are expected to share more than 67% of the total BEV market in the year 2020, whereas only four countries (Germany, France, Italy, and the UK) are expected to share more than 52% of the total market share of PHEVs [66]. According to Pike Research forecast, almost 1.8 million of BEVs, 1.2 million of PHEVs, and 1.7 million of HEVs are expected on Europe's roadways by 2020 [67–70].

While discussing about HEVs, a natural question that arises is about its affordability. The government of every country is taking interest to search for an alternative method of transportation which is accessible to the public. As such, there is a rapid deployment of well-managed infrastructure to supplement electric technology. Since the production of batteries has increased, costs have shaved by approximately 50% and are expected to be lesser than $200/kWh by 2020. Improvement in battery technology will reduce the cost of the hybrid vehicle and make it accessible to more people. It is expected that EVs can be made affordable by 2022 even if the conventional cars improve their fuel efficiency by 3.5% a year. The analysis uses the US government's projected oil price of $50–$70 (£36–£50) a barrel in the 2020s. If the price is $20, the tipping point is pushed back between 3 and 9 years. Consequently, EVs will reduce the revenue from ICE vehicle, but it will compensate via the revenue generated from this new window of opportunity for car manufacturers, for charging infrastructure companies, and for battery manufacturers. Solid-state batteries will be the key to the enhancement in battery performance as they are 2.5 times denser than lithium-ion batteries [5, 63–74].

Recent years have witnessed 10 future plans for launching EV from varied automakers. Based on these plans it is expected that about 25 million units would be on the road by 2025. Tesla Gigafactory is currently 35% operational and aims to produce 50 GWh of batteries in 2018. The EV charging station is a big hindrance which has yet not been focused and needs global attention. Currently, charging stations are present in limited areas where the sale of EVs is higher.

According to the 2018 statistics, the total cost of ownership of a Ford Fusion Hybrid would amount to ~ $35,606. This HEV lies comfortably in the midsize market range. In contrast to this, a Honda Accord, which is a traditional vehicle, would cost ~ $35,709. By comparing other car models, it is evident that HEVs are now as price competitive as ICE vehicles. When looking into the compact market, it is seen that a HEV would be about ~ $9,000–$10,000 more expensive. The difference of the pricing in the two segments can be understood due to the higher maintenance and fuel consumption of a midsize IC-driven vehicle compensating for a more expensive battery and base price. In a compact vehicle, fuel consumption is also lower. Apart from that, the depreciation seen in a compact HEV is surprisingly higher which could be due to its lesser demand and rapid technology changes in the field. It is evident that HEVs in the midsize market are already price competitive and are affordable choices [5, 71–74].

In summary, HEVs are rapidly emerging as a potential alternative to the existing state of transportation due to their lower petroleum consumption and toxic emission. Strict CO2 emission laws and increased public awareness will propel HEVs to be the future of road transportation. Penetration of PHEVs in the market will change the operations of electric grid substantially, and efforts are being made to provide a two-way communication between the user and the grid. Based on the literature review, it is found that the complex hybrid architecture will provide greater efficiency, trading off on higher costs and more complex designs. As inverters are needed to interface the motor engine with ESS, their selection is of prime importance and the q-ZSI is found to be a promising candidate. To extend the battery life, it is suggested to combine UC with battery which will further improve the fuel efficiency and performance during varying ambient conditions. The use of hybrid energy storage on HEV and EV is discussed in great detail in Chapter 4.

The literature indicates that there is a growing interest in developing advanced traction motors for hybrid vehicles and many traction motors are available in the market. However, considering the trade-off based on performance, robustness, reliability, and cost, the choice is often between the induction motor and permanent magnet AC motor. PVHEVs are still in the early stage and are being explored to minimize gasoline consumption and maximize the usage of renewable energy. Various MPPT algorithms tuned with artificial intelligence techniques like FL, ANN, and PSO are also being explored for PVHEV applications [6]

A comparison of various existing hybrid vehicles is provided in Table 3.5 which will serve as a guide to choose the best option.

In examining economics of HEV/EV, other markets that are influenced by them need to be examined. In 2018, the inverter power electronics market was worth roughly US$55 billion, coming from various power applications such as wind turbines, solar inverters, transport, UPS, and other industrial applications. Motor drives was still the larger part of the inverter market with about 40 percent market share in 2018. These are quite mature markets, where technology innovation is no longer the main driver of system performance, and the supply chain is very well stabilized. Today, however, the introduction of other markets such as EV/HEV is boosting technology innovation and market expansion into new applications. Over the last years, we have seen a big drive and increased EV/HEV sales, leading to an expected 28 percent CAGR 2017-2023 [2, 6, 8, 63–74].

EV/HEV market is driven by various incentive mechanisms which might still be subject to sudden changes, such as, for example, the recent decision by the Chinese government to cut EV/HEV subsidies and eventually abolish them completely by 2020. However, there are sustainable drivers, such as various governments' CO_2 emission reduction targets and the need for cleaner air in cities. These drivers, together with improving battery and power electronic technologies, reducing battery costs, as

TABLE 3.5
Comparison of various existing hybrid vehicles [2, 6, 8]*

Parameters	ICV	HEV	PHEV	PVHEV	BEV
Driving range	H	M	H	H	L
Fuel Type	Gasoline	Gas + Elec.	Gas +Elec.	Gas + Elec	Electric
Overall cost	H	M	M	M	L
Efficiency	L	L	H	H	H
Structure	S	M	C	C	S

* L = Low, M==Medium, H = High, S== Simple, C== Complex

well as constructive engagement by numerous automotive manufacturers will continue to drive the EV/HEV market. More than US$300 billion of investment into EV/HEVs in the coming years has been announced by the leading car manufacturers, with the biggest portion coming from European companies. The "dieselgate" affair over CO2 emissions measurements has further accelerated the car makers' strategic decisions to release more electrified car models earlier. To reach CO_2 emission reduction targets, strengthened in 2018 in Europe, car makers had to focus on increased electrification of their vehicle fleets, i.e. to full-hybrid, plug-in hybrid and full electric vehicles, which emit less CO2 compared to mild-hybrid electric vehicles.

The huge sales of Tesla' full-electric cars have given an additional impetus to the customer's and car maker's perceptions about the future of full electric cars. The full electric car with large battery capacity together with a rapid deployment of DC fast charging infrastructure has significantly reduced customer concerns about the limited driving range and long battery charging time. The electric vehicle has also become a symbol of an exciting driving experience. The electrified car is clearly evolving toward full electric car and the full electric car is evolving toward a car with higher traction power, larger battery capacity, and more functionality.

A. **Costs and Affordability**

Ever since its inception, the cost of HEVs has always been much higher than its traditional counterparts in the same segment. Studying just its model price would be misleading, so it is necessary to realize the various costs, such as maintenance, repair, depreciation and fuel to get the complete picture. Comparisons between various costs involved in vehicle are as follows [2, 6, 8, 63–74]:

1. ***Fuel costs:*** The main difference between EVs and ICVs is their fuel source; ICVs run on gasoline, while EVs run on electricity. According to a study of the University of Michigan's Transportation Research Institute, the operating cost of EV is less than half of gas-powered cars. The average operating cost of an EV is $485/year as against $1,117 for gasoline-powered vehicles in the USA. This figure is subject to the rates of gas and electricity.
2. ***Purchase cost:*** The base price for an EV in general is seen to be higher in comparison with traditional alternatives. Higher costs can be attributed to increased complexity and the number of components in an HEV and specialized parts.
3. ***Maintenance costs:*** In ICVs, vehicle maintenance costs can be very high due to engine maintenance which is a huge money sink, which further increases as the car ages. Changing the engine oil, coolant, transmission fluid, and belts can add up in value over time. Since an EV does not have these parts, such repair costs are averted. The universal vehicle expenses, i.e., tire and brake changes, insurance and structural repair, are part of owning any vehicle. EVs are also not free of expenses. The highest maintenance cost associated with EVs is due to its battery which is unlike normal batteries. EV has large

complex rechargeable batteries, which are quite resistant to any defect but degrade with time, and their replacement is quite expensive.

4. ***Depreciation cost***: HEVs have seen a higher rate of depreciation compared to ICVs. Depreciation is judged by its resale ability. One possible reason for the higher depreciation cost could be linked to the rapid advancement in HEV technology. A HEV which was once the state of the art could become much inferior within a short period due to the recent stress in research in this field.
5. ***Electric car rebates and incentives:*** A great reason which attracts consumers to go for EVs is the country and state incentives available in the form of various subsidies and policies. These rebates help offset the typically higher cost of an electric car to make "going electric" more financially feasible [66].

EVs are not for every lifestyle, but when compared to the myriad costs surrounding ICV purchase and maintenance, choosing an EV can be an intelligent fiscal decision. While the purchase price for various hybrid cars are higher than conventional ICE car, low operating cost makes up for high initial cost in few years. Currently, only EV, BEV and fuel cell cars appear to be more expensive in terms of total cost of ownership compare to conventional ICE cars. This will, however, change as technology for EV, BEV, and fuel cell car improves and their market grows [63–74].

3.10 FUEL CONSUMPTION AND ENVIRONMENTAL IMPACT OF HYBRID CARS

As mentioned before hybrid cars in general save fuel and emit less CO_2 in the environment. Some data for PHEV hybrid cars reported in the literature are described in Table 3.6

TABLE 3.6
Petroleum Use and CO_2 Emission of PHEV [13]

PHEV well-to-wheels Petroleum energy use and greenhouse gas emissions for an all-electric range between 10 and 40 miles (16 and 64 km) with different on-board fuels. (as a % relative to an internal combustion engine vehicle that uses fossil fuel gasoline)

Analysis	Reformulated gasoline and Ultra-low sulfur diesel	E85 fuel from corn and switchgrass	Fuel cell hydrogen
Petroleum energy use reduction	40–60%	70–90%	more than 90%
GHG emissions reduction(2)	30–60%	40–80%	10–100%

Source: Center for Transportation Research, Argonne National Laboratory (2009). with PHEV model year 2015. (2) No direct or indirect land-use changes included in the WTW analysis for bio-mass fuel feedstocks.

TABLE 3.7
Saving in Fuel Consumption in Some Top Models (Data from Oak Ridge Laboratory [21])

Technology	Nonhybrid/nonelectric base model (BEE* fuel efficiency star rating)	Hybrid/electric model (BEE fuel efficiency star rating)	Gasoline equivalent fuel consumption reduction over base model
Diesel-based mild hybrid	Maruti Ciaz VDI (5 star)	Maruti Ciaz VDI-shvs (5 star)	7%
Diesel-based mild hybrid	Maruti Ertiga VDI (4 star)	Maruti Ertiga VDI-shvs (5 star)	15%
Gasoline-based strong hybrid	Toyota Camry at 2.5 l (2 star)	Toyota Camry hybrid (5 star)	32%
Battery-operated electric	Mahindra Verito d2 (4 star)	Mahindra E-verito d2 (5 star)	68%
Battery-operated electric	–	Mahindra e2o (5 star)	–

The average fuel consumption of production hybrid HEVs was also determined from fuel economy data supplied by Oak Ridge National Laboratory [21]. These are illustrated in Table 3.7. Some other published data on fuel consumptions and CO_2 emissions by various hybrid cars are also illustrated in Table 3.8.

For several specific models, these data were compared to data from conventional (nonhybrid) vehicles of approximately similar performance and physical specifications. A significant contribution to the fuel consumption benefit of hybrid vehicles is due to modifications to the engine, body, and tires. For example, the fuel economy of the Prius is significantly influenced by engine improvements and optimized operating area. The 2007 model-year version of the Saturn Vue hybrid, which used a BAS design, exhibits a 25 percent improvement in fuel economy on the FTP cycle, but approximately half of that improvement is due to vehicle modifications, including a more aggressive torque converter lockup and fuel cutoff during vehicle deceleration.

The hybrid vehicle typically achieves greater fuel economy and lower emissions than conventional internal combustion engine vehicles (ICEVs), resulting in fewer emissions being generated. These savings are primarily achieved by three elements of a typical hybrid design:

1. Relying on both the engine and the electric motors for peak power needs, resulting in a smaller engine size more for average usage rather than peak power usage. A smaller engine can have less internal losses and lower weight.
2. Having significant battery storage capacity to store and reuse recaptured energy, especially in stop-and-go traffic typical of the city driving cycle.

TABLE 3.8

Typical Comparison of CO_2 Emission of Several Cars on the Road [13]

Comparison of tailpipe and upstream CO_2 emissions estimated by EPA for the MY 2014 plug-in hybrids available in the US market as of September 2014

Vehicle	EPA rating combined EV/ hybrid (mpg-e)	Utility factor(2) (share EV miles)	Tailpipe CO_2 (g/mi)	Tailpipe + Total Upstream CO_2		
				Low (g/mi)	Avg (g/mi)	High (g/mi)
BMW i3 Rex	88	0.83	40	134	207	288
Chevrolet Volt	62	0.66	81	180	249	326
Cadillac ELR	54	0.65	91	206	286	377
Ford C-Max Energi	51	0.45	129	219	269	326
Ford Fusion Energi	51	0.45	129	219	269	326
Honda Accord Plug-in Hybrid	57	0.33	130	196	225	257
Toyota Prius Plug-in Hybrid	58	0.29	133	195	221	249
BMW i8	37	0.37	198	303	351	404
Porsche Panamera S E-Hybrid	31	0.39	206	328	389	457
Average gasoline car	**24.2**	**0**	**367**	**400**	**400**	**400**

Notes: (1) Based on 45% highway and 55% city driving. (2) The utility factor represents, on average, the percentage of miles that will be driven using electricity (in electric only and blended modes) by an average driver. (3) The EPA classifies the i3 REx as a series plug-in hybrid

3. Recapturing significant amounts of energy during braking that are normally wasted as heat. This regenerative braking reduces vehicle speed by converting some of its kinetic energy into electricity, depending upon the power rating of the motor/generator.

Other techniques that are not necessarily 'hybrid' features, but that are frequently found on hybrid vehicles include:

1. Using Atkinson cycle engines instead of Otto cycle engines for improved fuel economy.
2. Shutting down the engine during traffic stops or while coasting or during other idle periods.

3. Improving aerodynamics (part of the reason that SUVs get such bad fuel economy is the drag on the car. A box shaped car or truck has to exert more force to move through the air causing more stress on the engine making it work harder). Improving the shape and aerodynamics of a car is a good way to help better the fuel economy and also improve vehicle handling at the same time.
4. Using low rolling resistance tires (tires were often made to give a quiet, smooth ride, high grip, etc., but efficiency was a lower priority). Tires cause mechanical drag, once again making the engine work harder, consuming more fuel. Hybrid cars may use special tires that are more inflated than regular tires and stiffer or by choice of carcass structure and rubber compound have lower rolling resistance while retaining acceptable grip, and so improving fuel economy whatever the power source.
5. Powering the a/c, power steering, and other auxiliary pumps electrically as and when needed; this reduces mechanical losses when compared with driving them continuously with traditional engine belts.

These features make a hybrid vehicle particularly efficient for city traffic where there are frequent stops, coasting and idling periods. In addition noise emissions are reduced, particularly at idling and low operating speeds, in comparison to conventional engine vehicles. For continuous high speed highway use these features are much less useful in reducing emissions. Hybrid vehicle emissions today are getting close to or even lower than the recommended level set by the EPA (Environmental Protection Agency). The recommended levels they suggest for a typical passenger vehicle should be equated to 5.5 metric tons of CO. The three most popular hybrid vehicles, Honda Civic, Honda Insight and Toyota Prius, set the standards even higher by producing 4.1, 3.5, and 3.5 tons showing a major improvement in carbon dioxide emissions. Hybrid vehicles can reduce air emissions of smog-forming pollutants by up to 90% and cut carbon dioxide emissions in half. More fossil fuel is needed to build hybrid vehicles than conventional cars but reduced emissions when running the vehicle more than outweigh this.

Electric hybrids reduce petroleum consumption under certain circumstances, compared to otherwise similar conventional vehicles, primarily by using three mechanisms:

1. Reducing wasted energy during idle/low output, generally by turning the ICE off,
2. Recapturing waste energy (i.e. regenerative braking),
3. Reducing the size and power of the ICE, and hence inefficiencies from under-utilization, by using the added power from the electric motor to compensate for the loss in peak power output from the smaller ICE.

Any combination of these three primary hybrid advantages may be used in different vehicles to realize different fuel usage, power, emissions, weight and cost profiles. The ICE in a HEV can be smaller, lighter, and more efficient than the one in a conventional vehicle, because the combustion engine can be sized for slightly above *average* power demand rather than *peak* power demand. The drive system in a vehicle is required to operate over a range of speed and power, but an ICE's highest efficiency is in a narrow range of operation, making conventional vehicles inefficient. On the contrary, in most HEV designs, the ICE operates closer to its range of highest efficiency more frequently. The power curve of electric motors is better suited to variable speeds and can provide substantially greater torque at low speeds compared with internal-combustion engines. The greater fuel economy of HEVs has implications for reduced petroleum consumption and vehicle air pollution emissions worldwide. Many hybrids use the Atkinson cycle, which gives greater efficiency, but less power for the size of the engine.

Individual technology contributes to the efficiency in fuel consumption in hybrid cars. In city mode, hybrid car uses only about 55% of fuel compared to full ICE car. This reduction can be attributed to engine improvement, engine operating conditions, idle stop, and regeneration. In highway mode a HEV consumes 85–90% of fuel required for ICE car. Here the reduction is largely due to engine improvement with minor additions from other areas. Progress towards fuel efficiency by different types of hybrids is illustrated in Table 3.4.Several other relevant references [50, 75–79] also address the fuel efficiency and CO_2 emissions of various types of hybrid cars. All of these references indicate the superior performance of various types of HEV/EV cars for fuel efficiency and CO_2 emission compared to conventional gasoline use in ICE cars.

A. **Environmental Impact of Hybrid Car Battery**

Though hybrid cars consume less fuel than conventional cars, there is still an issue regarding the environmental damage of the hybrid car battery. Today most hybrid car batteries are one of two types: 1) nickel metal hydride, or 2) lithium ion; both are regarded as more environmentally friendly than lead-based batteries which constitute the bulk of petrol car starter batteries today. There are many types of batteries. Some are far more toxic than others. Lithium ion is the least toxic of the two mentioned above.

There are mixed reports in the literature on the toxicity levels and environmental impact of nickel metal hydride batteries—the type currently used in hybrids—compared to lead acid or nickel cadmium batteries. Some claims that recycling and disposing nickel metal hydride batteries safely is difficult. In general, various soluble and insoluble nickel compounds, such as nickel chloride and nickel oxide, have known carcinogenic effects in chick embryos and rats. The main nickel compound in NiMH batteries is nickel oxyhydroxide (NiOOH), which is used as the positive electrode.

The lithium-ion battery has attracted attention due to its potential for use in hybrid electric vehicles. Hitachi is a leader in its development. In addition to its smaller size and lighter weight, lithium-ion batteries deliver performance that helps to protect the environment with features such as improved charge efficiency without memory effect. The lithium-ion batteries are appealing because they have the highest energy density of any rechargeable batteries and can produce a voltage more than three times that of nickel–metal hydride battery cell while simultaneously storing large quantities of electricity as well. The batteries also produce higher output (boosting vehicle power), higher efficiency (avoiding wasteful use of electricity), and provide

TABLE 3.9
Comparison of Emission from Conventional and Different Types of Electric Vehicles [2, 6, 8, 21]

Parameters	Conventional	EV	Series hybrid	Parallel hybrid
NOx (g/km), CO (g/km)	High	NA	Medium	Low
HC	High	NA	Low	Medium
Fuel consumption (km/L)	High	NA	Medium	Low

NA not applied

excellent durability, compared with the life of the battery being roughly equivalent to the life of the vehicle. Additionally, the use of lithium-ion batteries reduces the overall weight of the vehicle and also achieves improved fuel economy of 30% better than petro-powered vehicles with a consequent reduction in CO_2 emissions helping to prevent global warming.

There are two different levels of charging. Level one charging is the slower method as it uses a 120 V/15 A single-phase grounded outlet. Level two is a faster method; existing Level 2 equipment offers charging from 208 V or 240 V (at up to 80 A, 19.2 kW). It may require dedicated equipment and a connection installation for home or public units, although vehicles such as the Tesla have the power electronics on board and need only the outlet. The optimum charging window for Lithium ion batteries is 3–4.2 V. Recharging with a 120 V household outlet takes several hours, a 240 V charger takes 1–4 hours, and a quick charge takes approximately 30 minutes to achieve 80% charge. The charging details depend on three important factors—distance on charge, cost of charging, and time to charge. In order for the hybrid to run on electrical power, the car must perform the action of braking in order to generate some electricity. The electricity then gets discharged most effectively when the car accelerates or climbs up an incline. In 2014, hybrid electric car batteries can solely run on electricity for 70–130 miles (110–210 km) on a single charge. Hybrid battery capacity currently ranges from 4.4 kWh to 85 kWh on a fully electric car. On a hybrid car, the battery packs currently range from 0.6 kWh to 2.4 kWh representing a large difference in use of electricity in hybrid cars. A comparison of emission for different driving cycle is illustrated in Table 3.9 [2, 6, 8, 21].

REFERENCES

1. Badin F, Scordia J, Trigui R, Vinot E, Jeanneret B. Hybrid electric vehicles energy consumption decrease according to drive train architecture, energy management and vehicle use. In: IET-The Institution of Engineering and Technolgy Hybrid Vehicle Conference, Coventry, UK; 2006. pp. 213–23. https://doi.org/10.1049/cp:20060610.
2. De Lucena S. A survey on electric and hybrid electric vehicle technology. An open access intech paper. 2011. DOI: 10.57772/18046.

3. Thomas CES. Transportation options in a carbon-constrained world: Hybrids, plug-in hybrids, biofuels, fuel cell electric vehicles, and battery electric vehicles. *International Journal of Hydrogen Energy*. 2009;34:9279–96. https://doi.org/10.1016/j.ijhydene.2009.09.058.
4. Panday A, Bansal HO. A review of optimal energy management strategies for hybrid electric vehicle. *International Journal of Vehicular Technology*. Vol. 2014:article id 160510 (pp. 1–19). (2014) https://doi.org/10.1155/2014/160510.
5. Panday A, Bansal HO. Energy management strategy for hybrid electric vehicles using genetic algorithm. *Journal of Renewable and Sustainable Energy*. 2016;8:015701. https://doi.org/10.1063/1.4938552.
6. Singh KV, Bansal HO, Singh D. A comprehensive review on hybrid electric vehicles: Architectures and components. *Journal of Modern Transportation*. 2019;27:77–107. https://doi.org/10.1007/s40534-019-0184-3.
7. Eshani M, Gao Y, Gay S, Emadi A. *Modern Electric, Hybrid Electric and Fuel Cell Vehicles*. 2nd ed. Boca Raton, FL: CRC Press.
8. Hybrid vehicle. *Wikipedia, The Free Encyclopedia*. 2020 [last visited May 18, 2020] (2020).
9. Solar Vehicle. *Wikipedia, The Free Encyclopedia*. 2020 [last visited April 2, 2020] (2020).
10. Solar Car. *Wikipedia, The Free Encyclopedia*. 2020 [last visited April 22, 2020] (2020).
11. Hybrid electric vehicle. *Wikipedia, The Free Encyclopedia*. 2020 [last visited May 19, 2020](2020).
12. Electric car. *Wikipedia, The Free Encyclopedia*. 2020 [last visited May 23, 2020] (2020).
13. Plug in electric vehicle. *Wikipedia, The Free Encyclopedia*. 2020 [last visited May 23, 2020](2020).
14. Battery electric vehicle. *Wikipedia, The Free Encyclopedia*. 2020 [last visited May 16, 2020](2020).
15. Hybrid electric bus. *Wikipedia, The Free Encyclopedia*. 2020 [last visited May 15, 2020](2020).
16. Hybrid vehicle drive train. *Wikipedia, The Free Encyclopedia*. 2020 [last visited March 21, 2020](2020).
17. Brslica V. Plug in hybrid vehicles. Intech paper. 2011. DOI: 10.5772/21528. http://www.intechopen.com.
18. Dearborn DD. Modular Solar Photovoltaic Canopy System for Development of Rail Vehicle Traction Power. US20100200041A1. August 12, 2010 [February 6, 2019].
19. Soma A. Trends and hybridization factor for heavy duty working vehicles. Open access intech paper. 2017. DOI: 10.5772/intechopen.68296.
20. Shah YT. *Energy and Fuel Systems Integration*. New York: CRC Press, Taylor and Francis group; 2015.
21. Davis SC, Williams SE, Boundy RG, Moore S. 2016 vehicle technologies market report. Oak Ridge, TN: Oak Ridge national laboratory; 2017. Report No.: ORNL/TM-2017/238.
22. Fuel cell vehicle. *Wikipedia, The Free Encyclopedia*. 2020 [last visited April 18, 2020].
23. Aziz M. Advanced charging system for plug-in hybrid electric vehicles and battery electric vehicles. Intech open access paper. 2017. DOI: 10.57772/intechopen.68287.
24. Shah YT. *Modular Systems for Energy Usage Management*. New York: CRC Press; 2020.
25. Oeftering RC, Kimnach GL, Fincannon J, Mckissock BI, Loyselle P, Wong E. Advanced modular power approach to affordable, supportable space systems. June 2013. Available from: https://ntrs.nasa.gov/search.jsp?R=201400003332019-07-20T17:10:08+00:00Z. Report No.: NASA/TM-2013-217813, AIAA-2013-5253.

26. Andermann. Lithium-ion batteries for hybrid electric vehicles: Opportunities and challenges. In: Presentation to the National Research Council Committee on the Assessment of Technologies for Improving LightDuty Vehicle Fuel Economy, October 25, Washington, DC; 2007.
27. Lithium-ion battery. *Wikipedia, The Free Encyclopedia*. 2020 [last visited June 22, 2020].
28. Nelson P, Amine K, Yomoto H. Advanced lithium-ion batteries for plug-in hybrid-electric vehicles. In: Paper presented at 23rd International Electric Vehicle Symposium, December, Anaheim, CA; 2007.
29. Rousseau A, Shidore N, Carlson R, Nelson P. Research on PHEV battery requirements and evaluation of early prototypes. In: Paper presented at the Advanced Automotive Battery Conference, May 17, Long Beach, CA; 2007.
30. NRC (National Research Council). *Transitions to Alternative Transportation Technologies: A Focus on Hydrogen*. Washington, DC: National Academies Press; 2008.
31. Rowe DM. *CRC Handbook of Thermo- Electrics*. 1st ed. New York: CRC Press; 1995. 701 p. ISBN 978-0849301469.
32. Ensescu D. Thermoelectric energy harvesting: Basic principles and applications. InTech open access paper. 2019. DOI: 10.5772/intechopen.83495. http://www.intech.com.
33. Riffat SB, Ma X. Thermoelectrics: A review of present and potential applications. *Applied Thermal Engineering*. 2003;23:913–35.
34. Patil D, Arakerimath RR Dr. A review of thermoelectric generator for waste heat recovery from engine exhaust. *International Journal of Research in Aeronautical and Mechanical Engineering*. 2013; 1: 1–9.
35. Ismail BI, Ahmed WH. Thermoelectric power generation using waste-heat energy as an alternative green technology. *Recent Patents on Electrical Engineering (Continued as Recent Advances in Electrical & Electronic Engineering)*. 2009;2(1):27–39.
36. Shah YT.*Thermal Energy: Sources, Recovery and Applications*. New York: CRC Press, Taylor and Francis Group; 2018.
37. Zheng XF, Liu CX, Yan YY, Wang Q. A review of thermoelectrics research—Recent developments and potentials for sustainable and renewable energy applications. *Renewable and Sustainable Energy Reviews*. 2014;32:486–503.
38. Taguchi T. US20070193617. 2007.
39. Hicks W. *News Release: NREL Reveals Potential for Capturing Waste Heat via Nanotubes*. Golden, CO: National Renewable Energy Laboratory; 2016 [April 4, 2016].
40. Smith K, Thornton M. Feasibility of thermoelectrics for waste heat recovery in conventional vehicles. Golden, CO: National Renewable Energy Laboratory; 2009. P. 4. Technical Report NREL/TP540-4424.
41. Thacher EF, Helenbrook BT, Karri KA, Richter CJ. Testing of an automobile exhaust thermoelectric generator in a light truck. *Proceedings of the Institution of Mechanical Engineers, Part D: Journal of Automobile Engineering*. 2007;221(1):95–107. DOI: 10.1243/ 09544070JAUTO51.
42. Hsiao YY, Chang WC, Chen SL. A mathematic model of thermoelectric module with applications on waste heat recovery from automobile engine. *Energy*. 2010;35(3):1447–54. DOI: 10.1016/j.energy.2009.11.030.
43. Hsu CT, Yao DJ, Ye KJ, Yu B. Renewable energy of waste heat recovery system for automobiles. *Journal of Renewable and Sustainable Energy*. 2010;2:013105. DOI: 10.1063/ 1.3289832.
44. Hsu CT, Huang GH, Chu HS, Yu B, Yao DJ. Experiments and simulations on low-temperature waste heat harvesting system by thermoelectric power generators. *Applied Energy*. 2011;88(4):1291–7. DOI: 10.1016/j.apenergy. 2010.10.005.

45. Tian H, Sun X, Jia Q, Liang X, Shu G, Wang X. Comparison and parameter optimization of a segmented thermoelectric generator by using the high temperature exhaust of a diesel engine. *Energy*. 2015;84:121–30. DOI: 10.1016/j.energy.2015.02.063.
46. Meng JH, Wang XD, Chen WH. Performance investigation and design optimization of a thermoelectric generator applied in automobile exhaust waste heat recovery. *Energy Conversion and Management*. 2016;120:71–80. DOI: 10.1016/j.enconman.2016.04.080.
47. Mazumder SK, Rathore AK. Primary side converter assisted soft switching scheme for AC/AC converter and cycloconverter. *IEEE Transactions on Power Electronics*. 2011;58(9).
48. Samson D, Kluge M, Becker T, Schmid U. Energy harvesting for autonomous wireless sensor nodes in aircraft. *Procedia Engineering*. 2010;5:1160–63. DOI: 10.1016/j.proeng.2010. 09.317.
49. Kousksou T, Bedecarrats J-P, Champier D, Pignolet P, Brillet C. Numerical study of thermoelectric power generation for an helicopter conical nozzle. *Journal of Power Sources*. 2011;196:4026–32. DOI: 10.1016/j.jpowsour.2010.12.015.
50. Ricardo, Inc. A study of potential effectiveness of carbon dioxide reducing vehicle technologies. Prepared for the U.S. Environmental Protection Agency. Arbor, MI: EPA; 2008. EPA420-R-08-004. Contract No. EP-C-06–003. Work Assignment No. 1–14.
51. Drobnik J, Jain P. Electric and hybrid vehicle power electronics efficiency, testing and reliability. *World Electric Vehicle Journal*. 2013;6(3):719–30.
52. Drobnik J, Bernoux B. Power electronics study for a small urban. In: EVS25, Shenzhen; November 2010.
53. Jennings PA, Jones RP, McGordon A. Generalised fuzzy-logic-based power management strategy for various hybrid electric vehicle powertrain architectures. UKACC International Conference on CONTROL, Coventry, UK; 2010. pp. 197–202. https://doi.org/10.1049/ic.2010.0280.
54. Zulkifli SA, Mohd S, Saad N, Aziz ARA. Operation, power flow, system architecture and control challenges of split-parallel through-the-road hybrid electric vehicle. In: 2015 10th Asian Control Conf Emerg Control Tech a Sustain World (ASCC), Kotea Kinabalu, Malaysia. 2015. https://doi.org/10.1109/ascc.2015.7244637.
55. Miller JM. Hybrid electric vehicle propulsion system architectures of the e-CVT type. *IEEE Transactions on Power Electronics*. 21:756–67. https://doi.org/10.1016/j.healun.2006.03.007.
56. Cheng R, Dong Z. Modeling and simulation of plug-in hybrid electric powertrain system for different vehicular application. In: IEEE Vehicle Power and Propulsion Conference (VPPC), Montreal, QC, Canada; 2015. https://doi.org/10.1109/vppc.2015.7352976.
57. Cao D, Lei Q, Peng FZ. Development of high efficiency current-fed quasi-Z-source inverter for HEV motor drive. In: Twenty-Eighth Annual IEEE Applied Power Electronics Conference and Exposition (APEC), Long Beach, CA; 2013. pp 157–164. https://doi.org/10.1109/apec.2013.6520201.
58. Tolbert LM, Peng FZ, Habetler TJ. Multilevel inverters for electric vehicle apps. In: WPET'98, MI; October 1998.
59. Sun J. Small signal methods for AC power distributed systems. *IEEE Transactions on Power Electronics*. 2009;24(11):2545–54.
60. Steinke JK. Control strategy for a three phase with 3-level GTO PWM inverter PESC. 1998.
61. Le Gouic B. Market and technology analysis, power electronics and related markets. Yole Développement. Date 08/01/2010.

62. Sakai Y, Ishiyama H, Kikuchi T. Power control unit for high power hybrid system. In: Society of Automotive Engineers World Congress, April 16–19, 2007, Detroit, MI; 2007. SAE Paper 2007-01-0271.
63. Frost and Sullivan. Global electric vehicle market outlook. 2018 [Accessed 5 Oct 2018]. https://store.frost.com/global-electric-vehicle-market-outlook-2018.html.
64. Dhawan R, Gupta S, Hensley R, Huddar N, Iyer B, Mangaleswaran R. The future of mobility in India: challenges & opportunities for the auto component industry. 2017 [Accessed October 5, 2018]. Available from: https://www.mckinsey.com/~/media/mckinsey/industries/automotive%20and%20assembly/our%20insights/the%20future%20of%20mobility%20in%20india/the-future-of-mobility-in-india.ashx.
65. Alford, J., Irving, V., Zlatnicka, E. and Dembele, F. Sustainability compendium: MS views on sustainability topics a report by morgan stanley' s sustainability research team. New York, NY: Morgan Stanley Inc.; (Nov. 2, 2017). pp. 1–62.
66. IEA. Hybrid and electric vehicles, the electric drive establishes a market foothold. 2009 [Accessed October 5, 2018]. Available from: http://www.ieahev.org/assets/1/7/2008_annual_report.pdf.
67. Frost & Sullivan. Global market analysis of plug in hybrid electric vehicles. 2007 [Accessed October 5, 2018]. Available from: http://www.emic-bg.org/files/Global_Market_Analysis_of_Plug_in_Hybrid_Electric.pdf.
68. Singh S. Global electric vehicle market looks to power up in 2018. In: Forbes. 2018 [Accessed October 5, 2018]. Available from: https://www.forbes.com/sites/sarwantsingh/2018/04/03/global-electric-vehicle-market-looks-to-fire-on-all-motors-in-2018/#6bf966142927.
69. Electric cars to reach price parity by 2025. Bloomberg NEF [Accessed October 6, 2018]. Available from: https://about.bnef.com/blog/electric-cars-reach-price-parity-2025/.
70. Electric cars vs gas cars: What do they cost?. EnergySage [Accessed October 5, 2018]. Available from: https://www.energysage.com/electric-vehicles/costs-and-benefits-evs/evs-vs-fossil-fuel-vehicles/.
71. Electric cars "will be cheaper than conventional vehicles by 2022". Environment The Guardian [Accessed October 5, 2018]. Available from: https://www.theguardian.com/environment/2016/feb/25/electric-cars-will-be-cheaper-than-conventional-vehicles-by-2022.
72. Propfe B, Redelbach M, Santini DJ, Friedrich H. Cost analysis of plug-in hybrid electric vehicles including maintenance & repair costs and resale values. *World Electric Vehicle Journal*. 2012;5(4):886–95. https://doi.org/10.3390/wevj5040886.
73. Efficiency compared: Battery-electric 73%, hydrogen 22%, ICE 13% [Accessed October 5, 2018]. https://insideevs.com/efficiency-compared-battery-electric-73-hydrogen-22-ice-13/.
74. Panday A, Bansal HO. Green transportation: Need, technology and challenges. *International Journal of Global Energy Issues*. 2014;37:304. https://doi.org/10.1504/IJGEI.2014.067663.
75. Kalhammer FR, Kopf BM, Swan DH, Roan VP, Walsh MP. Status and prospects for zero emissions vehicle technology. Report presented to ARB Independent Expert Panel, April 13, State of California Air Resources Board, Sacramento; 2017.
76. Ragatz AC, Burton JL, Miller ES, Thornton MJ. Investigation of emissions impacts from hybrid powertrains. Golden, CO: National Renewable Energy Laboratory; January 2020. Report No.: NREL/TP-5400-75782, Contract No. DE-AC36-08GO28308.
77. Bennion K, Thornton M. Fuel savings from hybrid electric vehicles. Golden, CO: National Renewable Energy Laboratory; March 2009.Technical Report No.: NREL/TP-540-42681.

78. Simpson A. Cost-benefit analysis of plug-in hybrid electric vehicle technology. Paper presented at the 22nd International Battery, Hybrid and Fuel Cell Electric Vehicle Symposium and Exhibition (EVS-22), October 23–28, 2006, Yokohama, Japan; 2006.
79. Short W, Denholm P. A preliminary assessment of plug-in hybrid electric vehicles on wind energy markets. Golden, CO: National Renewable Energy Laboratoryl April 2006. Technical Report No.: NREL/TP-620-39729.

4 Hybrid Energy Storage

4.1 INTRODUCTION

There are two applications of energy storage: power and heating and cooling. Sometimes, like in building, both applications simultaneously occur. The architecture and design of energy storage for heating and cooling needs are relatively simple, and most time, either thermal or chemical storage systems are used. Thermal storage can also be used for power application, since thermal energy can be converted to power with relative ease. A detailed description of various energy storage systems is given in my previous book [1]. In this chapter, we assess the energy storage for power application. While we briefly assess the pluses and minuses of individual storage systems [1–19], we mainly focus on the need, options, and other fundamentals such as architecture, management, control, and applications of hybrid energy storage systems.

Electric energy consumption has steadily risen since its industrial introduction in the second half of the 19th century. In fact, the world's total electrical energy production in 2009 was about 20,000 TWh, which is equivalent to the generated (and consumed) power of around 2.3 TW on average. This level of average power consumption is achieved by a combination of electricity generation stations, including heat engines fueled by chemical combustion or nuclear fission, kinetic energy of flowing water and wind, solar photovoltaics, and geothermal processes. Fossil fuels (coal, gas, and oil in that order) account for 67%, renewable energy (mainly hydroelectric, wind, solar, and biomass) for 16%, nuclear power for 13%, and other sources for 4% of all electrical energy produced worldwide. In recent years, the contribution of renewable energy is rapidly increasing. Emissions of pollutants and greenhouse gases from fossil fuel-based electricity generation are responsible for a significant portion of world greenhouse gas emissions. Although solar PV generation is advertised as environmentally friendly, it should be noted that fabrication of PV cells utilizes large amounts of water in addition to releasing toxic chemicals such as phosphorus and arsenic [1–19].

Electricity is the key to the proper functioning of modern human society. Ever-increasing electricity consumption gives rise to recent regulations and significant endeavors to improve the energy efficiency in all kinds of human activity from manufacturing to commerce, from transportation to digital communication, and from entertainment to laptops and portable devices. An important technology for helping reduce energy consumption is the ability to store any excess electrical energy for long periods of time and efficiently retrieve the stored energy. This is very important at present times when serious efforts are being made to increase the contributions of renewable sources of energy in power generation mix. The benefits and challenges of storage in all parts of electricity value chain are illustrated in Figure 4.1 [2].

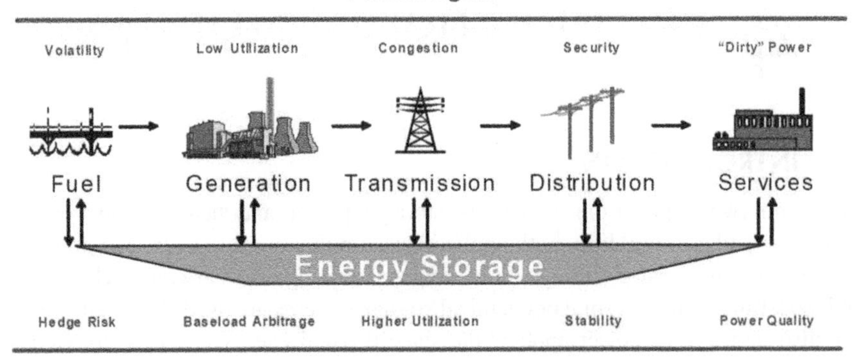

FIGURE 4.1 Benefits of ESS along the electricity value chain [2].

Reliable supply of electric energy is also an important issue. Power outage is regarded as a public emergency as people take the availability of uninterrupted power supply for granted. Electrical energy consumption in a system changes over time due to changes in the power requirements of load devices as well as the users' behaviors. Load-following power plants (for example, fossil fuel power plants) are intended to handle rapid changes in power demands on the power grid. In addition, the grid requires a certain level of operating reserve, which is made up of spinning and non-spinning reserves, in order to prevent blackouts and brownouts. Spinning reserve denotes the online extra generating capacity to deal with the peak power demand that can arise for a short period of time. Non-spinning reserve, on the other hand, refers to the off-line additional generating capacity that can be turned on and connected to the power grid after a short delay. Both the spinning and non-spinning reserves require extra capital investment by the utility companies for their generation facility setup and operation. Reserve power generation is generally more costly than the normal operation on the power grid. Some countries have only small reserve margin during peak hours, which threatens the power supply and demand match and gives rise to risky operating reserve guard banding. This can be remedied by building extra power plants. However, construction of new power plants requires large capital investment and has social and environmental costs.

To tackle the high demand for electric power and reduce the power plant overprovisioning, electrical energy storage systems (ESS) have been proposed [1, 8]. An ESS performs operating reserve management, which is performed by expensive, environmentally unfriendly load-following power plants. In addition, the ESS effectively enhances the power grid stability as well as the availability of renewable power sources such as windmills and photovoltaic (PV) panels. Renewable power sources have unreliable power generation characteristics; the level of power generation of the renewable power sources, such as PV cells and windmills, is heavily dependent on environmental factors (for example, the solar irradiance level or climate conditions).

The ESS also resolves the mismatch between the power generation and power consumption times in case of renewable power sources [19].

4.2 ENERGY STORAGE TECHNOLOGIES

Energy storage encompasses a wide range of technologies and resource capabilities, and these differ in terms of cycle life, system life, efficiency, size, and other characteristics. Although battery technology has attracted a great deal of industry attention in recent years, pumped hydro technology still supplies the vast majority of grid-connected energy storage (>95%) (see Table 4.1). The remaining categories combined comprise only 5% of installed capacity. As described below, there are several storage technologies in demonstration or development stages.

Homogeneous ESS are composed of a single type of energy storage elements. Majority of large-scale ESS that are currently deployed are homogeneous. Main advantages of homogeneous ESS are that the homogeneity of energy storage elements gives rise to relatively low design and operation complexities. Selecting the type of energy storage element is most critical for the homogeneous ESS design, whereas other decisions such as provisioning for the required energy and power capacities are naturally determined by performance requirements for the homogeneous ESS and characteristics of the energy storage elements [1, 19].

There are various types of energy storage systems. These can be listed as:

1. ***Electrochemical battery storage***: This class of energy storage includes advanced lead-acid, lithium-ion, sodium-based, nickel-based, and flow batteries. Technologies are further divided into subcategories based on the specific chemical composition of the main components (anode, cathode, separator, electrolyte, etc.). Each class and subcategory is at a different stage of commercial maturity and has unique power and energy characteristics that make it more or less appropriate for specific grid support applications [9–20]. In recent years, lithium-ion batteries are extensively used for both static and mobile applications. These different types of batteries and their specific characteristics are extensively described in my previous book [1].

TABLE 4.1
Energy Storage Distribution among Different Methods of Storage [1, 7]

Pumped hydro	95%	23.4 GW
Others (remaining 5% is distributed as)		1.2 GW
Thermal storage	36%	431 MW
Compressed air	35%	423 MW
Battery	26%	304 MW
Flywheel	3%	40 MW

2. ***Electrochemical capacitors***: This class of energy storage includes capacitors and super or ultracapacitors. A capacitor is a passive two-terminal electronic component that stores electrical energy in an electric field. The effect of a capacitor is known as capacitance. While some capacitance exists between any two electrical conductors in proximity in a circuit, a capacitor is a component designed to add capacitance to a circuit. Supercapacitors, also called electrochemical double-layer capacitors and ultracapacitors, bridge the gap between batteries and traditional capacitors; they store energy electrostatically. Supercapacitors are characterized by low internal resistance, which allows rapid charging and discharging, very high power (HP) density (but low energy density) and high specific energy (30 Wh/kg), and corresponding high cost per kWh. In recent years, electrostatic double-layer supercapacitors are widely used for both static and mobile applications. Different types of capacitors are extensively described in my previous book [1].
3. ***Mechanical storage systems***: This class of energy storage includes compressed air and flywheel storage systems. Compressed air energy storage (CAES) devices compress air and store it in a reservoir, typically underground caverns or aboveground storage pipes or tanks. Underground facilities are considered less expensive than aboveground and can operate for between 8 and 26 hours; however, siting underground compressed air storage facilities requires detailed research and time-consuming permission requirements. The low-pressure compressed air storage (five bar max) facilitates the use of isentropic relations to describe the system behavior and practically eliminates the need for heat removal considerations necessary in higher pressure systems to offset the temperature rise. The maximum overall system efficiency is around 97.6%, while the system physical footprint is less than 0.6 m^3 (small storage room). This provides a great option for storage in remote locations that operate on wind energy to benefit from a nonconventional storage system.

   Flywheels are the other mechanical energy storage technology. They accelerate a rotor (flywheel) to a very high speed in a very low-friction environment. The spinning mass stores potential energy to be discharged as necessary. Flywheels are modular and can range from 22 kW in size (Stornetic's EnWheel) to 160 kW (Beacon Power). Flywheels are best for short-duration, high-power, and high-cycle applications. They also have a much longer cycle life than other storage alternatives. Flywheels are less heat sensitive than batteries, and they last longer (up to 20 years guaranteed performance). More details are given in my previous book [1].
4. ***Magnetic energy storage***: This class of energy storage includes superconducting magnetic energy storage (SMES). SMES is a superconducting coil that stores energy in the form of magnetic field. SMES is capable of taking/delivering power from/to power system with excellent characteristics, for example, fast responding time, high efficiency, high power density, and high cycle life. SMES does not require any step-up or step-down interfaces;

its control is relatively simple, and the capabilities of SMES could be fully exploited in power systems [1, 20]. However, implementing a SMES system necessitates the refrigeration mechanism which is very costly and requires complex maintenance. Also, the special site requirements of the SMES have limited its applications to some stationary applications such as renewable power generation sites and railway supply substations.

5. ***Thermal storage***: This class of energy storage comes in many forms; the most well-known bulk thermal storage device is molten salt. Paired with solar thermal generation plants, molten salt thermal storage is used to improve the dispatchability of concentrated solar power (CSP) facilities. The stored energy powers steam turbines to continue generation after the solar day has ended. Other forms of thermal storage are more distributed in nature. These primarily interact with building heating and cooling systems and support demand-side services such as demand response. Some technologies, such as direct load control of water heaters, have already demonstrated deployment in electrical and heating networks. Thermal energy storage (TES) also involves storage of heat and cold. Besides thermal storage devices outlined above, storage of heat and cold also involves sensible heat storage such as ice chillers, ice box, phase change materials, and thermochemical heat storage [1].
6. ***Bulk gravitational storage***: This class of storage device includes technologies such as pumped hydro and gravel in railcars. Changing the altitude of solid masses can store or release energy via an elevating system driven by an electric motor/generator. Methods include using rails and cranes to move concrete weights up and down, using high-altitude solar-powered buoyant platforms supporting winches to raise and lower solid masses, and using winches supported by an ocean barge for taking advantage of a 4 km (13,000 ft) elevation difference between the surface and the seabed. Efficiencies can be as high as 85% recovery of the stored energy [1].

   The gravel/railcar storage method operates in a similar manner to pumped hydro. Off-peak power is used to move railcars filled with gravel or another heavy material up a slope. When power is needed, the railcar moves down the slope, converting gravitational energy into electricity as it moves down. Unlike pumped hydro, railcar/gravel energy storage does not require reservoirs to function. Rather, it requires a long slope of existing or new railroad track. This makes it potentially easier to site than pumped hydro, although it is still not suitable for urban areas, nor is it suitable for railroad segments where there is existing traffic.

   Pumped hydro is a mature technology used throughout North America and in other parts of the world. Off-peak power is used to pump water from a lower reservoir to a higher reservoir; then the water is released to generate electricity during peak periods. Because pumped hydro facilities require aboveground reservoirs, specific land configurations are needed. Pumped hydro projects are rarely located close to urban centers, and permission can take many years due to their large environmental impact.

7. ***Hydrogen storage***: Hydrogen is stored conventionally as a gas in steel cylinders at high pressures (e.g., 2,000 psi) and at lower pressures as a liquid in insulated containers. Both methods of storage require comparatively bulky storage containers. In addition to their unwieldy size, such containers are inconvenient due to the high pressure required for gas storage in cylinders and the ever present danger of gaseous hydrogen evolving from boiling-off of the liquid form. In recent years, considerable attention has been focused on the storage of hydrogen as a metallic compound, or hydride, or various substances. Metal hydrides can store large amounts of hydrogen at low- and even sub-atmospheric pressures in relatively small volumes. This is a subject of intensive research.

All of these storage technologies are extensively described in my previous book, and readers are encouraged to refer to my book [1].

4.2.1 Comparison of Energy Storage Devices

Not all ESS technologies are ready for large-scale and widespread deployment. As of 2009, only four energy storage technologies (sodium-sulfur batteries, pumped hydro, CAES, and thermal storage) have a total worldwide installed capacity that exceeds 100 MW [21]. The main reason is that, in spite of the large variety of ESS technologies, no technology offers sufficient performance in respect of key figures of merit needed of an electrical energy storage medium. The performance matrix needed for ESS to handle HRES is described in detail in section 4.4. For instance, a high-performance ESS should exhibit high cycle efficiency, high power and energy storage capacity, low cost, high volumetric and/or gravimetric density, efficiency of charge/discharge cycle, and long cycle life. The breakdown of various storage technologies based on its function as energy management or power quality and reliability is illustrated in Figure 4.2 [2]. The ESS technology of choice for many applications (especially those requiring high volumetric and/or gravimetric density) is battery storage. However, as shown in Table 4.2, different battery technologies have different storage characteristics. Again, no single storage technology can simultaneously achieve all the desired characteristics of a high-performance ESS mentioned above

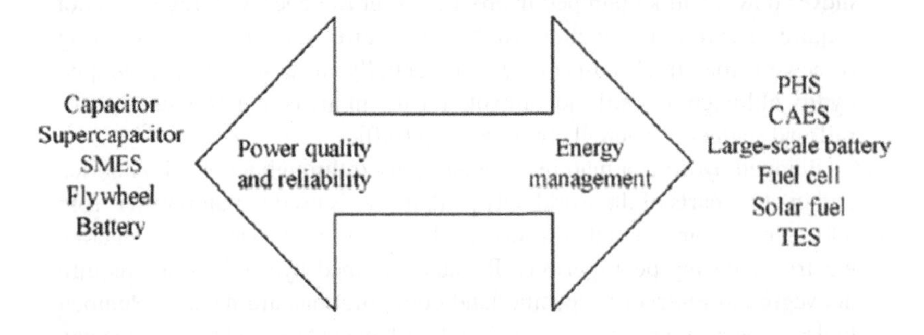

FIGURE 4.2 Energy storage classification with respect to function [2].

TABLE 4.2
ESS Technologies Properties [2–5, 22, 23]

System	Rating		Density		Lifetime	Efficiency	Self-discharge per Day
Storage Device	**Power Rating (MW)**	**Typical Discharge Time**	**Power Density (W/l)**	**Energy Density (Wh/l)**	**Years**		
Pumped hydro	100–5,000	1–24 h	0.1–0.2	0.2–2	450	70–80	Very small
Compressed air	5–300	1–24 h	0.2–0.6	2–6	425	41–75	Small
Flywheel	0–0.25	s–h	5,000	20–80	15–20	80–90	100
Fuel cell	0–50	s–24 h	0.2–20	600 (200 b)	10–30	34–44	0
Supercapacitor	0–0.3	ms–1 h	$(4–12)*10^4$	10–20	4–12	85–98	20–40
SMES	0.1–10	ms–8 s	2,600	0.5–10	–	75–80	10–100
Batteries							
Lead acid	0–20	s–h	90–700	3–15	3–15	75–90	0.1–0.3
NiCd	0–40	s–h	75–700	5–20	5–20	60–80	0.2–0.6
Li-on	0–0.1	min–h	1300–10,000	5–100	5–100	65–75	0.1–0.3
NaS	0.05–8	s–h	120–160	10–15	10–15	70–85	10–20
VRB	0.03–3	s–10 h	0.5–2	5–70	5–20	60–80	Small (0.1%–0.4%/day)
ZnBr	0.05–2	s–10 h	1–25	5–10	5–10	65–75	Small

Storage Device	Short Term (<1 min)	Mid Term (<1 m,<2d)	Long Term (>2d)	Reaction Time	Installation Costs (Euro/kW)	Installation Costs (Euro/kWh)
Batteries						
Lead acid		XXX	X	3–5 ms	150–200	100–250
Li-ion	X	XXX		3–5 ms	150–200	300–800
NaS		XX	X	3–5 ms	150–200	500–700

(Continued)

TABLE 4.2 (CONTINUED)
ESS Technologies Properties [2–5, 22, 23]

Storage Device	Short Term	Mid Term	Long Term	Reaction Time	Installation Costs	Installation Costs
	(<1 min)	(<1 m,<2d)	(>2d)		(Euro/kW)	(Euro/kWh)
VRB	X	XX	XX	>1s	1,000–1,500	300–500
Pumped hydro		XX	XXX	>3 min	500–1,000	5–20
Flywheel	XXX	X		>10 ms	300	1000
Fuel cell		X	XXX	10 min	1,500–2,000	0.3–0.6
Supercapacitor	XXX			<10 ms	150–200	10,000–20,000
SMES	XXX			1–10 ms	high	High
CAES		XX	XX	3–10 min	700–1,000	40–80

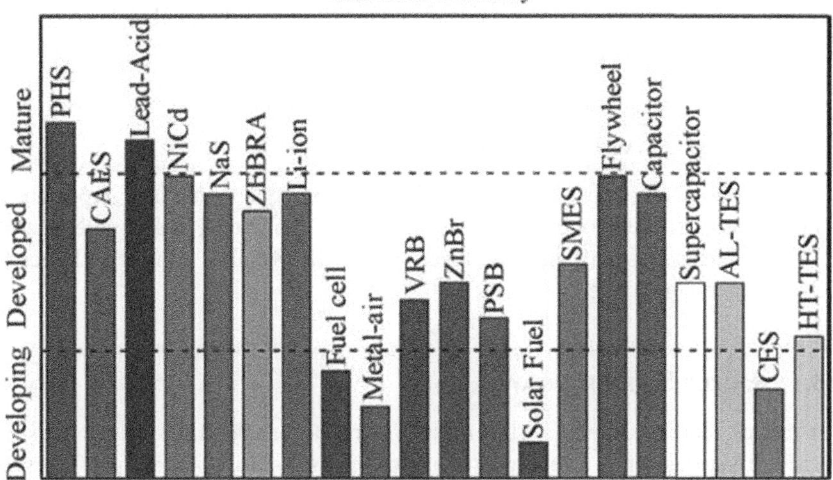

FIGURE 4.3 Technical maturity of EES systems [2].

(see Table 4.2). Furthermore, as shown in Figure 4.3, different storage devices have different degrees of readiness for commercialization. When considering a particular energy storage device, as mentioned above, a number of property requirements need to be considered, which include the technical and commercial maturity of the storage technology. While individual storage system may satisfy some of these elements, just like sources for power generation, each storage device possesses some drawbacks. Despite numerous research efforts in order to improve ESS capabilities over the past decade, a perfect ESS technology that copes the drawbacks in terms of all aspects is not to be expected to develop in the near future.

As mentioned in the earlier chapters, hybrid energy generation has become important because no single energy source satisfies all the criteria needed for the sustainable energy generation. The same principle applies for energy storage. As shown in Table 4.2, no single energy storage device has all the favorable properties for the storage. This has led to the concept of hybrid energy storage. A hybrid energy storage system comprised of heterogeneous types of energy storage elements organized in a hierarchical manner so as to hide the weaknesses of each storage element while eliciting their strengths has become important. The hybrid storage system also offers challenges that one faces when dealing with the optimal design and runtime management of a hybrid energy storage system, targeting some specific application scenario, for example, grid-scale energy management, household peak power shaving, mobile platform power saving, and more. In order to increase the range of advantages that a single ESS technology can offer, and at the same time, enhance its capabilities without fundamental development of the storage mechanism and only via complementary use of the existing ESS technologies, more than one ESS technologies can be hybridized. A hybrid energy storage system (HESS) is composed of two or more heterogeneous ESS technologies, with matching characteristics, and combines the power outputs of them in order to take advantage of each individual technology.

4.3 ENERGY STORAGE, POWER CONVERTERS, AND STORAGE CONTROLLER

Before we move onto various aspects of hybrid storage systems, it is important to examine basic architecture of energy storage, power converters, and storage controller of a homogeneous ESS, because heterogeneous ESS follow similar architecture with a greater degree of complexity.

4.3.1 Energy Storage Array

The energy storage element is the most important consideration for ESS design. There are various types of energy storage elements as discussed earlier. The type and capacity of energy storage element is determined according to the application and requirement of the ESS with restrictions on volume, weight, cost, and so forth. When a single energy storage element does not meet the energy or power requirement, which is usual for a large-scale ESS, multiple energy storage elements are connected together to form an array of energy storage elements. An energy storage array is a set of multiple identical energy storage elements that are connected in series and/or parallel, forming a regular matrix. The dimension of the energy storage array is determined by power and energy capacity requirement and the maximum voltage rating of the energy storage array. The regular array structure makes it possible to maintain the same states of charge (SoC) and state of health (SoH) of all the elements in the array.

Some types of energy storage elements require balancing between serially connected cells. Even though all the energy storage elements are of the same type, manufacturing variation in practice may result in an imbalance of characteristics such as capacity and internal resistance that causes imbalanced SoC during operation and even damages to the elements [24]. Supercapacitors and lithium-ion batteries require external cell balancing circuits. Cell balancing is an active research area, which is critical for large-scale ESS [25–28]. The literature categorizes cell balancing methods into three categories: charging methods (steering), active methods, and passive methods (bleeding) [29]. The charging method selectively bypasses fully charged cells. The active cell balancing methods moves energy from more charged cells to less charged cells by means of capacitors or power converters. The passive cell balancing methods simply dissipate energy from overcharged cell as heat.

Configuration of energy storage elements, that is, the number of series and parallel connections, may be dynamically adjusted. For example, with four supercapacitors, three configurations, 1-by-4, 2-by-2, and 4-by-1, are feasible. Supercapacitors, which have wide voltage variation according to SoC, benefit from the dynamic reconfiguration [30]. A supercapacitor array maintains roughly a constant voltage by increasing (or decreasing) the number of series connections as the voltage of each energy storage element decreases (or increases). Dynamic reconfiguration circuit proposed by Fang et al. [31], Kim et al. [32], Uno [33, 34], and Uno and Toyota [35] aims at maintaining marginally constant supercapacitor array voltage. It is shown that explicitly considering the power converter efficiency further improves the energy efficiency than simply maintaining a constant voltage [32].

Fault resilience is another benefit of the dynamic energy storage array reconfiguration. Energy storage elements are subject to manufacturing variations. Each energy storage element shows different aging-induced battery capacity degradations [30, 36]. Dynamic reconfiguration may prolong the lifetime of the energy storage array when short-circuit faults or open-circuit faults occur in the supercapacitor array [30]. It is more practical to operate the partly degraded supercapacitor array rather than replacing the whole array when most supercapacitors are healthy. Energy storage array reconfiguration improves dependability, efficiency, and scalability of large-scale battery arrays. The real-time controller proposed in [37, 38] makes decisions and rearranges in series or parallel while bypassing faulty batteries [30].

4.3.2 Power Converters

The energy storage array is not directly connected to the power source or load because of its varying terminal voltage depending on its SoC, load current, temperature, and so on. Therefore, a power converter is placed between the energy storage array and power source or load and generates regulated voltage or current for charging or discharging the energy storage array. Batteries and supercapacitors are DC energy storages, and so storing energy from the AC power grid and supplying power from and to the AC load electronics require AC-to-DC and DC-to-AC conversions. An AC-to-DC power converter is often referred to as a rectifier, and a DC-to-AC power converter is often referred to as an inverter [30].

Power conversion is not free. Converting the voltage level or between AC and DC involves non-zero amount of power loss. The overall power loss includes conduction losses by parasitic resistances of circuit components, switching losses by parasitic capacitances of switching devices, power consumption of the controller circuit, and so on. The power conversion efficiency η_c is defined as a ratio of power output P_{out} to power input P_{in} levels of the power converter. P_c (difference between input and output power) is the power loss in the power converter. This power loss is not constant, but varies depending on the input and output voltage and the amount of power that is transferred through the converter, and so the power conversion efficiency also varies. The power conversion efficiency is a critical factor, which determines the energy efficiency of the ESS together with the cycle efficiency of the energy storage elements. Note that the fact that the cycle efficiency of the supercapacitor is nearly 100% does not guarantee high energy (HE) efficiency of the ESS composed of supercapacitors only because the terminal voltage of such an ESS varies in a very wide range depending on its SoC, which makes it difficult to design an energy-efficient power converter. The efficiency of the LTM4609 buck-boost converter from Linear Technology [30] shows a wide range of variation depending on input voltage, output voltage, and output current.

Many research efforts have focused on analyzing the power conversion efficiency. Understanding and enhancing the efficiency of power converters is important for the optimal design and operation of the ESS. The modulation method (pulse width modulation (PWM) versus pulse frequency modulation (PFM)) of a power converter has a significant impact on its conduction and switching losses [30]. The PWM and PFM DC-to-DC converter efficiencies are studied and specified as functions of the input voltage, output voltage, and output current by Choi et al. [39]. Inverters, which generate AC power out of DC energy storage elements or DC power sources, are also significant. Many different topologies for the power converters are explored by Nergaard et al. [40] and Xue et al. [41].

The primary task of power electronics is to process and control the flow of electric energy by supplying voltages and currents in a form that is optimally suited for user loads. Modern power electronic converters are involved in a very broad spectrum of applications like switched-mode power supplies, active power filters, electrical machine motion control, renewable energy conversion systems, distributed power generation, flexible AC transmission systems, vehicular technology, etc.

Power electronic converters can be found wherever there is a need to modify the electrical energy. This is a different form of classical electronics in which electrical currents and voltage are used to carry information, whereas with power electronics, they carry power. Some examples of uses for power electronic systems are DC/DC converters used in many mobile devices, such as cell phones or PDAs (personal digital assistants), and AC/DC converters in computers and televisions. Large-scale power electronics are used to control hundreds of megawatt of power flow across our nation. Some of those converters are briefly described below.

A. **Dual converter**

Dual converter is a combination of a rectifier and an inverter in which the conversion of AC to DC happens and is followed by DC to AC, where the load lies in between. A dual converter can be of a single phase or a three phase. A dual converter consists of two bridges consisting of thyristors in which one is used for rectifying purpose where alternating current is converted to direct current which can be given to load. Other bridge of thyristors is used for converting DC to AC [4].

A single-phase dual converter uses a single phase as the source which is given to converter 1 of the dual converter for rectification which follows to load. An AC input is given to converter 1 for rectification. In this process, a positive cycle of input is given to the first set of forward biased thyristors, which gives a rectified DC on positive cycle, and a negative cycle is given to a set of reverse biased thyristors, which gives a DC on negative cycle, completing a full wave rectified output which can be given to load. During this process, converter 2 is blocked using an inductor. As the thyristor only starts conducting when current pulse is given to gate, it continues conducting until the supply of current is stopped. Output of thyristor bridge can be given to different loads. As a dual converter also consists conversion of DC to AC, to make it work, converter two is blocked, and DC inputs become load to DC power source conversion [4].

To make thyristors conduct, a trigger pulse must be given to its gate simultaneously, along with line voltage. A separate gate drive circuit must be added to dual converter thyristor bridges. Gate drive circuit must be equally synchronized with source voltage; any delay causes zero cross jitter and zero frequency fluctuates. To prevent these, circuits must be included with phase lock loops and comparators.

A single-phase dual converter can be used in controlling speed and direction of rotation interfacing with microcontroller. Combination of four SCRs is placed on either side of the motor. These thyristors can be triggered through an optocoupler which is connected to a port of the microcontroller. Rotation of motor can be initialized using optocoupler by setting a set of thyristor to trigger which is placed at one side, and change in the direction of the motor can be achieved by triggering another set of thyristor. Variation in speed of motor can be achieved by delayed firing angle of SCR. Mode selection and speed selection are microcontroller-interfaced switches, and using these switches speed and rotation can be selected [4].

B. **Single-phase three-leg AC/AC converter**

Power electronics is the application of electronics for power conversion. A subcategory of power conversion is the AC to AC conversion. An AC to AC voltage controller is a converter which controls the voltage, current, and average power delivered to an AC load from an AC source. There are two types of AC voltage controllers: single- and three-phase AC controller. A single-phase AC/AC converter is a converter which converts from a fixed AC input voltage into a variable AC output voltage with a desired frequency. They are used in practical circuits like light dimmer circuits, speed controls of induction motors and traction motor control, etc. There are many existing technologies in single-phase AC/AC converters; they are single-phase two legs, three legs, and four legs. The single-phase two- and four-leg converters have some demerits like they need large number of power devices, large control circuitry, more switching, and losses are reduced only half to control the 50% of the output. So, to overcome these demerits present in the conventionally used converters, a better approach is the use of single-phase three-leg AC/AC converter [4].

A single-phase three legs consists of three legs and six switches. A leg is common for both grid side and load side. A leg performs the rectifier operation and a grid performs the inverter operation. And in this, one uses pulse width modulation (PWM) techniques for controlling the converter output. The advantages and applications of a three-leg converter include:

1. The DC output voltage across the capacitor is almost doubled compared to the four-leg converter.
2. The power rating and voltage of the circuit can be improved.
3. Same output can be obtained with reduced losses and switches. Hence the efficiency and the power factor can be improved.

This converter is used in uninterruptable power supply (UPS) circuits and in power electronics for getting four quadrant operations of the drives [4].

4.3.3 System Control

Control of the ESS is a not trivial problem. Even a homogeneous ESS requires elaborate system control schemes beyond simple charging and discharging of power converters in order to maximize benefits of the ESS. Objectives of the ESS control typically include enhancing the energy efficiency, cycle life, and reliability.

The basic functionality of the ESS control is management of the energy storage elements. The ESS controller monitors the voltage, current, and temperature of each energy storage element and prevents the energy storage elements from operating in an unsafe range. Distributed ESS controllers communicate with each other or an external device for monitoring and control through a communication network. Control area network (CAN) bus communication is one of the most widely used communication networks for system control [30, 42], which can be used here as well.

An ESS also provides application-specific system controls. ESS controllers for renewable power sources perform MPPT (maximum power point tracking) in order to maximize power generation from the power sources [30, 43, 44]. The controller is responsible for finding and maintaining the energy-optimal operating point that dynamically varies depending on the environmental conditions. They also should determine the amounts of charge and discharge currents from the power grid considering the power generation and power demand. The optimization objectives include minimizing the daily energy cost considering the time-of-use pricing and maximizing the battery lifetime [30, 45]. ESS controllers for EV/HEV applications determine the direction and amount of power flow among components such as internal combustion engine (ICE), ESS, traction motor, and alternator depending on the operation modes. Optimal system control is critical not only for traction performance but also for energy efficiency and battery lifetime [30, 46].

4.4 PERFORMANCE MATRIX FOR ENERGY STORAGE FOR HRES

As the level of renewable energy sources in power generation increases, the role of storage devices becomes more important to handle the intermittency of solar and wind power generation. Matching renewable generation intermittency to the demand in an electricity supply system requires the introduction of the energy storage system (ESS) technologies in the power systems [23, 47]. Besides storing and smoothing renewable power, there are numerous advantages related to the advent of ESSs in the power systems. ESSs can increase power system operation and planning resiliency and efficiency by means of many applications as shown by Kim et al. [30] and others [23, 47, 48]:

- Energy time-shift supply capacity load following
- Area regulation
- Fast regulation
- Supply spinning reserve
- Voltage support

- Transmission congestion relief
- Transmission and distribution upgrade deferral power quality
- Renewable energy time-shift
- Renewable capacity firming
- Renewable energy smoothing
- Service reliability
- Black start

In addition, the rapid growth of electric and hybrid vehicle technology opens a new era of ESS. However, ESS solutions are not quite commercially or technologically mature in many features, causing obstacles to their extensive utilization. The ESS technologies are different in terms of cost and technical properties such as [23, 30, 47, 48]:

- Energy and power rating
- Volumetric and gravimetric energy density, volumetric and gravimetric power density, and discharge time
- Response time
- Operating temperature
- Self-discharge rate
- Round-trip efficiency
- Lifetime (years and cycles)
- Investment power and energy cost
- Spatial requirement
- Environmental impact
- Recharge time
- Memory effect (batteries)
- Maintenance requirements and recyclability
- Technical maturity
- Transportability
- Cumulative energy demand

Based on the above requirements for energy storage of power, the following performance matrix is required for energy storage devices:

A. **Cycle efficiency and internal resistance**

Cycle efficiency of an energy storage element is defined by the round-trip energy efficiency, that is, ratio of the amount of energy output during discharging to the energy input during charging. In other words, the cycle efficiency is the product of charging efficiency and discharging efficiency. Here, the charging efficiency is the ratio of energy stored in an energy storage element after charging to the total energy supplied to that element during the entire charging process, and discharging efficiency is the ratio of energy extracted from an energy storage element during discharging to the total energy before the discharging process begins.

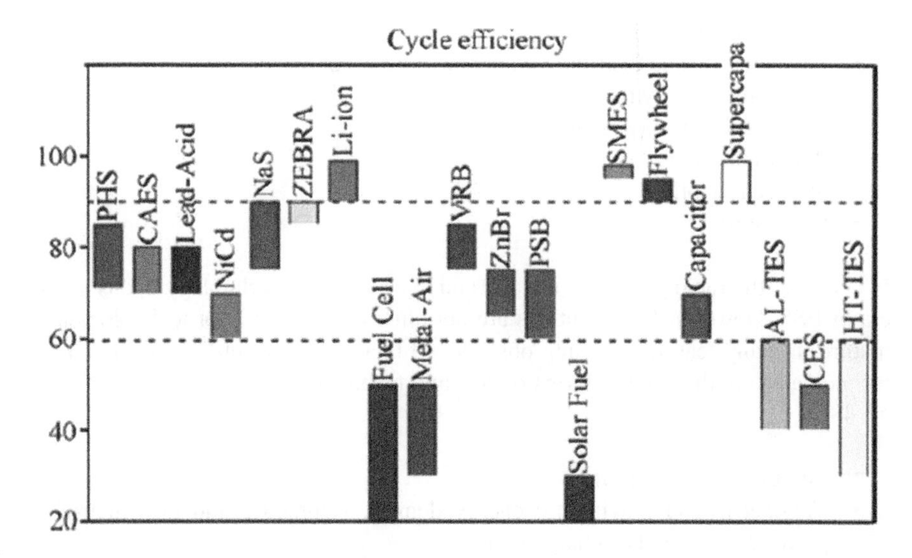

FIGURE 4.4 Cycle efficiency of EES systems [2].

Power loss during the charge and discharge cycles is mostly due to internal resistance of the energy storage elements. Cycle efficiency is drastically affected by charging/discharging profiles, that is, the magnitude and shape of the charging/discharging current. For example, due to the well-known rate capability effect, the total energy delivered by a battery goes down with the increase in load current, resulting in lower discharge efficiency. At the same time, the recovery effect of battery which recovers the terminal voltage during idle periods between current pulses also affects the cycle efficiency of the battery [49]. Cycle efficiency of various storage devices is illustrated in Figure 4.4 [2].

Supercapacitor and flywheel have a very high cycle efficiency, which is close to 100%, and is less affected by the charging/discharging profile. A high cycle efficiency means less energy loss during charging and discharging processes, which leads to low operational cost per each cycle. Therefore, it is wise to use a high cycle efficiency energy storage element such as a supercapacitor for frequent charging/discharging applications.

B. **Rate capability**

Rate capability (also referred to as the power capability and sometimes called the rate capacity) signifies the capability to provide high power without degradation of the total amount of energy. A battery's available capacity decreases as the discharging rate increases. Peukert's law describes the phenomenon that the delivered capacity normalized to the rated capacity has an exponential relationship to the discharge current normalized to the rated current [50]. That is,

$$I^k \cdot t = \text{fixed capacity} \tag{4.1}$$

where I is the discharge current, t is discharge time, and k is the Peukert constant, which is typically a value between 1 and 2. Generally, the Peukert constant varies according to the age of the battery, commonly increasing with age. In an ideal battery where the total delivered capacity is independent of the discharge current, k = 1. For a lead-acid battery, k is between 1.1 and 1.3. In contrast, supercapacitors have a very good rate capability and can withstand very high discharge rates with virtually no loss of available capacity. Note that if I is described as the actual discharge current relative to 1 ampere, then $I^k \cdot t$ will give the battery capacity at a 1-ampere discharge rate. This is another common form of Peukert's law.

C. **Energy density versus power density**

Power density is defined as the rated output power divided by volume (W/L) or mass (W/kg) of the energy storage element. Similarly, energy density is the stored energy divided by the volume (Wh/L) or mass (Wh/kg). Generally speaking, power density is related to the amount of instantaneous power that an ESS element can provide; on the other hand, energy density is related to the time duration that the energy storage element can last while supplying a certain amount of power.

Supercapacitor and flywheel have a very high power density, whereas typical lithium-ion batteries have a marginally high power density. High power density energy storage elements are suitable as temporary energy buffers to deal with short-duration high-power demand of some applications. Metal-air batteries and fuel cells have a high energy density, which is multiple orders of magnitude higher than those of typical batteries. High energy density energy storage elements are suitable as long-term energy storage means. Supercapacitor is one of the energy storage elements that has the worst energy density, and so it is not economical to use them alone as a large-scale energy storage means.

A HESS can rely on high energy density energy storage elements for long-term storage and a relatively small amount of high power density energy storage elements for high output power rating. Battery-supercapacitor hybrid is a representative HESS, which exploits the high energy density of the battery and high power density of the supercapacitor. Other suitable combinations for energy and power supply are illustrated in Table 4.3.

D. **Discharge time or self-discharge rate**

Self-discharge rate is a measure of how quickly a storage element loses its stored energy when there are no charging and discharging currents. It is heavily dependent on how the energy storage element stores the energy inside, as well as ambient conditions such as the temperature and humidity. Supercapacitor and flywheel, which have a high cycle life, have the highest self-discharge rate; that is, they typically lose all their stored energy within a few days, or even hours. One should thus not store energy in a supercapacitor if it is expected that the energy will not be used for a long time. On the other hand, electrochemical batteries store the energy with stable chemicals and do not lose so much energy by themselves.

TABLE 4.3
Suitable Storage Combinations for Energy and Power Supply and Their Applications [23, 30, 47, 48]

Energy Supplier	Power Supplier	HESS Application
Battery	Supercapacitor	General MG, hybrid PV-wind, PV, EV, wind, battery-powered EV
	SMES	General MG, wind, wave
	Flywheel	General MG, wind
Fuel cell	Supercapacitor	General MG, PV, wind, fuel cell-powered EV
	SMES	General MG
	Battery	General MG, hybrid PV-wind, PV
	Flywheel	General MG
Flow battery	Supercapacitor	General MG, PV, wind
	SMES	General MG
	Battery	General MG, hybrid, PV-wind, PV
CAES	Supercapacitor	General MG
	SMES	General MG
	Flywheel	General MG
	Battery	General MG

The choice of an energy storage device depends on its application in either the current grid or in the renewables/ vehicle to grid (VG)-driven grid; these applications are largely determined by the length of discharge. Energy storage applications are often divided into three categories, based on the length of discharge. Table 4.4 indicates the three regimes of energy storage applications commonly discussed [21].

The first two categories of energy storage applications in Table 4.4 correspond to a range of ramping and ancillary services, but do not typically require continuous discharge for extended periods of time. In the case of renewables-driven applications, this could require discharge times of up to about an hour to allow fast-start thermal generators to come online in response to forecast errors. Bridging power typically refers to the ability of a storage device to "bridge" the gap from one energy source to another. The third category (energy management) corresponds to energy flexibility or the ability to shift bulk energy over periods of several hours or more. Figure 4.5 shows the discharge time for various storage technologies [5]. It should

TABLE 4.4
Three Classes of Energy Storage [21]

Common Name	Example Applications	Discharge Time Required
Power quality	Transient stability, frequency regulation	Seconds to minutes
Bridging power	Contingency reserves, ramping	Minutes to ~1 hour
Energy management	Load leveling, firm capacity, T&D deferral	Hours

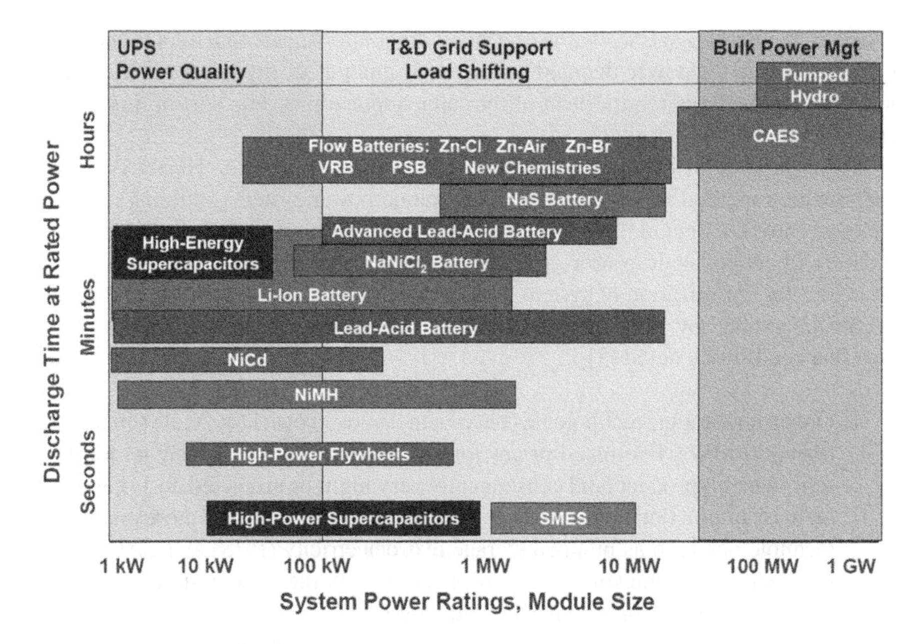

FIGURE 4.5 Discharge time characteristics of energy storage devices [5].

be noted that this figure does not include thermal energy storage, which would cover a power range of a few kilowatts (kW) for thermal energy storage in buildings to more than 100 MW in CSP plants, with a discharge time of a few minutes to several hours [5, 21].

E. **Capital cost**

Capital cost, which is an important consideration in the design and implementation of an ESS, is typically represented in the forms of cost per unit of delivered energy ($/Wh) or per unit of output power ($/W). The capital cost determines how much money should be invested in order to build an ESS with a certain amount of energy capacity or power capacity. Some batteries are superior to others in terms of both their energy and power densities (for example, lithium-ion battery is better than lead-acid in both respects), yet the cost of these batteries is quite different from lead-acid having much lower cost. One should thus carefully determine the portion of expensive (but higher performance) energy storage elements and inexpensive (but lower performance) energy storage elements for a given monetary budget. In fact, the capital cost gap between energy storage elements is a key motivation for the hybrid storage approach because one cannot afford to use an unlimitedly large amount of good, yet expensive, energy storage elements to meet the ESS requirements. For example, one can use a very large amount of supercapacitors to meet both the high-output power requirement and the high-energy storage requirement of some applications, but the cost of such an ESS will be impractically high. Instead, consider an alternative ESS that uses a small amount of supercapacitors to meet the high instantaneous power demand and a small amount of lithium-ion batteries to meet the high storage requirement of the same applications. The cost of this hybrid combination will be affordable.

Some previous researches attempted to minimize the cost of HESS. A sizing optimization method for stand-alone photovoltaic power systems utilizing a HESS was presented by Li et al. [51]. The optimization goal is to find a cost-optimal combination of photovoltaic power generation module, fuel cell, and battery. Control strategies for optimization of system performance and cost in a battery and fuel cell HESS are studied by Vosen and Keller [52]. There are other factors regarding the cost that needs to be noted [21]:

1. Only a few storage technologies have been deployed on a large scale (greater than 100 MW). Estimated prices for emerging technologies may be for a semi-custom product (and consequently very high) or projected costs based on mass production (and perhaps overly optimistic). Even with more mature technologies such as pumped storage hydroelectricity (PHS) and CAES, it has been some time since either has been built in the United States, so the cost of the next plant is somewhat uncertain.
2. As with any generation technology, large variations in prices occur from year to year due to commodity prices and the global economy. Therefore, cost estimates of storage technologies from different years may reflect market conditions as opposed to real differences.
3. Storage technologies offer different classes of services and are comprised of an energy component and power component. The total cost of a storage device includes both components, with the limits of the target application. As a result, a direct comparison of a PHS device with a flywheel, for example, has limited value.

F. **Cycle life and state of health (SoH)**

State of health (SoH) of an energy storage element is a measure of its age. It captures the general condition of the energy storage element and its ability to store and deliver energy compared to its initial state (that is, when it was a fresh device out of the manufacturing line). During the lifetime of the energy storage element, its capacity (or "health") gradually deteriorates due to irreversible physical and chemical processes that take place along with usage. Cycle life is the maximum number of charging and discharging processes that an energy storage element can perform before its capacity drops to a specific percentage (60%–80% typically) of its initial capacity. It is one of the key performance parameters, which gives an indication of the expected working lifetime of the energy storage element. The cycle life of an energy storage element is closely related to its replacement period and thus to the full cost of the element according to a life cycle analysis. Figure 4.6 illustrates the average energy storage duration versus the power capacity for various storage devices.

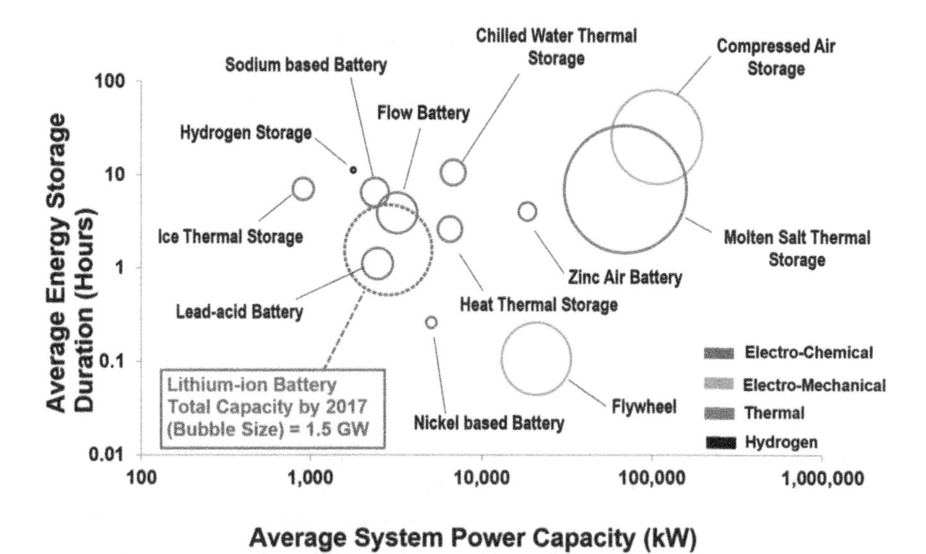

FIGURE 4.6 Average energy storage duration versus power capacity for various storage devices [3].

Cycle life of an energy storage element heavily depends on the usage pattern of the element, especially to the depth of discharge (DoD) which is defined as the ratio of used capacity to the initial full capacity. For lead-acid batteries, the number of available charging/discharging cycles increases when lowering the DoD [53]. Similar conclusions hold for most electrochemical batteries.

Typically, energy storage elements whose operation principles are based on electrical, mechanical, or thermal technologies, such as supercapacitor, flywheel, thermal energy storage (TES), and cryogenic energy storage (CES), typically have long cycle lives. In contrast, the cycle lives of electrochemical batteries are not that high due to unavoidable chemical deterioration of the electrodes during their operation.

G. **Efficiency**

While efficiency data for various storage technologies are reported in Table 4.2, the following factors about efficiency should be noted [21]:

1. The standard measure of an electricity storage device's efficiency in the grid is the AC to AC round-trip efficiency, or AC kWh_{out}/kWh_{in}. However, this is not always the value reported, especially for devices that store DC energy such as batteries and capacitors. In some cases, the DC-DC round-trip efficiency may be reported, and additional losses in power conversion efficiencies must be considered if the device is to provide applications in the grid.
2. Reported round-trip efficiencies may not include "parasitic" loads. These include heating and cooling of batteries and power-conditioning equipment.

These parasitic loads can vary considerably depending on the use, climate, and length of each storage cycle.

3. The round-trip efficiency of several technologies cannot be directly compared. Thermal storage provides some, but not all, of the services of a "pure" electricity storage device, while compressed air energy storage is a hybrid device that requires both electricity and natural gas. These factors limit the value of a direct comparison.

H. **Environmental impacts**

In recent years, the importance of environmental friendliness of energy storage elements is being emphasized. Typically, electrochemical batteries have a negative impact on the environment due to their reliance on toxic metals such as lead and cadmium that can be quite harmful if not disposed of properly. They should be recycled properly for both reducing environmental impact and conserving scarce resources [54]. The environmental impacts of an energy storage element are closely related to its expected cycle life. In other words, we can reduce the negative impacts of using hazardous energy storage elements by extending their cycle lives. Note that the supercapacitor and flywheel have very small or almost negligible impacts on the environments, not only because they do not contain harmful materials, but also because of their long cycle lives. A comparison of various storage elements toward the above described performance matrix is illustrated in Tables 4.2.

Furthermore, as shown in Table 4.1, battery energy storage is only a small portion of total energy storage currently used. Pumped hydro currently stores about 95% of our electrical energy storage. The distribution among various energy storage elements is illustrated in Table 4.1. The maturity of these and other technologies is illustrated in Figure 4.3. The largest percentage of these storage technologies are used for capacity between 100 kW and 1 MW. Just like sources of power, as shown in Table 4.2, these storage technologies have different strengths and weaknesses.

It is a practically promising solution to develop system-level design methodology that enhances the storage system performance and lifetime through efficient use of the current energy storage technologies. A hybrid ESS (HESS) consists of multiple heterogeneous energy storage elements so as to exploit the unique advantages of each energy storage element while hiding their unique shortcomings by introducing novel storage system architecture and hierarchy, along with sophisticated charge management policy and means [23, 30, 55]. The HESS concept is derived from an analogy to the computer memory hierarchy employed in computer systems and is used to provide low-latency, yet low-cost, access to program and data storage.

However, designing the optimal HESS is not a trivial problem. Simply mixing different types of energy storage elements does not automatically guarantee to make up a high-performance ESS. HESS architectural design is a multivariable multi-objective optimization, and the heterogeneity of the energy storage elements explodes the design complexity. It includes both continuous and discrete design parameters and many complex nonlinear models. Management policies of HESS involve in another highly complicated runtime optimization. Therefore, computer-aided design and optimization is a must for the optimal design and operation of the HESS with reasonable time and efforts.

An ESS performs many useful functions such as load leveling, contingency service, voltage stabilization, maximum power point tracking (MPPT) for renewable power sources, and more, for a wide range of applications including portable devices, household appliances, EV/HEV, and even power grid. Just like hybrid energy generation, hybrid storage device can give energy system more reliability, flexibility, cost-effectiveness and sustainability.

4.5 MOTIVATION, BASIC FEATURES, AND PRINCIPLES OF HESS

As discussed above, homogeneous ESSs are inherently subject to the limitations/weaknesses of the energy storage elements that constitute them. For example, battery-based ESSs have a limited number of operation years, which is determined by the cycle life of the batteries. A supercapacitor has a very high cost per energy, and so it is not very economical to use in low-power high-energy ESS. HESS is an emerging technology that overcomes the limitations of homogeneous ESS by utilizing heterogeneous energy storage elements. For example, the battery-supercapacitor HESS aims at achieving high energy capacity of the battery and high power capacity of the supercapacitor at the same time. However, a HESS requires a nontrivial initial design and runtime control in comparison with a homogeneous ESS. Not only the types but also the proportions of each energy storage element should be carefully determined at design time. Energy flow to each energy storage element during operation is also important. The benefits of the HESS are many and include high energy efficiency and long lifetime, but these benefits are achievable only through careful optimization and management strategies [30].

Recently HESS has attracted attention from the research community due to its need for systematic optimization and careful charge management at runtime [56–58]. For example, lifetime maximization for a battery-supercapacitor HESS is discussed by Mirhoseini and Koushanfar [58, 59], with systematic approaches utilized for the estimations of distributed algorithms (EDA) problems. More examples of power management of power sources and load devices when considering fuel cell-battery hybrid systems are given in references [60–62].

Battery-supercapacitor hybrid is an exemplary HESS. Research interest in the battery-supercapacitor hybrid is widespread and includes hybridization architecture design, control method, amortized system cost minimization, and so on. EVs/HEVs are representative applications of the HESS [23, 30, 46]. Many researches related to HESS design and control in EV/HEV focus on improving the energy efficiency of vehicles. A power control method proposed by Moreno et al. [63], for example, optimizes the supercapacitor current by using a neural network and achieves more than 20% improvement in km/kWh. A power control method introduced by Thounthong et al. [64] is for a HEV, with a shared bus composed of a fuel cell, battery, and supercapacitor. A simple passive parallel connection of battery and supercapacitor mitigates voltage ripple with limited volume and weight of the EV/HEV [46]. An optimization method by Romaus et al. [65] solves the multi-objective optimization of minimizing the energy loss and maximizing the power reserve for HEV. Economic viability of using the HESS for EV/HEV is analyzed by Miller et al. [66]. A genetic algorithm-based revenue maximization method for a HESS used in energy and regulation markets is presented by Jin et al. [67].

Low-power sensor nodes like the ones mentioned by Jiang et al. [53] and Park and Chou [68] employ a battery-supercapacitor hybrid. Due to very limited capability to produce power from energy harvesting devices such as PV cells, reducing the power loss during charge/discharge cycles is important. They take advantage of the high cycle efficiency of the supercapacitor while using the battery for low-leakage long-term energy storage. Similarly, a compressed air energy storage (CAES) may utilize supercapacitors to keep track of the maximum efficiency operating point [69]. A HESS introduced by Ise et al. [70] takes advantage of the high power density of superconducting magnetic energy storage (SMES) and high energy density of batteries. A fuzzy control logic determines the power split between the SMES and batteries.

HESS is an emerging approach for achieving performance improvement of the ESS with current nonideal energy storage element technologies relying on optimal design and control. Hybridization offers opportunities to take advantage of each energy storage element while hiding their shortcomings/drawbacks. Expected benefits include enhancement of energy efficiency, cycle life, power and energy capacity, and so on.

As pointed out by Kim et al. [30], the hybrid approach for the HESS has the same motivations as those for designing the memory subsystem in a computer system. For computer memory devices, the SRAM (static random access memory) has the lowest latency and highest throughput but is expensive and has low density. On the other hand, DRAM (dynamic random access memory) is inferior to the SRAM in terms of latency and throughput, but is cheap and has high density. Mass storage devices such as an HDD (hard disk drive) and NOR/NAND flash memory have even lower cost, higher capacity, non-volatility, but are subject to limited random access capability and write count. Composing the required memory space with the SRAM only is infeasible due to its high cost, except for supercomputers where cost is not a primary issue. On the other hand, using HDD or flash memory only cannot meet the latency and throughput requirements of the CPU core and suffers from poor random access capability and limited write count. Computer architects, therefore, have remedied this problem by building a hierarchy of different types of memory devices. In a typical memory hierarchy L1 (level 1), the SRAM cache provides the best latency and throughput but the smallest capacity, whereas in L2 (level 2), the SRAM cache has a bit poor latency and throughput but is larger than the L1 cache. They both generally reside on a chip to provide fast access speed. The DRAM main memory is placed off chip, or sometimes on chip, for better latency and provides a much larger capacity with a higher latency and lower throughput than the L1/L2 SRAM caches [30]. Finally, the slowest HDD and flash memory are used to provide the largest storage space. There are many policies to utilize this memory hierarchy efficiently, but generally speaking, we use a faster memory to store frequently accessed data and/or code in order to take advantage of its high speed. We overcome the capacity limitation of fast memory by moving less frequently accessed data down to a slower memory. As a result, this memory hierarchy enables the CPU to exploit the low latency of the L1 SRAM cache and the large capacity of the HDD at the same time [30].

A HESS aims at similar benefits by using multiple heterogeneous energy storage elements. Instead of relying on a single type of energy storage element, the HESS exploits distinct advantages of multiple heterogeneous energy storage elements and hides their drawbacks. For instance, the EV/HEV exhibits frequent charge and discharge cycles with a short period and a large amount of current. Conventional batteries make it difficult to maintain high efficiency and a longer cycle life in such an operational environment. Use of supercapacitors instead can be a huge upgrade in terms of efficiency and cycle life. However, the current supercapacitor technologies have serious disadvantages in terms of energy density and cost, which makes it hard to completely replace the batteries in an EV with supercapacitors (this is despite a test bed supercapacitor-only EV [71]). Use of supercapacitors in a complementary manner reinforces the drawback of the battery through high power density, long cycle life, and high efficiency [46, 63].

Kim et al. [30] pointed out that in a typical HESS-connected system, the HESS is comprised of a number of energy storage banks and is connected to external power sources and load devices. For example, the HESS can be comprised of a supercapacitor, a lithium-ion battery, and a lead-acid battery. Similar to the computer memory hierarchy, the HESS exploits different features of these three energy storages to its own benefit: the high power density and long cycle life of the supercapacitor and the relatively low cost and high energy density of lithium-ion battery and lead-acid battery. The charge transfer interconnect (CTI) internally connects the energy storages, external power sources, and external load devices through appropriate power converters [30]. In spite of the similarity of the HESS and computer memory subsystem, of course, there are differences as well. For example, there is some energy loss during transfer, but no data is lost during data transfer. Also, selection of storage elements does not cause any coherence or invalidation issues that a cache memory suffers from.

Employing the HESS concept comes with additional design considerations. Deployment of a HESS does not always guarantee better performance unless proper design consideration is given. Designer should thus carefully determine the selection of energy storage elements, the proportion of each energy storage elements, system architecture, and the management policy, among others, in order to maximize the benefits of the HESS over the homogeneous ESS [30].

4.6 HYBRIDIZATION CRITERION AND OPTIONS FOR HYBRID ENERGY STORAGE

The ESS technologies can be classified into four categories based on the type of stored energy, including mechanical, electrical, chemical, and thermal. Each technology encompasses several different properties which should be taken into account in different applications. As listed in Table 4.2, typical properties for different technologies consist of power rating, discharge time, power density, energy density, and lifetime in both years and cycles, among others [8, 10].

As illustrated earlier, ESS technologies can be classified into two main categories including high power and high energy technologies. High power storage systems supply energy at very high rates, but characteristically for short time periods. Out of all ESS technologies, superconducting magnetic energy storage (SMES), supercapacitor, flywheel, and high power batteries belong to this category. Alternatively, high-energy storage systems can supply energy for longer time periods. The remainder of the ESS technologies belong to this category, i.e., pumped hydro, compressed air energy storage (CAES), fuel cell, and high energy batteries. Table 4.3 categorizes ESS technologies based on this classification. It should be noted that battery technologies can be employed either in high-power or high-energy devices due to their extensive characteristics range. In this context, the hybridization of high-energy devices in conjunction with high-power devices seems to yield a more functional ESS. The desired process of such a hybrid system known as hybrid energy storage system (HESS) is as follows: the high-power device should supply short-term power needs, while the high-energy device should meet the long-term energy needs. Based on this idea, possible combinations are listed in Table 4.3. It should be noted that all combinations shown in the table are not feasible in terms of technical and some system-level limitations.

Battery energy storage systems are typically configured in one of two ways: a *power* configuration or an *energy* configuration, depending on their intended application. This is accomplished by adjusting the ratio of inverters to batteries in the system. A simple way to envision this is to imagine a bathtub; the volume of water in the bathtub would represent the batteries and the drain(s) in the tub would represent the inverter(s). For a fixed level in the bathtub, several drain lines can be incorporated, resulting in a rapid discharge (a power configuration), or a single drain line can be incorporated, resulting in a slower discharge (an energy configuration). In each case, the system has the same amount of water (stored energy in the battery), but the discharge rate is varied.

A. **Power configuration**

In a power configuration, the batteries are used to inject a large amount of power into the grid in a relatively short amount of time. There is a high inverter to battery ratio required to accomplish this. A typical application would be to simulate a turbine ramp up for frequency regulation, spinning reserve, or black start capability. This means that there will be a high converter to battery ratio, fast discharge, C rate of MW/MW-hr >1, low cost/MW, and required space more than that for energy configuration, and its typical application includes spinning reserve, black start, and fast start.

B. **Energy configuration**

In an energy configuration, the batteries are used to inject a steady amount of power into the grid for an extended period of time. This application has a low inverter to battery ratio and is typically be used for addressing issues such as the California "duck curve," in which power demand changes are occurring over a period of as long as several hours or shifting curtailed PV production to a later time of the day. This means that there will be low inverter to battery ratio, slow discharge, C rate of Mw/MW-hr <1, low cost/MW-hr, and required space less than that for a power configuration, and typical applications include "duck curve" power makeup, energy shifting, and curtailed energy capture.

Just like batteries, all other types of energy storage systems possess different levels of power and energy density. This is illustrated in Table 4.2. A plot of energy density versus power density is commonly called Ragone plot. This type of plot was illustrated earlier in Chapter 1 (Figure 1.5). Another way to demonstrate the difference in different storage elements is to illustrate average energy storage duration versus average system power capacity as shown in Figure 4.6.

4.6.1 Options for Hybrid Energy Storage

Just like a single battery system can accommodate power or energy configurations, combining different types of storage systems into hybrid systems to reap the benefits has always been an attractive prospect. In the past years, successful projects have come online for both solar-plus-storage and wind-plus-storage—the resiliency of *battery energy storage* combined with the financial boost from power generation. *Hybrid energy storage systems* (HESS) can refer to several different types of setup; the point in common is that two or more types of energy storage are combined to form a single system.

There is no single *energy storage* solution that is ideal for every grid-scale application. As explained by Greentech Media, they

> are typically designed for high-power applications (i.e., "sprinter" mode that provides lots of power in short bursts) or energy-dense applications (i.e., "marathon" mode that provides consistent lower power over long durations), and there are lifetime, performance, and cost penalties for using them in unintended ways.

A HESS typically combines both "sprinter" and "marathon" storage solutions to fulfill applications that have diametrically opposed requirements, e.g., fast response versus peak shaving. Hybrid storage offers other avenues for cost reductions; two or more systems can share much of the same power electronics and grid-connection hardware, reducing both upfront and maintenance costs. Several possible hybrid combinations are described below. Some of them are commercially installed. Battery-supercapacitor combination has found most applications.

A. **Power-to-heat/battery**

A utility in Bremen, Germany, has contracted AEG Power Solutions to design and build a 20-MW hybrid battery/power-to-heat storage solution for frequency regulation. Storing energy as heat is much cheaper than storing electrochemically, which translates into a much smaller required battery capacity; AEG estimates up to 50% smaller.

B. **Battery/battery**

Many arguments have been had as to which technology will ultimately prevail for the stationary energy storage—flow batteries or lithium-ion? Flow batteries' low cost and long lifespan against the fast response and energy density of lithium-ion. UK vanadium flow battery manufacturer RedT thinks the answer is to combine them. In October 2017, it supplied a 1-MW hybrid energy storage system to Australia's Monash University. RedT envisaged a system where the vanadium-flow "workhorse" provides 70%–80% of energy, while lithium-ion provides bursts of power for demand surges.

C. **Battery/thermal**

In the United States, electricity consumption accounts for about 61% of the total energy consumed in the commercial building sector, and cooling and refrigeration alone accounts for more than 25% of the electricity consumption [72]. Most of the electricity consumption for building cooling happens during the peak operating hours between 10 am and 5 pm when the solar gain into the building is very high. During these peak operating times, the AC units typically run close to their full load during hot summer days when their efficiency is low and carbon emissions are very high. When this scenario is aggregated over thousands of commercial buildings, the load on the electricity grid is so significant that utility companies charge three to four times higher electricity rates during peak operating times and also charge additional demand electricity charges in order to lower the demand on the grid. Currently, the utility companies have time of usage (TOU) electricity rates for electricity consumption and also demand charges for peak power consumption. The electricity rates during peak hours between 9 am and 5 pm are three to four times higher than off-peak times and the demand charges are about $15–$18/kW [73]. This will result in significant electricity costs that will encourage commercial building owners to adopt more energy-efficient technologies and renewable energy sources to partially meet the electricity demands or shift the electricity consumption to off-peak duration. Thus, solar photovoltaic (PV) energy storage systems have gained a lot of attention in recent years to reduce peak electricity consumption in commercial and residential sectors. Such energy storage systems typically use battery systems like lead-acid or lithium-ion battery to store the PV electrical energy and shift the building electricity load completely or partially during the peak operating hours [1].

D. **Modular thermal energy storage tanks using modular heat batteries**

This invention by Laverman [74] relates to the storage and extraction of thermal energy. More particularly, this invention is concerned with a thermal energy storage apparatus which employs a plurality of heat batteries of modular design located in an insulated tank. With the recent interest in energy conservation and efficiency, and with the development of solar energy technology, there has been considerable interest in the application of thermal energy storage. Many applications, both industrial and commercial, have been evaluated for possible use of thermal energy storage concepts. The temperature range, over which it is desired to store thermal energy, varies considerably with these different concepts and ranges from approximately 200°F to temperatures in excess of 1,000°F.

Many of the energy storage systems already proposed to employ a storage tank to contain the heated material. However, major design problems are involved with such tanks because of the thermal movements associated with placing the storage tank in service and with the normal temperature cycles through which it operates. These temperature variations cause considerable difficulty in the load bearing insulation and foundation of the storage tank. A need accordingly exists for improved thermal energy storage apparatus. According to the invention by Laverman [74], there is a provision of a thermal energy storage apparatus comprising a thermally insulated tank having a bottom, sidewall, and roof; a plurality of spaced apart modular heat batteries inside the tank supported on load bearing thermal insulation on the tank bottom; each heat battery constituting an enclosed metal shell containing a bed of solid objects and around which objects as liquid can flow; conduit means to feed a hot or cold liquid from outside of the tank to the top of each bed in each battery; and conduit means to withdraw a hot or cold liquid from the bottom of each bed in each battery and deliver it to a destination outside of the tank. Each heat battery can be a vertical, circular, cylindrical shell with a flat bottom. Furthermore, the solid objects in each battery can be rocks. The tank desirably has a flat bottom and a vertical, circular, cylindrical sidewall which supports the roof. The sidewall thermal insulation can comprise a layer of granular insulation supported between the sidewall and a thin gauge metal barrier which is suspended from the sidewall by a plurality of horizontal rods, and the bottom insulation can comprise concrete load bearing insulation, such as in the form of blocks. The roof thermal insulation can be provided by a layer of granular insulation supported by a metal deck suspended by rods from the roof.

The apparatus can also include means to flood the tank with an inert gas for safety purposes. For the same reason, means can be included to supply each battery with a blanket of an inert gas. Each battery bottom is desirably provided with means to maintain it axially stationary while permitting radial expansion and contraction with temperature change. Each battery is preferably supported on load bearing thermal insulation. According to a second aspect of the invention, there is a method provided for storing thermal energy which includes distributing a flowing hot liquid to the top of a plurality of heat storage batteries containing a bed of cold solid objects; allowing the hot liquid to flow downwardly in a trickle-flow manner over the bed of solid objects contained in the heat storage batteries; allowing a vertical temperature gradient, including a downwardly moving thermocline heat transfer zone, with the flowing hot liquid above the zone and the flowing cold liquid below the zone to develop; and removing the cold liquid from below the thermocline heat transfer zone from the bottom of the heat storage batteries. According to the invention, there is a provision for recovering stored thermal energy for distributing a flowing cold liquid to the top of a plurality of heat storage batteries containing a bed of hot solid objects; allowing the cold liquid to flow downwardly in a trickle-flow manner over the bed of solid objects contained in the heat storage batteries; allowing a vertical temperature gradient, including a downwardly moving thermocline heat transfer zone in which the cold liquid is located above the zone and hot liquid is located below the zone to develop in the heat storage batteries; and removing the hot liquid from below the thermocline heat transfer zone from the bottom of the heat storage batteries [74].

E. **Battery/supercapacitors**

Similar in concept to the battery/battery HESS, supercapacitors or ultracapacitors can be used to provide fast-responding bursts of electricity to complement the batteries that provide lower intensity and longer duration power. Supercapacitors can survive an order of magnitude of more cycles than most batteries and have faster charging times. In 2016, Duke Energy, the largest investor-owned utility in the United States, installed a hybrid ultracapacitor/battery system which includes a 100-kW/300-kWh Aquion Aqueous Hybrid Ion "saltwater" battery bank. The hybridization of batteries and supercapacitors has been, however, actively proposed and studied by many researchers over the past several decades [75–77]. Supercapacitors, also called in some literature electrochemical double-layer capacitors (EDLC), are energy storage devices with high specific power, typically above 10 kW kg^1, and low specific energy, typically below 10 Wh kg^1 [78, 79]. Also they possess a very high cycle life typically above 500,000 [22]. On the other hand, rechargeable batteries, also entitled as secondary batteries, are a type of energy storage devices typically with high specific energy. However, batteries possess low specific power in contrast to supercapacitors, which typically does not exceed 1 kW kg^1. In addition, battery cycle life is much lower than supercapacitors [79–82].

In general, the hybridization of the batteries and supercapacitors can be carried out according to the various methods. In [76], these methods are classified into external and internal in terms of hybridization level. Each of the internal and external hybridization levels is divided into serial and parallel methods. In other word, a battery-supercapacitor HESS can possess the form of internal serial, internal parallel, external series, and external parallel. Among the abovementioned methods, the external parallel hybridization has been studied by many researchers in specific applications such as electric powered vehicles [22, 83], intermittent renewable energy systems [22], and general pulse power systems [22].

The main part of the drive train in an electric or hybrid vehicle is its power source, which typically comprises rechargeable batteries and fuel cells [79]. The battery lifetime in terms of charge/discharge cycles is affected by the rate at which the current is quieted from or entered into the battery [22], for example, during vehicle accelerations or decelerations. In addition, during high power demands, and depending on the state of charge, the battery voltage might decrease below the minimum admissible voltage of the system [77]. In order to increase the battery lifetime, and in addition to preserve the system voltage above a minimum value, supercapacitors are usually hybridized with batteries in battery-powered electric vehicles [77].

The majority of the studies concerning the battery-supercapacitor HESS are about the utilization of this technology in electric and hybrid vehicles. The studies have been verified that the performance, lifetime, and cycle life of the battery for electric or hybrid vehicles can be increased significantly by hybridizing with supercapacitors. Many studies in the context of battery-supercapacitor HESS have investigated the use of supercapacitors with batteries to create an optimized energy storage system in order to flatten the power variations in renewable energy systems including: photovoltaic systems, wind energy systems, hybrid wind-solar systems, and general microgrids [22].

Other studies are concerned with the behavior of this HESS for general pulsed loads. For example, Kuperman and Aharonin have reviewed the battery-supercapacitor system for pulsed power applications [84]. In [85], the authors have claimed that they "analytically demonstrated that a battery-supercapacitor HESS can supply a pulsed load with higher peak power, smaller internal losses and greater discharge life of the battery." The considered benefits with respect to the load profile parameters include the pulse duty ratio, the pulse rate, and the supercapacitor bank size and configuration. Also in [86], a straightforward and generic model for the optimal hybridization of battery and supercapacitor is proposed. In [87], battery against battery-supercapacitor HESS is investigated and reviewed in terms of various hybridization architectures. Other studies have been evaluated for the usage of HESS and claimed that this technology decreases the stress on the batteries, while extending their life, and additionally enhances the pulse power supply functionalities of the system [88]. At last, some of the researches in this context concerned with the power electronic converters used with [89] and related control techniques [90, 91] and also optimal sizing and energy management strategies [86].

Today, supercapacitors are found in many electronic applications. In automotive applications alone, they are used in start-up systems, energy recovery solutions, and fast charge-discharge systems, to name a few. Supercapacitors can charge/discharge quickly without losing energy storage capacity over time. They also have very high power density. In contrast, batteries can store larger amounts of energy, but they have a defined cycle life. A combination of batteries and supercapacitors results in a hybrid energy storage system that could help meet the needs of myriad renewable energy applications [1].

Supercapacitors can assist in delivering peak power while improving the performance of batteries in energy storage systems. Many supercapacitor manufacturers, utility companies, and researchers are developing hybrid capacitor-battery energy storage systems for future projects. Some are already using them in case studies and pilot projects, such as the one built by Duke Energy at its Rankin Substation in Gaston County, North Carolina, mentioned earlier. Duke Energy partnered with Aquion Energy, Maxwell Technologies [1], and others to build a hybrid energy storage system (HESS) project. The hybrid system uses Maxwell's ultra capacitors (UCAPs) to help manage solar smoothing events in real-time—particularly when the solar power on the grid fluctuates due to cloud cover or other weather circumstances. The Aquion batteries are used to shift solar load to a time that better benefits the utility. The hybrid energy storage system integrates patented energy management algorithms [1]. The invention by Ibok [92] addresses the limitation of singular energy storage devices by integrating multiple modular storage units interfaced with each other such that the electrical energy content of one unit can be throttled into a connecting unit when the energy content of the said unit is detected to be below a predetermined threshold.

F. **Flow batteries and flow capacitors**

A flow battery generates electricity inside a reactor by using dissolved electroactive species stored in external tanks. This separation of reactor and tank decouples the energy density (limited by tank size) from the power density (limited by reactor size). Flow batteries have advantages of long cycle life, environmental friendliness, quick charging by electrolyte replacement, and so on. However, they require complicated components and complex circuitry such as sensors and actuators to run the system. Flow batteries are generally considered for large-scale stationary applications rather than portable applications. Vanadium redox battery and zinc-bromine battery are examples of flow batteries. Lead-acid flow battery is also considered as a replacement of the conventional lead-acid batteries [1].

A recent technology called the electrochemical flow capacitor has the advantages of both supercapacitors and flow batteries [1, 93]. It provides rapid charging and discharging, high cycle life, and high cycle efficiency like a supercapacitor. It also provides and high energy capacity, decoupled with the power capacity, like a flow battery. A flowable carbon-electrolyte mixture, called a slurry, captures or releases charged ions while flowing through a flow cell during a charging or discharging process.

G. **Battery-SMES**

The nominal voltage of a single cell in the supercapacitor is usually limited to a very low range. In practice, series connection of several units is used in order to provide higher voltage levels. But, connecting cells in series decreases the overall capacitance. Moreover, some protection circuits are needed to balance voltage. Usually, step-up and step-down converters are needed in order to adjust output voltage where these interfaces may cause voltage fluctuations in the power system in turn. The output voltage in a supercapacitor changes with its charge and discharge and is proportional to the stored energy, which means that if the stored energy is low, its output voltage is low [20]. On the contrary, the superconducting magnetic energy storage (SMES) does not require any step-up or step-down interfaces, its control is relatively simple and the capabilities of SMES could be fully exploited in power systems [20]. SMES for regenerative braking on board of the hybrid vehicle has been proposed by Morandi et al. [94], but it is not practically implemented as yet. A small number of applications of the battery-SMES HESS are investigated in the literature, as described below.

The first battery-SMES HESS has been proposed in [95] where Suzuki et al. proposed to install the SMES in Shinkansen railway substations in order to supply high-frequency component of the load fluctuation. In the proposed system, a flow battery supplies long-term and low-frequency power variations. An innovative control strategy has been proposed for the control of the HESS based on the measured data of a real high-speed railway substation. Simulation results have been shown that the proposed HESS offers a considerable load leveling effect along with a lower cost of the system, in spite of a compact system.

Ise et al. [70] proposed a power-sharing method for each storage device by using a fuzzy control method and filters. Simulations have been done for the compensation of railway loads and the power of wind turbines. Simulation results have been shown that utilizing proposed HESS results in longer life and higher efficiency for a battery due to reduced power and less charging and discharging cycles. At the same time, space and cost reduction of the SMES can be expected due to a reduced energy required for SMES.

The utilizing battery-SMES HESS in order to condition the output power from the direct drive linear wave energy converters have been proposed by Nie et al. [20]. Simulation results have been shown that the proposed system can successfully address both the frequent and the slow-varying power fluctuations and maintaining stable and dispatchable electrical power supply to the grid or local load. Shim et al. [96] have proposed a synergistic control of the SMES and the battery energy storage in order to increase the dispatchability of the renewable energy sources. They have been concluded that their designed HESS operation strategy can help minimize the size (reduce the investment in the planning) and sustain or even prolong the designed lifespan of the energy storage system as well.

Finally, the study conducted by Wang et al. [97] proposed a mathematical model and the topology of the battery-SMES HESS. The sophisticated control methodology is designed with both device and system levels. Besides, in order for efficient power sharing between ESS devices, a new system-level control strategy is proposed. In addition, a genetic algorithm-optimized fuzzy logic controller is used in accordance with the control objectives.

H. **Battery-flywheel**

A flywheel (FW) stores energy in the form of mechanical energy in a motor-generator set. Like a SMES, an FW possesses high cycle life and efficiency and can be charged and discharged at high power rates for many cycles without losing efficiency. On the other hand, and unlike SMES, battery, and supercapacitors, FW technology does not require chemicals and is a green solution. The FW is well matched to short-duration and high-power applications. Also, the discharge duration could reach up to tens of minutes employing magnetic levitation bearings [98]. The battery-flywheel HESS is first utilized by Allen Windhorn in [99], where a UPS system has been proposed using a single-phase inverter driving a flywheel consisting of a motor-generator set. It has been shown that the proposed system has advantages over other systems. UPS and standard rotary UPS systems provide reduced output impedance, reduced distortion with nonlinear loads, higher reliability, and better isolation. Also, it is reported that practical installation of the proposed system at a large computer facility has resulted in considerable improvement under conditions including brownouts and several outages of various durations with no interruption of power [1].

Beaman and Rao [100] have described the potential combined battery/flywheel energy storage system for aerospace applications. They claimed that battery charging control schemes and solar array regulation can be augmented with a flywheel system to improve spacecraft performance and allow an alternate energy storage source for single battery systems, reducing the size of the solar array and consequently the weight of the spacecraft. Briat et al. [101] discussed design and the integration of a flywheel into drivetrain of the electric heavy-duty vehicles, with discontinuous mission profiles like garbage collection. The authors proposed a solution to improve vehicle performances via maintaining the battery power within rated levels for charges and discharges with a flywheel which provides the energy during acceleration or braking, respectively. In order to validate the simulation results, the authors create an experimental test bench [1].

Lee et al. [102] proposed to hybridize battery with flywheel in order to balance the power output of Heangwon wind farm in Cheju Island in Korea. The considered flywheel in this study is a permanent magnet synchronous motor/generator set where superconducting bearings are being utilized to eliminate the thermal losses due to the friction in the pivot bearings. By using site-measured output data, the optimal power and energy combination of the battery and superconductor flywheel energy storage (SFES) is determined. As the authors have stated, an SFES has a higher response rate and efficiency, and adequately using it with a battery is a more efficient and effective method for stabilizing the output of the wind farms.

A number of flywheels for trackside energy recovery systems have been demonstrated by URENCO and Calnetix [1]. VYCON's flywheel, known as Metro's Wayside Energy Storage Substation (WESS), can recover 66% of the braking train energy [1]. The collected data, after 6 months of operation, showed 20% energy savings (approximately 541 MWh), which is enough to power 100 average homes in California [1]. A total of 190 metro systems operating in 9,477 stations and approximately 11,800 km of track has been reported globally [1]. The introduction of energy storage into rail transit for braking energy recovery can potentially reduce 10% of the electricity consumption, while achieving cost savings of $90,000 per station [1].

Flywheels are also used in roller coaster launch systems to accumulate the energy during downhill movements and then rapidly accelerate the train to reach uphill positions, using electromagnetic, hydraulic, and friction wheel propulsion. Since the late 2000s, the use of flywheel hybrid storage systems in motorsports has seen major developments, beginning with Formula 1, followed by the highest class of World Endurance Championship (WEC). In public transport, city buses are an ideal application for electric flywheel hybridization, due to their higher mass and frequent start-stop nature. The technology can save fuel and reduce greenhouse gas emissions by up to 30% [1]. Williams Hybrid Power (WHP) started developing flywheel energy storage for use in buses for the Go-Ahead Group in March 2012.

Flywheels find applications in space vehicles where the primary source of energy is the sun, and where the energy needs to be stored for the periods when the satellite is in darkness. For the past decade, the NASA Glenn Research Centre (GRC) has been interested in developing flywheels for space vehicles. Initially, the designs used battery storage, but now, FESSs are being considered in combination with or to replace batteries. The combined functionality of batteries and flywheels improves the efficiency and reduces the spacecraft mass and cost [1]. It has been shown that the flywheel offers a 35% reduction in mass, 55% reduction in volume, and a 6.7% area reduction for solar array [1]. FESS is the only storage system that can accomplish dual functions, by providing satellites with renewable energy storage in conjunction with attitude control [1].

In the military, a recent trend has been toward the inclusion of electricity in military applications, such as in ships and other ground vehicles, as well as for weapons, navigation, communications, and their associated intelligent systems. Hybrid electric power is essential for future combat vehicles, based on their planned electrically powered applications. Flywheels appear as an appropriate energy storage technology for these applications. They are combined with supercapacitors to provide power for high-speed systems requiring power in less than 10 μs.

Flywheels are also likely to find applications in the launching of aircraft from carriers. Currently, these systems are driven by steam accumulators to store the energy; however, flywheels could replace these accumulators to reduce the size of the power generating systems that would otherwise be sized for the peak power load [1]. A FESS is integrated into a microgrid serving the US Marine Corp in California, to provide energy storage applications throughout the entire distributed generation at the base [1]. The purpose of the project is to provide energy security to military facilities using renewable energy. It is a network of interconnected smaller microgrids that are nested into a 1.1-MW bigger-scale microgrid that includes solar PV systems, diesel generators, batteries, and 60-kW/120-kWh FESS [103]. The flywheel storage is intended to decrease the dependency on diesel generators by about 40% and provide peak shaving applications by mainly supplying high power loads such as elevators. In addition to extending the lifespan of the batteries, the FESS is estimated to work for 50,000 cycles and have a lifespan of 25 years [1]. Flywheels can assist in the penetration of wind and solar energy in power systems by improving system stability. The fast-response characteristics of flywheels make them suitable in applications involving RES for grid-frequency balancing. Power oscillations due to solar and wind sources are compensated for by storing the energy during sunny or windy periods and are supplied back when demanded [1]. Flywheels can be used to rectify the wind oscillations and improve the system frequency; whereas, in solar systems, they can be integrated with batteries to improve the system output and elongate the battery's operational lifetime [1]. Several other applications of flywheel storage are described in my previous book [1].

Finally, in [104–106], efforts have been made in order to commercialize flywheel and to lower the cost through hybridizing with battery. In this work, simulations have been conducted on a typical house in Naxos Island in Greece due to its high renewable energy potential. It is assumed that the considered system utilizes solar and wind energy simultaneously and uses a stack of batteries in combination with flywheel for the plant energy storage. Authors have proven that using the flywheel in combination with classic batteries is feasible and could supply the required load. Moreover, they observed that while the initial costs of the battery-only system are much lower, hybridizing it with flywheel can be economical.

I. CAES-supercapacitor

A compressed air energy storage (CAES) system consumes off-peak electric energy to compress air into underground cavities (salt cavern, abandon mines, rock structures, etc.) or surface vessel. Three types of air reservoirs are usually considered including naturally occurring aquifers (such as natural gas storage reservoirs), solution-mined salt caverns, and mechanically formed reservoirs in rock formations. The main limitation to construct a CAES system is related to the reservoir achievement. The stored energy as compressed air in the reservoir is combined with one of the conventional fuels to drive a turbine generator set during peak demand periods, to produce electric energy. The fundamental issue for a CAES system is that the reservoir has to be air tight and very large. Utilizing ground storage tanks in small CAES units is usually limited in their energy storage capacity to only a few hours.

Lemofouet and Rufer conducted several researches concerning hybridizing CAES and supercapacitor [69, 107, 108]. Lemofouet and Rufer [69] proposed a HESS composed of the CAES and supercapacitor energy storage systems. The CAES is the heart of this system and plays the role of the energy supplier. The considered CAES system benefits from a maximum efficiency point tracking system (MEPT). On the other hand, supercapacitor acts as a filter and smooths the output power. The authors have proposed the system concept, power electronic interface systems, the MEPT algorithm, and finally the utilized strategy to change the output power. Besides works of Lemofouet and Rufer, authors in [69, 107, 108] have proposed a similar HESS system. In these works, in addition to the dynamic modeling and the control strategy, an efficient charge and discharge strategy under maximum power conditions is proposed.

J. CAES-SMES

Theoretically, a CAES system can be hybridized with a SMES system to provide more flexibility and capability. In such a HESS, CAES provides long-term and slow response energy storage needs while SMES supplies short-term (fast response) and high-power storage requirements. This HESS technology has not yet been proposed in the literature.

K. CAES-flywheel

The study by Okou et al. [109] is the only work which has studied CAES-flywheel HESS so far. In this study, a HESS composed of adiabatic compressed air energy storage (A-CAES) system, and a flywheel energy storage system (FESS) is proposed in order to smooth the wind turbine output fluctuations. First, the design and the thermodynamic analysis of the proposed system is carried out. The A-CAES system operates in variable cavern pressure and constant turbine inlet pressure mode, whereas the FESS is controlled by constant power strategy. Then, the off-design analysis of the proposed system is performed. Meanwhile, a parametric analysis is also performed to investigate the effects of several parameters on the system performance, including the ambient conditions, inlet temperature of compressor, storage cavern temperature, and maximum and minimum pressures of the storage cavern.

L. **CAES-battery**

Although as yet not proposed in the literature, hybridizing CAES with batteries can yield a HESS with excellent characteristics. Slow response and low power density of the CAES can be compensated complementarily by the batteries. With regard to the fast and new advancements in the batteries technology, hybridizing them with any energy supplier ESS, particularly CAES, will be a promising tool to construct a functional HESS.

M. **CAES-thermal**

Design of a hybrid CAES and thermal energy storage has been reported in the literature [110]. The share of renewable energy sources in the power grid is showing an increasing trend worldwide. Most of the renewable energy sources are intermittent and have generation peaks that do not correlate with peak demand. The stability of the power grid is highly dependent on the balance between power generation and demand. Compressed air energy storage (CAES) systems have been utilized to receive and store the electrical energy from the grid during off-peak hours and play the role of an auxiliary power plant during peak hours. Using thermal energy storage (TES) systems with CAES technology is shown to increase the efficiency and reduce the cost of generated power [110]. In the study by Lakeh et al. [110], a modular solid-based TES system is designed to store thermal energy converted from grid power. The TES system stores the energy in the form of internal energy of the storage medium up to 900 K. A three-dimensional computational study using commercial software (ANSYS Fluent) was completed to test the performance of the modular design of the TES. It was shown that a solid-state TES, using conventional concrete and an array of circular fins with embedded heaters, can be used for storing heat for a high-temperature hybrid CAES (HTH-CAES) system [110].

N. **Fuel cell-supercapacitor**

The hybridizing fuel cell (along with a hydrogen tank) with a supercapacitor energy storage unit in order to constitute a HESS has been proposed in various literature, whereas applications are different from electric vehicles to renewable resources integration. Thounthong et al. have conducted many researches in this context [93, 111]. Thounthong et al. [111] first proposed a control concept in order to combine proton exchange membrane (PEM) fuel cell and supercapacitors for utilization in electric vehicle applications. The proposed scheme is based on DC-link voltage (VDC) regulation while fuel cell and supercapacitor operate in steady-state and transient regimes, respectively. The proposed system structure is realized by analogical current loops and digital voltage loops and the experimental results with a 500-W PEM fuel cell pointed out the fuel cell starvation problem when operating with dynamic load. Also, the results confirmed that the supercapacitor can improve system performance for hybrid power sources.

Wu and Gao [112] have minimized the cost of the fuel cell and supercapacitor in a fuel cell electric vehicle. Their proposed optimal design includes cost, volume, and weight of the fuel cell stack and supercapacitor bank of the powertrain. With the purpose of application in residences, Uzunoglu and Alam [113] presented a dynamic model, design, and simulation of a combined PEM fuel cell and supercapacitor system. A comparative study by Bauman and Kazerani [114] was conducted between fuel cell-battery, fuel cell-supercapacitor, and fuel cell-battery-supercapacitor vehicles. Besides abovementioned works, various researches have been conducted regarding hybridizing fuel cell and supercapacitor including control strategies [93]; power and energy management [115]; modeling, simulation, analysis, and experimental results [116, 117]; and converter design [118].

O. **Fuel cell-SMES**

The hybridizing hydrogen storage and superconducting magnetic energy storage (SMES) have been considered first in the literature by Louie and Strunz [119]. The authors with analogy to the computer systems have proposed to combine fast-response (low capacity) storage devices with slow response (high capacity) ones. Like data cache in computers, the former act as an access-oriented storage and serves as a cache for the capacity-oriented latter storage. The authors entitled their proposed scheme as "energy cache control" to reflect its similarity to the process of access to the stored data in the computer systems. In the proposed HESS, SMES acts as access-oriented storage, and a hydrogen-electric conversion and storage plant composed of an electrolyzer, a fuel cell, and tanks plays the role of capacity-oriented storage in order to compensate for power fluctuations of the DC bus of a distributed generation (DG) system. Through modeling and simulation of the proposed system, it has been shown that the energy cache control will compensate fast and slow power variation through SMES and hydrogen storage, respectively.

The other work which has been proposed to use fuel cell with SMES is reported by Sander et al. [120] where liquid hydrogen (LH_2) is integrated with SMES, named as LIQHYSMES. The LH_2 plays the role of the long-term high-power storage needs, and in the other side, short-term fluctuations will be suppressed by the SMES. The proposed HESS system combines the regenerative H_2 liquefaction part, the LH_2 tank, and the SMES. Authors have shown that their proposed system will be capable of handling strong variations on time scales of hours, minutes, and seconds. Also, it is claimed that the proposed system helps to integrate large-scale renewable resources into the grid through providing desirable combination in terms of operational safety, efficiency, and cost.

P. **Fuel cell-battery**

The hybridizing fuel cell with battery is one of the most studied HESS and it is safe to say that it is the first proposed HESS configuration [121]. The fuel cell and its applications have been the focus of many researches in the 1970s when lightweight fuel cells were being developed at that time [121]. At the end of this decade, the idea of combining a fuel cell with the battery emerged. The background of this idea goes back to 1970 when Frysinger and Wrublewski proposed to combine a 30-W fuel cell and a nickel-cadmium battery for powering lightweight portable devices like mobile military communications and surveillance equipment in order to reduce the weight of the equipment [121]. The secondary battery supplies the peak power pulses and offers the required power density, while the fuel cell efficiently converts the chemical fuel into electrical energy at a continuous steady rate to provide extremely high energy densities over extended missions. About 18 years later in 1988, Beyer et al. proposed to combine a fuel cell with lead-acid batteries for application in stand-alone wind-solar systems [122]. In this work, the authors proposed a system configuration, optimized it with respect to energy investment, and analyzed the effect of the hydrogen system cycle efficiency on the storage configuration. In [123], Ledjeff proposed to use a similar configuration and performed detailed calculations for achieving design data to use the storage system as part of the "Self-sufficient Solar House 2000" project which was intended to be built in Freiburg, Germany. After that, various researches were conducted regarding this HESS application in the renewable systems, including photovoltaic [123], hybrid wind-PV [124], microgrid, and general distributed generation systems [125].

Besides renewable energy system, hybridizing fuel cell with batteries has been proposed in many literatures in order to enhance powertrain functionalities in an FC-powered vehicle [126]. Specially, in [127] and [128], the batteries have been compared with supercapacitors in order to hybridize with fuel cell in the FC-powered vehicles. Other studies in this field are concerned with general topics such as control, power and energy management, and optimal sizing [129].

Q. **Pumped hydro—power supplier ESS**

Pumped hydraulic energy storage system is the only storage technology that is both technically mature and widely installed and used. These energy storage systems have been utilized worldwide for more than 70 years. This large-scale ESS technology is the most widely used technology today where there are about 280 installations worldwide. In a pumped hydraulic ESS system, during off-peak periods or equivalent periods with surplus cheap electricity, pumps are used to reposition water to a reservoir at a higher altitude than the original water source. In this situation, consumed electric energy to pump water upward is stored as the potential energy in the water in the reservoir. Then, during peak periods, the stored water in the reservoir is released to move down to its base position where it drives hydraulic turbines to produce electricity. Two main obstacles to construct a pumped hydraulic ESS system are lack of suitable location and high costs. Perhaps even more significant than the site limitations and cost of the system is the perceived environmental concerns, including flooding of valleys to create reservoirs and damage to wildlife habitats. Other ESS technologies do not have environmental issues as do pumped hydraulic systems. Despite technical maturity and worldwide use, environmental impacts of the pumped hydro ESS units limit their expansion to the HESS systems. Although a pumped hydro system as an energy supplier ESS unit can be hybridized with any power supplier ESS to form a HESS, this hybridization is not yet been proposed in the literature.

4.7 HYBRIDIZATION ARCHITECTURES

In HESS design, hybridization architecture has a considerable effect on the control and energy management strategy as well as a variety of characteristics, for instance, modularity, flexibility, efficiency, and especially, cost. Architectures with more flexibility provide more choices in implementing control and energy management strategies and offer most possible enhanced performance at the expense of complexity and cost. To date, the most proposed hybridization architectures are modified for a specific control and energy management strategy. Numerous architectures have been proposed to hybridize ESS technologies ranging from simple with low cost to complicated with high cost. Generally, the proposed architectures can be classified into three types including passive parallel, cascade, and active parallel. [130–132]:

A. **Passive parallel**

In passive parallel architecture, also known as direct parallel architecture, two ESS technologies are directly connected to power electronic device without any power-conditioning circuit (power electronic device) between them [133]. This topology requires that the output voltage of the two ESSs be equal. Benefits of this architecture are simplicity, easy implementation, and lack of control or power electronic converters. All these advantages lead to earning the merit of the lowest cost architecture. However, it encompasses several drawbacks, including [130–132]: (a) the system is not protected against ESSs faults; then, the failure of one ESS unit may affect another and involves the system totally; (b) the current distribution between ESSs is uncontrolled and determined just by the factors which vary with voltage; (c) there is no flexibility in the selection of ESSs nominal voltage; and (d) the output voltage changes because of the system charge and discharge. The amount of current that is extracted from each ESS is restricted by the voltage swing of the other.

B. **Cascade**

More efficient but more expensive architecture is to place an additional converter (power electronic device, PED) between the ESSs. This cascade architecture provides decoupling of the ESSs which allows active energy management by use of additional power electronic device between ESSs in turn. In other words, first, the PED controls the power output of the first energy storage device, allowing its voltage to vary, while the second energy storage delivers the remaining power requirement of the load. Usually, the ESS with higher voltage variations is operated as the first energy storage and other one as the second energy storage [30]. Also, the more sensitive ESS may be located in the position of first energy storage to extend the lifetime of the overall system by conditioning the power output of it. One of the most noticeable drawbacks of this architecture is the lack of freedom in the control policy. In addition, the cascade architecture is restricted in terms of scalability because it suffers from more conversion losses as the number of power conversion steps increases [132].

C. **Active parallel**

Finally, in active parallel architecture, each energy storage device is connected to its devoted PED [30]. This architecture shows clearly the highest level of flexibility compared to the previously mentioned two architectures. Having dedicated power electronic device for each ESS offers various advantages, including [130]: (a) scalability is higher because the number of power conversion steps between any ESS and load is always two, and the power conversion loss does not increase as the heterogeneity increases; (b) each ESS can operate at its specific voltage, which allows the specific power and specific energy be optimized using the best available technology; in other words, maximum power point tracking can be implemented for each source; (c) the stability is also improved since a failure of one source still allows the operation of the other; and (d) a variety of control and energy management strategies can be implemented. Although multiple input converters [134–136] have been proposed in order to decrease the cost of the overall system, the only demerit of this architecture is high cost due to employing two independent PED for each ESS [4].

A general conceptual diagram of a HESS includes multiple energy storage banks, which are connected through a network called the CTI (charge transfer interconnect). Each energy storage bank is similar to an independent homogeneous ESS, but it performs energy storage array management as determined by system-level HESS management. CTI is connected to power source, load, and power grid through power converter.

D. **Energy storage bank**

A HESS is a hybridization of heterogeneous energy storage banks, each of which is homogeneous. An energy storage bank is composed of a homogeneous energy storage array and a power converter as described earlier, for homogeneous ESS architecture. In addition to the individual control of energy storage array, an energy storage bank in a HESS supports system-wide bank management. The energy storage banks communicate with a central controller, which sets the system-level management. The controller receives the current status of each bank, such as voltage, current, SoC, SoH, and so on and sends out charge management commands to each bank [30].

E. **Charge transfer interconnect**

Kim et al. [30] showed how the energy storage elements are connected starting from a simple battery-supercapacitor HESS. There are three representative hybrid architectures of battery and supercapacitor. A basic hybrid approach is a passive direct connection of a battery and a supercapacitor in parallel [137]. The supercapacitor handles high current demands by suppressing large battery voltage variations, but this architecture cannot be generally applied for different combinations of heterogeneous energy storage elements. Cascaded converter architecture puts a converter in between the battery and supercapacitor to control the power. A constant-current regulator-based hybrid architecture for a battery and supercapacitor in [138, 139] improves the energy efficiency compared with the direct parallel connection. The parallel connection and cascaded converters architecture, however, are not suitable for three or more energy storage banks [30].

A general hybrid architecture, which can accommodate any type of energy storage elements in a symmetric manner, utilizes a DC bus. In the DC-bus architecture, each energy storage element is connected to the DC bus through a power converter (voltage regulator or current regulator). A HESS introduced in [64] is comprised of a fuel cell, a battery, and a supercapacitor connected to a fixed-voltage DC bus. The control method regulates a fast energy storage element with a slow energy storage element in order; it controls the DC-bus voltage, supercapacitor voltage, and battery voltage by controlling the supercapacitor current, battery current, and fuel cell current, respectively. Another HESS is comprised of a photovoltaic panel, a fuel cell, and a supercapacitor, and its basic operational principle is the same [140]. This simple and intuitive control method, however, cannot achieve energy optimality because (i) the DC-bus voltage is fixed, and so the conversion efficiency is not always the maximum, and (ii) current distribution among the energy storage elements is determined by other energy storage elements and cannot be optimized in a holistic manner. True energy optimality requires consideration of not only the dynamic response, but also other factors such as rate capability, residual energy, and load demand. For example, we may want to reserve energy in the supercapacitor and use the battery instead when a high current demand is expected in a near future, but this kind of intelligent energy control is not supported in [64, 140]. Another DC-bus-based HESS proposes a control method that increases the supercapacitor current as the battery current increases [141].

Heterogeneous energy storage elements in a HESS play distinct roles depending on their unique characteristics, but may or may not have a hierarchy among the elements. Abbey et al. [142] developed a knowledged based approach for control of two-level energy storage for wind energy systems. A hierarchy may be physical (explicit hierarchy that is exposed by the CTI architecture) or logical (implicit hierarchy achieved by system management policies). For example, the cascaded converters architecture has an explicit hierarchy, whereas the parallel connection and DC-bus architecture do not. However, the DC-bus architecture may have a logical hierarchy between the battery and supercapacitor by actively controlling their contribution in various aspects such as power and energy [30].

A HESS with a larger number of energy storage banks requires a more complicated connection than a passive parallel connection or cascaded converter architecture as the CTI. Connecting heterogeneous energy storage elements in a HESS is more than expanding power and energy capacities, but involves complicated energy transfer among the elements. Therefore, the CTI architecture is more critical in a HESS than in a homogeneous ESS, though the CTI architectures for the HESS can be used for homogeneous ESS as well. Similar to on-chip communication networks, the network topology of the CTI is one of the important design considerations for energy efficiency in a large-scale HESS. It has a significant impact on the charge transfer efficiency and thus should be carefully designed in order to maximize the benefits of the HESS. A system-on-chip design is subject to a similar problem of determining a suitable interconnect architecture. The interconnect architecture in a system-on-chip affects its communication latency, throughput, power consumption, and so on. For both HESS and system-on-chip designs, the interconnect architecture should be selected by considering the scale of the system. The interconnect architecture becomes more critical as the number of nodes increases [30].

Kim et al. [30] point out that CTI architectures of HESS are similar to the system-on-chip interconnect architectures. There are four interconnect architectures for four nodes. A shared bus interconnect is simple to implement, but has limited scalability. Variances of the shared bus CTI with higher scalability include a segmented bus CTI and multiple bus CTI. Finally, the point-to-point interconnect provides independent paths between every pair of nodes, but its cost increases exponentially as the number of nodes increases. These architectures are well explored for the system-on-chips and are also applied for the HESS [143, 144]. As the number of energy storage banks in HESS further increases, the CTI architectures cannot provide efficient paths for charge transfers [30]. Higher energy efficiency may be achieved if the CTI network is able to provide more isolated paths to simultaneous charge transfers for the energy-optimal CTI voltage. This leads to a networked CTI architecture which is comparable to a typical network-on-chip architecture [30]. As the number of processing elements in a network-on-chip increases, the single-level on-chip bus architecture is no longer able to handle increased data exchanges between the processing elements. Similar to the network-on-chip that requires packet routing, a HESS with a networked CTI architecture requires routing of the charge transfers [30].

4.8 ENERGY MANAGEMENT AND CONTROL OF HESS

Hybridizing different ESS technologies does not mean improved functionalities, in general [30]. In order to extract the maximum advantages of each ESS in the HESS, developing a sophisticated system management plan is compulsory. Due to the heterogeneous types of ESSs, control and energy management strategy for a HESS is more complicated than a single ESS unit. Properties of the ESS technologies, power electronic devices, load, and time frames should be taken into account in order to take the advantages of heterogeneity. In addition, nonlinear and time-dependent characteristics, for instance, self-discharge (especially in supercapacitors) and memory effect (in batteries), power converter efficiency, and so on should be included in the control system. The HESS architecture is the first and foremost factor governing control strategy design and implementation. One of the properties of each architecture is its flexibility to plan various control strategies. This issue should be considered to design a proper and feasible control system. Generally, the ultimate goal of the control system is optimizing three tasks related to HESS utilization including charge allocation, charge replacement, and charge migration [145].

When HESS gets energy from the power source, an appropriate choice for the stored energy in terms of ESS technologies should be selected. Charge allocation term refers to optimizing the charging process via selecting the proper ESS, taking into account the technology's type and state of charge (SoC), voltage, current, and time characteristics of the power source [146]. Analogous to the charge allocation, charge replacement strategy determines the most efficient ESS technology which can be discharged during power needs. Charge replacement significantly depends on load characteristics and capabilities and also on the SoC of the ESS technologies [147].

Almost all of the ESS technologies may experience self-discharge. The amount of the self-discharge energy depends on the ESS technology and its remainder lifetime. Supercapacitors have the highest self-discharge rate and may be fully discharged without any load within a few days. This problem may necessitate charge migration between various ESS technologies, depending on the time characteristics of the load variations. If required, charge migration can be appropriately carried out by precise forecasting of the future load demand as well as self-discharge rate. Regarding the abovementioned matters, various control techniques such as fuzzy, robust, digital, differential flatness, model predictive, adaptive, sliding mode, and knowledge-based control methods have been proposed by researchers.

Despite numerous research efforts in order to improve ESS capabilities over the past decade [148], a perfect ESS technology that copes the drawbacks in terms of all aspects is not to be expected to develop in the near future. On the other hand, particular ESS applications require a combination of energy and power rating, charge and discharge time, life cycle, and other specifications that cannot be met by a single ESS technology. In order to increase the range of advantages that a single ESS technology can offer, and at the same time, enhance its capabilities without fundamental development of the storage mechanism and only via complementary use of the existing ESS technologies, more than one ESS technologies can be hybridized [149]. A hybrid energy storage system (HESS) is composed of two or more heterogeneous ESS technologies with matching characteristics and combines the power outputs of them in order to take advantage of each individual technology, and at the same time, hide their drawbacks.

The HESS requires more sophisticated management policies than a conventional ESS because of the heterogeneity of energy storage elements. Proper management policies are very crucial to achieving high energy efficiency in the HESS. The HESS management policies are system-level policies for maximizing the benefits of the HESS in energy efficiency, lifetime, etc., by exploiting its heterogeneity. The most important decision required in a HESS is energy distribution among heterogeneous energy storage banks. We need to select particular energy storage banks to charge or discharge and determine the CTI voltage and amount of current that maximizes the energy efficiency. Also, we may need to internally move energy from one energy storage bank to another in order to mitigate the self-discharge or prepare for the expected demand for energy/power capacity. These operations are called charge management, which is named after cache management in a computer memory hierarchy in [55]. The charge management includes (i) charge allocation for charging

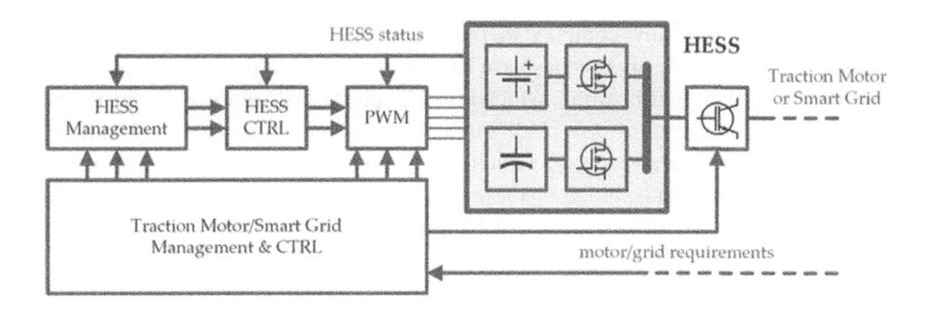

FIGURE 4.7 Schematic representation of an HESS management and control system [4].

energy storage banks [150], (ii) charge replacement for discharging energy storage banks [145], and (iii) charge migration for moving energy between the energy storage banks [143]. A SoH-aware charge management enhances battery lifespan by utilizing supercapacitors for high-frequency power [151].

As far as semi-active or active HESS configurations are concerned, the management and control system cover a fundamental role for exploiting the HESS at the maximum extent. Particularly, an appropriate selection of HESS management strategy is of paramount importance, even from the design stage, especially for sizing BESS and SCSS properly in accordance with target performances and technical and economic constraints. It is also fundamental for assuring HESS's efficiency, reliability, and durability. A schematic representation of HESS management and control system is illustrated in Figure 4.7.

In case of battery pack and ultracapacitor module as HESS, its management consists mainly of a sharing criterion for splitting the overall HESS energy flow between BESS and SCSS. Literature review reveals that several approaches have been proposed in order to exploit BESS and SCSS inherent features to the maximum extent, preventing them from unsuitable operation as well. In this regard, SCSS generally handles fast power fluctuations, whereas BESS copes with the average power demand. This principle is the basis of the simplest HESS management strategy known as frequency-based management (FBM); this consists of splitting the overall power demand into high- and low-frequency components, which have to be tracked by SCSS and BESS, respectively [152]. An alternative approach is the so-called rule-based management (RBM), which exploits the single ESSs in accordance with an appropriate order of priority by means of a preset of rules [153]. In this regard, it is worth noting that FBM and RBM may be combined to each other or with fuzzy logic algorithms in order to account for ESS constraints and to improve overall HESS performances [154].

Although FBM and RBM are intuitive, simple, and easy to implement, they generally do not lead to optimal solutions. For this reason, another popular approach is determining BESS and SCSS reference power profiles by minimizing suitable cost functions over a given time horizon. Hence, different optimal solving techniques can be used, such as model predictive control, mixed-integer/linear programming, nonlinear programming, and dynamic programming [155]. However, such solving techniques are generally complex to implement and are quite time-demanding. Consequently, heuristic approaches have been also proposed, like genetic algorithms and particle swarm optimization, which achieve suboptimal solutions but faster and with less computational efforts [156]. As a result, very complex and sophisticated cost functions can be considered, which can account for many system constraints and goals. The main advantage of these approaches consists of enabling HESS to provide multiple services in an optimal manner, by both economic and technical points of view; this aspect makes HESS very competitive, especially for smart grid applications.

In order to overcome the issues arising from both passive and active HESS configurations, a highly integrated solution has been proposed by Porru et al. [157], whose schematic representation is depicted in Figure 4.8. It consists of coupling a BESS (or BP in the figure) with an SCSS (or UM in the figure) through a multilevel converter, namely a three-level neutral-point-clamped converter (NPC). The key feature of the proposed highly integrated battery-ultracapacitor system (HIBUC) is the full integration of SCSS within the DC-link of the NPC, which decouples the overall DC-link voltage (VDC) from its energy content (EDC). As a result, HIBUC energy flow management is quite similar to that achieved with active HESS configurations without resorting to any DC/DC converter.

4.9 HESS OPTIMIZATION

The specification of HESS requires several types of design optimization. Here we summarize a few of them analyzed in an excellent review article by Kim et al. [30].

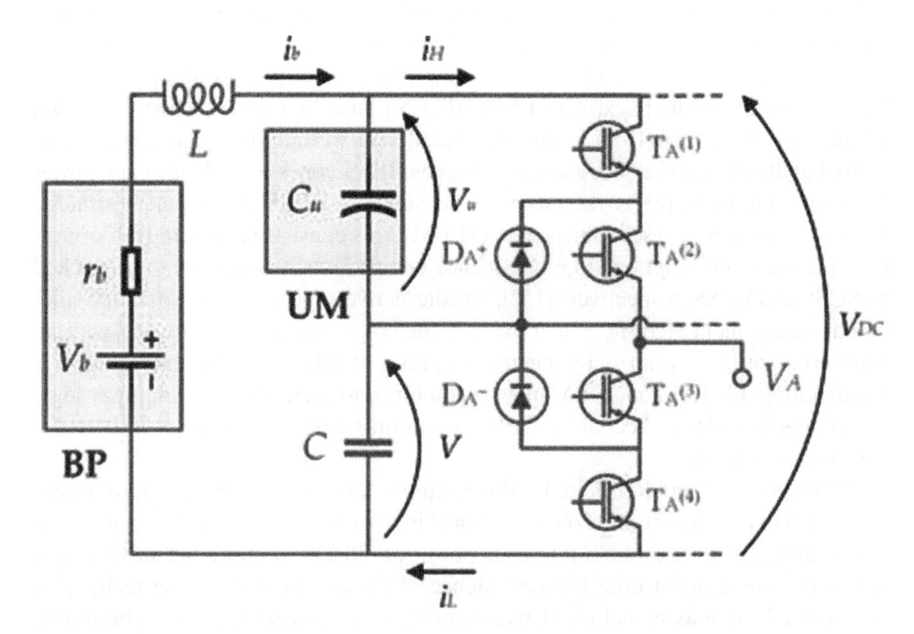

FIGURE 4.8 The highly integrated HESS configuration proposed in [4].

4.9.1 Design-time Optimization

The total energy capacity is one of the most important features in the HESS specification. However, HESS requires determination of the type of storage banks and individual storage bank capacities at the design time. Appropriately proportioned energy storage bank capacities can enhance HESS performance metrics, whereas smart control policies can create the illusion that the HESS consists of only the most suitable energy storage elements for the given load. Determination of the energy capacity of each energy storage bank in the HESS is a critical issue for minimizing the cost of the HESS. Not only the capital cost for purchase and disposal of the energy storage elements but also the operational cost for electricity energy is affected by the composition of energy storage bank types.

Kim et al. [30] considered a HESS that consists of a lithium-ion battery bank and a supercapacitor bank. The two energy storage elements have distinct characteristics, that is, lithium-ion batteries have an 80% round-trip efficiency, $1,000 per kWh cost, 0.1% self-discharge rate, and a cycle life of 2,000, whereas supercapacitors have a nearly 100% round-trip efficiency, $40,000 per kWh cost, 15% self-discharge rate, and 100,000 cycle life. They assumed that we are required to compensate for energy usage fluctuation. We have a small daily fluctuation and a large weekly fluctuation, and the total fluctuation is a superposition of them. A homogeneous ESS composed of only batteries will have a poor round-trip efficiency for daily storage and retrieval of energy. On the other hand, supercapacitor-only ESS will suffer from high leakage for the energy stored during weekend. By building a HESS with optimized composition of the energy storage elements, one can reduce both the capital cost and operational cost. Supercapacitors have near perfect round-trip efficiency and a long cycle life, and so they are suitable for compensating daily fluctuation. On the other hand, batteries, which have low round-trip efficiency and a short cycle life, are suitable for weekly energy storage. For an example of 500-kWh HESS, up to 15% cost reduction compared with homogeneous ESS has been shown [158].

Kim et al. [30] also pointed out that another important feature of the HESS specification is its overall power capacity, which in turn depends on the power capacity of each energy storage bank in the system, power capacity of the charger and the DC-DC converter, and the control policies employed. Kim et al. [30] defined the power capacity optimization problem as that of designing a HESS that satisfies a power capacity requirement while maximizing the total service time.

Suppose we have a high current pulsed load device such as a radio transceiver. Consider a connected two-cell lithium-ion battery of 350 mAh capacity used along with a supercapacitor in a HESS. The load device consumes 3.5 A of current for 10 seconds with a 10% duty ratio. We can enhance the total service time of the battery-supercapacitor HESS by the use of a constant-current charger circuit compared with a conventional hybrid architecture that simply connects the battery and supercapacitor in parallel. An optimization method introduced by Shin et al. [139] derives that a supercapacitor of 2.5°F capacity maximizes the total service time of the HESS. Experimental results of the real implementation of the constant-current charger-based HESS show 7.7% total service time improvement.

4.9.2 Design Optimization of a Residential HESS

Zhu et al. [146] investigated the design and control optimization of a HESS for a residential smart grid user in order to maximize the total profit. The study considered a HESS comprised of a lithium-ion battery bank and a lead-acid battery bank for a case study. Kim et al. [30] considered a simple time-of-day electricity pricing function as shown in reference [158]. Both energy storage banks will get charged during the low peak period and discharged during the high peak period under such electricity pricing function so as to reduce the daily electricity cost for the residential user. The study took into account the following factors in the design optimization framework to provide a more practical and accurate estimation of the amortized annual profit:

1. ***Characteristics of different battery banks,*** including energy capacity, power capacity, rate capacity effect, self-discharge rate, and so on. Power conversion efficiency variation of chargers and power converters.
2. ***The SoH degradation rate*, or the cycle life, of an energy storage bank** is strongly dependent on the DoD. The works of Xie et al. [150, 151, 159] and Millner [160] show that the battery cycle life increases (nearly) exponentially when the DoD decreases. The study determines the energy capacity of both energy storage banks subject to the following constraints.
3. ***System volume constraint***: The overall volume of the HESS must be limited since it targets residential usage, so

$$\text{Volume of Li-ion bank} + \text{Volume of lead-acid bank} \leq \text{Total volume constraint} \tag{4.2}$$

   Kim et al.'s [30] problem formulation uses the reciprocal of battery volumetric energy density, referred to as the unit volume, which is its volume divided by the maximum stored energy. The lead-acid battery's average unit volume is 12.5 L/kWh, much higher than that of lithium-ion battery, which is 2 L/kWh [55].
4. ***Monetary budget constraint:*** The total cost of the HESS consists of both the capital cost of energy storage banks and the maintenance cost. We know that lithium-ion batteries usually have three to four times the lifetime of lead-acid batteries. Normally different types of batteries do not reach their end of life together. It is uneconomical to discard or replace the whole HESS as soon as one battery bank reaches its end of life. Instead, replacing the aged battery bank with a new one can restart the HESS with lower extra cost. Apart from the purchase cost of new batteries, the replacement of devices also adds up to the total cost since it requires maintenance personnel to come in and restore the system. Moreover, we also consider the discount factor, which signifies the fact that there is a difference between the future value of a payment and the

present value of the same payment. The overall monetary budget constraint is given by:

$$\begin{aligned}&\text{Amortized cost of Li-ion bank} + \text{Amortized cost of lead-acid bank}\\&+ \text{Amortized maintenance cost} \leq \text{Monetary budget constraint}\end{aligned} \tag{4.3}$$

where the discount factor is reflected in the amortized costs of different parts.

The study proposed a unified framework for the optimal design and control of the HESS targeting at exploiting its potential for energy cost saving. First, we derive an effective HESS control algorithm to maximize the daily energy cost saving with a given specification of the HESS (in terms of types and capacities of different energy storage banks), under limitation on the DoD. This management algorithm properly controls the charging and discharging of each energy storage bank. For each possible design specification of the HESS under a monetary budget constraint and a total volume constraint, one can find its amortized annual profit based on the optimal control algorithm. The study found the optimal design and specification of the HESS that maximized the amortized annual profit. The study showed that the optimally designed HESS achieves an annual return on investment (ROI) of up to 60% higher than a lead-acid battery-only system or a lithium-ion battery-only system, under the same amount of investment. Based on their analysis, the study recommended that future work on the design optimization of HESS may include (but is not limited to) the following directions:

- Perform joint design optimization on both the PV module and the HESS in order to achieve the maximum amortized profit, under a given budget of the total investment.
- Study parameters for future energy storage technologies, such as their energy density, unit capital cost, and so on. In this way, we may explore the potential benefits of HESS in the future.

4.9.3 Runtime Optimization

Kim et al. [30] points out that runtime optimization involves manipulation of charge management, charge allocation, charge allocation efficiency, charge replacement, and charge migration. An operative framework for designing a HESS should enable holistic optimizations across different HESS components and structures through unified problem formulations and efficient solutions. Simultaneously, this framework should enable multi-objective optimizations with respect to cycle efficiency, cycle life, system cost, and so on.

An operative framework for designing a HESS should enable holistic optimizations across different HESS components and structures through unified problem formulations and efficient solutions. Simultaneously, this framework should enable multi-objective optimizations with respect to cycle efficiency, cycle life, system cost, and so on. It is crucial to provide policies and methods for charge allocation to the most suitable energy storage banks for a given incoming power level, and for charge replacement from the most suitable energy storage banks for a given load demand. Hence, the HESS runtime control comprises of policies that dynamically control the manner in which the system is used once it has been deployed. These policies include algorithmic/heuristic approaches for basic energy management operations like charge allocation, charge replacement, and charge migration [55]. These also include heuristics that predict the future energy needs [160], heuristics that predictively charge some energy storage banks [161], and so on. Even if the optimal charge allocation and replacement policies are put in place and executed, charge migration that moves charge from one energy storage bank to another is often necessary to improve the overall HESS efficiency and responsiveness. Charge migration can ensure the availability of the most suitable energy storage bank (in terms of its self-leakage, output power rating) to service a load demand (to the extent possible.) To achieve this goal, one must first invent an efficient low-cost charge migration architecture. One must also develop systematic solutions for multiple-source and multiple-destination charge migration considering the efficiency of charger rate capacity effect of the storage element, terminal voltage variation of the storage element as a function of the SoC, and so on. Kim et al. [30] described optimum processes for charge management, charge allocation, charge replacement, and charge migration with a focus on cycle efficiency enhancement.

4.9.4 Joint Optimization with Power Input and Output

One can expand the scope of optimization from HESS to the whole energy system that includes power input (power sources) and output (load devices). For both the homogeneous and hybrid ESS, joint optimization of the energy storage together with the power input and output design is curial to achieve the true system-wide energy optimality. It becomes more critical when the HESS is used for renewable power sources instead of the power grid with high reliability. Kim et al. [30] illustrated such a joint optimization with an example of solar renewable energy as a case study described below.

In a storage-equipped renewable energy harvesting system, it is possible and beneficial to generate the maximum amount of power from the renewable power sources, regardless of how much power is needed, consume as much as the load demands, and store the rest in the energy storage for the future use. The maximum power is achieved at a certain voltage and current, which is called the maximum power point (MPP). The MPP dynamically changes due to environmental condition change, and thus one needs to dynamically change the operating voltage and current to remain at the MPP which is adaptive to the environmental condition change. This technique is called the MPPT and is considered mandatory for renewable power systems [162].

Kim et al. [30] examined power generation, conversion (charger), and storage in a solar energy harvesting system. Typical MPPT techniques perform feedback control of the system so that the power from the PV cell P_{pv} is maximized. Perturb and observe technique, or hill-climbing method, makes a slight increase or decrease (perturbation) in the PV cell voltage V_{pv} to see which direction increases the P_{pv} (observe), and change V_{pv} in that way. Incremental conductance technique utilizes that $dP_{pv}/dV_{pv}=0$ at the MPP. It compares the incremental conductance (dI_{pv}/dV_{pv}) and the PV cell's instantaneous conductance (I_{pv}/V_{pv}) to determine whether to increase or decrease the V_{pv} to increase the P_{pv}. The common objective of these techniques is to maximize the P_{pv} in time-varying environmental conditions.

Kim et al. [30] pointed out that, in fact, the MPPT does not guarantee the maximum power production in real practices due to varying power loss P_c in the charger. The final amount of power delivered to the energy storage is $P_{charge}=P_{pv} - P_c$. The MPPT aims at drawing the maximum P_{pv} without consideration on P_c. However, P_c is not constant but variable by the input/output voltage and current of the charger. Consequently, P_{charge} may not be the maximum if increased P_c offsets the increment of P_{pv} even though the MPPT techniques maximize it. That is, consideration of both the maximum power extraction from the PV cell/module of the MPPT and minimization of the power loss in the converter are equally important to maximize PV system efficiency.

In this context, system-level energy efficiency should not be overlooked. Recently, some researches in design automation have been focusing on the significant impact of the power conversion efficiency on the system-wide energy efficiency of renewable energy harvesting systems [43]. It is shown that the PV cell MPP may not be the same as the system-level MPP in small-scale energy harvesting systems due to the significant impact of the power loss in the charger circuit. They presented an energy harvesting system that maximizes the output power of the charge circuit instead of the output power of the energy harvesting power source. Another solar energy harvesting system with charge pump charger is presented by Kim et al. [43] with the output power maximization consideration.

Systematic optimal design scheme which covers the power source, power converter, and energy storage was first introduced by Kim et al. [44]. Kim and Chou [163], in a more recent work, while attempting to find the optimal size and topology of the energy storage, examined several supercapacitors of different capacitances to find the best capacitance that maximizes energy efficiency. They also applied reconfiguration of the supercapacitors to dynamically change the terminal voltage and effective capacitance and took into account the reconfiguration overhead. A cross-layer optimization method by Mungan et al. [164] derives the optimal design parameters such as PV cell silicon thickness, PV module configuration, and charge pump stage and frequency. Wang et al. [161] consider a joint optimization method for the partial shading effect of PV cells for HESS optimization.

4.10 HESS FOR EV, MICROGRID, AND OFF-GRID APPLICATIONS

4.10.1 HESS for Electric Vehicle Applications

Electric energy storage systems (ESSs) are widely recognized as one of the most promising technology for enabling the transition toward a sustainable energy system [165]. Particularly, transportation electrification is pushing toward progressive improvements of ESS technologies, especially for light- and heavy-road electric vehicles: these have to rely on on-board ESSs for guaranteeing long mileage and short charging time [166]. Consequently, high efficiency, low costs, small volumes, and weights for ESS are desirable. The employment of ESSs is increasingly considered also for such systems that have been propelled electrically since a long time, such as railway and ships, in order to increase system efficiency and fuel economy, as well as ensuring reliable operation of on-board power systems [167]. In this context, both hybrid and all-electric ships are expected to be appealing for ESS use in the forthcoming future. Similarly, ESSs are a key point for designing more electric aircraft, in which pneumatic and hydraulic actuators are being replaced with electrical ones [168]. Therefore, on-board ESSs should start the engines, maintain DC-link voltage constant over dynamic operations, and, above all, guarantee emergency power supply. As shown later, ESSs can be employed successfully also for addressing several issues affecting modern power systems, such as reduced level of power quality, massive growth of distributed generation, and high penetration of renewable energy sources [169]. Particularly, integrating the massive and increasing share of photovoltaic and wind power plants installed all over the world is one of the main challenges for future power systems, which will be faced by resorting to smart grid, microgrid and, off-grid concepts. The development of solar electric cars will also require ESS. In this context, ESSs are the ideal solution for mitigating power fluctuations, storing overproduction, and releasing it when required, improving overall reliability and power quality.

An ESS consists of two main stages, i.e., the power conversion system and the energy storage unit, as shown in Figure 4.9. The power conversion system is generally represented by a power electronic converter, which has to regulate ESS voltage and current levels in order to match application requirements. Whereas, the energy conversion occurs within the energy storage unit, which exchanges electrical energy only, storing it into different forms (mechanical, chemical, magnetic, etc.).

There are several ESSs available on the market, which are generally classified in accordance with their energy/power density (Wh/l, W/l) or specific energy/power (Wh/kg, W/kg). Various characteristics of ESSs are depicted in Figure 4.10. High energy density ESSs are able to provide a large amount of energy over long

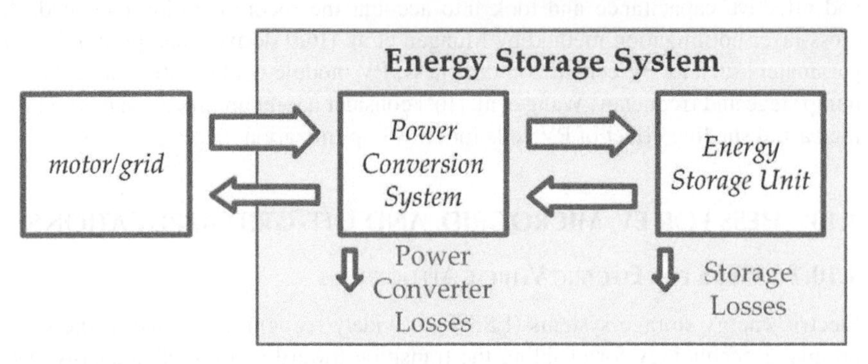

FIGURE 4.9 ESS schematic representation [4].

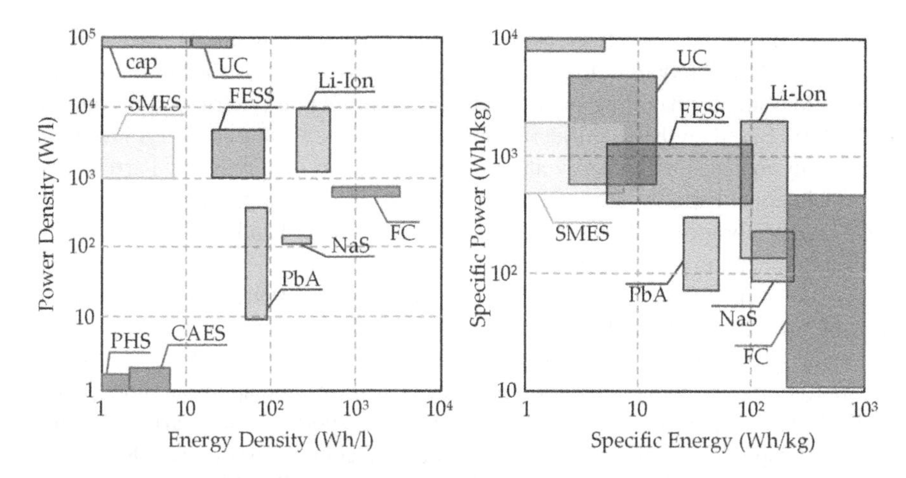

FIGURE 4.10 Energy/power density (left) and specific energy/power (right) of ESSs on Ragone plots: pumped hydroelectric storage system (PHS), compressed air energy storage system (CAES), and fuel cell (FC) [4].

time periods, as occurring for the majority of electrochemical batteries. These are the first ESSs introduced on the market and still represent the most widespread. Electrochemical batteries can be further classified based on the chemical reaction they exploit. Lead-acid (PbA) is probably the most known technology; it is being widely used on vehicles for starting, lighting, and ignition purposes. PbA batteries are currently employed when sizes and weights are not an issue, such as isolated power systems and UPSs [170], whereas, lithium-ion batteries (Li-ion) are surely the best solution for modern electric vehicles due to their very high energy density and specific energy. However, sodium-based batteries may be preferred when high capacity is required, namely, for load leveling and renewable energy sources integration. Despite their advantages, electrochemical batteries generally suffer from low power capabilities; thus, even if research is focused on improving power density, they are not yet the best solution for high-power applications [171].

Unlike electrochemical batteries, high power density ESSs can provide much little amount of energy but in very short times, which is the case of flywheel energy storage system (FESS), superconducting magnetic energy storage system (SMES), and ultra or supercapacitors (UCs or SCs). Particularly, a FESS is characterized by very high efficiency, long life expectancy, and low environmental impact. However, it presents a quite high self-discharge rate and also suffers from safety issues as far as high speeds are concerned [172]. Regarding SMES, a cryogenic system ensures the superconducting state of coils. Consequently, losses are due to power converters only, leading to a very high overall efficiency. Other advantages consist of fast response and wide power range, as well as a long lifespan. Nevertheless, SMES is very expensive due to the high costs of both superconductors and cryogenic system; thus, although it has been recently tested for power quality and voltage stabilization in both transmission and distribution systems, it is still employed mostly in military applications [173]. The main advantages of UCs are high power capability, quite long life cycle, and no memory effect, but they cannot store great amount of energy unless big and costly UC modules are used. Thus, UCs have been used for power quality applications and for handling small regenerative braking on electric propulsion systems [170]. Based on these considerations, it can be stated that a single ESS technology hardly matches both energy and power application requirements. Particularly, electrochemical batteries are very suitable for providing energy services, in which high-energy storage capability is mandatory. On the other hand, FESS, SMES, and UCs are more appropriate for power services.

As shown in Chapter 3, due to environment concerns, transportation industry is moving toward more hybrid and electrical transport. In the long term, use of oil and internal combustion engine will be phased out. Fuel cell which uses hydrogen is also gaining ground in the transportation industry. This transition applies to both public and private transportation. In the field of private transport vehicles, various alternatives are proposed to replace or improve the internal combustion engines (ICE), such as FC vehicles, hybrid electric vehicles (HEVs), plug-in HEVs (PHEVs), and battery EVs. All these alternatives are to replace ICE by various electric propulsion systems. The main element of the powertrain in electric or hybrid vehicles is the power source, which usually involves batteries or fuel cells [174]. The lifetime of the battery and fuel cell is influenced by the charge and discharge cycles and is limited. In addition, during high power needs with relatively low time periods, such as acceleration, the storage system may fail to supply the load. The use of HESS is one method for eliminating these issues.

In addition to private (road) transport, public (rail) transport also benefits from the HESS. The power consumption of the electrically powered railway systems seriously fluctuates because of the various accelerations and decelerations. In order to flatten these fluctuations, ESSs can be installed either at the railway substation or on the train itself. Besides smoothing out power variations, the ESS unit can be utilized to reuse the regenerative energy from vehicle braking. Therefore, in addition to the required constant power of the train, a large power can be drawn from the power source during train acceleration and similarly a large power feedback to the source during its deceleration.

These requirements greatly influence the energy storage characteristics which may not be provided by a single ESS unit. In this situation, a HESS unit can be the solution. Usually, high power density and high cycle devices such as supercapacitor and SMES are hybridized with batteries or fuel cells [175]. In this context, and with the purpose of application in electrified transport, supercapacitors (also called ultracapacitor), and sometimes SMES and flywheel because of their relatively high cycle life and power density, are often connected to the batteries or fuel cells in battery-powered and FC-powered vehicles, respectively. Also, fuel cell hybridization with batteries and/or supercapacitors as a functional HESS is proposed by many researchers to utilize in FC vehicles.

EV/HEV applications of HESS are in between portable small-scale applications and stationary large-scale applications. They can accommodate larger size and weight quota for the HESS, but still require high energy density because it directly affects the cruising distance of the vehicle. Cycle life is another critical factor for

EV/HEV applications because they suffer from deep DoD (depth of discharge). Low cycle life implies high maintenance cost for battery replacement. Lead-acid batteries have been used for traditional vehicles and EV such as golf carts and forklifts for low cost, but their heavy weight is the problem. Today's HEV such as Toyota Prius has a NiMH battery pack. Lithium-ion batteries are being actively researched for its use in EV/HEV applications to replace the NiMH batteries [176]. In large heavy-duty vehicles, supercapacitor/battery, battery/fuel cell, and flywheel/battery systems are also examined [177]. Some of HESS used in EV are described in Table 4.4.

There are a number of issues that need to be considered while using HESS for EV application. The most important ones are hybridization topologies and control methods and optimum storage sizing and improved storage efficiency and low cost. Battery-supercapacitor combination with Li-ion batteries is the most preferred one for EV vehicles. Cheng et al. [42] have developed test platform for hybrid electric vehicles with supercapacitor based energy storage. The objective of HESS in EV is to reduce electrical vehicular weight, improve driving range, and have efficient regenerative braking for battery recharging. HESS in EV are extensively covered in the literature [4, 8, 22, 23, 30, 48, 55] among others.

4.10.2 HESS for Microgrid Applications

In recent years, HESS has been extensively used for managing the increasing level of renewable energy sources in the grid. As shown in Chapter 5, microgrids which can be connected to a utility grid or can be operated in islanded mode have been found to be a very effective tool to harness distributed renewable sources in the grid. In microgrids, the integration of non-dispatchable sources like solar and wind cause numerous issues like grid regulations and fault support, grid stability, power quality, storage lifespan, etc. which need to be addressed particularly as the level of renewable sources increases significantly. The effective integration of HESS in the microgrid also requires proper capacity sizing of various storage elements, proper use of power converter topologies, and optimum management and control strategies so that HESS can improve the operation of microgrid for hybrid renewable energy sources. In this section, we address five issues related to the use of HESS for microgrid applications: (a) advantages of HESS in the performance of microgrid with hybrid renewable system which includes grid regulations and fault support; (b) methods of capacity sizing of various elements of HESS in microgrid (MG); (c) power converter topologies used in HESS in MG; (d) HESS energy management and control in MG; and (e) an overview on the control of microgrid with HRES and HESS. More detailed description of HESS for microgrids applications are given in several excellent reviews [22, 174, 178–180].

MGs and RESs suffer from problems like the intermittency nature, poor power quality, stability issues, frequency control, and unbalanced load. In typical MGs, the ESS usually experiences irregular and frequent discharging/charging patterns, which truncates the ESS lifespan, and therefore, the replacement cost of the ESS increases significantly [171]. HESS is an appropriate solution to overcome MGs and RESs challenges. Many studies have been done in recent years with the common goal of demonstrating the positive effects of HESS on the RESs [181]. Depending on the purpose of the hybridization, different energy storages can be used as a HESS. Generally, the HESS consists of high power storage (HPS) and high energy storage (HES), where the HPS absorbs or delivers the transient and peak power while the HES meets the long-term energy demand [182]. HESSs provide many benefits for MGs and RES including improving the total system efficiency, reducing the system cost, and prolonging the lifespan of the ESS [183]. Due to the various types of energy storages technologies with different characteristics, a wide range of energy storage hybridization can be formed. Earlier in Section 4.6, the combination of different storage technologies which can be used for various applications [184] were described. It can be seen that the SC/battery, SMES/battery, FC/battery, FC/SC, battery/flywheel, battery/CAES, and FC/flywheel HESS are commonly implemented in HRESs applications [185]. In recent years, however, the battery-supercapacitors combination has been preferred in many applications [22, 174, 179, 180]. Selection of appropriate HESS combinations depends on the variety of items, including storages hybridization targets, storage costs, geo-location, and storage space availability. Storage hybridization has many positive impacts on HRES in MG as outlined below.

4.10.2.1 Renewable System Intermittence Improvement

Most researchers have used HESS to improve the fluctuation of wind and solar power [152]. ESS can be integrated to alleviate some of the challenges associated with a fluctuation power production source such as solar and wind energy. Wind power has been composed from various frequency components with different amplitudes, and since the HESS includes both low- and high-speed responses, improved smoothing can be obtained in comparison to single ESS [186]. Jiang and Hong [187] presented a capacity configuration of HESS based on the wavelet transform algorithm to flatten short- and long-term fluctuations of wind power. A combination of batteries and SCs is used and dispatched, considering the frequency distribution of RES output power. A similar HESS (containing battery and SC) is also proposed by Mendis et al. [188] for managing the fluctuations of wind energy in a remote area power supply. Superconducting flywheel energy storages are proposed by Lee et al. [189] to improve the power fluctuations of the large-scale wind farms. Simulation results show that instead of only using the battery energy storage system, adequately mixing SFES with battery energy storage is a more efficient and effective solution to stabilize the output regime of wind farms.

Li et al. [190] also used SMES and battery for wind power generation connection to the grid, where a double-level control method is implemented. The system-level control performs power allocation between battery and SMES considering grid power demand. Meanwhile, the device-level control for converters aims at responding to the power sharing of the system-level control and DC-bus voltage regulation. A fuzzy logic controller optimized with the genetic algorithm is used to smooth the wind power fluctuations to satisfy the grid demand by Li et al. [190]. The output fluctuations of a solar power plant are appropriately compensated by the use of SMES and electrolyzer by Zhang et al. [191]. For HESS application for power smoothing in a grid-integrated wind and PV system, the HESS includes HPS and

HES that compensates both low- and high-frequency power fluctuations, respectively, in comparison to single ESS [191].

4.10.2.2 Storage Lifespan Improvement

Hajiaghasi et al. [178] described storage lifespan improvement by HESS in microgrid applications. One of the disadvantages of electrochemical energy storages such as battery and FC is their low lifespan. Avoiding frequent cyclic charging and discharging of battery prevents the degradation and improves its lifespan [192]. Battery lifespan in HESS structure can be increased by smoothing the battery power profile and preventing fluctuation supplied by battery, i.e., reducing the number of battery-involved charge and discharge actions [193, 194]. Ongaro et al. [195] and Gee et al. [196] proposed a power management strategy that utilizes both lithium battery and SC for battery life extension where SC provides the high-frequency demand. The results show the battery lifespan is increased by 19%. A control strategy is proposed by Li et al. [192] to contribute the instantaneous power between the battery and SMES. In this control method, the battery discharges and charges as a function of the SMES current rather than directly providing the power fluctuations.

As shown by Zhuo et al. [61], FC lifespan is dependent on the fuel consumption of the FC and power fluctuations. Mane et al. [197] and Aouzellag et al. [198] showed that the FC lifespan can be improved by combining FC with high power density storage. Most researchers have used SC/FC hybridization to improve the FC lifespan. The use of SC and FC with appropriate energy management is presented by Mane et al. [197]. In this method, instantaneous power is supplied by the SC trying to improve FC power leveling and consequently its lifespan. Aouzellag et al. [198] used hybridization of FC with SC to prevent the oxygen starvation and pressure oscillation of FC, caused by cyclic and frequent charging/discharging of FC.

4.10.2.3 Power Quality Improvement

Integration of RES (such as solar and wind) into the AC grid is growing very rapidly to meet the high and reliable load demand. This leads to the introduction of unique new power quality challenges with further penetration and growing of the microgrid structure within the distribution network. Nevertheless, it is mandatory to maintain the voltage and frequency within the prescribed limits under any condition of the RES and load demand. Hence, the integration of HESS becomes extremely important to regulate RES and load accordingly. The sharing of power among grid, RES, and HESS plays a very vital role in the stable operation of a grid-integrated microgrid system. The stable operation of HESS, therefore, becomes a common denominator that maintains the power quality and system stability during the transitional operation of the overall grid [178].

HESSs are used for various power quality purposes, such as frequency regulation, stability improvement, harmonic compensation of unbalanced loads, supply of pulse loads, and DC-bus voltage regulation in MGs [178]. Lahyani et al. [199] performed an investigation to identify the potential of HESS as a backup uninterruptible power supply (UPS) during short grid failure intervals. Based on this study, it has been observed that HESS can effectively ensure compensation for the whole load requirement during an outage lasting less than 10 seconds. However, considering the high cost of SCSS, for this case scenario, it should be pointed out that the system conceived would be efficient with at least four times BESS lifetime extension. Accordingly, a fault ride-through control scheme was presented by Georgious et al. [200], with BESS for DC-link voltage regulation and SCSS supporting large transience in the system. This topology consists of two parallel bidirectional converters, for BESS and SCSS, respectively, with the additional inclusion of switches, making it suitable for short-circuit fault tolerant, and further proposed that the control scheme for the converter is designed as such to reduce the losses due to these additions of switches. A PI-based DC voltage regulation scheme is proposed by Cabrane et al. [201] that aims to ensure continuity of power supply regardless of the load behavior and variation of solar irradiation. Maintaining the SoC, the duty ratios of HESS is controlled, via, the power flow variation projected as voltage variation at the DC-link.

Control strategies for DC voltage regulation in case of grid-connected and islanded operation are presented by Dong et al. [202]. In grid operation, renewable sources supply power to the loads and DC voltage is controlled by the inverter maintaining optimum power flow, while the AC bus is controlled by the utility. However, EMS in islanded mode/utility fault conditions maintains magnitude and frequency of the AC zone by controlling storage to regulate DC bus and inverter as a voltage source to regulate magnitude and frequency of the AC bus [178]. Hence, to maintain power quality of the power supplied from RES, the voltage of HESS is regulated to supply power in both DC and AC zones. This strategy maintains low power flow between utility and microgrid, reducing the influence of the load-generation fluctuant into the utility and regulating DC-bus voltage under any load profile. A capacity optimization of HESS is proposed by Zhou and Sun [203], aiming to reduce the one-time investment and operational costs of the system. Based on utilization rate and reliability constraints, optimal HESS capacity is obtained through a simulated annealing particle swarm optimization (SAPSO) algorithm. Therefore, the designed HESS capacity has the potential to fully absorb maximum surplus power of the fluctuant RES and can maintain a continuity of power supply to sensitive loads under unstable weather conditions through maximum power loss process. Wang et al. [204] investigated the operation of HESS combined with micro combined heat and power (CHP) generation for uninterruptible power supply to the heat loads and the electrical loads. In this investigation, a novel control scheme was introduced where CHP is dispatched to meet the heating load and the HESS reconciles any mismatch between the electrical load and CHP generations. Similarly, implementation of HESS in hybrid RES system consisting of PV-wind-diesel paves the way for optimal utilization of various energy sources and hence reduces the fuel consumption.

Similarly the study by Sanjeev et al. [205] provides a descriptive investigation to achieve an uninterruptible power supply by implementing the HESS as power backups in DER systems; with BESS providing the base DC loads and SCSS acts as storage buffer to cope with large transient of the system. Pan et al. [206] used maximum power point tracking (MPPT) control, using extreme search control (ESC) principle combined with sliding mode control scheme (SMC) (to avoid the need for wind speed forecasting) for extraction of maximum power from the stand-alone wind resource and over-generation and under generation of wind power is coped with fully active HESS employing a simple current-controlled power-sharing scheme based on the current level of the charge/discharge current between BESS and SCSS, that is, BESS absorbs all the excess power if the current level is lower than the predefined threshold current value. Further, if the power generated is low, SCSS discharges when the discharging current is too high; thus achieving rapid response to maintain the power balance in a wind power system with variations in both wind velocity and load power. The work has been further extended to a grid-tied system by Hassan et al. [207] (to simulate an uninterrupted power supply) with (wind turbine generator (WTG) as a primary source of power supply. Here MPPT is achieved through optimal torque (OT)-based control. An experimental study employing a coordinated band control strategy is presented by Wu et al. [208], for reliable supply to both AC and DC loads in an autonomous PV-wind-diesel generation system. The objective of the system is to maintain the voltage at the distributed constrained power control center DCPCC), thereby maintaining the overall system stability, and so under normal conditions with no significant power variation, DC voltage is regulated by SCSS, whereas if the state capacity of SCSS is not under its specified low/high operation bands, BESS replaces SCSS under a specified voltage band for voltage stabilization [178]. Further, in cases where the power supply remains to be continuously lower than the power demand, the diesel generator acts as a backup supply to maintain the DC-link voltage.

A. **Voltage regulation**

As indicated by Hajiaghasi et al. [178], the study by Cabrane et al. [209] provides an economical and investigative framework for DC-bus voltage regulation using HESS. In this investigation, a detailed graphical representation depicts the variability and sensitivity of power allocation between BESS and SCSS through reference current generation. Furthermore, in combination with a distinctive cost analysis, the study establishes an optimal number of supercapacitor in harmony with the filter cut-off constant, which reduces the overall system losses. Tummuru et al. [210] presented a voltage-based energy management scheme. Using small signal control gains for designing current and voltage control loops, SCSS is utilized solely to respond to quick load fluctuation, and BESS covering average load demand and a faster DC-link voltage regulation, with optimal current level in both BESS and SCSS, is achieved. The research described by Zheng et al. [211] formulates the methodology for making a PV-HESS system work as a classical generator. A systematic control strategy is proposed for a grid-tied PV system, where the HESS coordinately controls the DC bus while an active/reactive power control strategy is implemented on the DC/AC inverters. In this setup, a smooth power flow is maintained between the main grid and the RES grid owing to the characteristics of the HESS that actively overshadows the negative effects of power transience in the grid and hence allows a sustainable PV-HESS participation in the voltage regulation of the grid. In this regard, the work by Han et al. [212] strives to overcome the power fluctuations and LVRT (low voltage ride through) for a grid-interactive WTG farm. In this study, an enhanced active/reactive EMS is propositioned for a modular-based power conditioner that consists of an integrated STATCOM-HESS on the DC side of the H-bridge, via a bidirectional DC-DC chopper. Hence, by using a universal three-stage decoupling control technique, a cascaded STATCOM provides reactive power support, even in fault conditions and active power management, that is, wind power fluctuation is smoothened by HESS. Nikhil and Mishra [213] presented a study to highlight the contribution of RES-HESS for mitigating power imbalance and abrupt frequency variations in a hybrid AC/DC grid. For achieving this, suitable reference values are generated for RES, power converters of HESS, and the grid-connecting voltage source converter (VSC), using a control algorithm. Therefore, HESS mitigates the power imbalance in the DC grid, reducing its negative impact on the VSC, and hence, enhancing the VSC operational support to help compensate the frequency variations in the AC grid [178].

These characteristics can be further implemented in the case of implementing dual voltage source inverter (DVSI) grid inverters where the properties of BESS-SCSS HESS to maintain a dynamic power balance between different energy sources and sinks, empowers the DVSI to maintain the power quality of the power shared between the AC and the DC zones. Moreover, in an islanded mode of operation, DVSI scheme regulates a perfect sinusoidal voltage at point of common coupling (PCC) through auxiliary inverters, improving power quality at the PCC, which is a very essential aspect in a microgrid, while the main inverter of DVSI ensures power transfer between AC and DC zones. Further, in this perspective, the operation of dynamic voltage restorer (DVR) can be enhanced to mitigate voltage sag and voltage swell at the grid side, with the DVR being regulated by MPC and powered by the DC microgrid consisting of RES-BESS-SCSS. A nonlinear flatness-control strategy is presented by Benaouadi et al. [214], to manage energy flow in domestic autonomous hybrid power systems. The advantage of the proposed strategy is that, without solving differential equations, the output variables can be known, and hence, the desired trajectory can be planned to maintain a flat output. Considering voltage profiles of SCSS and the DC-link voltage, SCSS can provide/absorb the power needed to control its voltage and hence the DC-link voltage under permissible limits and the power flow can be maintained to compensate the power between inconsistent load and generation [178].

In accordance, a new unified power quality conditioner (UPQC) configuration is presented by Ismail and Mishra [215], mitigating voltage interruption, taking into account current harmonics, current unbalance, load reactive power, voltage sag, voltage swell, and voltage fluctuation. The UPQC consists of a shunt inverter that ensures a balanced sinusoidal and unity power factor at the PCC through a current injection under normal condition and operating in voltage control mode to maintain balanced sinusoidal load voltage under voltage interruption. Moreover, a series inverter injects voltage to ensure balanced and sinusoidal load voltage, isolating the inverter under voltage interruptions. The HESS is integrated into the shunt and series inverters, with BESS delivering average load power during voltage interruption and SCSS providing load peak power fluctuation. The study by Zhang et al. [216] investigates

DC microgrid with HESS and posits a novel power management strategy for the integration of unstable RES DC grid to traditional grid power supply network. The DC-bus voltage is employed as a carrier to represent operation modes of the system based on critical DC voltage conditions, the control strategy automatically judges and switches within the specified seven ranges of power operation state formed by six critical DC-bus voltages.

Further, a unified adaptive energy management system is proposed by Kotra and Mishra [217] for a residential grid-integrated PV-based hybrid microgrid with BESS-SCSS energy storage system. The proposed EMS ensures reliable and continuous power supply to various local loads and enables the bidirectional real power transfer between microgrid and utility grid while maintaining the grid standards at the PCC through VSC. Smooth mode transfer of grid, hierarchical load shedding, off-MPP operation in excess mode, and fast DC-link voltage regulation are the considered features of the proposed EMS. In addition, the EMS minimizes the power drawn from the grid and maximizes the power injection during peak pricing and can effectively operate without forecasting and voltage measurement of the HESS [178].

For stand-alone MG, a common DC bus is the preferred choice due to various reasons. DC-bus voltage regulation being as fast and accurate as possible is one of the important issues in MGs. Manandhar et al. [218] and Kllimala et al. [90] used battery-SC storages for rapid DC-bus voltage restoration and effective power allocation between the SC and the battery in an isolated system. In some researches, battery-SC are used for DC-bus regulation improvement in grid-connected mode [154].

B. **Frequency regulation and inertial support**

The use of HESS for frequency control can be divided into two main parts: frequency control in off-grid systems [219] and in on-grid systems [220]. The high penetration of RESs in power system reduces system inertia, which could jeopardize the power system frequency, and it may cause blackouts and equipment damage. A new design framework to system frequency regulation using HESS (coupled battery and SC) is presented by Akram and Khalid [221]. Moreover, an efficient coordinated operation strategy in electricity markets is proposed for frequency regulation. Fang et al. [222] used battery and SC to achieve the power management of the virtual synchronous generator. Particularly, the SC is employed to emulate the inertia of a virtual synchronous generator and cope with high-frequency power fluctuations [178].

As indicated by Hajiaghasi et al. [178], the battery energy storages have good features for regulating frequency in the off-grid MG systems. However, for frequency regulation, the battery charges and discharges at a high rate, which reduces its lifespan. Moreover, the battery needs to deal with the sudden power changes in the primary frequency control, which will also accelerate the battery degradation process. To solve the mentioned problems, Li et al. [219] proposed a novel concept of primary frequency control by combining the SMES with the battery, thereby achieving both the frequency regulating function and the battery life-service extension.

The study by Anzalchi et al. [223] demonstrates the contribution of HESS to the inertial response in power systems with high penetration of RES. To demonstrate this, a virtual inertia emulator has been formulated essentially consisting of a three-phase inverter and an output load controller (LC) filter. In combination with the characteristic of HESS, an efficient virtual synchronous generator can be formulated that actively contributes to the inertia of the overall system. Huiyu et al. [224] studied the introduction of the working principle of virtual synchronous generators (VSG) with the configuration of HESS for enhancing the frequency stability of the system. A novel control strategy is proposed by which BESS provides the low-frequency part of the frequency deviation, and SCSS the high-frequency deviation part for the VSG system.

The study by Zhou et al. [225] exploits the BESS-SCSS HESS characteristics to enhance the primary frequency control (PFC) and the inertial response in a power grid. In this study, a configuration in which SCSS enhances inertia response and BESS support toward PFC has been presented on a two-area system model in which one area is only composed of conventional energy sources and other with increased penetration of RES has been tested. Modeling and sizing of SCSS have been performed based on the rate of change of frequency (RoCoF) whereas BESS is done to meet the steady-state frequency deviation limit. Gu et al. [226] proposed the concept of frequency-coordinating virtual impedance for autonomous control of a DC microgrid. The control strategy enables an effective power frequency split for BESS and SCSS for a seamless transition, even in grid-integrated systems. Furthermore, this methodology ensures more operational flexibility in comparison to conventional droop control methods through coordination of multiple converters in dual timescale and power-scale. Chirkin et al. [227] proposed a frequency regulation strategy in a grid-connected microgrid with high penetration of wind farms. The strategy is based on maintaining a power balance between the generation and the load. Based on the power imbalance caused by the wind farm, the power is divided into small, medium, and large power fluctuations, which are compensated and smoothened by BESS, SCSS, and conventional grid generators. Based on this power frequency split methodology and a proposed computational technique for choosing optimal cut-off frequencies for the low and medium frequency, a sizing methodology of HESS is proposed.

C. **Pulse loads**

Pulse loads require a high instantaneous power with low average power [138]. Once a single energy source is used to supply pulsed loads, thermal and power disturbance problems can be created. If a high power density storage is integrated to the system, various benefits such as the elimination of thermal issues, less volume and weight of the system, and reducing voltage deviation and frequency fluctuation can be achieved [228]. Farhadi and Mohammed [229] proposed a real-time control strategy for a stand-alone DC MG with heavy pulse load and high redundancy. Battery and SC are used as HESS. The SC bank is used to supply the pulsed load and support the grid during transient periods. The results show that using this control method prevents the frequency fluctuation of the generator and improves system performance. Pulsed loads have significantly negative impacts on battery service life. The impact of pulsed load on battery lifespan is investigated by Lahyani et al. [199]. Two scenarios are considered to evaluate battery lifespan. In the first scenario, the pulsed load power is only supplied with battery, whereas in the latter, battery and SC as a hybrid system supplied the pulsed load. Finally, a substantial gain of 17.6% in the lifespan cost of the hybrid system was concluded [178].

D. **Unbalanced load and harmonics**

High-power quality supply for consumers is one of the important issues for MGs. Nonlinear and unbalanced load conditions can be dealt with using HESS. Using the negative-sequence voltage control method, the MG voltage quality can be enhanced. Nevertheless, the unbalanced current challenges remain unsolved for the majority of converters [230]. The use of battery/SC combination for MG performance improvement under unbalanced load was proposed by Hajiaghasi et al.[178, 230]. The results show that HESS utilization leads to fast and accurate voltage regulation under unbalanced load conditions. Zhu et al. [231] proposed a coordination control strategy that uses HESS to improve the power quality for MG under unbalanced load conditions. Tabart et al. [232] presented the power management of a HESS composed of a lithium battery and a vanadium redox battery. The four-leg three-level neutral-point clamped inverter is used to interface the HESS with the MG, due to its low total harmonic distortion, its ability to manage unbalanced AC loads through the fourth leg, and its high efficiency [178].

E. **Stability**

The MG stability is generally divided into three categories: stability of the rotor angle, voltage stability, and frequency stability. The stability of the rotor angle is due to the stability that generators can keep their synchronization in the face of turbulence. It's the stability between the electromagnetic torque and the mechanical torque of the generator and rotor. Frequency stability refers to the stability that the power grid can maintain a constant frequency under different conditions. This kind of stability is an equilibrium between the production and the power dissipation and loss of loads. Voltage stability rests on the stability that maintains a constant voltage in all buses after confronting turbulence. This type of stability is the balance between load demand and power supply per bus. Stability in the MG can be divided into the stability of the MG in the grid-connected mode as well as islanded from the grid [178].

Energy storages can be used for transient stability issues in MG applications [233]. Chang et al. [234] proposed a new method based on active damping to overcome instability problems in DC MG, which is created by constant power loads. This method is based on existing energy storages in DC MG. These storage units also have an additional duty, such as setting the system's damping rate to deal with the problem of instability caused by constant power loads. Using the proposed method, known as the energy storage method by the SC, the constant power loads in the system are reduced virtually and the resistive loads are increased virtually. Therefore, the undesirable effects of constant power loads on the stability of the grid are virtually eliminated and MG stability is improved. Chen et al. [235] employed a battery/

SMES HESS to improve the transient performance of a PV-based MG under various faults. The results of this study have shown that the HESS presents a better performance in comparison of HES to the timely handling of transient fault issues of the MG. The HESS is capable of offering a rapid power injection for the MG at the initial stage of the fault feeding. About the articles reviewed, it can be concluded that the use of energy storage devices increases MG's stability margin, and with the use of HESS, this improvement is further than conventional storage [178].

As pointed out by Hajiaghasi et al. [178], in contrast to the complicated restoration process of conventional power systems that is carried out according to predefined guidelines, the whole restoration process is much simpler in case of distributed generation microgrid due to the reduced number of controllable variables (micro-sources, loads, and switches). However, due to the unpredictability and transience of the micro-sources, they are unsuitable for direct connections to the microgrid. In such cases, different power control switch (PCS) such as DC/AC, AC/DC/AC converter interface is required. Furthermore, considering a worst-case scenario in which fully controllable synchronous generators that would control the voltage and frequency and balance the load and generation conditions of the grid are absent, the operations of the energy source converter interface hence plays a key role to emulate a synchronous generator operational support to allow islanded mode microgrid operation. In these situations, ESS coupled with PCS usually injects active power into the grid proportional to the frequency deviation [178].

The role of ESS is very important in order to maintain the stability of the system as they handle the transient requirement of the system [236], usually modeled as constant DC voltage sources, coupled with voltage source inverter (VSI) in cases of gird-connected microgrid, and inject active power into the grid in proportion to the frequency deviation at least during the initial stages of black start process. Correspondingly, the amount of power to be connected should consider the capacity of the ESS in order to avoid any significant voltage and frequency deviation. Therefore, during the black start process, auxiliary power requirements are compensated by the BESS and transient power by SCSS [237]. The study by Gao et al. [238] presents a simulation evaluating the performance of BESS-SCSS HESS in various black start conditions of 55-kW asynchronous motors. The rating of BESS and SCSS are 250 kW/250 kWh with V/F (voltage and frequency) control and 50 kW/10 seconds with P/Q (active power P and reactive power Q) control, respectively. Accordingly, at the start-up, the SCSS is positioned to compensate for the inrush current generated at the start process of the asynchronous generator. Hence, it has been observed that BESS-SCSS HESS can successfully support the power quality requirements during direct start, star delta start, and frequency conversion start conditions and maintain continuous power supply during these conditions [178].

4.10.3 HESS Capacity Sizing

One of the most important issues in HESS applications is to determine the appropriate storages capacity [178]. Various methods have been proposed for storage capacity sizing. Some methods are developed to determine the HESS capacity of a particular technology, and some other, regardless of technology, can be used for sizing all types of storages. Yang et al. [239] reviewed battery sizing methods and its applications in various RESs. Günther et al. [240] pointed out that in HESS sizing procedure, total cost and the reliability of the system should be considered. The HESS sizing methods based on the purpose of HESS application may be different. As shown by Hajiaghasi et al. [178], storage capacity sizing techniques can be classified into the analytical methods (AM), statistical methods (SM), search-based methods (SBM), pinch analysis method (PAM), and Ragone plot method (RPM). Analytical methods are most commonly used for HESS sizing. Analytical methods are based on analyzing a series of power system configurations with the system elements varied being those that need to be optimized against performance criteria. Various analytical methods have been proposed to determine the HESS capacity and the overall trend of these methods is similar. Compared to analytical methods, statistical methods have more flexibility to determine the energy storage's capacity in some applications. A statistical method for HESS capacity sizing is suggested by Abbassi et al. [241]. A statistical method using Monte Carlo simulation is proposed by Jia et al. [242] to determine the battery and SC capacity in a hybrid system [178]. Search-based methods are subdivided into heuristic methods (HM) and mathematical optimization methods (MOM). Due to the nonlinearity of the objective function in the sizing problem, some researchers have used the heuristic optimization methods. An improved simulated annealing particle swarm optimization algorithm is proposed by Zhou and Sun [203] to solve the HESS optimization problem to reach the minimum cost. A genetic algorithm to determine hybrid storages capacity including battery and SC is proposed by Masih-Tehrani et al. [243]. Wen et al. [244] presented a new strategy based on DFT analysis for optimizing the size of a HESS. Moreover, a cost analysis is done using a particle swarm optimization (PSO) algorithm to optimize the size of various types of ESSs. Ghiassi-Farokhfal et al. [245] proposed a method that returns a Pareto-optimal frontier of the sizes of the underlying storage technologies. This frontier can be used to find the optimal operating point of any application with a linear objective function that is non-decreasing in the vector of storage sizes [178].

Pinch analysis is a simple and flexible methodology for the minimum energy points determination in a utility heat exchanger network. This method is the low-burden computational tool which can be used in renewable MGs. Janghorban-ESfahani et al. [246, 247] used the PAM in HESS applications. A generic HESS sizing method for an islanded MG based on PAM and design space approach is proposed by Jacob et al. [248]. This method is based on variation in the production and load and discharges time of energy storages. Ragone plot is used to compare the performance characteristics of different energy storages. This theory separates energy storage based on power density and energy density. The use of Ragone theory to design HESS is proposed by Zhang et al. [249] and Tankari et al. [250]. Zhang et al. [249] considered the maximization of the storage life cycle as the objective function, and the Ragone plots of energy storages are added to the problem as a constraint. Their analysis indicated that a deeper insight into the problem and more exact solution can be reached with simultaneous consideration of both capacity sizing and real-time operation strategy. Sizing of storage in a wind/HESS system is studied by Cao et al. [251], considering the employed fuzzy-based control and state of charge (SoC), reflecting the real-time operation of system [178].

4.10.3.1 Comparison of Different Capacity Sizing Methods

The advantages and limitations of various sizing methods are compared by Hajiaghasi et al. [178]. The use of appropriate methods depends on the different parameters such as available data of generation and load and the linearity or nonlinearity of the problem, taking into account the dynamics of generations and load, incorporating dynamic characteristics of HESS, and different constraints. Depending on the importance of these factors, different capacity sizing methods can be adopted. AM method is very straightforward and simple and easy to understand. However, the method has a high computational time, and if data resolution is not sufficient, the optimal global solution may not be reached. Numerical approximation of system components is required. SM method is more flexible for engineers to choose the optimum capacity according to the needs of the operation. Synthetic data can be generated when weather data are incomplete. The developed system by this method will be less sensitive to the variation of the parameters. PAM method has a low-burden computational tool, and it is simple and flexible. However, it does not optimize the PV module slope angle and PV area and wind turbine swept parameter.

RPM method is rigorously defined for any kind of energy storages, and it readily displays the two parameters with cost impact. However, it is unable to represent the dynamic performance of the hybrid energy system. SBM methods are mostly used due to their capabilities such as considering different objective functions and solving nonlinear problems. SBM (MOM) can find the solution in a limited number of steps. Also professional software is available for solving optimization problems such as GAMS and MATLAB. However, when the problem becomes more complex, especially for nonlinear programming cases, these tools face difficulties in converging to an optimum solution. SBM (HM) can avoid complicated derivatives, especially for nonlinear optimization problems using reasonable computation time and memory. However, it is unable to represent the dynamic performance of the hybrid energy system and the optimum solution may not be achieved in some problems. On the whole, SBM methods are mostly used due to their capabilities for considering different objective functions and solving nonlinear problems [178].

4.10.4 Hybrid Energy Storages Power Converter Topologies

A HESS can be connected to an MG through different topologies. Different topologies can be employed to combine HPS and HES [252]. A comprehensive review of HESS topologies is presented in [180]. Power converter topologies can be classified as active, semi-active, and passive [178].

In passive topology, two storages with same voltage are simply connected together, that is, efficient, simple, and cost-effective topology [88]. Since the terminal voltage of the storages is not regulated, the power distribution between HPS and the HES units is mainly determined by internal resistances and their voltage-current characteristics. As a result, the available energy from the HPS is very limited, and it acts as a low pass filter for the HESS.

In the semi-active topology, a power converter is inserted at the terminals of one of storages, while the other storage is directly connected to the DC bus [253]. Although the use of a converter requires extra installation space and increases the cost, this class of topology offers more controllability and dispatch capability. There are different semi-active HESS topologies that are reviewed in [132, 253]. The use of the extra converters in this topology makes a better working range of the HESS [132].

Active HESS topologies consist of two or more energy storages in which each energy storage unit is connected to the system by the separate power converter. The active HESS topology can be further broken down as series active topology, parallel active topology, isolated active topology, and multilevel active topology, depending on how high energy storage and high power storage are connected to the two or more converters and eventually to the microgrid. Although the complexity, losses, and cost of the system increase, this class of topology has certain advantages. The advantage of this configuration is that all powers of storages can be controlled actively. Sophisticated control schemes can be implemented in a fully active HESS which is the most commonly used configuration [254]. The parallel active hybrid topology utilizes two converters for power control of the storage 1 and the storage 2. While in the traditional parallel topologies, the energy is exchanged via the common DC bus and the two cascaded converters that negatively affects the overall efficiency of the system. To solve the mentioned problem, a reconfigurable topology is proposed in [255]. Compared to conventional active topologies, the reconfigurable topology has advantages including reduction in the DC-bus capacitor size, increased efficiency during the energy exchange modes, and the capability of reconfiguration. In some researches, multilevel converters are used as hybrid storage power converters [256]. Using a multilevel structure, the system reliability and power quality can also be enhanced. Meanwhile, connecting multiple energy storage in one converter reduces costs and coordination control complexity. However, the number of power electronics switches and capacitors in the multilevel converters are high and their control is more complicated. Isolated multi-port converters for HESS have been used in [257]. Each structure has its own advantages and disadvantages, but in recent years, the use of the active structure has been considered due to its high capabilities. For HESS connection to MG, the various class of power converter can be used [258]. To select a power converter, various parameters such as efficiency, cost, reliability, and flexibility should be considered [178].

4.10.4.1 Comparison of Different Topologies

The HESS topology directly affects energy management strategy. In the passive topology, there is no direct control of the storage power. In the semi-active topology, the output power of one of the storages is uncontrollable and the voltage of the other should be same as the DC bus. The active structure controls the output or input power of both storages by a rational control strategy, in cost of lower efficiency. Appropriate topology should be selected regarding different factors such as costs, efficiency, controllability, complexity, and flexibility. Table 4.5 compares the

TABLE 4.5
Comparison of Different Hybrid Storages Topologies [178]

Parameter	Passive Topology	Semi-active Topology	Active Topology
Cost	Low to medium	Medium	High
Efficiency	High	Medium	Low to medium
Flexibility	Low to medium	Medium	High
Complexity	Low to medium	Medium	High
Controllability	Low to medium	Medium	High

HESS topologies from different operational aspects. The passive topology is simple and cost-effective but noncontrollable. In terms of controllability and flexibility and taking into account different constraints, such as SoC, the active topology exhibits the best performance, but its cost and complexity is high. The semi-active topology introduces limited controllability with lower cost [178].

4.10.5 HESS Energy Management and Control in MG

Designing and implementing a proper control system is the most important issue in HESS. The choice of the appropriate control method for HESS depends on different parameters: the purpose of the use of HESS (such as storage life extension, power quality, intermittency improvement, etc.), the type of system (DC MG, AC MG, grid-connected), the cost of the control method, the control method response time, the hybridization architecture, etc. To achieve the safe, stable, and efficient operation of HESS, a reasonable power-sharing strategy is essential.

Generally, the energy management and control system of HESS can be classified into two parts: (i) the underlying control unit that controls the current or power flowing of HESS elements based on the reference signal generated by energy management control unit, and (ii) the energy management unit which performs power allocation between the HESS storages to enhance system dynamics, reach high overall efficiency, monitor SoC, reduce the loss and cost of system, and minimize operational cost of the system and frequency regulation.

4.10.5.1 HESS Energy Management System

Power allocation is a major concern in HESS. In literature, various methods have been proposed for allocating power in HESS. In general, the energy management methods can be categorized into intelligent control methods and classical control methods. Various HESS control methods are proposed in literatures and these are described in reference [178].

4.10.5.1.1 Classical Energy Management Control Strategies

The classical control methods can be divided into the rules-based controller (RBC), droop-based control (DBC), and filtration-based controller (FBC). These control methods are sensitive to changes in parameters, and the exact mathematical model of the system is required. In RBC, power allocation of HESS is performed according to the predefined rules. RBC can be divided into thermostat, state machine, and power follower control methods [259]. The RBC controller is simple and easy to implement and is an effective method for real-time energy management. Nevertheless, the sensitivity to parameters variation is the disadvantage of this method.

In FBC, a filter is used to decompose net power to high and low frequencies. In most MG applications power sharing between hybrid storages is based on FBC control. In this method, the storages' reference powers are determined based on the filter parameters. Scheduling the reference current among ESs is characterized by different ramp rates, low pass filters, and high-pass filters which are normally utilized to share power demand into low/high-frequency components. Power sharing in energy storages is based on DC-bus voltage regulation and filtration method. Implementing this method is easy and cost-effective. But in cases where the number of DGs or ESs increases, this method is not suitable because the power sharing between the storage devices is less accurate [178].

When HESSs are used in MG, the hybrid storages power is determined by active coordination, and this power is divided by the energy management unit between different storages. Active coordination can further be classified as centralized, distributed, and decentralized ones. However, both the distributed and the centralized controls are vulnerable to system communication failures. In this regard, the decentralized control methods are the most reliable one among the MGs control strategies. The DBC methods are one of the effective decentralized control methods.

The various types of droop control in MGs applications are reviewed and compared in [260]. However, the conventional droop control methods cannot be directly used to HESS control. In recent researches about HESS control, various types of DBC methods such as high-pass filter-based droop control, extended droop control strategy, adaptive droop-based control, integral droop (ID) control, virtual capacitance droop, virtual resistance droop, virtual capacitance droop with SoC recovery, adaptive droop, secondary voltage recovery droop, and virtual impedance droop are proposed [178]. Lin et al. [261] used an ID for a group of energy storages with high ramp rates. Conventional voltage power (V-P) droop control method contains the power split between the hybrid storages with ID. The smooth change of load power components is supplied by the storages with V-P droop, whereas the storages with ID respond to high-frequency power demand. Additionally, the coefficient of ID can be designed based on the nominal ramp rate of the energy storages with slow response, which leads an improved lifespan of HESS. Each of the droop control methods has some advantages and disadvantages. All droop methods have plug "n" play capability. However, droop control methods with SoC recovery loop provide better storage performance. Extended droop can realize SoC of SC automatic recovery only when the line impedance can be ignored. Virtual capacitance droop is not suitable for DC.

In MGs with multiple HESSs, if the SC leakage current is considered, the performance of virtual capacitance droop with SoC recovery strategy will deteriorate. To solve the problems associated with droop control, Shi et al. [262] proposed an advanced secondary voltage recovery strategy using consensus algorithm based on the primary virtual impedance droop for multiple HESSs. In this strategy, secondary voltage recovery control for battery restore voltage deviation caused by virtual impedance droop control, and secondary voltage recovery for SC recover the SoC of the SC after discharging and charging, in spite of the existence of SC leakage current [178].

To improve the dynamic response of HESSs, the deadbeat control method (DCM) is presented by Wang et al. [263]. This strategy generates the optimal duty ratio in one control cycle, so that it can respond faster to the disturbance than various PI-based methods. The fast dynamic response of SC relieves the stress on the battery to extend its lifetime, and the bus voltage can be maintained around its reference tightly. In addition, it is implemented with a reduced hardware cost by eliminating extra current sensors compared to the other model-based control approaches. However, this method is dependent on the system model, and in cases where the exact model of the system does not exist and the system parameters change, it does not respond appropriately [178].

Hierarchical control is another control method used for HESS control [260, 261]. In comparison with other energy management strategy, the hierarchical method not only handles the uncertainty of loads and RESs by embedding the robust/stochastic energy management, but also reduces such uncertainty through using the intraday forecast data with higher precision. For a secure and economic operation of MG with HESS under load and RESs uncertainty, a hierarchical energy management strategy is presented in [260], which consists of real-time scheduling and hour-ahead scheduling. The results confirm that the control strategy is effective in providing a safe generation and economic plan. The calculation speed is sufficiently fast to satisfy online operation requirements. Moreover, the proposed control method of HESSs increases the lifespan of battery by 10.48%. To enhance DC MG reliability, a hierarchical strategy, which is composed of both distributed and centralized control, is proposed in [264]. In this strategy, the traditional HESS control has been refined with implementations of secondary voltage regulation, autonomous SoC recovery, and online iteration. Using autonomous SoC recovery, the SoC violation of storage is reduced; moreover, DC-link voltage deviation is minimized and power quality improved. However, the accuracy of the system has decreased compared to the central control methods. A hierarchical control method to manage and control a shipboard power system is presented in [261], which is composed of the diesel generator, HEESs, and loads. To overcome the limitation controllability of HESS, a new inverse droop control method is presented, for which the power sharing is according to DG characteristic, rather than their power rating. The results show that compared with conventional control strategy, the method has advancement in bus voltage regulation and real-time fuel efficiency.

4.10.5.1.2 Intelligent Energy Management Control Strategies

As pointed out by Hajiaghasi et al. [178], to control nonlinear and complex systems, optimization-based methods have become more popular. These methods can be categorized as the artificial neural network (ANN), Fuzzy logic control (FLC), and evolutionary algorithms such as genetic algorithm, dynamic programming, linear programming (if the system is convex and could be mathematically represented via a set of linear functions), and model predictive control (MPC). The FLC method is one of the effective methods for controlling complex systems. This control method is simple and does not require the exact mathematical model of the system. In some references such as [265–268], FLC has been used for HESS control. An FLC strategy for power allocation of HESS with three modes is proposed by Feng et al. [265]. One mode is dedicated to exchange power between SCs and batteries to prevent overcharging and discharging of the energy storages. The results of control implementation show that SoC values of storages maintain within the acceptable range, and the power compensation efficiency is improved. To minimize battery peak current, authors in [269] optimized the membership functions of FLC by PSO algorithm. A HESS and relevant control strategy for DC-bus voltage regulation are proposed in [209]. The DC-bus voltage is controlled according to the principle described in [209]. FLC and FBC are used for energy storage management. A low pass filter is applied to the DC-bus current to divert sudden power variation into the SCs. A PI controller is used to calculate the energy storages reference currents so that the bus voltage is maintained at the reference value. The main purpose of using the FLC is to control the power of storage devices considering the limits of battery and SC charging within acceptable limits. An adaptive fuzzy logic-based energy management strategy to split the power requirement between the battery and the SC is proposed in [270]. The key issues of this strategy are to improve the system efficiency, to reduce the battery current variation, and to minimize SC state-of-charge difference [178].

The MPCs provide another alternative to control the HESSs. The MPC in HESS control can be used either for power allocation in HESS and reference determination of storages, with consideration of different constraints [155, 271], or for direct control of power converter by controlling modulation index [272]. Moreover, the combination of both the mentioned approaches can be incorporated. Naseri et al. [273] used the MPC control to power sharing in an MG containing FC/batteries/SC. The different degradation challenges of energy storages related to start-up cycles, shut-down cycles, and load fluctuation are investigated and minimized. The block diagram for the MPC controller is described in [271]. The controller receives as a reference power scheduled by the economic dispatch of the MG for the batteries, the electrolyzer, and the FC, as well as the energy exchange with the main grid. Garcia-Torres and Bordons [155] proposed the MPC controller to solve the optimization problem in a grid-connected MG containing various energy storages. In the control design, various constraints including degradation issues, operational limitations, and operational control associated with HESS are considered. An example of the power converter modulation index control by MPC is described by Garcia-Torres [155] and Garcia-Torres et al. [271]. In this example, MPC predicts the modulation indices m_{cap} and m_{batt} for the SC and the battery DC/DC converter, respectively, so that the total power matches the required total power subjected to specified constraints. The purpose of the control system is to allocate the fast-current changes to SC and slow-current changes to the battery [272]. ANN approaches do not require an exact system model, and pattern recognition capabilities of these methods make them attractive for renewable energy and MG applications. In literatures, ANNs have been used to predict the wind velocity and solar radiation, and few articles have used the ANN for energy management in HESS. Nevertheless, the ANN-based control methods require historical data for the tuning and learning process, which makes it difficult to implement [178].

An ANN-based control system for investigating the behavior of a grid-connected hybrid AC/DC MG including PVs modules, a wind turbine generator, solid oxide FC, and a battery energy storage system is proposed by Yin et al. [270]. This strategy tracks the maximum power point of renewable energy generators and controls the power exchanged between the front-end converter and the electrical grid. A grid-connected hybrid MG which consists of a PV system, a battery energy storage, a wind turbine generator, an FC, and the AC and DC loads is presented by Chettibi et al. [274]. A feed-forward ANN is used for the DC-bus voltage regulation. Two Elman neural networks-based controllers are designed to ensure the control of the bidirectional flow of the active power as well as the compensation of the AC load reactive power. Moreover, an FLC-based power management system is adopted to minimize the energy purchased from the electrical grid. The performance of the mentioned control system has been evaluated for different situations: variable load demands, perturbed grid, and variable climate conditions. The results show that it is possible to control complicated nonlinear systems without the need for a precise model. Furthermore, the mentioned methods are adaptable, flexible, and easy to implement in real-time applications with low computational costs.

4.10.5.2 Underlying Controller

The task of underlying control is to tracking HESS references power generated by energy management control system. Concerning the HESS control, existing control methods can be classified into the modern control method and the classical control method [275]. In classical control strategies, the controller based on the linearized system is designed in the frequency domain. The main problem of the classical methods is that the system nonlinearities are not considered. Moreover, in some applications, the controller performance may not be acceptable. In this regard, some researchers have used nonlinear control methods to control HESS. There are various types of nonlinear control methods. In some of the references, sliding mode control is used to control energy storages [276, 277]. The comparison between the classical PI controller and the sliding mode controller for a HESS is investigated by Etxeberria et al. [276]. The simulation results show that the sliding mode controller has higher stability and robustness compared to the PI controller. However, the conventional sliding mode controller has some drawbacks for ESS converters such as variable and high switching frequency that increases system complexity and power losses. To solving the mentioned problem, a fixed-frequency sliding mode control method is proposed by Wickramasinghe et al. [277] for HESS control. In comparison with the traditional method, the SM controller was able to achieve better output capacitor reference tracking. Classical control methods are most common to the power or current reference tracking in HESS. However, due to the nonlinear properties of HESSs, the use of nonlinear controllers will lead to better results [178].

4.10.5.3 Comparison of Various Control Methods

Choosing an energy management approach and control strategy do not have a specific flowchart or a predetermined process. Different items such as the purpose of the control system, response time, control system cost, energy storages characteristics, and reliability can be considered. Advantages and limitations of different control methods of HESS are presented by Hajiagashi et al. [178]. Due to less computational complexities classical control methods are more suitable for real-time applications. However, the need for the mathematical model of the system makes it difficult to use in some cases. One of the main disadvantages of intelligent control methods is the high computational burden of these strategies, wherein real-time applications may affect the performance of the system and requires hardware systems with high computational capabilities. These methods are flexible, where different cost functions or power quality parameters can be adopted.

4.10.6 Overview on Control of Microgrid with HRES and HESS

Generally, the control system must be placed at different levels of the system, and a consistent communication between several control units is required since there is a continuous change of power production in the DGs and the load demand in fluctuation with time. MG central controller (MGCC) installed at the medium-/low-voltage substation, which has a supervisory task of centrally controlling and managing the MG, integrates with the main grid. The MGCC includes several key functions, such as economically managing functions and control functionalities, and is the head of the hierarchical control systems and communicates between network operators. The MG is intended to operate in the following two different operating conditions: the normal interconnected mode with a distribution network and the emergency mode in islanding operation via a central switch, which must also implement the synchronization between both power systems.

The overall architecture of a microgrid consists of a low-voltage (LV) network on the consumer load side (both critical and noncritical loads), both noncontrollable and controllable power generators, energy storage units, and a hierarchical energy management. Controlling and monitoring each DG and loads and managing energy system require communication infrastructure to support the control scheme so that the microgrid central control (MGCC), the center for the hierarchical control system, follow sequential low control level, like local controllers (LOCs) of loads and DGs which exchange information with the MGCC for managing the whole MG operation by providing set-points to LCs. The common relevant data in exchange includes mainly information about MG switch orders that are sent by the MGCC to LC and the sensed voltage/current information to MGCC from each local capture; power and frequency reference setting for each source and the state charge and discharge of the energy storage; and the protection device conditions the system in the case of fault happening to isolate the abnormal zone of the system.

The typical single-line structure of a microgrid control system is described in Figure 4.11 [278]. The radial power line arrangement is connected to the main utility grid at the point of common coupling (PCC) through a separation device, usually a

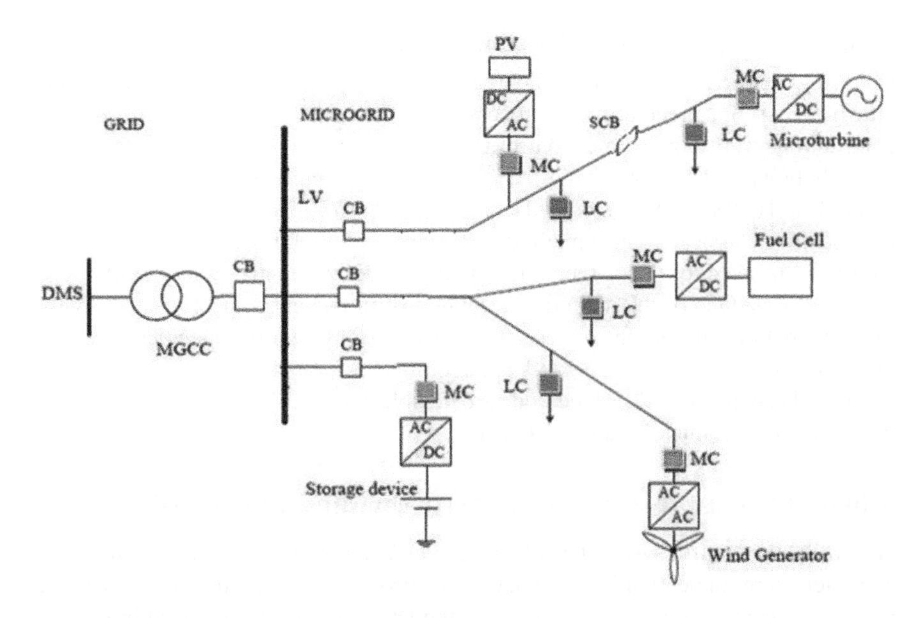

FIGURE 4.11 The microgrid control architecture [278].

static switch, having circuit breaker and a power flow controller for each feeder line [133]. The limited capacity of distribution generators can result in the development of several microgrids, which are interconnected to each other and operate with or without the main grid. It is clear that a direct connection of the microgrid LV line to DGRs (PV, wind generator, microturbine) and to the electrical grid network is not possible and so power electronic interfaces (DC/AC or AC/DC/AC) are required due to the characteristics of the energy produced. Inverter control and circuit protection is thus an important concern in MG operation. In the microgrid control system, there are main parts including: micro-source controllers (MCs) on the consumer production side and load controllers (LCs) on the consumer demand side; microgrid system central controller (MGCC) on the middle of the main grid; and microgrid structures and distribution management system (DMS) in the grid network side.

The different DG sources and energy storage devices are connected to the low feeder lines through the micro-source controllers (MCs). MC has a function of controlling the power flow and bus voltage profile of the micro-sources according to the load changes or any other disturbances. These feeders are also supplied with several sectionalizing circuit breakers (SCBs) which help in isolating a part of the microgrid as needed in case trouble. Power electronics interfaces and inverters (AC/DC, AC/AC, DC/AC) are important means for controlling and monitoring the loads using load controllers (LCs).

The overall operation and management in both the modes (isolated and grid-tied) is controlled and coordinated with the help of micro-source controllers (MCs) at the local level and microgrid system central controller (MGCC) at the global level; there is a point of common coupling (PCC) through the circuit breakers (CBs) between the microgrid and the medium voltage-level utility grid. The MGCC is responsible for the overall control of microgrid operation and protection, like maintaining specified bus voltages and frequency of the entire microgrid and energy optimization for the microgrid. On the utility side, there is a distribution management system (DMS) having several feeders including several microgrids and functions for distribution area management and control. So there are two parts in control tasks: the first one is the microgrid-side controller (MC and LC) to take the maximum power from the input source, and the protection of input-side converter must be considered. The second part is the grid-side controller (MGCC and DMS), which has the following main tasks: (a) input active power control derived for network; (b) control of the reactive power transferred between network and microgrid; (c) DC-link voltage control; (d) synchronization of network; and (e) assurance of power quality injected to the network [279].

The major element in the second level of the hierarchical control is the power converter control for the sources. In an AC grid, the power converter can be controlled using the following methods:

PQ control: In this mode, the active power (P) and reactive power (Q) output from the converter are controlled by controlling the direct and quadrature axis current of the converter output. The reference PQ values will be set by MGC. In this mode, the converter is controlled as a current-controlled voltage source [280].

VSI control: In this the converter operates as voltage source inverter (VSI), mimicking the operation of a synchronous generator. The droop characteristic is used for obtaining the P and Q reference [281].

DC current control: The DC converter control for the storage devices and the sources are done in a similar way to the PQ control. The P reference generates the current reference which will result in the converter working as current-controlled voltage source as suggested by Thounthong et al. [93].

4.10.6.1 Overview of Control under Islanded Mode for Microgrid

The microgrids as discussed above can operate in the grid-connected mode and islanded mode during the event of the faults. The main challenge in the islanded mode of operation is ensuring the system's stability. In the absence of synchronous machines, the converters have to emulate the droop characteristics and ensure frequency stability. The VSI control scheme is suited for this. There are two main control strategies under the islanded mode of operation as shown in Lopes et al. [280]:

- ***Single master operation (SMO)***: It is where one converter is connected to source or the storage element will operate in VSI mode. This converter will therefore set a voltage and frequency reference for the grid. The other converters will work in PQ control mode.
- ***Multi-master mode (MMO)***: In this mode, more than one converter will work in VSI mode, setting the voltage and frequency reference, while others work in the PQ control mode.

Another control strategy for microgrid control under island mode is presented in Ding et al. [282]. Here the authors treat the microgrid as an AC subgrid formed by the AC sources and its load along with a DC subgrid formed by DC sources and its load in the main microgrid. The two subgrids are controlled through their individual droop characteristics. An interconnecting converter between subgrids enables the power transfer and stable operation under island mode. Most of the controls in the island mode utilize the droop characteristics for the stabilizing the grid.

4.10.7 HESS for Off-grid Applications Using HRES

The main advantages of a HESS in off-grid applications are [30, 178]:

1. Reduction of total investment costs compared to a single storage system (due to decoupling of energy and power, the HE unit only has to cover average power demand).
2. Increase of total system efficiency (due to operation of the HE unit at optimized, high-efficiency operating points and reduction of dynamic losses).
3. Increase of storage and system lifetime optimized operation and reduction of dynamic stress of the HE unit.

To take full advantage of the characteristics of the complementary ESS requires an energy management and control system, which in advanced systems also controls the energy flows in the HRES. In essence, the HESS has moved from a subsidiary part to become the heart of the HRES. To understand the concept of the HESS, it is necessary to consider hybridization principles and proposed topologies, power electronics interface architectures, as well as control and energy management strategies.

The current HRES system design places a number of functions on the HESS, in addition to basic energy storage, particularly in stand-alone or islanded mode. These functions are performed by the ESS management system using stored energy. The two main active functions of the HESS are [30, 178]:

- ***Frequency control and frequency support:*** This requires fast response and a high-power capability. Battery energy storage has good features for regulating frequency in off-grid MG, stand-alone, and HRES systems. However, for frequency regulation, the battery charges and discharges at a high rate, which reduces its lifespan. Moreover, the battery needs to deal with the sudden power changes in the primary frequency control, which will also accelerate the battery degradation process. To solve the problems, a novel concept of primary frequency control by combining the HP unit (e.g., supercapacitor) with the battery, achieves both the frequency regulating function and the battery life-service extension. The supercapacitor (SC) is employed to emulate the inertia of a virtual synchronous generator and cope with high-frequency power fluctuations.
- ***Reactive power and PF control (voltage control)***: These require long-term action and are supplied using the energy in the HE unit.

Initially HESS systems are operated on a simple parallel connection of the high power (HP) and high energy (HE) storage systems. Subsequent research has shown that improved performance can be achieved by adopting other configurations. Literature showed that there are different ways for the coupling of the energy storages in a HESS. A simple approach is the direct DC coupling of two storages (passive topology). The main advantage is the simplicity and cost-effectiveness. Moreover, the DC-bus voltage experiences only small variations. The main disadvantage is the lack of possibilities for power flow control and energy management and a resulting ineffective utilization of the storages (e.g., in a SC/B HESS); with direct coupling, only a small percentage of the SC capacity can be utilized when operated within the narrow voltage band of the battery. In series active system, high energy storage is coupled to DC bus by bidirectional converter while high power storage system is directly connected to the DC bus. Finally, in parallel hybrid active technology, both high energy and high power storage systems are connected to DC bus using bidirectional DC/DC converters. Both power flow control and energy management are the best in the last most complex system [30, 178].

REFERENCES

1. Shah YT. *Modular Systems for Energy Usage Management.* New York: CRC Press; 2020.
2. Ibrahim H, Ilinca A. Techno-economic analysis of different energy storage technologies. Intec open access paper. 2013. DOI: 10.57772/52220.
3. Fu R, Remo T, Margolis R.U.S. utility-scale photovoltaics- plus-energy storage system costs benchmark. Golden, CO: National Renewable Energy Laboratory; 2018 [November 2018]. Technical Report No.: NREL/TP-6A20-71714, Contract No. DE-AC3608GO28308.
4. Serpi A, Porru M, Damiano A. A novel highly integrated hybrid energy storage system for electric propulsion and smart grid applications, Chapter 4. INTEch Publication; 2018. http://dx.doi.org/10.5772/intechopen.73671, http://www.intechopen.com.
5. Smart grid system report, report to Congress 2018. Washington, DC: Department of Energy; Novermber 2018.
6. Supercapacitors. *Wikipedia, The Free Encyclopedia.* 2020 [last visited May 14, 2020].
7. Shah YT. *Thermal Energy: Sources, Recovery and Applications.* New York: CRC Press, Taylor and Francis Group; 2018.
8. Wagner L. Overview of energy storage technologies, Chapter 27. In: Letcher TM, editor. Future Energy: Improved, Sustainable and Clean Options for Our Planet. 2nd ed. Amsterdam: Elsevier; 2014. pp. 613–31. DOI: 10.1016/ B978-0-08-099424-6.00027-2.
9. Energy storage technology: An overview. ScienceDirect Topics. 2015. Available from: http://www.sciencedirect.com/topics/engineering/energy-storage-technology. A website report.
10. Energy storage. *Wikipedia, The Free Encyclopedia.* 2019 [last visited July 25, 2019].
11. Energy storage technologies. Washington, DC: Energy Storage Association; 2019. Available from: http://energystorage.org/energy-storage/energy-storage-technologies. A website report.
12. Gur T. Review of electrical energy storage technologies, materials and systems: Challenges and prospects for large-scale grid storage. *Energy and Environmental Science.* 2018;11:2696–767. DOI: 10.1039/C8EE01419A.

13. MedinaP, Bizuayehu AW, Catalao JPS, Rodrigues EMG, Contreras J. Electrical energy storage systems: Technologies' state-of-the-art, techno-economic benefits and applications analysis. In: Proceedings of the 47th Hawaii International Conference on System Sciences, 6–9 January, Waikoloa, HI, USA; 2014. pp. 2295–304.
14. Chen H, Cong TN, Yang W, Tan C, Li Y, Ding Y. Progress in electrical energy storage system: A critical review. *Progress in Natural Science*. 2009;19:291–312.
15. Hadjipaschalis I, Poullikkas A, Efthimiou V. Overview of current and future energy storage technologies for electric power applications. *Renewable and Sustainable Energy Reviews*. 2009;13:1513–22.
16. Sabihuddin S, Kiprakis A, Mueller M. A numerical and graphical review of energy storage technologies. *Energies*. 2014;8:172–216.
17. EPRI (Electric Power Research Institute). *Electric Energy Storage Technology Options: A White Paper Primer on Applications, Costs, and Benefits*. Palo Alto, CA: EPRI; 2010.
18. SBC Energy Institute. *Leading the Energy Transition, Factbook, Electricity Storage*. Gravenhage, Netherlands: SBC Energy Institute; 2013.
19. Khan N, Dilshad S, Khalid R, Kalair A, Abas N. Review of energy storage and transportation of energy. *Energy Storage*. 2019;1(3):e49. https://doi.org/10.1002/est2.49.
20. Nie Z, Xiao X, Kang Q, Aggarwal R, Zhang H, Yuan W. SMES-battery energy storage system for conditioning outputs from direct drive linear wave energy converters. *IEEE Transactions on Applied Superconductivity*. 2013;23(3):5000705.
21. Denholm P, Ela E, Kirby B, Milligan M. The role of energy storage with renewable electricity generation. Golden, CO: National Renewable Energy Laboratory; January 2010. Technical Report No.: NREL/TP-6A2-47187.
22. Khalid M. A review on the selected applications of battery-supercapacitor hybrid energy storage systems for microgrids. *Energies*. 2019;12:4559. DOI: 10.3390/en12234559. http://www.mdpi.com/journal/energies.
23. Hemati R, Saboori H. Emergence of hybrid energy storage systems in renewable energy and transport applications: A review. *Renewable and Sustainable Energy Reviews*. 2016;65:11–23. http://dx.doi.org/10.1016/j.rser.2016.06.029.
24. Barsukov Y. Battery cell balancing: What to balance and how. Dallas, TX: Texas Instruments; 2009. Technical Report.
25. Altaf F, Johannesson L, Egardt B. Performance evaluation of multi-level converter based cell balancer with reciprocating air flow. In: Proceedings of the Vehicle Power and Propulsion Conference (VPPC), Seoul, Korea; 2012. pp. 706–13.
26. Chanhom P, Sirisukprasert S, Hatti N. DC-link voltage optimization for SOC balancing control of a battery energy storage system based on a 7-level cascaded PWM converter. In: Proceedings of the International Conference on Electrical Engineering/Electronics, Computer, Telecommunications and Information Technology (ECTI-CON), Phetchaburi, Thailand; 2012. pp. 1–4.
27. Einhorn M, Roessler W, Fleig J. Improved performance of serially connected li-ion batteries with active cell balancing in electric vehicles. *IEEE Transactions on Vehicular Technology*. 2011;60(6):2448–57.
28. Imtiaz A, Khan F, Kamath H. A low-cost time shared cell balancing technique for future lithium-ion battery storage system featuring regenerative energy distribution. In: Proceedings of the Applied Power Electronics Conference and Exposition (APEC), Fort Worth, TX; 2011. pp. 792–9.
29. Moore SW, SchneiderPJ. A review of cell equalization methods for lithium ion and lithium polymer battery systems. In: Proceedings of the SAE 2001 World Congress, Detroit, MI; 2001. pp. 2001–01–0959.

30. Kim Y, Wang Y, Chang N, Pedram M. Computer-aided design and optimization of, hybrid energy storage systems. *Foundations and Trends in Electronic Design Automation*. 2013;7(4):247–338. DOI: 10.1561/1000000035.
31. Fang X, Kutkut N, Shen J, Batarseh I. Analysis of generalized parallel- series ultracapacitor shift circuits for energy storage systems. *Renewable Energy* 2011;36:2599–2604.
32. Kim Y, Park S, Wang Y, Xie Q, Chang N, Poncino M, Pedram M. Balanced reconfiguration of storage banks in a hybrid electrical energy storage system. In: Proceedings of the International Conference on Computer- Aided Design (ICCAD), San Jose, CA; 2011. pp. 624–31.
33. Uno M. Cascaded switched capacitor converters with selectable intermediate taps for supercapacitor discharger. In: Proceedings of the TENCON. 2009. pp. 1–5.Singapore
34. Uno M. Series-parallel reconfiguration technique for supercapacitor energy storage systems. In: Proceedings of the TENCON. 2009. pp. 1–5.Singapore
35. Uno M. Toyota H. Supercapacitor-based energy storage system with voltage equalizers and selective taps. In: Proceedings of the power electronics specialists conference, Rhodes, Greece; 2008. pp. 755–60.
36. Li K, Wu J, Jiang Y, Hassan Z, Lv Q, Shang L, Maksimovic D. Large-scale battery system modeling and analysis for emerging electric-drive vehicles. In: Proceedings of the International Symposium on Low-Power Electronics and Design (ISLPED), Austin, TX; 2010. pp. 277–82.
37. Kim H, Shin K. On dynamic reconfiguration of a large-scale battery system. In: Proceedings of the Real-Time and Embedded Technology and Applications Symposium (RTAS), San Francisco, CA; 2009. pp. 87–96.
38. Kim H, Shin K. DESA: Dependable, efficient, scalable architecture for management of large-scale batteries. *IEEE Transactions on Industrial Informatics*. 2012;8(2):406–17.
39. Choi Y, Chang N, Kim T. DC–DC converter-aware power management for low-power embedded systems. *IEEE Transactions on Computer-Aided Design of Integrated Circuits and Systems*. 2007;26(8):1367–81.
40. Nergaard T, Ferrell J, Leslie L, Lai JS. Design considerations for a 48 v fuel cell to split single phase inverter system with ultracapacitor energy stor- age. In: Proceedings of the Power Electronics Specialists Conference (PESC), Queensland, Australia; 2002. Vol. 4, pp. 2007–2012.
41. Xue Y, Chang L, Kjaer SB, Bordonau J, Shimizu T. Topologies of single-phase inverters for small distributed power generators: an overview. *IEEE Transactions on Power Electronics*. 2004;19(5):1305–14.
42. Cheng Y, Joeri V, Lataire P. Research and test platform for hybrid electric vehicle with the super capacitor based energy storage. In: Proceedings of the European Conference on Power Electronics and Applications, Aalborg, Denmark; 2007. pp. 1–10.
43. Kim S, No KS, Chou P. Design and performance analysis of super capacitor charging circuits for wireless sensor nodes. *IEEE Journal on Emerging and Selected Topics in Circuits and Systems*. 2011;1(3):391–402.
44. Kim Y, Wang Y, Chang N, Pedram M. Maximum power transfer tracking for a photovoltaic-supercapacitor energy system. In: Proceedings of the International Symposium on Low Power Electronics and Design (ISLPED), Austin, TX; 2010. pp. 307–12.
45. Glavin M, Hurley W. Optimisation of a photovoltaic battery ultracapacitor hybrid energy storage system. *Solar Energy*, 2012;86(10):3009–20.
46. Frenzel B, Kurzweil P, Rönnebeck H. Electromobility concept for racing cars based on lithium-ion batteries and supercapacitors. *Journal of Power Sources*. 2011;196(12):5364–76.
47. Battery power for your residential solar electric system. Golden, CO: National Renewable Energy Laboratory; 2002. Technical Report.

48. Ericson SJ, Rose E, Jayaswal H, Cole WJ, Engel-Cox J, Logan J, et al. Hybrid storage market assessment: A JISEA white paper. Golden, CO: National Renewable Energy Laboratory; October 2017. Report No.: NREL/MP-6A50-70237, Contract No. DE-AC36-08GO28308.
49. LaFollette RM, Bennion DN. Design fundamentals of high power density, pulsed discharge, lead-acid batteries. 2. Modeling. *Journal of the Electrochemical Society*. 1990:3701–07.
50. Doerffel D, Sharkh SA. A critical review of using the peukert equation for determining the remaining capacity of lead-acid and lithium-ion batteries. *Journal of Power Sources*. 2006;155(2):395–400.
51. Li C-H, Zhu X-J, Cao G-Y, Sui S, Hu M-R.Dynamic modeling and sizing optimization of stand-alone photovoltaic power systems using hybrid energy storage technology. *Renewable Energy*. 2009;34(3):815826,.
52. Vosen A, Keller J. Hybrid energy storage systems for stand-alone electric power systems: Optimization of system performance and cost through control strategies. *International Journal of Hydrogen Energy*. 1999;24(12):1139–56.
53. Jiang X, Polastre J, Culler D. Perpetual environmentally powered sensor networks. In: Proceedings of the International Symposium on Information Processing in Sensor Networks (IPSN). 2005. pp. 463–68. Los Angeles, US: University of California.
54. Yoda S, Ishihara K. The advent of battery-based societies and the global environment in the 21stcentury. *Journal of Power Sources*. 1999;81–82:162–9.
55. Pedram M, Chang N, Kim Y, Wang Y. Hybrid electrical energy storage systems. In: Proceedings of the International Symposium on Low-Power Electronics and Design (ISLPED). 2010. pp. 363–8. Austin, Texas
56. Koushanfar F. Hierarchical hybrid power supply networks. In: Proceedings of the Design Automation Conference (DAC), Anaheim, CA; 2010. pp. 629–30.
57. Koushanfar F, Mirhoseini A. Hybrid heterogeneous energy supply networks. In: Proceedings of the International Symposium on Circuits and Systems (ISCAS), Rio de Janeiro, Brazil; 2011. pp. 2489–92.
58. Mirhoseini A, Koushanfar F. Learning to manage combined energy supply systems. In: Proceedings of the International Symposium on Low-Power Electronics and Design (ISLPED), Fukuoka, Japan; 2011. pp. 229–34.
59. Mirhoseini A, Koushanfar F. HypoEnergy. hybrid supercapacitor- battery power-supply optimization for energy efficiency. In: Proceedings of the Design, Automation and Test in Europe Conference and Exhibition (DATE), Grenoble, France; 2011. pp. 1–4.
60. Zhuo J, Chakrabarti C, Chang N. Energy management of DVS-DPM enabled embedded systems powered by fuel cell-battery hybrid source. In Proceedings of the International Symposium on Low-Power Electronics and Design (ISLPED), Portland, Oregon; 2007. pp. 322–27.
61. Zhuo J, Chakrabarti C, Chang N, Vrudhula S. Maximizing the life- time of embedded systems powered by fuel cell-battery hybrids. In Proceedings of the International Symposium on Low-Power Electronics and Design (ISLPED), New York, NY; 2006. pp. 424–9.
62. Zhuo J, Chakrabarti C, Lee K, Chang N, Vrudhula S. Maximizing the lifetime of embedded systems powered by fuel cell-battery hybrids. *IEEE Transactions on Very Large Scale Integration (VLSI) Systems*. 2009;17(1):22–32.
63. Moreno J, Ortuzar M, Dixon J. Energy-management system for a hybrid electric vehicle, using ultracapacitors and neural networks. *IEEE Transactions on Industrial Electronics*. 2006;53(2):614–23.
64. Thounthong P, Ra äel S, Davat B. Energy management of fuel cell/battery/supercapacitor hybrid power source for vehicle applications. *Journal of Power Sources*. 2009;193(1):376–85.

65. Romaus C, Bocker J, Witting K, Seifried A, Znamenshchykov O. Optimal energy management for a hybrid energy storage system combining batteries and double layer capacitors. In: Proceedings of the Energy Conversion Congress and Exposition (ECCE), San Jose, CA; 2009. pp. 1640–7.
66. Miller J, Deshpande U, Dougherty T, Bohn T. Power electronic enabled active hybrid energy storage system and its economic viability. In: Proceedings of the IEEE Applied Power Electronics Conference and Exposition (APEC). 2009. pp. 190–8.
67. Jin C, Lu S, Lu N, Dougal R. Cross-market optimization for hybrid energy storage systems. In: Power and Energy Society General Meeting. 2011. pp. 1–6. IEEE.
68. Park C, Chou P. AmbiMax: Autonomous energy harvesting platform for multi-supply wireless sensor nodes. In: Proceedings of the Communications Society Conference on Sensor, Mesh and Ad Hoc Communications and Networks. 2006. pp. 168–77.
69. Lemofouet S, Rufer A. A hybrid energy storage system based on compressed air and supercapacitors with maximum efficiency point tracking (mept). *IEEE Transactions on Industrial Electronics*. 2006;53(4):1105–15.
70. Ise T, Kita M, Taguchi A. A hybrid energy storage with a SMES and secondary battery. *IEEE Transactions on Applied Superconductivity*. 2005;15(2):1915–1918.
71. Carter P, Baxter J, Newill T, Erekson T. An ultracapacitor-powered race car update. In: Proceeding of the Electrical Insulation Conference and Electrical Manufacturing Expo, Indianapolis, IN; 2005. pp. 267–74.
72. EIA (United States Department of Energy Information Administration). [May 1, 2016]. Retrieved from http://www.eia.gov/energyexplained/index.cfm?page=us_energy_commercial.
73. Southern California Edison Commercial Time of Usage Electricity Rates. Available from: http:www.//sce.com/NR/rdonlyres/6B523AB1-244D-4A8F-A8FE-19C5E0EFD095/0/090202-Business-Rates-Summary.pdf.
74. Laverman RJ. Modular Thermal Energy Storage Tanks Using Modular Heat Batterie [June 25, 1985]. US4524756A.
75. Cericola D, Novák P, Wokaun A, Kötz R. Hybridization of electrochemical capacitors and rechargeable batteries—An experimental analysis. *Journal of Power Sources*. 2011;196:10305–13.
76. Cericola D, Kötz R. Hybridization of rechargeable batteries and electrochemical capacitors—Principles and limits. *Electrochimica Acta*. 2012;72:1–17.
77. Bolborici V, Dawson FP, Lian KK. Hybrid energy storage systems: connecting batteries in parallel with ultracapacitors for higher power density. *IEEE Industry Applications Magazine*. 2014;20(4):31–40.
78. Burke A. Ultracapacitors: why, how, and where is the technology. *Journal of Power Sources*. 2000;91(1):37–50.
79. Burke AF. Batteries and ultracapacitors for electric, hybrid, and fuel cell vehicles. *Proceedings of the IEEE*. 2007;95(4):806–20.
80. Corson D. High power battery systems for hybrid vehicles. *Journal of Power Sources*. 2002;105:110–3.
81. Nelson R. Power requirements for batteries in hybrid electric vehicles. *Journal of Power Sources*. 2000;91:2–26.
82. Karden E, Shinn P, Bostock P, Cunningham J, Schoultz E, Kok D. Requirements for future automotive batteries—A snapshot. *Journal of Power Sources*. 2005;144:505–12.
83. Baisden AC, Emadi A. ADVISOR-based model of a battery and an ultra-capacitor energy source for hybrid electric vehicles. *IEEE Transactions on Vehicular Technology*. 2004;53(1):199–205.
84. Kuperman A, Aharon I. Battery-ultracapacitor hybrids for pulsed current loads—A review. *Renewable and Sustainable Energy Reviews*. 2011;15:981–92.

85. Dougal RA, Liu S, White RE. Power and life extension of battery-ultracapacitor hybrids. *IEEE Transactions on Components and Packaging Technologies.* 2002;25(1):120–31.
86. Henson W. Optimal battery/ultracapacitor storage combination. *Journal of Power Sources.* 2008;179:417–23.
87. Holland C, Weidner J, Dougal RA, White RE. Experimental characterization of hybrid power systems under pulse current loads. *Journal of Power Sources.* 2002;109:32–7.
88. Zheng JP, Jow TR, Ding MS. Hybrid power sources for pulsed current applications. *IEEE Transactions on Aerospace and Electronic Systems.* 2001;37(1):288–92.
89. Gao L, Dougal R, Liu S. Power enhancement of an actively controlled battery/ultracapacitor hybrid. *IEEE Transactions on Power Electronics.* 2005;20(1):236–43.
90. Kollimalla SK, Mishra MK, Narasamma NL. Design and analysis of novel control strategy for battery and supercapacitor storage system. *IEEE Transactions on Sustainable Energy.* 2014;5(4):1137–44.
91. Jung H, Wang H, Hu T. Control design for robust tracking and smooth transition in power systems with battery–supercapacitor hybrid energy storage devices. *Journal of Power Sources.* 2014;267:566–75.
92. Ibok EE. Modular, portable and transportable energy storage systems and applications thereof [September 19, 2013]. US 20130244064 A1.
93. Thounthong P, Rael S, Davat B. Control strategy of fuel cell and super capacitors association for a distributed generation system. *IEEE Transactions on Industrial Electronics.* 2007;54(6):3225–33.
94. Morandi A, Trevisani L, Negrini F, Ribani PL, Fabbri M. Feasibility of super-conducting magnetic energy storage on board of ground vehicles with present state-of-the-art superconductors. *IEEE Transactions on Applied Superconductivity.* 2012;22(2):5700106.
95. Suzuki S, Baba J, Shutoh K, Masada E. Effective application of superconducting magnetic energy storage (SMES) to load leveling for high speed transportation system. *IEEE Transactions on Applied Superconductivity.* 2004;14(2):713–6.
96. Shim JW, Cho Y, Kim S, Min SW, Hur K. Synergistic control of SMES and battery Energy storage for enabling dispatchability of renewable energy sources. *IEEE Transactions on Applied Superconductivity.* 2014;23(3):5701205.
97. Wang S, Tang Y, Shi J, Gong K, Liu Y, Ren L, et al. Design and advanced control strategies of a hybrid energy storage system for the grid integration of wind power generations. *IET Renewable Power Generation.* 2015;9(2):89–98.
98. Doucette RT, McCulloch MD. A comparison of high-speed flywheels, batteries, and ultracapacitors on the bases of cost and fuel economy as the energy storage systemin a fuel cell based hybrid electric vehicle. *Journal of Power Sources.* 2011;196:1163–70.
99. Windhorn A. A hybrid static/rotary UPS system. *IEEE Transactions on Industry Applications.* 1992;28(3):541–5.
100. Beaman BG, Rao GM. Hybrid battery and flywheel energy storage system for LEOspacecraft. In: Proceedings of the thirteenth annual battery conference on applications and advances, Long Beach, CA; 1998. pp. 113–6.
101. Briat O, Vinassa JM, Lajnef W, Azzopardi S, Woirgard E. Principle, design and experimental validation of a flywheel-battery hybrid source for heavy-duty electric vehicles. *IET Electric Power Applications.* 2007;1(5):665–74.
102. Lee H, Shin BY, Han S, Jung S, Park B, Jang G. Compensation for the power fluctuation of the large scale wind farm using hybrid energy storage applications. *IEEE Transactions on Applied Superconductivity.* 2012;22(3):5701904.
103. Truong LV, Wolff FJ, Dravid NV. Simulation of flywheel electrical system for aerospace applications. In: Proceedings of the 35th Intersociety Energy Conversion Engineering Conference and Exhibition, July 24–28, Las Vegas, NV, USA; 2000. Vol. 1, pp. 601–8.

104. Easy street ramps up data center operations, deploys additional VYCON flywheel systems to protect its green data center. [Accessed February 1, 2017]. Available from: http://www.calnetix.com/newsroom/press-release/easystreet-ramps-data-center-operations-deploys-additional-vycon-flywheel.
105. Tarrant C. Kinetic energy storage wins acceptance. [Accessed February, 12, 2017]. Available from: www.railwaygazette.com/news/single-view/view/kinetic-energy-storage-wins-acceptance.html.
106. VYCON technology allows Los Angeles metro to be first transit agency in U.S. using flywheels to achieve nearly 20 percent in rail energy savings. [Accessed February 1, 2017]. Available from: http://www.calnetix.com/newsroom/press-release/vycon-technology-allows-los-angeles-metro-be-first-transit-agency-us-using.
107. Lemofouet S, Rufer A. Efficiency considerations and measurements of a hybrid energy storage system based on compressed air and supercapacitors. In: Proceedings of the 12th power electronics and motion control conference (EPE-PEMC). 2006. pp. 2077–81.
108. Rufer A, Lemofouet S. Energetic performance of a hybrid energy storage system based on compressed air and supercapacitors. In: Proceedings of the international symposium on power electronics, electrical drives, automation and motion (SPEEDAM). 2006. pp. 469–74.
109. Okou R, Sebitosi AB, Khan A, Pillay P. The potential impact of small- scale flywheel energy storage technology on Uganda's energy sector. *Journal of Energy in Southern Africa*. 2009;20:14–9.
110. Lakeh RB, Villazana IC, Houssainy S, Anderson KR, Kavehpour HP. Design of a modular solid-based thermal energy storage for a hybrid compressed air energy storage system. In: ASME 2016 10th International Conference on Energy Sustainability collocated with the ASME 2016 Power Conference and the ASME 2016 14th International Conference on Fuel Cell Science, Engineering and Technology, Volume 2: ASME 2016 Energy Storage Forum, June 26–30, Advanced Energy Systems Division, Solar Energy Division, Charlotte, NC, USA; 2016. ISBN: 978-0-7918-5023-7.
111. Thounthong P, Rael S, Davat B. Control strategy of fuel cell/supercapacitors hybrid power sources for electric vehicle. *Journal of Power Sources*. 2006;158:806–14.
112. Wu Y, Gao H. Optimization of fuel cell and supercapacitor for fuel-cell Electric vehicles. *IEEE Transactions on Vehicular Technology*. 2006;55(6):1748–55.
113. Uzunoglu M, Alam MS. Dynamic modeling, design, and simulation of a combined PEM fuel cell and ultracapacitor system for stand-alone residential applications. *IEEE Transactions on Energy Conversion*. 2006;21(3):767–75.
114. Bauman J, Kazerani M. A comparative study of fuel-cell–battery, fuel-cell–ultra-capacitor, and fuel-cell–battery–ultracapacitor vehicles. *IEEE Transactions on Vehicular Technology*. 2007;57(2):760–9.
115. Payman A, Pierfederici S, Meibody-Tabar F. Energy management in a fuel cell/supercapacitor multisource/multiload electrical hybrid system. *IEEE Transactions on Power Electronics*. 2009;24(12):2681–91.
116. Maclay JD, Brouwer J, Samuelse GS. Dynamic modeling of hybrid energy storage systems coupled to photovoltaic generation in residential applications. *Journal of Power Sources* 2007;163:916–25.
117. Todorovic MH, Palma L, Enjeti PN. Design of a wide input range DC-DC converter with a robust power control scheme suitable for fuel cell power conversion. *IEEE Transactions on Industrial Electronics*. 2008;55(3):1247–55.
118. Tao H, Duarte JL, Hendrix MAM. Line-interactive UPS using a fuel cell as the primary source. *IEEE Transactions on Industrial Electronics*. 2008;55(8):3012–21.

119. Louie H, Strunz K. Superconducting magnetic energy storage (SMES) for energy cache control in modular distributed hydrogen-electric energy systems. *IEEE Transactions on Applied Superconductivity.* 2007;17(2):2361–4.
120. Sander M, Gehring R, Neumann H, Jordan T. LIQHYSMES storage unit – hybrid energy storage concept combining liquefied hydrogen with superconducting magnetic energy storage. International Journal of Hydrogen Energy. 2012;37: 14300–6.
121. Frysinger GA, Wrublewski F. Fuel cell-battery power sources. In: *Research and Development in Non-Mechanical Electrical Power Sources*, Amsterdam: Elsevier; 1970. pp. 567–581.
122. Beyer HG. Combined battery hydrogen storage for autonomous wind/solar systems. *Advances in Solar Energy Technology.* 1988:413–7.
123. Ledjeff K. Comparison of storage options for photovoltaic systems. *International Journal of Hydrogen Energy.* 1990;15(9):629–33.
124. Ghosh PC, Emonts B, Stolten D. Comparison of hydrogen storage with diesel-generator system in a PV–WEC hybrid system. *Solar Energy.* 2003;75:187–98.
125. Vosen SR. A design tool for the optimization of stand-alone electric power systems with combined hydrogen-battery energy storage. Livermore, CA: Sandia National Labpratory; 1997. Report No.: UC-406:SAND97-8601J.
126. Nadal M, Barbir F. Development of a hybrid fuel cell/battery powered electric vehicle. *International Journal of Hydrogen Energy.* 1996;21(6):497–505.
127. Thounthong P, Chunkag V, Sethakul P, Davat B, Hinaje M. comparative study of fuel-cell vehicle hybridization with battery or supercapacitor storage device. *IEEE Transactions on Vehicular Technology.* 2009;58(8):3892–904.
128. Bruni G, Cordiner S, Mulone V. Domestic distributed power generation: Effect of sizing and energy management strategy on the environmental efficiency of a photovoltaic-battery-fuel cell system. *Energy.* 2014;77:133–43.
129. Han J, Park ES. Direct methanol fuel-cell combined with a small back-up battery. *Journal of Power Sources.* 2002;112(2):477–83.
130. Laldin O, Moshirvaziri M, Trescases O. Predictive algorithm for optimizing power flow in hybrid ultracapacitor/battery storage systems for light electric vehicles. *IEEE Transactions on Power Electronics.* 2012;28(8):3882–95.
131. Amjadi Z, Williamson SS. Prototype design and controller implementation for a battery-ultracapacitor hybrid electric vehicle energy storage system. *IEEE Transactions on Smart Grid.* 2012;3(1):332–40.
132. Cao J, Emadi A. A new battery/ultracapacitor hybrid energy storage system for electric, hybrid, and plug-in hybrid vehicles. *IEEE Transactions on Power Electronics.* 2012;27(1):122–32.
133. Kifle Y, Khan B, Singh J. Designing and modeling grid connected photovoltaic system: (case study: EEU building at Hawassa city). *International Journal of Convergence Computing.* 2018;3(1):20–34.
134. Dobbs BG, Chapman PL. A multiple-input DC–DC converter topology. *IEEE Power Electronics Letters.* 2003;1(1):6–9.
135. Matsuo H, Wenzhong L, Kurokawa F, Shigemizu T, Watanabe N. Characteristics of the multiple-input DC-DC converter. *IEEE Transactions on Industrial Electronics.* 2004;51(3):625–31.
136. Solero L, Lidozzi A, Pomilio J. Design of multiple-input power converter for hybrid vehicles. *IEEE Transactions on Industrial Electronics.* 2005;20(5):1007–16.
137. Kim Y, Park S, Chang N, Xie Q, Wang Y, Pedram M. Networked architecture for hybrid electrical energy storage systems. In: Proceedings of the Design Automation Conference (DAC), San Francisco, CA; 2012. pp. 522–8.

138. Shin D, Kim Y, Seo J, Chang N, Wang Y, Pedram M. Battery- supercapacitor hybrid system for high-rate pulsed load applications. In: Proceedings of the Design, Automation and Test in Europe Conference and Exhibition (DATE), Grenoble, France; 2011. pp. 1–4.
139. Shin D, Kim Y, Wang Y, Chang N, Pedram M. Constant-current regulator-based battery-supercapacitor hybrid architecture for high-rate pulsed load applications. *Journal of Power Sources*. 2012;205:516–24.
140. Thounthong P, Chunkag V, Sethakul P, Sikkabut S, Pierfederici S, Davat B. Energy management of fuel cell/solar cell/supercapacitor hybrid power source. *Journal of Power Sources*. 2011;196(1):313–324.
141. Zhang Y, Jiang Z, Yu X. Control strategies for battery/supercapacitor hybrid energy storage systems. In: Proceedings of the IEEE Energy 2030 Conference, Atlanta, Georgia; 2008. pp.1–6.
142. Abbey C, Strunz K, Joos G. A knowledge-based approach for control of two-level energy storage for wind energy systems. *IEEE Transactions on Energy Conversion*. 2009;24(2):539–57.
143. Wang Y, Kim Y, Xie Q, Chang N, Pedram M. Charge migration efficiency optimization in hybrid electrical energy storage (HEES) systems. In: Proceedings of the International Symposium on Low Power Electronics and Design (ISLPED), Fukuoka, Japan; 2011. pp. 103–8.
144. Wang Y, Xie Q, Pedram M, Kim Y, Chang N, Poncino M. Multiple-source and multiple-destination charge migration in hybrid electrical energy storage systems. In: Proceedings of the Design, Automation and Test in Europe Conference and Exhibition (DATE), Dresden, Germany; 2012.
145. Xie Q, Wang Y, Kim Y, Shin D, Chang N, Pedram M. Charge replacement in hybrid electrical energy storage systems. In: Proceedings of the 17th Asia South Pacific Design Automation Conference, Sydney, Australia; 2012. pp. 627–32.
146. Zhu D, Wang Y, Yue S, Xie Q, Pedram M, Chang N. Maximizing return on investment of a grid-connected hybrid electrical energy storage system. In: Proceedings of the 18th Asia South Pacific Design Automation Conference, Yokohama, Japan; 2013. pp. 638–43.
147. Xie Q, Kim Y, Wang Y, Kim J, Chang N, Pedram M. Principles and efficient implementation of charge replacement in hybrid electrical energy storage systems. *IEEE Transactions on Power Electronics*. 2014;29(11):6110–23.
148. Beaudin M, Zareipour H, Schellenberglabe A, Rosehart W. Energy storage for mitigating the variability of renewable electricity sources: An updated review. *Energy for Sustainable Development*. 2010;14:302–14.
149. Strunz K, Louie H. Cache energy control for storage: Power system integration and education based on analogies derived from computer engineering. *IEEE Transactions on Power Systems*. 2009;24(1):12–9.
150. Xie Q, Wang Y, Kim Y, Chang N, Pedram M. Charge allocation for hybrid electrical energy storage systems. In: Proceedings of the International Conference on Hardware/Software Codesign and System Synthesis (CODES+ISSS), Taipei, Taiwan; 2011. pp. 277–84.
151. Xie Q, Lin X, Wang Y, Pedram M, Shin D, Chang. State of health aware charge management in hybrid electrical energy storage systems. In: Proceedings of the Design, Automation and Test in Europe Conference and Exhibition (DATE), Dresden, Germany; 2012. pp. 1060–1065.
152. Mendis N, Muttaqi KM, Perera S. Management of battery-supercapacitor hybrid energy storage and synchronous condenser for isolated operation of PMSG based variable-speed wind turbine generating systems. *IEEE Transactions on Smart Grid*. 2014;5(2):944–53.

153. Zeng A, Xu Q, Ding M, Yukita K, Ichiyanagi K. A classification control strategy for energy storage system in microgrid. *IEEJ Transactions on Electrical and Electronic Engineering*. 2015;10(4):396–403.
154. Tummuru NR, Mishra MK, Srinivas S. Dynamic energy management of renewable grid integrated hybrid energy storage system. *IEEE Transactions on Industrial Electronics*. 2015;62(12):7728–37.
155. Garcia-Torres F, Bordons C. Optimal economical schedule of hydrogen-based microgrids with hybrid storage using model predictive control. *IEEE Transactions on Industrial Electronics*. 2015;62(8):5195–207.
156. Herrera V, Gaztanaga H, Milo A, Saez-de-Ibarra A, Etxeberria I, Nieva T. Optimal energy management of battery-supercapacitor based light rail vehicle using genetic algorithms. *IEEE Transactions on Industry Applications*. 2016; 1: 99–106.
157. Porru M, Serpi A, Marongiu I, Damiano A. A novel hybrid energy storage system for electric vehicles. In: Proceedings of the 41th Annual Conference of the IEEE Industrial Electronics Society (IECON, Yokohama, Japan; 2015. pp. 3732–7.
158. Consolidated Edison Company of New York, Inc. *Service Classification No. 1—Residential and Religious*.
159. Xie Q, Wang Y,Kim Y, Pedram M, Chang N. Charge allocation in hybrid electrical energy storage systems. *IEEE Transactions on Computer-Aided Design of Integrated Circuits and Systems*. 2013;32(7):1003–16.
160. Millner A. Modeling lithium ion battery degradation in electric vehicles. In: Proceedings of IEEE Conference of Innovative Technologies for an Efficient and Reliable Electricity Supply (CITRES), Waltham, MA: 2010. IEEE.
161. Wang Y, Lin X, Kim Y, Chang N, Pedram M. Enhancing efficiency and robustness of a photovoltaic power system under partial shading. In: Proceedings of the International Symposium on Quality Electronics Design (ISQED), Santa Clara, CA; 2012.
162. Bhatnagar P, Nema R. Maximum power point tracking control techniques: State-of-the-art in photovoltaic applications. *Renewable and Sustainable Energy Reviews*. 2013;23:224–1.
163. Kim S, Chou P. Size and topology optimization for supercapacitor- based sub-watt energy harvesters. *IEEE Transactions on Power Electronics*. 2013;28(4):2068–80.
164. Mungan ES, Lu C, Raghunathan V, Roy K. Modeling, design and cross-layer optimization of polysilicon solar cell based micro-scale energy har- vesting systems. In: Proceedings of the International Symposium on Low Power Electronics and Design (ISLPED), Redondo Beach, CA; 2012. pp. 123–8.
165. Huggins RA. *Energy Storage*. Boston, MA: Springer; 2010.
166. Apsley JM, Gonzalez Villasenor A, Barnes M, Smith AC, Williamson S, Schuddebeurs JD, et al. Propulsion drive models for full electric marine propulsion systems. *IEEE Transactions on Industry Applications*. 2009;45(2):676–84.
167. Arboleya P, Bidaguren P, Armendariz U. Energy is on board: Energy storage and other alternatives in modern light railways. *IEEE Electrification Magazine*. 2016;4(3):30–41.
168. Li R, Zhou F. *Microgrid Technology and Engineering Application*. Oxford, UK: Elsevier; 2015. 200 p.
169. Gao DW. *Energy Storage for Sustainable Microgrid*. Oxford, UK: Academic Press; 2015.153 p.
170. Luo X, Wang J, Dooner M, Clarke J. Overview of current development in electrical energy storage technologies and the application potential in power system operation. *Applied Energy*. 2015;137:511–36.
171. Boicea VA. Energy storage technologies: The past and the present. *Proceedings of the IEEE*. 2014;102(11):1777–94.

172. Gayathri NS, Senroy N, Kar IN. Smoothing of wind power using fly wheel energy storage system. *IET Renewable Power Generation.* 2017;11(3):289–298.
173. Liu Y, Tang Y, Shi J, Shi X, Deng J, Gong K. Application of small-sized SMES in an EV charging station with DC bus and PV system. *IEEE Transactions on Applied Superconductivity.* 2015;25(3):1–6.
174. Khaligh A, Li Z. Battery, ultracapacitor, fuel cell and hybrid energy storage systems for electric, hybrid electric, fuel cell and plug-in hybrid electric vehicles: State of the art. *IEEE Transactions on Vehicular Technology.* 2010;59(6):2806–14.
175. Kamali SK, Tyagi VV, Rahim NA, Panwar NL, Mokhlis H. Emergence of energy storage technologies as the solution for reliable operation of smart power systems—A review. *Renewable and Sustainable Energy Reviews.* 2013;25:135–65.
176. Araújo RE, DeCastro R, Pinto C, Melo P,. Freitas D. Combined sizing and energy management in EVs with batteries and supercapacitors. *IEEE Transactions on Vehicular Technology.* 2014;63:3062–76.
177. Chatzivasileiadi A, Ampatzi E, Knight I. Characteristics of electrical energy storage technologies and their applications in buildings. *Renewable and Sustainable Energy Reviews.* 2013;25:814–30.
178. Hajiagashi S, Salemnla A, Hamzeh M. Hybrid energy storage system for microgrids applications: A review. *Journal of Energy Storage.* 2019;21:543–70.
179. Jing W, Lai CH, Wong SHW, Wong MLD. Battery-supercapacitor hybrid energy storage system in standalone DC microgrids: A review. *IET Renewable Power Generation.* 2017;11:461–9.
180. Zimmermann T, Keil P, Hofmann M, Horsche MF, Pichlmaier S, Jossen A. Review of system topologies for hybrid electrical energy storage systems. *Journal of Energy Storage.* 2016;8:78–90. DOI: 10.1016/j.est.2016.09.006.
181. Sharma Rk, Mishra S. Dynamic power management and control of PV PEM fuel cell based standalone AC/DC microgrid using hybrid energy storage. IEEE Transactions on Industry Applications. 2018;54(1):526–538. https://doi.org/10.1109/TIA.2017.2756032.
182. Chia YY, Lee LH, Shafiabady N, Isa D. A load predictive energy management system for supercapacitor-battery hybrid energy storage system in solar application using the support vector machine. *Applied Energy.* 2015;137:588–602. https://doi.org/10.1016/J.APENERGY.2014.09.026.
183. Pimm AJ, Garvey SD. The economics of hybrid energy storage plant. *International Journal of Environmental Studies.* 2014;71:787–795. https://doi.org/10.1080/00207233.2014.948321.
184. Chemali E, Preindl M, Malysz P, Emadi A. Electrochemical and electrostatic energy storage and management systems for electric drive vehicles: State-of-the- Art review and future trends. *IEEE Journal of Emerging and Selected Topics in Power Electronics.* 2016;4:1117–34. https://doi.org/10.1109/JESTPE.2016.2566583.
185. Mahlia TMI, Saktisahdan TJ, Jannifar A, Hasan MH, Matseelar HSC. A review of available methods and development on energy storage; technology update. *Renewable and Sustainable Energy Reviews.* 2014;33:532–45. https://doi.org/10.1016/J.RSER.2014.01.068.
186. Zhao P, Wang J, Dai Y. Capacity allocation of a hybrid energy storage system for power system peak shaving at high wind power penetration level. *Renewable Energy.* 2015;75:541–9. https://doi.org/10.1016/J.RENENE.2014.10.040.
187. Jiang Q, Hong H. Wavelet-based capacity configuration and coordinated control of hybrid energy storage system for smoothing out wind power fluctuations. *IEEE Transactions on Power Systems.* 2013;28:1363–72. https://doi.org/10.1109/TPWRS.2012.2212252.

188. Mendis N, Muttaqi KM, Perera S. Management of low- and high-frequency power components in demand-generation fluctuations of a DFIG-Based wind- dominated RAPS system using hybrid energy storage. *IEEE Transactions on Industry Applications.* 2014;50:2258–68. https://doi.org/10.1109/TIA.2013.2289973.
189. Lee H, Shin BY, Han S, Jung S, Park B, Jang G. Compensation for the power fluctuation of the large scale wind farm using hybrid energy storage applications. *IEEE Transactions on Applied Superconductivity.* 2012;22:5701904. https://doi.org/10.1109/TASC.2011.2180881.
190. Li J, Liu Y, Wang S, Shi J, Ren L, Gong K, et al. Design and advanced control strategies of a hybrid energy storage system for the grid integration of wind power generations. *IET Renewable Power Generation.* 2015;9:89–98. https://doi.org/10.1049/iet-rpg.2013.0340.
191. Zhang Z, Miyajima R, Sato Y, Miyagi D, Tsuda M, Makida Y, et al. Characteristics of compensation for fluctuating output power of a solar power generator in a hybrid energy storage system using a Bi2223 SMES coil cooled by thermosiphon with liquid hydrogen. *IEEE Transactions on Applied Superconductivity.* 2016;26:1–5. https://doi.org/10.1109/TASC.2016.2529565.
192. Li J, Gee AM, Zhang M, Yuan W. Analysis of battery life time extension in a SMES-battery hybrid energy storage system using a novel battery lifetime model. *Energy.* 2015;86:175–85. https://doi.org/10.1016/J.ENERGY.2015.03.132.
193. Li J, Xiong R, Mu H, Cornélusse B, Vanderbemden P, Ernst D, et al. Design and real-time test of a hybrid energy storage system in the microgrid with the benefit of improving the battery lifetime. *Applied Energy.* 2018;218:470–8. https://doi.org/10.1016/J.APENERGY.2018.01.096.
194. Weitzel T, Schneider M, Glock CH, Löber F, Rinderknecht S. Operating a storage-augmented hybrid microgrid considering battery aging costs. *Journal of Cleaner Production.* 2018;188:638–54. https://doi.org/10.1016/J.JCLEPRO.2018.03.296.
195. Ongaro F, Saggini S, Mattavelli P. Li-ion battery-supercapacitor hybrid storage system for a long lifetime, photovoltaic-based wireless sensor network. *IEEE Transactions on Power Electronics.* 2012;27:3944–52. https://doi.org/10.1109/TPEL.2012.2189022.
196. Gee AM, Robinson FVP, Dunn RW. Analysis of battery lifetime extension in a small-scale wind-energy system using supercapacitors. *IEEE Transactions on Energy Conversion.* 2013;28:24–33. https://doi.org/10.1109/TEC.2012.2228195.
197. Mane S, Mejari M, Kazi F, Singh N. Improving lifetime of fuel cell in hybrid energy management system by Lure–Lyapunov-based control formulation. *IEEE Transactions on Industrial Electronics.* 2017;64:6671–9. https://doi.org/10.1109/TIE.2017.2696500.
198. Aouzellag H, Ghedamsi K, Aouzellag D. Energy management and fault tolerant control strategies for fuel cell/ultra-capacitor hybrid electric vehicles to enhance autonomy, efficiency and life time of the fuel cell system. *International Journal of Hydrogen Energy.* 2015;40:7204–13. https://doi.org/10.1016/J.IJHYDENE.2015.03.132.
199. Lahyani A, Venet P, Guermazi A, Troudi A. Battery/supercapacitors combination in uninterruptible power supply (UPS). *IEEE Transactions on Power Electronics.* 2013;28:1509–22.
200. Georgious R, Garcia J, Navarro-Rodriguez A, Garcia-Fernandez P. A study on the control design of non-isolated converter configurations for hybrid energy storage systems. *IEEE Transactions on Industry Applications.* 2018;54:4660–71.
201. Cabrane Z, Ouassaid M, Nationale E, Maaroufi M. Management and control of storage photovoltaic energy using battery-supercapacitor combination. In: Proceedings of the 2014 Second World Conference on Complex Systems (WCCS), November 10–13, 2014, Agadir, Morroco; 2014.

202. Dong B, Li Y, Zheng Z. Control strategies of DC-bus voltage in islanded operation of microgrid. In: Proceedings of the 2011 4th International Conference on Electric Utility Deregulation and Restructuring and Power Technologies (DRPT), 30 August–1 September 2011, Birmingham, UK; 2011. pp. 1671–1674.
203. Zhou T. Sun W. Optimization of battery–Supercapacitor hybrid energy storage station in Wind/Solar generation system. *IEEE Transactions on Sustainable Energy*. 2014;5:408–15. https://doi.org/10.1109/TSTE.2013.2288804.
204. Wang X, Yu D, LeBlond S, Zhao Z, Wilson P. A novel controller of a battery-super capacitor hybrid energy storage system for domestic applications. *Energy and Buildings*. 2017;141:167–74.
205. Sanjeev P, PrasadPadhy N, Agarwal P. Effective control and energy management of isolated DC microgird. In: Proceedings of the IEEE Power & Energy Society General Meeting, July 16–20, 2017, Chicago, IL, USA; 2017.
206. Pan TL, Wan HS, Ji ZC. Stand-alone wind power system with battery/supercapacitor hybrid energy storage. *International Journal of Sustainable Engineering*. 2014;7:103–10.
207. Hassan SZ, Kamal T, Awais M. Stand-alone/grid-tied wind power system with battery/supercapacitor hybrid energy storage. In: Proceedings of the 2015 International Conference on Emerging Technologies (ICET), December 19–20, 2015, Peshawar, Pakistan; 2015. pp. 1–6.
208. Wu G, Ono Y, Alishahi M. Development of a resilient hybrid microgrid with integrated renewable power generations supplying DC and AC loads. In: Proceedings of the 2015 IEEE International Telecommunications Energy Conference (INTELEC), October 18–22, 2016, Osaka, Japan; 2016.
209. Cabrane Z, Ouassaid M, Maaroufi M. Battery and supercapacitor for photovoltaic energy storage: A fuzzy logic management. *IET Renewable Power Generation*. 2017;11:1157–65. https://doi.org/10.1049/iet-rpg.2016.0455.
210. Tummuru NR, Mishra MK, Srinivas S. Dynamic energy management of hybrid energy storage system with high-gain PV converter. *IEEE Transactions on Energy Conversion*. 2015;30:150–60.
211. Zheng Z, Wang X, Li Y. A control method for grid-friendly photovoltaic systems with hybrid energy storage units. In: Proceedings of the 2011 4th International Conference on Electric Utility Deregulation and Restructuring and Power Technologies (DRPT), July 6–9, 2011, Weihai, Shandong, China; 2011. pp. 1437–40.
212. Han Y, Xu Y, Li Y. Research on application of STATCOM/HESS in wind power integrated to the grid. In: Proceedings of the IEEE 2nd International Future Energy Electronics Conference (IFEEC), November 1–4, 2015, Taipei, Taiwan; 2015.
213. Nikhil K, Mishra MK. Application of hybrid energy storage system in a grid interactive microgrid environment. In: Proceedings of the IECON 2015—41st Annual Conference of the IEEE Industrial Electronics Society, November 9–12, 2015, Yokohama, Japan; 2015. pp. 2980–5.
214. Benaouadj M, Aboubou A, Ayad MY, Becherif M. Nonlinear flatness control applied to supercapacitors contribution in hybrid power systems using photovoltaic source and batteries. *Energy Procedia*. 2014;50:333–41.
215. Ismail NM, Mishra MK. Control and operation of unified power quality conditioner with battery-ultracapacitor energy storage system. In: Proceedings of the 2014 IEEE International Conference on Power Electronics, Drives and Energy Systems (PEDES), December 16–19, 2014, Mumbai, India; 2014. pp. 1–6.
216. Zhang F, Yang Y, Ji C, Wei W, Chen Y, Meng C, Jin Z, Zhang G. Power management strategy research for DC microgrid with hybrid storage system. In: Proceedings of the

2015 IEEE First International Conference on DC Microgrids (ICDCM), May 24–27, 2015, Atlanta, GA, USA; 2015. pp. 62–8.

217. Kotra S, Mishra MK. Control algorithm for a PV based hybrid microgrid. In: Proceedings of the 2015 Annual IEEE India Conference (INDICON), December 17–19, 2015, New Delhi, India; 2015. pp. 4–9.
218. Manandhar U, Tummuru NR, Kumar S, Ukil A, Beng GH, Chaudhari K. Validation of faster joint control strategy for battery and supercapacitor based energy storage system. *IEEE Transactions on Industrial Electronics.* 2017;1:1–10. https://doi.org/10.1109/TIE.2017.2750622.
219. Li J, Xiong R, Yang Q, Liang F, Zhang M, Yuan W. Design/test of a hybrid energy storage system for primary frequency control using a dynamic droop method in an isolated microgrid power system. *Applied Energy.* 2017;201:257–69. https://doi.org/10.1016/J.APENERGY.2016.10.066.
220. Tamura S. Economic analysis of hybrid battery energy storage systems applied to frequency control in power system. *Electrical Engineering in Japan.* 2016;195:24–31. https://doi.org/10.1002/eej.22816.
221. Akram U, Khalid M. A coordinated frequency regulation framework based on hybrid battery-ultracapacitor energy storage technologies. *IEEE Access* 2018;6:7310–20. https://doi.org/10.1109/ACCESS.2017.2786283.
222. Fang J, Tang Y, Li H, Li X. A battery/ultracapacitor hybrid energy storage system for implementing the power management of virtual synchronous generators. *IEEE Transactions on Power Electronics.* 2018;33:2820–4. https://doi.org/10.1109/TPEL.2017.2759256.
223. Anzalchi A, Pour MM, Sarwat A. A combinatorial approach for addressing intermittency and providing inertial response in a grid-connected photovoltaic system. In: Proceedings of the 2016 IEEE Power and Energy Society General Meeting (PESGM), July 17–21, 2016, Boston, MA, USA, 2016; pp. 1–5.
224. Huiyu M, Chenyu Z, Fei M, Yun Y, Jianyong Z. A novel control strategy for hybrid energy system in virtual synchronous generator. In: Proceedings of the 2018 13th IEEE Conference on Industrial Electronics and Applications (ICIEA), 31 May–2 June 2018, Wuhan, China; 2018. pp. 2244–9.
225. Zhou X, Dong C, Fang J, Tang Y. Enhancement of load frequency control by using a hybrid energy storage system. In: Proceedings of the 2017 Asian Conference on Energy, Power and Transportation Electrification (ACEPT), 30 October–1 November 2017, Singapore; 2017. pp. 1–6.
226. Gu Y, Li W, He X. Frequency-coordinating virtual impedance for autonomous power management of DC microgrid. *IEEE Transactions on Power Electronics.* 2015;30:2328–37.
227. Germanovich Chirkin V, Yurievich Lezhnev L, Anatolyevich Petrichenko D, Arkadyevich Papkin I. A battery-supercapacitor hybrid energy storage system design and power management. *International Journal of Pure and Applied Mathematics.* 2018; 119:2621–4.
228. Farhadi M, Mohammed OA. Performance enhancement of actively controlled hybrid dc microgrid incorporating pulsed load. *IEEE Transactions on Industry Applications.* 2015;51: 3570–8. https://doi.org/10.1109/TIA.2015.2420630.
229. Farhadi M, Mohammed O. Adaptive energy management in redundant hybrid DC microgrid for pulse load mitigation. *IEEE Transactions on Smart Grid.* 2015;6:54–62. https://doi.org/10.1109/TSG.2014.2347253.
230. Hajiaghasi S, Salemnia A, Hamzeh M. Hybrid energy storage for microgrid performance improvement under unbalanced load conditions. *Journal of Energy Management and technology.* 2018;2:32–38. https://doi.org/10.22109/JEMT.2018.109536.1065.

231. Zhu Y, Zhuo F, Wang F. Coordination control of lithium battery-spercapacitor hybrid energy storage system in a microgrid under unbalanced load condition. In: 16th European Conference on Power Electronics and Applications, Lappeenranta, Finland; 2014. pp. 1–10. IEEE. https://doi.org/10.1109/EPE.2014.6910727.
232. Tabart Q, Vechiu I, Etxeberria A, Bacha S. Hybrid energy storage system mi- crogrids integration for power quality improvement using four-leg three-level NPC inverter and second-order sliding mode control. *IEEE Transactions on Industrial Electronics.* 2018;65:424–35. https://doi.org/10.1109/TIE.2017.272.
233. Alafnan H, Zhang M, Yuan W, Zhu J, Li J, Elshiekh M, et al. Stability improvement of DC power systems in an all-electric ship using hybrid SMES/Battery. *IEEE Transactions on Applied Superconductivity.* 2018;28:1–6. https://doi.org/10.1109/TASC.2018.2794472.
234. Chang X, Li Y, Li X, Chen X. An active damping method based on a super- capacitor energy storage system to overcome the destabilizing effect of instantaneous constant power loads in DC microgrids. *IEEE Transactions on Energy Conversion.* 2017;32:36–47. https://doi.org/10.1109/TEC.2016.2605764.
235. Chen L, Chen H, Li Y, Li G, Yang J, Liu X, et al. SMES-battery energy storage system for the stabilization of a photovoltaic-based microgrid. *IEEE Transactions on Applied Superconductivity.* 2018;28:1–7. https://doi.org/10.1109/TASC.2018.2799544.
236. Thale, S, Agarwal V. A smart control strategy for the black start of a microgrid based on PV and other auxiliary sources under islanded condition. In: Proceedings of the 2011 37th IEEE Photovoltaic Specialists Conference, June 19–24, 2011, Seattle, WA, USA; 2011. pp. 002454–002459.
237. Zhu Y, Zhuo F, Shi H. Power management strategy research for a photovoltaic-hybrid energy storage system. In: Proceedings of the 2013 IEEE ECCE Asia Down Under, June 3–6, 2013, Melbourne, Australia; 2013. pp. 842–8.
238. Gao ZQ, Xie ZJ, Fan H, Meng L. Hybrid energy storage system for GSHP black start. *Advanced Materials Research.* 2014;953–954:748–51.
239. Yang Y, Bremner S, Menictas C, Kay M. Battery energy storage system size determination in renewable energy systems: A review. *Renewable and Sustainable Energy Reviews.* 2018;91:109–125. https://doi.org/10.1016/J.RSER.2018.03.047.
240. Günther S, Bensmann A, Hanke-Rauschenbach R. Theoretical dimensioning and sizing limits of hybrid energy storage systems. *Applied Energy.* 2018;210:127–37. https://doi.org/10.1016/J.APENERGY.2017.10.116.
241. Abbassi A, Dami MA, Jemli M. A statistical approach for hybrid energy storage system sizing based on capacity distributions in an autonomous PV/Wind power generation system. *Renewable Energy.* 2017;103:81–93. https://doi.org/10.1016/J. RENENE.2016.11.024.
242. Jia H, Mu Y, Qi Y. A statistical model to determine the capacity of batter- y–supercapacitor hybrid energy storage system in autonomous microgrid. *International Journal of Electrical Power & Energy Systems.* 2014;54:516–24. https://doi.org/10.1016/J.IJEPES.2013.07.025.
243. Masih-Tehrani M, Ha'iri-Yazdi M-R, Esfahanian V, Safaei A. Optimum sizing and optimum energy management of a hybrid energy storage system for lithium battery life improvement. *Journal of Power Sources.* 2013;244:2–10. https://doi.org/10.1016/J.JPOWSOUR.2013.04.154.
244. Wen S, Lan H, Yu DC, Fu Q, Hong YY, Yu L, et al. Optimal sizing of hybrid energy storage sub-systems in PV/diesel ship power system using frequency analysis. *Energy.* 2017;140:198–208. https://doi.org/10.1016/J.ENERGY.2017.08.065.
245. Ghiassi-Farrokhfal Y, Rosenberg C, Keshav S, Adjaho M-B. Joint optimal design and operation of hybrid energy storage systems. IEEE Journal on Selected Areas in Communications. 2016;34:639–50. https://doi.org/10.1109/JSAC.2016.2525599.

246. Janghorban Esfahani I, Ifaei P, Kim J, Yoo C. Design of hybrid renewable energy systems with battery/hydrogen storage considering practical power losses: A MEPoPA (modified extended-power pinch analysis). *Energy*. 2016;100:40–50. https://doi.org/10.1016/J.ENERGY.2016.01.074.
247. Janghorban Esfahani I, Lee S, Yoo C. Extended-power pinch analysis (EPoPA) for integration of renewable energy systems with battery/hydrogen storages. *Renewable Energy*. 2015;80:1–14. https://doi.org/10.1016/J.RENENE.2015.01.040.
248. Jacob AS, Banerjee R, Ghosh PC. Sizing of hybrid energy storage system for a PV based microgrid through design space approach. *Applied Energy*. 2018;212:640–53. https://doi.org/10.1016/J.APENERGY.2017.12.040.
249. Zhang Y, Tang X, Qi Z, Liu Z. The Ragone plots guided sizing of hybrid storage system for taming the wind power. *International Journal of Electrical Power & Energy Systems*. 2015;65:246–53. https://doi.org/10.1016/J.IJEPES.2014.10.006.
250. Tankari MA, Camara MB, Dakyo B, Lefebvre G. Use of ultracapacitors and batteries for efficient energy management in wind–diesel hybrid system. *IEEE Transactions on Sustainable Energy*. 2013;4:414–24. https://doi.org/10.1109/TSTE.2012.2227067.
251. Cao J, Du W, Wang H, McCulloch M. Optimal sizing and control strategies for hybrid storage system as limited by grid frequency deviations. IEEE Transactions on Power Systems. 2018;33(5):5486–5495. https://doi.org/10.1109/TPWRS.2018.2805380.
252. Ju F, Zhang Q, Deng W, Li J. Review of structures and control of battery-supercapacitor hybrid energy storage system for electric vehicles. In: *Advances in Battery Manufacturing, Service, and Management Systems*. Hoboken, NJ: Wiley; 2016. pp. 303–18. https://doi.org/10.1002/9781119060741.ch13.
253. Song Z, Hofmann H, Li J, Han X, Zhang X, Ouyang M. A comparison study of different semi-active hybrid energy storage system topologies for electric vehicles. *Journal of Power Sources*. 2015;274:400–11. https://doi.org/10.1016/j.jpowsour.2014.10.061.
254. Cohen IJ, Wetz DA, Heinzel JM, Dong Q. Design and characterization of an actively controlled hybrid energy storage module for high-rate directed energy applications. IEEE Transactions on Plasma Science. 2015;43:1427–33. https://doi.org/10.1109/TPS.2014.2370053.
255. Momayyezan M, Abeywardana DBW, Hredzak B, Agelidis VG. Integrated reconfigurable configuration for battery/ultracapacitor hybrid energy storage systems. *IEEE Transactions on Energy Conversion*. 2016;31:1583–90. https://doi.org/10.1109/TEC.2016.2589933.
256. Mo R, Li H. Hybrid energy storage system with active filter function for shipboard MVDC system applications based on isolated modular multilevel DC/DC converter. *IEEE Journal of Emerging and Selected Topics in Power Electronics*. 2017;5:79–87. https://doi.org/10.1109/JESTPE.2016.2642831.
257. Piris-Botalla L, Oggier GG, García GO. Extending the power transfer capability of a three-port DC–DC converter for hybrid energy storage systems. *IET Power Electronics*. 2017;10:1687–97. https://doi.org/10.1049/iet-pel.2016.0422.
258. Khosrogorji S, Ahmadian M, Torkaman H, Soori S. Multi-input DC/DC converters in connection with distributed generation units—A review. *Renewable and Sustainable Energy Reviews*. 2016;66:360–79. https://doi.org/10.1016/J.RSER.2016.07.023.
259. Bocklisch T. Hybrid energy storage approach for renewable energy applications. *Journal of Energy Storage*. 2016;8:311–9. https://doi.org/10.1016/J.EST.2016.01.004.
260. Tayab UB, Roslan MAB,Hwai LJ, Kashif M. A review of droop control techniques for microgrid, *Renewable and Sustainable Energy Reviews*. 2017;76:717–27. https://doi.org/10.1016/J.RSER.2017.03.028.

261. Lin P, Wang P, Xiao J, Wang J, Jin C, Tang Y. An integral droop for transient power allocation and output impedance shaping of hybrid energy storage system in DC microgrid. *IEEE Transactions on Power Electronics*. 2018;33:6262–77. https://doi.org/10.1109/TPEL.2017.2741262.
262. Shi M, Chen X, Zhou J, Chen Y, Wen J, He H. Advanced secondary voltage recovery control for multiple HESSs in a droop-controlled DC microgrids. *IEEE Transactions on Smart Grid*. 2019;10:3828–3839. https://doi.org/10.1109/TSG.2018.2838108.
263. Wang B, Manandhar U, Zhang X, Gooi HB, Ukil A. Deadbeat control for hybrid energy storage systems in DC microgrids. *IEEE Transactions on Sustainable Energy*. 2019;10(4):1867–1877. https://doi.org/10.1109/TSTE.2018.2873801.
264. Xiao J, Wang P, Setyawan L. Hierarchical control of hybrid energy storage system in DC Microgrids. *IEEE Transactions on Industrial Electronics*. 2015;62:4915–24. https://doi.org/10.1109/TIE.2015.2400419.
265. Feng X, Gooi HB, Chen SX. Hybrid energy storage with multimode fuzzy power allocator for PV systems. *IEEE Transactions on Sustainable Energy*. 2014;5389–97. https://doi.org/10.1109/TSTE.2013.2290543.
266. Shi J, Huang W, Tai N, Qiu P, Lu Y. Energy management strategy for microgrids including heat pump air-conditioning and hybrid energy storage systems. *The Journal of Engineering*. 2017;2017(13):2412–16. https://doi.org/10.1049/joe.2017.0762.
267. Sikkabut S, Mungporn P,Ekkaravarodome C, Bizon N, Tricoli P, Nahid- Mobarakeh B, et al. Control of high-energy high-power densities storage devices by Li-ion battery and supercapacitor for fuel Cell/Photovoltaic hybrid power plant for autonomous system applications. *IEEE Transactions on Industrial Applications*. 2016;52:4395–4407. https://doi.org/10.1109/TIA.2016.2581138.
268. Wu G, Lee KY, Sun L, Xue Y. Coordinated fuzzy logic control strategy for hybrid PV array with fuel-cell and ultra-capacitor in a microgrid. *IFAC-Papers On Line*. 2017;50(1):5554–9. https://doi.org/10.1016/J.IFACOL.2017.08.1098.
269. Chong LW, Wong YW, Rajkumar RK, Isa D. An optimal control strategy for standalone PV system with battery-supercapacitor hybrid energy storage system. *Journal of Power Sources*.2016;331:553–565. https://doi.org/10.1016/j.jpowsour.2016.09.061.
270. Yin H, Zhou W, Li M, Ma C, Zhao C. An adaptive fuzzy logic-based energy management strategy on battery/ultracapacitor hybrid electric vehicles. *IEEE Transactions on Transportation Electrification*. 2016;2:300–11. https://doi.org/10.1109/TTE.2016.2552721.
271. Garcia-Torres F, Valverde L, Bordons C. Optimal load sharing of hydrogen-based microgrids with hybrid storage using model-predictive control. *IEEE Transactions on Industrial Electronics*. 2016;63:4919–28. https://doi.org/10.1109/TIE.2016.2547870.
272. Hredzak B, Agelidis VG, Jang M. A model predictive control system for a hybrid battery-ultracapacitor power source. *IEEE Transactions on Power Electronics*. 2014;29:1469–79. https://doi.org/10.1109/TPEL.2013.2262003.
273. Naseri F, Farjah E, Ghanbari T. An efficient regenerative braking system based on battery/supercapacitor for electric, hybrid, and plug-in hybrid electric vehicles with BLDC motor. *IEEE Transactions on Vehicular Technology*. 2017;66:3724–38.
274. Chettibi N, Mellit A, Sulligoi G,. Massi Pavan A. Adaptive neural network-based control of a hybrid AC/DC microgrid. *IEEE Transactions on Smart Grid*. 2016;1:1–13. https://doi.org/10.1109/TSG.2016.2597006.
275. Song Z, Hou J, Hofmann H, Li J, Ouyang M. Sliding-mode and Lyapunov function-based control for battery/supercapacitor hybrid energy storage system used in electric vehicles. *Energy*. 2017;122:601–12. https://doi.org/10.1016/J. Energy.2017.01.098.

276. Etxeberria A, Vechiu I, Camblong H, Vinassa J-M. Comparison of sliding mode and PI control of a hybrid energy storage system in a microgrid application. *Energy Procedia.* 2011;12:966–74. https://doi.org/10.1016/J.EGYPRO.2011.10.127.
277. Wickramasinghe Abeywardana DB, Hredzak B, Agelidis VG. A fixed-frequency sliding mode controller for a boost-inverter-based battery-supercapacitor hybrid energy storage system. *IEEE Transactions on Power Electronics.* 2017;32:668–80. https://doi.org/10.1109/TPEL.2016.2527051.
278. Yeshalem M, Khan B. Microgrid integration. An intech open access paper. 2018. http://dx.doi.org/10.5772/intechopen.78634, Chapter 4.
279. Mohamed Y, ElSaadany EF. Adaptive decentralized droop controller to preserve power sharing stability of paralleled inverters in distributed generation microgrids. *IEEE Transactions on Power Electronics.* 2008;23:2806–16.
280. Lopes JAP, Hatziargyriou N, Mutale J, Djapic P, Jenkins N. Integrating distributed generation into electric power systems: A review of drivers, challenges and opportunities. *Electric Power Systems Research.* 2007;77(9):1189–203.
281. Katiraei F, Iravani MR. Power management strategies for a micro-grid with multiple distributed generation units. *IEEE Transactions on Power Systems.* 2006;21(4):1821–31.
282. Ding G, Gao F, Zhang S, Loh PC, Blaabjerg F. Control of hybrid AC/DC microgrid under islanding operational conditions. *Journal of Modern Power Systems and Clean Energy.* 2014;2(3):223–32.

5 Hybrid Microgrids

5.1 INTRODUCTION

During the last century, electricity transport mainly occurred by centralized utility grid which transmitted electricity by high-voltage transmission line to medium-voltage distribution network and finally to the customers by low-voltage distribution lines. When power is generated by coal, gas, or nuclear energy, the grid itself acted as a storage device to handle peak demand or some unexpected surges in the power need. The utility grid worked well in an urban environment, but it did not satisfy the need of rural environment or isolated places, where power lines were not accessible. As shown in Chapter 2, in recent years, this utility grid has gone through substantial changes and has become hybrid due to (a) increased need to improve efficiency, (b) reduced CO_2 emission, and (c) increased desire to use renewable sources for power generation. As shown in that chapter, the recent and future trend is to make the utility grid more and more hybrid having more than one source of power (or energy in general) generation with the additional use of energy storage device to handle intermittent sources for power like solar and wind energy. The chapter indicated that in order to handle greater penetration of low-density and distributed renewable energy sources, two main strategies are to be adopted: one is to make the utility grid more hybrid with strong insertion of energy storage device like batteries, and second is to create microgrid to handle distributed energy resources which is connected to a utility grid or it can operate in an islanded mode in case the utility grid fails for some reasons. Unlike coal-, gas- and nuclear-based power plants, renewable sources tend to be more distributed and smaller in size. Solar and wind energy can be directly connected to users such as in home power need. Microgrids are ideally suited to capture power from these sources. Chapter 2 also showed that future utility grid will be more complex and smarter and will be digitally controlled by intelligent management and control systems and operate in a bidirectional way where consumers will become prosumers.

As pointed out in Chapter 1, one of the defining elements of hybrid energy system is the hybrid grid transport, largely created to harness distributed and low-density renewable energy sources (like biomass, waste, geothermal, hydro, solar, and wind) in a larger amount and to satisfy the needs of the rural and remote communities. In Chapter 1, we examined recent developments of hybrid grid operation which include three levels of grid operation and associated management and control systems; centralized utility grid with high-voltage transmission followed by medium-voltage distribution and low-voltage supply to the customer; low- to medium-voltage microgrids to supply largely renewable energy to local customers and off-grid energy systems which include mini, nanogrid operations, and stand-alone systems for very isolated customers. The off-grid hybrid energy system is described in detail in Chapter 6.

Generally, microgrids can be either connected to a utility grid or operated in the islanded mode. Also, generally, mini- or nanogrid or off-systems operations are not connected to a macrogrid. A combination of a significant number of mini- or nanogrids, however, can result in a microgrid with subsequent connection to a macrogrid.

As the need for distributed renewable energy expands, it is the microgrid that will grow fastest in the future. It should be noted that for each grid operation, the power generations can be hybrid, involving multiple sources for power or heat generations with or without energy storage. The most complex situation arises when hybrid microgrid (utility grid-connected or in islanded mode) contains also hybrid storage systems. In general, for microgrids, both homogeneous and hybrid storage devices are very important, particularly when intermittent sources like solar and wind are used to generate power. In this chapter, because of its importance, we examine various aspects of hybrid microgrids involving renewable energy sources. The chapter examines the role of microgrids to harness distributed energy resources, basic features of microgrids, its power architecture, benefits, control systems, management strategies, main problem areas and issues for the use of microgrids to harness hybrid renewable resources, and finally, value-added propositions by microgrids with HRES. The last subject is illustrated with sample case studies.

Hybrid microgrids carry hybrid power systems. As mentioned before, there are many advantages of hybrid power generation systems. These advantages can be listed as [1–26]:

- The possibility to combine two or more renewable energy sources, based on the natural local potential of the users.
- Environmental protection especially in terms of CO_2 emission reduction.
- Low operating cost: Wind energy and also solar energy can be competitive with nuclear, coal, and gas, especially considering possible future cost trends for fossil and nuclear energy.
- They provide diversity and security of supply.
- They provide rapid deployment—modular and quick to install.
- The fuel is abundant, free, and inexhaustible.
- Costs are predictable and not influenced by fuel price fluctuations, although fluctuations in the price of batteries will be an influence where these are incorporated.

As the use of hybrid energy grows, so does the development of hybrid microgrids. Hybrid microgrids evolution has been directly dependent on the evolution of distributed energy resources.

It has been noted that the world's electricity systems are starting to "decentralize, decarbonize, and democratize," in many cases from the bottom up [27]. These trends, also known as the "three Ds," are driven by the need to rein in electricity costs, replace aging infrastructure, improve resilience and reliability, reduce CO_2 emissions to mitigate climate change, and provide reliable electricity to areas lacking electrical infrastructure. While the balance of driving factors and the details of the particular solution may differ from place to place, microgrids have emerged as a flexible architecture for deploying distributed energy resources (DERs) that can meet the wide ranging needs of different communities from metropolitan New York to rural India.

In industrialized countries, microgrids must be discussed in the context of a mature "macrogrid" that features gigawatt-scale generating units, thousands or even hundreds of thousands of miles of high-voltage transmission lines, minimal energy storage, and carbon-based fossil fuels as a primary energy source. Today's grid is not a static entity, though; we are traveling a historic arc that began with small-scale distributed generation (recognized as the original DC microgrids), pioneered by Thomas Edison in the late 19th century, that underwent consolidation and centralization driven by growing demand, and that is now experiencing the beginning of a return to decentralization. From the 1920s through the 1970s, the increased reliability afforded by connecting multiple generating units to diverse loads, decreased construction costs per kilowatt (kW), and ability to draw power from distant large generating resources like hydropower drove the development of the grid we see today [28, 29]. However, those advantages seem to have reached their limits and are increasingly undermined by environmental and economic concerns. Driven by utility restructuring, improved DER technologies, and the economic risks that accompany the construction of massive generating facilities and transmission infrastructure, companies that generate electricity have been gradually shifting to smaller, decentralized units over time [30]. This transition is driven by a range of DER benefits that have been studied in detail [31, 32], such as deferral of generation, transmission, and distribution capacity investments; voltage control or volt amperes reactive (VAR) power supply, ancillary services, environmental emissions benefits, reduction in system losses, energy production savings, enhanced reliability, power quality improvement, combined heat and power, demand reduction, and standby generation. These benefits accrue not only to small, dispatchable fossil-fueled plants—many also accompany deployment of intermittent renewable generating sources, as shown by a foundational study of a 500 kW distributed generation PV plant in California [32, 33]. The challenge of radically decreasing greenhouse gas emissions to avoid catastrophic climate disruption has also led to governmental policies that incentivize deployment of carbon-free generating sources, many of which lend themselves to distributed applications.

Systematic research and development programs [34, 35] began with the Consortium for Electric Reliability Technology Solutions (CERTS) effort in the United States [36] and the MICROGRIDS project in Europe [37]. Formed in 1999 [38], CERTS has been recognized as the origin of the modern grid-connected microgrid concept [39]. It envisioned a microgrid that could incorporate multiple DERs yet present itself to the existing grid as a typical customer or small generator, in order to remove perceived challenges to integrating DERs [36, 40, 41]. Emphasis was placed on seamless and automatic islanding and reconnection to the grid and on passive control strategies such as reactive power versus voltage, active power versus frequency, and flow versus frequency [42]. The goals of these strategies were: (1) to remove reliance on high-speed communications and master controllers, yielding a "peer-to-peer" architecture, and (2) to create a flexible "plug-and-play" system that would not require extensive redesign with the addition or removal of DERs, in order to lower system costs and provide the freedom to locate cogeneration facilities near thermal loads. The CERTS microgrid concept has been deployed in a test bed setting [43, 44] and in real-world microgrid projects [45, 46]. While the initial motivation of CERTS was to improve reliability rather than to reduce greenhouse gas emissions, per se, CERTS microgrids can incorporate renewable microgeneration sources. The European Union MICROGRIDS project explored similar technical challenges such as safe islanding and reconnection practices, energy management, control strategies under islanded and connected scenarios, protection equipment, and communications protocols [37]. Active research continues on all of the topics pioneered in these early studies [47].

In recent years, the rapid increase of distributed generation (DG) installation, especially those based on renewable energy sources, is expected to address the concerns on greenhouse gas emissions, energy sustainability, energy security, etc. Among various kinds of renewable energy-based DGs, wind and photovoltaic (PV) power generation are relatively mature and are the fastest growing DG technologies. Since mid-1990s, wind power installation has been growing by 25% per year. By 2011, China has the highest installed wind power capacity with more than 60 GW [48], and PV installation follows a similar trend, but with an even faster growth rate around 48% each year in recent years, making it the fastest growing energy technology in the world. As of 2011, PV installation has reached around 70 GW with Germany as the leader [28]. The global cumulative wind and PV installed capacity in 2017 were 539 GW and 401 GW, respectively [28]. Large investments made by a number of countries in renewable energy is illustrated in Figure 5.1 [9].

In addition to renewable energy sources, alternative energy sources or microsources such as fuel cells and microturbines have also been used increasingly in recent years for power generation. Fuel cells produce electrical power directly from chemical energy contained in a fuel, which can be hydrogen, natural gas, methanol, gasoline, etc. These power generations are inherently modular in nature, and their capacity can be added easily as loads grow. On the other hand, microturbines had been originally designed for aircrafts applications [29]. Modern microturbines using advanced components such as heat exchangers, power electronics, communications, and control systems are becoming more popular for DG applications [29].

As a part of upgrading traditional utility grid, the traditional power systems are changing globally, and a large number of distributed generation (DG) units are integrated into distribution power grid driven by the environmental concerns and economical factors [48]. However, both utility grid and power consumers can suffer severe power quality problems as the consequence of high-penetration DG units integrated without autonomous control capability. In order to overcome the inherent limitation of distributed generation concept, microgrid as a new concept has then been proposed to well manage the local DG units and loads. RESs have been extensively used to supply the electrical energy demands and reduce greenhouse gas emission with an increasing trend. The intermittency nature of the clean energy sources influences the power generation adversely, becoming a challenge for the uninterrupted and regular supply of power to the consumer and endangering grids operation in terms of different

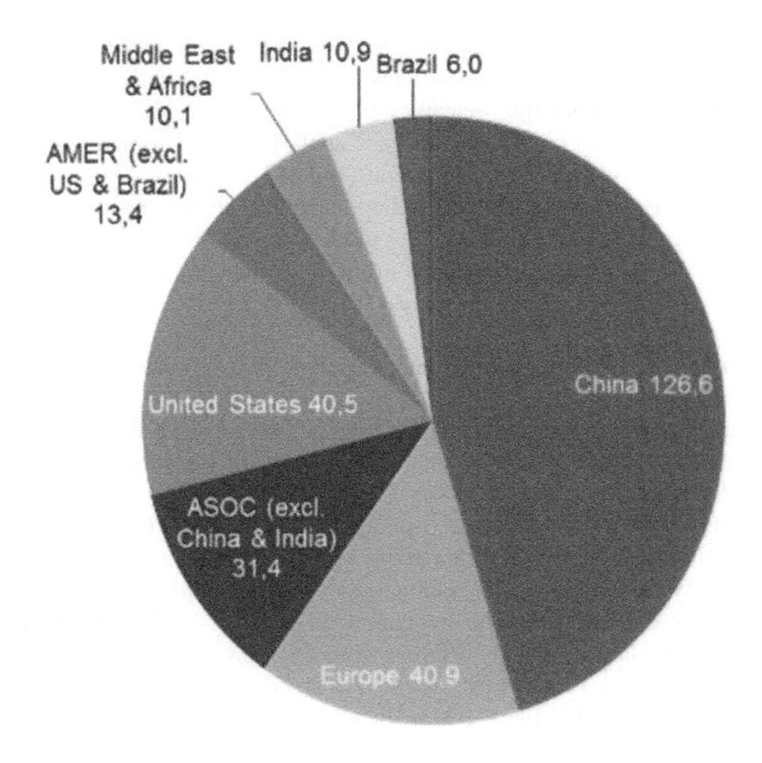

FIGURE 5.1 Global investment in renewable energy in billions [9].

operational and technical aspects. Microgrid (MG) as a cluster of loads and distributed generations (DGs) is proposed to take maximum benefits of RES which can be operated in both islanded and grid-connected modes. An ESS could contribute to integration of RESs into the MG by flattening the RESs fluctuations, power quality improvement, and contributing in frequency and other ancillary services [29].

While much has been written about the concept and promise of microgrids, much can also be learned from examples of real, operating microgrids. For an exhaustive list of existing, experimental, and simulated microgrid systems, the reader is recommended to consult a recent review by Mariam et al. [48]. According to Navigant Research, which has tracked microgrid deployment since 2011, the United States has been the historical leader in deployed capacity; today, though, the United States and Asia have roughly the same capacity of operating, under development, and proposed microgrids, each with 42% of the market. Europe trails with 11%, Latin America with 4%, and the Middle East and Africa currently have just a 1% share. Total capacity was approximately 1.4 GW in 2015 and is expected to grow to roughly 5.7 GW (considered a conservative estimate) or 8.7 GW (under an "aggressive" scenario) by 2024 [49]. More details on microgrid market are given in Section 5.5. Navigant breaks the microgrid market into the following segments (with percentage of total deployed power capacity as of Q1 2016): remote (54%), commercial/industrial (5%), community (13%), utility distribution (13%), institutional/campus (9%), and military

(6%) [26]. It should be noted that Navigant Research does not track purely diesel generator-based remote microgrid systems; to be considered, they must include at least one renewable generating source.

Microgrids are used for a wide variety of applications—distribution grids, electric transportation such as vessels or aircraft, isolated grids, etc.—and therefore can have very different features. The research has identified six key types of microgrids or market segments where microgrids would best apply. These key microgrid categories include: (a) campus/institutional, (b) remote, (c) commercial/industrial, (d) community/utility, (e) military and (f) residential.

1. **Campus environment/institutional microgrids**

The focus of campus microgrids is aggregating existing on-site generation with multiple loads that are co-located in a campus or institutional setting (e.g., industrial park) of tight geography in which the owner can easily manage them. Deploying on-site generation, especially in a combined cooling, heating, and power (CCHP, also known as "trigeneration") application with multiple loads collocated on a campus owned by a single entity, has been a successful model so far and typically includes the largest microgrids to date, with capacities ranging from 4 to over 40 MW [50]. Santa Rita Jail, located in Alameda County, California, is a real institutional microgrid proof-of-concept employing the CERTS concept [45]. The microgrid includes a 1-MW fuel cell, 1.2 MW of solar PV, two 1.2-MW diesel generators, a 2-MW/4-MWh lithium iron phosphate electrical storage system (chosen because this chemistry features high AC-AC round trip efficiency and offers improved thermal and chemical stability compared to other battery technologies, despite some sacrifice in energy density), a fast static disconnect switch, and a power factor correcting capacitor bank. The CERTS protocol allowed all of these distributed energy resources to work together during grid-connected and island modes without requiring a customized central controller. The ability of an institutional microgrid to deliver peak load reduction, and the trade-offs between optimizing net load shape for the facility versus for grid needs, has been demonstrated using Santa Rita Jail as an example, using DER-CAM software to determine optimal equipment scheduling and dispatch [51].

2. **Remote "off-grid" microgrids**

These microgrids never connect to the macrogrid and instead operate in an island mode at all times. Examples of this type of microgrid include the remote village power systems on islands that usually include diesel power or wind generation. Microgrid is being developed with the goal of reducing diesel fuel by integrating solar photovoltaic (PV), distributed wind, and/or run-of-the river hydropower.

Typically, an "off-grid" microgrid is built in areas that are far distant from any transmission and distribution infrastructure and, therefore, have no connection to the utility grid. Studies have demonstrated that operating a remote area or islands off-grid microgrids that are dominated by renewable sources, will reduce the levelized cost of electricity production over the life of such microgrid projects.

Large remote areas may be supplied by several independent microgrids, each with a different owner (operator). Although such microgrids are traditionally designed to be energy self-sufficient, intermittent renewable sources and their unexpected and sharp variations can cause unexpected power shortfall or excessive generation in those microgrids. This will immediately cause unacceptable voltage or frequency deviation in the microgrids. To remedy such situations, it is possible to interconnect such microgrids provisionally to a suitable neighboring microgrid, to exchange power and improve the voltage and frequency deviations. This can be achieved through a power electronics-based switch after a proper synchronization or a back-to-back connection of two power electronic converters and after confirming the stability of the new system. The determination of the need to interconnect neighboring microgrids and finding the suitable microgrid to couple with can be achieved through optimization or decision-making approaches.

More than 1 billion people in developing and underdeveloped countries currently lack access to reliable electricity—or to any electricity at all. Often, the limited electricity that is available is generated using expensive diesel fuel. In particular, for rural areas in these countries, electricity is a key resource for meeting basic human needs, and microgrids may be the best way to deliver that electricity [52, 53]. Remote microgrids combining clean generation and storage, in some cases facilitated by innovative mobile payment platforms, can provide a lifeline to those people, allowing children to study at night, medical systems to provide reliable service, and entrepreneurs to improve their livelihoods. These remote microgrids are leveraging the same advances in power electronics, information and communications technologies, and distributed energy resources that are driving changes in the grid in industrialized countries, allowing developing nations to potentially leapfrog to a world of smart microgrids, in the same way that mobile communications allowed them to connect to each other and the outside world without building up extensive landline networks.

The so-called "hybrid" microgrids [49] that incorporate renewable energy sources, often as an add-on to diesel generator-based systems, show great potential to diversify generation and lower microgrid operating costs in island communities that rely on expensive imported oil for generating electricity and in remote areas far from existing electricity infrastructure [54–59]. Remote microgrids need not use a one-size-fits-all approach to system design; with careful resource evaluation and understanding of demand profiles, projects can be optimized to fit local conditions [60, 61]. However, careful attention needs to be paid to the impact of resource variability on level of service as well as the level of maintenance required to keep the system running or to restore service in the case of generator failure. Examples of research featuring remote microgrids include Huatacondo Island in Chile [62], Xingxingxia in Xinjiang, China [63], and Lencois Island in Brazil [64]

3. **Military base microGrids**

These microgrids are being actively deployed with focus on both physical and cyber-security for military facilities in order to assure reliable power without relying on the macrogrid. This segment also includes mobile military microgrids for forward operating bases.

Cost-effective energy security, "the ability of an installation to access reliable supplies of electricity and fuel and the means to use them to protect and deliver sufficient energy to meet critical operations during an extended outage of the local electrical grid [55]," is the main driver for grid-connected military microgrids (off-grid solutions for operational deployment are also being developed). A good example of military microgrid research and demonstration efforts is the Smart Power Infrastructure Demonstration for Energy Reliability and Security (SPIDERS) [65] Joint Capability Technology Demonstration (JCTD) [56], a three-phase program, with the scope and complexity growing with each phase. Phase 1 took place at Joint Base Pearl Harbor-Hickam, Hawaii, in 2012 and 2013, featuring a single distribution feeder, two electrically isolated loads, two diesel generators, and a PV array. Phase 2 took place in 2013 and 2014 at Fort Carson, Colorado, and included three distribution feeders, seven building loads, three diesel generators, a 1-MW PV array, and five bidirectional electric vehicle chargers. The final phase 3, at Camp Smith, Hawaii, finished in late 2015; it used new and existing generation sources to support the loads of the entire base. A more detailed description of SPIDERS [65], including the project's cybersecurity components and comparisons to other military microgrids are available in the literature [, 66, 67]. Microgrids are also imbedded in the department of defense. Examples of types of common DoD microgrids are illustrated in Figure 5.2.

4. **Commercial and industrial (C&I) microgrids**

These types of microgrids are maturing quickly in various countries; however, the lack of well-known standards for these types of microgrids limits them globally.

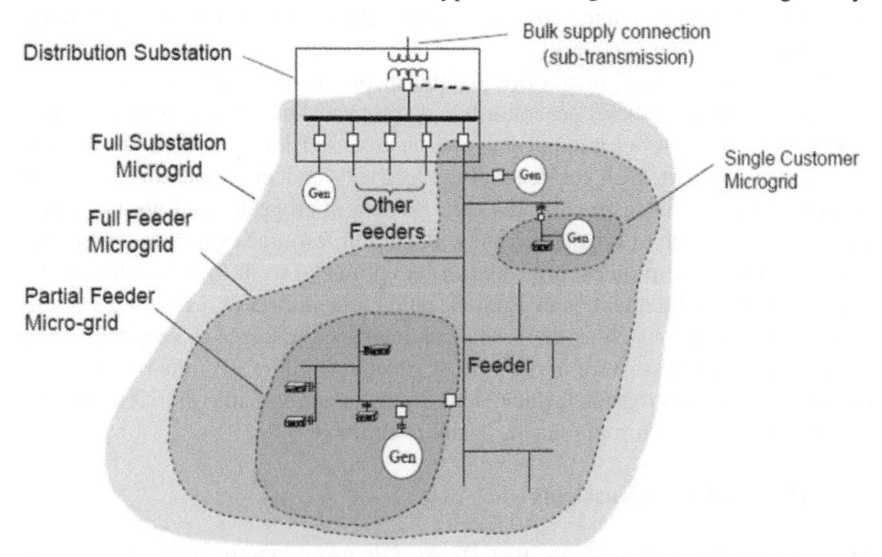

FIGURE 5.2 Examples of types of common DoD microgrids [25]. Source: EPRI, 2016.

Therefore they are a "type" of microgrid but without clear characteristics. The major reasons for the installation of an industrial microgrid are power supply security and its reliability. There are many manufacturing and commercial processes in which an interruption of the power supply may cause high revenue losses and long start-up time. Industrial microgrids can be designed to supply circular economy (near-)zero-emission industrial processes and can integrate combined heat and power (CHP) generation, being fed by both renewable sources and waste processing; energy storage can be additionally used to optimize the operations of these subsystems.

One area where the use of commercial microgrid is being explored is the energy supply to the large data centers. In recent years, significant interest has arisen to use renewables for energy needs for mega data centers. Interest in renewables is driven by two key requirements: environmental sustainability and energy savings. Power consumption within a data center environment is substantial, accounting for almost half of total operating expenses.

There are numerous efforts being made on the use of renewables for data centers. Google sites its data centers based on a number of factors related to reliable service delivery, in locations that may not offer the best potential for renewable power. As a result, it is using renewable energy in the form of wind and solar by engaging in power purchase agreements (PPAs) to power over 35% of its operations, an approach that encourages the development of renewables on the part of utility providers. Facebook is building facilities in Iowa that will be 100% powered by a local wind project that the social network helped to develop as well as a data center in the cool Swedish climate to take advantage of local hydropower. Microsoft, for its part, has built in Quincy, Washington, using 100% hydropower, and is experimenting with powering a 200 kW data center in Wyoming with biogas from a local Cheyenne wastewater treatment facility. A third approach is demonstrated by companies like Apple, which has built into its Maiden, NC facility, a 100-acre, 20-MW solar plant that enables total reliance on renewables, or Verizon Communications, which announced plans in 2013 to invest $100 million in the installation of solar panels and fuel cells at more than a dozen facilities that will use and generate the clean resource [68].

Renewables of choice for data centers include rooftop solar, wind, geothermal, and waste heat reclamation. Solar has limitations in light of the high cost of photovoltaic solar arrays, climatic requirements and space restrictions. However, it has become one of the more widely used approaches in the data center environments since rooftop real estate, when available, is virtually free, reducing the cost of implementation. In recent years, the price of solar cell has also come down significantly. Wind turbines are less widely used, largely because of real estate needs and cost, although interest in this resource is growing. Projects are also underway for geothermal (particularly in the US Midwest) and in waste heat recovery. Battery storage is also top of mind in renewable management discussions, as storage can mitigate issues with intermittency. While the technology is viable, large-scale implementation is out of reach for many enterprises from financial and space perspectives. Once pricing comes down, battery storage promises to play an important role in overcoming some of the reliability concerns that are inherent in renewable power [68].

In decision-making around renewables, there are a number of dynamics that come into play. The first question to ask is how the data center consumes electricity and to what extent consumption is controllable. In other words, can renewables play a role in delivering controllable, interruptible load and can they be integrated in a way that enables them to provide added value to the data center? And can control systems be installed that would allow interface with the utility for sale back of excess energy? Recognizing the existing limitations of renewables, data center operators can still realize their benefits by introducing microgrid platforms that integrate energy from multiple resources, including power from the grid, diesel generation, and renewables. This hybrid energy approach is proving especially useful in regions where grid-delivered power is costly or unreliable. Microgrids can also send back excess energy to the utility grid, generating some profit for the data centers. There are other platforms such as the one proposed by Li et al. [69]. They proposed Greenworks, a power management framework for green high-performance computing data centers powered by renewable energy mix. Greenworks features a hierarchical power-coordination scheme tailored to the timing and capacity of different renewable energy sources. The platform proposed by Li et al. [69] is once again similar to the microgrid concept with a scheme that allows green data centers to achieve better design trade-offs.

Use of hybrid microgrids makes sense for data centers because the reality is that renewable options typically struggle to address 100% of the enterprise data center's power needs. There are few scenarios in which on-site or even rooftop solar PV (photovoltaic) would fully supply a data center's energy needs simply because there is not enough space available to house all the solar arrays that would be needed. In addition, weather conditions can have a dramatic impact on generation capacity: ultimately solar is not dispatchable power—i.e., capable of reliably delivering the required level of power when it is needed. Energy production can be extremely variable, depending on the time of day and weather conditions: a cloud passing over the sun could drop power generation from 5 MW to 100 kW in a matter of seconds. And while rooftop solar may have a natural "home" that does not add to operational costs, wind generation would require substantial real estate to generate energy at any meaningful levels. An underground source of energy, geothermal also comes with significant and costly logistics and infrastructure requirements.

The challenge for many data center facilities interested in on-site generation is that renewable technologies, as they stand today, are rarely a whole solution for the supply of energy 24/7 due to intermittency (in the case of wind and solar), real estate limitations, and implementation costs or all of the above. But combining various resources such as rooftop solar, biofuel, battery, and fuel cells in a hybrid system with the help of microgrids and backing them up with diesel generation that is on-site for redundancy purposes can help to leverage and extend the role of renewables in the data center. The choice of resources depends on a number of variables, including climate, local utility provider fees, and available access to renewable resources.

5. **Community/utility microGrids**

These deployments do not meet the classic definition of a microgrid because they do not "island." Community microgrids can serve up to a few thousands of customers and support the penetration of local energy (electricity, heating, and cooling). In a community microgrid, some houses may have some renewable sources that can supply their demand as well as that of their neighbors within the same community. The community microgrid may also have a centralized or several distributed energy storages. Such microgrids can be in the form of an AC and DC microgrid, coupled together through a bidirectional power electronic converter.

6. Residential microgrids

The question of optimal aggregation scale is an open one in the microgrid literature and an active area of investigation. For example, is it better to integrate detached home residential customers into large community microgrids or to deploy microgrid technology at the level of individual homes? The advantages of a fully decentralized building-integrated microgrid approach [58, 70] include control over energy resources by customers and the fact that individual homes are already connected to the electrical distribution network, so that any changes performed behind the utility meter to add microgrid capabilities will likely not introduce significant legal or regulatory complications beyond what is already encountered for the interconnection of rooftop solar installations today. At the same time, this fully decentralized approach, especially if it includes islanding capability, forfeits cost-saving economies of scale and the generation and load diversity that comes with networking multiple generators and loads. For example, the cost of interconnection protection can add as much as 50% to the cost of a micro-source (i.e., serving an individual home or small building) project so that it may be better to site multiple micro-sources behind a single utility interface [71]. Some authors envision a nested system where energy management systems at the building level communicate with each other and use neighborhood-level master controllers to coordinate distributed energy resources, including shared community energy resources and loads like street lighting [72]. The building-integrated microgrid deployment model [70] would likely benefit from innovative financing (akin to solar leasing models) due to the expense of generating resources, controllers, power electronics, and integration with existing building systems. Literature exploring the so-called "customer microgrids" examines the technical feasibility and economic viability of this mode of broad decentralized residential deployment [73, 74]. Many of these studies are motivated by the question of whether it is feasible and or/desirable to cost-effectively gain full autonomy from the electrical grid using PV and battery storage [73, 75].

One appealing residential microgrid application combines market-available grid-connected rooftop PV systems, electrical vehicle (EV) slow/medium chargers, and home or neighborhood energy storage system (ESS). During the day, the local ESS will be charged by the PV, and during the night, it will be discharged to the EV. The effect is twofold: (1) feed-in tariff schemes are not necessary since little power needs to be exchanged with the main grid and (2) the voltage quality at the PCC is improved [76]. The inclusion of the ESS alleviates over voltages during the day due to excess PV power generation and under voltages during the night caused by the huge current drained to charge the vehicle.

Besides the breakdown of different types of microgrids based on its applications outlined above, one of the most common approaches is to classify various types of microgrids based on the nature of their current [29]: AC, DC, or hybrid AC/DC. Typically, researchers have focused most part of the research activity related to microgrids on AC systems, as they are the most straightforward solution based on the current infrastructure. The knowledge gained over the years with the electric grid can be directly applied for the development of AC microgrids. Therefore, these systems are characterized by efficient modification of voltage levels with transformers and by advanced fault management capabilities with optimally designed devices. As shown later, in recent years, DC and, more importantly, AC/DC hybrid current systems have become more important. These different types of microgrids and their control and management strategies are described in detail in Sections 5.6–5.8.

In recent years, the microgrid has been very instrumental in increasing the use of hybrid renewable energy sources for power generation. The increasing level of RES creates problems for utility grid. Microgrid is a solution to increase level of RES without destabilizing utility grid. The value-added proposition offered by microgrid for HRES is treated in detail in Section 5.10. This subject is supported by several case studies.

5.2 EVOLUTION OF DISTRIBUTED ENERGY RESOURCES

Cutting-edge technologies are being combined to deliver game-changing disrupters as the grid expands from an electricity supplier to an energy-service platform and consumers are becoming prosumers. The word "transformational" keeps popping up in the literature. These extraordinary technologies are transforming electric utilities' conventional business models into unconventional business strategies that are blurring the distinction between producers, distributors, and customers. They are transforming the generation resources mix, too, by using less coal and nuclear as well as more wind, solar, and natural gas [1–26].

In turn, more emphasis on renewable resources and the need to serve isolated and rural communities opened the door for distributed energy resource (DER) systems, which many claim as the biggest accelerator of the grid modernization undertaking. The growth in renewable sources worldwide is indicated by the global investments depicted in Figure 5.1. The original DER concept was simple. Small energy sources—such as gasoline or diesel generators, rooftop solar, and fuel cells—were placed on the customer's side of the meter to keep their lights on or reduce their electric bill. From that starting point, DER has grown over the years and morphed into a wide array of digital technologies now referred to as DER systems.

These systems include generation devices, monitoring systems, control schemes, and energy storage systems (ESS), which enable greater real-time responsiveness in connecting power producers and consumers. When these individual schemes are combined, they become more than the sum of their parts. They also become more complicated. Although it is not rocket science, it does require a better understanding of the technology and what the user is trying to accomplish. It could be said that what makes DER systems valuable in the grid modernization process is that they can be tailored to meet highly specific requirements and applied exactly where needed.

The 2018 Black & Veatch strategic direction report for energy industry contained several topics, and grid modernization was one of them. Black & Veatch reported,

> "The majority of utility respondents (76%) are busy developing grid modernization plans or already have a strategy for electric distribution in place. Of these, 40% already have a grid modernization plan in place and are implementing it, while 36% are in the process of putting theirs together. Of the 24% not currently engaged in grid modernization, more than half are considering it".

Grid modernization is analogous to building complex hybrid grid that can satisfy both urban and rural and isolated community needs for the power. Black & Veatch's study confirms that grid modernization is important to utilities and operators of the grid due to their increasing interests in DER systems [77]. A recent survey revealed 45.8% of the responding utility executives selected the integration of DER systems as a driving element of their plans. The inclusion of DER to grid is a key factor in the process. When the penetration of DER technology increases, it needs to be integrated into the power system to ensure the grid operates as a single entity in real time. Navigant Research published a report that projects that the installed global DER system capacity will reach 530.7 GW by 2024. The report stated that, in many markets, DER deployments are challenging incumbent grid-operating models, requiring a more dynamic network with advanced communications and orchestration to ensure stability, efficiency, and equality [26]. DER types, owners, and grid characteristics are illustrated in Table 5.1 [7].

Several organizations and agencies report that the real strength of DER systems is their ability to enable a wide variety of different resources to be matched locally to the load. Being matched locally is significant because authorities like the US Department of Energy (DOE) have been saying, for years, that 90% of all electrical power outages take place on the distribution system. Localization of DER/ESS schemes has proven to reduce outages for individual users, which has attracted attention on the distribution side of the grid. It is a short step to apply DER systems to the bulk power side of the grid for that same purpose. Supersizing DER applications requires the door to open to grid-scale DER schemes as well as the transmission-distribution interface changes [1–26].

There are many reasons for this transformation. Renewable energy has helped to shift the boundaries, as has the swing from consumers to prosumers, which many of our customers have become. The desire for building zero-energy buildings propels this cause. Extreme weather events also have had an impact [78]. Interestingly, it is the influence of weather that has gotten more attention from regulators and customers. When ice storms and hurricanes knock out the power, everyone is upset. This is especially true after a great deal of resources were spent on storm hardening of the power system.

TABLE 5.1
DER Types, Owners, and Grid Characteristics [7]

DER Type	Primary Owner	Characteristics
Photovoltaic (PV) System	Utility, merchant, or customer	Provides electricity to customers, microgrids, and/or utility grids; power output depends on the intensity of solar irradiation
Energy storage system	Utility, merchant, or customer (aggregator may be involved)	Consumes, stores, and delivers electricity to customers, microgrids, and/or utility grids; often used to enhance system flexibility
Combined heat and power systems	Utility, merchant, or customer	Provides district heating (steam) and electricity to customers, microgrids, and/or utilities
Energy efficiency	Customer (sometimes aggregator involved)	Use of energy efficient technology to reduce electricity consumption; often promoted in utility programs
Demand response	Customer (often aggregator involved)	Often associated with utility programs where customers are compensated for reducing demand (load) during peak periods of electricity usage
Variable rates	Utility/customer	Utilities may impose variable rates to customers to incentivize behavior that reduces overall energy usage or demand during peak periods
Electric vehicles	Customer	Consumes electricity and methods for delivering electricity back to the grid are being investigated
Building energy management system	Customer	Optimizes energy use for the building owner
Microgrid	Utility, customer, or merchant	A grid system providing electricity services to a set of customers or buildings (e.g., a university campus); optimizes energy use within its domain, provides backup power, and offers energy or other services (e.g., frequency regulation) back to the utility grid

Most groups understand that there is only so much technology can do in the face of tornados, fires, floods, and other disasters, but they also notice when localized DERs/ESSs keep the lights on in these ravaged areas. These small devices have provided critical infrastructure with power that enabled them to continue operation after electricity from the grid went down, which has increased the awareness of the DER technology. As shown in Figure 5.1, in recent years, global investment in renewable technologies (which largely imbedded in DER family) has significantly increased, and these numbers in all the countries are going to go up as global warming takes the front seat in all global policies. Granted, the devices that have gotten most of the attention are microgrids [79], but they are an important part of the DER family [80,

81]. Microgrids have been defined as smaller scale versions of the power grid, complete with distributed generation (usually wind or solar), advanced control systems, islanding capabilities, and energy storage facilities. In the past, these devices have been small systems located behind the meter, sized for individual customers, but that is changing. The current crop of these schemes has moved to the front of the meter, and they are getting larger in scale, which has helped them to gain acceptance by grid owners and operators [77].

There are some practical real-life cases of taking advantage of DER system through microgrids which are worth noting. The Kauai Island Utility Cooperative is almost saturated with midday solar power because of the number of solar facilities installed on the island. Consequently, the cooperative installed a 52-MWh lithium-ion (Li-ion) battery on its 13-MW solar farm to enable storage of the solar power. The cooperative recently announced it will be adding another 19.3 MW of solar generation in conjunction with 70 MWh of battery storage, adding to its DER system of residential solar, batteries, and controllable loads. This grid-scale DER system provides a time-shifting service for the solar-generated electricity and stabilizes voltage shifts.

The Connecticut Municipal Electric Energy Cooperative (CMEEC) and its partner SolarCity installed several solar facilities, totaling 13 MW with a 1.5-MW/6-MWh Li-ion battery storage system in Bozrah, Norwich, and Groton, Connecticut. According to press releases, "The battery storage to be set in Norwich will be integrated with SolarCity's GridLogic Control technology to allow CMEEC to remotely dispatch stored energy to manage load spikes and optimize the operational efficiency of the network." These grid-connected DER/ESS projects provide electrical support, flexibility, and reliability to CMEEC's network and make the power supply more secure [77].

The shifting of the transmission-distribution interface is not new. It has been taking place for many years. One of the first attention-getters was San Diego Gas & Electric Co.'s (SDG&E's) Borrego Springs microgrid project. Borrego Springs sits on the end of a radial line that was taken down by a wildfire in 2007. SDG&E built a microgrid for the town, funded by the DOE and the California Energy Commission. The system interconnects with customers' rooftop solar installations and includes two 1.8-MW diesel generators, with a 500-kWh/1,500-kWh battery and three smaller 50-kWh batteries. The system was put to the test in 2013 when severe weather knocked out the transmission line. The microgrid picked up the load for about 24 hours while the restoration took place. The DER system was modified in 2015 to connect to a 26-MW solar facility. SDG&E says the project is frequently used as a poster child of energy innovation [77].

The original Borrego Springs microgrid has been the subject of modifications over the years for its changing role on the grid. Another project providing hands-on experience with large-scale DER projects is the Ameren Technology Applications Center DER project. Adjacent to the University of Illinois campus in Champaign, Illinois, the project made history in December 2016 as the first utility-scale microgrid to be placed in service in North America. Ameren's microgrid operates between 4 kV and 34.5 kV, with multiple levels of control. The Ameren DER project interconnects a solar farm, a wind turbine generator, natural gas generators and battery storage to produce 1,475 kW of power. Ameren's DER system is controlled by Schneider Electric's EcoStruxure software. This cloud-based software platform simplifies the integration of DER systems into the grid. The Ameren system uses a 100-kW Northern Power Systems wind turbine, a 125-kW Yingli Solar solar array, two 500-kW Caterpillar Inc. natural gas generators, and a 250-kW S&C Electric Co. battery storage system that can supply two hours of electric power [1–26].

The Ameren microgrid interconnects a wind turbine, natural gas generators, a solar farm, and a battery storage. The microgrid can deliver power directly to local customers, route power to the grid, or store the power. Recently, Xcel Energy Inc. announced it was installing a large-scale DER system in conjunction with Panasonic Corp. and the Denver International Airport. The project includes a battery storage system located on a feeder with about 30% solar penetration. According to a project spokesperson,

> "The four primary components of the Panasonic project include a 1.6-MW DC carport solar installation, a 259-kW DC rooftop photovoltaic system at Panasonic's new corporate facility, a 1-MW/2-MWh Li-ion battery, and switching and control systems to operate the energy storage system and microgrid functionality."

The battery system contains Younicos GmbH's Y.Cubes, controlled by Younicos' intelligent Y.Q. storage control software. The DER system provides frequency regulation, solar-grid integration through ramp control, and solar time-shifting that includes grid peak shaving, energy arbitrage, and backup power. Xcel Energy has a unique public/private partnership with Panasonic and the Denver International Airport to build a large-scale DER system demonstration project [77].

These front-of-the-meter DER systems projects are gaining momentum as manufacturers such as ABB, Eaton, General Electric, S&C Electric, Siemens AG, and Schneider Electric introduce more advanced DER systems. The result of these efforts is an increasing focus on redefining grid boundaries. An Electric Power Research Institute (EPRI) report on integrating DER systems into the grid said, "Through a combination of technological improvements, policy incentives and consumer choices in technology and service, the role of DER is likely to become more important in the future." In the report, EPRI stated the real values of DER technologies are the services it offers, but it is made difficult by regulatory policies, procedures, and rules. Fortunately, this is changing. Regulators and other stakeholders better understand what DER technologies offer the grid in resilience, security, demand response, and efficiency [1–26].

The use of microgrids has also been implemented in developing countries. A wirelessly managed microgrid is deployed in rural Les Anglais, Haiti. The system consists of a three-tiered architecture with a cloud-based monitoring and control service, a local embedded gateway infrastructure, and a mesh network of wireless smart meters deployed at 52 buildings. Nontechnical loss (NTL) represents a major challenge when providing reliable electrical service in developing countries, where it often accounts for 11%–15% of total generation capacity [47]. An extensive data-driven simulation on 72 days of wireless meter data from a 430-home microgrid deployed in Les Anglais, Haiti, has been conducted to investigate how to distinguish NTL from the total power losses, which helps energy theft detection [77].

A community-based diesel-powered microgrid system was set up in rural Kenya near Mpeketoni called the Mpeketoni Electricity Project. Due to the installment of these microgrids, Mpeketoni has seen a large growth in its infrastructure. Such growth includes increased productivity per worker with an increase of 100% to 200% and income levels increase of 20%–70% depending on the product.

The friendlier regulatory climate started several years ago when policy-makers began to change interconnection rules and provide incentives for deploying DER systems. The Federal Energy Regulatory Commission (FERC) has been issuing orders for several years that recognize the value of a DER system with ESS. With orders 890 and 719, FERC allowed companies installing energy storage technologies to receive compensation. Then in 2011, FERC recognized the value of fast-responding technologies (for example, batteries and flywheels) with order 755. A few years later, in 2013, FERC's order 784 broadly opened the ancillary service markets to DER/ESS technology. In late 2016, FERC announced a notice of proposed rulemaking to integrate electric storage resources more effectively into organized markets. However, real progress was made in 2018. On February 15, FERC voted unanimously to remove barriers to the participation of electric storage resources in the capacity, energy, and ancillary services markets operated by regional transmission organizations and independent system operators. This opened up the door for the development of hybrid microgrids with hybrid energy storage as described in detail in Chapter 4.

In the European Union, the central agencies have no specific regulations or policies formulated on the usage or deployment of DER systems. It has been left to the individual countries to establish their own strategies, which is taking place now. The UK, Ireland, Germany, and Italy are working on establishing a climate friendly technology, with directives and framework programs focused on renewable energy penetration and microgrid developments. In China, the government has developed favorable policies. Its goal is to encourage large-scale development and deployment of renewable energy and distributed generation to promote Chinese DER systems. The infusion of massive DG in the electrical system is illustrated in Figure 5.3 [3]. This global effort has not gone unnoticed by analysts like Navigant Research. It is predicted that the distributed energy storage annual power capacity will rise from 683.9 MW in 2017 to 19,699 MW by 2026. According to the Navigant Research, "Recent market analysis indicates the largest global markets in 2017 (latest available data) would be the U.S., China, Germany, Japan and the U.K." The growing interest in a more decentralized power-delivery system and the maturing DER systems have had a positive influence on the acceptance of technology. Xcel Energy's 2-year 54-kW Community Energy Storage project will test behind-the-meter batteries and utility-scale batteries on the feeder to test battery storage technology [3, 77].

This acceptance increases as DER technologies expand in capabilities and performance abilities, but it is important to know how these systems are evaluated and integrated into the grid. As the technology's complexity increases, it seems the application of the technology gets harder. However, it still revolves around basic

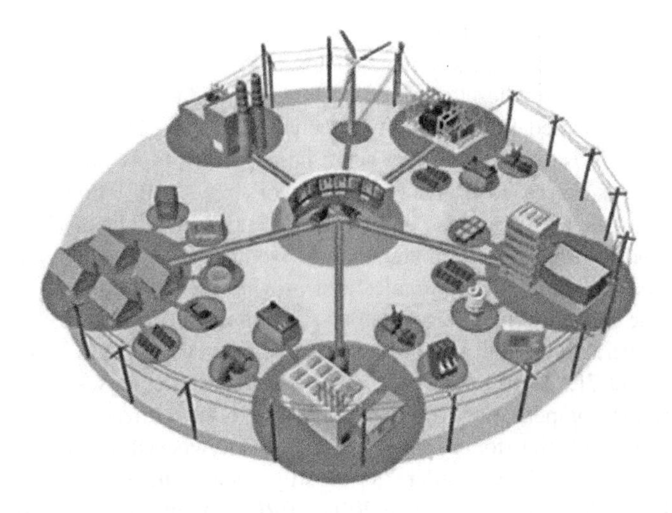

FIGURE 5.3 Massive DGs in the electrical network [3].

engineering practices. Not all DER systems are created equal. Is a fast response needed or is it better to have a lasting power supply? DER systems can provide energy for a few seconds or many hours. Are large amounts of megawatts needed or would a short burst fit the need best? DER systems can supply a few watts or megawatts.

Therefore, it is critical to understand the problem that needs to be addressed and the abilities of the technology. DER systems supply services to support the grid and provide the customer with dependable power. They can improve the power quality and make possible the ability to time-shift generation resources. The industry needs the successful integration of DER systems into the grid. The technology is changing the transmission-distribution interface and moving the industry toward decentralization of the grid. However, it is important to understand that DER systems and the power grid are not competing but rather complementary technologies. The evolution of DER and its management will also change the nature of grid power supply as shown in Figure 5.4 [22].

5.3 BASIC FEATURES, DRIVERS, BENEFITS, AND RESOURCE OPTIONS FOR HYBRID MICROGRIDS

A microgrid, a local energy network, offers integration of distributed energy resources (DER) with local elastic loads, which can operate in parallel with the grid or in an intentional island mode to provide a customized level of high reliability and resilience to grid disturbances. This advanced, integrated distribution system addresses the need for application in locations with electric supply and/or delivery constraints, in remote sites, and for protection of critical loads and economically sensitive development [82]. In principle, a microgrid is any small or local electric power system that is independent of the bulk electric power network. For example,

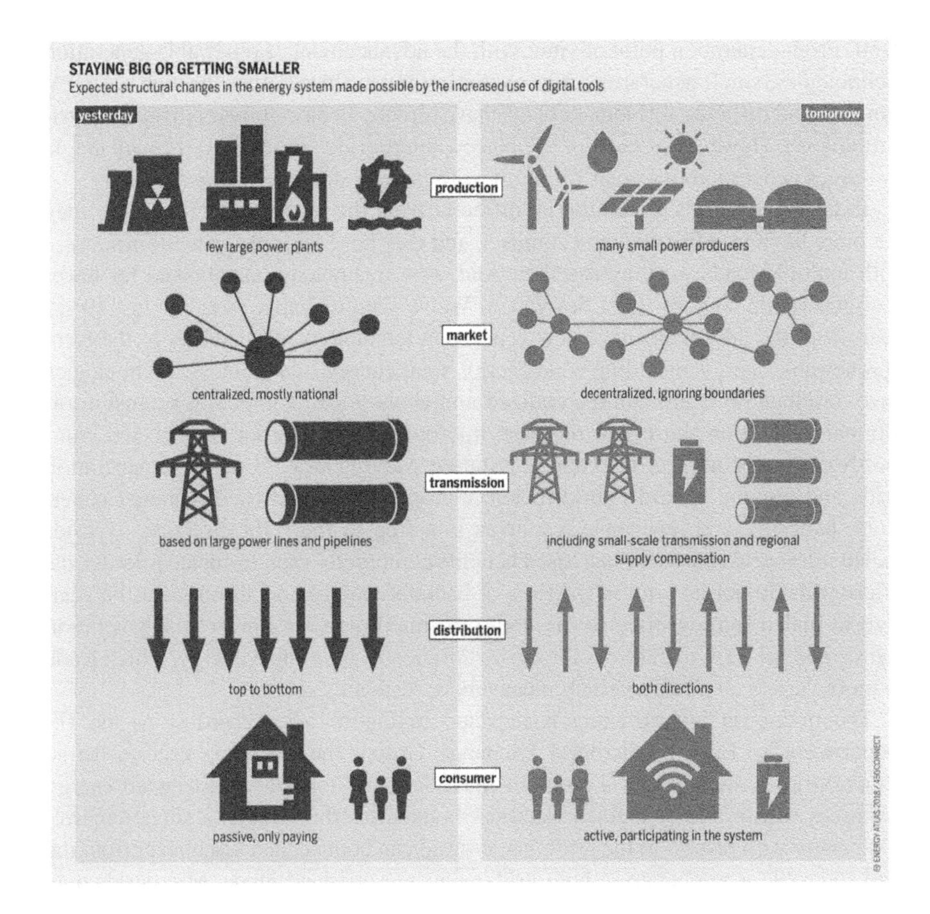

FIGURE 5.4 Characteristics of a smart grid (right) versus the traditional system (left) [22].

it can be a combined heat and power system based on a natural gas combustion engine (which cogenerates electricity and hot water or steam from water, which is used to cool the natural gas turbine), diesel generators, renewable energy, or fuel cells. As mentioned before, a microgrid can be used to serve the electricity needs of data centers, colleges, hospitals, factories, military bases, or entire communities (i.e., "village power") [78]. A true microgrid is much more than a backup power system. It also has to include real-time, on-site controls to match the microgrid's generation and storage capacity to power use in real time, as well as have some way to interact with the grid [83].

The microgrid can be realized through utilizing the potential of distributed renewable energy resources where various small power systems working as independent "microgrids" may be established which can cater to several consumers' loads through small-size distributed energy resources. In other words, microgrids are modern, small-scale version of centralized electricity systems which generates, distributes, and regulates the flow of electricity to a set of consumers at the local level itself. From generation point of view, with the advancement of renewable generation technologies, small generating units exploiting renewable sources through solar PV panels, wind turbines, or biomass plants have already been commercialized at distribution level. However, at most of the places, microgrids are realized through single resources generation like solar PV along with battery storage systems.

Today's microgrids are a little bit different from the microgrids of the past: they are more flexible, advanced, and compact, and they have multiple generation sources with multiple loads working together. And, new technology has allowed for lower cost-generation sources to be developed. We've seen a change over the last 10–15 years with the advancement of new generation technology, particularly in the form of renewable energy, including wind energy, solar energy, and biomass. Technologies have developed and are commercialized and economical, competing against a traditional generation source. In the past, microgrids looked a lot simpler and commonly used only one type of generation source. Moving away from single-generation grids and moving toward hybrid-fueled microgrids powered by distributed generators, batteries, and renewable resources is a step in the right direction. Hybrid-fueled microgrids allow for increased benefits compared to single-fueled microgrids. Microgrids do not just run in the time of backup, they run all throughout the year. Hybrid-fueled microgrids give the ability to maximize the lowest cost-generation source and, in turn, that allows for the installation of renewable energy, which leads to more variety, and more variety means more resiliency.

The microgrid infrastructure is depicted in Figure 5.5. According to the US Department of Energy Microgrid Exchange Group, the following criteria define a microgrid: a microgrid is a group of interconnected loads and distributed energy resources within clearly defined electrical boundaries that acts as a single controllable entity with respect to the grid. A microgrid can connect and disconnect from the grid to enable it to operate in both grid-connected or island mode. Microgrids may

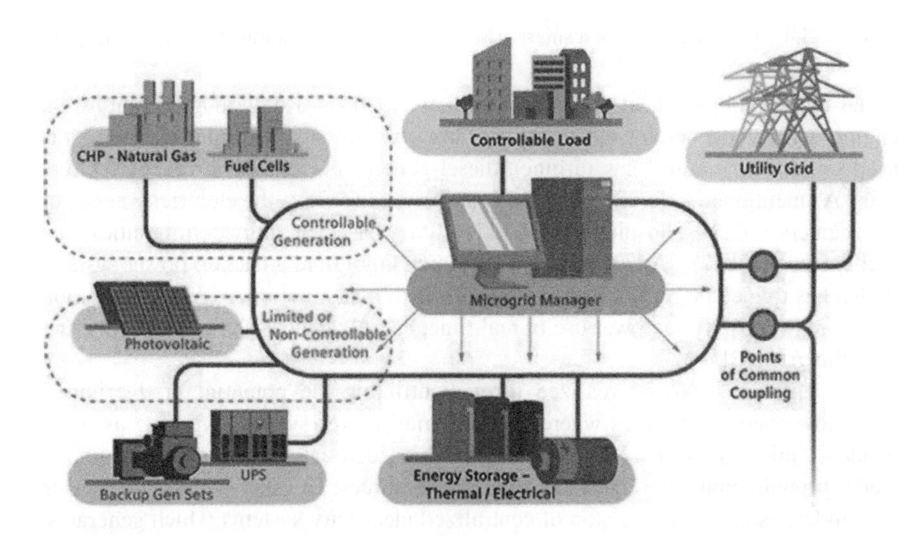

FIGURE 5.5 Microgrid infrastructure [3].

involve a combination of resources, sometimes a quite complex one. This description includes three requirements: (1) that it is possible to identify the part of the distribution system comprising a microgrid as distinct from the rest of the system; (2) that the resources connected to a microgrid are controlled in concert with each other rather than with distant resources; and (3) that the microgrid can function regardless of whether it is connected to the larger grid or not. The definition says nothing about the size of the distributed energy resources or the types of technologies that can or should be used. While as mentioned before, there are different types of microgrids based on its application, microgrids can also be defined in two major categories. These include microgrids wholly on one site, akin to a traditional utility customer, which are usually called customer microgrids or true microgrids (μgrids), and microgrids that involve a segment of the legacy regulated grid, often called milli grids (mgrids).

The operation of microgrids offers distinct advantages to customers and utilities, i.e., improved energy efficiency, minimization of overall energy consumption, reduced environmental impact, improvement of reliability of supply, network operational benefits such as loss reduction, congestion relief, voltage control, or security of supply, and more cost efficient electricity infrastructure replacement. There is also a philosophical aspect, rooted in the belief that locally controlled systems are more likely to make wise, balanced choices, such as between investments in efficiency and supply technologies. Microgrids can coordinate all these assets and present them to the mega grid in a manner and at a scale that is consistent with current grid operations, thereby avoiding major new investments that are needed to integrate emerging decentralized resources. Microgrids have been proposed as a novel distribution network architecture within the smart grids concept, capable of exploiting the full benefits from the integration of large numbers of small-scale distributed energy resources into low-voltage electricity distribution systems. At the highest level, the smart grid has three components [1–26]:

- Improved operation of the legacy high-voltage grid, e.g., through use of synchrophasers.
- Enhanced grid-customer interaction, e.g., by smart metering or real-time pricing.
- New distributed entities that have not existed previously, e.g., microgrids and active distribution networks.

There are several drivers for microgrids, and these include:

- Microgrid advocates that reliability and power quality can be dramatically improved at the local distribution level through systematic application of microgrid technologies.
- To meet local demand.
- To enhance grid reliability.
- To ensure local control of supply.
- Lower frequency responses include enhancing supply reliability, reducing energy cost, and enhancing grid security.

- Microgrid as a foundational building block in the ultimate smart grid.
- Microgrid provides reliability and integration of distributed energy resources (DER) and energy storage assets through improved system intelligence.
- To reduce the physical vulnerabilities of the electric grid during natural disasters.

These drivers fall into three broad categories: energy security, economic benefits, and clean energy integration, as described below.

A. **Energy security**

There are three areas very important for energy security: damage by severe weather, cascading outages, and cyber and physical attacks. In each case, microgrids provide some benefits.

1. **Severe weather**
   There is a growing concern that weather-related disruptions [78] will become more frequent and more severe over time across the United States due to climate change, lending a sense of urgency to addressing grid resilience. Grid outage costs from severe weather, in the United States alone, from 2003 to 2012, averaged $18 billion–$33 billion per year due to lost output and wages, spoiled inventory, delayed production, and damage to the electric grid [84]. Microgrids can provide power to important facilities and communities using their distributed generation assets when the main grid goes down.
2. **Cascading outages**
   Because electrical grids are run near critical capacity, a seemingly innocuous problem in a small part of the system can lead to a domino effect that takes down an entire electrical grid [85]. Microgrids alleviate this risk by segmenting the grid into smaller functional units that can be isolated and operated autonomously if needed. The United States Northeast Blackout of August 2003 impacted 50 million people and 61,800 MW of load [86].
3. **Cyber and physical attacks**
   The grid today increasingly relies on advanced information and communications technologies, making it vulnerable to cyberattack [87]. The centralized grid also contains large, complex components that are expensive and slow to replace if damaged. Microgrids, through their decentralized architecture, are less vulnerable to attacks on individual pieces of key generation or transmission infrastructure. Natural [88, 89] or man-made [90–93] electromagnetic pulse (EMP) events could also have potentially catastrophic results. Examples of cyberattacks are cyberattacks on Ukraine [94] in 2015 and Israel in 2016 (which was successfully thwarted) [95]. Large transformers were physically attacked at a major California substation in 2013 [96, 97].

B. **Economic benefits**

Microgrids can also have some economic benefits. They largely lie in three areas of infrastructure cost savings, fuel savings, and new possibilities for ancillary services.

1. **Infrastructure cost savings**

   Investment in the US electricity grid has not kept pace with generation. As a result, the capacity is constrained in many areas and components are quite old, with 70% of transmission lines and transformers now over 25 years old. The average power plant age is over 30 years [78, 98]. Microgrids could avoid or defer investments for replacement and/or expansion like deferred construction of a $1 billion substation in the Brooklyn and Queens area of New York [99]. It costs $40,000–$100,000 per mile (depending on design, terrain, and labor costs) to build new primary distribution systems [28].
2. **Fuel savings**

   Microgrids offer several types of efficiency improvements including reduced line losses; combined heat, cooling, and power; and transition to direct current distribution systems to avoid wasteful DC-AC conversions. Use of absorption cooling technology in a combined heat and power application could help address summer critical peak electrical demand [35]. Transmission and distribution losses waste between 5% and 10% of gross electricity generation [28, 29]. If the supply of reused waste heat is well matched to the thermal loads, efficiencies of combined heat and power systems can reach 80%–90% [28], much higher than the average efficiency of the existing US grid (which is only ~30%–40%) [28, 100].
3. **Ancillary services**

   Traditional ancillary services include congestion relief; frequency regulation and load following; black start; reactive power and voltage control; and supply of spinning (due to their ability to mimic the inertia of traditional generation), non-spinning, replacement reserves [101, 102], and power quality (reactive power and voltage harmonics compensation). When discussing microgrids, intentionally islanded operation should be added to this list [39]. Recent rulings 755 and 784 from the US Federal Energy Regulatory Commission (FERC) mandate that fast-responding reserves like those used in microgrids be compensated based on their speed and accuracy, opening new revenue possibilities [103, 104].

C. **Clean energy integration**

1. **Need to firm variable and uncontrollable resources**

Important clean energy sources to address climate change like solar PV and wind are variable and noncontrollable, which can cause challenges like overgeneration [105], steep ramping [105, 106], and voltage control [107, 108] problems for the existing grid if deployed in large quantities. Microgrids are designed to handle variable generation, using storage technologies to locally balance generation and loads. In locations with high renewable penetration like California, Texas, and Germany, electricity prices have occasionally gone negative, reflecting an imbalance between supply and demand [109, 110].

The main driver of microgrid development in the United States has been their potential to improve the resiliency (the ability to bounce back from a problem quickly) and reliability (the fraction of time an acceptable level of service is available) of "critical facilities" such as transportation, communications, drinking water and waste treatment, health care, food, and emergency response infrastructure. One major area of activity is the Northeastern United States, where aging infrastructure and frequent severe weather events [78] have led to billions of dollars of losses in recent years.

2. **Need for larger clean renewable energy generator for climate change**

In Europe, climate change and the need to integrate large amounts of clean renewable energy generation into the grid have been more significant drivers spurring microgrid activity. Climate scientists have concluded that to avoid a global average temperature rise exceeding 2°C over pre-industrial levels, currently accepted as the threshold between "safe" and "dangerous" climate change, human society needs to reduce the proportion of electricity produced by burning fossil fuels from 70% (in 2010) to under 20% by 2050 [111]. Many of the energy resources scaling up to fill this gap are decentralized, intermittent, and non-dispatchable, making them a challenge to integrate into a legacy grid designed for a one-way flow of electricity from centralized generating plants to customer loads. Deploying intermittent renewables in with co-located flexible loads and storage technologies in microgrids allows for local balancing of supply and demand, which makes widespread distributed renewable deployment more manageable. Rather than having to track and coordinate thousands or millions of individual distributed energy resources, each microgrid appears to the distribution utility as a small source or consumer of electricity with the ability to modify the net load profile in ways that benefit the main grid [36].

Despite differences in the priority given to resilience and emissions in the United States and Europe, microgrid fuel savings and ancillary grid services are important components of the business case in both areas. Extensive research is now underway to design microgrids using advanced analytical approaches in order to maximize these benefits across a broad range of criteria, including land use, water use, employment, CO_2 emissions, investment costs, and cost of electricity, among others [112–114].

5.3.1 Generation, Storage, and Other Resource Options for Hybrid Microgrid

The hybrid microgrid may consist of solar photovoltaic array (PV), wind turbine (WT), biomass gassifier (BG), and battery energy storage (BES). In grid-interfaced hybrid microgrid architecture systems, two types of control, i.e., sources control and load control should be designed. Source control is achieved primarily through generation control of biomass gassifier or battery energy storage systems which serves as flexible generation whereas solar PV and wind, considering its nature of resources, is never controlled or backed down. At the time of supply exceeding demand, flexible resources like biomass are asked to back down/shut down and the balance excess energy is used to charge the batteries for utilization in times of lower generation by other resources like wind/solar. Thus energy storage batteries provide an economical and/or logistical advantage by making better use of off-peak hours to supply the daily energy needs in peak hours. In case of demand exceeding the generation, batteries provide immediate power response dynamically to the net load fluctuations as biomass gassifier gives comparatively slower responses, but longer duration support. In real terms, design of microgrid should be based on self-sufficiency criterion, which means in a normal scenario, energy is not drawn from the main grid. In general, the power generating technologies can be:

- Internal combustion engines (10 kW–10 MW)
- Mini- to small-size combustion turbines (0.5–50 MW)
- Microturbines (20–500 kW)
- Fuel cells (1 kW–10 MW)
- Photovoltaic systems (5 W–5 MW)
- Wind turbines (30 W–10 MW)

Not all of them are available at all locations. However, ideal microgrid generation technologies should have the following characteristics:

- Modular design (scalable from 1 kW up to 10 MW)
- Low capital cost (preferably much less than $1000/kW installed)
- Low operation and maintenance cost
- Suitable for residential, commercial, and industrial permitting constraints
- Low emissions (meet or exceed those of large modern fuel-fired power plants)
- High efficiency over broad range of loading conditions (at least 40%)
- Usable waste heat (higher exhaust temperature is better)
- High power quality (low harmonics, good voltage, and frequency regulation)
- Good load-following characteristics (for large load steps and transient motor starts)
- Rapid start-up (from cold start and standby conditions)
- Good energy density (high power/weight and high power relative to footprint area required)
- High reliability and dispatchability
- Resistant to damage by power system voltage and current anomalies (surges, voltage unbalances, and so on)
- Operate on fuel that can be easily delivered or transported to the site
- A mature technology with excellent support infrastructure

As the renewable resources generally suffer from the limitations of intermittency and variability, use of more than one distributed resource, i.e., resource diversity, improves reliability and security of power supply. Thus hybrid microgrid system is preferred over the use of single resource microgrid system, but the challenge lies in the integration of all such resources-based generations to meet consumer demand with reliability, security, and best of quality.

The power generated by a PV system and wind turbine [6] is highly dependent on the weather conditions. Batteries are used as a long-term energy storage system to use solar and wind energy resources more efficiently and economically [8]. Ultracapacitors (UCs) as short-term energy storage are used to compensate for transient conditions because they can be charged and recharged rapidly in the condition of uncertainty and sudden interruption. The distributed generators (DG) in microgrid are generally inverter-based that has very low inertia. Unlike, rotating DGs, the low inertial inverter-based DGs have a tendency to respond very quickly causing large transients. These transients of high magnitude are not favorable for stable operation of microgrid. However, due to their fast response, the inverter-based sources ensure the supply of dynamic load regardless of slow rotating machines which require seconds of response time to transients [35]. Therefore, this aspect needs to be taken care of while designing the architecture of microgrid.

Several multidisciplinary studies cover a wide variety of distributed energy resources that can be deployed in microgrids [48, 115–117]. Some examples of the options available for generation and storage today, including their advantages and disadvantages, are provided in Tables 5.2 and 5.3. In general, microgrids are somewhat "technology agnostic," and design choices will depend on project-specific requirements and economic considerations. While not strictly required, incorporating some energy storage helps prevent microgrid faults [118]. Since most microgrid generating sources lack the inertia used by large synchronous generators, a buffer is

TABLE 5.2
Advantages and Disadvantages of Various Generation Options for Microgrids [1–26]

Generation Option	Advantages	Disadvantages
Diesel and spark ignition reciprocating internal combustion engine	Dispatchable, quick start-up, load following, can be used for CHP	GHG emissions, noise generation, nitrous oxide and particulate emissions
Microturbines	Dispatchable, multiple fuel options, low emissions, CHP capable, mechanical simplicity	GHG emissions
Fuel cells (including solid oxide, molten-carbonate, phosphoric acid, alkaline, and low-temperature proton exchange membrane or PEM)	Dispatchable, zero on-site pollution, CHP capable, higher efficiency available compared to microturbines	Relatively expensive (high cost/kwh), limited lifetime, unproven durability
Renewable generation (solar photovoltaic cells, small wind turbines, and mini-hydro)	Zero emissions, zero fuel costs	Variable and not controllable, not dispatchable without storage

TABLE 5.3
Advantages and Disadvantages of Storage Options for Microgrids [1–26]

Storage Options	Advantages	Disadvantages
Batteries (including lead-acid, sodium-sulfur, lithium ion, and nickel-cadmium)	Long history of research and development, flexible	Limited number of charge-discharge cycles
"Flow batteries," also known as "regenerative fuel cells" (including zinc-bromine, polysulphide bromide, vanadium maximum load and complete redox)	Decouple power and energy storage, able to support continuous operation at discharge without risk of damage	Relatively early stage of deployment
Hydrogen from hydrolysis	Clean	Relatively low end to end efficiency, challenge to store hydrogen
Kinetic energy storage (flywheels)	Fast response, high efficiency, high charge-discharge cycles	High standing losses, limited discharge time

needed to mitigate the impact of imbalances of electricity generation and demand. Microgrids also lack the load diversity of larger geographical regions, and so they must deal with much greater relative variability. The array of technologies for energy storage currently under development that could potentially play a role in microgrids is extensive [119, 120] Much of the attention is focused on storage of electricity; however, storage of thermal and mechanical energy should be kept in mind where appropriate. The ability of storage technologies to provide ancillary services like voltage control support, spinning reserves, load following, and peak shaving, among others, has also been analyzed [119] A power architecture of hybrid AC/DC microgrid is illustrated in Figure 5.6 [6].

All microgrids have certain basic components which include:

A. **Local generation**

A microgrid presents various types of generation sources that feed electricity, heating, and cooling to user (see Figure 5.7). These sources are divided into two major groups—thermal energy sources (e.g., natural gas or biogas generators or micro combined heat and power) and renewable generation sources (e.g., wind turbines and solar).

B. **Energy consumption**

In a microgrid, consumption simply refers to elements that consume electricity, heat, and cooling which range from single devices to lighting, heating system of buildings,

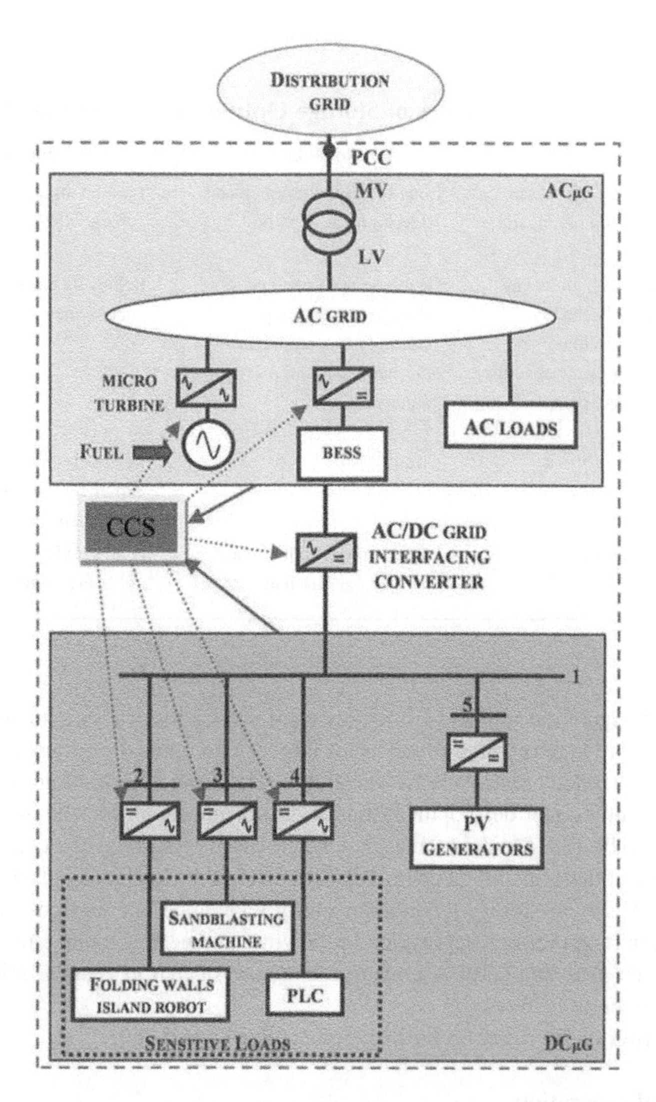

FIGURE 5.6 Scheme of the hybrid AC/DC MG. [6].

commercial centers, etc. In the case of *controllable loads*, the electricity consumption can be modified in demand of the network.

C. **Energy storage**

In a microgrid, the energy storage is able to perform multiple functions, such as ensuring power quality, including frequency and voltage regulation, smoothing the output of renewable energy sources, providing backup power for the system, and playing a crucial role in cost optimization. It includes all of chemical, electrical,

FIGURE 5.7 The solar settlement, a sustainable housing community project in Freiburg, Germany [4].

pressure, gravitational, flywheel, and heat storage technologies. When multiple energy storages with various capacities are available in a microgrid, it is preferred to coordinate their charging and discharging such that a smaller energy storage does not discharge faster than those with larger capacities. Likewise, it is preferred a smaller one does not get fully charged before those with larger capacities. This can be achieved under a coordinated control of energy storages based on their state of charge [23]. If multiple energy storage systems (possibly working on different technologies) are used and they are controlled by a unique supervising unit (an energy management system—EMS), a hierarchical control based on a master/slaves architecture can ensure best operations, particularly in the islanded mode.

D. **Point of common coupling (PCC)**

It is the point in the electric circuit where a microgrid is connected to a main grid. Microgrids that do not have a PCC are called isolated microgrids, which are usually presented in the case of remote sites (e.g., remote communities or remote industrial sites), where an interconnection with the main grid is not feasible due to either technical or economic constraints.

5.3.2 Benefits and Technology Requirements

Microgrids provide efficient, low-cost, and clean energy and enhance local resiliency and improve the operation and stability of the regional electric grid. Microgrids offer many benefits from various different perspectives including [1–26]:

- From the electric grid's perspective, the primary advantage of a microgrid is that it can operate as a single collective load within the power system, reduce grid "congestion" and peak loads, and improve the operation and stability of the regional electric grid. It has critical infrastructure that increases reliability and resilience.

- Customers benefit from the quality of power produced and the enhanced reliability versus relying solely on the grid for power. It provides efficient, low-cost, and clean energy.
- Distributed power production using smaller generating systems—such as small-scale combined heat and power (CHP), small-scale renewable energy resources can yield energy efficiency, and therefore, environmental advantages over large, central generation. It enables highly efficient CHP, reducing fuel use, line losses, and carbon footprint. It integrates CHP, renewables, thermal and electric storage, and advanced system and building controls.
- Blackouts and power disturbances are either eliminated or substantially minimized.
- Economically, the microgrid's improved reliability can significantly reduce costs incurred by consumers and businesses due to power outages, brownouts, and poor power quality. When properly designed, a regional power grid that combines both large central plants and distributed microgrids can be built with less total capital cost, less installed generation, higher capacity factor on all assets, and higher reliability. It uses local energy resources and jobs.
- Microgrids can also generate revenue for constituent consumers and businesses by selling the microgrid power back to the grid/utility when not islanded.
- Microgrids have the flexibility needed to use a wider range of energy sources such as wind, solar, fuel cells, etc. Using electric and thermal storage capabilities, a microgrid can provide local management of variable renewable generation, particularly on-site solar. Energy storage options and capabilities are a very weak link in the success of microgrid operations.
- Microgrids makes RTO markets more competitive. It offers grid services including energy, capacity, and ancillary services and supports places of refuge in regional crises and first responders. Microgrid diversifies risk rather than concentrates.

Microgrid is capable of operating in grid-connected and stand-alone modes and of handling the transition between the two. In the grid-connected mode, ancillary services can be provided by trading activity between the microgrid and the main grid. Other possible revenue streams exist. In the island mode, the real and reactive power generated within the microgrid, including that provided by the energy storage system, should be in balance with the demand of local loads. Microgrids offer an option to balance the need to reduce carbon emissions while continuing to provide reliable electric energy in periods of time that renewable sources of power are not available. Microgrids also offer the security of being hardened from severe weather and natural disasters by not having large assets and miles of aboveground wires and other electric infrastructure that needs to be maintained or repaired following these events.

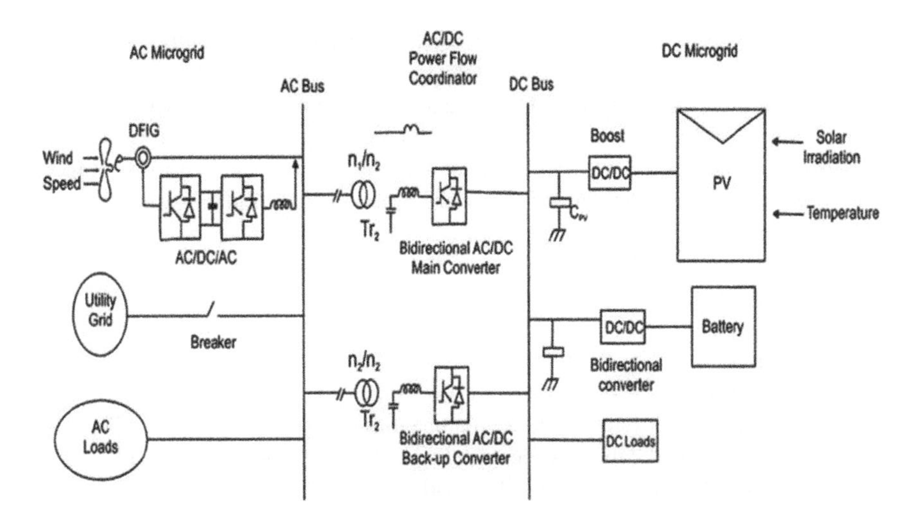

FIGURE 5.8 Hybrid microgrid [9].

Microgrid may transition between grid-connected and islanded modes because of scheduled maintenance, degraded power quality or a shortage in the host grid, faults in the local grid, or for economical reasons. By means of modifying energy flow through microgrid components, microgrids facilitate the integration of renewable energy generation such as photovoltaic, wind, and fuel cell generations without requiring redesign of the national distribution system. Modern optimization methods can also be incorporated into the microgrid energy management system to improve efficiency, economics, and resiliency.

Hybrid microgrid (see Figure 5.8) systems combine the best from grid-tied and off-grid systems. These systems can either be described as off-grid with utility backup power, or grid-tied with extra battery storage. If one owns a grid-tied system and drive a vehicle that runs on electricity, you already kind of have a hybrid setup. The electrical vehicle is really just a battery with wheels. In recent years, hybrid microgrid solar PV systems are used for solar home and zero-energy buildings. The advantages of hybrid solar systems are:

1. ***Less expensive than off-grid systems***

   Hybrid systems are less expensive than off-grid systems. One does not really need a backup generator, and the capacity of your battery bank can be downsized. Off-peak electricity from the utility company is cheaper than diesel.
2. ***Smart grid holds a lot of promise***

   The introduction of hybrid systems has opened up many interesting innovations. New inverters let homeowners take advantage of changes in the utility electricity rates throughout the day. Solar panels happen to output the most electrical power at noon—not long before the price of electricity peaks. Home and electrical vehicle can be programmed to consume power during off-peak hours (or from your solar panels). Consequently, one can temporarily store whatever excess electricity your solar panels in batteries and put it on the utility grid when you are paid the most for every kWh.

Smart solar holds a lot of promise. The concept will become increasingly important as we transition toward the smart grid in the coming years. Typical hybrid solar systems are based on the following additional components:

a. Charge controller
b. Battery bank
c. DC disconnect (additional)
d. Battery-based grid-tie inverter
e. Power meter

Hybrid solar systems utilize battery-based grid-tie inverters. These devices combined can draw electrical power to and from battery banks, as well as synchronize with the utility grid.

5.4 COMPARISON OF HYBRID MICROGRIDS WITH TRADITIONAL POWER SYSTEM APPROACHES

In making a comparison between microgrids and conventional source of power systems, some key areas that are clearly important include:

1. Cost
2. Efficiency
3. Reliability
4. Potential for ancillary services

1. ***Costs of integrated microgrid systems compared to traditional T&D systems***

   In comparing the cost of microgrid systems to conventional T&D systems, it is important to analyze the components that make up the cost of energy delivered via the traditional bulk power system and compare those to the costs of energy delivered via microgrids. The cost elements of the traditional power system include the bulk generation, transmission, sub-transmission, distribution substation, primary feeder, and secondary system. For a microgrid, the cost is the distributed generation cost added to the cost of the power system that makes up the microgrid. If it is a low-voltage microgrid, the only cost is basically that of the low-voltage wiring infrastructure with all of its controls and protection. There is no high-voltage infrastructure. Table 5.4 is an example showing a hypothetical comparison between a utility grid and microgrid delivering the same amount of power [121].
2. ***Efficiency of microgrid systems***

   There are two sides to the distributed generation and microgrid efficiency argument. One school of thought considers that distributed generation is extremely efficient compared to the bulk utility system, and the

TABLE 5.4
Comparison of Costs for Traditional Power System and Hybrid Microgrids [121]

Conventional Delivered Power			Low Voltage Microgrid Application Running Full Time Off-Grid
Cost of bulk generation	3 cents /kw	Distributed generation	7.5 cents /kwh
Transmission/Sub (including value of CHP)			
Transmission	3 cents /kwh	Secondary services	1.5 cents/kwh
Cost of substation	2 cents /kwh	Special microgrid controls	1.0 cents/kwh
Primary feeder	1 cent/kwh		
Secondary service	1.5 cents /kwh		
Total	10.5 cents/kwh		10 cents/kwh

other school considers that it is less efficient than a current central-station power plant. The efficiency of distributed generation can certainly be greater than the bulk power system, but it is important to recognize that this is not universally guaranteed for all applications, many applications are less efficient when they do not employ the correct elements needed for efficiency success.

3. ***Reliability of microgrid compared to conventional power***

   The reliability of average power distribution circuits in the United States is about 99.98%, which means that power is available for 99.98% of the year or there are about 2 hours of cumulative interruption time each year. This is far better than typical individual distributed generators such as ICE units or CT units, which have availability in the range of 95% to 98%, depending on how they are maintained and operated. To achieve 99.98% with ICE units, it requires more than one unit, and they need to be sized so that, if one should fail, the others can pick up the load.

4. ***Potential for ancillary services***

   Microgrids offer the potential for two key ancillary services that may be able to generate a revenue stream for the owners/operators of the microgrid. The two key ancillary services are: thermal energy and reliability. These services really cannot be offered with a conventional power system because generation is located a long distance from loads and because it is very difficult to have targeted high power quality/reliability on the conventional system. So these are the service areas where the microgrid with distributed generation really has an advantage over a conventional power system.

5.5 MARKET FOR MICROGRIDS

Microgrids are small-scale localized power stations that have their own storage resource, generating plant, and defined boundaries. The system can be powered by batteries, renewable sources, and distributed generators based on the fuel requirement, application area, and easy availability of generation source. In addition, the market estimates and forecast are in line with data represented by various regulatory authorities including DOE, IEA, IDEA, USAID, and associated company presentations [20, 26, 50].

Accelerating demand for reliable and uninterrupted power supply along with favorable measures toward integration of clean energy sources will stimulate the microgrid market growth. Rapid industrial development in remote areas will increase penetration of captive and renewable power generating sources. In addition, a positive outlook toward electrification programs backed by government aided schemes and an upsurge in awareness toward security of supply will provide opportunities for business expansion. Furthermore, declining component prices along with increasing investments toward integration of sustainable energy technologies will positively influence the industry dynamics. Rapid technological advancements toward the development of advanced energy storage systems, distributed power generating systems, and software will complement the business outlook.

5.5.1 Globally Microgrid Market is Thriving Due to a Number of Reasons Including:

- In order to ensure uninterrupted power supply through microgrid, various technological advancements have been made to enhance the performance of battery inverters. Other than this, improved battery technologies have also been developed with higher power inputs and outputs for a longer period of time. Such features and technologies are driving the growth of the microgrid market.
- The demand for microgrids is on a rise as they allow consumers and developers to meet the environmental objectives by using renewable energy in the form of power generation source. In line with this, governments of several nations are taking initiatives to establish bio power, solar, and wind energy farms. Other than this, countries like the United States have implemented regulations like the Clean Power Plan (CPP) rule which are aimed toward reducing carbon dioxide emissions.
- Nowadays, dependence on modern communication technologies, such as wireless cloud computing, is increasing which makes power systems susceptible to cyberattacks and hackers. Therefore, some of the sectors like military and research labs require a secure network with round-the-clock power supply, which is provided by microgrids as they are capable of functioning under "island mode" and independent of all external power and data transmissions. This has contributed toward their augmenting demand across the globe.

Another force that has been proactive in maintaining the growth of the global microgrid market is low transmission losses. Microgrids generate power locally and reduce dependence on long-distance transmission lines and thus cut transmission.

Microgrids, which are popping up like mushrooms globally, continue to be one of the biggest disruptors to the traditional electrical grid. The Microgrid Deployment Tracker 2Q17, in its 12th edition, reinforces this point. As shown earlier. It covers six major microgrid market segments and six principal geographies. As of this update, some 1,842 projects (some of which represent country or company portfolios of projects) are listed in the *Tracker*, with a total capacity of 19,279.4 MW. There are 135 countries represented across all seven continents.

Navigant Research has been tracking the microgrid market since 2009, publishing its first Microgrid Deployment Tracker in 2011. While a dynamic process—due to the lack of reporting requirements for microgrids of any kind and the lingering ambiguity of what is and is not a microgrid—this is the only tally of global deployments microgrids published by any research or government organization. The data collected during this process provides insights into the most common application segments, regional differences, preferred resource mixes, and emerging company partnerships.

5.5.2 Microgrids by Segment

In terms of overall breakout by market segment, there is a notable shift in market share in this update of the *Tracker* due in large part to the addition of the sizable commercial and industrial (C&I) portfolio (see Table 5.5). C&I market share increased from 6% to 16% from the 4Q 2016 *Tracker*, thanks to an additional 2,140.4 MW of capacity. (This increase corresponds with the recent Navigant Research report, C&I Microgrids, that projects this segment will be the fastest growing microgrid market segment over the next decade.) Remote projects added 212.7 MW for a total of 8,708.1 MW and still hold the largest share at 45%. In the US military, the microgrid sector is also expanding. The microgrid market size was valued at over $10 billion in

TABLE 5.5
Total Microgrid Power Capacity Market Share by Segment, World Markets (2Q, 2017) [26, 50]

Remote	45%
Commercial/industrial	16%
Utility distribution	15%
Community	10%
Institution/campuses	9%
Military	5%
Direct current	<0.1%

(*Source:* Navigant Research)

2019 and installation is anticipated to exceed 13 GW worth $38 billion by 2026. This gives CAGR of about 23% between 2020 and 2026. Solar PV microgrid industry will witness rapid growth over 2020–2026. Also, shifting trends of gas-based power generating systems will significantly increase gas use in microgrid market.

The research states that governments across several countries are actively working toward higher adoption of microgrids for power generation. Several initiatives have been taken by various governments, including Smart Power Infrastructure Demonstration for Energy Reliability and Security (SPIDERS) program, which has augmented the large-scale deployment of microgrids in the US military sector and the Clean Power Plan would further strengthen the deployment of microgrids in the United States. In China and India, the respective governments have given emphasis to the adoption of microgrids for electricity generation with the launch of programs such as China's 13th Five-Year Plan for Electricity Development and National Smart Grid Mission of India.

According to the report, one of the major benefits of microgrids is the reduction in energy loss due to the long-distance transmission of electricity as power is generated locally. As a result, the demand for microgrids is increasing across several segments. Healthcare application segment is expected to grow at a faster rate over the coming years on account of the requirement of uninterrupted power supply for critical life-support systems and equipment [26, 50].

5.5.3 Microgrids by Region

Looking at the total microgrid capacity in the *Tracker*, Asia Pacific moved ahead of North America after the region added 1,703.6 MW of total capacity from 4Q 2016, compared to 493.8 MW for North America. North America added 16 more projects compared to Asia Pacific, but several of the Asia Pacific additions were large projects, including a couple as large as 300 MW. Europe remains in a distant third place among the six regions with 1,911.3 MW of total capacity, representing a total increase of only 68.3 MW from 4Q 2016. The Middle East and Africa pushed ahead of Latin America in total microgrid capacity with 801.8 MW compared to 776.9 MW for Latin America (see Table 5.6). The Middle East and Africa had a strong

TABLE 5.6
Total Microgrid Power Capacity Market Share by Region, World Markets [26, 50]: 2Q 2017

Asia Pacific	42%
North America	40%
Europe	10%
Middle East and Africa	4%
Latin America	4%
Antarctica	<0.1%

(*Source:* Navigant Research)

update with 334.9 MW of new capacity between project updates and newly added projects. While the strong showings in Asia Pacific and Middle East and Africa are not all from newly installed projects, it does show that these regions that commonly lack robust electrical grid infrastructures are hotspots for microgrid development. Microgrid is well known in North America and Europe and is used in those developed countries; however, there will be positive progress in less developing country to build their electricity and power infrastructure in a futuristic microgrid and smart grid model—one knows that there are big financial obstacle and skill gap in those countries—unless there is a willingness from the government to transform their development plan into small-scale microgrid in power and heat demand [20, 26, 50].

Many vendors claim to have large project portfolios, but due to client sensitivity and a desire to remain opaque in business development efforts, they did not reveal project details for publication in the Navigant report. For example, Power Secure—recently purchased by Southern Company—claims to have a portfolio equaling 1,500 MW in North America. With the new utility ownership, the company promises to be more forthcoming with project data, but was unable to respond in time for this update. Numerous microgrid developers and integrators have nondisclosure agreements in place. Nevertheless, the microgrid market is inching its way into the mainstream. In fact, Navigant Consulting predicts that by 2026, commercial and industrial electricity customers will add nearly 5,000 MW of on-site, hybrid generation systems—that is, microgrids—to help power their organizations to lower costs and better payoff.

Microgrids are on-site generation and storage resources that serve a localized load and can disconnect from the larger electric grid in the case of outages or instability. By localized load, think of a hospital, a big manufacturing plant, a school campus, or any other sizable facility that pays dearly for electricity and can't afford to go without it for long. Hybrid systems incorporate generation assets—often renewables—and battery-based energy storage.

Demand for such hybrid systems is up largely because costs have come down significantly. According to the US Department of Energy (DOE), the average price of utility-scale solar is now 6¢/kWh, a target DOE had hoped to reach in 2020. Efficiency is on the rise too. As recently as 2015, solar panel efficiency hovered around 18%, but last year, the National Renewable Energy Lab (NREL) clocked it at nearly 30%. Average prices for lithium-ion batteries fell 53% between 2012 and 2015, according to analysts at IHS, a global market research firm. They've further come down some 18% in the past year. While solar-plus-storage installations can take an organization far, they may not make it through the night. That's why many on-site generation systems include some form of fossil fuel-based generation. In fact, 40% of distributed generation systems contain fossil fuel generator sets (gensets), according to Navigant's Microgrid Deployment Tracker.

Policy transitions toward the integration of efficient and sustainable power generation sources will augment the natural gas-powered microgrid market demand. Sturdy increase in the global and domestic gas demand, primarily for power generation, has instituted a positive business scenario. In addition, strengthening of the economic and social rationale for the energy transition driven by concerns over fuel diversification has further reinforced an expansive industrial platform. Moreover, increasing competition from the US shale gas boom and its expanding gas industry have further led nations to pursue investments in shale and tight gas projects, which, in turn, will nourish the technological adoption.

Ongoing expansion of utility-based grid infrastructure along with extensive deployment of sustainable energy technologies including solar, hydro, and wind energy has enhanced the development of grid-tied systems. Low cost of energy storage, improved efficiency, and resiliency and reduction in emissions are some of the key features which will boost the microgrid market share. These systems are being deployed in synchronization with main power grid and are being used to provide electricity across manufacturing, healthcare, and commercial applications. Moreover, Asia Pacific and Europe are witnessing a significant growth on account of development of Tier II and III cities and shifting industrial base across developing nations. Extensive research and development activities along with technological innovations backed by government funded schemes will positively influence the business outlook [26, 50].

The regulated European markets have numerous parameters that enhance the development of these grids. The major focus in the region has been on large-scale renewable technology deployment including offshore wind that requires significant investment in transmission infrastructure. The industry has been interwoven and the variability of renewable energy including solar and wind primarily leans toward the cross-border energy trading. The deployment of distributed energy technologies on account of ongoing initiatives including FIT's, tax rebate and other promotional schemes have further provided favorable opportunities for business expansion. In addition, technological advancement across clean energy sector along with stringent measures to reduce the dependency on conventional power generating systems will upsurge the microgrid market growth.

5.5.4 Microgrids by Energy Source

The microgrid market has been segmented on the basis of energy sources which mainly include natural gas, combined heat and power, solar photovoltaic (PV), diesel, fuel cell, and others. Among these, natural gas is the most popular source of energy generation used by microgrids as it produces less carbon emissions as compared to oil- or coal-fired generation. Based on the UK microgrid market statistics, in 2019, about 30% of the market is for diesel generators, 25% for natural gas, 20% for solar PV, 13% CHP, and 12% others.

Incorporating renewable energy, such as PV, into a microgrid may offer a variety of benefits over a microgrid served by fossil fuels alone. Renewable energy can prolong operation of a facility during a long-term grid outage, especially in a scenario where the renewable energy system incorporates battery storage. For example, an analysis of a modeled telecommunications facility using a hybrid solar, storage, and diesel system was estimated to survive 1.8 days longer without grid electricity than a diesel system operating alone. Renewable energy also produces zero emission electricity that can provide environmental and public health benefits. This electricity generation may also provide an economic benefit to participating customers if surplus electricity is sold to an electric service provider or systems are used to provide grid services such as demand response during normal operations.

Prolonged operation during a grid disruption relates to the resilience value offered by PV. Resilience in the context of energy systems can be defined as "the ability to anticipate, prepare for, and adapt to changing conditions and withstand, respond to, and recover rapidly from disruptions through adaptable and holistic planning and technical solutions" [122]. There are a wide variety of resilient energy solutions that can be incorporated at critical infrastructure such as microgrids or conventional diesel generator sets. Despite this technological diversity, each resilience solution has a few key components: on-site electricity generation, energy supplies during a grid outage, it must have local electricity generation that can provide islanding controls, black start capability, and energy storage solutions.

The system integration companies play a critical role as they assemble, integrate, and deliver the final substation to the installers. These companies provide high-end services to their customers by providing more competitive prices for the integration and assembly of the system. The industry has been facing large investments from both public and private players to develop microgrids across the regions. Key players operating across the industry include Siemens, Exelon, Schneider, Homer Energy LLC, ABB, Lockheed Martin, Caterpillar, Honeywell, Viridity Energy, EnSync, General Electric, Eaton, Exelon Corporation, and Advanced Microgrid Storage.

The report by Research and Markets.com says that the market is expected to witness significant growth over the coming years, owing to the ongoing shift from conventional power stations toward distributed energy. Rising demand for uninterrupted power supply and resilient power infrastructure has paved the way for the growth of the microgrid market globally. Also, microgrids use the renewable energy mix for power generation that would help the governments and enterprises across the globe to achieve the target to reduce global carbon footprint.

5.5.5 Market Structure and Degree of Market Decentralization

The EU "More Microgrids" project [123] presented four different scenarios of microgrid resource ownership including: ownership by the distribution system operator (DSO), where the DSO owns the distribution system and is responsible for retail sales of electricity to the end customer; ownership by the end consumer or even consortium of *prosumers* (entities that both import and export electricity); ownership by an independent power producer; or, ownership by an energy supplier in a free market arrangement. According to Navigant Research [26], the majority of grid-tied microgrids today are owned and financed by facility owners, especially in the campus/institutional category. It is important to recognize that microgrids, especially community microgrids, can utilize the existing distribution system infrastructure, radically reducing their costs [20].

Three models have been proposed for integrating energy prosumers into the grid—peer-to-peer, prosumer-to-grid, and prosumer community groups—and identified barriers to their adoption [124, 125]. In the peer-to-peer model, perhaps the farthest from today's centralized grid model, the underlying platform would support the ability of electricity producers and consumers to directly buy and sell electricity and other services from each other, with a fee going to the manager of the distribution grid for providing distribution services [126]. Pilot projects of this type are starting to appear in places like Brooklyn, New York, the Netherlands, and the United Kingdom. Researchers, practitioners, and even large European energy companies, for applications like electric vehicle charging, are starting to apply secure peer-to-peer platforms like blockchain-based distributed ledgers to peer-to-peer energy markets [127, 128].

One focus area is the market for voltage control in distribution networks with microgrids. Some researchers propose that each microgrid in a future multi-microgrid network act as a virtual power plant—i.e., as a single aggregated distributed energy resource—with each microgrid's central controller (assuming a centralized control architecture) bidding energy and ancillary services to the external power system, based on the aggregation of bids from the distributed energy resources in the microgrid (responsive loads, microgenerators, and storage devices) [129]. They conceive of the distribution system operator running a day-ahead market for reactive power, which is required for the flow of power from large generators to customers across a radial transmission and distribution network, and propose a mechanism for optimal market settlement. This vision is similar to that presented in New York's Distributed Energy Resource Roadmap [130], which proposes to open the state's wholesale electricity market to DER aggregators [20].

Innovative business models such as power purchase or energy services agreements and design-build-own-operate-maintain (DBOOM) will likely play a big role in the ability of microgrids to scale [26]. Once microgrid design and procurement becomes more streamlined, power purchase agreements (PPAs) are poised to play a larger role in the microgrid market [26]. The PPA is currently a very successful business model in the US residential and commercial solar PV markets because it can be used to capture tax and other related incentives while avoiding large upfront capital costs for the facility hosting the system. The infrastructure in a PPA is owned by a third party and is leased to customers to provide electricity and related services to end customers. In the case of microgrids, improved security, reliability, and sustainability can be marketed along with economic benefits like energy cost savings. In the case of combined cooling, heat, and power projects, thermal energy can be bundled in the PPA along with electricity. It is reasonable to expect that operations and maintenance will be included in the PPA, since PPA revenues depend on systems performing to their potential [20].

5.6 MICROGRID POWER ARCHITECTURES

Microgrid consists of group of multiple distributed sources with interconnected demands. Microgrid can be defined as a system or a subsystem, which incorporates single, or multiple sources, controlled demands, energy storage systems, security and supervision system. These elements and subsystems make microgrid operational in utility integrated or isolated mode. Here, the main function of the utility grid is to maintain system frequency and bus voltage by supplying deficient power instantly [4]. Microgrid consists of bidirectional connections that means it can transmit and receive power from utility grid. Wherever any fault occurred on utility grid, microgrid switched to stand-alone mode [5]. Even though emerging power electronic (PE) technologies and digital control systems make possible to build advanced microgrids capable of operating independently from the grid and integrating multiple distributed energy resources, there are a lot of challenges in integration, control, and operation of microgrid to a whole distribution system. Microgrid is not designed to handle the large power being fed by the utility distribution feeders. Further, the characteristic of large microgrid components possesses big challenges. The issues related to the integration of microgrid raise the challenges to operation and control of the main utility grid.

It is a distribution network which is supplied through low- and medium-voltage distribution lines. Various self-sufficient and independent distributed energy sources, i.e., PV, wind, fuel cell, microhydro, etc., and storage devices such as battery storage, flywheel storage, etc., along with demands, are incorporated and grouped inside a microgrid structure. Different distributed energy sources are integrated in microgrids by its corresponding bus bars, equipped with power electronics converter. Point of common coupling (PCC) is the point where microgrid is connected to the upstream network.

In grid-interfaced mode of operation, PCC is closed and microgrid is linked with utility grid. Whenever there is any disturbance in utility grid or microgrid, PCC is opened and a microgrid is disconnected to the main grid, then the microgrid is operated in stand-alone mode. While many renewable energy systems (RES) are large scale and are connected directly to the transmission system, there are small-scale and interconnected distributed energy resources located near consumption points within low-voltage electric distribution to achieve efficient and economical requisites. DG should be provided at strategic points in the microgrid system. These strategic points may be load centers so that these sources provide voltage and capacity support, reduce line losses, and improve stability [20].

The renewable energy production is further classified into dispatchable and non-dispatchable production. Dispatchable production is able to change their power production upon demand and by the request of grid operators. They are microhydro and mega hydropower, ocean/marine current power and wave power, geothermal and ocean thermal energy conversion, biofuel biomass, etc. [1–26]. Non-dispatchable renewable energy-based generators are wind and solar, which cannot be obtained on demand. They are, however, more accessible in most parts of the world.

Microgrids often include technologies like solar PV (which outputs DC power) or microturbines (high-frequency AC power) that require power electronic interfaces like DC/AC or DC/AC/DC converters to interface with the electrical system. Inverters can play an important role in frequency and voltage control in islanded microgrids as well as facilitating participation in black start strategies [131]. The static disconnect switch (SDS) is a key microgrid component for islanding and synchronization; it can be programmed to trip very quickly on overvoltage, undervoltage, over frequency, underfrequency, or directional overcurrent [45].

The interface with the main grid can be a synchronous AC connection or an asynchronous connection using a DC-coupled electronic power converter [118]. The former approach has the advantage of simplicity, while the latter isolates the microgrid from the utility regarding power quality (frequency, voltage, harmonics) and is a natural match with DC-only microgrid strategies. Since most distributed energy resources (including fuel cells, solar PV, and batteries) provide or accept DC electricity and since many end loads, including power electronics, lighting, and variable speed drives for heating, ventilation, and air conditioning, use direct current internally, all DC microgrids have been proposed to avoid losses from converting between DC and AC (and often again back to DC) power [28, 132–136]. These losses can waste from 5% to 15% of power generation depending on the number of back-and-forth conversions. Additionally, faults in DC systems can be isolated with blocking diodes and issues of synchronization, harmonic distortion, and problematic circulating reactive currents are alleviated [135]. Furthermore, a grid-tied DC-based nonsynchronous architecture simplifies interconnection with the AC grid and permits straightforward plug-and-play capabilities in the microgrid, allowing addition of components without substantial reengineering [26].

In simplistic terms, there are two types of microgrids power system available. They are AC microgrid and DC microgrid, which depend on distributed sources and required demands. DC grid has the advantage of easier control. Further, it does not require DC-AC or AC-DC converters; therefore, it provides lower cost and better efficiency. On the other hand, AC grid has the advantage of full utilization of available AC grid technologies but it requires synchronization and stability from the reactive power point of view [12].

Generally, the AC microgrid with high penetration of DG units and storage devices [28–30] has more capacity and control flexibilities to be connected to the conventional AC power systems under grid-tied and islanding operation conditions. However, during the last two decades, DC microgrids are arising as an interesting alternative due to the inherent advantages they provide over AC ones. Some new energy sources in the microgrid, such as photovoltaic (PV) panels, batteries, and fuel cells are DC sources in nature, which unavoidably require inverters to be connected to grid. Meanwhile, more DC loads, e.g., LED lighting and computers, are rapidly growing at the end users, which need the power factor correctors (PFCs) to convert the standard AC voltage to a desired DC voltage. Intuitively, establishing a DC power supply network to connect the DC sources and DC loads directly using high-efficient DC/DC converters could reduce the unnecessary power conversion circuit and simplify the control complexity. Doing so, the DC grid can demonstrate its significant characteristics with high efficiency and low power conversion cost. The increment of DC-based systems or devices that require a DC stage to operate, the lack of reactive power circulation or the fact that there is no need for synchronization, among other features, is clearing the path toward DC-operated distribution networks.

The DC microgrid was proposed [3, 20] to integrate various distributed generators. In practice, several DC microgrid pilot projects, such as low-voltage DC microgrids at the Italian Research Center on Energetic Systems and the Lawrence Berkeley

National Laboratory, have been established and tested with 10% energy savings for their data center compared to a very efficient AC baseline case. Compared with AC microgrid, DC microgrid has other advantages of less energy conversion links, lower line losses, and higher system efficiency. In addition, the DC microgrid does not need to track the phase and frequency of AC voltage unlike the traditional grid-tied inverters, which greatly influence the controllability and the reliability of AC microgrid. Therefore, DC microgrid is more suitable for the integration of distributed renewable energy sources.

However, for a comprehensive microgrid, where the various sources as complementary should be integrated to overcome the environmental influence and reduce the interruption maintenance time, a pure DC grid would be deemed inappropriate. Therefore, a hybrid AC/DC microgrid should be assumed to fully demonstrate the advantages of AC and DC distribution networks in view of easier renewable energy integration, higher power conversion efficiency, less energy storage capacity, and higher reliability. For example, a hybrid AC/DC microgrid project in Dongfushan Island, built by State Grid Corporation of China, integrated multiple distributed generation resources and energy storage and improved the power supply reliability of whole island. It is worth noting that while the success of promising initiatives like "DC homes," i.e., low-voltage DC grids for residential applications, has been limited by a lack of DC appliances and the need for large grid-connected AC-DC converters, DC or hybrid AC/DC microgrids have flourished in maritime applications, data centers, and the so-called mini-grids utilizing PV solar generation and batteries to charge electronic devices like laptops or cellphones.

In general, microgrid system is a configuration of single or multiple renewable energy sources with even nonconventional sources as main energy generation mechanism, so that the capacity shortage of power from one source will substitute by other available sources to provide sustainable power. Additionally, it incorporates energy storage and power electronics circuitry [3]. Some of the components produce direct current (DC) power and other alternating current (AC) power directly with no use of converter. Microgrid is configured based on the following technical topologies to couple the available renewable sources and to meet the required load. Here, the voltage and the load demand are the determinant factors. Power system configurations can be grouped in a number of different ways as described below [3, 20].

A. **AC microgrid**

In this configuration, various renewable sources and the energy storage system are linked at the AC bus with the demands. For this type of configuration, two subcategories, centralized and decentralized, are available. AC microgrid is currently the main form, and radiation type is the basic structure. Sometimes ring type structure is also used. According to the application of microgrid, load type, and capacity size, the AC microgrid is also divided into three categories: system level microgrid, commercial and industrial microgrid, as well as rural microgrid.

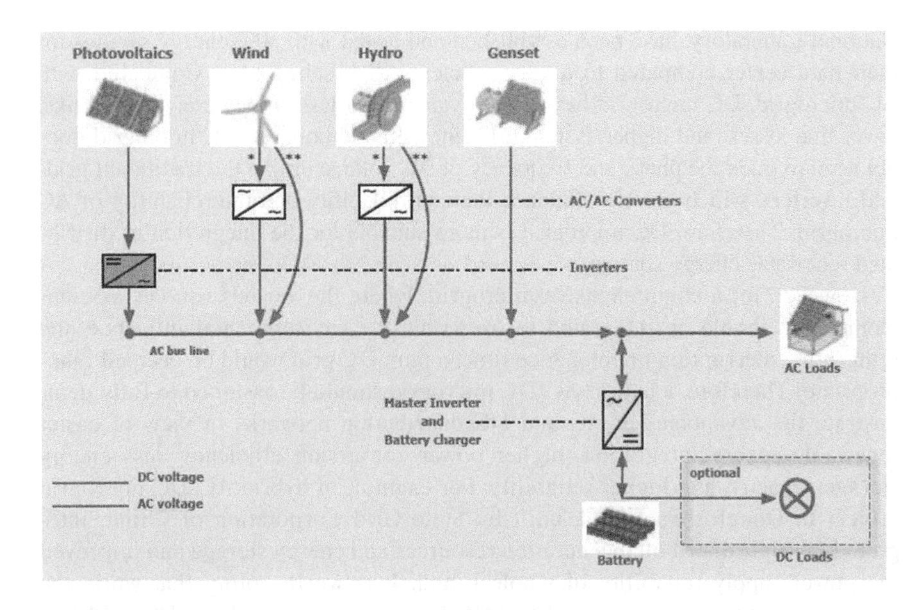

FIGURE 5.9 AC-coupled centralized microgrids [3].

1. **Centralized AC-coupled microgrid**

   In this case (see Figure 5.9), all the elements are linked to the AC bus. AC-power producing elements are connected to AC line in direct manner or with the help of AC/AC converter, for getting even component coupling topology. For controlling the energy flow to the battery and from the battery to the load, the master inverter is required. Furthermore, DC electricity can be provided from battery if needed. All the elements are linked to the AC bus. AC power producing elements are connected to AC line in direct manner or with the help of AC/AC converter, for getting even component coupling topology. For controlling the energy flow to the battery and from the battery to the load, the master inverter is required. Furthermore, DC electricity can be provided from battery if needed.
2. **Decentralized AC-coupled microgrid**

   In this type of architecture (Figure 5.10), all the technologies are not connected to any of the bus, rather they individually connect to the load directly. The energy sources may not be situated in one location or close to one another, and they can connect to the load from anywhere the renewable resources are available. The merit of such configuration is that the power generating components can install from the location where renewable resource is available. But it has a disadvantage due to the difficulty of power control of the system. Thus, comparing the two configurations, the centralized system is better due to its controllability than the distributed system [3, 20].

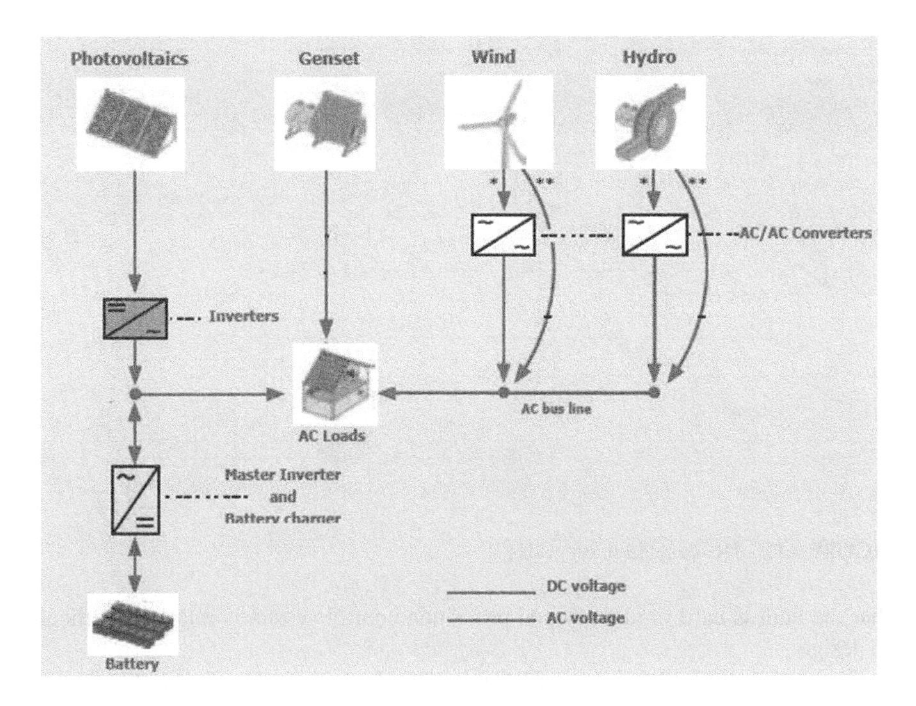

FIGURE 5.10 Decentralized AC-coupled microgrids [3].

B. **DC microgrid**

Due to the problems which AC microgrid has, such as voltage, frequency, and synchronization, DC microgrid has been rapidly developed (see Figure 5.11). According to the different quality requirements for electricity, there have been several typical DC grid structures in recent years, such as ring type, radiation type, feeder type, etc.

In the direct current (DC) combination, all the energy sources are linked to the DC bus prior to connection to the AC bus. All AC power sources are converted to DC and then linked to the AC demand by using converter. The merit of DC-coupled topology is that the demand is met with no cutoffs. Despite the advantage of this, it has disadvantages of low conversion efficiency and no power control of diesel generator. Wind turbine and diesel generator produce AC voltage and need AC/DC converter to supply appropriate load to the DC bus. Charge controller is also employed to protect the deep discharge and overcharge of the battery. If required, the AC load can be supplied using an inverter. Recently, the number of DC loads has been increased, and the urban distribution grid has been developed.

When the trunk lines adopt the DC bus completely in the microgrid (Figure 5.11), the distributed generation and AC/DC loads can access to the DC bus through the power electronic converter, and then DC microgrid can be connected with the large systems through the bidirectional converters. Small low-voltage DC microgrid accesses the medium-voltage DC bus voltage via the DC transformer. The advantage of the ring structure is the high reliability of supply, while the disadvantages are

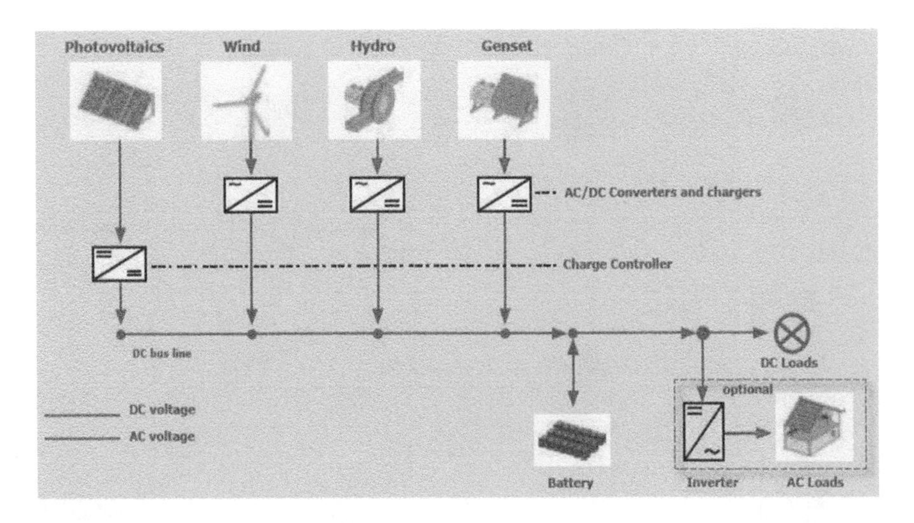

FIGURE 5.11 DC-coupled microgrids [3].

that the fault is hard to identify and protection control system is relatively difficult to design.

Sometimes the radiation type of DC microgrid is connected to the AC distribution network through a bidirectional converter. Each distributed generation and loads are connected to a DC bus directly or indirectly, and then buses with different voltage level are connected together through the DC transformer. The disadvantage of the radial structure is the low reliability of power supply, while the advantage is that the ability of fault identification is reliable and the design of the protection system is relatively easy.

Another type is the feeder type structure of microgrid. Compared with the ring type, there is only one bidirectional converter which connects the microgrid with distribution network. This type of microgrid is suitable for rural and other remote areas, and its advantages are as follows: system design is simple, the control system is relatively simple than the ring and radiation type, besides, and fault identification and protection is relatively simple. The disadvantage is the low reliability of power supply.

C. AC-DC microgrid architecture

There is a possibility to join AC- and DC-coupled microgrid systems (Figure 5.12). This type is called mixed-coupled microgrid system [8, 9]. In this kind of topology, some renewable are linked with battery storage at DC bus, while others are linked with DC at AC bus. The distributed generation and AC/DC loads can access to the DC bus through the power electronic converter, and then AC microgrid can be connected with the large systems through the transformers. A new AC-DC hybrid microgrid structure is the ideal choice. Using the AC-DC hybrid microgrid, the power supply will meet the reliability, economy, and flexibility requirements.

There are a number of internal inductive loads in AC-DC microgrid, such as transformers, and asynchronous motors. Also there are a large number of microgrid

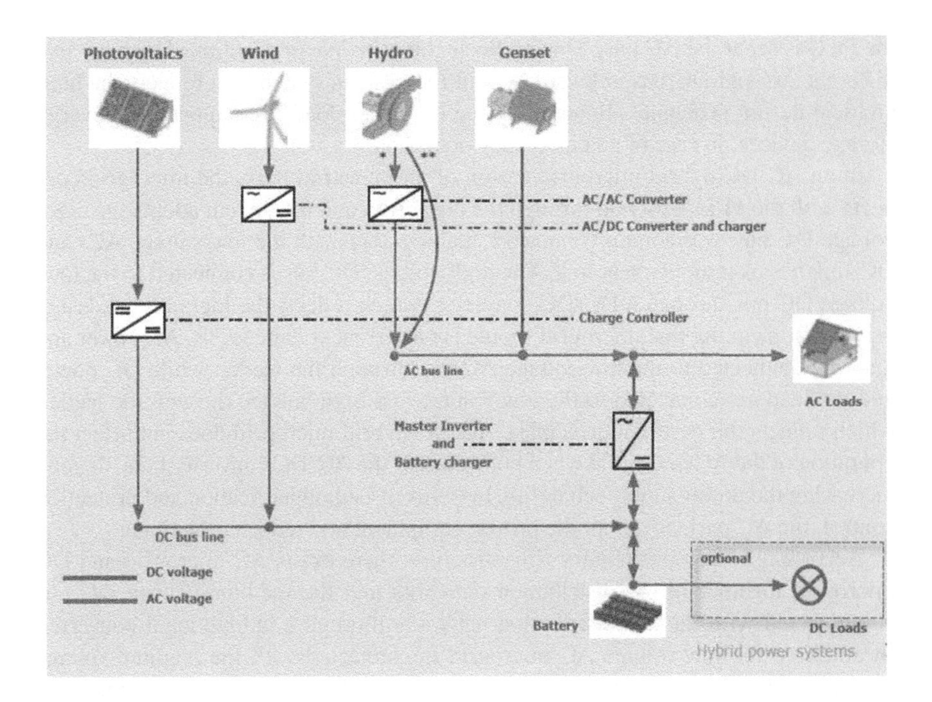

FIGURE 5.12 Mixed microgrids [3].

power electronic devices (such as rectifiers, inverters, chopper circuits), and these nonlinear devices need to consume large amounts of reactive power, while these devices will generate a lot of harmonics, which need to absorb some reactive power as well. Therefore, in the internal of the microgrid, it must be configured reasonably with static or dynamic reactive power compensation devices [3, 20].

The research of AC and DC microgrid topology has been more mature, but still less for AC-DC hybrid microgrid topology research literature. In [15], an AC-DC hybrid microgrid structure is presented. In this study, the microgrid consists of two parts, AC part is connected with DC part through a bidirectional converter, and then the AC bus is connected with the distribution network through a transformer. AC and DC load access to microgrid through appropriate power electronics devices. The system can run on a variety of operating status by controlling the bidirectional converter and isolating switch at the PCC, such as: (a) grid-connected operating status; (b) AC parts operate on grid-connected status, while the DC parts operate on islanding status, and the bidirectional converter stays off; (c) AC and DC parts operate on the islanding status, PCC at isolating switch off; (d) AC and DC parts operate independent, and isolating switch off at PCC, while the bidirectional converter turns off at the same time.

According to the design principles of the microgrid, the literature [3, 4] has analyzed the structure-1. In this structure, the system AC and DC microgrid are in two separate areas—the AC power supply and AC load combine together—while the DC power supply and DC load combine together, and the two microgrids are connected via converters. The hybrid microgrid is connected with the distribution network, and the PCC is set at the AC bus, which reflects the principle of partition. There are two different AC and DC bus voltage levels in the system, which also reflects the hierarchical design principles. In addition, the important load is equipped with energy storage devices, in case of an emergency operation.

In an AC-DC hybrid microgrid design of an industrial park, the microgrid connects with the 11-kV AC grid through the converter, and the system adopts the high-voltage DC bus as the main framework and together with the low-voltage AC- and DC-link bus as a supplement to it. The high-voltage DC bus is connected to the low-voltage DC bus through a DC/DC converter, which reflects the hierarchical design principles, meaning that microgrid should set different voltage levels. AC power and load are connected to the low-voltage AC bus through the feeder, while DC power and the load are connected to the low voltage and high voltage through the feeder, which reflects the partition principles. AC/DC hybrid microgrid does not affect the operation of the AC microgrid when both sides of the AC/DC converter fails, thereby increasing the power supply reliability. In terms of fault identification and protection control, the AC part can adopt the proven equipment and reduce investment.

According to complementary ring structure, there exists AC microgrid and DC microgrid together in the distribution networks and the medium-voltage DC bus is connected with the AC distribution networks through a bidirectional converter. In addition, the low-voltage AC microgrid is connected with the medium-voltage AC bus through the step-up/down transformer. The AC/DC hybrid complementary structure can greatly improve the reliability of power supply. Compared with the pure DC microgrid and AC microgrid, using the AC/DC hybrid approach can save a lot of AC/DC converters and reduce the investment as well.

In addition, compared with the ring type DC microgrid, hybrid complementary ring structure can reduce AC/DC inverter capacity. With much higher reliability of power supply compared to the ring type distribution network, fault handling capability can be greatly enhanced. If the main AC/DC converter fails, DC microgrid can be easy to form an island. Due to the lack of internal microgrid power, some of the loads have to be removed, resulting in a certain impact on the user.

Hybrid AC/DC microgrids are an interesting alternative, as they would enable the integration of DC-based systems through a DC network while maintaining the AC infrastructure [4]. A schematic of hybrid AC/DC is depicted in Figure 5.13. This way, the advantages of AC as well as DC grids can be combined, facilitating the shift to a more distributed electric grid composed by DG and ESSs. In this case, hybrid microgrids are composed of AC and DC subgrids, which are linked by one or more interface converter. The integration of these converters not only enables the power exchange between the AC and DC subgrids (Figure 6.14) but they also increase the degrees of freedom regarding the management of the grid, as they can provide other ancillary services [5].

In general, the advantage of hybrid AC/DC microgrid can be summarized as follows [3, 12]:

1. The elimination of unnecessary DC-AC or AC-DC-AC power conversion circuits installed in the power supplies, meaning the significant reduction of power conversion losses.

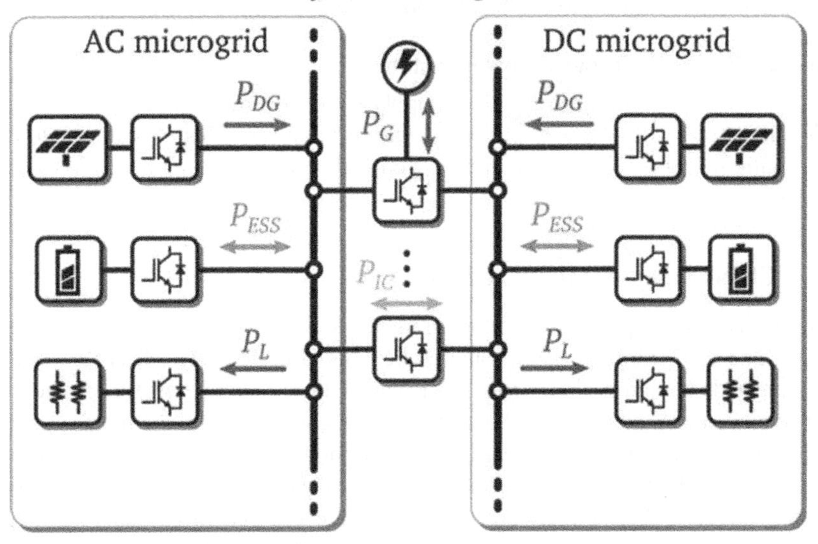

FIGURE 5.13 Structure of the AC/DC hybrid microgrid [12].

2. The elimination of embedded PFCs for powering DC loads in the traditional AC grid, meaning the significant cost and loss reduction of power electronics equipment at end users.
3. The improved power quality in AC grid since the DC loads will not directly generate harmonics pollutions and the interlinking converters with full controllability can significantly enhance the power quality.
4. The improved unsymmetrical current control capability since negative and zero sequence current problems caused by the unbalanced loads in AC grid can be handled by the DC grid.

5.6.1 Series/Parallel Microgrid Power System

Microgrid systems are also categorized on the basis of the type of supply provided to the demands from renewables and diesel generators [3, 12, 20]. Series and parallel hybrid microgrid are the two configurations and their brief details are given below.

5.6.1.1 Series Microgrid Power System

In this configuration, all the generated DC power is supplied to the battery. Therefore, the energy produced by the PV, wind, and diesel generator is utilized for charging battery storage. Hence, charge controller is equipped with each component, other than diesel generator. Diesel generator is equipped with a rectifier. Afterward, inverter converts the DC power into standard AC power and feed to the AC demands. Overcharging of the battery storage by PV/wind is prevented by charge controllers. Similarly, deep discharging of the battery bank is also prevented by the charge controller. This topology is also called centralized DC bus configuration, because all the sources are linked to DC bus and load is fed through a single point [3].

5.6.1.2 Parallel Microgrid Power System

In this type of configuration, a part of supply demand is directly fed by the renewable sources and diesel generator. This configuration is further classified into two sub-configurations: DC coupled and AC coupled, which are already discussed earlier [3].

5.7 CONTROL OF MICROGRIDS

Microgrids feature special control requirements and strategies to perform local balancing and to maximize their economic benefits [131, 137–141]. There is general agreement that microgrid controls must deliver the following functional requirements: present the microgrid to the utility grid as single self-controlled entity so that it can provide frequency control like a synchronous generator [137]; avoid power flow exceeding line ratings; regulate voltage and frequency within acceptable bounds during islanding; dispatch resources to maintain energy balance; island smoothly; and safely reconnect and resynchronize with the main grid [142].

Although microgrids are gaining a lot of interest, especially during the past decades, most of the challenges still reside in their control and management, especially when they operate islanded from the main grid. Even if the control techniques employed at AC, DC, and hybrid AC/DC microgrids can be considerably different, their concept of operation is usually very similar. Inspired by the classical AC grid, the management of microgrids is most of the times carried out by employing a hierarchical structure. Each control layer is responsible for certain functions, such as the voltage/frequency control or the management of the islanding/reconnection process.

The main difference for microgrid control is that, at microgrids, conventional synchronous generators are replaced by converter-interfaced generation and storage systems, and therefore, the inertia of the grid is drastically reduced. This is one of the main challenging tasks in the management of microgrids, as lower inertia in the grid means that their voltage and/or frequency is significantly deteriorated under power variations. Consequently, microgrids become more susceptible to failures—especially when operating in the islanded mode—and hence more advanced control strategies need to be adopted to replace the lack of inertial behavior and ensure a stable operation [3, 12].

One of the most common solutions is to configure the converters associated to DG, ESSs, and even loads to contribute to the voltage and frequency regulation of microgrids. This is usually carried out by primary controllers, which are most of the times integrated locally in each device. Here, the lack of inertial response is partially replaced by the primary control of converter-interfaced devices connected to the microgrid. In the literature, there is a wide collection of this type of primary techniques that can be integrated at AC, DC, and hybrid microgrids. A review of some of the most relevant strategies can be seen in numerous literature reports [1–26]. The subject of inverters is also briefly discussed in the next section.

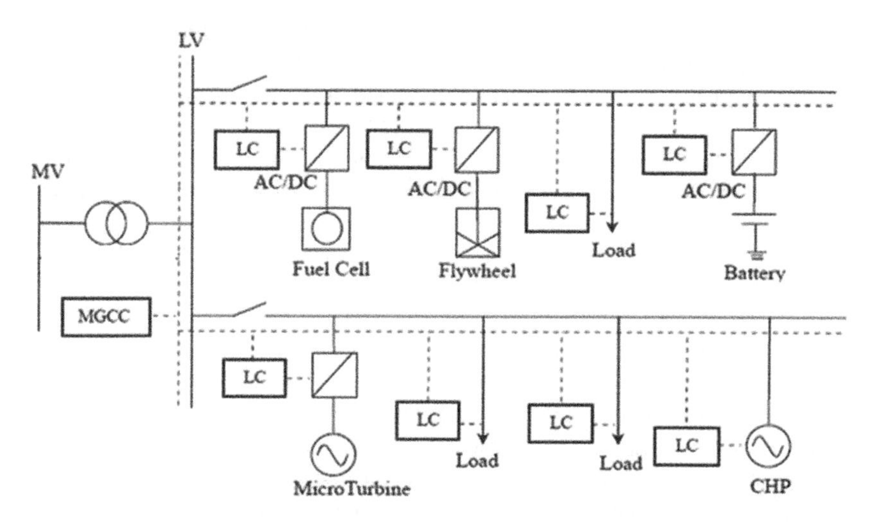

FIGURE 5.14 Basic microgrid architecture with an MGCC [3].

The factor that has made the greatest difference between modern concepts and the older versions of MG is the development of purpose-designed controllers for microgrids. Several manufacturers are offering microgrid controllers in a range of sizes and configurations. The second major development in microgrids is communication between elements of the microgrid, so that a data or IT network exists in parallel with the electricity network.

A typical microgrid control scheme is illustrated in Figure 5.14. In a typical MG control infrastructure, the microgrid central controller (MGCC) is installed at the LV side of the MV/LV substation. The MGCC interfaces the MG and the distribution network and has several vital functions. At a second hierarchical level, each MS and storage device is locally controlled by a micro-source controller (MC) and each electrical load is locally controlled by a load controller (LC). In order to achieve a good performance of the control scheme, an efficient communications infrastructure is necessary. Inverter control is used to ensure the accuracy of the power sharing while controlling or regulating the microgrid frequency and voltage. The MGCC coordinates a hierarchical control scheme. In order to perform load-frequency control, the MGCC receives and stores data from the load controllers (load levels) and the micro-source controllers (active power levels) as well as taking frequency measurements. The main function of the MGCC is to perform load or frequency control [3, 4, 143] by:

1. **Balancing load and generation by increasing or decreasing generation in response to load variations:** The MGCC can be set up to dispatch generation according to the most economical combination of MS at the time of dispatch. Using the load profile created based on historical power usage and adjustments made on real-time monitoring, the MCGG can optimize the use of the available generation to provide power in the most economical means

possible. Using frequency deviation as input and cost-functions associated with each micro-source and economic set points for the micro-sources, the frequency control function implemented at the MGCC specifies active power set points that are sent back to the MCs in order to adjust the production levels and consequently correct the frequency offset [3, 4, 143].

2. **Load shedding or DSM where the load exceeds generation capacity:** Using the same frequency deviation as input and load criticality functions associated with each load, the MGCC disconnects or reduces loads to achieve the required balance.

Some MGCC can do generation and load forecasting, based on historical data and prevailing conditions or real-time data. In addition, the MGCC will provide the voltage and frequency reference to the master controller in the case of islanding, as well as controlling connection to and disconnection from the grid. A critical security trait for MG operation is to ensure that they can run into islanded operation following an unexpected event without collapse due to imbalance of loads and generations. Depending on the MG operating conditions, such as local load, local generation profile and MS availability for active power or frequency regulation, high amounts of power may be required to be injected or absorbed in the MG in the first moments subsequent to islanding [3, 4, 143].

5.7.1 Frequency and Voltage Reference and Control

In a network consisting of many generation units, a reference frequency is necessary to ensure that all units remain in synchronization. For grid-connected units, the grid frequency is used as the reference but, when the grid is unavailable, a secondary frequency source is necessary to ensure that the microgrid remains within the frequency and voltage limits and to resynchronize the MG to the main grid before reconnecting to the network when power is restored. Many MS (microgrid source of power) make use of inverters to convert DC to AC, which can be used to provide frequency and voltage control. In the simplest case, a single inverter operating in voltage source mode is chosen as the reference and is supplied with voltage and frequency information from the controller. Other inverter units are operated in PQ (active/reactive power) mode and are slaved to the frequency and voltage generated by the master inverter [3, 4, 143].

In the case of off-grid systems, there is usually only a single source of generation, with sometimes a prime mover added for backup. In hybrid MGs, where synchronous generators (SG) are run full time, the SG can be used as the voltage and frequency source, and the other inverter output MS are slaved to the SG. In cases where the SG is not run full time, a more complex control system is required which allows one of the inverters to act as the voltage and frequency source. In these cases, either the inverter of the prime source becomes the frequency reference or an independent frequency reference is used.

A problem exists with grid-connected microgrids in that, in the event of disconnection from the grid, the MSs may not be able to supply the instantaneous load, causing frequency and voltage deviations. This problem is solved by incorporating energy storage devices, either collocated with the MS or centrally located within the MG. If a synchronous generator is included in the MG, this would need to be running at the time of changeover to island operation. If the SG needs to be started up, the same situation will exist until it reaches full speed and full power. Most MGs rely on some form of storage to facilitate frequency and voltage control during islanding, and because of the rapid response characteristics, the storage unit is generally used as the master VSI unit. An independent reference frequency and voltage source is nonetheless still required to control the master unit [5].

5.7.2 Architecture of Microgrid Control

In regard to the architecture of microgrid control, there are two different approaches that can be identified: centralized and decentralized. A fully centralized control relies on a large amount of information transmittance between the involving units and then the decision is made at a single point. Hence, it will present a big problem in implementation since interconnected power systems usually cover extended geographic locations and involve an enormous number of units. On the other hand, in a fully decentralized control, each unit is controlled by its local controller without knowing the situation of others. A compromise between those two extreme control schemes can be achieved by means of a hierarchical control scheme consisting of three control levels: primary, secondary, and tertiary.

Microgrids can essentially be controlled in the same way as the main grid, i.e., by using a three-level hierarchical control [137, 144] (see Figure 5.15). Control of frequency and voltage—the so-called primary and secondary control—can be achieved either under the guidance of a microgrid central controller (MGCC) that sends explicit commands to the distributed energy resources [145] or in a decentralized manner, like CERTS, in which each resource responds to local conditions. In addition, microgrids generally include a tertiary control layer to enable the economic and optimization operations for the microgrid, mainly focused on managing battery

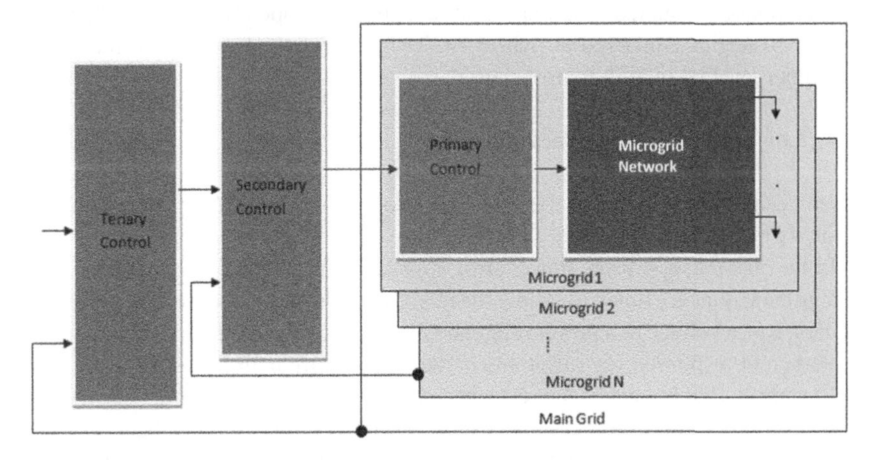

FIGURE 5.15 Hierarchical control [4].

storage, distributed generation scheduling and dispatch, and managing import and export of electricity between the microgrid and the utility grid [131, 139, 144, 146]. Hierarchical control architectures (see Figure 5.15) that manage power within a microgrid and mediate exchanges with the main grid have been deployed using a "multi-agent system" approach in two European microgrids, one in the Greek island of Kythnos and another in the German "Am Steinweg" project [79]. Increasingly, microgrid research and development is focusing on adding "intelligence" to optimize operational controls and market participation [42, 72, 76, 79, 137, 138, 147–152].

A. **Primary control**

The primary control (Figure 5.15) is designed to satisfy the following requirements:

- To stabilize the voltage and frequency
- To offer plug-and-play capability for DERs and properly share the active and reactive power among them, preferably, without any communication links
- To mitigate circulating currents that can cause overcurrent phenomenon in the power electronic devices

Primary control techniques employed at microgrids are usually composed of two main stages: the lower level stage usually contains the faster regulation loops, which are responsible for the current and/or voltage regulation of converters for DER. These control loops are commonly referred to as zero-level control [4]. Similarly, the upper level of the primary control, which is slower than the previous one, determines the reference value where the converter should be controlled, e.g., the active or reactive power. When there is a variation in the grid, the lower level control stage primarily defines the transient response of the converter. In addition, the upper level stage determines the steady-state operation point of the converter. However, depending on how these stages are designed, their effect in the transient as well as steady-state response is coupled. It is important to analyze the operation modes of primary control strategies employed at hybrid AC/DC microgrids, focusing on their lower and upper level (primary) control stages.

A-1 **Lower level control operation**

When designing the lower level control stages of primary regulators, one can usually follow two main approaches.

In the classical approach, this control stage is composed by one or more cascaded PI regulators that are tuned to follow the reference value provided by upper level controllers, e.g., a voltage or a power reference. In this case, the regulators do not provide any inherent response over variations in the grid and are mainly designed to control the system so that it reaches the reference value as fast as possible.

On the other hand, in the last decades, a different approach has been proposed where the lower level regulators are designed to participate in the transient regulation of the voltage and frequency of the network. These regulators are designed to emulate the behavior of classical synchronous generators with power converters associated with DG, ESSs, and loads connected to microgrids. In the literature, these techniques have been widely employed for different applications and are also known as virtual synchronous machines (VSMs) or synchronverters [8–15]. Following the main AC grid configuration—where synchronous generators are directly connected—most of these techniques are usually employed for devices connected to AC microgrids. However, recently, similar approaches have been developed to reproduce an analogous behavior at DC systems. For instance, the study carried out in reference [16] shows that a similar response can be emulated at DC systems by employing virtual impedances in the control strategy, for example, as virtual capacitors. In this case, instead of controlling the frequency as in VSMs, the variable controlled is the bus voltage [12].

A-2 **Upper level (primary) control operation modes**

Due to the dispersed nature of microgrids, primary controllers are usually autonomous and operate based on local measurements of the device they are controlling. Whether the device contributes to the frequency/voltage regulation or not, their upper level regulator is responsible for defining the steady-state point of operation of the converter. Similar to lower level regulators, upper level ones are mainly classified into two different types.

On the one hand, there are certain devices that do not contribute in the frequency/voltage regulation of the microgrid and operate based on the reference provided by another control level (e.g., the secondary) or based on a reference internally calculated to, for example, extract as much energy as possible from the energy source they are connected to—a maximum power point tracking (MPPT) technique. These units are also named grid-following devices, as they do not regulate the bus but rather they "follow" their frequency and/or voltage [12].

On the other hand, systems that contribute to the regulation of the bus are known as grid-forming or grid-supporting systems. These devices share the power variations occurring in the microgrid to decrease the variations of the bus frequency/voltage [12]. One can design generation, storage systems, and loads connected via a power converter to operate differently, for example, depending on the bus voltage or frequency level. These systems usually include both types of upper level controllers and change their mode of operation based on some external condition. This approach is also known as mode-adaptive control, and one of the most interesting methods is to adapt the behavior of each controller based on the level of the bus frequency or voltage. Different types of mode-adaptive control strategies for microgrids, where the devices connected to the microgrid adapt their characteristics based on the frequency or voltage of the grid, have also been proposed [12].

B. **Secondary control**

Secondary control has typically seconds to minutes sampling time (i.e., slower than the previous one), which justifies the decoupled dynamics of the primary and the secondary control loops and facilitates their individual designs. The set point of the primary control is given by the secondary control in which, as a centralized controller, it restores the microgrid voltage and frequency and compensates for the deviations caused by variations of loads or renewable sources. The secondary control can also be designed to satisfy the power quality requirements, e.g., voltage balancing at critical buses [4].

C. **Tertiary control**

Tertiary control is the last (and the slowest) control level which considers economical concerns in the optimal operation of the microgrid (sampling time is from minutes to hours) and manages the power flow between microgrid and main grid. This level often involves the prediction of weather, grid tariff, and loads in the next hours or day to design a generator dispatch plan that achieves economic savings. In case of emergency like blackouts, tertiary control could be utilized to manage a group of interconnected microgrids to form what is called "microgrid clustering" that could act as a virtual power plant and keep supplying at least the critical loads. During this situation, the central controller should select one of the microgrid to be the slack (i.e., master) and the rest as PV and load buses, according to a predefined algorithm and the existing conditions of the system (i.e., demand and generation), in this case, the control should be real-time or at least high sampling rate [4].

D. **IEEE 2030.7**

A less utility influenced controller framework has been designed in the latest microgrid controller standard from the Institute of Electrical and Electronics Engineers (IEEE), the IEEE 2030.7. The concept relies on four blocks: (a) device level control (e.g., voltage and frequency control), (b) local area control (e.g., data communication), (c) supervisory (software) controller (e.g., forward looking dispatch optimization of generation and load resources), and (d) grid layer (e.g., communication with utility).

E. **Elementary control**

A wide variety of complex control algorithms exist, making it difficult for small microgrids and residential distributed energy resource (DER) users to implement energy management and control systems. Especially, communication upgrades and data information systems can make it expensive. Thus, some projects try to simplify the control via off-the-shelf products and make it usable for the mainstream (e.g., using a Raspberry Pi) [4].

5.7.3 Control Scheme of Hybrid AC/DC under Islanding Conditions

The study by Ding et al. [143] reviews the control schemes of hybrid AC/DC microgrid under islanding operational condition. The study starts with a discussion about the operational mode classification of microgrid according to the power flow in hybrid AC/DC microgrid. Based on the classification, operational features and detailed control schemes of AC sub-microgrid, DC sub-microgrid, and interlinking converter in hybrid AC/DC microgrid are reviewed and discussed.

A. **Classification of different operation modes in a hybrid AC/DC microgrid**

As indicated earlier, generally, a hybrid AC/DC microgrid consists of three main parts: (1) AC sub-microgrid, (2) DC sub-microgrid, and (3) power electronics interfaces between AC and DC buses. In a general architecture of hybrid AC/DC, a microgrid is connected to a utility AC grid, where the DC sub-microgrid is connected to AC grid through an interlinking converter. Since the AC microgrid can be directly connected to utility grid through a simple circuit breaker, the AC sub-microgrid is generally dominant in the hybrid AC/DC microgrid to provide a stable voltage. The AC power generators, such as wind turbine and small diesel generator, and the AC loads, such as AC motors and traditional lighting, can connect to the AC sub-microgrid. On the other hand, DC power sources such as photovoltaic panels, fuel cells, and batteries can be connected to DC sub-microgrid through simple DC/DC converters. Besides, the AC loads have variable frequency operation requirements, such as adjustable speed motors could be connected to DC sub-microgrid. Either the energy storage can be installed in AC sub-microgrid or DC sub-microgrid or it can be inserted into the interlinking converter, whose installation location should be optimized by considering load types, power flow, operational reliability, and cost [143, 153].

Considering the configuration of hybrid AC/DC microgrid, three different operational modes and their power flow patterns are listed in Table 5.7. As shown, when AC sub-microgrid operates independently, the control targets are to properly control the AC current or AC voltage to manage the real and reactive power flow during grid-tied or islanding operational conditions. While DC sub-microgrid operates independently, the DC voltage or DC current should be carefully controlled to dispatch the power flow between DC sources and DC loads. At last, the joint operation of AC and DC sub-microgrids can properly manage the power flow between AC and DC networks to optimize the operational efficiency and stability.

TABLE 5.7
Operations of Hybrid AC/DC Microgrid [143]

Operation Modes	Microgrid Sources and Power Flow	Control System
Independent AC sub-microgrids	AC-sub-microgrids AC sources to AC loads and grid	Control of AC voltage/AC current for AC load sharing
Independent DC sub-microgrids	DC-sub-microgrids DC sources to DC loads and grid	Control of DC voltage/DC current for DC load sharing
Joint AC-DC sub-microgrids	Both AC and DC sub-microgrids AC and DC sources to AC and DC microgrid	Control of AC and DC voltage for AC and DC load sharing

B. Control schemes of AC sub-microgrid

The distributed energy sources can connect to the AC sub-microgrid through the interfacing converters. Therefore, the DGs can be treated as AC voltage sources or current sources operated in parallel. The load power sharing among the parallel converters under islanding operation condition has been an active research topic [154–156]. In general, to fully utilize the capacity of DGs, the output power of DGs should be proportional to their rated power. To achieve this control goal, many power sharing control schemes have been proposed.

The centralized control scheme proposed by Martins et al. [157] used a central controller to produce and deliver the current reference of each DG so that all DGs can generate proper real and reactive powers simultaneously to maintain the grid stable. Master-slave control method proposed by Chen and Chu [158] contained a combination of one master converter with voltage control capability and several slave converters with only current control capability like the traditional grid-tied converters. In specific, the voltage-controlled converter acts as the main converter to establish the voltage reference for other current-controlled converters, while the slave converters track the voltage reference to inject the dispatched real and reactive power. The master converter should have a relatively large capacity to fast establish the grid voltage during the grid transient interval when the current-controlled converters stop injecting power to microgrid due to the loss of stable voltage. An improved method called the circular-chain-control (3C) strategy was presented by Wu et al. [159], where the parallel converters were cascade connection, and each converter generated the current reference for its adjacent converter. Doing so, the reliability of whole control system will be improved. The average current control scheme was proposed by Sun et al. [160], where the current references were generated and delivered to each converter by means of communication. Although the above-summarized control schemes can achieve the steady and dynamic performances of AC microgrid under islanding operational condition, the stability of the control system highly relies on the effective communications, which would reduce the operational reliability and increase the maintenance cost.

Considering the distributed source configuration in microgrid, the decentralized control method without mutual communication is more reliable for the islanding operation. Therefore, the droop control method [161] known as the independent, autonomous, and wireless control scheme is studied due to the elimination of intercommunication links among distributed converters. The main idea of droop control is to imitate the behavior of a synchronous generator, which reduces the frequency when the real power increases [162, 163]. This control scheme can regulate the frequency and the magnitude of voltage reference in order to achieve proper power sharing.

When assuming the droop control scheme, the line impedance between distributed converter and point of common coupling (PCC) will influence the accuracy of power sharing. He and Li [164] implied that a virtual output impedance loop added in the droop control system was assumed to solve the problem. In principle, the added virtual impedance will force the equivalent impedance to be highly inductive so that the traditional droop control theory can still work effectively. However, when directly implementing the droop control scheme in grid-tied operation condition, it will produce the frequency deviation resulting in the loss of synchronization. A hierarchical control scheme was proposed by Guerrero et al. [161], where, as mentioned before, three control levels were defined: primary control, secondary control, and tertiary control. In specific, the primary control is indeed a droop control method, including an output virtual impedance loop, and the secondary control allows the restoration of deviations produced by the primary control and the tertiary control manages the power flow between the microgrid and the utility grid. The hierarchical control architecture needs communication between distributed sources and central controller, and it can be implemented in both grid-tied and islanding operational conditions. Because the droop control method is assumed in the bottom-level controller, the controller reliability can still be guaranteed.

C. **Control schemes of DC sub-microgrid**

With the rapid increase of DC power generators and DC loads, the DC microgrid has been investigated for many years. The DC sub-microgrid can be treated as a pure DC microgrid without power flow between AC and DC buses in the hybrid AC/DC microgrid to simplify the analysis and control schemes of DC sub-microgrid. Usually, DC microgrid should equip with energy storage components, otherwise, the variable DC sources would lead to unstable power supply. However, even equipped with energy storage, the DC sources still need to share their output power to stabilize the DC grid voltage. Therefore, the proper power sharing control schemes implemented in DC microgrid have so far been deeply investigated, which will be reviewed in this section to facilitate the derivation of control schemes for hybrid AC/DC microgrid under islanding operational condition.

Power sharing control scheme for DC microgrid can be generally classified as centralized control scheme and distributed control scheme. The centralized control scheme captures the system information by a central controller and generates the corresponding operational commands. Nevertheless, the method depends on the communication between distributed sources and central controller, which definitely will induce additional investment on the communication infrastructure and reduce the reliability of control system [165, 166]. As a consequence, the distributed methods are proposed to eliminate the disadvantage of centralized control scheme. A typical example of distributed control scheme is DC bus signaling (DBS) method [167–169], in which the bus voltage is regarded as global indicator to determine the operational modes of converters according to the defined voltage thresholds. Droop control has been proved to be an effective method to merge multiple sources and storages for proportionally sharing load according to their ratings, using only local information. Because DC microgrid does not have reactive power and frequency, the droop control method used in DC sub-microgrid only needs to control the real power sharing by carefully considering converter output voltage and line resistance. In order to achieve proper power sharing, the droop control scheme regulates the output voltage magnitude of distributed converters.

In order to make the DC sub-microgrid more flexible and expandable, the hierarchical control strategy presented by Guerrero et al. [161] can also be assumed. In specific, hierarchical control principle can be divided into three levels. The primary level control is responsible for DC bus voltage stability. When the local loads are connected to the DC bus, DC droop controller will obtain equal or proportional DC load current sharing. The secondary control is aiming to eliminate the DC bus voltage deviation introduced by the droop controller. After guaranteeing the performance of DC bus operation by means of primary and secondary level controllers, the tertiary level control is employed to realize the connection to external DC systems. Being similar, the hierarchical control scheme for DC microgrid still needs communication to achieve the superior operational features of accurate power sharing and enhanced reliability.

D. **Control schemes of hybrid AC/DC microgrid**

Different control methods of hybrid AC/DC microgrids have been presented in the literatures [144, 161]. Droop control scheme was employed by Kurohane et al. [153], where the controllable loads with different capacities were taken into account. A coordinate control method for a hybrid microgrid composed of various kinds of renewable energy sources was proposed by Liu et al. [170], where detailed models of PV modules, batteries, and wind turbines were derived and the energy management strategy for the whole system was developed. A configuration with both DC- and AC-links and the corresponding control method were presented by Jiang and Yu [171], where the DC-link could integrate the local converters with DC couplings and connect to the common AC-link through DC-AC interfacing converters. A power quality enhancement method was proposed by Shahnia et al. [172], where the unbalanced and nonlinear loads were taken into account and the control strategies were developed for a multi-bus microgrid. The above-summarized methods can effectively enhance the performance of a hybrid microgrid. But they mainly focus on the AC side performance and the corresponding control schemes.

For the islanding operation of hybrid AC/DC microgrid, the control schemes mentioned in earlier sections allow the distributed sources to share the load demands in their respective sub-microgrid. But sharing the load demands in both AC and DC sub-microgrids simultaneously cannot be simply realized by means of droop-controlled distributed sources. The power sharing in both sub-microgrids would heavily depend on the control strategy of interlinking converter. Jin et al. [173], Loh and Blaabjerg [174], and Loh et al. [175, 176] proposed an autonomous control scheme for the interlinking converter whose responsibility was to link AC and DC sub-microgrids together to form the hybrid AC/DC microgrid, where the distributed AC and DC sources were classified into two consolidated sources, tied to the same interlinking converter. Generally, the interlinking converter consists of a standard DC-DC boost converter and a standard DC-AC inverter.

At the common DC-link, either capacitor or energy storage can be added for buffering, filtering, or storage. According to the configuration of interlinking converter, the control scheme of interlinking converter can be classified as general control scheme, control scheme with DC-link capacitor, and control scheme with energy storage.

D-1 General control scheme

In AC sub-microgrid, the real power sharing is influenced by the controlled frequency as already addressed; however, the real power sharing in DC sub-microgrid is affected by the controlled voltage as mentioned earlier. It is obvious that the droop variables assumed in two sub-microgrids are totally different. Therefore, both droop variables should be properly merged before using them to control the real power across the interfacing converter. The details of this merging control scheme are outlined by Ding et al. [143].

D-2 Control scheme with DC-link capacitor

In the case of a DC-link capacitor added to the interlinking converter the same as the general AC-DC conversion circuits, the above general control scheme should be further specified. A proportional-integral (PI4) controller is introduced to the general control scheme to keep the DC-link voltage constant by generating a real current reference for compensating the losses in the power conversion circuit. This current reference can be added to the DC current reference of DC side drawn by the DC-DC boost converter from the DC sub-microgrid. Meanwhile, the current reference for AC terminal of the interlinking converter remains unchanged [143].

D-3 Control scheme with energy storage

For case where the energy storage instead of a capacitor is added to the DC-link of interlinking converter, the control scheme of interlinking converter should be revised to fully utilize the energy storage capacity. It means that the energy storages should be charged only when the sources can generate additional energy after meeting the load demands. On the other hand, it should discharge only when the sources cannot fully satisfy the load demands. The two operation modes require sensing the additional generation capacities of sources. More details on control schemes with hybrid energy storage are given in Chapter 4 [143].

ESSs are one of the most important agents in the microgrid regulation because they serve as an energy "buffer" to compensate generation and demand deviations during normal operation. One must design and size these systems in order to cope with the most severe conditions of the system; otherwise, a poor regulation would cause the malfunction or disconnection of devices. Under a balanced operation, the ESS does not exchange any power with the microgrid. This hysteresis range must be carefully determined in order to avoid an excessive cycling and therefore ageing of the ESS, but also to prevent the system from entering into an unstable point of operation.

As mentioned above, when the generated power is higher than the demanded one, the voltage or frequency of the microgrid increases over the hysteresis upper value, and therefore, the ESS absorbs power according to the charging droop slope. This slope depends not only on the charging capabilities of the ESS but also on the sizing of its power converter. When the voltage/frequency of the microgrid increases above the preestablished value or the state of charge (SoC) of the ESS is too high, the device turns to a mode where the power absorbed from the microgrid is kept constant although the v/f deviation keeps increasing.

A similar behavior is reproduced when there is an excessive demand in the microgrid and the voltage or frequency decrease below the hysteresis lower value. At this point, the ESS supplies power to the microgrid according to its discharging droop slope. We must note that we could design the droop slopes differently for the ESS charging or discharging process, which means they do not have to be symmetric. The ESS remains in this mode until the voltage or frequency reach their minimum value or the SoC of the ESS is too low. At some point, when the minimum SoC is reached, the ESS will not supply more power to the microgrid [3, 12, 143].

5.8 ENERGY MANAGEMENT OF MICROGRIDS FOR RENEWABLE ENERGY-BASED DISTRIBUTED GENERATION

A common feature of the renewable energy-based or micro-sources-based DG systems is the power electronics interfaces that are required to convert the energy sources output to the grid-ready voltages [177]. These power electronics also provide enhanced flexibility for the DG operation and energy management. In order to better organize these DG systems, the concept of microgrid has been developed, which has higher capacity and more control flexibility than a single DG system. For proper operation of a microgrid, energy management strategies are important to regulate the output powers of each DG as well as the voltage and frequency of the microgrid systems [79, 178, 179]. Additionally, to achieve improved power quality in a microgrid, as mentioned above, proper design and control of the DG interfacing converters to provide the ancillary services are important. This is particularly true when considering that most renewable energy-based DG systems are not always operating at their rated power, and the available converter rating can be used to facilitate grid frequency and voltage regulation, harmonics control, unbalance voltage compensation, etc. [164, 180].

Interface AC-DC converters located along the microgrid provide an extra degree of freedom in the management of microgrids, as we can control them to perform diverse operations. The most typical approach would be to employ these converters to compensate the v/f deviations in the AC and DC subgrid of the hybrid microgrid [3, 12, 164, 181, 182]. The converters would transfer power from one subgrid to the other in order to equalize the excess of generated or demanded power at both interface converters, which carry out the power transfer between different subgrids with a droop controller. Unlike classical approaches, this droop is based on the difference between the frequency deviation in the AC subgrid and the voltage deviation in the DC part. Whenever there is a mismatch between these deviations, AC-DC converters located in the microgrid will transfer power to balance them. Various techniques can be employed in the control of interface converters integrated at hybrid AC/DC microgrids such as the state of charge balancing of ESSs located in the AC and DC subgrid of the system.

The simulations reported in the literature [3, 4, 12, 143] demonstrate that a mode-adaptive control is very useful to take advantage of DG systems and loads. DG systems—which are usually based on RES—are only taken out of their MPP when the voltage or frequency of the bus reaches a very high boundary, maximizing the energy produced. Similarly, loads only contribute to the regulation of the system when the voltage or frequency reaches a very low level, which avoids any possible malfunction of these devices. The most optimal approach would be to design a bus-regulating strategy for the entire operation range composed by a combination of ESSs, DG systems, loads, and converters, taking advantage of their different dynamic characteristics.

5.8.1 Interfacing Converter Topologies

Since the output of renewable energy sources and micro-sources are mainly DC or nonutility-grade AC, power electronic converters are critical to interface the energy source to the grid in these DG systems. In the DG interfacing power electronic converters, the requirements related to the energy source characteristics, the energy storage system, the distribution system configuration, voltage levels, power quality, etc. need to be considered. The patented system configuration and control enable powering of DC loads (such as lighting and ventilation) directly from PV arrays with only a single conversion stage between the source and load, so that the PV energy is delivered with minimal losses. In contrast, a conventional AC system requires two conversions. Figure 5.16 shows the basic DCMG configuration alongside a conventional AC system. In the DC system, only the DC light-emitting diode (LED) driver

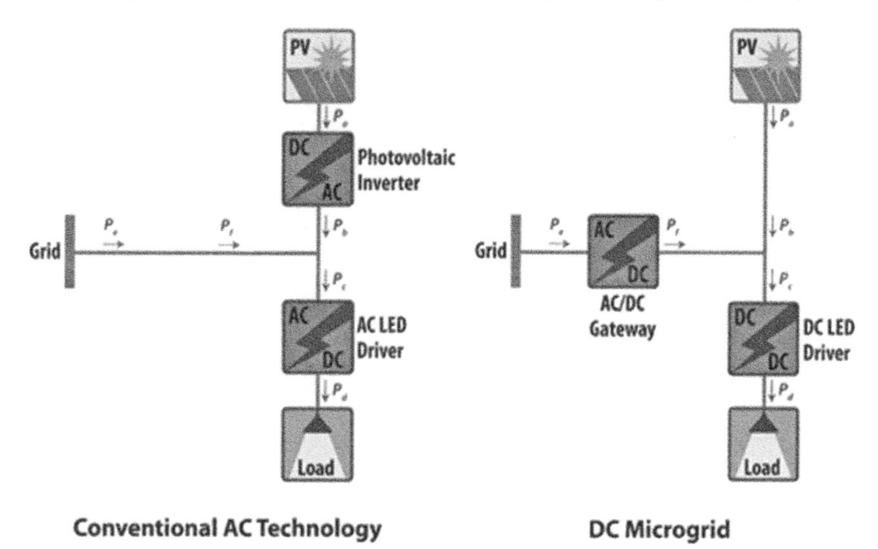

FIGURE 5.16 Configurations of conventional AC and Bosch DC microgrid systems [183].

appears in the PV-to-load path. In the basic configuration, the AC/DC gateway is a rectifier. The PV is sized such that it does not normally exceed the load, so an inverter is not needed.

In the Bosch DCMG, maximum power point tracking (MPPT) is performed by the voltage regulation of the AC/DC gateway converter, which is not directly in the PV-to-load path but rather supplies the balance of power required to operate the load. This patented configuration enables higher efficiency and greater reliability than other DC systems which use a dedicated MPPT converter through which all the PV power passes. More advanced configurations include ventilation fans, forklift chargers, flow batteries, ultracapacitors, and bidirectional AC/DC gateways. A bidirectional converter enables the PV system to be sized larger than the load power for more flexibility. Lighting and ventilation are the preferred load types because their profiles are predictable and align well with PV generation profiles.

An unbiased comparison of energy performance between conventional AC technology and an equivalent DCMG system requires that the definitions of energy inputs and outputs are consistent between the two system types. Figure 5.17 presents

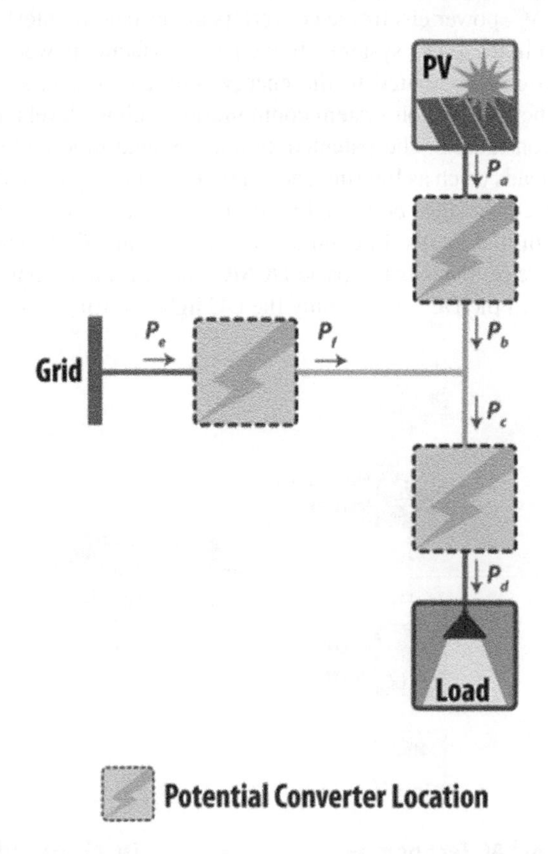

FIGURE 5.17 Conceptual representation of a building electric power distribution network with a local PV generation, DC load, and grid interconnection [183].

a generic conceptual representation of a building's electric power distribution network that contains a PV source, a DC load, and a grid interconnection. This conceptual model is useful for structured comparison of AC and DC distribution systems.

Three possible converter locations are identified in Figure 5.17: at the PV source, at the load, and at the grid interconnection. The wiring between these converters may be either AC or DC depending on the system configuration. The figure also identifies six power measurement points, P_a – P_f, one on each side of each of the three converters. Depending on how converters are selected and configured, the generic network of Figure 5.17 may represent either a conventional AC system or the Bosch DCMG. In the AC case, there is no central AC/DC converter and $P_e = P_f$. The PV system uses a conventional PV inverter and the load uses an AC driver. In the DC case, there is instead no converter at the PV source and therefore $P_a = P_b$. A DC driver matches voltage between the PV source and the lighting load, and a central, or gateway, AC/DC converter provides the balance of power. If the available generation exceeds the combined load demand and export capacity of the gateway inverter, then PV curtailment occurs. The model represents the curtailed energy conceptually as a system loss between points a and b, such that $P_b < P_a$.

The single stage with less power electronics converters and power conversion process is more popular in recent years. However, its drawbacks include compromised control flexibility and limited operation range. Moreover, this topology needs an overrated inverter and high DC output voltage from DGs [184]. Multilevel converters have been increasingly used in the single-stage topology with better DC voltage utilization and output power quality. In DGs with AC output power such as wind turbine or microturbine, the power electronic interfaces can be classified as double-stage or multistage converters. In the double-stage converters, the front-end PWM rectifier is usually used, and DC bus voltage is controlled by this rectifier. While in multistage topology, lower cost diode rectifier can be used, and DC-DC converter controls DC bus voltage. From the cost point of view, the multistage topology could be more cost-effective although it has low efficiency in comparison to two-stage topology [177]. With the development of power electronic technologies, multi-port interfacing power converters are becoming more attractive in microgrid. Multi-port power converters are used to connect various power sources (DGs or energy storage elements) to the grid and load through a single converter structure with lower cost and compact size [185–189].

These multi-port converters can be classified as electrically coupled and magnetically coupled structures. The electrically coupled types are usually implemented with non-isolated topologies such as buck, boost, and buck-boost switching cells [185, 186, 188]. In these topologies, in order to avoid large buck-boost conversion ratios and to effectively handle the ports, the operating voltage of different power ports need to be close to each other. This constraint is the main drawback of electrically coupled multi-port converters [189]. In magnetically coupled topologies, the energy sources are coupled through magnetic coupling (mainly high-frequency AC-link), which provides electrical isolation between the power ports [185, 187]. In these topologies, because of the use of multi-winding transformers, DGs with different operating voltage levels can be connected to power ports. These converters can handle quite different operation levels of DGs and energy storage units, but with relatively complex structures and control systems.

Finally, different interfacing converter topologies such as Z-source converters [190], multilevel converters (neutral point clamp (NPC) [191, 192], cascaded H-bridge [193], multi-modular converter [194, 195], etc.), soft-switching converters, matrix converters, etc. have been used in microgrids as DGs interfacing power electronics.

5.8.2 Energy Management Schemes

For sound operation of a microgrid in both grid-connected mode and stand-alone mode, proper energy management strategies are very important. These energy management schemes determine output powers and/or voltages of each DG source, which are then fed to the control system of interfacing converters as the control tracking references. In general, power management schemes in microgrids can be classified into communication-based and communication-less schemes.

5.8.2.1 Communication-Based Energy Management Schemes

In the communication-based energy management schemes, the system information (current, voltage, power, etc.) is communicated in the microgrid to determine the operation point of each DG. These schemes take the full advantage of intelligence in the integration of the computing and communications technologies in order to determine the output powers of each DG. Considering the distances of power sources, level of security, cost, and available technologies, the appropriate communication method is determined. The communication methods can be based on fiber optics, microwave, infrared, power line carrier (PLC), and/or wireless radio networks (GSM and CDMA) [196, 197]. In these schemes, combination of Internet Protocol (IP) with existing industry protocols and standards are used to communicate over the grid.

In general, the communication-based energy management schemes can be divided into centralized and decentralized energy management schemes [196, 197]. These schemes are explained in the following sections.

5.8.2.1.1 Centralized Energy Management Scheme

This strategy is also known as supervisory energy management. In general, master-slave control and central mode strategies belong to centralized energy management strategies. In this scheme, one centralized system or control center makes decisions and determines operation points of DGs. This control center receives all the measured signals of all energy units in microgrid and sets the operating points of DGs based on the objectives and constraints, which can be minimizing system operation and maintenance costs, environmental impact (carbon footprint), maximizing system efficiency, etc. [138, 198–200]. These objective functions along with the constraints can be conflicting, and sometimes solving these problems is difficult (if not impossible). After making decisions, the control signals are sent to the DG local control systems. The DG local control is mainly realized by controlling the DG's power electronics converters. In centralized energy management scheme, the DG units include both the energy sources and DG-grid interfacing power electronic converters.

An example of such supervisory energy management scheme is for microgrids based on PV-wind-battery-FC (fuel cell) input power sources [201]. In this work, the measured microgrid data are sent to the central system, and the objective function in the center controller is to provide the load power with high reliability.

The advantage of this centralized control scheme is that the central system receives all the data of system, and then based on the available information, the multi-objective energy management system can achieve global optimization. However, heavy computation burden is one of its main drawbacks. Another drawback of this system is the reliability concern, as a failure in the communication system may cause an overall shutdown in the system.

5.8.2.1.2 Decentralized Energy Management Scheme

In decentralized energy management scheme, all the local controllers are connected through a communication bus. This bus is used to exchange data among DGs' controllers. In this energy management system, each local control system knows the operation point of other converters. This information is used to determine the DGs' operating points according to different optimization objectives [138, 198, 202]. In these systems, intelligent algorithm has been often used to find optimal operation point [203].

This strategy has some advantages over centralized strategy. For example, it is easy to extend the control system to newly added energy sources with the plug-and-play feature. Moreover, computation requirement of each controller is reduced, and the redundancy and modularity of the system is improved [138, 198]. However, failure in the communication link can still cause problem in the system (although unlike the supervisory control where a communication failure may collapse the system). Also, potential complexity of its communication system is still a concern of this strategy.

Multi-agent system (MAS) can be the best example of decentralized energy management system [138]. In MAS, autonomous computational agents make decisions based on goals within an environment, and they communicate information about their goal achievement to other independent agents [204–206]. These systems are mainly used for large and complex microgrids, and artificial intelligence-based methods such as neural network or fuzzy systems are used to determine each DG's operation point while improving the overall performance of the microgrid [204–206].

In addition to centralized and decentralized energy management schemes, a combination of these schemes will produce a hybrid centralized and decentralized scheme. In this hybrid strategy, DGs are divided into groups. In each group, centralized scheme is used, which is responsible for local optimization within the group. Among different groups, decentralized energy management scheme is utilized for global optimization. Such a hybrid strategy could be suitable for large systems with interconnected microgrids, where centralize control of each microgrid and decentralized coordination among microgrids could improve reliability and resilience of the system. The recently proposed hierarchical energy management scheme can be considered as a hybrid centralized and decentralized energy management scheme [161, 207, 208].

5.8.2.2 Communication-less Energy Management Schemes

The main idea of communication-less energy management strategy is that every DG unit must be able to operate independently when communication is too difficult or costly. In these methods, each energy source has its own local controller without having communication links with the other controllers.

Droop control method is probably the most popular strategy in communication-less energy management [178, 209, 210]. This method emulates the behavior of synchronous generator, where the voltage and frequency vary with the DG output real and reactive power. The droop control is based on the assumption that the output impedance of DG is mainly inductive, and it uses droop characteristics of the voltage amplitude and frequency of the each DG to control its output [210]. In other words, the virtual communication link here is the microgrid voltage amplitude and frequency.

This strategy has obvious advantages: there is no communication requirement, so the control strategy is more reliable. Also, the control system is expandable with true plug-and-play function. However, there are some potential issues. First, in this method, nonlinear loads are not considered and the nonlinear current sharing among DG units cannot be addressed directly. In addition, in low-voltage microgrid systems, high line impedance may lead to real and reactive power coupling and stability concerns [211, 212]. Also, the mismatched DG output can cause power sharing error. Recent works have been done to improve droop control by adjusting the output voltage bandwidth, adding virtual impedance, or implementing the droop in virtual frames [211]. However, without a central control/optimization algorithm, optimal operation of the microgrid system is still difficult with the communication-less-based control strategy.

Other than the droop control method, if all DGs work at the MPPT mode, it is not necessary for communication between DGs. As a result, this method can also be considered as a communication-less energy management strategy. However, in such a system, energy storage devices are essential in stand-alone operation mode to provide the microgrid voltage and frequency regulation.

Considering the drawbacks of communication-based and communication-less energy management schemes, a combination of the droop control with communication-based control could have both improved reliability and control performance, and may be a good option for future microgrid systems. In such a combination strategy, with the help of communication-based energy management, the DG operation point in both grid-connected and stand-alone modes can be determined more accurately. Also with the droop control as the backbone, the communication requirement (such as speed and bandwidth) can be reduced and the failure of the communication links will not cause system collapse.

5.8.3 Interfacing Converter Control Strategies

The previously discussed energy management strategies determine operating point (such as power references) for each DG, and at the same time, guarantee voltage and frequency regulations, load demand matching, etc. In the DG interfacing converter control system, the reference real and reactive powers are controlled through DG output current and voltage regulations. Therefore, the DGs output power control strategies are generally categorized as current control mode (CCM) and voltage control mode (VCM). These strategies are explained in the following sections.

5.8.3.1 CCM-Based Power Flow Control Strategy

In the CCM scheme, both active and reactive powers, are tracked in the closed-loop manner. The real power control loop and reactive power control loop generate reference currents. Note that the real power reference could be from energy management scheme or from a MPPT scheme. With these reference currents, the DG output current can then be controlled in the synchronous frame or in the stationary frame. In this control strategy, the grid-voltage angle information from phase-locked loop (PLL) is used to synchronize the inverter output current with the grid voltage.

The *real power control loop* reference current can also be generated by DC-link voltage control. This condition occurs mainly in two-stage converters (in either DC-DC+DC-AC or AC-DC+DC-AC), where real power is controlled by first stage (DC-DC or AC-DC converters). In other words, output real power of the inverter is controlled to regulate DC-link voltage where the power difference between the input stage and inverter output can be used to charge or discharge the DC-link capacitor [213, 214].

In general, the CCM-based power flow control strategy is popularly used in grid-connected operation mode, where the AC bus frequency and voltage are determined by the grid. However, in stand-alone operation of a microgrid, the CCM-based method cannot directly regulate the microgrid voltage and frequency, and therefore the VCM-based control strategy of at least one or more large DG units or energy storage units in a microgrid is necessary.

5.8.3.2 VCM-Based Power Flow Control Strategy

In this control strategy, output voltage of DG is controlled to regulate the DG output power, and the DG behaves like a synchronous generator. The droop control can be easily implemented on VCM-based DG units [215].

In this strategy, the output voltage phase angle is determined by active power controller, and the output voltage magnitude is controlled by the reactive power controller. The DG output three-phase voltages are regulated on their reference values with a closed-loop control system. In this strategy, the voltage closed-loop control system can have an inner current loop for transient and stability performance improvement [216]. In this control scheme, the active and reactive power controllers can be proportional controllers. More complex controller can also be used here to closely mimic the synchronous generator with excitation and torque dynamics [217].

Compared to CCM-based control, the main advantage of VCM-based control is that it can be used in both grid-connected and stand-alone operation modes, which makes the operation mode transition easy and smooth. Possible issues when utilizing this method are mainly related to the lack of direct control of DG output current, especially during fault- or grid-voltage disturbances. These problems can be avoided by implementing virtual impedance control at the DG output [164, 212].

5.8.4 Ancillary Services

Ancillary services for DG systems are becoming an important issue that may further improve the cost-effectiveness of DG systems. This is a promising idea, especially considering that many renewable energy-based DG systems (such as PV and wind) do not operate at the maximum rating all the time (PV systems are simply idle during the night). As a result, the available ratings from these DGs' interfacing converters can be utilized to provide ancillary services such as flicker mitigation [218], unbalance voltage compensation [219, 220], harmonic control [221], power factor correction, etc. Here the harmonics compensation and unbalance voltage compensation are briefly discussed.

5.8.4.1 Harmonics Compensation

The power electronics interfaced DGs can be controlled like active power filters at the harmonic frequencies to mitigate system harmonics. As mentioned earlier, there are two types of control strategies in DG systems: CCM and VCM. The CCM-based control strategy is widely adopted in active power filters to mitigate harmonics [222]. As a result, CCM-based DGs can be easily controlled as shunt active power filters to absorb harmonic currents produced by nonlinear loads. To do this, DGs can be controlled to act as virtual resistances at the selected harmonic frequencies. In VCM-based DG systems, the current-controlled harmonic compensation schemes mentioned before are not applicable, as they cannot directly control the DG output current.

5.8.4.2 Unbalance Voltage Compensation

Using the DG interfacing converters to compensate the grid-voltage unbalance can be an important ancillary service for the utility, where the unbalanced loads could cause serious unbalanced voltage resulting in poor power quality and even protection responses. For unbalance compensation, DG mitigates/reduces voltage sag and unbalances by injecting additional negative sequence current. Therefore, the DG injected current contains both positive sequence and negative sequence components, where the positive sequence component can help to improve the power factor or voltage support as discussed earlier, while the negative sequence component could reduce negative sequence of voltage at PCC [219].

Finally, other than the abovementioned ancillary services, the DG systems or the microgrid as a whole can be used to improve the power system operation by providing the reserve functions [223]. For these reserve functions, the DG or microgrid can be controlled with frequency or voltage droop control and help the grid frequency and voltage regulations. This can be done by the DG systems alone or collectively with both the DG and load response control. With more controllability and flexibility in a microgrid system, valuable ancillary functions can be provided for better grid operation and better grid power quality.

Microgrid is becoming an important aspect of future smart grid, which features great control flexibility, improved reliability, and better power quality. The important aspects of the microgrid are the grid integration and energy management strategies, which enables sound operation of the microgrid in both grid-connected mode and stand-alone mode. The recent research trend on the DG interfacing converter is focused on better efficiency, reduced size, multi-port, and modular design. For the energy management strategy, a hybrid combination of communication-based and communication-less energy management technologies could be a good balance of system optimal operation, reliability, and resilience. The interfacing converter control schemes show that VCM-based methods are gaining more attention due to its ability to mimic the behavior of a synchronous generator. Finally, the ancillary service is becoming a promising topic to further assist the grid control, enhance the grid power quality, and, at the same time, to improve the cost-effectiveness of power electronic-based DGs and microgrids.

5.8.5 Generation Systems

Generation systems mainly operate in two different modes: maximum energy extraction/constant power operation and droop regulation. During normal operation, as most DG systems are based on renewable energy sources (RESs), the converters attached to generation systems are controlled to absorb as much energy as possible from the energy sources. In the case of other types of DG systems such as diesel generators, secondary level controllers determine their constant power reference.

When the voltage or frequency of the microgrid increases above the preestablished level, DG systems shift out of their MPP to reduce the power amount they supply to the system. In this mode, DG systems contribute in the v/f regulation of the microgrid through a droop slope. This means that depending on the available power that can be absorbed from the RES, the controllers will have to adapt their operation characteristics to meet the grid codes predefined by the system operator. Classically, most converters associated with DG systems have been configured to exclusively operate on the MPP. However, the transition toward a more decentralized electrical system requires the participation of these generators in the regulation of the grid [3, 12, 224].

5.8.6 Demand Response

Just like for distributed generation systems, for microgrids in demand response mode, loads absorb the power required by the attached device. When generation systems are producing all the power they can and no power can be absorbed from the main grid and energy storage systems are not able to provide more power, the voltage or frequency decreases below the predefined level and the power consumed by loads is consequently decreased. In the literature, this type of operation is a part of the so-called demand-side management, as the loads actively participate in the regulation of the microgrid by reducing their consumed power when required. A high research activity has been carried out in the past years highlighting the importance of the participation of loads in the management of different types of electric systems [3, 12, 225].

5.8.7 Connection to the Main Grid

Depending on the topology and type of microgrid, one can follow different approaches with respect to the connection to the main grid. On a classical approach, the connection to the main grid can be employed to contribute to the regulation of the microgrid for the entire voltage range [3, 12, 226]. Another solution would be to use the link to the main grid at specific cases where the voltage or frequency levels are above the maximum or below minimum levels, avoiding the malfunction or disconnection of other devices.

5.9 TECHNICAL AND OPERATIONAL CHALLENGES AND ISSUES OF HYBRID MICROGRID

Microgrids, and the integration of DER units in general, introduce a number of operational challenges that need to be addressed in the design of control and protection systems, in order to ensure that the present levels of reliability are not significantly affected and the potential benefits of distributed generation (DG) units are fully harnessed. Some of these challenges arise from assumptions typically applied to conventional distribution systems that are no longer valid, while others are the result of stability issues formerly observed only at a transmission system level. There are also regulatory and grid-connection issues. The most relevant challenges in microgrid protection and control include [3, 4, 20, 78, 84, 143]:

1. ***Bidirectional power flows***: The presence of distributed generation (DG) units in the network at low-voltage levels can cause reverse power flows that may lead to complications in protection coordination, undesirable power flow patterns, fault current distribution, and voltage control.
2. ***Stability issues***: Interactions between the control systems of the DG units may create local oscillations, requiring a thorough small-disturbance stability analysis. Moreover, transition activities between the grid-connected and islanding (stand-alone) modes of operation in a microgrid can create transient instability. Recent studies have shown that direct current (DC) microgrid interface can result in a significantly simpler control structure, with more energy efficient distribution and higher current carrying capacity for the same line ratings.
3. ***Modeling***: Many characteristics of traditional schemes such as the prevalence of three-phase-balanced conditions, primarily inductive transmission

lines, and constant power loads, do not necessarily hold true for microgrids, and consequently, models need to be revised.

4. ***Low inertia***: Microgrids exhibit a low-inertia characteristic that makes them different to bulk power systems, where a large number of synchronous generators ensures a relatively large inertia. Especially if there is a significant proportion of power electronic-interfaced DG units in the microgrid, this phenomenon is more evident. The low inertia in the system can lead to severe frequency deviations in island mode operation if a proper control mechanism is not implemented. Synchronous generators run at the same frequency as the grid, thus providing a natural damping effect on sudden frequency variations. Synchronverters are inverters which mimic a synchronous generator to provide frequency control. Other options include controlling battery energy storage or a flywheel to balance the frequency.
5. ***Uncertainty***: The operation of microgrids involves addressing much uncertainty, which is something the economical and reliable operation of microgrids relies on. Load profile and the weather are two of these uncertainties that make this coordination more challenging in isolated microgrids, where the critical demand-supply balance and typically higher component failure rates require solving a strongly coupled problem over an extended time horizon. This uncertainty is higher than those in bulk power systems, due to the reduced number of loads and highly correlated variations of available energy resources (the averaging effect is much more limited).
6. ***Operation support in grid-tied and islanded conditions***: In grid-tied condition, the DG has to work in the PQ mode where the source controller of DG controls the output current. In this condition, the voltage and frequency of the microgrid is governed by grid parameters, and therefore, DGs have to follow the superior grid. In islanded condition, when the microgrid is in isolated state from the utility grid, the DG has to transition into PV mode, controlling voltage and frequency of its own autonomous mode. In hybrid system, where more than one DG is present, all DGs are operated in parallel in master-slave configuration. A microgrid following master-slave strategy needs only one energy source to operate in PV mode when the utility grid is absent.
7. ***Integration of microgrid to the main grid***: Most of the small-scale DG sources in the load side are integrated at medium- or low-voltage network as low penetration fashion where they are connected as passive systems, and they are not involving grid-voltage controlling, frequency controlling, and stability activities. Still in the case of high penetration, the interfaces can be modified to work as active generators so that DER can participate in the frequency, voltage, and system stability control activities of the grid. Power electronic is used to interface between the grid and the renewable power source of microgrid so that there are not any negative influences in reliability, stability, and power quality of the supply after the interconnection DERs to the grid. Numerous components and constraints are involved

in the integration of DER to the utility grid [18]. The integration of varying intermittent renewable sources like solar and wind energy conversion systems to the grid can provide a technical relief in the form of reduced losses, reduced network flows, and voltage drops. However, there are also several undesirable impacts due to high penetration of these variable DERs which include voltage swell, voltage fluctuations, reverse power flow, changes in power factor, injection of unwanted harmonics, frequency regulation issues, fault currents, and grounding issues and unintentional islanding [18]. Advanced protection system should be included in the DG units to disconnect the units in case of fault or unfavorable grid conditions.

Grid integration of distributed renewable sources are classified depending on the resource availability, load demand, and existing electrical power system into three categories namely low penetration with existing grid, high penetration with existing grid, and high penetration with future smart-microgrid configuration.

A. **Low penetration with existing grid**

In low-penetrated networks, the distributed generator units are not involving in frequency control activities and voltage control activities of the PCC point. Grid operator is responsible for managing the overall system stability, and DG operators can send the maximum available power to the main grid and local loads without major consideration of grid constraints.

The DG operators have to deliver the power based by grid synchronization via PLL systems with correct phase sequence. Whenever grid frequency exceeds the allowable limit, the inverters are required to disconnect from the grid. And it operates in power factor (PF) correction mode, where PF keeps closer to unity. Most of PV units and wind generators can inject the maximum available active power into the grid; most existing voltage source converters (VSC) operate in power factor correction mode (zero reactive power).

The network operators face a real problem when DG sources are connected to low-voltage lines since microgrids have dispersed generation units; sizes of the DGs are very small and have low-inertia characteristic, especially frequency deviations. The amount of DG units connecting to a particular distribution network is limited by the voltage control margins of that distribution network; to overcome these challenges, static synchronous compensator (STATCOM), voltage source converter (VSC), automatic tap control transformers, and special control mechanism are used by operators to control the network voltage [3].

B. **High penetration with existing grid**

When growing the renewable energy source, penetration causes complication in the system constraints due to the intermittency of RES; that the percentage of the renewable power injected into the existing grid is relatively high as compared to the power assigned to the conventional power plant. Therefore, in such type of situation, intermittent power sources cannot work as passive generators, but they have to actively participate in grid-frequency and voltage-control activities. In addition to grid synchronization with phase sequence matching and protection system, controls and inverters should be more intelligent.

The grid operator cannot transfer the energy to or from the main grid in the case of islanded power systems with a significant penetration of RES power, so the isolated system has to deal with intermittency issue. Since the amount of power delivered considerably effects the grid stability, phase-balanced operations and proper VSC inverter connection strategies have to be implemented in the system. Voltage control loop can be included in VSC inverters to provide the required reactive power to the grid; in this way, the VSC will intelligently respond to the grid conditions. On the other hand, inverters have to operate within a defined power factor range, not unit power factor, so that the VSC will have the capability to control the grid voltage at the PCC point [3].

C. **High penetration with smart grid concepts**

The combination of different renewable energy generation resources (such as micro hydropower, photovoltaic arrays, geothermal, and wind turbine generators) in a microgrid can be integrated into the grid and increase the penetration of renewable energies to change the whole system into a smart grid with advanced technologies. Upcoming smart grid networks will provide a real-time, multidirectional flow of energy and information. Smart intelligent equipment with modern digital controls is used in an entire electricity grid, from central control office to end customer levels [3].

However, maintaining the stability and reliability of the network becomes a problem when the contribution from DGs is maximizing, then the solution may be using smart grid concepts such as microgrids, large-scale energy storage with advanced energy management systems, smart homes with demand response management, etc. This will help in better communication and coordination between all the participants in the electricity business such as power plant operators, network operators, end consumers, and government.

8. ***Black start capability***: Black start phenomenon is to start the microgrid system from a complete shutdown state. It is a challenging task to black start a microgrid system because it requires a complete analysis of the system's state. Moreover the start of DGs and the connection of loads require a certain procedure to be followed. It involves step-by-step connection of DGs and loads to the LV grid on the basis of overload capacity. However, in comparison to conventional power restoration, the microgrid restoration process is much simpler due to reduced number of variables (switches, DGs, and loads) [6]. The capability of black start is most needed in remote areas where utility grid is absent and where utility grid outages are very frequent. In a microgrid system, a bottom-up approach to black start is more preferred as it reduces the restoration times [3, 12].

9. ***Protection strategy***: The protection scheme of a microgrid is much more complex and challenging than the conventional power system because of the requirements of both grid-interfaced and grid-isolated modes in a microgrid and also the existence of the two types of faults—internal and external faults [3].
10. ***Implementation needs***: Despite their many advantages, microgrids face significant barriers to widespread implementation. As a fundamental complication, microgrids face often-conflicting regulation at the federal, state, and sometimes local levels. The key capability and feature of a microgrid is its ability to island itself from a utility's distribution system during blackouts. However, in order to have an operational microgrid that can perform in the manner expected—both online and islanded—it requires the use of the following technologies:
    - Distributed generation (DG)
    - Islanding and bidirectional inverters
    - Smart meters
    - Distribution automation (DA)
    - Substation automation
    - Smart transfer switches
    - Advanced energy storage
    - Microgrid control systems
    - Energy management systems
    - Distribution management systems
    - Communication technologies and sensors
11. ***Market need***: The design of a microgrid also requires the knowledge of the number of customers served, whether or not the microgrid is used full time or part time, physical length of circuits and types of loads to be served, voltage levels to be used, feeder configuration (looped, networked, or radial), AC or DC microgrids, heat recovery options, and desired power quality and reliability levels [3, 26, 50].
12. ***Legal and regulatory uncertainty***: There are two key legal issues that impact microgrids: first, whether they are deemed to be electrical distribution utilities and are therefore subject to oversight by state regulatory agencies, and second, even if they are exempt from state regulation as utilities, do they fit into existing legal frameworks governing the sale and purchase of electricity and rights to generate and distribute electricity? A clear legal identity for microgrids is needed to achieve the regulatory certainty required to make microgrid projects "bankable"—otherwise the potential costs are too high and benefits too uncertain to justify investing time and money [227]. Several states in the United States have evaluated microgrids in the context of the current legal and regulatory framework pertaining to electricity generation, transmission, and distribution. The resulting reports are a good starting point for understanding the issues the states are wrestling with regard to the future of their electrical distribution systems [227–230].

13. ***Interconnection policy:*** One fundamental source of legal uncertainty centers on the laws regulating connection of distributed energy resources to the grid. Following deregulation in the United States in the late 1990s, there were no nationwide standardized requirements for small independent power producers to connect their equipment to the grid. Manufacturers and project developers had to deal with a patchwork of requirements that varied from utility to utility [231], adding substantial cost and time to the microgrid development process. The development of IEEE 1547 (released in 2003) was an important step toward a consistent set of rules for integrating distributed energy resources (< 10 MVA) to the grid in a safe manner [232]. Until recently, though, the main focus of interconnection policy for distributed energy resources, including IEEE 1547, was on ensuring that those resources would disconnect in the case of grid failure (a so-called "unintentional islanding" situation) to protect the safety of line workers. It wasn't until the IEEE approved standard 1547.4 in 2011 that standardized protocols became available for safe *intentional* islanding and reconnection of microgrid systems. IEEE 1547.4 includes guidance for planning, design, operation, and integration of distributed resource island systems with the larger utility grid. It covers the functionality of microgrids including operation in grid-connected mode, the transition to intentionally islanded mode, operation in islanded mode, and reconnection to the grid, specifying correct voltage, frequency, and phase angle. Finally, IEEE 1547.4 also covers safety considerations, protection, monitoring, communications, control, and power quality. California's Rule 21 also addresses interconnection requirements to help remove barriers put in place by legacy utility providers, by establishing standardized technology- and size-neutral requirements, a clear review process, testing and certification procedures, set fees, and a streamlined application process. Interconnection is of paramount importance: if microgrids are not able to connect to the utility grid, they must operate permanently in an islanded mode, forfeiting the opportunity to derive revenue from grid services that they could otherwise provide and crippling their business case.
14. ***Utility regulation***: A microgrid is likely to be considered an electric corporation if it intends to serve multiple, otherwise unrelated, retail customers, cross a public way with power lines, and/or obtain a franchise from a local authority. The reasons for this conclusion are discussed below in more detail. If a state utility regulatory agency decides that services provided by microgrids qualify them as utilities, that body can regulate the rates charged for electricity and decide whether to approve the facility construction, among other powers, all of which have major implications for microgrid developers and owners. In the event that the microgrid *is* deemed to be a distribution utility, it may assume an obligation to serve, meaning that it would be required to provide service upon the written or oral request of a potential retail customer.

    All microgrids that intend to use public ways to distribute electricity to customers (for example, sending thermal energy or electricity across a public street) require permission from the local municipal authority [227]. This permission can be in the form of a "franchise" or other "lesser consent." A microgrid's ability to obtain this permission depends in large degree on whether a preexisting electric utility has been given an exclusive franchise, effectively blocking out competitors. In New York, for example, if the existing franchise is nonexclusive, the state law still mandates that a competitive process be used to determine the franchise grantee, allowing incumbents and other service providers to compete against the microgrid developer for the franchise.

Due to their small scale and limited scope of services, it is unlikely in most cases that a microgrid would require a franchise and therefore, that most microgrids would *not* be under the jurisdictional authority of the utility regulatory agency; however, these cases are being decided on a project-by-project basis in the courts. In addition, microgrids selling to retail customers may have to comply with various consumer protection laws. Finally, regardless of their status as a distribution utility, microgrids that produce power through combustion (such as microturbines or diesel generators) are subject to federal and state laws governing emissions and will require a permit under certain conditions. The choice of business or ownership model will also impact the degree to which utility franchise or lesser consent come into play; these considerations are discussed in more detail below.

Today's regulations governing electric utilities in the United States reflect a process referred to as "restructuring," and colloquially as "deregulation," that occurred in the mid- to late 1990s in many states in the United States, following the example of deregulation in other major industries like airlines, railroads, telecommunications, and others [233]. In general, restructuring introduced a separation between the generation, transmission, and distribution functions of what were previously vertically integrated monopolies. In the case of New York, generators can sell electricity into competitive wholesale markets or directly to local distribution utilities or retailers for resale to customers. A system operator (in the case of New York, the NYISO) is responsible for maintaining a balance between supply and demand at all times. The ecosystem of players in the restructured New York electricity market includes smaller generating companies called independent power producers (IPPs). Microgrids, as such, do not fit neatly into the classes of market participant defined by restructuring, perhaps because they transcend the categories of generation, transmission, and distribution. As a result, further work is needed to incorporate them into the regulatory legal structure.

15. ***Utility opposition***: Although grid-tied microgrid customers will likely stay connected to the grid for the foreseeable future, only islanding in the case of utility grid failure, self-consumption of microgrid generated energy could erode the revenue base that has traditionally paid for utility infrastructure investments. There is also still reluctance to add large amounts of distributed

energy resources to the grid because of perceived management, safety, and protection challenges. As a result, many utilities are seeking to impose additional fees on DER owners and threatening to halt net metering programs. Market restructuring, like that proposed in New York's "Reforming the Energy Vision (REV)" effort, will be required to move from a situation where microgrids are viewed as a threat to one in which distributed energy resource services are valued by the utility grid and fairly compensated [234]. As part of this restructuring, utility regulators will fully unbundle generation, transmission, and distribution services and will allow independent power producers to compete in wholesale (and potentially retail) markets. Real-time or time-of-use (ToU) electricity prices will become the norm so that microgrids receive the economic signals they need to manage their DERs to provide grid services like frequency regulation, black start, and congestion relief and to maximize their own revenues. However, utility restructuring has not been a universal phenomenon, and progress slowed dramatically following the challenges experienced in California in the early 2000s [233].

Even for deregulated utilities, the structure of electricity markets and the manner in which investor-owned utilities are paid for providing service (using so-called "cost of service" accounting) still represent impediments to distributed energy resource adoption in general, including microgrids. Decoupling electric company revenues from electricity sales, which is already done in 14 states in the United States, is a major step toward removing utility resistance to microgrids based on concerns about a so-called "utility death spiral," where widespread self-generation leads to demand reduction for the grid's electricity, which in turn leads to higher electricity costs for traditional customers, fueling additional uptake of self-generation to the point that utilities cannot cover their costs.

A potential path forward is to move from the traditional cost-of-service paradigm to a performance-based approach [235] that recognizes that the utility grid is being asked to provide functions that are much different from those they have historically been responsible for, such as resilience, security, and clean generation. In this new paradigm, utilities would be incentivized to invest in upgrading infrastructure and improving efficiency as opposed to selling the maximum number of kilowatt hours. Several states in the United States have taken it upon themselves to commission or formulate their own plans for how to modernize their grids and electricity markets to provide more reliable, efficient, and clean electricity to their customers [236–238]. Countries like Great Britain are also formulating plans for the evolution of the grid to a more clean, secure, and distributed energy future and examining the social, legal, and regulatory factors that help or hinder that transition [239].

Utilities are also coming around to the view that they may be well positioned, if allowed by regulators, to provide microgrid services to their existing customers since they have extensive knowledge, distribution infrastructure already in place, and franchise rights from local authorities. Electrical utilities have begun testing microgrid concepts in laboratory-type settings. One example is Duke Energy, which maintains two test microgrid facilities: one in Gaston County, North Carolina [240], and one in Charlotte, North Carolina [241]. The first installation focuses on interoperability and building partnerships with manufacturers; the second, originally built to test virtual power plant capabilities, is a solar PV and storage microgrid serving a fire station. The partnership between the CERTS team and American Electric Power (AEP) to develop a CERTS test bed represents a productive partnering model between the industry and the government [43]. Other utility companies [80], like Arizona Public Service, Consolidated Edison, Commonwealth Edison, Green Mountain Power, NRG Energy, San Diego Gas and Electric, and Southern California Edison [242] are also exploring microgrids as a way to provide additional services to customers, defer capital investments, improve overall reliability, and manage potential disruption to their business model.

16. ***Competing smart grid paradigms***: While it has been argued that microgrids are a better approach to contain and manage local problems [243] and could even serve as a possible pathway to a "self-healing" smart grid of the future [244], it is possible that society will find grid architecture paradigms like "smart supergrids" [245, 246] or "virtual power plants" [144], [81, 247]—which do not feature local balancing of generation and loads or isolating segments of the grid—to be more compelling architectures. Smart supergrids rely on improved fault detection, isolation, and restoration capabilities to alleviate congestion, route power around faults, and shorten recovery time from outages. Virtual power plants rely on software and analytics to manage widely dispersed distributed energy resources, although grid-connected microgrids can also function as virtual power plants, as mentioned above. New information and communications developments, broadly known as the "Internet of Things (IoT)," are also facilitating the emergence of a decentralized, so-called "transactive" energy market platform, where individual distributed energy resources and loads can bid to buy and sell electricity from each other [248]. Whether microgrids become the dominant strategy to deploy large amounts of intermittent renewables and improve resilience depends on whether the benefits are perceived to be great enough in relation to the costs, when compared to the alternative smart grid paradigms. It is possible that—even in situations where there is a low value placed on islanding for resilience and reliability—it will be deemed advantageous to collocate virtual power plant assets in microgrid-like architectures [20].
17. ***Legal and social response***: Whether microgrids remain a niche application or become ubiquitous depends on two main factors: (1) to what degree regulatory and legal challenges can be successfully surmounted, and (2) whether the value they deliver to property owners and communities in terms of power quality and reliability (PQR) and other economic benefits outweigh

any cost premiums incurred to capture those benefits. These questions are now being answered in court rooms and commercial markets around the globe as electricity grids evolve to address social and economic concerns and incorporate 21st-century technology to update Thomas Edison's original vision of the grid [20].

5.10 SUCCESSFUL EXAMPLES OF HRES IN MICROGRIDS

Hybrid energy systems are becoming attractive to supply electricity to rural areas in all aspects like reliability, sustainability, cost, and environmental concerns. The advances in renewable energy technology like solar and wind provide reliable power supply with improved system efficiency and significant cost reduction, especially for communities living far in areas where grid extension is difficult. Besides this, the demand for renewable energy source in large urban cities is increasing, and their integration to the existing conventional grid has become more fascinating challenges. So the future requires stable and reliable integration of renewable distributed generators to the grid, where the local loads are close to distributed generators. Microgrid allows everyone to be a producer and consumer of energy at the same time; by doing so, everyone can become energy independent from the overcharging utility companies. Microgrids are not a replacement for traditional utility infrastructure. From the utility viewpoint, microgrids allow the transmission and distribution cost to be lowered, along with the reduction in line losses, network congestion, and load shedding; improvement in power quality and reliability; and reduction in infrastructure investment needs. HRES in microgrids provides many value-added propositions including its effects on economics and carbon emission. In this section, we analyze these benefits with the help of a few successful case studies.

5.10.1 Economic Value of Microgrids with Energy Storage for HRES

An economic value of hybrid microgrid was analyzed in an excellent article by Saury and Tomlinson [84]. They considered hybrid systems of diesel generator and solar and wind systems to analyze the effects of contribution of renewable sources on the overall economics. Here we summarize their findings.

Aided by sharp declines in the cost of wind and solar energy, as well as lower energy storage costs relative to the price of fuel, hybrid microgrids using renewable resources are well suited to a host of applications, including individual buildings, resorts, mine sites, remote villages, small islands, and others. The most promising applications are those with total power demand from 100 kW to 20 MW [84]. The basic concept of HRES impact on economics and environment is simple. Wind or solar energy reduces reliance on power produced from generator sets, saving fuel and, to a lesser extent, maintenance costs. The generator sets, along with renewable sources, follow the load in hybrid setting, and sophisticated digital controls tie the system together. Energy storage enhances system economics and helps the generator sets respond smoothly to significant fluctuations in output from the renewable resources, while maintaining consistent voltage and frequency [84].

This concept of hybrid energy has become increasingly attractive as the costs of energy from wind and solar photovoltaic generations have declined. Conversely, the cost of diesel fuel—usually the most available fuel for remote locations—has risen. In 2000, the levelized life cycle energy cost of wind generation was similar to that of diesel, while solar energy was nowhere near being competitive (nearly five times higher). Since then, diesel fuel prices have mostly trended upward, while wind power prices trended slightly down, and solar photovoltaic prices fell dramatically. Conservative projections place the price of wind energy at $0.09 per kWh by 2020, and the price of solar energy only slightly higher. In 2020, while the levelized cost of diesel fuel can vary anywhere between $0.2 and $0.7 per kWh, this gives renewable energy a meaningful, long-term price advantage over diesel-generated power. In addition, advances in energy storage, system control, power conditioning, and connected load-side management have helped drive down the total cost of ownership of hybrid microgrid systems [84].

The hybrid microgrid combines the benefits of renewable and conventional power generation while offsetting their weaknesses. The basic cost equation indicates that there is a balance between high capital cost for renewables compared to diesel generator versus low operating cost for renewables versus fuel for generator. Low operating cost for renewables pays off high capital cost more rapidly, higher the percentage contribution of renewables in hybrid energy. Thus, a hybrid microgrid delivers lower long-term operating cost and a lower total cost of ownership than pure conventional power generation. In a hybrid microgrid, renewable energy capacity can account for any percentage of the total peak load. In general, the greater the contribution from renewables, the greater the potential fuel and operating cost savings (see Table 5.8).

Saury and Tomlinson [84] indicated that in a hypothetical system with a 3-MW peak load served by three 1-MW generator sets, a conservative approach would limit the renewable components to a small share of the capacity—for example, 300 kWp (10%). In this scenario, the renewables lower the fuel cost, but by far less than 10%, because the intermittent nature of renewable energy can only offset a small share of generator set runtime. An intermediate approach would increase the renewables to 1 MWp (33%). Here, one generator set could shut down while the renewables produced at full capacity, saving significant fuel. If the renewable output declined suddenly, such as when a cloud shades a photovoltaic array, energy storage could provide

TABLE 5.8
Microgrid Renewable Component versus Fuel Savings [84]

Fuel Savings	Renewable Energy Capacity	Energy Storage Function
30%–100%	Greater than generator set capacity	Power stability and energy time-shifting
10%–30%	Equal to generator set capacity	Power stability only
5%–10%	<15%–20% of generator set capacity	No energy storage
Baseline	Diesel or gas generator sets used for prime power	

short-duration intermittent power, achieving good system stability. Optimally, renewables would total the full 3 MW or more. In that event, under ideal wind or sun conditions, the renewables could carry the entire load, and at times, inject the excess energy into storage for use later when wind or sun conditions are less favorable. This would result in long periods where no fuel is consumed from running the generator sets [84]. This analysis indicates the flexibility and corresponding cost savings associated with hybrid energy system with renewable sources and storage.

Each component of a hybrid microgrid brings advantages that strengthen the system as a whole. Wind and solar, while requiring higher capital and significant space, have minimal operating and maintenance costs once installed. They displace greenhouse gases and other pollutants and contribute strongly to sustainability initiatives. Two basic types of solar photovoltaic systems can serve microgrids. The frameless thin-film technology has an energy conversion rate of about 20% and is well suited for hot, humid, and dusty environments found in Australia, Africa, and the Middle East, where its performance degrades at a lower rate than silicon panels. More costly monocrystal silicon technology is more efficient, delivering up to 25% energy conversion under ideal light and temperature conditions. It also requires less space than a thin-film system. However, more crystal silicon and polycrystal silicon cells degrade rapidly under hot, humid, and dusty conditions, rendering their performance advantage in the real world moot. Reciprocating-engine generator sets deliver reliable energy and are fully dispatchable while renewable sources are offline or produce at less than rated capacity. Units in ratings from 350 kW to 15 MW and larger can be added in a modular fashion to create systems of substantial size. Multiple units add ample flexibility for variable load conditions [84].

Generator sets offer high power density, simple-cycle electrical efficiencies from 40% to 48%, high part-load efficiency and excellent capability to follow loads. They tolerate a wide range of ambient temperatures and high altitude without derating. Diesel units in particular accept load rapidly, with start times to full load as fast as 10 seconds and ramp-up time from 25% to 100% load in as little as 5 seconds. They tolerate unlimited starts and stops throughout the day with limited impact on service life or maintenance requirements. The technology is thoroughly proven and reliable, with hundreds of gigawatts of capacity installed and qualified service technicians readily available worldwide [84].Generator sets also offer the potential for combined heat and power (CHP). Indeed, settings where microgrids are economically attractive—communities, resorts, and industrial facilities—tend to have significant thermal requirements. Heat captured from an engine exhaust or cooling circuits can be converted to steam, hot water, and chilled water (by way of absorption chillers) or used in water desalination plants. In each case, the project economics can be greatly enhanced. In temperate climate zones, it is very attractive to combine CHP and solar energy because of their complementary nature. CHP capabilities are typically used during cooler times of the year, when the solar contribution lessens. Energy storage helps to balance any surplus.

Energy storage is a key enabler of hybrid microgrids due to rapidly advancing technology. The conventional energy storage system consists of banks of deep-cycle lead-acid, nickel-metal hydride batteries, flywheels, or lithium ion. However, two other energy storage technologies—ultracapacitors and rechargeable metal-air—are now gaining favor [84]. Metal-air energy storage originated with hearing-aid batteries, providing long life in a safe and nontoxic package. More recently, a rechargeable capability has been developed for the zinc-air energy storage technology, already common in applications such as backup power for cellular communication towers. The batteries can be 95% discharged and can be recharged with no cycle limit. Rechargeable zinc-air provides the most economical electricity storage, and it includes integrated controls and monitoring at the cell level.

According to Saury and Tomlinson [84], zinc-air batteries do not overheat or discharge dangerous concentrations of hazardous gases, and they operate in a range of freezing temperatures to 122°F (50°C) without degradation. Life expectancy is at least twice as long as lead-acid batteries. The next generation of zinc-air storage will be offered in capacities at megawatt scale, well suited for the hybrid microgrid concept, and will provide an attractive total cost of ownership The range of energy storage technologies outlined above are used in microgrid applications today. A fully flexible offering enables a combination of these technologies (hybrid energy storage systems), depending on the application. The use of hybrid energy storage in microgrids is also rapidly expanding. Role of energy storage (particularly hybrid energy storage) is discussed in detail in Chapter 4.

The key question is determining whether a hybrid microgrid is appropriate to a given site. Analytical tools are available that make it relatively easy to check economic feasibility. An initial high-level analysis requires little more than basic information about these factors [84]:

- The load profile of the community or facility to be served
- The site latitude and longitude and historic solar and wind conditions
- The cost of fuel for the primary power unit generator sets
- The cost of capital

Saury and Tomlinson [84] points out that the results of this type of analysis will indicate whether a deeper investigation is warranted or whether the project should be abandoned. HOMER microgrid analysis software can be used to perform a much more rigorous analysis for making a final decision. This software simulates one year of system performance, uses site-specific solar and wind energy data, and predicts annual hours of operation and fuel use for generator sets. The resulting data can then be used to develop an operating protocol that enables financial optimization of the system [84].

5.10.2 Case Studies Demonstrating Microgrids Value

A. *Case Study 1: Tropical Island*

This case study was outlined by Saury and Tomlinson [84]. A power company serving a tropical island wanted a hybrid microgrid with photovoltaic energy and a diesel-fueled generation system to help reduce its costs. The existing power generation equipment consisted of generator sets, fueled by No. 2 diesel distillate, supplied by Caterpillar. Island power generation system before microgrid had total generation capacity of 10.7 MW with average demand of 1.6 MW, and peak demand of 4.5 MW, [84].

In order to identify the optimal system design for the island's needs, planners used the NASA EOSWEB database for global solar irradiation data, combined with the HOMER software tool. This enabled engineers to generate an estimate of the performance of a microgrid system incorporating renewable energy. The NASA data estimated the average annual solar irradiation to be 5.32 kWh/m^2/day. The analysis evaluated the site with a 670-kW solar photovoltaic system distributed over various rooftops. The results for fuel savings were obtained with rooftop PV with grid stabilization and rooftop PV with grid stabilization and energy storage. In both cases, the PV capacity was the same at 670 kW. The results showed that in the first case, fuel savings was 5% (53,373 gallons per year), with a simple payback period of 6.4 years by combining PV with diesel. In the presence of 250 kW energy storage system, the fuel savings were 7% (69,927 gallons per year), with a simple payback period of 4.9 years even though that system would result in a higher total capital expense.

The system demonstrated the value of the microgrid concept in combining the benefits and offsetting the limitations of both renewable and conventional power generation. The system with integrated energy storage delivered rapid economic benefits, along with enhanced power system stability and availability. The system included rooftop-mounted photovoltaic panels, since ground space was too scarce and too valuable for that purpose. It included 2,316 solar modules and the racking structure needed to roof-mount the panels in various locations. String inverters were also included.

The energy storage system consisted of one Cat® BDP250 bidirectional energy storage inverter and two strings of batteries. The inverter interacted with the power system frequency and voltage to add power when it was needed to compensate for energy fluctuations. The grid stabilization effect of integrating energy storage helped the overall power plant maintain system frequency and voltage in a narrower control band. The system readily compensated for fluctuations in output from the solar panels. During cloudy periods, solar energy generation can vary from 50% to 100% of the array capacity over short periods of 10 seconds or less. A sharp drop in photovoltaic output causes a sharp rise in the power required from the diesel generator sets. This causes a drop in system frequency and voltage as the diesel generators respond to the load.

The benefit of energy storage was demonstrated when the bus frequency dipped. The energy storage inverter outputs powered to assist the generator set in responding to the transient, thus reducing the magnitude of frequency and voltage dips. The energy storage inverter can provide 250 kW continuously, up to the storage capacity of the battery. In transient operation, the inverter has higher power capability—150% for 30 seconds or 200% for 3 seconds. The battery capacity in this example can power the inverter for 10 minutes at 250 kW. An additional benefit of the energy storage inverter is the overall improvement in diesel power efficiency. The energy storage system can supply operating reserve or spinning reserve for the power plant. As the power plant load rised, additional generator sets were usually started when the operating load reached 80% to 90% of the running generator sets' capacity. This provided extra capacity to respond to transient loads. Using the stored energy as operating reserve allowed the system operator to wait to dispatch additional generator sets until the power system load is 95% to 100% of the running units' capacity. The overall average load of the diesel generator sets was therefore higher, resulting in higher overall power plant efficiency [84].

The Cat microgrid control system can be used with or without energy storage in an existing power plant to control the network and integrate with all energy sources to provide coordinated control of the diesel generator sets, wind power energy sources, energy storage, and photovoltaic panels. This is important because high reliance on renewable generation may cause voltage on some power feeder circuits to rise. The microgrid control system automatically adjusts to maintain system voltage and power quality within specifications. The controller monitors generator set operation and maintains minimum load levels on the diesel generator sets in response to set points determined by the operator. Where fuel consumption information is available, the microgrid controller can be used to dispatch the most efficient generator set or a combination of generator sets for a particular operating mode [84].

B. **Case study 2: Norvento case**

Norvento, a company in UK, helped a client needing support in making an investment decision for their power system. The site facility was running for several years powered by diesel gensets at a high cost of electricity. The client currently has a diesel price equivalent to £0.58 per liter, which includes the costs of delivery to its remote location. Now they face the requirement to either invest in a new energy system based on the same technology, or finding an alternative that may help them to save OPEX (operating expenses)as well as help them comply with carbon emission targets [249].

At the start of the design process, Norvento analyzed the available renewable resources at the site, as well as collecting data on the typical electrical demand of the operation. Solar PV was considered the optimal primary generation source at the site due to the favorable solar irradiation conditions. In order to size the microgrid appropriately, data were collected on the typical daily electrical demand at the site. The client operated in two shifts 7 days a week, consuming electricity in an approximately constant pattern all year round. Simulations allowed the company to compare demand with generation across a range of solar PV systems of different sizes as well as a diverse set of storage capacities, taking into considerations the seasonality existing for this resource. The final specification that Norvento arrived at for the microgrid sought to address the priorities of the client. The main priorities were to significantly reduce the operational cost of electricity generation on site, but also to achieve a relevant CO_2 emissions reduction to comply with carbon emission targets [249].

In order to fairly assess the economics of two separate electrical generation systems, Norvento used the levelized cost of electricity (LCOE) as a way to measure the cost of energy used by the client, regardless of the source it comes from. It is the net present value of the unit cost of electricity over the lifetime of a generating asset, meaning that it considers not only the costs and the energy delivered but also when they happen. LCOE was used to compare the current operational electrical generation for the client with a renewable-driven microgrid provided by Norvento.

Norvento used a discount rate of 8% (which was indicated by the client) when calculating the LCOE over a 20 year period, which is meant to represent the cost of capital in real terms; neither inflation nor fuel price change have been accounted to calculate OPEX of the plant [249].

B-1 **Business case**

Initial assessment of the mining facility showed electrical needs to be met by 2 MW of old diesel generators with one backup generator of 1 MW. These deliver 8,500 MWh of electricity each year at an operational cost of £42,134 for maintenance and £1,674,261 for diesel. The replacement of the existing generators required a CAPEX of £422,400. The diesel generators would consume a total of 2,880,658 L of diesel per year, producing 7,736 tons of CO_2. A lifetime of 10 years is predicted when utilizing the generators to meet total electricity demand at the site, at which time they would need to be replaced [249].

Assessment by Norvento engineering arrived at an alternative option to the diesel-only system. They proposed a microgrid comprised of 5.3 MWp of solar PV and 6 MWh of battery storage, retaining the requirement for 2 MW of diesel generator as a failsafe; maintaining the gensets makes sense as the CAPEX associated to the gensets is just a fraction of the total costs in their lifetime. The operation of the gensets in the microgrid will also be much less intense than in the base case, leading not only to a much lower diesel cost but also to an increased lifespan, as well as reduced maintenance and overhaul needs [249].

A comparison of the parameters of each system is shown in Table 5.9. Figures are converted into GBP where appropriate.

The results in Table 5.9 demonstrate the economic feasibility of a renewable microgrid as an alternative to diesel generation, especially in remote areas. It is important to comment on coverage of 75% of electrical demand with renewable generation, rather than 100%. The relationship between cost and renewable penetration is not linear.

For example, it may be feasible to cover 90% of a particular operation's demand with renewables and battery storage, but to cover the final 10% may cause the system cost to increase by three or four times. In order to meet the client's priorities, Norvento reached a balance between system costs and results, arriving at a 75% renewable penetration. It is for this reason that Norvento combined renewable generation and storage, and in the case study above, retained the diesel generation. The optimal control of the system should also ensure that the generators are used in the most efficient manner possible with minimal diesel consumption and minimal wear of the equipment, by limiting the number of hours in operation and the number of start-ups required.

TABLE 5.9
Comparative Data for Norvento Study: Comparison of Diesel Generation versus a Renewable Microgrid with 75% Renewable Penetration over 20 year Lifespan [249]

***Base case*: 2 MW installed diesel generation.**

System: Two main gensets (expected lifespan 10 years with overhaul after 5 years).
1 MW for continuous operation with one backup genset 1 MW for emergency and maintenance.
Capital cost: Total £985,000 with £42,200 three gensets, £281,800 replacement, and £140,800 overhaul.
Operating costs: Fuel cost £1,674,261 and maintenance cost £42,134, both per year: levelized cost of energy £210 per kWh.
Emissions: CO_2 emission of 7,736 tons per year and emission of NOx, SOx, and particles.

Renewable case: 75% renewable penetration.

System: Three base case gensets used half of the hours used in the base case; 5.3 MWp solar PV, 6 MWh/2 MW lithium-ion battery storage system; expected lifespan for solar PV, batteries, and genset 20 years, genset overhaul 10 years.
Capital investment: Three gensets £422,400, main gensets overhaul £140,800, solar PV £3,731,200, storage battery £2,100,000, and balance of system design £88,000 with total £6,482,400.
Operating cost: Fuel: £448,159 per year, maintenance cost: £117,156 per year with total £548,802 per year. Levelized cost £148 per kWh
Emissions: CO_2 emission down by 73% to 2,071 tons per year. Significant reduction in local pollution.

By switching to a renewable microgrid, Norvento demonstrated that the economic benefits to the business could be more than £1 million per year in operational costs and a LCOE that is 29% lower. By utilizing the generators as a backup in the microgrid system, their life can be greatly extended and operate with lower maintenance costs, as indicated in Table 5.9. As the LCOE calculation considers the initial capital expenditure in both projects, it demonstrates that a renewable microgrid can deliver operational efficiencies, with a significantly lower cost of energy. Clearly important also is the saving in CO_2 emissions as a result of the transition to renewables; 5,665 tons/year of savings (about 73%) is significant and is to be considered of high value. The microgrid system also provides future flexibility to the business due to the simplicity of adding further generation or storage modules. In addition, relying on renewable power generation and battery storage will allow the business to hedge against future volatility in diesel prices.

While the numbers reported here can substantially change in future due to price changes in solar panels and diesel fuels, batteries cost along with other ancillary costs, the final conclusions will remain unchanged. For many years we have witnessed a steep and consistent fall in solar PV and battery prices; a trend which is set to continue. The likely LCOE of microgrid projects are even likely to be comparable with main grid power in the close future. Assuming the above project was connected to the UK power grid rather than powered purely by diesel, the LCOE would be around £0.11 per kWh (data from Department of Business, Energy and Industrial Strategy published in 2018, excluding taxes and levies). It is clear to see that the proposed system in this case study is not far away from this reality; grid parity will soon arrive for microgrid systems [249].

C. **Case study 3: California Energy Commission study**

A recently published report by the California Energy Commission profiled 26 microgrid case studies. The case studies include several demonstration projects sponsored by governments, public utilities, and universities, but the majority feature private industry projects for commercial hosts. The portfolio was focused on California and North America, but also included projects in India, China, Singapore, Japan, Mozambique, and Denmark [250]. The report was quite extensive and illustrated DER mix of seven global case studies (Figure 5.18) and comparison of DER mix by region (California, North America, and global) (Figure 5.19). The report examined nine case studies as outlined in Table 5.10 [250]. Table 5.11 outlines some of the value propositions obtained in one of the case studies [250].

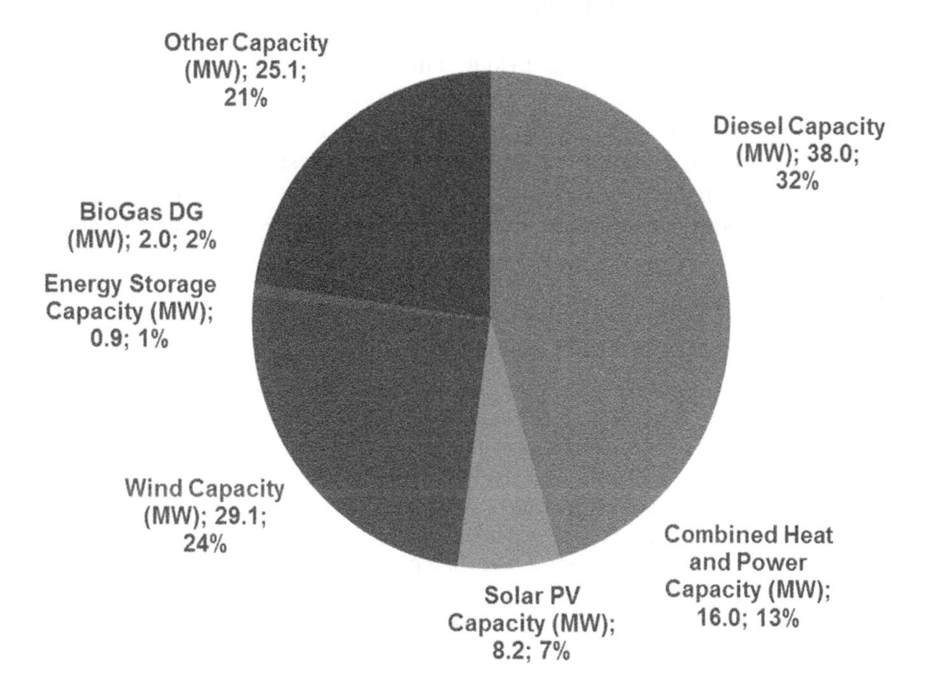

FIGURE 5.18 DER mix of seven global case studies, by capacity. Source: Asmus, Peter, Adam Forni, and Laura Vogel. Navigant Consulting, Inc. 2017. Microgrid Analysis and Case Study Report. California Energy Commission. Publication Number: CEC-5002018-022 [250].

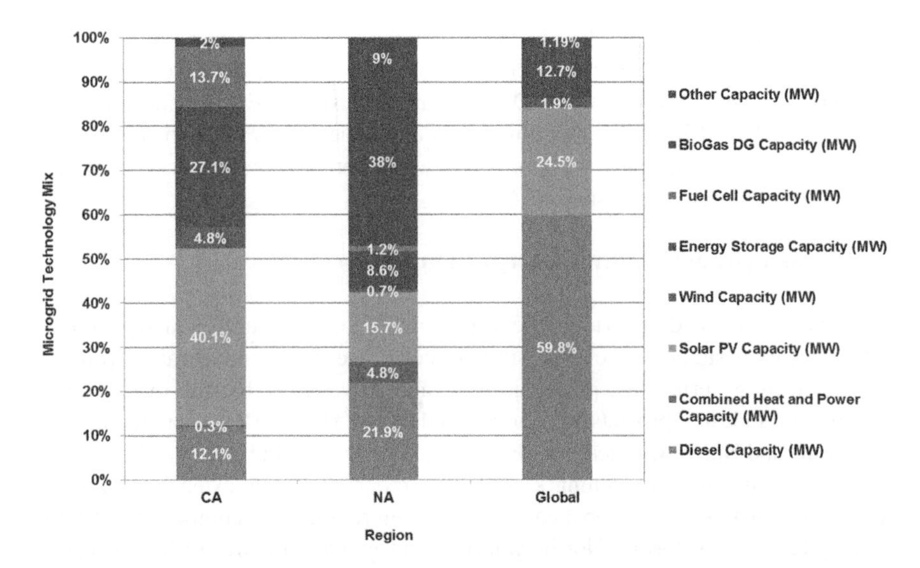

FIGURE 5.19 Comparison of DER mix, by region. Source: Asmus, Peter, Adam Forni, and Laura Vogel. Navigant Consulting, Inc. 2017. Microgrid Analysis and Case Study Report. California Energy Commission. Publication Number: CEC-5002018-022 [250].

A key aspect of the analysis was to look at the distributed energy resources (DER) deployed in each region's microgrids. Solar PV was a clear focus in California compared to the rest of the world. While Navigant Research purposefully targeted case studies that included renewable energy, results still demonstrate that North America and global case studies more often include fossil fuel generation (diesel and combined heat and power) to achieve resilience and reliability benefits through the ability

TABLE 5.10
Case Studies Examined

1. Inland Empire Utilities Agency, San Bernardino County
2. Mission Produce Facility, Oxnard
3. 2500 R Midtown Development, Sacramento
4. San Diego Zoo Solar-to-EV Project, San Diego
5. Alpha Omega Winery, Napa Valley
6. Stone Edge Farm, Sonoma
7. US Marine Corps Base Camp Pendleton, Oceanside
8. Thousand Oaks Real Estate Portfolio, Thousand Oaks
9. The Thacher School, Ojai

Source: Asmus, Peter, Adam Forni, and Laura Vogel. Navigant Consulting, Inc. 2017. Microgrid Analysis and Case Study Report. California Energy Commission. Publication Number: CEC-5002018-022) [250]

TABLE 5.11
Inland Empire Utilities Agency Value Proposition Ranking

Category	Not Important	Somewhat Important	Very Important	Essential
Reliability		●		
Resiliency			●	
Bill savings/ demand charge abatement				●
Future transactive energy revenue		●		
Provision of energy and capacity services			●	
Provision of ancillary services				●
Reduction of carbon footprint			●	
Renewable energy integration				●
Services beyond electricity (thermal, water, etc.)	●			
Linkage to virtual power plant			●	
Other: N/A	●			

Source: Navigant

Source: Asmus, Peter, Adam Forni, and Laura Vogel. Navigant Consulting, Inc. 2017. Microgrid Analysis and Case Study Report. California Energy Commission. Publication Number: CEC-5002018-022) [250]

to completely island from the grid. One factor contributing to this result is a focus primarily on projects not dependent on government subsidies. Projects that relied on at least 50% nongovernmental funding were the focus of this analysis [250].

C-1 The multidimensional value propositions of microgrids

The case studies also explored what value propositions are driving commercially viable microgrids. Despite the diversity of use cases and microgrid applications, there are common value propositions driving microgrid adoption in California, nationally, and globally. Ten value propositions were provided to project developers and owners, who then ranked the importance of each value proposition. Average results for all case studies are shown in the Table 5.12 [250].

The drivers for microgrids in California, as shown by the value proposition rankings, appear to align with state policy goals for renewable energy and carbon

TABLE 5.12
Value Proposition for Various Parameters by Microgrids (0=not important, 3=essential)

Topic	California	North America	Global
Renewable energy integration	2.7	2.2	2.7
Resiliency	2.3	2.6	2.0
Bill savings/demand charge abatement	2.3	1,7	1,7
Reduction of carbon footprint	2.3	1,5	2.6
Reliability	2.0	2.4	2.1
Provision of energy and capacity services	1.9	1.5	1.7
Provision for ancillary services	1.9	1.7	1.6
Linkage to virtual power plant	1.6	1.3	1.6
Future transactive energy revenue	1.3	1.2	1,4
Non-electricity services (thermal, water, etc.)	0.4	0.3	1.0

Source: Navigant Consulting, Inc.
Source: Asmus, Peter, Adam Forni, and Laura Vogel. Navigant Consulting, Inc. 2017. Microgrid Analysis and Case Study Report. California Energy Commission. Publication Number: CEC-5002018-022) [250]

reduction. The rankings also show the value placed on resiliency and microgrid economics, particularly for electricity bill savings and demand charge abatement. This is a driver that microgrids share with energy storage systems for commercial customers in California, where electricity rates and demand charges are quite high. In comparison, most of the microgrids profiled from North America are in states with relatively low electricity costs. The North American projects score higher on reliability and resiliency, likely due to the impact of extreme weather events in the East Coast [250].

Global microgrids reported value propositions more like California than the rest of North America, signaling a strong interest in renewable energy integration and carbon reduction (despite reliance on legacy fossil fuel generation). The relatively high scores for the provision of ancillary services and energy and capacity services speak to a recent trend of exploring ways for microgrids to offer services to the larger grid in general.

The cost of a microgrid varies greatly depending upon size, the DER resource mix, legacy assets being incorporated into the microgrid, and number and types of vendors involved. The best measure of the cost of microgrids is the cost per unit capacity ($/MW). Acknowledging the limited dataset available from these case studies (only 26 microgrid projects overall), North America is the most expensive microgrid market, followed closely by California, which is identified as a separate market. As key enabling technologies such as solar PV and energy storage experience additional price decrease, microgrid costs are expected to continue to decline over time. This trend is already apparent. The average cost of all the case study microgrids is just over $3 million/MW down significantly from data provided in the past by vendors showing average costs exceeding $4 million/MW. Global adjusted unit cost (%/MW) is just over $2 million. These costs are certainly going to drop significantly in the future [250].

5.10.3 Additional Perspectives on the Cost of Hybrid Microgrid

Diesel is the fuel of choice for standby gensets. It's reliable, easy to obtain, and can be stored on site. It also has high power density, meaning that it produces more energy per gallon than many other liquid fuels. That's one reason diesel generators are the most cost-effective to operate and maintain over long periods of time. Added to a combined photovoltaic (PV) solar and battery storage installation, diesel gensets provide the reliability communities and organizations need while enhancing the performance of all assets in the on-site power system [84, 249, 250].

The one optimization that goal hybrid systems can help C&I customers meet is demand charge reduction. "Demand ratchets" represent the extra costs businesses pay based on peak demand, which is often measured by the top 15 minutes of usage during any billing period. After examining some 10,000 utility rate structures, researchers at NREL found demand charges typically represent between 30% and 70% of a C&I customer's total electricity bill. The NREL research team also determined that there's a solid business case for the storage benefits of hybrid systems in states nationwide, not just the "first-mover" states of New York and California. Likewise, GTM Research found that where demand charges reach or exceed $15/kW, implementing 1-hour or 2-hour storage systems begins to make sense for demand charge management.

Time-based rates are another billing bane for C&I customers. Under such pricing models, researchers at the Rocky Mountain Institute, an energy think tank, found that peak rates typically run as high as seven times more than off-peak rates. With a hybrid system on site, any asset—solar, storage, or the diesel genset—can kick in when rates skyrocket. Along with high electricity costs, C&I customers face business interruption from unplanned outages. According to the ASCE (American Society of Civil Engineers), most of the nation's power lines have a 50-year life expectancy that's already been reached. High-voltage lines are at full capacity. The ASCE's report card says, "Without greater attention to aging equipment, capacity bottlenecks, and increased demand, as well as increasing storm and climate impacts, Americans will likely experience longer and more frequent power interruptions."

An on-site hybrid microgrid can minimize disruption, but running it for the greatest gain and lowest possible operating expense takes the right optimization and control software. With that software, owners can limit battery degradation by maximizing the optimal state of charge, run the generator at its sweet spot for efficiency, and use the storage as a fast-ramping spinning reserve because storage systems can inject or absorb power in sub-seconds. Such coordination lowers the operating costs for all devices, because the system runs each for maximum asset life and efficiency.

This type of coordination requires operational integration. The objective is to ensure that each resource gets dispatched when conditions warrant, that is, the software should prioritize goals—whether it's maximizing use of renewables, slashing demand charges, avoiding time-based rates, prolonging equipment life, or raising revenue by bidding into power markets. Each of these objectives can be achieved with hybrid systems that combine renewables, storage, and diesel gensets. On the island of Graciosa in Portugal, power suppliers are reducing the island's dependence on diesel. The hybrid system implemented, which includes a 4-MW battery system combined with 4.5 MW of wind energy and 1 MW of PV solar power, has enabled up to 100% renewable power penetration and dramatically decreased the levelized cost of energy for the island by replacing around two-thirds of fossil fuel-based generation with cheap renewable energy [84, 249, 250].

Kodiak Island in Alaska has a similar success story. There, a 3-MW battery storage system allowed grid operators to incorporate 9 MW of wind capacity while cutting diesel purchases by 8 million gallons since commissioning in November 2012. Diesel gensets continue to backup demand when the wind doesn't blow for long periods of time [84, 249, 250].

REFERENCES

1. Donahue E. Microgrids: Applications, solutions, case studies and demonstrations. An open access intech paper. 2019. DOI: 10.57772/intechopen.83560.
2. Hajiagashi S, Salemnla A, Hamzeh M. Hybrid energy storage system for microgrids applications: A review. *Journal of Energy Storage*. 2019;21:543–570.
3. Yeshalem M, Khan B. Microgrid integration. An intech open access paper, http://dx.doi.org/10.5772/intechopen.78634, Chapter 4. 2018. DOI:10.5772/intechopen.7863.
4. Microgrid. Wikipedia, The Free Encyclopedia. 2020 [last visited May 21, 2020].
5. Distributed generation. Wikipedia, The Free Encyclopedia. 2020 [last visited April 15, 2020].
6. Alfieri L, Bracale A, Caramia P, Carpinelli P. Distributed energy resources to improve power quality and to reduce costs of a hybrid AC/DC microgrid. An intech paper. 2017. DOI: 10.5772/intechopen. 68766.
7. Smart grid system report. A 2018 report to Congress by U.S. Department of Energy. Washington, DC; 2018.
8. Rethinking Energy-2017-accelerating global energy transformation. A report by IRENA. Abu Dhabi; 2017.
9. Kannan N. Microgrid. An open access intech report. 2019. DOI: 10.57772/intechopen.88812.
10. Department of Energy and National Energy Technology Laboratory. Environmental impacts of smart grid. Washington, DC; 2010 [Jan 10, 2010]. DOE/NETL-2010/1428. Available from: http://www.netl.doe.gov/File%20LibraryResearch/Energy%20Analysis/Publications/Envimpact_SmartGrid.pdf.
11. Chupka MW, Earle R, Fox-Penner P, Hledik R. Transforming America's power industry: The investment challenge 2010–2030. The Edison Foundation November [2008]. http://www.eei.org/ourissues/finance/Documents/Transforming_Americas_Power_Industry_Exec_Summary.pdf.
12. Unamuno E, Barrena J. Hybrid AC/DC microgrid mode-adaptive controls. An open access intech paper. 2017. DOI: 10.57772/intechopen.69026.

13. Ibrahim M, Khair A, Ansari S. A review of hybrid renewable energy systems for electric power generation. *International Journal of Engineering Research and Applications.* 2015;5(8): 42–8. www.ijera.com.
14. Dykes K, King J, DiOrio N, King R, Gevorgian V, Corbus D, Blair N, Anderson K, Stark G, Turchi C, Moriarty P. Opportunities for research and development of hybrid power plants. National Renewable Energy Laboratory; 2020. Technical Report NREL/TP-5000-75026, Contract No. DE-AC36-08GO28308.
15. Denholm PL, Eichman JD, Margolis RM. Evaluating the technical and economic performance of PV plus storage power plants. Golden, CO: National Renewable Energy Laboratory; 2017. NREL/TP-6A20-68737. Available from: https://www.nrel.gov/docs/fy17osti/68737.pdf.
16. Grid connected renewable energy sourced building. Washington, DC: Department of energy; 2020. A Website Report.
17. Tiwari P, Manas M, Jan P, Nemec Z, Radovan D, Mahanta P, Trivedi G . A review on microgrid based on hybrid renewable energy sources in south-asian perspective. *Technology and Economics of Smart Grids and Sustainable Energy.* 2017;2:10. https://doi.org/10.1007/s40866-017-0026-5.
18. García Vera YE, Dufo-López R, Bernal-Agustín JL. Energy management in microgrids with renewable energy sources: A literature review. *Applied Sciences.* 2019;9:3854. DOI: 10.3390/app9183854. www.mdpi.com/journal/applsci.
19. Hybrid renewable energy system. Wikipedia, The Free Encyclopedia; 2020 [last visited February 27, 2020].
20. Hirsch A, Parag T, Guerrero J. Microgrids: A review of technologies, key drivers and outstanding issues. *Renewable and Sustainable Energy Reviews.* 2018;90:402–11.
21. Mariam L, Basu M, Conlon MF. A review of existing microgrid architectures. *Journal of Engineering.* 2013;2013:937614. 8 p. https://doi.org/10.1155/2013/937614.
22. Smart grid. Wikipedia, The Free Encyclopedia; 2020 [last visited May 14, 2020].
23. Renewable energy. Wikipedia, The Free Encyclopedia; 2020 [last visited May 25, 2020].
24. Carpintero-Rentería M, Santos-Martín D, Guerrero JM. Microgrids literature review through a layers structure. *Energies.* 2019;12:4381. DOI: 10.3390/en12224381. www.mdpi.com/journal/energies.
25. Booth S, Reilly J, Butt R, Wasco M, Monohan R. Microgrids for energy resilience: A guide to conceptual design and lessons from defense projects. Golden, CO: NREL; 2020 [Revised January 2020]. Technical Report NREL/TP-7A40-72586.
26. Asmus P, Mackinnon L. *Emerging Microgrid Business Models.* Chicago, IL: Navigant Research; 2016..
27. Green M. Community power. *Nature Energy.* 2016;1:16014. http://dx.doi.org/10.1038/nenergy.2016.14.
28. Barker P, Johnson B, Maitra A. *Investigation of the Technical and Economic Feasibility of Micro-Grid Based Power Systems.* Palo Alto, CA: EPRI; 2001.
29. . Lovins AB, Rocky Mountain Institute (eds). *Small is Profitable.* 1st ed. Snowmass, CO: Rocky Mountain Institute; 2002.
30. Gumerman EZ, Bharvirkar RR, LaCommare KH, Marnay C. *Evaluation Framework and Tools for Distributed Energy Resources.* Berkeley, CA: Lawrence Berkeley Natl Lab; 2003. 73 p.
31. Iannucci JJ, Cibulka L, Eyer JM, Pupp RL. DER *Benefits Analysis Studies: Final Report.* Golden, CO: National Renewable Energy Laboratory; 2003.
32. Farmer BK, Wenger H, Hoff TE, Whitaker CM. Performance and value analysis ofthe Kerman 500 kW photovoltaic power plant. In: Proceedings of Am Power Conference, Chicago, IL; 1995.

33. Hoff TE, Wenger HJ, Farmer BK. Distributed generation: An alternative to electric utility investments in system capacity. *Energy Policy* 1996;24:137–47.
34. Ton DT, Smith MA. The US department of energy's microgrid initiative. *The Electric Journal.* 2012;25:84–94.
35. Marnay C, Chatzivasileiadis S, Abbey C, Iravani R, Joos G, Lombardi P, et al. Microgrid evolution roadmap. In: Proceedings of International Symposium on Smart Electric Distribution Systems and Technologies (EDST), Vienna, Austria; 2015. pp. 139–44. IEEE.
36. Lasseter R, Akhil A, Marnay C, Stephens J, Dagle J, Guttromson R, et al. The CERTS microgrid concept. *White Paper for Transmission Reliability Program, Office of Power Technologies, US Department of Energy.* 2002;2:30.
37. Hatziargyriou N, Jenkins N, Strbac G, Lopes JAP, Ruela J, Engler A, et al. Microgrids–large scale integration of microgeneration to low voltage grids. In: CIGRE, Paris, France; 2006.
38. McGowan E. *Why Two Grids Can Be Better than One: How the CERTS Microgrid Evolved from Concept to Practice.* Washington, DC: US Department of Energy.
39. Lopes JAP, Madureira AG. Moreira CCLM. A view of microgrids. *Wiley Interdisciplinary Reviews: Energy and Environment.* 2013;2:86–103. http://dx.doi.org/10.1002/wene.34.
40. Lasseter R, Piagi P. Microgrid: *A Conceptual Solution.* Germany: Aachen; 2004.
41. Marnay C, Bailey OC. *The CERTS Microgrid and the Future of the Macrogrid.* Berkeley, CA: Lawrence Berkeley National Laboratory; 2004.
42. Piagi P, Lasseter RH. Autonomous control of microgrids. In: Proceedings of IEEE Power Engineering Society General Meeting, Montreal, Quebec; 2006. 8 pp. IEEE.
43. Eto J, Lasseter R, Schenkman B, Stevens J, Klapp D, VolkommeRr H, et al. *CERTS Microgrid Laboratory Test Bed.* Berkeley, CA: Lawrence Berkeley National Laboratory; 2009. pp. 1–8.
44. Lasseter RH, Eto JH, Schenkman B, Stevens J, Vollkommer H, Klapp D, et al. CERTS microgrid laboratory test bed. *IEEE Transactions on Power Delivery.* 2011;26:325–32.
45. Alegria E, Brown T, Minear E, Lasseter RH. CERTS microgrid demonstration withlarge-scaleenergy storage and renewable generation. *IEEE Transactions on Smart Grid.* 2014;5:937–43. http://dx.doi.org/10.1109/TSG.2013.2286575.
46. Panora R, Gehret J, Furse M, Lasseter RH. Real-world performance of a CERTS microgrid in Manhattan. *IEEE Transactions on Sustainable Energy.* 2014;5:1356–60.
47. Parhizi S, LotfiH, Khodaei A, Bahramirad S. State of the art in research on microgrids: A review. *IEEE Access.* 2015;3:890–925. http://dx.doi.org/10.1109/ACCESS.2015.2443119.
48. Mariam L, Basu M, Conlon MF. Microgrid: Architecture, policy and future trends. *Renewable and Sustainable Energy Reviews* 2016;64:477–89. http://dx.doi.org/10.1016/j.rser.2016.06.037.
49. Nema P, Nema RK, Rangnekar S. A current and future state of art development ofhybrid energy system using wind and PV-solar: A review. *Renewable and Sustainable Energy Reviews.* 2009;13:2096–103. http://dx.doi.org/10.1016/j.rser.2008.10.006.
50. Asmus P, Lawrence M. Market data: Microgrids;a report by Navigent Research, Chicago, IL; 2016.
51. DeForest N. Microgrid dispatch for macrogrid peak-demand mitigation. In: Proceedings of 2012 ACEEE Summer Study on Energy Efficiency in Buildings, Asilomar Conference Center, August 12–17, 2012, Pacific Grove, CA; 2013.
52. Illindala M, Siddiqui A, Venkataramanan G, Marnay C. *Localized Aggregation of Diverse Energy Sources for Rural Electrification Using Microgrids.* Berkeley, CA: Lawrence Berkeley National Laboratory; 2006.

53. Williams NJ, Jaramillo P, Taneja J, Ustun TS. Enabling private sector investment in microgrid-based rural electrification in developing countries: A review. *Renewable and Sustainable Energy Reviews.* 2015;52:1268–81. http://dx.doi.org/10.1016/j.rser.2015.07.153.
54. Bunker K, Hawley K, Morris J. *Renewable Microgrids: Profiles from Islands and Remote Communities across the Globe.* Basalt, CO: Rocky Mountain Institute; 2015.
55. Nayar CV. Recent developments in decentralised mini-grid diesel power systems in Australia. *Applied Energy.* 1995;52:229–42.
56. Schnitzer D, Lounsbury D, Carvallo J, Deshmukh R, Apt J, Kammen D. United Nations Foundation; 2014. Available from: Micro-grids_for_Rural_Electrification-A_critical_review:of_best_practices_based_on_seven_case_studies.pdf.
57. Ustun TS, Ozansoy C, Zayegh A. Recent developments in microgrids and example cases around the world—A review. *Renewable and Sustainable Energy Reviews.* 2011;15:4030–41. http://dx.doi.org/10.1016/j.rser.2011.07.033.
58. Hazelton J, Bruce A, MacGill I. A review of the potential benefits and risks of photovoltaic hybrid mini-grid systems. *Renewable Energy.* 2014;67:222–9. http://dx.doi.org/10.1016/j.renene.2013.11.026.
59. Kuang Y, Zhang Y, Zhou B, Li C, Cao Y, Li L, et al. A review of renewable energy utilization in islands. *Renewable and Sustainable Energy Reviews.* 2016;59:504–13. http://dx.doi.org/10.1016/j.rser.2016.01.014.
60. Domenech B, Ferrer-Martí L, Lillo P, Pastor R, Chiroque J. A community elec-trification project: Combination of microgrids and household systems fed by wind, PV or micro-hydro energies according to micro-scale resource evaluation and so-cial constraints. *Energy for Sustainable Development.* 2014;23:275–85. http://dx.doi.org/10.1016/j.esd.2014.09.007.
61. Khatib T, Mohamed A, Sopian K. Optimization of a PV/wind micro-grid for rural-housing electrification using a hybrid iterative/genetic algorithm: Case study of Kuala Terengganu, Malaysia. *Energy and Buildings.* 2012;47:321–31. http://dx.doi.org/10.1016/j.enbuild.2011.12.006.
62. Palma-Behnke R, Ortiz D, Reyes L, Jimenez-Estevez G, Garrido N. A social SCADAapproach for a renewable based microgrid—The Huatacondo project. In: Proceedings of Power and Energy Society General Meet, Detroit, MI; 2011. pp. 1–7. IEEE.
63. Yangbo C, Jingding R, Kun L. Construction of multi-energy microgrid laboratory. In: 4th International Conference on Power Electronics Systems and Applications, Hongkong, China; 2011. p. 1–5. IEEE.
64. De Bosio F, Luna AC, Ribeiro L, Graells M, Saavedra OR, Guerrero JM. Analysis and improvement of the energy management of an isolated microgrid in Lencois island based on a linear optimization approach. In: IEEE Energy Conversion Congress and Exposition (ECCE), Milwaukee, WI; 2016. IEEE.
65. Naval Facilities Engineering Command. Technology transition final public report: Smart power infrastructure demonstration for energy reliability and security (SPIDERS). U.S. Department of Defense; 2015.
66. Van Broekhoven SB, Judson N, Nguyen SV, Ross WD. Microgrid study: Energy security for DoD installations. DTIC Document; 2012.
67. Van Broekhoven S, Judson N, Galvin J, Marqusee J. Leading the charge: Microgrids for domestic military installations. *IEEE Power and Energy Magazine.* 2013;11:40–5. http://dx.doi.org/10.1109/MPE.2013.2258280.
68. ABB Article. Adapting renewable energy to the data center. Carey, NC: ABB; 2015. Available from: library.e.abb.com › public › Adapting Renewable Energy.
69. Li C, Wang R, Li T, Qian D, Yuan J. Managing green datacenters powered by hybrid renewable energy systems. In: Proceedings of the11th International Conference on Autonomic Computing (ICAC, Philadelphia, PA; June 18–20, 2014.

70. Sechilariu M, Wang B, Locment F. Building-integrated microgrid: Advanced local energy management for forthcoming smart power grid communication. *Energy and Buildings*. 2013;59:236–43. http://dx.doi.org/10.1016/j.enbuild.2012.12.039.
71. Lasseter RH. Microgrids and distributed generation. *Journal of Energy Engineering*. 2007;133(3):144–9.
72. Bahramirad S, Khodaei A, Svachula J, Aguero JR. Building resilient integrated grids: One neighborhood at a time. *IEEE Electrification Magazine*. 2015;3:48–55. http://dx.doi.org/10.1109/MELE.2014.2380051.
73. Khalilpour R, Vassallo A. Leaving the grid: An ambition or a real choice? *Energy Policy*. 2015;82:207–21. http://dx.doi.org/10.1016/j.enpol.2015.03.005.
74. Suryanarayanan S, Rietz RK, Mitra J. A framework for energy management incustomer-driven microgrids. In: IEEE PES General Meeting, Basalt, CO; 2010. pp. 1–4. IEEE. http://dx.doi.org/10.1109/PES.2010.5589873.
75. Bronski P, Creyts J, Guccione L, Madrazo M, Mandel J, Rader B. et al. *The Eco-nomics of Grid Defection: When and Where Distributed Solar Generation Plus Storage Competes with Traditional Utility Service*. Basalt, CO: Rocky Mountain Institute; 2014.
76. Rodriguez-Diaz E, Vasquez JC, Guerrero JM. Intelligent DC homes in future sus-tainable energy systems: When efficiency and intelligence work together. *IEEE Consumer Electronics Magazine*. 2016;5:74–80. http://dx.doi.org/10.1109/MCE.2015.2484699.
77. Black and Veatch strategic directions: Smart cities and utilities report. Virginia Beach, VA: Black and Veatch; 2018. A Website Report.
78. Campbell R. Weather related power outages and electric system resiliency. A report by congressional research service [August 28, 2012]. Available from: www.crs.gov R42696. Report No.: 7-5700.
79. Hatziargyriou N, Asano H, Iravani R, Marnay C. Microgrids. *IEEE Power and Energy Magazine*. 2007;5:78–94.
80. Asmus P. Microgrids: Friend or foe? *Public Utilities Fortnightly*. 2015:18–23.
81. Asmus P. Microgrids, virtual power plants and our distributed energy future. *The Electricity Journal*. 2010;23:72–82. http://dx.doi.org/10.1016/j.tej.2010.11.001.
82. Myles P, Miller J, Knudsen S, Grabowski T. *430.01.03 Electric Power System Asset Optimization*. Morgantown, WV: National Energy Technology Laboratory; 2011.
83. John JS. How microgrids helped weather hurricane sandy. November 20, 2012 [Accessed November 24, 2012]. Available from: http://www.greentechmedia.com/articles/read/how-microgrids-helped-weather-hurricane-sandy.
84. Saury F, Tomlinson C. Hybrid microgrids: The time is now a website report by Caterpillar; February 2016.
85. Newman D. Right-sizing the grid. *Mechanical Engineering*. 2015;137:34.
86. U.S.-Canada Power System Outage Task Force. Final report on the August 14, 200 3blackout in the United States and Canada: Causes and recommendations. Washington, DC: U.S. Department of Energy; 2004.
87. Wang W, Lu Z. Cyber security in the smart grid: Survey and challenges. Computer Networks. 2013;57:1344–71. http://dx.doi.org/10.1016/j.comnet.2012.12.017.
88. Baker D, Balstad R, Bodeau JM, Cameron E, Fennell J. Severe space weather events–understanding societal and economic impacts: A workshop report. Washington, DC: National Academies Press; 2008.
89. Maize K. The Great Solar Storm of 2012? POWER Magazine; 2011 [Accessed September 13, 2016]. Available from: http://www.powermag.com/the-great-solar-storm-of-2012/.
90. Foster Jr JS, Gjelde E, Graham WR, Hermann RJ, Kluepfel HM, Lawson RL. et al. Report of the commission to assess the threat to the united states from electro-magnetic pulse (EMP) Attack. Volume 1: Executive Report. DTIC Document; 2004.

91. Maize K. EMP: The biggest unaddressed threat to the grid. Power Magazine; 2013 [Accessed May 30, 2016]. Available from: http://www.powermag.com/emp-the-biggest-unaddressed-threat-to-the-grid/.
92. Miller CR. Electromagnetic pulse threats in 2010. DTIC Document; 2005.
93. Wilson C. High altitude electromagnetic pulse (HEMP) and high power microwave (HPM) devices: Threat assessments, DTIC Document; 2008.
94. Perez EUS. Investigators find proof of cyberattack on Ukraine power grid. CNN; 2016 [Accessed September 13, 2016]. http://www.cnn.com/2016/02/03/politics/cyberattack-ukraine-power-grid/index.html.
95. Times of Israel. Steinitz: Israel's electric authority hit by severe cyber-attack; 2016 [Accessed May 22, 2016]. Available from: http://www.timesofisrael.com/steinitz-israels-electric-authority-hit-by-severe-cyber-attack/.
96. Smith R. Assault on California power station raises alarm on potential for terrorism. A website report by *Wall Street Journal.* (Feb. 6, 2014).
97. Smith R. How America could go dark. A website report by *Wall Street Journal.* July 14, 2016.
98. President's Council of Economic Advisers, U.S. Department of Energy Office of Electricity Delivery and Energy Reliability. Economic benefits of increasing elec-tric grid resilience to weather outages. Washington, DC: Executive Office of the President, U.S. Department of Energy; 2013.
99. Tweed K. Con. (editor) Looks to batteries, microgrids and efficiency to delay $1Bsubstation build. 2014 [Accessed September 18, 2016]. Available from: http://www.greentechmedia.com/articles/read/con-ed-looks-to-batteries-microgrids-and-efficiency-to-delay-1b-substation.
100. U.S. Department of Energy, U.S. Environmental Protection Agency. Combined heat and power–A clean energy solution. Washington, DC: .S. Environmental Protection Agency; 2012.
101. Braun M. Technological control capabilities of DER to provide future ancillary services. International Journal of Distributed Energy Resources. 2007;3:191–206.
102. Morris GY, Abbey C, Joos G, Marnay C. A framework for the evaluation of the costand benefits of microgrids. In: CIGRÉ International Symposium, The Electric Power System of the Future-Integrating Supergrids and Microgrids, Bologna, Italy, September 13–15, 2011; 2012.
103. Bhatnagar D, Currier A, Hernandez J, Ma O, Kirby B. Market and policy barriers to energy storage deployment. Washington, DC: Sandia National Laboratories/Office of Energy Efficiency and Renewable Energy; 2013.
104. Byrne RH, Concepcion RJ, Silva-Monroy CA. Estimating potential revenue from electrical energy storage in PJM. In: 2016 IEEE Power and Energy Society General Meeting (PESGM), Boston, MA; 2016. pp. 1–5. IEEE.
105. Denholm P, O'Connell M, Brinkman G, Jorgenson J. Overgeneration from solar energy in California: A field guide to the duck chart. Golden, CO: National Renewable Energy Laboratory; Nov 2015. Techical Report No.: NRELTP-6A20-65023.
106. California Independent System Operator. What the duck curve tells us about managing a green grid;a website report by California ISO, shaping a renewed future 2016.
107. Liu E, Bebic J. Power system planning: Emerging practices suitable for evaluating the impact of high-penetration photovoltaics. Golden, CO: National Renewable Energy Laboratory; 2008.
108. Hoke A, Hambrick RBJ, Kroposki B. Maximum photovoltaic penetration levels on typical distribution feeders. Golden, CO: National Renewable Energy Laboratory; 2012.

109. Martin R. Texas and California have a bizarre problem: Too much renewable en-ergy. MIT Technology Review; 2016 [Accessed September 18, 2016]. Available from: https://www.technologyreview.com/s/601221/texas-and-california-have-too-much-renewable-energy/.
110. Martin R. Loading up on wind and solar is causing new problems for Germany. MIT Technology Review; 2016 [Accessed September 18, 2016]. Available fron: https://www.technologyreview.com/s/601514/germany-runs-up-against-the-limits-of-renewables/.
111. Pachauri RK, Allen MR, Barros VR, Broome J, Cramer W, Christ R. et al. Climatechange: 2014: Synthesis Report. Contribution of working groups I, II and III to the fifth assessment report of the intergovernmental panel on climate change. Geneva, Switzerland: Intergovernmental Panel on Climate Change; 2014.
112. Shmelev SE, van den Bergh JCJM. Optimal diversity of renewable energy alter-natives under multiple criteria: An application to the UK. *Renewable and Sustainable Energy Reviews.* 2016;60:679–91. http://dx.doi.org/10.1016/j.rser.2016.01.100.
113. Gamarra C, Guerrero JM. Computational optimization techniques applied to microgrids planning: A review. *Renewable and Sustainable Energy Reviews.* 2015;48:413–24. http://dx.doi.org/10.1016/j.rser.2015.04.025.
114. Shmelev SE. *Ecological Economics: Sustainability in Practice.* New York: Springer; n.d.
115. Akorede MF, Hizam H, Pouresmaeil E. Distributed energy resources and benefits tothe environment. *Renewable and Sustainable Energy Reviews.* 2010;14:724–34. http://dx.doi.org/10.1016/j.rser.2009.10.025.
116. Bayindir R, Hossain E, Kabalci E, Perez R. A comprehensive study on microgrid technology. International Journal of Renewable Energy Research. 2014;4:1094–107.
117. El-Khattam W, Salama MM. Distributed generation technologies, definitions and benefits. *Electric Power Systems Research.* 2004;71:119–28. http://dx.doi.org/10.1016/j.epsr.2004.01.006.
118. Abusharkh S, Arnold R, Kohler J, Li R, Markvart T, Ross J, et al. Can microgrids make a major contribution to UK energy supply? *Renewable and Sustainable Energy Reviews.* 2006;10:78–127. http://dx.doi.org/10.1016/j.rser.2004.09.013.
119. Díaz-González F, Sumper A, Gomis-Bellmunt O, Villafáfila-Robles R. A review of energy storage technologies for wind power applications. Renewable and Sustainable Energy Reviews. 2012;16:2154–71. http://dx.doi.org/10.1016/j.rser.2012.01.029.
120. Suvire GO, Mercado PE, Ontiveros LJ. Comparative analysis of energy storage technologies to compensate wind power short-term fluctuations. In: EEE/PES Transmission and Distribution Conference and Exposition: Latin America (T&D-LA), Sao Paulo, Brazil; 2010. pp. 522–8. IEEE. http://dx.doi.org/10.1109/TDC-LA.2010.5762932.
121. Giraldez Miner JI, Flores-Espino F, MacAlpine S, Asmus P. Phase i microgrid cost study: Data collection and analysis of microgrid costs in the United States. Golden, CO: National Renewable Energy Laboratory; 2018. Technical Report NREL/TP-5D00-67821.
122. Hotchkiss E, Cox S, Resilient energy platform power sector resilience technical solutions. Golden, CO: National Renewable Energy Laboratory; May 2019. Available from: www.resilient-energy.org, www.nrel.gov/usaid-partnership.
123. Schwaegrl C, Tao L, Lopes JAP, Madureira A, Mancarella P, Anastasiadis A, et al. Report on the technical, social, economic, and environmental benefits provided by Microgrids on power system operation; 2009.
124. Parag P. Beyond energy efficiency: A "prosumer market" as an integrated platform for consumer engagement with the energy system. In: ECEEE 2015 Summer Study on Energy Efficiency. 2015. pp. 15–23.
125. Parag Y, Sovacool BK. Electricity market design for the prosumer era. *Nature Energy.* 2016;1:16032. http://dx.doi.org/10.1038/nenergy.2016.32.

126. Linnenberg T, Wior I, Schreiber S, Fay A. A market-based multi-agent-system for decentralized power and grid control. In: Proceedings of the 16th Conference On Emerging Technologies and Factory Automation (ETFA 2011), Toulouse, France; 2011. pp. 1–8. IEEE.
127. Lacey S. The energy blockchain: How bitcoin could be a catalyst for the distributed grid. 2016 [Accessed February 28, 2016]. Available from: http://www.greentechmedia.com/articles/read/the-energy-blockchain-could-bitcoin-be-a-catalyst-for-the-distributed-grid.
128. Mihaylov M, Razo-Zapata I, Rădulescu R, Nowé A. Boosting the renewable energy economy with NRG coin; Series: Advances in Computer Science Research, Proceedings of ICT for Sustainability-Online Report 2016. https://doi.org/10.2991/ict4s-16.2016.27.
129. Madureira AG, Peças Lopes JA. Ancillary services market framework for voltage-control in distribution networks with microgrids. Electric Power Systems Research. 2012;86:1–7. http://dx.doi.org/10.1016/j.epsr.2011.12.016.
130. Distributed energy resources roadmap for New York's wholesale electricity markets. New York Independent System Operator; 2017.
131. Lopes JAP, Vasiljevska J, Ferreira R, Moreira C, Madureira A. Advanced architectures and control concepts for more microgrids; 2009, a website report by European Commison CORDIS, Contract No: SES6-019864.
132. Dragicevic T, Vasquez JC, Guerrero JM, Skrlec D. Advanced LVDC electrical power-architectures and microgrids: A step toward a new generation of power distribution networks. *IEEE Electrification Magazine*. 2014;2:54–65. http://dx.doi.org/10.1109/MELE.2013.2297033.
133. John JS. A current in every ceiling. 2008 [Acceesed May 4, 2016]. Available from; http://www.greentechmedia.com/articles/read/a-current-in-every-ceiling-5278.
134. Justo JJ, Mwasilu F, Lee J, Jung J-W. AC-microgrids versus DC-microgrids with distributed energy resources: A review. *Renewable and Sustainable Energy Reviews*. 2013;24:387–405. http://dx.doi.org/10.1016/j.rser.2013.03.067.
135. Che L, Shahidehpour M. DC Microgrids: Economic operation and enhancement of resilience by hierarchical control. *IEEE Transactions on Smart Grid*. 2014;5:2517–26. http://dx.doi.org/10.1109/TSG.2014.2344024.
136. Patterson BT. DC, come home: DC microgrids and the birth of the "enernet." *IEEE Power and Energy Magazine*. 2012;10:60–9.
137. Guerrero JM, Loh PC, Lee T-L, Chandorkar M. Advanced control architectures for intelligent microgrids Part II: Power quality, energy storage, and AC/DC micro-grids. IEEE Transactions on Industrial Electronics. 2013;60:1263–70. http://dx.doi.org/10.1109/TIE.2012.2196889.
138. Katiraei F, Iravani R, Hatziargyriou N, Dimeas A. Microgrids management. *IEEE Power and Energy Magazine*. 2008;6:54–65. http://dx.doi.org/10.1109/MPE.2008.918702.
139. Madureira AG, Pereira JC, Gil NJ, Lopes JAP, Korres GN, Hatziargyriou ND. Advanced control and management functionalities for multi-microgrids. *European Transactions on Electrical Power*. 2011;21:1159–77. http://dx.doi.org/10.1002/etep.407.
140. Shuai Z, Sun Y, Shen ZJ, Tian W, Tu C, Li Y, et al. Microgrid stability: Classification and a review. *Renewable and Sustainable Energy Reviews*. 2016;58:167–79. http://dx.doi.org/10.1016/j.rser.2015.12.201.
141. Olivares DE, Mehrizi-Sani A, Etemadi AH, Canizares CA, Iravani R, Kazerani M, et al. Trends in microgrid control. *IEEE Transactions on Smart Grid*. 2014;5:1905–19. http://dx.doi.org/10.1109/TSG.2013.2295514.
142. Palizban O, Kauhaniemi K, Guerrero JM. Microgrids in active network manage-ment–Part II: System operation, power quality and protection. *Renewable and Sustainable Energy Reviews*. 2014;36:440–51. http://dx.doi.org/10.1016/j.rser.2014.04.048.

143. Ding G, Gao F, Zhang S, Loh PC, Blaabjerg F. Control of hybrid AC/DC microgrid under islanding operational conditions. *Journal of Modern Power Systems and Clean Energy*. 2014;2:223–232. https://doi.org/10.1007/s40565-014-0065-z.
144. Palizban O, Kauhaniemi K, Guerrero JM. Microgrids in active networkmanagement—Part I: Hierarchical control, energy storage, virtual power plants, and market participation. *Renewable and Sustainable Energy Reviews*. 2014;36:428–39. http://dx.doi.org/10.1016/j.rser.2014.01.016.
145. Kaur A, Kaushal J, Basak P. A review on microgrid central controller. *Renewable and Sustainable Energy Reviews*. 2016;55:338–45. http://dx.doi.org/10.1016/j.rser.2015.10.141.
146. Meng L, Sanseverino ER, Luna A, Dragicevic T, Vasquez JC, Guerrero JM. Microgrid supervisory controllers and energy management systems: A literature review. *Renewable and Sustainable Energy Reviews*. 2016;60:1263–73. http://dx.doi.org/10.1016/j.rser.2016.03.003.
147. Roche R, Blunier B, Miraoui A, Hilaire V, Koukam A. Multi-agent systems for grid energy management: A short review. In: Proceedings of IECON 2010-36th Annual Conference on IEEE Industrial Electronics Society; 2010, pp. 3341–6. IEEE.
148. Kantamneni A, Brown LE, Parker G, Weaver WW. Survey of multi-agent systems for microgrid control. *Engineering Applications of Artificial Intelligence*. 2015;45:192–203. http://dx.doi.org/10.1016/j.engappai.2015.07.005.
149. Dimeas AL, Hatziargyriou ND. Agent based control of virtual power plants. In: Proceedings of ISAP 2007 International Conference on Intelligent Systems Applications to Power Systems, Kaohsiung, Taiwan; 2007. pp. 1–6. IEEE.
150. Chatzivasiliadis SJ, Hatziargyriou ND, Dimeas AL. Development of an agent based intelligent control system for microgrids. In: Proceedings of IEEE Power and Energy Society General Meeting-Conversion and Delivery of Electrical Energy in the 21st Century, Pittsburgh, Pennsylvania; 2008; 2008. pp. 1–6. IEEE.
151. Colson CM, Nehrir MH, Wang C. Ant colony optimization for microgrid multi-objective power management. In: Proceedings of EEE/PES Power Systems Conference and Exposition (PSCE09), Seattle, Washington; 2009. pp. 1–7. IEEE.
152. Serna-Suárez ID, Ordóñez-Plata G, Carrillo-Caicedo G. Microgrid's energy management systems: A survey. In: Proceedings of the 12th International Conference on European Energy Market (EEM), Lisbon, Portugal; 2015. pp. 1–6.
153. Kurohane K, Senjyu T, Uehara A, Yona A, Funabashi T, Kim CH. A hybrid smart AC/DC power system. *IEEE Transactions on Smart Grid*. 2010;1(2):199–204.
154. De Brabandere K, Bolsens B, Van den Keybus J, Woyte A, Driesen J, Belmans R. A voltage and frequency droop control method for parallel inverters. *IEEE Transactions on Power Electronics*. 2007;22(4):1107–15.
155. Prodanovic M, Green TC. High-quality power generation through distributed control of a power park microgrid. *IEEE Trans Ind Electron*. 2006;53(5):1471–82.
156. Marwali MN, Jung JW, Keyhani A. Control of distributed generation systems: Part II. Load sharing control. *IEEE Transactions on Power Electronics*. 2004;19(6):1551–61.
157. Martins AP, Carvalho AS, Araujo AS. Design and implementation of a current controller for the parallel operation of standard upss. In: Proceedings of the 1995 IEEE IECON, 21st International Conference on Industrial Electronics, Control, and Instrumentation, vol. 1 Orlando, FL; Nov 6–10, 1995. pp. 584–9.
158. Chen JF, Chu CL. Combination voltage-controlled and current-controlled PWM inverters for UPS parallel operation. *IEEE Transactions on Power Electronics*. 1995;10(5):547–58.

159. Wu TF, Chen YK, Huang YH. 3C strategy for inverters in parallel operation achieving an equal current distribution. *IEEE Transactions on Industrial Electronics.* 2000;47(2):273–81.
160. Sun X, Lee YS, Xu DH. Modeling, analysis, and implementation of parallel multi-inverter systems with instantaneous average-current-sharing scheme. *IEEE Transactions on Power Electronics.* 2003;18(3):844–56.
161. Guerrero JM, Vasquez JC, Matas J, De Vicuña LG, Castilla M. Hierarchical control of droop-controlled ac and dc microgrids: A general approach toward standardization. *IEEE Transactions on Industrial Electronics.* 2011;58(1):158–72.
162. Guerrero JM, Matas J, De Vicuna LG, Castilla M, Miret J. Wireless-control strategy for parallel operation of distributed generation inverters. *IEEE Transactions on Industrial Electronics.* 2006;53(5):1461–70.
163. Zhang Q. Robust droop controller for accurate proportional load sharing among inverters operated in parallel. *IEEE Transactions on Industrial Electronics.* 2013;60(4):1281–90.
164. He JW, Li YW. Analysis, design and implementation of virtual impedance for power electronics interfaced distributed generation. *IEEE Transactions on Industry Applications.* 2011;47(6):2525–38.
165. Xianghui C, Jiming C, Yang X, Youxian S. Building-environment control with wireless sensor and actuator networks: Centralized versus distributed. *IEEE Transactions on Industrial Electronics.* 2010;57(11):3596–605.
166. Wang BC, Sechilariu M, Locment F. Intelligent DC microgrid with smart grid communications: Control strategy consideration and design. *IEEE Transactions on Smart Grid.* 2012;3(4):2148–56.
167. Sun K, Zhang L, Xing Y, Guerrero JM. A distributed control strategy based on DC bus signaling for modular photovoltaic Generation systems with battery Energy storage. *IEEE Transactions on Power Electronics.* 2011;26(10):3032–45.
168. Schonberger J, Duke R, Round SD. DC-bus signaling: A distributed control strategy for a hybrid renewable nanogrid. *IEEE Transactions on Industrial Electronics.* 2006;53(5):1453–60.
169. Zhang L, Wu T, Xing Y, Sun K, Gurrero JM. Power control of DC microgrid using DC bus signaling. In: Proceedings of Twenty-Sixth Annual IEEE Applied Power Electronics Conference and Exposition, Fort Worth, TX; March 6–11, 2011. pp. 1926–32. IEEE.
170. Liu X, Wang P, Loh PC. A hybrid AC/DC microgrid and its coordination control. *IEEE Transactions on Smart Grid.* 2011;2(2):278–86.
171. Jiang Z, Yu X. Hybrid dc- and ac-linked microgrids: towards integration of distributed energy resources. In: Proceedings of the IEEE Energy 2030 Conference, November 17–18, 2008, Atlanta, GA; 2008. 8p.
172. Shahnia F, Majumder R, Ghosh A, Ledwich G, Zare F. Operation and control of a hybrid microgrid containing unbalanced and nonlinear loads. *Electric Power Systems Research.* 2010;80(8):954–65.
173. Jin J, Loh PC, Wang P, Mi Y, Blaabjerg F. Autonomous operation of hybrid ac–dc microgrids. In: Proceedings 2010 IEEE International Conference on Sustainable Energy Technologies (ICSET), Kandy Sri Lanka; December 6–9, 2010. pp. 1–7.
174. Loh PC, Blaabjerg F. Autonomous control of distributed storages in microgrids. In: Proceedings of the IEEE 8th International Conference on Power Electronics and ECCE Asia, 30 May–3 Jun 2011, Jeju, South Korea; 2011. pp 536–42.
175. Loh PC, Li D, Chai YK, Blaabjerg F. Autonomous operation of hybrid microgrid with ac and dc subgrids. *IEEE Transactions on Power Electronics.* 2013;28(5):2214–23.

176. Loh PC, Li D, Chai YK, Blaabjerg F. Autonomous control of interlinking converter with energy storage in hybrid ac–dc microgrid. *IEEE Transactions on Industrial Applications.* 2013;49(3):1374–82.
177. Carrasco JM, Franquelo LG, Bialasiewicz JT, Galván E, PortilloGuisado RC, Prats MM, et al. Power-electronic systems for the grid integration of renewable energy sources: A survey. *IEEE Transactions on Industrial Electronics.* 2006;53(4):1002–16.
178. Li YW, Vilathgamuwa DM, Loh PC. Design, analysis and real-time testing of controllers for multibus microgrid system. *IEEE Transactions on Power Electronics.* 2004;19(5):1195–204.
179. Katiraei F, Iravani MR. Power management strategies for a microgrid with multiple distributed generation units. *IEEE Transactions on Power Systems.* 2006;21(4):1821–31.
180. Li Y, Nejabatkhah F. Overview of control, integration and energy management of microgrids. *Journal of Modern Power Systems and Clean Energy.* 2014;2:212–22. https://doi.org/10.1007/s40565-014-0063-1.
181. Loh PC, Li D, Chai YK, Blaabjerg F. Autonomous operation of hybrid microgrid with ac and dc subgrids. *IEEE Transactions on Power Electronics.* 2013;28:2214–23. DOI: 10.1109/ TPEL.2012.2214792.
182. Baharizadeh M, Karshenas HR, Guerrero JM. Control strategy of interlinking converters as the key segment of hybrid AC-DC Microgrids. *IET Generation, Transmission & Distribution.* 2016;10:1671–81. DOI: 10.1049/iet-gtd.2015.1014.
183. Fregosi D, Ravula S, Brhlik D, Saussele J, Frank S, Bonnema E, et al. A comparative study of DC and AC microgrids in commercial buildings across different climates and operating profiles. To be presented at the IEEE First International Conference on DC Microgrids, Atlanta, Georgia, May 24–27, 2015;2015. Conference Paper NREL/CP-5500-63959 April 2015, Contract No. DE-AC36-08GO28308. Golden, CO: National Renewable Energy Laboratory.
184. Xue Y, Chang L, Kjaer SB, Bordonau J, Shimizu T. Topologies of single-phase inverters for small distributed power generators: An overview. *IEEE Transactions on Power Electronics.* 2004;19(5):1305–14.
185. Jiang W, Fahimi B. Multiport power electronic interface—Concept, modeling, and design. *IEEE Transactions on Power Electronics.* 2011;26(7):1890–1900.
186. Chen YM, Liu YC, Hung SC, Cheng CS. Multi-input inverter for grid-connected hybrid PV/wind power system. *IEEE Transactions on Power Electronics.* 2007;22(3):1070–7.
187. Sarhangzadeh M, Hosseini SH, Sharifian MBB, Gharehpetian GB. Multiinput direct DC–AC converter with high-frequency link for clean power-generation systems. *IEEE Transactions on Power Electronics.* 2011;26(6):1777–89.
188. Nejabatkhah F, Danyali S, Hosseini SH, Sabahi M, Niapour SM. Modeling and control of a new three-input DC–DC boost converter for hybrid PV/FC/battery power system. *IEEE Transactions on Power Electronics.* 2012;27(5):2309–24.
189. Tao H, Duarte JL, Hendrix MAM. Multiport converters for hybrid power sources. In: Proceedings of the IEEE Power Electronics Specialists Conference (PESC'08), June 15–19, 2008, Rhodes; 2008. pp 3412–18. IEEE.
190. Peng FZ. Z-source inverter. *IEEE Transactions on Industrial Applications.* 2003;39(2):504–510.
191. Alepuz S, Busquets-Monge S, Bordonau J, Gago J, González D, Balcells J. Interfacing renewable energy sources to the utility grid using a three-level inverter. *IEEE Transactions on Industrial Electronics.* 2006;53(5):1504–11.
192. Yazdani A, Iravani R. A neutral-point clamped converter system for direct-drive variable-speed wind power unit. *IEEE Transactions on Energy Conversion.* 2006;21(2):596–609.

193. Villanueva E, Correa P, Rodriguez J, Pacas M. Control of a single-phase cascaded H-bridge multilevel inverter for grid-connected photovoltaic systems. *IEEE Transactions on Industrial Electronics*. 2009;56(11):4399–4406.
194. Akagi H. Classification, terminology, and application of the modular multilevel cascaded converter (MMCC). *IEEE Transactions on Power Electronics*. 2011;26(11):3119–30.
195. Islam MR, Guo Y, Zhu J. A high-frequency link multilevel cascaded medium-voltage converter for direct grid integration of renewable energy systems. *IEEE Transactions on Power Electronics*. 29(8):4167–82.
196. Islam SZ, Mariun N, Hizam H, Othman ML, Radzi MA, Hanif M, Abidin IZ. Communication for distributed renewable generations (DRGs): A review on the penetration to smart grids (SGs). In: Proceedings of the 2012 IEEE International Conference on Power and Energy (PECON, Kota Kinabalu; December 2–5, 2012: pp. 870–5.
197. Bouhafs F, Mackay M, Merabti M. Links to the future: Communication requirements and challenges in the smart grid. *IEEE Power and Energy Magazine*. 2012;10(1):24–32.
198. Nehrir MH, Wang C, Strunz K, Aki H, Ramakumar R, Bing J, et al. A review of hybrid renewable/alternative energy systems for electric power generation: Configurations, control, and applications. *IEEE Transactions on Sustainable Energy*. 2011;2(4):392–403.
199. Colet-Subirachs A, Ruiz-Alvarez A, Gomis-Bellmunt O, Alvarez-Cuevas-Figuerola F, Sudria-Andreu A. Centralized and distributed active and reactive power control of a utility connected microgrid using IEC61850. *IEEE Systems Journal*. 2012;6(1):58–67.
200. Tan KT, Peng XY, So PL, Chu YC, Chen MZ. Centralized control for parallel operation of distributed generation inverters in microgrids. *IEEE Transactions on Smart Grid*. 3(4):1977–87.
201. Wang C, Nehrir MH. Power management of a stand-alone wind/photovoltaic/fuel-cell energy system. *IEEE Transactions on Energy Conversion*. 2008;23(3):957–67.
202. Cheng YJ, Sng EKK. A novel communication strategy for decentralized control of paralleled multi-inverter systems. *IEEE Transactions on Power Electronics*. 2006;21(1):148–56.
203. Colson CM, Nehrir MH. Comprehensive real-time microgrid power management and control with distributed agents. *IEEE Transactions on Smart Grid*. 2013;4(1):617–27.
204. Ren F, Zhang M, Sutanto D. A multi-agent solution to distribution system management by considering distributed generators. *IEEE Transactions on Power Systems*. 2013;28(2):1442–51.
205. Zhao P, Suryanarayanan S, Simoes MG. An energy management system for building structures using a multi-agent decision-making control methodology. *IEEE Transactions on Industrial Applications*. 2013;49(1):322–30.
206. Ko HS, Jatskevich J. Power quality control of wind-hybrid power generation system using fuzzy-LQR controller. *IEEE Transactions on Energy Conversion*. 2007;22(2):516–27.
207. Dou CX, Duan ZS, Liu B. Two-level hierarchical hybrid control for smart power system. *IEEE Transactions on Automation Science and Engineering*. 2013;10(4):1037–49.
208. Jiang Z, Dougal RA. Hierarchical microgrid paradigm for integration of distributed energy resources. In: Proceedings of the Power and Energy Society General Meeting—Conversion and Delivery of Electrical Energy in the 21st Century (PES'08), Pittsburgh; 20–24, 2008. 8 p.
209. Chandorkar MC, Divan DM, Adapa R. Control of parallel connected inverters in standalone AC supply systems. *IEEE Transactions on Industrial Applications*. 1993;29(1–1):136–43.
210. Guerrero JM, De Vicuña LG, Matas J, Castilla M, Miret J. A wireless controller to enhance dynamic performance of parallel inverters in distributed generation systems. *IEEE Transactions on Power Electronics*. 2004;19(5):1205–13.

211. Li Y, Li YW. Power management of inverter interfaced autonomous microgrid based on virtual frequency-voltage frame. *IEEE Transactions on Smart Grid.* 2011;2(1):30–40.
212. Li YW, Kao CN. An accurate power control strategy for power-electronics-interfaced distributed generation units operating in a low-voltage multibus microgrid. *IEEE Transactions on Power Electronics.* 2009;24(12):2977–88.
213. Blaabjerg F, Teodorescu R, Liserre M, Timbus AV. Overview of control and grid synchronization for distributed power generation systems. *IEEE Transactions on Industrial Electronics.* 2006;53(5):1398–1409.
214. Yazdani A, Iravani R. *Voltage-sourced converter in power systems: Modelling, control, and application.* New York: Wiely; 2010.
215. Pogaku N, Prodanovic M, Green TC. Modeling, analysis and testing of autonomous operation of an inverter-based microgrid. *IEEE Transactions on Power Electronics.* 2007;22(2):613–25.
216. Li YW. Control and resonance damping of voltage-source and current-source converters with LC filters. *IEEE Transactions on Industrial Electronics.* 2009;56(5):1511–21.
217. Zhong QC, Weiss G. Synchronverters: inverters that mimic synchronous generators. *IEEE Transactions on Industrial Electronics.* 2011;58(4):1259–67.
218. Marei MI, Abdel-Galil TK, El-Saadany EF, Salama MM. Hilbert transform based control algorithm of the DG interface for voltage flicker mitigation. *IEEE Transactions on Power Delivery.* 2005;20(2–1):1129–33.
219. Camacho A, Castilla M, Miret J, Vasquez JC, Alarcón-Gallo E. Flexible voltage support control for three-phase distributed generation inverters under grid fault. *IEEE Transactions on Industrial Electronics.* 2013;60(4):1429–41.
220. Li YW, Vilathgamuwa DM, Loh PC. Microgrid power quality enhancement using a three-phase four-wire grid-interfacing compensator. *IEEE Transactions on Industrial Applications.* 2005;41(6):1707–19.
221. He J, Li YW, Munir MS. A flexible harmonic control approach through voltage-controlled DG–grid interfacing converters. *IEEE Transactions on Industrial Electronics.* 2012;59(1):444–56.
222. Lee TL, Cheng PT, Akagi H, Fujita H. A dynamic tuning method for distributed active filter systems. *IEEE Transactions on Industrial Applications.* 2008;44(2):612–23.
223. Vandoorn TL, Vasquez JC, De Kooning J, Guerrero JM, Vandevelde L. Microgrids: Hierarchical control and an overview of the control and reserve management strategies. IEEE Industrial Electronics Magazine. 2013;7(4):42–5.
224. Zhong Q-C. Virtual synchronous machines: A unified interface for grid integration. *IEEE Power Electronics Magazine.* 2016;3:18–27. DOI: 10.1109/MPEL.2016.2614906.
225. Palensky P, Dietrich D. Demand side management: Demand response, intelligent energy systems, and smart loads. *IEEE Transactions of Industrial Informatics.* 2011;7:381–8. DOI:10.1109/TII.2011.2158841.
226. Unamuno E, Barrena JA. Hybrid ac/dc microgrids—Part I: Review and classification of topologies. *Renewable and Sustainable Energy Review.* 2015;52:1251–9. DOI: 10.1016/j. rser.2015.07.194.
227. Center for Energy, Marine Transportation and Public Policy at Columbia University. Microgrids: An assessment of the value, opportunities and barriers to deployment in New York State. New York: New York State Energy Research and Development Authority; 2010.
228. Kema DNV. Microgrids–benefits, models, barriers and suggested policy initiatives for the commonwealth of Massachusetts; a website report by Massachusetts Clean Energy Center 2014.

229. Burr MT, Zimmer MJ, Meloy B, Bertrand J, Levesque W, Warner G. et al. Minnesota microgrids; a website report FINAL REPORT: September 30, 2013 Prepared by Microgrid Institute for the Minnesota Department of Commerce
230. Maryland Energy Administration. Maryland resiliency through microgrids task force report. Maryland: Maryland Energy Administration2014.
231. DeBlasio D. Cycle of innovation. Fortnightly; 2013 [Accessed September 20, 2016]. Available from: https://www.fortnightly.com/fortnightly/2013/07/cycle-innovation.
232. Basso TS. IEEE 1547 and 2030 standards for distributed energy resources inter-connection and interoperability with the electricity grid. Golden, CO: National Renewable Energy Laboratory; 2014.
233. Borenstein S, Bushnell J. *The US Electricity Industry after 20 Years of Restructuring.* Cambridge, MA: National Bureau of Economic Research; 2015.
234. Romankiewicz J, Marnay C, Zhou N, Qu M. Lessons from international experience for China's microgrid demonstration program. *Energy Policy.* 2014;67:198–208. http://dx.doi.org/10.1016/j.enpol.2013.11.059.
235. Malkin D, Centolella P. Results-based regulation. Fortnightly; 2014 [Accessed October 19, 2014]. Available from: https://www.fortnightly.com/fortnightly/2014/03/results-based-regulation.
236. Energy Future Coalition. Maryland Utility 2–0 Pilot Project, a website report by Maryland Utility, Maryland; 2013.
237. Massachusetts Department of Public Utilities. Investigation by the department of public utilities on its own motion into the modernization of the electric grid, a website report, The Commonwealth of Massachusetts, Department of Public Utlities, Maryland; 2014.
238. New York Public Service Commission. Reforming the Energy Vision: NYS Department of Public Service Staff Report and Proposal. 2014. Case 14-M–0101.
239. Barton J, Emmanuel-Yusuf D, Hall S, Johnson V, Longhurst N, O'Grady A. et al. Distributing power. A transition to a civic energy future, a website report from University of Bath, UK; 2015.
240. Cohn L. Playing nice: What the duke energy microgrid test bed teaches. MicrogridKnowl. 2016 [Accessed February 28, 2016]. Available from: http://microgridknowledge.com/duke-energy-microgrid-test-bed/.
241. John JS. How Duke's solar and battery microgrid is weathering disruptive gridevents. 2016 [Accessed September 21, 2016]. Available from: http://www.greentechmedia.com/articles/read/how-dukes-solar-and-battery-powered-microgrid-is-weathering-grid-events.
242. Montoya M, Sherick R, Haralson P, Neal R, Yinger R. Islands in the storm: Integrating microgrids into the larger grid. *IEEE Power and Energy Magazine.* 2013;11:33–9. http://dx.doi.org/10.1109/MPE.2013.2258279.
243. Walton R. Former FERC chair says microgrids are key to grid security. Utility Dive; 2014 [Accessed September 13, 2014]. Available from: http://www.utilitydive.com/news/former-ferc-chair-says-microgrids-are-key-to-grid-security/327814/.
244. DeBlasio D. Toward a self-healing smart grid. Fortnightly; 2013 [Accessed September 20, 2016]. Available from: https://www.fortnightly.com/fortnightly/2013/08/toward-self-healing-smart-grid.
245. Marnay C, Asano H, Papathanassiou S, Strbac G. Policymaking for microgrids. *IEEE Power and Energy Magazine.* 2008;6:66–77. http://dx.doi.org/10.1109/MPE.2008.918715.
246. Minkel JR. The 2003 northeast blackout–five years later. *Scientific American*, A website report. (Aug. 13, 2008.)

247. Soshinskaya M, Crijns-Graus WHJ, Guerrero JM, Vasquez JC. Microgrids: Experiences, barriers and success factors. *Renewable and Sustainable Energy Reviews*. 2014;40:659–72. http://dx.doi.org/10.1016/j.rser.2014.07.198.
248. Gridwise Architecture Council. Gridwise transactive energy framework version 1.0. Prepared by The GridWise Architecture Council January 2015 Transactive Energy Framework PNNL-22946 Ver1.0.
249. Renewable microgrids: A case study. Nprvento, UK: Norventoenerxfa; December 2018. A Website Report.
250. Asmus P, Forni A, Vogel L. Navigant Consulting, Inc. Microgrid Analysis and Case Study Report. Sacramento, CA: California Energy Commission; 2017. Publication Number: CEC-5002018-022.

6 Off-grid Hybrid Energy Systems

6.1 INTRODUCTION

In Chapter 2, we examined the basics of grid-connected hybrid energy systems. This chapter mainly focused on utility grid-scale hybrid energy systems which are becoming more and more norms as the large-scale fossil fuel and nuclear-based power plants are becoming problematic. Utility grid developed over the last century is still the best way to supply power in the urban environment. The chapter points out three reasons for the use of hybrid energy in a utility grid. The first one is to increase the energy efficiency via either by cogeneration or by generating extra power with thermoelectricity using waste heat. The second reason is to generate additional power from carbon dioxide using technology such as CO_2 fuel cell. And the third reason is to increase the use of renewable energy sources like solar and wind to reduce the use of harmful fossil fuels or nuclear energy. The renewable energy sources will generally require the insertion of an energy storage device (either homogeneous or hybrid) in a utility grid in order to provide high-quality and uninterruptable power. The third approach is very useful for obtaining zero-energy buildings and making the customers prosumers rather than just consumers. Thus, the chapter illustrated the use of hybrid energy (power and/or heat) by multiple sources (including renewables) in a number of different ways which can include energy storage. The chapter also articulated new system design, intelligent management, and the control needed to manage hybrid power systems. The chapter articulates new perspectives on the concepts outlined in my previous book [1].

As pointed out in Chapter 1, in recent years, in order to capture distributed energy resources, hybrid microgrids are being rapidly developed which either can be connected to the utility grid or can operate in an islanded mode. These microgrids are very useful in the further expansion of the use of distributed renewables for power generation and they directly connect to consumers at low- to medium-voltage distribution network. The use of microgrid either as a subset of macrogrid or operated in an islanded mode makes harnessing of hybrid renewable sources easier. Chapter 5 described in detail the fundamentals of microgrids, its management and control strategies, market future, implementation issues, and value-added propositions with the help of case studies. It was pointed out that the use of hybrid microgrids will make electricity transport more smarter, safer, and efficient. As the use of distributed renewable energy sources will increase, so will the use of hybrid microgrids.

As pointed out in Chapter 1, the third layer of hybrid energy and hybrid grid system is off-grid energy systems. While utility grid serves well urban communities and the microgrid harnesses well the distributed energy sources in both urban and rural environment, it is the off-grid energy that will serve the needs of rural and isolated communities. The small scale of off-grid energy will allow its implementation of electrical system to be bottom-up and more economical. It will also allow for a more efficient use of local renewable sources. While by definition, off-grid hybrid energy systems are not connected to the grid; in the case of nanogrid for individual homes, it can be connected to a utility grid to allow consumers to be prosumers. In general, off-grid systems include independent mini- and nano (or even pico) grids and stand-alone systems, and they serve the needs of rural and isolated communities or single buildings or equipment.

The literature is somewhat confusing in the definitions of mini, micro, nano, and pico grids based on the size of the power. While microgrids are in principle smaller than mini-grids, the latter are used more for rural and remote area electrification. In this book, we consider mini-grid, nano (or pico) grid, and stand-alone systems as parts of off-grid energy systems. As pointed out in Chapter 5, microgrid (irrespective to its size) is defined as the one that harnesses distributed energy sources and can be connected to a mega grid or operate in an islanded mode. Pico, nano, and mini-grids in rural and remote environments are bottom-up approaches for grid formulations, and they are independent of the mega grid, irrespective of their sizes. It is possible that interconnected nano or mini-grids can eventually become large enough to connect to micro or mega grids. However, for the present discussion, they are considered as independent grids. As pointed out in Chapter 1, while micro and mini-grids serve the communities in a similar and smaller scale, microgrids can be attached to the utility grid while mini- or nanogrids are operated independently. In general, differentiation of grids based on size is somewhat irrelevant. The IRENA, however, breaks down capability and complexity of various grids as follows [1]:

Stand-alone systems-------0–0.1 kW.

Pico grids------- 0–1 kW—single controller.

Nanogrids-----0–5 kW----Single voltage, single price, controller negotiates with other across gateways to buy or sell power. Both grid-tied and remote systems, preference for DC systems. Typically serving single building or single load, single administrator.

Microgrids---5–100 kW----Manage local energy supply and demand, provide variety of voltages. Provide variety of quality and reliability options, optimize multiple-output energy systems, incorporate generation, varying pricing possible.

Mini-grids---0–100,000 kW Optimize multiple-output energy systems, local generation satisfying local demand, transmission limited to 11 kW, interconnected customers.

The IEA's [2, 3] definition of access to electricity is at the household level and includes a minimum level of electricity consumption, ranging from 250 kWh in rural areas to 500 kWh in urban settings per household per year. The electricity supplied must be affordable and reliable. The initial level of electricity consumption should increase over time, in line with economic development and income levels, reflecting the use of additional energy services. Sometimes, a mini-grid is defined as a stand-alone AC grid (of undetermined size) while a microgrid often deals exclusively with DC power. In this chapter, we choose to use the term mini-grid to refer to any grid that is not linked to the main central grid in the country or territory in which it is located. Thus, in this chapter we define:

Mini-grids: A power source of a typical capacity ranging from a few kWs to a few MWs, supplying electricity to consumers in a remote location through a local distribution grid, justified by the population density in the concerned location. The power source could be a diesel-powered generator, a renewable energy power plant, or a hybrid between the two technologies including energy storage.

Nanogrid and pico grid: These are terms that are sometimes used to differentiate different kinds of mini-grids with size thresholds, capability, and complexity as some of their defining characteristics. A pico PV system is a small PV system with a power output of 1–10 W, mainly used for lighting and thus able to replace unhealthy and inefficient sources such as kerosene lamps and candles. Depending on the model, small ICT applications (e.g., mobile phone charger and radio) can also be added.

Off-grid systems: A collective term to refer to stand-alone systems for individual appliances/users and also to mini-grids and nanogrids of different sizes (serving multiple customers) that are not connected to a larger centralized grid.

Clean energy mini-grids (CEMGs) utilize one or several renewable energies (solar, hydro, wind, and biomass) to produce electricity with storage. Backup power can be supplied by electricity stored in, for example, batteries or otherwise by diesel generator. The storage provides or absorbs power to balance supply and demand and to counteract the moment-to-moment fluctuations in customer loads and unpredictable fluctuations in generation. All mini-grids are fundamentally hybrid in nature.

The second and equally competitive option is to use a nanogrid. This grid is generally used for an individual house, equipment, or a mobile system like automobile, airplane, etc. As shown later, many nanogrids can combine with each other to form a mini-grid. The third off-grid energy system is stand-alone systems. Stand-alone systems are small electricity systems, which are not connected to a central electricity distribution system and provide electricity to individual appliances, homes, or small productive uses such as a small business. Sometimes, stand-alone systems for homes are connected to the grid to make a consumer into a prosumer, but generally this is not possible in a rural environment. They thus serve the needs of individual customers, while utilizing locally available renewable resources. Due to the price drop, stand-alone off-grid systems, powered by biomass, small wind, small hydro, and small solar power, are becoming more and more common. Both nanogrids and stand-alone systems using renewable energy are generally hybrid because they require some energy storage and often multiple sources [1, 4, 5].

To extend the time of use and stability of power supply, energy storage systems have become more popular. Storage is typically implemented as a battery bank. Power drawn directly from the battery is often of extra low voltage (DC), and this is used especially for lighting as well as for DC appliances. An inverter is used to generate AC low voltage which powers standard appliances [1, 4, 5]. Stand-alone systems can be differentiated into pico, home, and productive systems. Pico systems are used to power individual appliances like lights, TV, and radio. Home systems are used to power individual households. Productive systems are used to power a small business, clinic, hotel, factory, etc.

Mini- or nanogrids and stand-alone systems are in the vast majority of cases more cost-competitive than extension of the national grid network. As rural areas in developing and emerging countries are often located far away from the national grid in difficult terrain or on islands, extending the national grid to rural areas is normally extremely costly and technically difficult, whereas off-grid systems are flexible, easy to use, and adaptable to local needs and conditions. With appropriate training, they can also be operated by local technicians, which in turn leads to local employment. Mountainous and forest areas as well as small islands, for instance, with difficult access to machinery, require more time and resources to install transmission lines, whereas off-grid systems are easier and less costly to implement and can use local renewable energy sources to provide electricity.

6.2 METHODOLOGICAL ISSUES AND DRIVERS FOR OFF-GRID HYBRID ENERGY

One of the fastest growing markets for hybrid energy is the off-grid systems. The hybrid renewable energy deployment in off-grid systems is growing steadily in both developed and developing countries, but there are only limited data available on their scope and extent. With declining costs and increasing performance for small hydro installations, solar photovoltaics (PV) and wind turbines, as well as declining costs and technological improvements in electricity storage and control systems, off-grid renewable energy systems could become an important growth market for the future deployment of hybrid renewable systems [1, 4, 5]. In the short to medium term, the market for off-grid renewable energy systems is expected to increase through the hybridization of existing diesel grids with wind, solar PV, biomass gasification, and small hydropower (SHP), especially on islands and in rural areas. Furthermore, renewables in combination with batteries allow stand-alone operations, and batteries are now a standard component of solar PV lighting systems and solar home systems. The impact of off-grid renewable energy systems will not only be measured in terms of their usage or reduced costs for electricity consumption in rural areas, but also in the context of their effect on the lives of some 1.16 billion people who today are totally without access to electricity.

Off-grid hybrid renewable energy systems are not only urgently needed to connect this vast number of people with a source of electricity, but are also most appropriate due to geographical constraints and costs for grid extension. At the same time, off-grid systems could become an important vehicle to support the development of renewables-based grids. In developed countries, mini-grids are increasingly considered an option to improve energy security, power quality, and reliability, as well as to avoid power blackouts due to natural disasters (e.g., hurricanes) or even deliberate attacks. Furthermore, declining costs for solar PV and wind, together with reduced costs for battery storage, make this option attractive for households and small communities to intentionally disconnect from the grid and start producing and consuming their own electricity. The ability to cover residential electricity demand with local renewables is already commercially interesting in some regions of the world and could become more widespread in the next 5 years. Australia, Denmark, Germany, Italy, Spain, and the United States (Hawaii) are expected to comprise the most promising early markets [6].

Despite the growing attention and market opportunities, there are to date only limited data available and only inadequate definitions of what exactly constitutes an off-grid renewable energy system. In this book, we have defined off-grid systems as mini, nanogrids, and stand-alone systems that are not connected to the macrogrid. As shown here, these definitions are not always followed in the literature. For example, in China, many mini-grids are connected to the macrogrid. In many cases, mini- and nanogrids are used interchangeably. Furthermore, data sources are scarce and inconsistent across countries and regions.

Based on the application and their key components, however, IRENA [1, 5, 7, 8] has made a classification for stand-alone and various grid systems. According to them, there are two types of stand-alone DC systems; solar lighting kits and DC solar home systems. The first one is used for lighting in residential and community areas, and its key components are generation, storage, lighting, and cell charger. The second one is used for lighting and appliances for residential and community usages, and its key components are generation, storage, and DC special appliances. There are also stand-alone AC solar home systems and single-facility AC systems. They are used for lighting and appliances in community and for commercial purposes. Its key components are generation, storage, lighting, regular AC appliances, and building wiring but no distribution system.

There are also AC/DC pico, nano, micro, and mini-grid systems. The pico and nanogrids are used for lighting, appliances, and emergency power for community and commercial usages. The key components for these systems are generation plus single-phase distribution. The micro and mini-grids can be applied to all usages for community, industry and commercial purposes. Their key components are generation plus three-phase distribution and controller. Finally full AC grid can be used for all purposes, and its key components are generation plus three-phase distribution and transmission.

To address the challenge of proper reporting, IRENA has identified a number of key areas where methodological improvements are needed. These methodological improvements include: (1) an overview of systems; (2) a categorization of off-grid renewable energy systems based on their application and system design; (3) categorization based on size, capability, and complexity; (4) consistent indicators to differentiate, evaluate, compare, and aggregate data on off-grid renewable energy systems, including hybrid systems; and (5) measures to compile existing data sources, identify their limitations, and create consistency. While these recommendations are not always followed, based on these methodological suggestions, the key results of this status update are summarized below [1, 4–7, 9]:

1. A significant number of households—almost 26 million households or an estimated 100 million people in all—are served through off-grid renewable energy systems: some 20 million households through solar home systems, 5 million households through renewables-based mini-grids, and 0.8 million households through small wind turbines.
2. There is a large market to replace diesel generators with renewable energy sources in off-grid systems. Currently, there are ~400 GW of diesel capacity (>0.5 MW) in operation, either in the form of industrial facilities and mines operating remotely, as backup units where electricity supply is unreliable, or as community mini-grids. Some of the diesel generators are single machines with more than 10 MW capacity. Of the total installed diesel capacity, 50–250 GW could be hybridized with renewables, of which around 12 GW is located on islands [1, 4–6].
3. China has led the development of renewables-based mini-grids in recent decades. However, Chinese mini-grids are not off-grids because they are mostly integrated into the centralized grid, while generation still takes place on a decentralized basis. There are about 50,000 such decentralized production units, mainly small hydropower (SHP). The power capacity of SHP by continent is described in Table 6.1 [9].
4. Total global SHP capacity amounted to 75 GW in 2011/12, complemented by 0.7 GW of small wind turbines (SWT).
5. There are a few thousand mini-grids in operation that are not connected to the main grid.

Bangladesh, Cambodia, China, India, Morocco, and Mali are among those countries with more than 10,000 solar PV village mini-grids. India has a significant number

TABLE 6.1
Small Hydropower Operational Capacities below 10 MW for 2011/2012 [9]

Region	Capacity (MW)
Africa	525
Caribbean	124
America	10,177
Asia	45,972
Europe	12,018
Oceania	310
Pacific islands	102
Total	~ 70,000

of rice husk gasification mini-grids. More than 6 million solar home systems are in operation worldwide, of which 3 million are installed in Bangladesh. Moreover, nearly 0.8 million small wind turbines are installed. A significant, but not precisely quantified, number of solar street lights and other electronic street signs are in operation. More than 10,000 telecom towers have been fitted with renewable electricity systems, especially with solar PV. In 2013, Africa had around 8 million solar lighting systems were installed to provide outside (public) lighting.

Interest in off-grid renewable energy systems is on the rise, both in developed and developing countries. A decentralized system requires less land than a utility-scale renewable project, experiences less distance-related transmission losses (as it serves only a local customer or area), and provides electricity like a traditional grid connection. Furthermore, off-grid systems are able to support the integration of decentralized renewable power generation into the grid and provide power reliability and stability, evidenced in reduced outages and their respective massive economic impacts upon related economic activities.

For *developing* countries, providing and maintaining energy access is an important driver for off-grid renewable energy systems [1, 5, 7, 8]. An estimated 1.16 billion people (17% of the world's population) currently live without access to electricity; an estimated 615 million of them in Asia and the majority of those in India (306 million), Indonesia (66 million), and the Philippines (16 million). Off-grid renewable energy systems are in many cases the most economical solution for these population groups [10, 11]. The energy access industry, excluding grid extension, is currently estimated as a $200–$250 million annual industry [12]. However, fewer than 5 million households (i.e., 25 million people) had energy access through off-grid solutions in 2013 [12]. Rapid expansion is projected with 500 million households being provided with access to electricity through mini-grids and solar home systems by 2030 [12]. Until recently, mini-grids have been a stepping stone toward grid expansion. However, with improved control systems and declining renewable energy technology costs, it is currently not clear whether this evolution will continue or if mini-grids can become an alternative to main transmission grid extensions.

For *developed* countries, off-grid systems consist of two types: (1) mini-grids for rural communities, institutional buildings, and commercial/industrial plants and buildings and (2) self-consumption of solar PV power generation in residential households. The latter category is relatively small and most residents still rely on the grid for part of their load, but Germany and Japan are currently providing subsidies for electricity storage technologies for renewable self-consumption. In Germany, 10,000 rooftop solar PV systems are already coupled to battery storage systems. With increasing grid parity of solar PV systems expected in a number of countries, this could be an important development. Furthermore, in a number of countries, businesses have entered the market and are leasing solar PV systems coupled to battery storage technologies.

Mini-grids are particularly relevant for island states, both developed and developing. It is estimated that there are more than 10,000 inhabited islands around the world and an estimated 750 million islanders. The United States, Canada, and Chile are the countries with the largest number of islands, while Indonesia, the Philippines, and

China are the countries with the largest population of islanders [13]. Many of these islands, especially those in the range of 1,000–100,000 inhabitants each, rely on diesel generators for their electricity production and spend a considerable percentage of their gross domestic product (GDP) on the import of fuels. In most cases, renewables are already a cost-effective replacement for these diesel generators [1, 5, 7, 8], creating an important market for off-grid renewable energy systems.

Although off-grid renewable energy systems are not new, there is still only limited information on them. Some overviews exist for solar home systems or mini-grids, but data sources are inconsistent or difficult to compare and aggregate. As renewables increasingly become cost-effective resources for off-grid systems, it is important to have a clear picture of their current status, their future prospects, and the methodological improvements required to track their progress [14].

According to IRENA [1, 5, 7, 8], there are four areas where methodological improvements are needed to provide a global status update on off-grid renewable energy systems. *First*, there is only a limited understanding of what aspects and designs comprise an *off-grid renewable energy system* and what applications, renewable energy resources, and technologies should be considered under this broad topic. *Second*, the definition of what constitutes an "off-grid system," including its application, users, and system components, is incoherent and inconsistent. *Third*, the indicators used to quantify and report off-grid systems differ, depending on the specific organization and country, so it is difficult to compare and aggregate data across the world. *Fourth*, data on renewables used in off-grid applications are often not captured in national energy statistics, and they often do not distinguish between installed capacity and operating capacity, especially considering that many off-grid systems have been in place over long periods of time or are used as backup systems. Methodological improvements are therefore necessary to increase the accuracy and comprehensiveness of data on renewable mini (or nano)-grids and off-grid or stand-alone systems.

There is, however, a broad agreement that off-grid systems are different from centralized or microgrids in two ways. *First*, centralized grids are larger in size. They can include several hundred megawatts or even a 1,000 GW of central generation capacity that can cover countries or even continents. Such grids include transmission at medium and high voltage (above 11 kV) by microgrids to transport electricity over long distances, generated largely by distributed energy sources. In contrast, off-grid systems are smaller in size, and the term "off-grid" itself is very broad and simply refers to "not using or depending on electricity provided through main grids and generated by main power infrastructure". *Second*, off-grid systems have a (semi)-autonomous capability to satisfy electricity demand through local power generation, while centralized grids predominantly rely on centralized power stations. In recent years, the use of solar energy for home is rapidly expanding. According to IRENA [1, 5, 7, 8], there were about 6 million off-grid solar home systems in the world, with Bangladesh leading with 3.2 million, followed by India with 861,654, China with greater than 400,000, Kenya with 320,000, Indonesia with 264,000, Nepal with 229,000, South Africa with 150,000, Sri Lanka with 132,000, Morocco with 128,000, Zimbabwe with 113,000, Mexico with 80,000, and Tanzania with 65,000. The last four are estimates.

The term "off-grid systems" covers mini-grids (serving multiple customers) nanogrids (serving single customers), and pico grids or stand-alone systems for individual appliances/users. Customers of off-grid systems can be residential users or they can be used solely for commercial purposes. Thus, the first step is to precisely define the extent to which off-grid systems are actually *renewable*. For off-grid systems that are based 100% on renewables, the capacity of the system can be taken to evaluate the capacity of renewables, but for hybrid systems, different indicators would be required, such as the capacity of the renewable power generation sources connected to the off-grid system. In off-grid systems that are partly sourced by renewable energy, the capacity or renewable energy production share can be taken to represent the ratio of renewables. Through recent smart grid technology developments, it is becoming easier to monitor and track the production of each component of hybrid systems [1, 8].

Considering these questions, Table 6.2 provides an overview of the possible areas for renewables-based mini-grids and off-grid deployment. The information is organized into two "dimensions": resource type and application. The application areas include mini-grids, including systems for owner consumption only, stand-alone residential systems, and systems for productive use. Renewable resources available for mini-grids and stand-alone systems are biomass, hydro, solar photovoltaics (PV), and wind. In a number of cases, geothermal energy is also used to provide base-load power generation to off-grid systems, mainly on (volcanic) islands. Ocean energy technologies could also provide base-load power generation, especially to islands, as they are scalable (1–10 MW plants exist today), but so far they have not been deployed commercially. This overview reflects a very diverse market that is not yet

TABLE 6.2
Renewables in Grid-connected, Mini-grid, and Stand-alone Systems [1, 5, 7, 8]

Gas	Grid connected – 1,500 GW
Diesel	Mini-grids – 5–10 GW, 50,000–100,000 systems
Hydro	Grid connected large >10 MW, 10,000–50,000 systems>1,000 GW
	Mini-grids small <10 MW, 100,000–150,000 systems 75 GW
	Stand-alone systems microhydro 0.1–1.0 MW-pico-hydro <0.1 MW
Wind	Grid connected 310 GW, 250,000 turbines
	Mini-grids-diesel-wind hybrid <1,000village/mining systems
	Stand-alone-small wind turbines-0–250 kW-806,000 turbines
Solar PV	Grid connected-50 GW/0.5 min large systems >50 kW
	Grid connected 80 GW/10–20 min rooftop systems 1–50 kW
	Mini-grids-diesel-PV hybrid <10,000 village systems
	Stand-alone-SHS <1 kW-5–10 min systems
Biogas/Biodiesel to power	Grid connected 14 GW-30,000–40,000 systems
	Mini-grids <100 kW biogas plants >1 million biogas systems, gasification/rice = 1,000–2,000 systems husks etc.

well documented. Table 6.2 also compares renewables, mini-grids, and stand-alone systems to grid-connected renewable energy resources and to fossil fuel-(mainly diesel) driven mini-grids and stand-alone systems. For each category, the figure summarizes existing data on the number and/or total installed capacity of renewables.

Table 6.2 shows that hydropower is the largest source of renewables and that most hydropower is grid connected. However, small hydropower systems (1–10 MW) constitute the largest source of renewables for mini-grids (75 GW). The second largest energy source for mini-grids is diesel generators (23 GW). Diesel generators are used as a backup power for renewables-based mini-grids and also as a backup for unreliable grids. Around 500 GW of diesel generators are used in industrial applications [15]. With declining costs for renewable energy and electricity storage technologies, the replacement of diesel generators could become an attractive market for accelerated renewables deployment. Solar PV and wind turbines are increasingly used as an energy source for mini-grid and off-grid systems; however, their total installed capacity is fairly limited.

Looking forward, three markets are projected: (1) the replacement of existing diesel generators, usually through hybridization of existing diesel grids; (2) the supply of electricity to consumers who were previously without it; and (3) the "islanding" of existing grids and/or extension of existing grids through mini- or nanogrids and stand-alone systems. In the latter two categories, mini-grids might remain independent but could also at some point be integrated into the grid while maintaining island operation capability on a community, industrial plant, or cluster level. According to the definition used in this book, this will then resemble to hybrid microgrids.

Off-grid systems are not new: there are millions of diesel and gasoline generators in use worldwide. These provide electricity where there is no grid or where the existing grid is unreliable. What has changed is that new renewable solutions have emerged that are cheaper, cleaner, and better able to produce electricity locally without reliance on imported/transported fuels. Due to this development, off-grid renewable energy systems are now seen as potential replacements for both traditional diesel-based off-grid systems and for existing grid systems.

Ideally, off-grid renewable energy systems would need to be characterized in a systemic way in order to compare and aggregate the applications and to provide a global perspective on the scale of deployment. However, in reality, there are many different kinds of off-grid renewable energy systems, ranging, for example, from single-home rooftop PVs, with and without battery storage, to solar lanterns, PV street lighting and traffic lights, PV pumping systems, PV telecom towers, small wind turbines, wind pumping systems, off-grid fridges/refrigeration systems, and mobile phone recharging systems.

Based on the available definitions of off-grid systems provided by international and national sources, Table 6.2 also proposes an overview to categorize different off-grid renewable energy systems in terms of their applications, users, and system components. Stand-alone systems tend to either be small and/or exist where the owner has no intention to connect them to a wider grid. These include residential solar home systems (SHSs) without grid connection, small wind turbines (SWTs), as well as solar lighting systems. Finally, there are the off-grid systems for productive use:

telephone towers, water pumps, street lighting, etc. Mechanical wind-driven water pumps could also theoretically be included as they offer the same function as solar PV-driven pump sets.

6.3 MINI-GRIDS

A *mini-grid* is an off-grid electricity distribution network, involving small-scale electricity generation. Often conflated with microgrids, a mini-grid is sometimes defined as having a power rating less than 11 kW and as being disconnected from utility-scale grids. The United Nations Framework Convention on Climate Change (UNFCCC) defines a mini-grid with a power rating below 15 MW and disconnected from larger electric grids. Mini-grids are used as a cost-effective solution for electrifying rural communities where a grid connection is challenging in terms of transmission and cost for the end user population density [4].

The electric grids of many developed, high-income countries once started out as mini-grids. These isolated electrical systems were then connected and integrated into a larger grid. Today, issues such as energy security, energy access, and depleting energy resources remain key issues worldwide. Mini-grids are functional at small scale and as potential future connections to a larger grid, making this technology attractive for ensuring the reliability of future energy systems. In many areas, centralized power is not feasible due to relatively small dispersed loads in remote areas, but a mini-grid provides an attractive alternative with enhanced stability when compared to stand-alone systems. Furthermore, transmission remains an issue for geographically isolated areas, making off-grid alternatives necessary in certain situations. Considering the large distances transmission lines must cover to reach rural areas from centralized generation, up to 30% of power can be lost before reaching the destination, significantly decreasing the efficiency of the overall electrical system. As such, localized mini-grids with on-site generation provide a reasonable alternative [4].

6.3.1 MINI-GRIDS FOR RURAL ELECTRIFICATION

Many rural communities remain isolated from larger, traditional grids due to geographic and economic constraints. The electrification of the global off-grid rural population remains a major task of many developing and developed countries, and according to the International Energy Agency in the 2013 World Energy Outlook, mini-grids represent the most cost-effective way to provide universal electricity access to these populations. Due to new technology innovations that have resulted in declining costs for both mini-grids and energy generation sources, specifically solar and wind power, mini-grids have the potential to electrify remote areas that would otherwise remain outside of a grid connection. Mini-grids are a cost-effective and timely solution for more isolated areas in which connection to the main electric grid is unavailable and represent a practical option for meeting the energy demand in sub-Saharan Africa, South and East Asia, and Small Island Developing States.

Millions of people remain without access to electricity today, and the U.N. Sustainable Development Goals commits to the global community to provide a

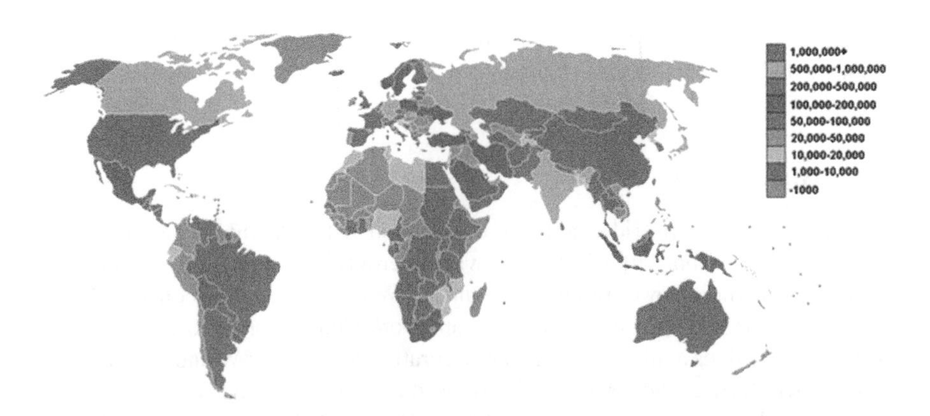

FIGURE 6.1 Electricity consumption per country in million kWh, from CIA Factbook, accessed April 2006 [4].

solution. The map shown in (Figure 6.1) demonstrates energy disparity between developed countries such as the United States, China, and Europe while South America, Africa, and Southeast Asia still have many communities that lack reliable, sustainable, and affordable energy. Mini-grids are currently being viewed as one of the most effective solutions to bringing energy to rural populations where the energy demands are such that individual stand-alone systems or nanogrids are impractical but where the population is large enough to require a larger grid system. Because a grid must balance the supply of energy with the demand, the mini-grid's larger size and flexibility allow for safer and more affordable power [4].

6.3.1.1 Technical Details

A vital component of a mini-grid electric system is on-site, reliable source of energy generation. Traditional mini-grid generation for remote areas came from diesel engine alternators, which incurred high running costs, low efficiency, high CO_2 emission, and high maintenance. To obtain the reliability of a fossil fuel–powered grid with greater sustainability, hybrid energy systems can be used to integrate renewable energy technologies with diesel generators, batteries, and inverters. The main concern with generation is the fluctuation in load demand that imposes varied power requirements from the generation system. These fluctuations can vary throughout a single day, from day to day, or even on the scale of weeks to months, which necessitates flexible mini-grid generation. In the case of limited power generation without a source of energy storage, peak loads can demand more power than the mini-grid generation is capable of supplying, which results in brownouts or blackouts [4].

There exist a variety of energy sources to provide on-site power to a mini-grid system. Recent developments to renewables provide a lucrative option due to the low cost and sustainable economic and environmental nature. As an example, the energy cost of solar PV decreased from \$4/W to \$0.55–\$0.65/W between 2007 and 2016, and this has been going down further in recent years. The common sources of mini-grid generation are solar photovoltaic, wind power, hydropower, biomass, or traditional fuel generators. Because of the intermittent nature of renewable energy sources, generators, fuel cells, or batteries are required to ensure the reliability of mini-grid renewable energy systems (MRES). Otherwise, communities relying on real-time generation experience power outages when renewable generation is not possible. Due to fluctuations in load demand, the energy storage system must be able to meet the peak demand, which can entail large and expensive battery or fuel cell systems. To balance cost with sustainability, energy storage can be combined with diesel power and introduced to a mini-grid in a series or switched hybrid system.

Hybrid mini-grid systems are a popular option to ensure mini-grid reliability, especially when considering renewable energy sources. A hybrid mini-grid is identified by diversified distributed energy resources (DERs), where the energy generation comes from a variety of sources such as solar PV, microhydro power plants, wind turbines, biomass, and small conventional generators. Series hybrid systems have both a renewable energy source and a diesel generator which are used in conjunction to maintain the charge of a battery bank, which is then converted to AC and fed to the load. This system allows for simple implementation, but has low efficiency and requires large battery capacity. In contrast, switched hybrid systems enable renewable energy plus storage to supply the base-load power supply while the diesel generator helps meet peak energy demand [4].

A mini-grid distribution system carries the energy produced by the generation source to the end users. It consists of transmission lines, transformers, and the infrastructure necessary to enable safe and effective energy distribution. Depending on the load requirements, a distribution system can be in AC or DC single- or three-phase power. AC has many benefits, as it allows for effective electricity transmission over distances, meets the requirements for consumer appliances, and is more widely used. However, AC also requires transformers to decrease high-voltage distribution network costs and decrease system loss, but is also generally more expensive than DC because of the enhanced power electronics.

A smart mini-grid (SMG) is an intelligent electricity distribution network that manages the various technical components of a mini-grid system. Often coupled with hybrid power generation, the smart mini-grid operates using smart controllers and advanced control techniques, accommodating various energy sources, energy storage, and distribution. The smart mini-grid relies on a management system which allows for the measurement, monitoring, and control of electric loads and can be coupled with automation to allow for remote operation, smart metering, load shedding, and optimized performance. Another key component is self-healing or the ability for the smart mini-grid to detect, respond, and restore itself immediately in case of disturbances or changes to the system [4].

There are many potential benefits of mini-grids ranging from technical and environmental to social and financial advantages. Mini-grids can be used in rural areas and are often more efficient and cost-effective than other types of power systems. They can also strengthen the community while having less impact on the environment. The technology used in mini-grids provides various benefits. Mini-grids are relatively quick and easy to implement in areas without electricity. They can also be used to improve existing electrical grids that are ineffective or unreliable by providing additional power or by replacing them completely. Most hybrid electrical systems contain at least some redundancy, but mini-grids are organized to prevent this and enhance productiveness. Mini-grids are also more efficient because they can provide a low load at night when less electricity is needed. Unlike conventional energy generation, mini-grids reduce the energy lost at nighttime when less energy is required by the community. Larger electrical systems such as diesel generators cannot offer this because they are inefficient at low loads and most often continue operating at higher loads regardless of the amount of electricity needed. The use of mini-grids also decreases the amount of time the generators are run at low loads, thereby increasing the efficiency of the entire system [4, 5].

An additional benefit that mini-grids provide is that they do not require a traditional fuel source (see Figure 6.2) as many larger-scale electric grids do. This means they can be easily implemented in areas without access to diesel or other fossil fuels. This reduces operating costs and reliance on often fluctuating fuel prices. Mini-grids also require less maintenance than larger electrical grids. Since they reduce the hours that diesel generators are used at low loads, generators last longer and do not need to be replaced as often. Because of the rural areas where mini-grids are typically used, there is often little access to supplies or technicians if system maintenance is needed.

Other than the reduced cost of fuel, mini-grids offer other financial advantages. Mini-grids can be run by a combination of energy sources, which means that they have a lower levelized cost of electricity. Mini-grids are also able to spread electrical storage across many users which reduces the cost when compared to off-grid or solar home systems where electrical storage is concentrated in one area. Mini-grids are also more profitable than other types of electric grids. Due to their improved electrical services and decreased malfunctions such as blackouts, customers are more satisfied overall and are thus willing to pay for the services mini-grids provide, leading to an increase in revenue [4, 5].

FIGURE 6.2 Solar panels are often used in mini-grids to reduce the need for diesel generators [4].

Mini-grids are much more environmentally friendly than other types of grids. Since they reduce the need for diesel generators, greenhouse gas emissions are greatly reduced. This also improves air and noise pollution in the areas mini-grids are used. In addition to their technical and economic advantages, mini-grids also benefit the people and communities they serve. In order for many businesses and organizations to function, they must have working and efficient electricity. Mini-grids provide the necessary services for businesses to succeed in developing areas. This leads to the creation of more jobs and an increase in income for the community. Improved electricity can also benefit healthcare technology and institutions in the areas and lead to a higher standard of living. The electricity that mini-grids provide also allows for more opportunities for social gatherings and events which strengthen the community. Improved electricity also creates the opportunity to construct more buildings and expand the community [4, 5].

6.3.1.2 Risks

Although mini-grids have many benefits, there are also some drawbacks. There are some risks associated with their technology and organization as well as risks to the community they are implemented in. One of the main technical risks associated with mini-grids is the load uncertainty. It is often difficult to estimate the load size, growth, and schedule, which can lead to the system running with lower efficiency and higher cost. It is also difficult to support loads that are constantly changing over time, as they typically are when using mini-grids with renewable sources like solar or wind energy. There is also a risk to power quality when using mini-grids. Integrating photovoltaic devices and batteries can be disruptive to the existing grid and can cause it to become unstable. Another technical drawback of using mini-grids is that failure of hardware in one part of the grid could affect the entire system. If one section of the grid is damaged, the rest of the grid could fail as well. This is a risk that exists with any type of grid; however, the regions where mini-grids are typically used are poor rural areas with less access to maintenance services so the effects are exacerbated. While helpful for energy storage, the batteries used in mini-grids also have risks of their own. They are usually expensive, and as they age, they have a large influence on the energy that is supplied to the grid. If the batteries are not replaced at the correct time, the energy provided by the whole grid could be decreased. As shown in Figure 6.3, most mini-grids are used in rural areas with little access to supplies [4, 5].

Because of their complex nature, there are a few organizational risks associated with using mini-grids. In order to be effective, mini-grids must have effective business models to support their operations. There needs to be a steady flow of revenue to keep the business up and running and in order to keep providing customers with electricity. Due to the remote and underdeveloped locations where mini-grids are typically implemented, it is difficult to transport supplies and skilled personnel to the areas they are needed. It is especially difficult when installing the system and when repairs are needed. Implementing a mini-grid into a community takes meticulous planning and cooperation between the people living in the area as well as the technicians installing the devices. There also needs to be communication among the

FIGURE 6.3 Most areas where mini-grids are used are rural and have little access to supplies [4].

community with regard to allotted energy quotas. Each user is typically assigned an energy quota to be used over a certain amount of time. If some users over-consume the electricity, this leaves a deficit for the other users and could disrupt the entire system. The community must work in cooperation in order for the mini-grid to work successfully [4, 5].

6.3.1.3 Economics

Mini-grids provide communities with a reliable source of energy as well as many benefits to their economy. It is often too expensive for government electrical companies to attempt to bring electricity to undeveloped areas, and there is less potential for profit in these areas with poor economies. Since mini-grids can operate separately from the larger national grids, private companies can implement them and provide rural communities with electricity more quickly than state-owned companies. In terms of market size, the consulting company Infinergia estimates that there were around 2,000 solar mini-grids installed in Africa in 2018, expecting them to reach 16,000 by 2023. Likewise, the analyst estimates 5,000 of them in Asia in 2018, expecting them to reach 15,000 by 2023.

A case study performed in the Leh District of India demonstrates the effects of mini-grids on the economy. Since the operational costs of mini-grids are less than those of diesel and hydro generators, the companies that run them are able to bring in more revenue. This increase in revenue means the companies can increase the salaries of their workers. In turn, the workers are able to spend more in the local businesses, and the economy is allowed to grow. Furthermore, mini-grids provide opportunities for the local economy to grow and improve. Businesses can provide more and better services with improved electricity and expand their organizations [4, 5].

6.3.2 Village-scale Mini-grids

Village-scale mini-grids can serve tens or hundreds of households in settings where sufficient geographical density allows economical interconnections to a central power generator. Rena [16] described few of them in her study. Traditionally, mini-grids in remote areas and on islands have been powered by diesel generators or small hydro. Generation from solar PV, wind, or biomass, often in hybrid combinations, can replace or supplement diesel power in these grids [17–19].

Most village-scale mini-grids have developed in Asia on the basis of small hydro, particularly in China, where more than 60,000 mini-grids exist, as well as in Nepal, India, Vietnam, and Sri Lanka, each with 100–1,000 mini-grids. In China, most mini-grids have resulted from government programs. More recently, rural entrepreneurs have built and run small hydro stations by borrowing from agricultural banks; revenue from just 3 years of electricity sales is apparently sufficient to repay such loans [20–23]. Standardization of the industry has also facilitated interconnection of multiple stations into county-level grids. These mini-grids thus do not form within the nomenclature of the off-grid used in this book. In Nepal, most mini-grids have been installed and managed by rural entrepreneurs. This Nepali entrepreneurial success story of the 1980s and 1990s has been attributed to several factors, including availability of credit from a public sector agricultural development bank, simplified licensing procedures to reduce transaction costs, unrestricted power tariffs, private financing from commercial banks, and capital cost subsidies from the government. Also, technical assistance by bilateral donors and NGOs led to technology development and manufacturing within Nepal's industrial base [24].

Very few hybrid mini-grids employing combinations of solar PV, wind, and diesel exist, perhaps on the order of 150 systems in developing countries. Such systems are still not yet economically competitive with conventional diesel power and must be financed at least partly with government or donor funds. China's roughly 80 PV/wind/diesel mini-grids (about half of which are PV-only systems), sized 10–200 kW, are installed mostly on islands along the coast and in the northern and western remote regions. In India, nine PV mini-grids (most 25 kW) and two biomass mini-grids serve 35 villages in West Bengal [20, 25–28].

Although electricity provides improvements in the quality of life through lighting, entertainment, and increased conveniences, it is the productive uses of this electricity that increase incomes and provide development benefits to rural areas. As incomes increase, rural populations are better able to afford greater levels of energy service, which can even allow for a greater use of renewable energy. The major emerging productive uses of renewable energy are for agriculture, small industry, commercial services, and social services like drinking water, education, and health care [17, 29].

6.4 NANOGRID

The "smart grid" has many different definitions, but all share a common "top-down" approach to understanding the problem and potential solutions. This assesses each component of the grid for how it can meet grid goals of reducing costs, increasing reliability, and gaining environmental advantages. Because the electricity grid is a single interconnected system, it is tempting to see this as the only way to improve how we distribute power.

An exception to this dominant paradigm is the topic of mini-grid in which methods of operation of electrical connectivity and control are significantly altered from normal grid operation, but only within a circumscribed domain of the mini-grid. Mini-grids have been around since before the "macrogrid" was created, though for the most part only existed in industrial facilities, large campuses, and off-grid houses. Mini-grid expansion has been hampered by real or imagined complexity in implementing them, and a lack of standard off-the-shelf technologies that can be readily, and cheaply, utilized.

There is now a "third way," which is not an alternative to the other two, but rather a useful complement—"nanogrids." Nanogrids take the general approach of mini-grids (and many design principles) and carry it considerably further. Nanogrids offer the possibility of attaining a critical mass of technology, affordability, and familiarity to enable nanogrids, and then mini-grids, to flourish. Each nanogrid is a single voltage, reliable, and administrative domain and can contain implementation details within it to enable interoperability with other grids. The information and control architecture for interconnecting nano and mini-grids should be independent of the physical layers within them. A key feature of nanogrids is their ability to be interconnected with each other, as well as implemented within mini-grids. Doing this requires interface standards that can be reliably implemented [30–32].

Nanogrids are already common today, in the form of USB-powered devices of a PC, power over Ethernet (PoE) distribution systems, and the electricity systems in cars and other vehicles. The fact that they are small and simple does not mean they are not useful and important. One notable absence in discussions around the smart grid is how it will help people who today have no access to electricity. Many countries are skipping the landline telephone phase that industrialized countries spent many decades embedded in, going directly to a much newer mobile technology. Similarly, many areas may skip the phase of a capital-intensive traditional grid for an economic and environmental advantage. They also may gain the services of electricity much earlier than if they waited for the traditional grid to reach them. This does not preclude their later joining a macrogrid, but possibly a leaner and different sort than we have today. The consequences of this possibility affect at least hundreds of millions of people in isolated environment and so should not be ignored.

A nanogrid is a single domain for voltage, reliability, and administration. It must have at least one load (sink of power, which could be storage) and at least one gateway to the outside for power source. Electricity storage may or may not be present. Electricity sources are not part of the nanogrid, but often a source will be connected only to a single nanogrid. Interfaces to other power entities are through "gateways." Nanogrids implement power distribution only, not any functional aspects of devices. Components of a nanogrid are a controller, loads, storage (optional), and gateways. Nanogrids can be highly dynamic, with the set of devices part of it changing over time. This includes loads, storage, and connections to other grids (particularly sources). The concept of nanogrids has been mentioned in several papers [30–32], but with somewhat different meaning and no reference to them already being widely deployed. There is a range of nanogrid functionality, and for convenience, the endpoints are called "minimal" and "full." Loads in a nanogrid can be any electrical device of any size, though generally they will be under 100 W, and sometimes under 1 W. They request power from the controller and relay changing customer needs. Loads can be electrical household appliances which are supplied power by local production via the nanogrid [33]. Some examples are loads such as a water heater, lighting, oven, and television.

The core of a nanogrid is the controller, which has the ability to provide or deny power to the loads, to negotiate with other grids through gateways, to set prices, and to manage internal storage. The controller is *the* authority in a nanogrid. In a full nanogrid, devices are always entitled to a minimum amount of power, to enable basic communication functions, and with this, can request more power from the controller. The controller can grant this request fully, grant it partially, or deny it. In addition, the controller can revoke a grant of power. Controllers may have knowledge of usage patterns from past operation and use that in decision-making. They also can have embedded preferences about their behavior such as how much storage capacity to try to always maintain under different circumstances. It is not necessary to standardize controller algorithms. The modular nature of the nanogrid delivers an opportunity to connect multiple nanogrids and then form a mini-grid or a microgrid with an eventual connection to a macrogrid [34].

Storage can be included internal to a nanogrid, or in a second attached nanogrid that may only contain storage. When storage is present, the controller will store or withdraw energy as needed, being cognizant of its technical characteristics. The energy storage is considered optional in a nanogrid structure but is usually present, as it adds stability. The energy storage most suited to nanogrids, due to capacity and residential location, is a battery bank. This makes all nanogrids hybrid energy systems.

Gateways can be one way or two way and have a capacity limit. Each gateway has two components: communication and power exchange. The communication should be generic, at higher layers, and will run across various physical layers. The power exchange will be defined for a variety of voltages and capacities; a challenge is to determine what the best sets of these is, but certainly the voltages already in use today—5 V, 12 V, 24 V, 48 V, 380 V, etc.—are good candidates. The nanogrid does not know what is on the other side of the gateway, just the basic price, capacity, and availability information passed across the interface from the counterpart gateway. The gateway also has the ability to disconnect from external power entities, allowing the nanogrid to operate in islanded mode. The gateway allows the nanogrid to purchase power from, and sell power to, connected power entities, increasing the financial benefit of owning distributed generation [35, 36].

6.4.1 Basic Structure of Nanogrid

One of the main features of a nanogrid is its ability to increase the efficient use of residential-sized distributed generation. These structures can support the integration of a variety of renewable and/or non-renewable energy sources. The typical renewable energy sources are solar and wind, whereas the non-renewable may be sources such as diesel generators or fuel cells [37, 38]. A full nanogrid uses price as a way for devices to express preferences about their relative importance. When more power is requested than is available, those with the lowest price have their allocation revoked or not granted. The nanogrid has a current price that may reflect the price of the marginal devices. This price also affects decisions to store or withdraw energy from storage and is exposed to the wider world through gateways.

Entities connected to gateways may do nothing (that is, exchange no power), offer to sell power to the nanogrid, or offer to buy it. This way, the nanogrid can seek an optimal behavior for the entire system. To account for losses through gateways, in wires and in possible voltage conversions (and between AC and DC), the buy and sell prices may be different, much as with currency exchanges. The actual price and other algorithms implemented in a nanogrid are internal to it and so do not need to be standardized; only the gateway definition and behavior needs to be interoperable.

Gateway connections and loads may come and go or rise and fall in size over time. The controller simply reassesses the situation each time such an event happens and adjusts its behavior as needed. Whether costs for exchanged energy is actually "paid" is not the point—there is no barrier to that, but within a single building, that may not be worth doing. Both gateways at a connection will track accumulated energy and costs, protecting nanogrids from malfunctioning nanogrids or nefarious other grids.

Compared to microgrids and macrogrids, nanogrids address a much reduced set of problems, though with much greater application potential; thus, they enable the development of standard technology that can quickly become widespread. A nanogrid seeks to provide only a single voltage and level of quality/reliability; they do not address systems with complex optimization (such as combined heat and power)—in fact, they do not address power sources at all, and they have only one entity that controls power distribution within it, and exchange of power with adjacent grids [30–32].

6.4.2 Nanogrid Format in Public and Private Domains

There are other forms of power connectivity like LoCal (x), USB port, Ethernet connectivity, power connectivity in vehicle, eMerge, Redwood Systems, Green plug, off-grid systems in household in developing countries, and Nextek etc. that can provide functions of nanogrid [30–32]. These are power connectivity in different formats (some are analogous to nanogrid) in public and private domains. LoCal (x) is a concept for how to interconnect electricity systems at various scales, first described several years ago. It has much in common with the nanogrids approach and is a significant inspiration. There are differences, though some may ultimately be as much a matter of presentation as of practical implementation. One difference between data networking (in the internet) and "grid networking" is that, in the former, it is necessary to have a consistent architecture across the entire network to enable end-to-end connectivity. LoCal envisions a hierarchy of Intelligent Power Switches (IPSs) in the network, eventually spanning the entire grid, like the end-to-end connectivity of the internet. For power connectivity in the nanogrid concept, connectivity is only needed between adjacent grids, as communication and knowledge only extends to adjacent grids. It seems likely that both approaches will be valuable going forward.

While LoCal (x) conceives an Intelligent Power Switch (IPS) as sitting equidistant among loads, generation, and storage, with nanogrids one controlling authority is tightly coupled to its loads, with the links between nanogrids much looser. Put another way, the internal links of a nanogrid are of a very different nature than the external links [30–32].

LoCal (x) begins from an overall concept that can eventually be applied to large-scale electricity systems and works down to specific implementations. Nanogrids start from *existing* technologies for local connectivity and explores how these could be connected to each other. Again, both approaches are needed. LoCal (x) presumes storage for the great benefits it offers in enabling the system to adjust to varying supply and demand conditions. Nanogrids do not assume storage, but can (and often do) include it. LoCal (x) and some other source [36] have at least the abstract concept of "packetizing" energy, much as data is packetized on the internet. Nanogrids lack this explicitly, though there is some expectation negotiated about the timing of changes in loads and exchanges of power, which provides a limited notion of "chunks" of energy rather than simply just continuous power [30–32].

Few USB cables and devices emanating from a PC is a minimal nanogrid. A USB port provides power, and the connected device is a load (if it wants to be). Multiple USB ports on the same PC or same hub are part of the same nanogrid. Unpowered hubs enable connecting more devices to a single port. A powered USB hub becomes its own nanogrid, independent (for power) from the upstream PC. Standard modern Ethernet cables are capable of carrying power according to IEEE 802 standards. The latest version 803.3 in [39] provides for up to 26 W and even 51 W. This can be accomplished by a "mid-span" device that sits between the network switch and the edge device, or, the entire switch can be capable of providing PoE power over some or all of its ports. While a mid-span device is just an external power supply, with a switch, we have a nanogrid. Many components of a car (lights, radio, etc.) are powered by the 12 V battery used to start and maintain its electrical stability. The cigarette lighter has long been a standard "outlet" in cars to plug in many accessory devices. Modern cars have an increasing amount of entertainment electronics, and sometimes provide Wi-Fi inside; these need high-speed communications wires, which may be able to also provide power. An increasing number of cars also have a 115-V AC outlet—essentially a second nanogrid. Aircraft and ships have a variety of non-standard AC and DC "grids" within them and so serve as important examples. They also already operate connected to the grid (at the gate or port), and off-grid [30–32].

Nordman [32] and Nordman and Christensen [30] point out that the eMerge Alliance is backing a technology which distributes 24-V DC power for use in commercial buildings, from external AC or DC sources [40]. It provides up to 100 W on each distribution channel. Lighting is a key application, but it is not limited. Some companies such as Redwood Systems [41] have technologies for distributing DC power and providing communications, also intended for commercial buildings. This system is beginning with lighting, but not limited to that. Nextek Power Systems sells devices that interface between the macrogrid, local renewables, local storage, and AC and DC building loads [35]. These thus implement several interconnected nanogrids, and the Nextek hardware serves as a controller for each with several gateways. Green Plug Inc. sells enabling technology for single and multiple outlet DC power strips that negotiate voltage and current to be delivered with the attached devices [42]. Ignoring the possibility of different voltages, these are a nanogrid. There is a proprietary protocol, *Green Talk*, which accomplishes the needed communications.

As nanogrids are already relatively inexpensive to purchase and install, they should see a quick uptake. This enables price reductions of components to make them even more accessible.

6.4.3 Interconnecting Nanogrids and Their Applications

While nanogrids are off-grid sources, they can be connected to a microgrid or a macrogrid (vehicles mostly an exception, but plug-in vehicles will change this). Usually this is only for power, not communication, and usually, power only flows into the nanogrid. If a nanogrid has non-dispatchable power (e.g., solar or wind), and all storage is full, then it can export any excess power, but this is a simplistic and limiting notion of when sharing power might make sense. By adding the price characteristic to electricity, connected nanogrids can share power when their offered and bid prices are compatible. Gateways between nanogrids have some economic cost to purchase and maintain. They also have some efficiency loss, for wire losses between nanogrids and for conversion if they are at different voltages. The purchaser of the gateway can ensure that there is a price difference between the selling and purchase price, so that it can be dedicated to covering these costs, ensuring that the system is fair. Such a price difference also inserts some "friction" in the system which should enhance stability.

Consider the example of an off-grid household in a developing country, with a car battery and a PV panel, and a number of devices of varying priority (lighting, TV, etc.). This nanogrid can operate in isolation or could connect to adjacent houses and other structures, e.g., a school or clinic. A school will have days off, in which case, its excess power can be readily sold to its neighbors. Any time a household has unexpected high demand, low demand, or equipment failure, the system can better serve the occupants than they could without any interconnection. Electricity production capacity expansion is also much more flexible with this system, with the easy sharing of any surplus power.

Also consider a village with dozens of nanogrids interconnected in some haphazard fashion. In principle, there could be a net flow across many "links" of the grid, with many nanogrids simultaneously buying and selling power on different "ports." The cost function of the transactions introduced by the price difference should help keep the amount of this that occurs to a manageable level. This raises the question of how the amount of power exchanged among the connected nanogrids should be determined. A central controller solution would impose costs, communication needs, and administrative burden and be a potential single point of failure for the whole system. A much better approach is one that is fully distributed, with each nanogrid periodically reconsidering its selling and buying based on its own needs, quantities available or desired, and prices [30–32].

Interoperation of nanogrids with each other as well as optimal operations internally all requires some forms of communication across the gateways. Some nanogrid technologies are built on data or network communications methods and so naturally have one available for internal use. As pointed out by Nordman [32] and Nordman and Christensen [30] for others, identifying a single standard for each for adding communications would be helpful. Data rates can be low and Internet Zero (IØ) [43, 44] deserves consideration. For interconnecting grids, it seems unlikely that a single physical layer could be agreed on, but the number of different ones in use should be kept as few as possible. In addition to having some means of exchanging data, it is necessary for interoperability to have standard higher layer protocols. Even if communication within a nanogrid is different from that between nanogrids, it would be helpful if they had common higher level concepts to minimize the difficulty in creating gateways between such domains. This argues for the creation of one or more "meta-standards" that define a nanogrid behavior in the abstract, with each particular technology implementing it in its own way. A key point is not to consider creating any standards for interoperation of products for functional purposes. There are already many standards for doing that and the difficulty of doing so would likely derail the process. Nanogrids need to be kept only for distributing power, and so the communication should be limited to what is needed for that purpose.

6.4.4 Types of Nanogrid Technology

The debate between alternating current (AC) and direct current (DC) power is not a new argument. As we know, for the national grid, AC emerged the victor, mainly due to the technical limitations at the time the grid was established [45]. With increased research into the benefits of distributed generation, where the supply and storage is often DC, the advantages of a DC grid are still regularly discussed. This is also a subject that frequently arises in mini- or microgrid and nanogrid literature, the reason being an increase in efficiency when distributing DC power [46].

The basic difference between a DC nanogrid and an AC nanogrid is that, in the former case, an AC/DC converter was used for the AC source past the gateway before linking it to DC link and DC/DC load. In the latter case, the DC source is converted to AC by a DC/AC converter and then past the gateway converted back to a DC load by an AC/DC converter. There are similarities between the two topologies at the source end of the power chain, these are as follows:

6.4.4.1 DC Source

Although there are no limitations as to what type of renewable/non-renewable resource is used to generate power, some are more practical than others (e.g., hydro is not often used in nanogrids as it requires access to a body of water, which most residential or commercial properties do not have). Commonly used resources are solar (photovoltaic modules (PV)), wind (small-scale wind turbines (SSWT)), which do generate AC but usually output DC (as the AC frequency varies), and battery storage (which is envisioned to include plug-in hybrid electric vehicles (PHEVs) in the future) [47, 48]. Diesel generators and fuel cells are also mentioned within the nanogrid literature, but not as regularly as SSWT, PV, and batteries [38, 49]. A SSWT or PV typically output voltage which is less than 50 V [47, 48]. There are a vast multitude of batteries available for storing charge which can be used during times of low SSWT or PV power production [50]. These typically come in denominations of 2 V which means, by creating a series string of multiple batteries, most values can be achieved (2 V, 4 V, 24 V, 48 V). Of course the SSWT and PV modules can also be arranged either in series, increasing voltage (five 24 V, 10 A PVs in series would make 120 V at 10 A), or parallel, increasing current (five 24 V, 10 A PVs in parallel would make 24 V at 50 A). The number of SSWT/PV modules selected would vary depending on the power requirements of the loads powered by the nanogrid [51, 52].

6.4.4.2 Source DC-DC Converter

A DC-DC converter is a circuit that takes an input voltage and either steps it up or down depending on the required output voltage. The source DC-DC converter can be used to fulfill a number of functions:

1. **Multiple source interface**
   Nanogrids can have a variety of sources at any one time, for example, a hybrid system may have a PV array, SSWT, and storage supplying power to the nanogrid. Each source has its own operating characteristics. In order to integrate the various sources into the nanogrid, each requires a DC-DC converter. The converter ensures regulation of the supply and provides protection [48].
2. **Bus voltage**
   The source DC-DC converter can also be used to convert the source voltage up to a DC-bus voltage of level of 380 V [34, 53]. This 380 V has become an industry-standard intermediate DC voltage level [47]. In the case of the AC topology, the voltage can then be rectified. The DC-bus voltage has the additional advantage of simplifying the control of the nanogrid [54].
3. **Maximum power point tracking**
   The behavior of the SSWT and PV is nonlinear. Under specific environmental conditions there is only one point of operation that ensures the maximum power output. This maximum power point is dynamic, and by utilizing sensors to observe the behavior of the renewable source/environmental conditions, this point can be tracked. This is done by varying the duty cycle controlling the source DC-DC converter, essentially presenting the source with a variable load. By creating the ideal load for the environmental condition, the source is forced to operate at its maximum power point [55]. The source DC-DC converter is usually of the boost or buck-boost variety as the source voltage typically needs stepping up [38]. The efficiency of these converters is greater than 85% and in some cases can

achieve can achieve a percentage in high 90s [56, 57]. As the DC source and load converter for both topologies are equal, at this point, the efficiencies are the same. From here, at the load end of the power chain, the two topologies differ.

At the other end of the power chain, from the DC source, is the load. The load has a DC-DC converter to interface it with the DC bus and the gateway requires AC power. The conversions are as follows:

A. **Load DC-DC converter**

This DC-DC converter is used to step down the bus voltage to a device (load) level. For the DC nanogrid, the conversion is performed by an external DC-DC converter such as a buck converter. Like the boost converter, the buck has an efficiency greater than 80% (in some cases, greater than 90%) [58]. The favored voltage levels for this stage are 24 V and 48 V which are the standard telecom voltage [59, 60]. Most existing DC loads are designed to run either at 12 V, 24 V, or 48 V. The range of DC loads available to purchase, in comparison to AC loads, are still extremely limited [61].

B. **Bidirectional AC-DC converter**

A bidirectional converter is needed to interface the national grid or other power entities with the local nanogrid [62, 63]. As the nanogrid functions on DC voltage and the grid AC, as power passes between them, it needs to be converted from AC-DC and vice versa. The reason a bidirectional converter is required is because when the nanogrid has excess power, it will sell the additional power to the grid (DC-AC) [64]. If the load requirements are greater than the local production, the nanogrid will need to purchase power from the grid (AC-DC). The efficiency of a bidirectional AC-DC converter should not be less than 80% (if designed properly), and well-designed converters can reach efficiencies in excess of 95% [56].

6.4.4.3 AC Nanogrid

When compared to the DC nanogrid, the AC has additional conversions that take place to ensure the correct power is supplied to the load. These additional conversions are where the AC nanogrid loses efficiency, and these conversions take place with [65]:

1. **DC-AC converter**

The DC-AC converter takes the DC voltage from the source converter and outputs 230 V AC (or 120 V AC depending on origin), which can be used by the majority of consumer loads sold today. This is also the voltage level supplied to a nanogrid from the national grid. This means if a converter is used that can synchronize to the grid's frequency of 50 Hz (60 Hz depending on origin), power can be shared easily between the power entities. With technologies like the inverter discussed in [66], this conversion can reach efficiencies in excess of 90%.

2. **Load AC-DC converter**

   The AC voltage is then converted to DC, and this conversion takes place in a "power adaptor" (also known as a wall wart) or in the device itself. For AC loads that draw less than 15 W of power (e.g., cell phones), the DC-DC conversion is often executed by a "linear power supply." The efficiency of these devices can vary from 20% to 75%. Loads that draw high power implement switch-mode power conversion which is more efficient, ranging from 50% to 90% [67]. The conversion efficiencies of some common households are illustrated in references [51, 52, 61].

6.4.4.4 DC Nanogrid/AC Nanogrid Comparison

There are a number of elements to consider when comparing nanogrid topologies (DC or AC), making it difficult to determine which is "superior." If efficiency is of utmost importance, the DC nanogrid has the advantage [68]. For both topologies, the source DC-DC converter has a similar efficiency. The DC-AC converter hardware also has the same efficiency for the two topologies but the operating conditions (how often power is passed through and the magnitude of power) may have an overall effect on the system's efficiency. The largest loss in the AC nanogrid comes at the device level AC-DC conversion, and this adds an average loss of approximately 14% to the system [61].

Currently most consumer products in homes, retail stores, and rolling out of production factories are still AC loads. To retrofit a current house with a DC power system will require either replacing AC loads with the limited DC compatible loads available to purchase or modifying the AC loads to function on DC power. So although the increased efficiency of a DC system would equate to a financial saving in the long run, replacing/retrofitting AC loads will increase the initial capital required to install distributed generation in a household. Of course this is only a relevant point if retrofitting, when the initial setup of a house is intended for DC, loads can be selected for DC compatibility [51, 52].

Within the DC nanogrid literature, protection also arises as an issue [54, 69]. The topic of this discussion focuses on protection against short-circuit line fault and ground fault [70]. These faults can occur at output terminals, loads, and switching devices and severely damage a DC system [71]. These faults can be mitigated by including fault protection such as traditional arcing-type circuit breakers or more advanced protection strategies as in [70] and [71].

6.4.5 Swarm Electrification via DC Nanogrids

Solar home systems (PV based) are widely used and are proven to be efficient, but their cost limits their capacity [72]. Only basic appliances and lights are being supported by most of the solar home systems due to their limited storage capacity. The cost of the storage system for a PV system is 29% of the total cost of the installation versus 20% for solar panels [73]. Therefore, swarm electrification home system in which first individual solar home system (nanogrid) is extended to multiple solar home system which are then interconnected and finally establishing household interconnection makes a lot of sense. This is the most efficient way to increase capacity of the system. This strategy known as "swarm electrification" in which a single solar home system is considered as a nanogrid [74]. It is defined as the smallest entity in the colony of interconnected generation points. Each nanogrid offers a single source, reliability, price and administrative domain [75, 76]. Considering the DC nature of PV solutions, they are referred to as DC nanogrids.

Swarm electrification concepts work on the principle that a unified community would overcome the lack of access to electricity. The DC nanogrid swarm electrification for off-grid operates on the principle of a household with a solar system (nanogrid), sharing its excess energy with its neighbor. This strategy has been successful in Bangladesh and India, since nanogrid owners earn money and the buyer gets energy at a low cost without investing in acquiring their own system [77].

The drawback of the strategy is, like that of a single nanogrid, that there is a limitation by the storage system of the amount of energy that can be harvested. Since storage is the expensive part of the system and that the sun shines for more than 10 hours in most of sub-Saharan Africa, there is a lot of solar energy that goes unharvested.

On the other hand, due to the limited capacity, few consumers would be able to tap on those nanogrids, and it would be difficult for commercial use of the harvested energy. Furthermore, though the nanogrid owner and customers might appreciate lights and appliances that those nanogrids can power, they would like to move up on energy ladder and be able to enjoy high-power appliance benefits, such as milling, sewing, and other small commercial activities which would improve the community in general. Hence, the proposed DC nanogrid-based energy bank would overcome the issue of storage capacity and increase total energy harvested by nanogrids [78, 79].

6.4.6 Energy Bank Concept

Swarm electrification concept described above can be further extended to energy bank concept. With energy bank concept, nanogrids won't need to have a storage system as they will be replaced by huge central storage system (called energy bank) which might be operated by a utility or an investor. So, during the day, all the nanogrids would feed both their local loads and the bank. The latter in turn can supply users for commercial use and to the nanogrids after the sun sets.

In this concept, all the nanogrid's PV panels in the network are interconnected to feed, at the same time, their local load and the central storage unit, the bank. The individual storage systems are removed. The storage capacity of the banks should be sized such that it won't get fully charged, as it should be continuously fed by the nanogrids as long as the sun shines. During the day, while being charged, it should also be able to supply commercial loads (commercial center and telecommunication tower) [30–32].

6.4.7 Operation and Metering

The operation of the setup consists of energy storing and retrieving in the bank. The latter can sell energy to other users and feed it back to the nanogrid at night. In comparison to a financial bank where there are charges on transactions, in this context, the transaction is the energy storing and retrieving. Here, the element of transaction is the energy and would be monitored by a smart meter. There would be two forms of transaction charge: for nanogrids and for customers [30–32].

- ***For a nanogrid:*** On the amount of energy stored, a certain percentage will be retained as transaction charge. This retained energy will be the one to be sold to customers in need of power, especially for commercial usage like for milling, sewing, small workshop, and telecom towers. This category of customers operates usually during the day.
- ***For customers:*** Those without their own PV system, monetary prepaid system would be used for the supplied energy. Their tariff will be calculated from the business model that will insure the return on investment of the bank as well as the soft cost related to the interconnection of nanogrids.

  The advantages of the above presented system will be [30–32]:
- Increased harvesting of solar energy as the removal of individual storage system which accounts up to nearly 29% of the total cost, will be allocated on acquiring more PV panels. Furthermore, given the drop in the total cost, more people will be able to afford a PV solar system. Thus, an increase in the number of nanogrids and more energy for the community.
- Moreover, the owner of a nanogrid will be relieved of the burden of storage system maintenance cost.
- Increased revenue generating services for the commercial use of energy with the community.
- Extended possibility, provided the required technology in place, would be the ability for the nanogrid owner to pay some services from the commercial client (commercial center, households, and telecommunication company) connected on the network using their energy stored in the bank. The energy will be a new form of payment within the community which might be an immune system against the money value volatility.

6.4.8 Stakeholders and Responsibilities

In most communities, people are able to afford their own panels but not an energy storage system, as these are expensive. Hence the proposal for multiple stakeholders: the nanogrid owner, investors, and customers. For the investor, such as a utility, instead of investing in grid extension, an investment in the energy bank would be cheaper and still generate revenues. Besides utilities, other investors could be telecommunications companies; instead of burning fuel to power the telecom tower, an investment in the bank would reduce telecom tower operation costs while shortening bank return on investment through energy selling.

Customers of the bank would benefit from those who use energy for commercial purposes, who, instead of investing in acquiring their own system, would rely on the bank and reallocate funds into growing their business. Lastly, domestic users who can't afford to own their own system, they would still be supplied by the bank through a prepaid system [30–32].

6.4.9 Nanogrid Control Types and Techniques

6.4.9.1 Nanogrid Control

The control of a nanogrid, implemented by the nanogrid controller, is what gives the system the ability to coordinate multiple sources and optimize power production and consumption. It is the "brains" of systems, and if implemented correctly, can increase the efficient operation of the nanogrid. Within a nanogrid structure, there are two categories for control, supply-side management (SSM) and demand-side management (DSM). Supply refers to the nanogrid's source of power, for example, photovoltaic modules, small-scale wind turbines, and grid. The demand is the consumption of power by the household loads, for example, refrigerator, television, and heater. Both the supply and demand are extremely dynamic, frequently changing from maximum to minimum consumption/production during a single day [80, 81]. Unfortunately, high-consumption/production times rarely coincide in an uncontrolled nanogrid system. It is for this reason that supply/demand-side management is an integral part of nanogrid control. Supply-side management is used to optimize the behavior of the nanogrid's power sources in order to best match power production to the consumption curve and utilize renewable energy sources. Demand-side management is used to optimize the consumption curve of the nanogrid's loads to match the power output of the nanogrid's sources.

There are a number of control topologies that can be used to implement SSM and DSM with various levels of success. Using nanogrid control topologies, implementation of supply-side management is presented in [82, 83]. Below is an explanation of how each topology is set out for both supply-side and demand-side management with the advantages/disadvantages of each system.

Central control consists of a central controller that acts on the information from sensors measuring the power production and consumption of the system (and in some cases, other variables such as temperature). In the central control topology, the communication lines are between the central controller and various sources, and DC supply power lines are between the central controller and various loads. As all control decisions are made from a central location, this topology has in-depth knowledge of the system dynamics and so has the ability to implement a cohesive control strategy. The central controller measures parameters in real time, making the system fast when implementing control. One disadvantage to this topology is its reliance on a high-bandwidth communications line for collecting data from its sensors in order to implement control in a timely fashion. Another disadvantage is by centralizing the control to a single controller, the system becomes susceptible to failure. If a communication line or the central controller itself is damaged, the system will no longer have the ability to implement control.

Decentralized control has a series of control nodes operating independently to sense the status of each local source or load. The information gathered by the node is then used to control the local source/load. Unlike central control, decentralized control does not require an extensive communication line, negating this reliance. As this topology has many independent controllers, it is also more robust than the central control. This makes the decentralized topology fast and reliable. However, as a control topology, the decentralized scheme is limited in its usefulness. This is due to the lack of communication between the system's nodes. Most control strategies rely on the ability to force a reaction within a power system, to an event that may only be sensed by a single node. This can only be implemented if communication between nodes exists, which in this case, it does not [82].

Distributed control takes the decentralized topology and adds communication between nodes via a communication line as in the central control [84]. This means the distributed system adopts certain characteristics of both systems. It remedies the shortcomings of the decentralized scheme by enabling each node to communicate its power status. As each control node stores segments of an overlying control scheme (pertaining to its own relationship to the system), the network as a whole then creates a cohesive control strategy. Distributed control, like decentralized control, has the advantage of multiple controllers reducing the likelihood of complete failure within in the system. However, like the central control, this topology is dependent on communication lines.

Hybrid distributed control is, as the name would suggest, a hybrid of the distributed control and decentralized control. As in the distributed topology, the nodes can communicate creating a cohesive control strategy. However, the hybrid system looks to improve on the distributed control topology by avoiding the need to use a communication line. It does so by utilizing the DC bus/supply lines to communicate between nodes, as used by the popular "droop control" [85]. This means the hybrid distributed control does not need to rely on a communications link, increasing the reliability of the system.

Hybrid central control is taken from the central control scheme which is combined with decentralized control. This creates a system with a central controller that communicates with decentralized control nodes. The control nodes implement the source/load-level control, whereas the central controller coordinates each node. The hybrid central control delivers a powerful and fast control system with an increased resilience to failure. However, the system relies on communication lines which can still make it vulnerable to faults.

6.4.9.2 Nanogrid Control Techniques

Within the nanogrid literature, there are a number of proposed nanogrid control techniques. The goal of nanogrid control is to optimize supply and consumption patterns to create a more efficient system. This is usually quantified by the financial savings gained from a controlled nanogrid over a nanogrid without control strategies in place. Listed below are a number of nanogrid control techniques with an explanation of their functionality [51, 52].

Ad hoc nanogrid is an implementation of the distributed control technique. It looks to ensure that the loads receive the required power, the sources are not overloaded, and power is transmitted via optimal paths. The ad hoc control technique has been developed for scenarios where the nanogrid does not have access to a national grid (islanded mode), for example, isolated rural areas. The goal of the system is to create a power structure that can be scaled, adaptive in dynamic conditions, and reliable with limited infrastructure planning and no central control. The system allows a single source or load to be connected to each node and uses an interconnected node structure to route power from the source to the load. The nodes communicate wirelessly during a series of phases in which the control algorithm (embedded in each node) selects the path for power flow. During the first phase, a node connected to a load requests power, first from local sources, then expands its enquiry until an available source is located. When an available source is located, the request is answered with an offer. The loads node then goes into a holding mode where it waits to collect replies from various sources at which point evaluates the cost of each to find the optimal path. A confirmation message is then sent via the selected path to allow power to flow from the source to the load. During the second phase, this newly formed connection between the source and load is monitored to ensure it remains active. This is done by periodically sending a confirmation message from the load's node to the source node and back. Another important aspect of the system is keeping track of which source is supplying certain loads as one source may supply multiple loads. This is taken care of in phase three where the amount of power coming in to a source node is accounted for. By then, observing the power entering connected nodes, the power flow can be determined and monitored. The last phase is the one where the nodes attempt to improve the flow of power by looking for alternative paths of less cost. This ensures that the nanogrid always finds the most efficient path for power to flow from the source to the load [86].

Biabani et al. [87] presented *cost function* as a control technique, which takes advantage of the fluctuations in the national grid's power price to implement demand-side management. The system uses central control, which gathers information from the nanogrid's hardware, a two-way communication system for smart devices, a smart grid connection, and an internet connection. This information is used to implement the control algorithm and send control signals to the converters which interface the nanogrid with the national grid, photovoltaics, and battery storage. This rule-based algorithm responds to fluctuations in power price by shifting or reducing loads. It also chooses to sell or make use of photovoltaic production based on the grids' buyback price. And although similar systems exist, they are often too complex to implement in home-based systems or do not make use of the grid connection [88]. The algorithm has three states of operation: automatic response, load curtailment, and islanding mode. In automatic response mode, considered the standard operating mode, the algorithm uses pricing information (from the smart grid) and weather information (from the internet) to control the charging/discharging of the batteries. The system then sets the batteries to discharge during the day's high-power price times and charges the batteries when the price is low. Depending on the grid's buyback price, the photovoltaics modules will either sell back to the grid (while high-buyback prices are offered) or power loads/charge batteries. During this mode, the system also implements some load shedding. If the power price is high, unnecessary loads will be removed and rescheduled for low-power price times. Load curtailment is a mode that can be requested by the utility. During this mode, a consumption limit, based on available resources, is set and an agreement is arranged with the utility that grid consumption of the nanogrid will be kept to a minimum. In islanding mode, the generation and consumption are monitored, and the power is rationed to ensure that essential loads receive power during an outage. The goal of the system is to reduce the payback time of a purchased nanogrid [30, 51, 52].

Predictive control makes decisions within the nanogrid based on accumulated load knowledge of the previous day's power demand. Mansour et al. [89] used this system to schedule the charge times of a plug-in electric vehicle (PHEV), creating a demand-side management algorithm. This algorithm uses a hybrid central control, where the central controller sends advisory messages to a node that responds by allowing/refusing the charging of a PHEV. This helps to flatten the demand curve of the nanogrid by charging the PHEV when excess power is available and delaying the charging process during times of high power consumption. The algorithm stores the aggregated power consumption of the previous day and forecasts the average consumption for each time slot of the next day. An upper and lower bound is calculated around this average consumption value to give three distinct areas (less than lower bound, between bounds, and greater than upper bound). It then compares real-time consumption data to the three areas and dispatches one of the three instructions depending on the result. If the real-time power use is less than the lower bound, the PEV is encouraged to charge. If the consumption is between bounds, a delay charging command is issued, and if above the upper bound, the system denies charging if possible. As the algorithm can cause deviations in the low-voltage distribution grid (when peak shaving is implemented), the nanogrid controller can deny actions if a 5% deviation is monitored. The overall result is a flatter consumption curve and a reduction in peak consumption. This, in turn, generates a number of benefits including financial savings [90].

Flattening peak electricity demand is a demand-side management technique used to reduce the difference between the peaks and troughs of a nanogrid's power consumption. This technique helps to reduce the quantity of power purchased from the national grid equating to a financial saving for the consumer. The system discussed in [91] employs the hybrid central control topology to implement a "Least Slack First" scheduling algorithm (derived from the "earliest deadline first" algorithm [92]). There are two categories of loads defined in the nanogrid system: the first are interactive loads and the second are background loads. Interactive loads are devices that are switched on or/and off by a consumer (e.g., television and computer). A background load is a load that has no direct interaction with consumers but instead switches on periodically to fulfill the load's required task (hot water tank, refrigerator, and air conditioning). The algorithm looks to schedule these background loads using "Slack," defined as the length of time each background load can be held off with minimal effect to load's output (off meaning disconnected from power). When a load's slack reaches zero, the power required by the load will be supplied regardless of the system's power level. The least slack first system then supplies power to the loads in ascending order of slack value, meaning the loads with the lowest slack value receive priority to power. To reduce peak power consumption, the background loads are scheduled on/off to reduce the number of loads drawing power simultaneously.

Droop control is a control technique that can be used for either demand-side management or supply-side management. Droop control is also used in AC and DC microgrid control referring to voltage-level control (in DC and AC systems) and frequency control (in AC systems) [93–97]. Within nanogrids, the control technique is used in the same manner to manage the supplies and loads. The DC-bus signaling (voltage level) approach to droop control is presented in [37, 98] and [99] as a supply-side management technique [100]. DC-bus signaling is implemented on a DC transmission network (DC bus) which has a defined voltage level between source DC-DC converters and load DC-DC converters. To control the nanogrid's sources, the DC-bus voltage is allowed to droop or reduce in magnitude as loading occurs. When the DC-bus voltage reaches certain predefined voltage levels, the controller connects alternative sources to the system. For example, if a nanogrid is comprised of photovoltaics, batteries, and a grid connection, initially the system may be supplied by just the photovoltaic modules. Under light load conditions, this may suffice, but as the loading increases the DC bus will begin to droop. When the first voltage threshold is reached, the batteries will be added to meet the load demands. If the loading continues to increase, reaching the second voltage threshold, the grid is required. In the same way, this control technique can be used for demand-side management. In [101, 102], the DC-bus voltage is also allowed to droop to predefined voltage thresholds. However, rather than triggering sources when each threshold is reached, the system sheds nonessential loads to balance power production and consumption. By assigning loads to certain voltage thresholds, a load hierarchy can be developed. If the above example is used in this instance, when the DC bus droops to the first voltage threshold, the lowest priority load will be switched off. If the voltage drops to the second threshold, the next load in the priority list will be switched off, and this will continue until the production and consumption are balanced. For an AC system, droop is a measure of the change in frequency from the fundamental nanogrid frequency. Although not readily discussed in nanogrid literature, frequency droop is presented in [102]. Much like the voltage droop, as the frequency droops from the fundamental, it reaches predefined thresholds at which either supply-side or demand-side management can be implemented. The supply/demand-side management works in the same manner as explained for voltage droop [30, 51, 52].

6.4.10 Nanogrid Hardware

There are a variety of technologies used with nanogrids, but the subject that dominates the nanogrid literature is converter topologies. Converters are responsible, within the nanogrid, for manipulating voltages to meet the requirements of a specific task. This is typically (but not limited to) interfacing the nanogrid's sources (both renewable and non-renewable), storage, national grid, and loads with the systems bus [103]. For example, within a DC nanogrid, the source converters would take the varying/low DC-input voltage and boost the output to the required 380 V for the DC bus. At the other end of the chain, the load converter will buck (reduce) the bus voltage (380 V) to a level the loads can use (12 V, 24 V, or 48 V). The common categories of converters used in nanogrids are, DC to DC, DC to AC, and AC to DC [54].

DC to DC converters accept a DC voltage at the input and output a modified DC voltage. The amplitude of the converter's output voltage can be smaller or larger than the input voltage. To achieve this change in amplitude, the converters use reactive components (capacitors and inductors) and switching components such as diodes, metal-oxide-semiconductor field-effect transistors (MOSFETs), and insulated-gate bipolar transistors (IGBTs) [28, 56]. By rapidly switching the converter between states, using pulse width modulation (PWM), the necessary output voltage can be attained. Moreover, by measuring the output (or input) voltage (or current) of the converter, the PWM can be altered to ensure the output voltage remains stable even when the input voltage varies. This property is essential when implementing nanogrid control strategies such as droop control. The most commonly used DC to DC converter topologies are buck style converters (output voltage is always less than the input), boost style converters (output voltage is always greater than the input), and buck-boost style converters (output voltage can be either greater or lesser than the input) [51, 52].

There are other tasks the converters can perform within the nanogrid, such as maximum power point tracking (MPPT) of a renewable source or charge controlling a battery bank. The goal of MPPT is to address the nonlinearities presented to a system by a renewable source (primarily photovoltaics but can also apply to wind turbines) [104]. A converter being used for MPPT is controlled with the PWM, according to the MPPT algorithm, to present an optimum impedance to renewable energy source. This is to ensure the source is always operating at its maximum power output [26]. Charge controllers make use of converters to regulate the speed of charging and ensure battery banks are not overcharged, which lengthens the life of a battery bank [27].

DC to AC converters accept a DC voltage at the input and output an AC voltage. This procedure is also implemented with reactive components, switching components, and PWM. DC to AC converters often have a DC to DC converter front end to ensure the DC input is of the right amplitude for the AC conversion. The DC to AC converter has similar characteristics to the AC to DC converter which accepts an AC voltage as an input and outputs a DC voltage. In fact, it is not rare to see bidirectional converters of this nature. As the name would suggest, these are converters that can work as an AC to DC converter, and in reverse, a DC to AC converter. This is ideal for grid-tied DC nanogrids which will in some circumstances feed the grid AC power, and in others, receive power from the AC grid which will need to be converted to DC for the nanogrid.

The research undertaken within the nanogrid literature is focused on increasing the efficiency of the converters used within the nanogrid. It also looks to improve power quality for the loads that require high power quality, reduce the physical size of the nanogrid system, and increase the ease of controllability within the nanogrid. These goals are achieved by researching multiple input/output converters, switching variations, galvanic isolation, and alternative topologies [105].

6.4.11 Nanogrid Network

One of the advantages of the bottom-up approach that nanogrids bring to power systems is their modular nature. This approach lends itself to creating larger power systems by interconnecting multiple nanogrids, forming a microgrid structure. It focuses on a hierarchical approach to power distribution from nanogrid to microgrid to the national grid. This hierarchy takes advantage of the semi-autonomous control structures within the nanogrid to alleviate some of the stresses caused by intermittent power production/consumption (DSM, SSM). These subsystems are then interfaced to form the microgrid, adding further intelligence and creating one bidirectional point of connection to the national grid. Power can then be shared between nanogrids via the network, decreasing the effects of an intermittent power supply and increasing the financial benefit to the consumer. From the microgrid, power can be sold to or bought from the national grid with further financial benefits and DSM/SSM advantages for the utilities. Although the theory behind nanogrid networks, in general, is still at a high level, exploration has begun and implementation is being pursued. The advantages and future research topics for nanogrid networks are outlined in [59, 106–109], and these are described in the following paragraphs.

Bidirectional power sharing is the main function of a nanogrid network. Houses/small buildings are almost guaranteed to have varied instantaneous power consumption curves and may also have varied power production capabilities. The diversity of electronic devices and consumer behavior is responsible for the variety of consumption curves. This is similar with power production, where the renewable energy source, capacity, and use of storage all play a role in creating varied production patterns. In the case of a nanogrid network, the diversity works in its favor to utilize the sharing of excess power [110]. As consumption/production peaks and troughs of individual nanogrids vary within the nanogrid network, it is likely that the demand can be met by the various connected nanogrids. By sharing power within the nanogrid network, the need to purchase power from an external source (national grid) is reduced. This equates to financial savings for the consumers within the network [51, 52].

Communication is an important aspect within the nanogrid network as it lies at the heart of information sharing, which is what creates an intelligent network [111]. There are multiple layers of communication within a nanogrid network, and as with any information sharing, there are a number of technical- and security-based considerations. The layers consist of internal nanogrid communication, which the nanogrid controller uses to gather data and implement control strategies pertaining to single house/building level power flows. The next layer is microgrid communication which organizes power offers/requests between individual nanogrids and may deal with the financial aspect of selling/buying power within the network and to the national grid. The national grid level will focus on DSM/SSM at the national level but would also be expected to have minimal involvement in day to day communication. This then creates a complex communication network where delicate information is shared. Although the technical aspects of a communication network fitting a nanogrid network still requires research/definition, it is not confined to novelty as appropriate data protocols such as ModBus, TCP/IP, and RS485 already exist [112, 113].

Financial benefits are a motivating factor for operating a controlled nanogrid and is also an incentive for the interconnection of multiple nanogrids (nanogrid network) [106]. By adding a financial cost to power shared within the nanogrid network, this motivation can be realized. Within the nanogrid network, a nanogrid can either be a source of power (if excess power is available from the nanogrid) or a load (if power is required by a nanogrid). If a source, the nanogrid can sell power either to another connected nanogrid at a negotiated price, or to the national grid at the set buyback price. As the price within the nanogrid network can be negotiated based on variables such as quantity of available excess power and grid buyback/purchase price, the cost of power can be customized to benefit both the buyer and seller. Meaning power can be sold within the network at price less than the grid purchase price but greater than the buyback price.

Withstanding power grid outages is important now our lives rely so heavily on power. The nanogrid network has the ability to island itself from the national grid in the case of a blackout. In islanded mode, the nanogrid network will be an individual power entity, servicing the connected nanogrids. This means the production and storage power within the nanogrid network, like a microgrid, can continue to power loads for a period of time [114]. The period of time is dependent on renewable energy availability (sun/wind) and storage capacity, but if well designed, the network should be able to withstand a lengthy blackout.

Gradual introduction is an advantage to the nanogrid network paradigm. As nanogrids operate at a single house level, it is envisioned that the introduction of small nanogrid networks can take place over an appropriate length of time [110]. This negates the need for investing large sums of money on replacing central power plants over a short time period. Instead, nanogrid networks can be integrated in to the existing national grid at a manageable rate.

Grid stability is another consideration, though not yet pursued within a nanogrid network context. It has been suggested that the nanogrid network has the ability to respond quickly to commands from the utility grid. This gives the nanogrid network the opportunity to participate in grid stabilization, voltage and frequency control, and real-time pricing at a national grid level [51, 52].

6.4.12 Overview on Nanogrids

In summary, a nanogrid is a power distribution system for a single house/small building, with the ability to connect or disconnect from other power entities via a gateway. It consists of local power production powering local loads, with the option of utilizing energy storage and/or a control system [30, 32, 51, 52]. Nordman, Christensen, Bumester, and others [30, 32, 51, 52] have pointed out the following features of nanogrids:

- Nanogrids have the potential to address the intermittent nature and the long payback time that are inherent with small-scale renewable energy sources. By using a nanogrid controller, demand-side and supply-side management techniques can be implemented to better match a building's supply and load curves. This helps to reduce the negative effects of intermittency. The controller, via a gateway, can also sell and buy power from other power entities such as the national grid or interconnected nanogrids, which shortens the payback time for RE. Nanogrids have a similar structure to microgrids;

however, as the name would suggest, nanogrids are smaller in capacity. Where a microgrid is large enough to power a community, a nanogrid focuses on a single house or small building. The two are not mutually exclusive; multiple nanogrids can be interconnected to form a microgrid.

- There are a number of aspects to consider when comparing DC and AC nanogrids. DC nanogrids are more efficient as they have fewer power conversions. This is because most RE sources are DC, as is storage and often electronic components within loads. This being said, most current household loads are created to operate with an AC power supply as the power supplied by the national grid is AC. So although DC is more efficient, to modify AC loads, or purchase new (limited) DC loads, will be a costly procedure. This is of course only an issue when retrofitting existing properties; if designing a new household/small building with the intention of creating a DC nanogrid, DC loads can be purchased specifically (though limited DC loads exist). A strong case can be made for both AC and DC nanogrids, with AC appealing to the near future/retrofitting and DC for the efficient home of the future.
- The topologies used to lay out the nanogrid and implement control are centralized, decentralized, distributed, and two hybrid topologies (centralized/decentralized and distributed/decentralized). Central control issues command from a central controller whereas decentralized and distributed implement control via a number of control nodes located close to the load or source they are controlling (decentralized nodes do not communicate with one another, distributed nodes do communicate). The hybrid systems address weaknesses in each topology. The distributed hybrid topology negates the need for a communication line by using the nanogrids bus to send control signals. The hybrid central topology still issues commands from a central location but uses decentralized nodes to control the nanogrid which increases its resilience to failure.
- The algorithm used to control the nanogrid is responsible for ensuring the success of the demand/supply-side management and ultimately the efficient operation of the nanogrid. The goal of the nanogrid control is to firstly ensure the loads are supplied with adequate power and the sources are not overloaded. It can then work to match the demand with the supply or vice versa.
- The nanogrid's hardware consists mainly of DC to DC, AC to DC, and DC to AC converters. The converters are used to manipulate the power produced by a source, to meet the demands of the nanogrid's bus, loads, and power exchanged with an external power entity (e.g., national grid). They can also carry out control tasks such as maximum power point tracking (RE source) or charge control (battery bank). These converters, if designed properly, can increase the efficient operation of the nanogrid.
- Multiple nanogrids can be connected, via gateways, to create a nanogrid network (microgrid). This facilitates the sharing of power between multiple nanogrids, delivering a number of advantages. Power can be sold and

purchased at a rate that can benefit both the buyer and seller, reducing payback time of RE sources. It offers the gradual introduction of carbon-free RE sources to the national grid. The community of connected nanogrids are equipped to withstand power outages, and it gives multiple nanogrids a single grid connection point offering benefits to the national grid.

6.5 STAND-ALONE HYBRID ENERGY SYSTEMS

A *stand-alone power system* (SAPS or SPS) (see Figure 6.4), also known as *remote area power supply* (RAPS), is an off-the-grid electricity system for locations that are not fitted with an electricity distribution system. Typical SAPS include one or more methods of electricity generation, energy storage, and regulation.

Electricity is typically generated by one or more of the following methods along with energy storage [116, 117]:

- Photovoltaic system using solar panels
- Wind turbine
- Geothermal source
- Micro combined heat and power
- Microhydro
- Diesel or biofuel generator
- Thermoelectric generator (TEGs)

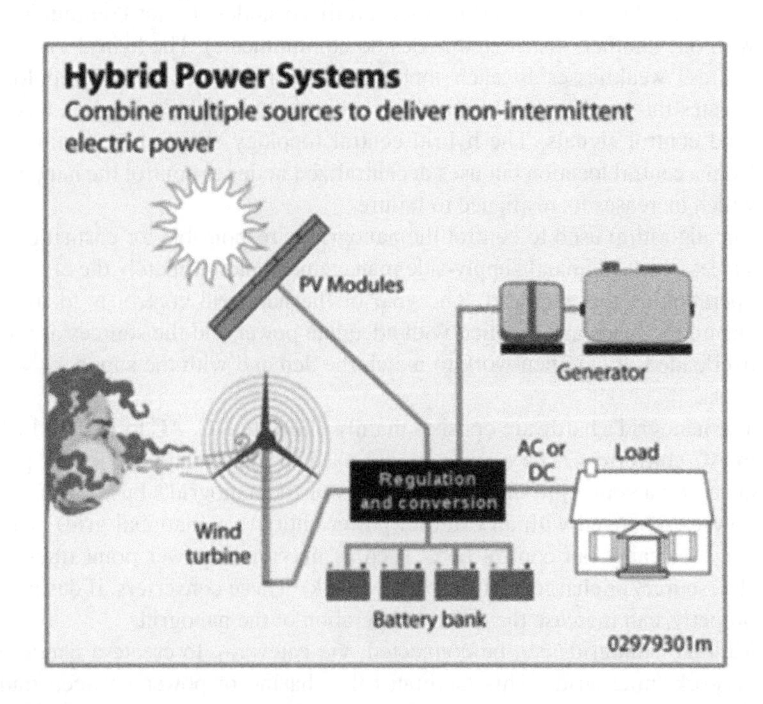

FIGURE 6.4 Schematics of a stand-alone hybrid system [115, 116, 117].

Storage is typically implemented as a battery bank, but other solutions exist including fuel cells. Power drawn directly from the battery will be direct current extra low voltage (DC ELV), and this is used especially for lighting as well as for DC appliances. An inverter is used to generate AC low voltage, which can be used with more typical appliances. Stand-alone photovoltaic power systems are independent of the utility grid and may use solar panels only or may be used in conjunction with a diesel generator, a wind turbine or batteries. The two types of stand-alone photovoltaic power systems are direct-coupled system without batteries and stand-alone system with batteries [116].

6.5.1 Direct-coupled System

The basic model of a direct-coupled system consists of a solar panel connected directly to a DC load. As there are no battery banks in this setup, energy is not stored, and hence it is capable of powering common appliances like fans, pumps, etc. only during the day. MPPTs are generally used to efficiently utilize the sun's energy, especially for electrical loads like positive-displacement water pumps. Impedance matching is also considered as a design criterion in direct-coupled systems. The hybrid system is preferred over direct-coupled system because it provides better quality and stable power.

In stand-alone photovoltaic hybrid power systems, the electrical energy produced by the photovoltaic panels cannot always be used directly (see Figure 6.5). As the

FIGURE 6.5 Schematic of a stand-alone PV system with battery and charger [116].

demand from the load does not always equal the solar panel capacity, battery banks are generally used. The primary functions of a storage battery in a stand-alone PV system are [115, 116]:

- **Energy storage capacity and autonomy:** To store energy when there is an excess available and to provide it when required.
- **Voltage and current stabilization:** To provide stable current and voltage by eradicating transients.
- **Supply surge currents:** To provide surge currents to loads, like motors, when required [6].

In general, the *hybrid power plant* is a complete electrical power supply system that can be easily configured to meet a broad range of remote power needs. There are three basic elements to the system—the power source, the battery, and the power management center. Sources for hybrid power include wind turbines, diesel engine generators, thermoelectric generators, and solar PV systems. The battery allows autonomous operation by compensating for the difference between power production and use. The power management center regulates power production from each of the sources, controls power use by classifying loads, and protects the battery from service extremes.

6.5.2 System Monitoring

Monitoring photovoltaic systems can provide useful information about their operation and what should be done to improve performance, but if the data are not reported properly, the effort is wasted. To be helpful, a monitoring report must provide information on the relevant aspects of the operation in terms that are easily understood by a third party. Appropriate performance parameters need to be selected, and their values consistently updated with each new issue of the report. In some cases it may be beneficial to monitor the performance of individual components in order to refine and improve system performance, or be alerted to loss of performance in time for preventative action. For example, monitoring battery charge/discharge profiles will signal when replacement is due before the downtime from system failure is experienced [7]. IEC (International Electrotechnical Commission) has provided a set of monitoring standards called the "Standard for Photovoltaic system performance monitoring" (IEC 61724). It focuses on the photovoltaic system's electrical performance, and it does not address hybrids or prescribe a method for ensuring that performance assessments are equitable [116].

Performance assessment involves:

- Data collection, which is a straightforward process of measuring parameters.
- Evaluation of that data in a manner that provides useful information.
- Dissemination of useful information to the end user.

The wide range of load-related problems identified are classified into the following types:

- **Wrong selection:** Some loads cannot be used with stand-alone PV systems.
- **House wiring:** Inadequate or low-quality wiring and protection devices can affect the system's response.
- **Low efficiency:** Low-efficiency loads may increase energy consumption.
- **Stand-by loads:** Stand-by mode of some loads waste energy.
- **Start-up:** High-current drawn by some loads during start-up. Current spikes during the start-up can overload the system temporarily.
- **Reactive power:** The circulating current can differ from the current consumed when capacitive or inductive loads are used.
- **Harmonic distortion:** Nonlinear loads may create distortion of the inverter waveform.
- **Mismatch between load and inverter size:** When a higher rated inverter is used for a lower capacity load, overall efficiency is reduced

6.5.3 GE Stand-alone Modular, Containerized Digitally Connected Hybrid Power Solution for Off-grid Electrification

According to their website publication [119], GE hybrid containerized power unit is designed to provide reliable electricity in off-grid applications, such as village electrification or remote commercial and industrial operations. The unit can also act as a backup to unreliable or inadequate grid supply. Pre-wired and configured, the hybrid energy power unit can be installed and commissioned within hours upon arriving at site. Its hybrid controller ensures that the lower cost power source is prioritized, automatically controlling the available sources to ensure that demand is met. Its remote monitoring and diagnostic capability eliminates the need for local supervision and control. The modular hybrid power unit also enables paralleling of multiple units, helping supply easily to adjust with changing demand.

A 20-foot enclosure with pre-configured diesel generator, energy storage, protection and control equipment, and power electronics can be connected to the distribution lines to begin generating power. According to their publication [119], the unit carries the following features:

- Easy external DC connection point for solar arrays
- Hybrid controller reduces operating costs by up to 40% when compared with traditional diesel generators
- Cloud-enabled monitoring and diagnostics using Predix™ allows for remote supervision and control across multiple installations
- Access doors to key components allowing easy service
- Temperature and moisture-controlled compartments ensure components remain at optimal operating conditions

- Quickly increase capacity either by paralleling multiple systems or adding incremental solar and energy storage
- Optional equipment includes grid-tied transfer switch, weather monitoring—integrated into the solution's remote monitoring system—as well as interior/exterior lighting

Benefits of the system include:

- Lower installation and commissioning time and expenditures
- Reduced operating cost and emissions versus diesel systems
- Quickly scale output to capture growing demand
- Achieve higher uptime, identifying issues before they cause unplanned downtime
- Enhanced ability to monitor and control multiple installations
- Operate reliably through a variety of environmental conditions

GE's hybrid power containerized unit for off-grid applications focuses on quality service to remote areas, supported by local service capability to facilitate increased uptime. The picture and the detailed characterization of various components of GE system are described in their website publication. There has been an extensive investigation on off-grid energy systems including stand-alone power systems for rural electrification over the past decade. This is largely driven by the need for supplying electricity to rural, remote, and poor areas of the world. The global landscape of off-grid energy systems is addressed in Section 6.7 of this chapter.

6.6 OFF-GRID HYBRID RENEWABLE ENERGY SYSTEMS (HRES)

For a long time, different conventional power generation systems have been adopted for on and off-grid operation, and these mainly include the hydroelectric power systems, thermal power systems, and distributed utilities (i.e., microturbines and combustion turbines) [120]. The constraints of limited water reserves, fast depleting behavior of conventional energy sources, very large maintenance cost of diesel generators, volatile prices of the fossil fuels, large capital investment, and the escalated concern about global warming and inability to satisfy power need in rural and remote environments have raised concerns about conventional power generation operations [121].

In recent years, the use of renewable energy source options and distributed energy generation has gained momentum particularly to satisfy the needs of areas where grids are not available. Renewable energy (RE) sources are attractive because they are non-depletable or dispatchable, nonpolluting, and monetarily cheap or available for free. The main RE sources and their different attributes are listed in Table 6.3. These RE sources often comprise of the systems of distributed generation either connected to grid or in stand-alone/off-grid formation. In fact, the deployment of the energy systems mentioned in Table 6.3 is increased by their exalted benefits, which include: better quality and reliability of power, lessened

TABLE 6.3
Important Characteristics of RE Sources*

Characteristics	Solar (PV)	Wind	Geothermal	Waves/ Tides	Microhydro	Biomass	Fuel cells
Efficiency (%)	12–22	40–50	15–20	85–90	85–90	35–40	40–60
Capacity factor (%)	20–25	25–35	85–95	30–50	51–55	80–85	90–95
Cost ($/kWh)	0.18–0.27	0.08–0.20	0.09–0.21	0.10–0.30	0.09–0.14	0.10–0.15	0.16–0.24
Emissions	No	No	No	No	No	Yes	No

* Sept., 2014 from [121–126, 148]

emission of carbon, and less/no maintenance and operational costs. During the past several decades, wide research in the area of RE has been done regarding the feasibility studies, control, modeling and simulation, and experimental work [127–138]. Electrification of the isolated and far-flung rural areas, where the grid supply is absent, has acquired much significance as a useful way for improving the living values. Based on the RE sources, a number of off-grid systems have been designed for this purpose [139].

In the use of renewable energy sources, the strategy of hybrid operation, containing one or more energy sources with energy storage, has captured most attention due to its ability to provide reliable and sustainable power supply of high quality. Different methods for storing energy are described in detail in Chapter 4. The hybridization makes the power generation operation cost-effective, and it also reduces the unavailability of harnessed power when one source is idle due to the unavailability of natural phenomenon which drives that source (i.e., at night, when solar is idle, wind will do the job) [140, 141]. In order to utilize the hybrid systems more effectively, the component systems which fabricate the hybrid system need proper sizing and combination leading to a system with improved performance, control, and dispatch [2, 78, 79, 142–147].

6.6.1 Design of Off-grid Hybrid Renewable Energy Systems (HRES)

In a design stage, renewable hybrid systems are synthesized, i.e., which types of generation technologies will be allocated and integrated to build a hybrid system. This is very crucial aspect in the design, since there are usually many alternative possibilities related to which individual components can be included in a hybrid energy system. The objectives of the design phase would be [121, 148]:

1. The type of renewable energy system to be included
2. The number and capacity of renewable energy units to be installed
3. Whether a backup unit, such as diesel generator and/or fuel cell, would be included in the system

4. Whether energy storage would be integrated into the system
5. Whether the system is stand-alone or grid connected

The selection of the technology depends on the availability of renewable resources for a particular site where the system is to be installed in which the local weather conditions play an important role for taking decision. Based on the weather statistics (hourly data), a feasibility study for different possible combinations of renewable sources is studied using optimization techniques to get the optimum configuration. Then the number and size of the selected components is optimized in order to get an economical, efficient, and reliable system. Component sizing is important and widely and extensively studied, e.g., [149–151]. Several factors or constraints directly influence the sizing of the system components, e.g., system economics, greenhouse emission requirements, and system reliability. Over sizing of the components may lead to high system cost, and therefore, the system may become economically unviable. On the other hand, under sizing will reduce the initial cost but one has to compromise with system reliability. For a particular load, different criteria constraints may be applied to the set of system components based on the objectives that have to be achieved. Some of the criteria constraints that are mostly considered while designing a hybrid energy system are [148]:

1. **Reliability criteria:** A number of methods are used to analyze the reliability of hybrid energy systems including, loss of load probability (LOLP), loss of power supply probability (LPSP), and unmet load (UL) (stand-alone) [152, 153].
2. **System cost criteria:** System cost criteria may include total energy systems costs, capacity costs, and societal costs.
3. **Operational optimization criteria:** This may include fuel savings, emissions reduction, reserve/backup capacity, and elimination of excess power generation.

Design of hybrid energy systems has been extensively studied and numerous optimization techniques, such as linear programming [154] and genetic algorithms [155] have been employed for the optimum economic and reliable design of hybrid systems. Recently, many software packages, such as HOMER [156], RETScreen [157], Hybrids [158], and HOGA [159], have been developed for the proper selection of suitable generation technologies and their sizes. These software packages have made the study of hybrid systems interesting and easier. Some of studies address only the reliability analysis for the design of hybrid systems [159–162], while others consider various types and sizes of the available generation resources for the reduction of investment costs and fuel costs and to improve system operations [163]. Optimization-based approaches that simultaneously minimizes the investment cost (installation and unit cost) and fuel cost while retaining the reliability and emission constraints have been implemented in [164]

These and other software simulation and optimization studies are illustrated in detail in Chapter 7.

Simulation and optimization studies reported in the literature indicate that the design of off-grid hybrid renewable energy systems (HRES) is quite complex due to the fact that it has to satisfy several constraints and to achieve several objectives that are conflicting in nature. In addition, the investment costs can become high, as the reliability constraints subjected to the system are stricter for a high degree of reliability. In general, RE sources have high capital costs but lower operating costs because there is no fuel requirement.

6.6.1.1 Integration Scheme

For stand-alone hybrid systems, various possible configurations that may be used for integration of the energy sources that form the hybrid system are shown in Figure 6.6 [121, 148].

1. **Series hybrid system:** This can be of two forms: centralized DC bus and centralized AC bus. In centralized DC bus, all the energy sources, storage devices, and loads are connected to a DC bus through appropriate electronic devices as shown in Figure 6.6a. The DC bus eliminates the need for frequency and voltage controls of individual sources connected to the bus, and the power supplied to the load is not interrupted when diesel generator starts [165] DC loads can be directly connected to the DC bus which reduces the harmonics from the power electronic devices. DC-bus configuration has low-efficiency limitation because in case of both source and load are AC, the power passes through two-stage conversions. Another limitation is that the inverter must be rated for the peak load requirements and in case of inverter failure results in complete power loss to the load. In centralized AC bus, all the energy sources, storage devices, and loads are connected to an AC bus through appropriate electronic devices as shown in Figure 6.6b. It is modular configuration, which facilitates the growth to manage the increasing energy needs. It offers major constraint in the synchronization of the inverters and AC sources to maintain the voltage and frequency of the system. The undesired harmonics introduced into the system by the use of inverters increases the level of power quality problems [149, 165].
2. **Parallel hybrid system:** In this configuration, the AC sources and loads are directly connected to AC bus. While the DC sources and loads are directly connected to DC bus, the bidirectional converter connects both the buses to permit the power flow between them (Figure 6.6c). The inverter rating required is less than that of series configuration and the efficiency is higher. In addition, for the same inverter rating as that used in series configuration, the power supplying capacity of the parallel configuration is much more [121, 148]. Thus, such a configuration arrangement increases the system reliability and ensures the supply continuity [166]. However, synchronization between the output voltage of the inverter and AC bus is needed.
3. **Switched hybrid system:** As shown in Figure 6.6d, in which the AC sources, such as diesel generator, can directly be connected to the load leading to

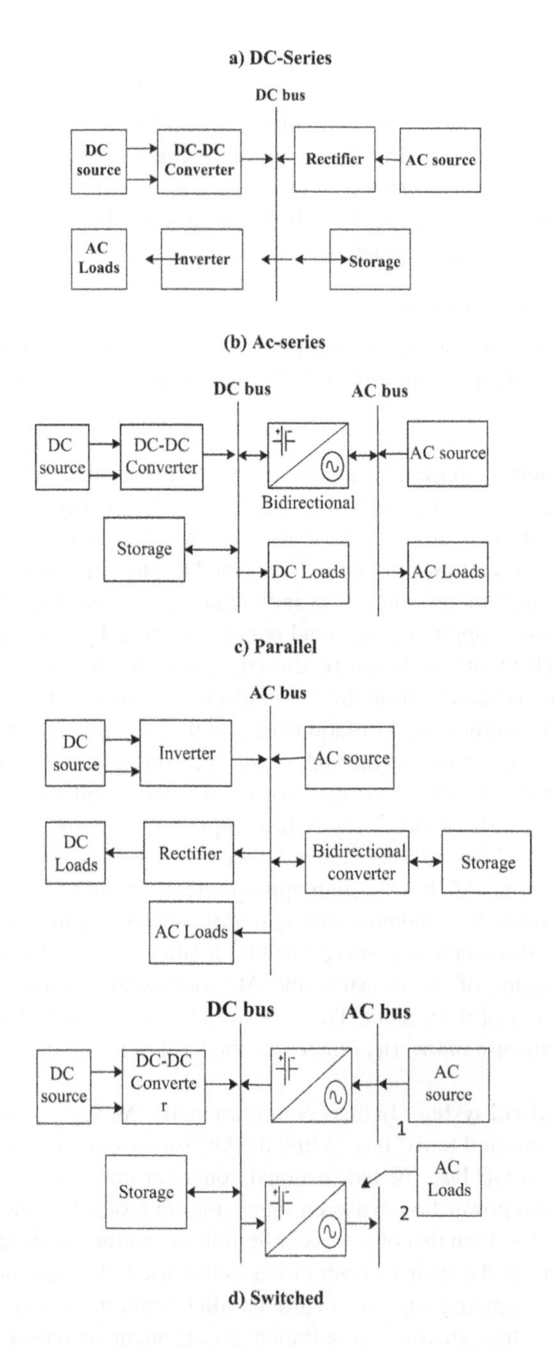

FIGURE 6.6 Various stand-alone hybrid system configurations [148].

higher efficiency and synchronization is not needed. This configuration, although popular, has several limitations that only one of the sources is connected to the load at a given instance. Furthermore, during switching between the sources, the power is interrupted [121, 148].

If stand-alone systems are grid connected, they take somewhat different configurations. Different grid-connected configurations are shown in Figure 6.7. The choice of the layout for particular location depends upon geographical, economical, and technical factors [148]:

1. *Centralized DC-bus architecture* is shown in Figure 6.7a. The AC energy sources, such as wind and diesel generator, firstly deliver their power to rectifiers to be converted into DC before being delivered to the main DC-bus bar. An inverter, main, takes the responsibility of feeding the AC grid from this DC bus. DC buses are used for the hybridization/integration of RE sources with ultimate DC output. RE sources with AC outputs need AC/DC converters (rectifiers) while those with DC outputs are either connected directly to DC buses or through DC/DC converters (if needed). AC loads are joined to DC buses through inverters (DC/AC converters) while DC loads are connected either directly or through DC/DC converters (if needed).

   In structure and control, hybridization of RE sources using DC bus is simple. Yet, it is categorized by its low efficiency and less reliability. All the power has to go through the AC/DC converter; therefore, the converter size is much bigger here than in other configurations. These negatives result into an increased system cost which, when added to the reduced efficiency and less reliability, result in an increased kWh cost of energy harnessed from the system. Also, there will be AC power loses, resulting from a malfunctioning of DC/AC converter, if the system is configured in the series DC-bus configuration. To avoid this, several low-power inverters have to be connected in parallel which will also result in the increased system cost. To obtain the desired distribution of load among different inverters, an appropriate control scheme of power sharing is needed [167, 168].
2. *Centralized AC-bus architecture* is shown in Figure 6.7b. In this case, the sources and the battery all are installed in one place and are connected to a main AC-bus bar, through appropriate power electronic devices, before being connected to the grid. This system is centralized in the sense that the power delivered by all the energy conversion systems and the battery is fed to the grid through a single point. AC buses are used for the hybridization/integration of RE sources with ultimate AC outputs. RE sources with DC outputs need DC/AC converters (inverters) while those with AC outputs are joined to AC buses either directly or linked through AC/AC converters (if needed). DC loads are joined to AC buses by AC/DC converters (rectifiers) while AC loads are connected either directly or through AC/AC converters (if needed). Further, AC buses can be of two types: (i) high-frequency AC (HFAC) buses and (ii) power-frequency AC (PFAC) buses.

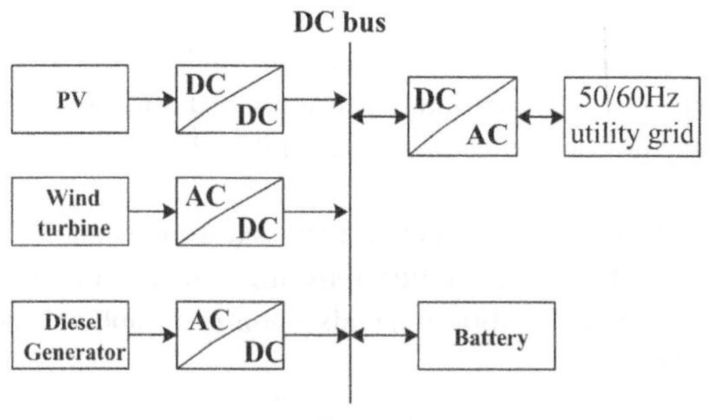

a) Centralized dc-bus

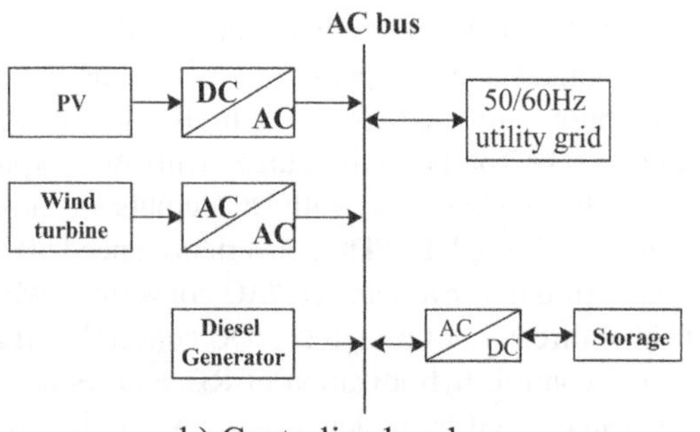

b) Centralized ac-bus

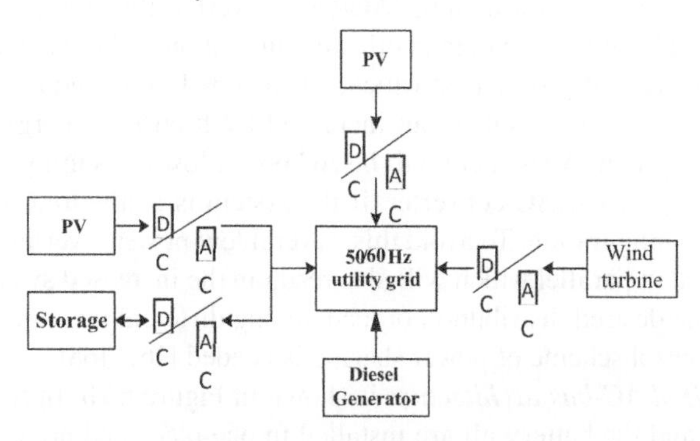

c) Distributed ac-bus

FIGURE 6.7 Various stand-alone grid-connected hybrid system configurations [148].

High-frequency (HF) AC power generation systems and HF loads are connected to HFAC buses while those of power frequency (PF) (i.e., 50–60 Hz) RE systems and PFAC loads are connected to PFAC buses. AC/DC rectification is available for obtaining DC power in case of both HFAC and PFAC configuration system. The HFAC may also have a PFAC bus and connection to a utility grid system (through a DC/AC or an AC/DC converter), to which AC loads can be connected [167].

3. *Distributed AC-bus architecture* is shown in Figure 6.7c. In this case the power sources do not need to be installed close to each other, and they do not need to be connected to one main bus. The sources are distributed in different geographical locations and connected to the grid separately. The power produced by each source is conditioned separately to be identical with the form required by the grid. Hybrid (AC-DC) bus is a combination of a DC bus and an AC bus with a master inverter in between. RE sources with AC outputs and AC loads are joined to the AC bus either directly or through AC/AC converter (if needed). Whereas, the RE sources with DC outputs and DC loads are connected to DC bus of the hybrid bus either directly or through DC/DC converter (if needed). Hybrid buses can be more energy-efficient and economical, but they can show more complexity in energy management and control when compared with the other AC and DC buses alone.

6.6.1.2 Advantages of Hybrid Renewable Energy Systems

Hybrid renewable energy systems provide many advantages [121, 148]:

1. *A hybrid energy system:* It can make use of the ***complementary nature of various sources***, which increases the overall efficiency of the system and improve its performance (power quality and reliability) [121, 148]. For instance, combined heat and power operation, e.g., MT and FC, increases their overall efficiency [166, 169–172] or the response of an energy source with slower dynamic response (e.g., wind or FC) can be enhanced by the addition of a storage device with faster dynamics to meet different types of load requirements [173–176].
2. *Lower emissions*: Hybrid energy systems can be designed to maximize the use of renewable resources, resulting in a system with lower emissions.
3. *Acceptable cost*: Hybrid energy systems can be designed to achieve desired attributes at the lowest acceptable cost, which is the key to market acceptance.
4. *Flexibility:* They provide *flexibility* in terms of the effective utilization of renewable source0s.

6.6.1.3 Issues with Hybrid Renewable Energy Systems

Though a hybrid energy system has a number of advantages, there are some issues and problems related to hybrid systems that have to be addressed [121, 148]:

1. Most of hybrid systems require storage devices, and batteries are mostly used. These batteries require continuous monitoring, and there's an increase in the cost, as the batteries life is limited to a few years. It is reported that the battery lifetime should increase to around a few years for its economic use in hybrid systems.
2. Due to dependence of renewable sources involved in the hybrid system on weather results, the load sharing between the different sources employed

for power generation, the optimum power dispatch, and the determination of cost per unit generation is difficult.

3. The reliability of power can be ensured by incorporating weather-independent sources like diesel generator or fuel cell. The use of diesel generator results in fuel cost and harmful emission. Fuel cell can be expensive.
4. As the power generation from different sources of a hybrid system is comparable, a sudden change in the output power from any of the sources or a sudden change in the load can affect the system stability significantly.
5. Individual sources of the hybrid systems have to be operated at a point that gives the most efficient generation. In fact, this may not occur, due to that the load sharing is often not linked to the capacity or ratings of the sources. Several factors decide load sharing like the reliability of the source, economy of use, switching require between the sources, and availability of fuel. Therefore, it is desired to evaluate the schemes to increase the efficiency to as high level as possible. This is a difficult task for weather- and time-dependent solar and wind energy.

The formation of a hybrid system is possible by any combination of different RE sources which may have an additional backup system. Sometimes hybridization is achieved by single source (like PV cell) with backup storage. As mentioned earlier, choice of the combination for a hybrid system considers different factors: feasibility study, site location, technical, sociocultural, and economic considerations.

6.6.2 Examples of Off-grid Renewable Hybrid Systems (HRES)

There are two types of renewable energy resources: dispatchable and non-dispatchable. The examples of dispatchable renewable sources are biomass (waste), geothermal, and hydro energy (some also consider gas because of its abundance as a part of dispatchable energy). These types of energy can be available continuously, but they are not available at all locations. The non-dispatchable renewable energy sources are solar and wind, both are intermittent but available in most parts of the world. These sources are not amenable for continuous power generation. Besides these natural renewable sources, in recent years, fuel cell (which runs on hydrogen and can be considered as power generation as well as storage device) has captured a lot of attention and development effort. The cost of energy, durability, and flexibility of required raw materials are researched, and in recent years, significant progress has been made [121, 148].

Based on all available data, the most prominent renewable energy resources that are presently considered for HRES for off-grid energy are solar, wind, biomass (waste), hydro, and fuel cell. In HRES, these can be considered individually with storage device, two together with/without storage device or in some extreme cases three together with or without storage devices. The storage device most commonly used is battery, although other candidates mentioned above can be considered for special situations. Most often three sources of power generation are not considered because they insert significant complexities in control, optimization, and management for rural electrification. Hybridization, provided with the presence of some storage device, will satisfy the load demand when the mismatch between demand and supply occurs [177]. Many HRESs are currently working successfully worldwide, and some of them are illustrated in this section. Electrification of different areas has been realized by the combination of different renewable sources, for instance, PV/wind/microhydro/fuel cell/biomass, etc. Dependency on the conventional systems is removed by the hybrid systems comprising of 100% renewable sources. However, costs of conventional sources and renewable sources are comparable at some places, and conventional sources act as a backup in order to optimize the power generation process such as diesel generator (DG) and gas generator. Main off-grid HRES power generation systems are based on available sources, along with the favorable environmental conditions, favorable location sites, and other attributes. A control system for hybrid renewable stand-alone system generally involves a multi-port converter unit, energy storage bus, multi-port inverter unit, isolation/protection and load network, analog and digital signal processing and conditioning network, control channels (depending on number of energy sources), and gate drive networks (depending upon number of sources) [121, 148].

The optimized design problem of hybrid power generation systems has emerged as a hot topic. This problem consists of meeting the load demand at minimum cost by carefully determining the optimal location, optimal configuration of the hybrid system, and optimal sizes of the units being installed [178]. There are many software available on websites of different universities and research laboratories that can be utilized for the optimum configuration of the hybrid systems based on the comparison of their performance and cost analysis [179]. There are many economic criteria dedicated to the cost analysis of power generation systems, such as initial capital cost (ICC), cost of energy (COE), net present cost (NPC), and life cycle cost (LCC) [180, 181]. Besides cost, there are other optimization objectives examined in the literature. All the issues related to simulation and optimization are examined in detail in Chapter 7.

6.6.2.1 HRES Involving Solar Energy

PV-wind hybrid system consists of the combined operation of a PV system and a wind generator [121]. This is the most prominent HRES in the world, largely because they are available in most parts of the world, they compensate each other intermittency to some extent, and the cost of both solar and wind energy operation has dramatically decreased over the past decade. Such systems are seasonal and perform depending upon the weather conditions and site selection. The presences of fluctuations in wind flow and variations in solar radiation may result in the failure to meet load demand by a single source alone. However, in some periods of the year, lone operation of PV system or wind generator can meet the load demand. Coastal areas with wind corridors and sun shine available for most of the day are preferable. For example, wind concentration is higher during the winter season, whereas PV concentration is less. Combined operation of these two sources minimizes the chances of power shortage, thereby minimizing the dependency of load on a single source alone [121]. One source compensates the load demand in the absence of the other source combined with batteries and a diesel generator backup. For example, at night, when PV does not work, load can enjoy power from the wind, DG, and batteries [182–189]. Different PV/wind hybrid power generation systems have been introduced in literature for different purposes. Roy et al. [190] presented a PV/wind hybrid power generation system for optimally meeting the remote load in Sagar Island, India. Nandi and Ghosh [191] sized and designed a PV/wind hybrid power generation system for the variable demand of a typical community in Sitakunda, Bangladesh. Shiroudi et al. [192] suggested a well-sized PV/wind hybrid system for the AC appliances of a distant area located in Taleghan, Iran. Beyene and palm [193] introduced a PV/wind hybrid system for the electrification of a distant area, comprising of 200 families, in Ethiopia. Also, Getachew and Gelma [194] designed a hybrid system comprising of photovoltaic–wind system for power generation for a remote community in Ethiopia. Sureshkumar et al. [195] sized, designed, and analyzed a PV/wind hybrid system for optimally meeting the load demand of remote village in India. PV-wind hybrid energy system offers a distinctive means to offset electrical bills of a home or otherwise can make it completely off of the power grid [196].

According to Aziz et al. [121], the proper management of energy flow of this HRES requires following strategies:

1. When load demand is less than the wind generation, the excess power of wind generator and the PV harvester is utilized in charging the backup batteries.
2. When load demand is higher than the wind generation, PV generation combined with the wind generation covers the load demand.
3. When combined generation of wind and PV cannot meet the load demand, batteries bank and diesel generator are utilized [121, 151, 182, 185].

They are ideal at smaller off-grid energy requirement such as for residential or single building purposes. PV-wind operation is nonpolluting and quieter with relatively higher output. The system has a long lifetime of about 15–30 years. The cost of electricity is expected to be in the range of $0.12–$0.41 per kWh [121]. Once the system recovers its initial investment, it is free of cost. Of all hybrid systems, PV/wind system is most popular because it can be available almost anywhere in the world. Solar and wind are most commonly used hybrid energy systems. There are several reasons for this popularity [121, 151, 182, 185]:

1. The times when solar and wind energy are at their best are the exact opposite of each other. Solar is best during daylight hours in the summer. Meanwhile, wind turbines tend to produce the most electricity during nighttime hours in the winter, especially in the case of offshore wind. This makes a wind turbine plus solar panel hybrid system a natural combination.
2. A hybrid energy system with solar and wind energy can produce a consistent source of electricity throughout the year, with the strengths of each resource balancing the other's weaknesses. As production from one

resource dwindles daily or seasonally, the other begins to pick up the slack with more generation.

3. Hybrid energy systems can be expensive, as you're relying on two separate types of electricity resources to solve a single need. However, there are certain instances such as solar and wind when a hybrid energy system makes a lot of sense financially.
4. For off-grid rural and remote areas, a stand-alone, hybrid wind plus solar energy system can be a great option, especially when paired with energy storage.
5. At a higher, grid-scale level, pairing solar and wind energy systems allows renewable developers to participate to a greater degree in deregulated electricity markets. By providing more electricity during more hours, as well as by ensuring production during both summer and winter hours of peak electricity usage, hybrid energy systems are a way to open access to greater levels of renewable energy on the macrogrid as well as microgrid.
6. The maintenance cost of the hybrid solar/wind energy systems is low compared to the traditional generators which use diesel as fuel. No fuel is used, and they do not require frequent servicing.
7. The hybrid solar energy systems work more efficiently than traditional generators which waste the fuel under certain conditions. Hybrid solar systems work efficiently in all types of conditions without wasting the fuel.
8. Unlike traditional generators, which provide high power as soon as they turned on, most of hybrid solar/wind power systems manage load accordingly. A hybrid solar system may have technology that adjusts the energy supply according to the devices they are connected to, whether it's an air conditioner requiring high power or a fan which requires less.

Hybrid solar/wind energy systems also have few disadvantages [121, 151, 182, 185]:

1. With different types of energy sources in use, the systems require some additional knowledge. The operation of different energy sources, their interaction and coordination must be controlled and it can become complicated.
2. Although the maintenance cost is low, the initial investment for the installation of a hybrid solar/wind energy systems is high. On large scale, land requirements can also be very high.
3. The batteries connected to the system may have a lower life as they are often exposed to natural elements like heat and rain.
4. The number of devices you can connect to a hybrid solar/wind energy system is limited and vary from system to system.

Most of the disadvantages mentioned above also apply to other hybrid systems outlined here. With these advantages and disadvantages, the hybrid solar/wind energy systems are becoming more and more popular around the world and are being installed for homes and offices.

All hybrid systems outlined here benefit from battery storage. The hybrid stand-alone system involving solar PV-wind-battery with or without diesel generator involves an AC bus and a DC bus and AC/DC converters because wind energy, diesel generator, and many loads (like home) require AC bus whereas solar energy and battery may require DC bus.

Several studies [197–202] on the use of hybrid PV/wind energy systems for rural electrification are worth noting. Sawle et al. [197–199] found PV and wind hybrid to be the most lucrative solution for diminishing the use of traditional energy sources. These renewable sources of the energy have many remarkable rewards like cost of energy, stability, and feasibility, and also their complementary nature makes them most attractive. The study gave the ideas about various configuration, control strategy, techno-economic analysis, and social effect of the hybrid system. The findings of comprehensive review by Sawle et al. [197–199] will help further improvements in hybrid system design and control and practical system implementation. The study also presents a case study of the remote area of Barwani, India, and results are compared using HOMER and PSO simulations. Objectives such as suitable location, assessment of present and future energy requirement, estimation of payback period, evaluation of LPSP, and economic along with energy reliability analysis are also reported in the literature [197–199]. The design of PV-based hybrid systems and PV-wind-based hybrid system [197–199] to assess the practical performance of the system using distinct methodology for a variety of geographical locations to meet the load demand are presented. The literature including the study of Sawle et al. [197–199] addresses the following goals:

a. Optimal sizing/configuration of hybrid system components
b. Reliable operation control of hybrid system
c. Minimization of the annual COE
d. Minimization of the LPSP for a given load
e. Satisfy the load demand by effective use of renewable sources
f. Decrease the pollutant emissions
g. Improve battery efficiency
h. Higher conversion efficiency
i. Reasonable price and ease of operation
j. Estimation of the loss of load probability (LLP)
k. Minimize costs of O&M of the PV sections and the battery
l. Minimize life cycle costs and ensuring reliable system operation by appropriate design and process control of HRES
m. Minimize the total annual cost including initial cost, operation cost, and maintenance cost

Various strategies for modeling and control of hybrid system using different methodologies and software have been reported [121]. The PV-wind HRES with battery unit and DG as a support are capable of electrifying the remote area population where it is uneconomical to extend the conventional utility grid. The merits of HRES can be achieved only when the system is designed and operated appropriately [80]. Boolean model in geographical information system software was used for identifying the location for best optimal solution of the HRES in the Middle East [121].

Babatunde et al. [201] carried out an optimal hybrid renewable off-grid energy system model that supplies a typical rural healthcare center across the six regions in Nigeria. A technical and economic evaluation was carried out to identify the optimal off-grid hybrid energy system combination based on photovoltaic (PV), wind, diesel generator, and battery. The study considered the effect of fuel subsidy removal by carrying out a sensitivity analysis on the fuel pump price. The impact of a change in diesel fuel pump price and interest rates on the economic performance criteria of the optimal configuration was explored. Results show that across all the locations considered, PV/diesel/battery system is the most economically viable with a net present cost and renewable fraction (RF) ranging between \$12 779 and \$13 646 and 70%–80%, respectively. The cost of energy (COE) was also estimated to range between \$0.507 and \$0.542 per kWh.

Querikiol and Taboada [200] evaluated the performance of a 1.5 kW micro off-grid solar power generator in a 2-hectare area of a 23-hectare agricultural farm located in Camotes Island, Cebu, Philippines. The area required at least 3,000 L of water every day to irrigate its plantation of passion fruit and dragon fruit; however, there is no water source within the immediate vicinity that can support such requirement. A 1/2 horsepower water pump was installed to provide the required irrigation. A 1.5 kW solar photovoltaic (PV) system consisting of 6 units of 250 W solar PV panel with corresponding 6 units of 200 Ah deep cycle batteries managed by a 3-kW industrial grade inverter provided the power for the water pump and supplied for the electricity demand of the farm. The actual energy usage of the farm was measured from the built-in monitoring of the charge controller, and the installed system was analyzed to determine its efficiency in meeting the actual load demand. The HOMER optimization tool was used to determine the optimal configuration for the micro off-grid system based on the actual load demand. Simulation results showed that the optimum configuration that could supply the actual load is a 2.63-kW all-PV system with 8-kWh batteries. Sensitivity analysis was done to consider (1) possible increase in electrical load when the current plantation expands either in progression or outright to its full-scale size of 23 hectares and (2) variations in fuel cost. This study can be considered a good model in assessing renewable energy needs of farms in the country, which can be operationalized for agricultural purposes.

Esan et al. [202] presented the use of a novel approach in assessing the generation reliability of a hybrid mini-grid system (HMS) based on the optimal design result obtained from the HOMER software. A typical Nigerian rural community—Lade II in Kwara State was used as a case study where the energy demand for the residential and commercial loads was 2.5 MWh/day and 171 kWh/day, respectively. The optimized hybrid mini-grid systems (HMGS) results from HOMER comprising of a solar photovoltaic (PV) array (1.5 MW), diesel generators (350 kW), and battery storage (1,200 units) have a combined least net present cost of \$4,909,206 and a levelized electricity tariff of \$0.396 per kWh. Contrasting the HMGS with a diesel-only system for the community, an approximate 97% reduction in all pollutant emissions was observed. Furthermore, fluctuations in diesel fuel prices, variations in average solar insolation, and variations in the solar PV's capital/replacement costs were utilized in conducting a sensitivity analysis for the HMS. The capacity outage probability table (COPT) was utilized in validating the reliability of the simulation results obtained from HOMER. The HMS was observed to experience a load loss of 0.769 MW, 0.594 MW, and 0.419 MW when zero, one, and two diesel generator(s), respectively, were operational for all of the solar PV's and batteries being off-line. The loss of load probability (LOLP), loss of load expectation (LOLE), and total expected load loss (ELL) obtained from the COPT were 5.76×10^{-8}, 5.0457×10^{-4} h/year, and 0.025344 W, respectively. The results show the reliability of the HMS and also depict a highly economical and feasible hybrid energy system.

Microhydro-PV hybrid system consists of the combined operation of a microhydro (MH) generator and a photovoltaic (PV) system. The presence of MH generation depends on the availability of water flow [121]. River bank, canal's bank, and natural water falls in many areas would be helpful. During the wet season, sufficient water flow is present to harness the required power from microhydro generator. During off season/dry season, the water is insufficient to harness the required power from microhydro generator. Therefore, its hybridization is required with a PV system [203]. The solar radiation, either obtained from direct observation or from meteorological measurements, at any site, does have certain limits. This limited solar radiation is not sufficient for harnessing enough power to meet the load demand when it increases during the dry season [121]. Therefore, an energy storage backup is necessary to play its role during peak hours [204]. Different MH/PV hybrid power generation systems have been introduced in literature for different purposes. Hossan et al. [205] sized, designed, and analyzed MH/PV hybrid system for optimized electrification of a distant area in Bangladesh. Qais et al. [206] sized, designed, and analyzed MH/PV hybrid system to optimally fulfill the domestic load demand in distant Malaysian area. Kenfack et al. [207] presented a well-sized MH/PV hybrid system for the electrification of a remote village in Cameroon. Kusakana et al. [208] introduced a MH/PV hybrid system to optimally meet the remote area load in Pretoria, South Africa. Nfah and Ngundam [209] presented the criteria for sizing, designing, and optimally meeting the load demand of a hostel located in Garoua, Cameroon. Ogueke et al. [210] examined the potential of a small hydro/photovoltaic hybrid system for electricity generation in FUTO, Nigeria. Major components of the hydro and PV systems were considered. The study area electricity load demand for a 10-year period, available hydrological data of the Otamiri river, and a 12-year solar radiation data of the study area were used in the study. The results show that at a net head of 3 m, the maximum available hydropower, was determined during October as 174.70 kW while the Kaplan turbine with a runner diameter not less than 4.67 m was found to be the most suitable. For the photovoltaic (PV) system, the maximum area, the number of PV modules, and the battery bank capacity were determined as 3,248 m^2, 3,025, and 98,521,098 Ah, respectively. Thus, a SHPV hybrid system was proven to be a viable source of power generation for the study area.

Syahputra and Soesanti [211] examined the feasibility of microhydro-PV hybrid system for rural areas of Central Java, Indonesia. The Indonesian government has paid great attention to the development of renewable energy sources, especially solar and hydropower. One area that has a high potential for both types of energy is the province of Central Java, located on the island of Java, Indonesia. In this study, authors conducted field research to determine the ideal capacity of solar and microhydro hybrid power plants, electricity load analysis, and optimal design of hybrid power plants. Data on the potential of microhydro plants were obtained by direct measurement on the Ancol Bligo irrigation channel located in Bligo village, Ngluwar district, Magelang regency, Central Java province, Indonesia. Data on solar power potential were obtained from NASA's database for solar radiation in the Central Java region. Hydropower potential data include channel length, debit, heads, and power potential in irrigation channels originating from rivers. These data were used to design an optimal hybrid power plant. The method used to obtain the optimal design of a hybrid power plant system is based on the analysis of capital costs, grid sales, cost of energy, and net present cost. Based on the parameters of the analysis, the composition of the optimal generator for the on-grid scheme to the distribution network was determined. The results showed that hybrid power plants were able to meet the needs of electrical energy in the villages around the power plant and that the excess energy could be sold to national electricity providers.

Presence of solar radiations and water availability are the key requirements for the system operation. The system operates in the following modes [121]:

1. During the wet season, sufficient water flow is present and the load demand is fulfilled by the MH generation. Batteries are charged through rectifier by the MH generator and also by the PV generation during the sunshine hours.
2. During the dry season, MH generation fails to meet the load demand due to insufficient water flow. Low-load demand is met by the fraction of PV generation directly.
3. Any mismatch between the load demand and the PV generation is entertained by the battery bank and DG. Battery bank either stores or supplies the power depending on the load demand [207, 212–217].

The principal advantage of microhydro-PV hybrid system is the enhanced reliability of the system by the elimination of drawbacks encountered in the alone operation of either of two sources. This hybrid system utilizes the freely available sources and has very small maintenance cost. Hence, the only cost of power production from this hybrid system is its initial capital cost, and after recovering its initial investment, it is free of cost. This makes it a monetarily reliable power generation system. The system is nonpolluting, and operation is noise free with a slightly improved output. Lifetime of this hybrid system is expected to be 30–40 years depending on the location of the project. The cost of electricity can be in the range \$0.15–\$0.38 per kWh [121].

PV-fuel cell hybrid system consists of the combined operation of a PV system and fuel cell power generation system [121]. PV system utilizes the solar radiation to directly produce electricity. Fuel cells utilize the hydrogen gas fuel to produce electricity. Availability of hydrogen supply, sufficient solar radiation fall are required. Areas with sunlight for most of the day and where the hydrogen supply is easy are preferable. Hydrogen, produced by the reformation of natural gas or electrolysis of water, can be stored in a hydrogen storage tank which then supplies to the fuel cell for power generation [218]. PV system cannot meet the load claim alone as it depends on the solar radiation which is not present at all times; i.e., at night, no radiations are present [121]. Therefore, using fuel cell generation combined with batteries as backup makes the power supply a reliable process [219–223]. Different PV/FCs hybrid power generation systems have been introduced in literature for different purposes. Lagorse et al. [224] sized, designed, and analyzed PV/fuel cell hybrid system to optimally meet the typical electric load requirement in Belfort-city, France. Krishna [225] reported a PV/fuel cell hybrid system to produce power for typical remote area load in Kondapalli, India. Bruni et al. [226] introduced a PV/fuel cell hybrid system for a house to fulfill the demand of typical variable load of the house appliances in Roma-Ciampino, Rome, Italy. Dufo-Lopez et al. [227] suggested PV/FCs hybrid system for electrification of typical area in Zaragoza-city, Spain. Cetin et al [228] designed PV/Fuel cell hybrid system to produce, distribute, and consume the DC energy in a house for household DC appliances, located in Denizli, Turkey.

Al-Shater et al. [229] examined hybrid PV/fuel cell system design and simulation. In this study, a hybrid photovoltaic (PV)-fuel cell generation system employing an electrolyzer for hydrogen generation is designed and simulated. The system is applicable for remote areas or isolated loads. Fuzzy regression model (FRM) is applied for maximum power point tracking to extract maximum available solar power from PV arrays under variable insolation conditions. The system incorporates a controller designed to achieve permanent power supply to the load via the PV array or the fuel cell, or both according to the power available from the sun. Also, to prevent corrosion of the electrolyzer electrodes after sunset, i.e., when its current drops to zero, the electric storage device is designed so as to isolate the electrolyte from the electrolysis cell.

Ghenai et al. [230] examined design, optimization, and control of a stand-alone solar PV/fuel cell hybrid power system for a residential community. The principal objective was to design a stand-alone power system to meet the desired electric load of a residential community (~150 houses) with high penetration of renewables in the energy mix, low excess power, low cost of energy, and low greenhouse gas emissions. Simulation and optimization were used to determine the performance and the cost of the hybrid power system architecture. The results showed that the solar PV/fuel cell system integrated with solar-based electrolyzer for hydrogen production and using cycle charging control strategy offers the best performance. The total power production from the hybrid system was 52% from the solar PV, and 48% from the fuel cell. From the total electricity generated from the photovoltaic hydrogen fuel cell hybrid system, 80.70% was used to meet all the AC load of the residential community with negligible unmet AC primary load (0.08%), 14.08% is the input DC power for the electrolyzer for hydrogen production, 3.30% was the losses in the DC/AC inverter, and 1.84% was the excess power (dumped). The proposed hybrid power system had 40.2% renewable fraction, was economically viable ($145/MWh), and was environmentally friendly (zero carbon dioxide emissions during the electricity generation).

According to Aziz et al. [121], the system operates in the following modes:

1. When the load demand is less than PV generation, PV alone supplies the load and excess power is used for hydrogen production through the electrolysis of water or reformation of natural gas.
2. When load is greater than PV generation or at night when there is no solar radiation, the stored hydrogen is used for power generation by the fuel cells.
3. When mismatch is large, batteries combined with fuel cell generation are used as backup [226, 227, 231–239].

This system has life expectancy of 20–35 years and cost of electricity can be in the range of $0.12–$0.47 per kWh. Hydrogen generation and availability of solar radiations are the key drivers for the system operation [121].

Solar-biomass hybrid system consists of the combined operation of a solar generation system and a biomass backup system [121, 240, 241]. Biomass consists of animal dung, crops residue, wood chips, and other organic wastes. Through anaerobic digestion, biomass can be utilized to produce biogas. Using this biogas, a turbine can be run for the production of electricity in the power production unit [242]. Solar generator is provided with the parabolic trough to increase the concentration of solar radiation. Areas with sunshine for most of the day, especially agricultural areas, are preferable for producing good amount of biomass [121]. The parabolic trough has an absorber tube filled with a heat transfer fluid (HTF), which produces super-heated steam to drive a turbine for electricity production in the power production unit [243]. The combined operation of solar and biomass generation eliminates the mismatch between power generation and load demand. Different solar/biomass hybrid power generation systems have been introduced in literature for different purposes. Mahalakshmi and Latha [244] reported solar/biomass hybrid system for meeting the demand of industrial load in distant area of Ethiopia. Jradi and S. Riffat [245] carried out modeling and testing of a hybrid solar-biomass ORC-based micro-CHP system. Pradhan et al. [246] suggested solar/biomass hybrid system to optimally meet the demand of a distinct load in Gunupur, Odisha, India. Frebourg et al. [240] introduced a solar/biomass hybrid system for meeting the variable load demand of a distant area in Thailand. Janardhan et al. [247] suggested that the solar/biomass hybrid system was feasible for electrification of the distant rural area in Gorakhpur, India. Nixon et al. [248] conducted a study on the feasibility on hybrid solar-biomass power plant implementation in India. Balamurugan et al. [249] sized, designed, and analyzed solar/biomass hybrid system for meeting the load demand of a university campus in India. Janardhan et al. [247] analyzed the electrical load requirement of Maulana Ganz village in Uttar Pradesh in India and analyzed feasibility of PV-biomass system to provide this load. The study concluded that PV-biomass system can successfully carry out the task.

Sou et al. [250] carried out the design of a photovoltaic/biomass hybrid electrical energy system for a rural village in Cambodia. A hybrid renewable energy system, consisting of a 1.27 kWp solar photovoltaic generator, a 15 kWe biomass gasification system, and a 7.28 kWh battery backup was designed for the electrification of a representative village, namely Chhouk Ksach in Cambodia, which is not currently connected to the electrical power grid and where car batteries are used for electrification. The hybrid system was designed to ensure that the load demand is supplied for 24 hours a day throughout the year. The photovoltaic generator allowed the gasification system to shut down when the load demand is small during daytime hours. The system had an electrical generation capacity of 34,376 kWh per year, of which 4% was generated by the photovoltaic system and 96% by the biomass gasification system. The annual load demand was 24,518 kWh, of which 90% was taken directly from the biomass gasification system, 3% was taken directly from the photovoltaic system, and 7% was taken from the backup battery. The annual net loss was estimated to be 2,878 kWh leaving an energy surplus of 6,980 kWh, which can be used for charging external batteries.

Zhang et al. [251] examined environment assessment of a hybrid solar-biomass energy supplying system. This study investigated the environmental performance of a hybrid solar-biomass energy supplying system by life cycle assessment method. The study examined a hybrid energy supply system driven by biomass and solar energy in northwest China. Based on the results obtained, the system construction stage contributed the most to environmental impacts compared to the operation stage and the disassembly stage. To reduce the environmental influence, it was significant to design the energy structure of the system, i.e., to adjust the proportion of solar energy and biomass energy supply according to local conditions. As is well known, the recycling of nutrients from digestate to local farms as fertilizer could create favorable benefits. Using livestock manure in the anaerobic digestion subsystem is advantageous in terms of AP (as purchased) and EP (edible portion), and meanwhile the greenhouse gas emissions from manure storage could be avoided. Analyzing the overall system process (construction, operation, and disassembly), as well as all the unit processes of the system, allowed the users to select optimal processes that minimize the environmental impacts. For example, the feedstock pretreatment including drying or compressing may reduce the emissions originated from transportation. According to the analyses, the hybrid solar-biomass energy supplying pattern is a promising solution to optimize the energy structure and promote the harmonious development of regional socioeconomic in northwest China, where the solar radiation and bioresources are relatively abundant, and the corresponding incentive policies could be provided to encourage the wide spread of the proposed energy supplying mode. The integrated energy supplying system significantly reduces non-renewable energy consumption, climate change impacts, acidification, as well as eutrophication effects due to the replacement of alternatives such as lignite coal and from fertilizer production. The hybrid solar-biomass energy supplying system not only produces clean thermal energy but also reduces the disposal of organic wastes and produces valuable agricultural products.

In all of these studies, biogas generation and solar radiation availability are the key drivers for the system operation. The system operates in the following modes [121]:

1. When load demand is lower than the solar generation, solar power satisfies the load demand alone. Any excess power charges the batteries bank.
2. When load demand is greater or when solar generation is not present (at night), biomass backup system operates to compensate for this mismatch. Any excess power charges the batteries bank.
3. When load demand is greater than the combined production of solar and biomass, batteries are used to compensate the mismatch. If needed, diesel generator provides the additional backup [121, 245, 247, 248, 252–258].

The main advantage of this system is that it utilizes biomass which is of no use and has negative cost. Slightly increased output and less noisy operation are the main attributes of this system. Long life of about 30–35 years is expected depending on the availability of biomass. Biomass system with capacity factor of 80%–85% makes this system more reliable for power supply. The cost of electricity is expected to be in the range of $0.12–$0.33 kWh [121].

6.6.2.2 HRES Involving Geothermal Energy

Geothermal power plants typically experience a decrease in power generation over time due to a reduction in the geothermal resource temperature, pressure, or mass flow rate. McTigue et al [259] from NREL explored methods to hybridize a double-flash geothermal plant with a concentrating solar power collector field. The solar field generated heat that was added to geothermal fluid and then recirculated through the steam turbine, thereby increasing the mass flow rate and pressure and consequently the power generation. The objective was to augment the geothermal plant power generation from its off-design operating condition at low cost.

A model of a double-flash geothermal power plant was developed, and results were validated against the operation of the Coso geothermal field, in China Lake, California. The concentrating solar system was based on linear Fresnel reflectors developed by Hyperlight Energy Ltd. Data for a wide variety of potential heat transfer fluids were collected. Thermal storage, in the form of two liquid storage tanks, was included with the objective of maximizing the electricity generated by the system. The most suitable fluid (in terms of thermal properties and cost) was determined to be either a mineral oil such as Xceltherm 600 or a synthetic fluid such as Therminol VP-1.

Several different methods of adding solar heat were considered, and a computer model was developed to compare the performance. Practicalities, such as the risk of mineral deposition in the pipes and heat exchangers, were also considered. The best method of integrating the solar heat involved extracting fluid from the first flash tank, heating it with the solar heat, and recirculating it into the high-pressure steam turbine. This configuration can achieve a solar-heat-to-electrical-work conversion efficiency of 24.3%.

Annual simulations were undertaken for this system to determine the optimal sizing of the solar field and the thermal storage. The system was sized to increase the power output from 22.2 MWe to 24.2 MWe and the levelized cost of electricity (LCOE) was calculated. The LCOE is not a suitable metric for assessing the optimal storage duration: storage merely shifts the time that the energy is produced, and as a result, the LCOE always increases with storage duration. (In this study, the power plant received a fixed price for electricity, regardless of the time and demand when it is dispatched.) The optimal storage duration was found by considering the storage capital cost per unit energy dispatched from the storage. For instance, large storage tanks have economies of scale (and low capital costs per unit energy *capacity*), but this is at the expense of the storage tank rarely being fully charged.

The study indicated that an optimal solar field size exists: increasing the solar field size produces more energy and reduces the LCOE. However, if the solar field size is increased further, not all the energy can be absorbed by the power block and must be curtailed, at which point the LCOE begins to increase. The optimal solar field size depends on the duration of thermal storage available: a solar multiple of 2 has an optimal storage duration of 3 hours, while for a solar multiple of 3, the optimal storage size is 10 hours.

Two different heat transfer fluids, a mineral oil and a synthetic oil, were compared. Mineral oils have lower maximum operating points, but low vapor pressures. On the other hand, synthetic fluids can operate at higher temperatures, thereby potentially reducing the storage size at the expense of higher vapor pressures, meaning that storage vessels should be pressurized. Results indicate that a diphenyl oxide/biphenyl-based synthetic fluid is currently a more cost-effective solution.

The LCOE results were compared to an equivalent photovoltaic array (PV) with battery energy storage (BES). Two annual energy generation cases 6.98 and 9.34 GWhe were examined. In the first case, 2 solar multiple for the concentrating solar field with storage duration of 0 and 3 hours were examined. The results indicated that for storage duration of 0 hours, the LCOE ($/kWhe) for hybrid plant was 0.067 compared to 0.062 for PV+BES system. For the storage duration of 3 hours, the LCOE for hybrid plant was 0.081 compared to the one for PV+BES of 0.112. Similarly for the second case, solar multiple of 3 for the concentrating solar field was considered. In this case, for the storage duration of 0 hour, the hybrid plant gave LCOE 0.076 compared to 0.062 for PV+BES system. For the second case and for the storage duration of 10 hours, LCOE for hybrid system was 0.091 compared to 0.172 for PV+BES system. Thus, the hybrid plant had an LCOE comparable to photovoltaics when there is no storage. However, the hybrid plant achieves lower LCOEs than PV+BES once storage was included, because thermal storage is relatively inexpensive compared to batteries. Furthermore, the replacement rate of the heat transfer fluid is low (and therefore low in cost) compared to the cost of replacing batteries that currently have a 10–15-year lifetime.

The main conclusions of NREL study were [259]:

- The thermodynamic performance of five different hybrid plant configurations was investigated. The configuration with the best performance and suitability involved reheating the unflashed brine at the exit of the first flash tank.
- An economic analysis of the hybrid plant was undertaken, and the LCOE was evaluated for plants with a range of thermal storage durations and solar field sizes. The results are described above.

- There is an optimal solar field size: larger solar fields generate more power, but some of this power may have to be curtailed.
- The LCOE is not a suitable metric for evaluating systems with storage. Instead, the storage capital cost per unit energy discharged was used to determine the optimal storage size.
- Using a 400°C hot store with a synthetic fluid was found to be more cost-effective than using a 300°C hot store with a mineral oil, even though the synthetic fluid is more expensive and requires pressurized stores. However, the health and safety and environmental risks of the synthetic fluid need to be carefully considered.
- As mentioned above, the LCOE of a hybrid plant with a solar multiple of 2 and a synthetic fluid storage duration of 3 hours was $0.081 ± $0.011 per kWh, while the LCOE of a comparably sized PV field with 3 hours of Li-ion battery storage was $0.112 ± $0.024 per kWh.
- Storage allows a power plant to dispatch power flexibly and provide value to the grid in several ways. If energy dispatched from the stores is valued at around 1.75 times the typical electricity price of $0.09 per kWh_e, then the hybrid plant including 4 hours of storage has an internal rate of return of 10%.

Geothermal energy has been successfully employed in Switzerland for more than a century for direct use but presently there is no electricity being produced from geothermal sources. Deep geothermal energy is a potential resource for clean and nearly CO_2-free electricity production that can supplant slowly phasing out nuclear power in Switzerland. Deep geothermal resources often require enhancement of the permeability of hot-dry rock at significant depths (4–6 km), which can induce seismicity. The geothermal power projects in the Cities of Basel and St. Gallen, Switzerland, were suspended due to earthquakes that occurred during hydraulic stimulation and drilling, respectively. Garapati et al. [260] presented an alternative unconventional geothermal energy utilization approach that uses shallower, lower temperature, and naturally permeable regions that drastically reduce drilling costs and induced seismicity. This approach uses geothermal heat to supplement a secondary energy source. Thus this hybrid approach may enable utilization of geothermal energy in many regions in Switzerland and elsewhere that otherwise could not be used for geothermal electricity generation. In this work, Garapati et al. [260] determined the net power output, energy conversion efficiencies, and economics of these hybrid power plants, where the geothermal power plant is actually a CO_2-based plant. Varied parameters include geothermal reservoir depth (2.5–4.5 km) and turbine inlet temperature (100–220°C) after auxiliary heating. The study found that hybrid power plants outperform two individual, i.e., stand-alone geothermal and waste-heat power plants, where moderate geothermal energy is available. Furthermore, such hybrid power plants are more economical than separate power plants. More details on this hybrid plant and simulation process are described by Garapati et al. [260].

6.6.2.3 HRES Involving Wind Energy

Besides PV-wind HRES mentioned earlier, wind energy can also be hybridized with other sources [121]. Microhydro-wind hybrid system consists of the combined operation of a MH generator and a wind generator. In many cases, the stand-alone wind power systems can be feasible for the electrification of far-flung and distant areas where the sufficient wind speeds are available, and no grid supply and other sources of electricity generation are available due to economic or some other reasons. In order to meet the escalated demand of the distant areas, by some means, this stand-alone power generation has to be expanded. Out of several, one viable choice is the addition of a MH generator with the wind generator where abundant water falls are available. Provided with some storage backup means, system becomes more reliable [261–264]. Coastal areas, bank of a river or canal, and natural water falls with natural wind flows are desirable. Different MH/wind hybrid power generation systems have been introduced in literature for different purposes [121]. Khan et al. [265] sized, designed, and analyzed a MH/wind hybrid system to optimally fulfill the domestic load demand in Tioman Island, Malaysia. Sadiqi et al. [266] introduced a MH/wind hybrid power generation system for optimally meeting the village area load demand in Bamiyan province, Afghanistan. Bekele and Tadesse [267] presented a MH/wind hybrid system for optimized electrification of a distant rural area in Ethiopia. Atiqur Rahman et al. [268] sized a MH/wind hybrid system for a remote area in Bangladesh and analyzed for optimum operation conditions. Lal et al. [269] suggested a MH/wind hybrid system to optimally meet the load demand of a typical community in Sundargarh district, Orissa state, India. Presence of wind flows and water availability are the key drivers for the system operation.

Bakele and Tadesse [267] examined the feasibility study of small hydro/PV/wind hybrid system for off-grid rural electrification in Ethiopia. In this work, feasibility of small-scale hydro/PV/wind-based hybrid electric supply system to the district was studied. Six sites (two on Taba stream, one on Bechet stream, two on Muga stream, and one on Suha stream) with small-scale hydropower potentials were identified. The hydro potentials were analyzed with the help of GIS and data obtained from the Ministry of Water Resource (MoWR) of Ethiopia. Meteorological data from National Meteorological Agency (NMA) of Ethiopia and other sources, such as NASA, was used for the estimation of solar and wind energy potentials. Electric load for the basic needs of the community, such as, for lighting, radio, television, electric baker, water pumps, and flour mills, was estimated. Primary schools and health posts were also considered for the community. HOMER energy was used for optimization and sensitivity analysis of the hybrid system. Taba (B) site was selected for a detailed study and, as a final result, different system types and their component sizes were identified having a cost of energy less than $0.16/kWh.

In general, wind-microhydro system operates in the following modes [121]:

1. When there is no mismatch between wind power generation and the load demand, the wind generator feeds the load alone.
2. When wind generation is not sufficient to entertain the load demand, micro-hydro generator starts working.

3. When the load decreases, the surplus power starts charging the backup batteries.
4. When the load increases or the wind-microhydro generation is not sufficient to meet the load demand, the DG starts supplying power to the load.
5. When the difference between the load demand and the generation is severe or long term, the backup batteries and DG can serve the load [270–281].

The enhanced reliability due to combined operation of MH and wind resources, pollution-free operation, less maintenance cost, and free fuel are the plus points of this hybrid system. Capacity factors of MH and wind are 51%–55% and 25%–35%, respectively. The lifetime of the system is 15–25 years. The cost of electricity can be in the range $0.10–$0.36 per kWh [121].

A wind-biomass hybrid system consists of the combined operation of a wind generator and a biomass backup system. Wind drives the wind generator to directly produce electricity. Biomass is gasified to obtain biogas. This biogas is utilized to drive turbine in a power conversion unit (PCU) for electricity production [282]. The security of wind power reliability depends on the wind flow availability and a lesser amount of fluctuations in it. Windy areas with good amount of biomass being produced are desirable. Wind flows keep changing during the year, thereby affecting the wind generator reliability in meeting the load demands [283]. The reliability of the wind power generation can be ensured by its combined operation with a biomass backup system [284]. Biomass backup system, combined with batteries and diesel generator backup, compensates the mismatch between predicted and real wind power production [282, 285]. Different wind/biomass hybrid power generation systems have been introduced in literature for different purposes. Kesraoui [286] sized and designed wind/biomass hybrid system for power generation to meet the variable load demand of an isolated pilot village in Algeria. Maherchandani et al. [287] suggested wind/biomass hybrid system for household and small industrial load of a remote location in Udaipur Rajasthan, India. Raheem et al. [284] reported that a well-sized wind/biomass hybrid system was feasible for electrification of small rural area community in Kallar Kahar, Punjab, Pakistan. Balamurugan et al. [288] designed a wind/biomass hybrid system suitable for the electrification of three distant villages in India. Biogas generation and wind flow availability are the key nuts and bolts for the system operation.

Balamurugan et al. [289] examined an optimal hybrid wind-biomass gasifier system for rural areas. Biomass gasifier-based power generation has a lot of potential for power generation in India. In this study, optimal sizing and operation of a wind-biomass gasifier-based hybrid energy system is proposed. A case study is done using real-time load demand data collected from a site. Economic analysis of wind-biomass gasifier hybrid energy system is also presented. The cost of energy is found to be Rs.3.499 ($0.078). The cost of energy of the wind-biomass system is also compared with the wind-diesel system. It is found that economically, the wind-gasifier system is a very good alternative for wind-diesel system.

Parez-Navarro et al. [282] investigated wind-biomass hybrid system to generate stable power. Wind generators are greatly affected by the restrictive operating rules of electricity markets because, as wind is naturally variable, wind generators may have serious difficulties on submitting accurate generation schedules on a day ahead basis, and on complying with scheduled obligations in real-time operation. In this study, an innovative system combining a biomass gasification power plant, a gas storage system, and stand-by generators to stabilize a generic 40 MW wind park is proposed and evaluated with real data. The wind park power production model is based on real data about power production of a Spanish wind park and a probabilistic approach to quantify fluctuations, and so, power compensation needs. The hybrid wind-biomass system is analyzed to obtain main hybrid system design parameters. This hybrid system can mitigate wind prediction errors and so provide a predictable source of electricity. An entire year cycle of hourly power compensation needs is simulated deducing storage capacity, extra power needs of the biomass power plant, and stand-by generation capacity to assure power compensation during critical peak hours with acceptable reliability.

Liu et al. [256] analyzed feasibility study of stand-alone PV-wind-biomass hybrid energy system in Australia. This study presents a feasibility study of photovoltaic (PV), wind, biomass, and battery storage-based hybrid renewable energy (HRE) system providing electricity to residential area in Australia. The system with load of 200 kWh/day is analyzed through the environmental and economic aspects. The study computes the net present cost (NPC), the cost of energy (COE, KW/h), and the emissions (kg/year) of greenhouse gas (GHG) of the HRE system running under the specific renewable energy resource mentioned above. The monthly daily mean global solar irradiance and wind speed data of the capitals of the seven regions of the six states and various territories of Australia (Queensland, Northern Territory, South Australia, Tasmania, Victoria, Western Australia, and New South Wales) are generated by the RETScreen Clean Energy Project Analysis Software produced by Natural Resources Canada. The long-term continuous implementation of the system is simulated. The software HOMER produced by the National Renewable Energy Laboratory is used as a simulating tool. Their cost and emissions are compared with each other among the systems. It is found that an off-grid PV-wind-biomass HRE system is an effective way of emissions reduction, and it does not increase the investment of the energy system.

According to Aziz et al. [121], the wind/biomass hybrid system operates in the following modes:

1. When load demand is lower than the wind generation, only wind generator supplies the required power. Any extra power charges the batteries.
2. When load demand is greater than the wind generation, power is supplied by the combined operation of wind generator and biomass backup system. Any extra power in the combined operation charges the batteries.
3. When mismatch between supply and load demand is so large that the combined operation fails to fulfill the load demand, DG and batteries bank are utilized as a backup [121, 256, 282].

The capacity factor of wind power system is 25%–35% while that of biomass is 80%–85%. When combined as a hybrid system, the capacity factor is improved which results in more reliable power supply operation. Emissions from this hybrid system are less as compared with the conventional sources, and therefore environmental impacts are limited. After compensating for the capital cost, this hybrid system is a free source of power as the fuel is almost free. The lifetime of this system is expected to be 20–35 years depending upon the quality of biomass and site location of the system. The cost of electricity can be in the range of $0.09–$0.33/kWh [121].

Wind-fuel cell hybrid system consists of the combined operation of a wind generator and fuel cell power generation system [290, 291]. Wind generator utilizes the wind flow to generate electric power [292]. Hydrogen is produced by the electrolysis of water or by the reformation of natural gas and stored in a storage tank [218]. Fuel cell utilizes this stored hydrogen to generate electric power [231]. Wind generation depends on the availability of wind flows. Hydrogen supply and wind flow greater than the cut-in speed of wind turbine should be available. Windy areas, especially of coastal lines, are preferable as the hydrogen supply is easily available. As wind flows are not constant, wind generation is not constant at all times, which may cause a mismatch between load demand and the generation [121]. Therefore, combining the wind generation with fuel cell generation can compensate this mismatch. For severe mismatch conditions, battery and diesel generator backup is necessary [293, 294]. Different wind/FCs hybrid power generation systems have been introduced in literature for different purposes. Leva and Zaninelli [295] designed wind/FCs hybrid power system to produce, distribute, and consume the energy in a remote house with a specific load located in Milano, Italy. Alam et al. [296] reported wind/fuel cell hybrid system for power production to entertain a typical load in Cookeville, USA. Panapakidis et al. [297] suggested wind/fuel cell hybrid system to meet the load requirements of a typical residence in Greece. Mills and Al-Hallaj [298] sized and designed wind/fuel cell hybrid system to optimally meet the variable load demand of a residence in Chicago, USA. Khan and Iqbal [299] introduced wind/fuel cell hybrid system to meet the typical household electric load requirement in Canada. Hydrogen generation and availability of wind flows are the key drivers for the system operation.

Sobotka [300] analyzed wind-fuel cell hybrid system. In this study a part of the energy produced by the wind turbine is stored in the form of hydrogen and is then delivered for consumption at variable power through a fuel cell. A model was developed to determine the key technical parameters influencing the operation of a wind energy system with hydrogen storage. The model incorporates the simulation results of a 600 kW wind energy system with a 100 kW proton exchange membrane fuel cell (PEMFC) and an electrolyzer. Dynamic modeling of various components of this small isolated system is presented for the month of January, 2006. In this way, the energy availability can be estimated and is presented for hybrid installations. The study presents the technology of the system for each particular element.

In general, wind-fuel cell hybrid system operates in the following modes [121]:

1. When load demand is less than wind generation, only wind generation is utilized by the load. Any excess power is utilized for hydrogen production.

2. When load demand is greater than the wind generation or in case of unavailability of wind flow, the stored hydrogen is used for power generation by the fuel cell.
3. When mismatch between load demand and supply is large, batteries and diesel generator are used as backup [121, 232, 236, 297, 298].

A wind generator has a capacity factor of 25%–35%, and a fuel cell has a capacity factor of 90%–95%. So, the overall capacity factor of the system is enhanced by transforming both systems into a single hybrid system. This results in an enhanced reliability of power supply system. This hybrid system is emission free and favorable to maintain a clean and healthy environment. The lifetime of the system is expected to be 30–45 years, but it depends on its location. The cost of electricity can be in the range of $0.12–$0.50 per kWh [121].

Wind-tidal hybrid system consists of the combined operation of a wind generator and a tidal generator. Wind generator utilizes the natural wind flows for power generation. Wind generation system performs depending upon the natural wind flow conditions and site selection [63]. Due to the existence of fluctuations in wind flows, stable power generation is not feasible by this single source to meet the load requirement. Oceanic areas receiving tides and bearing high wind flow are required. Tidal generator utilizes the periodic water currents (tides) for power generation. These tides result from the gravitational pull on sea water by the moon and sun. High water currents result in greater power generation from the tidal generator [301–304]. For the purpose of stable power output for the load, the fluctuations in the wind generation can be compensated by the tidal generation (wind-tidal hybrid system) [305] and a backup system (batteries and DGs) [306, 307]. Different wind/tidal hybrid power generation systems have been introduced in literature for different purposes. Lim [308] designed wind/tidal prototype hybrid system to produce energy for a small load. Sikder et al. [309] sized, designed, and analyzed wind/tidal hybrid system to meet the load requirements of a typical coastal area, consisting of 35,000 families, in Kutubdia Island, Bangladesh. Khan et al. [310] sized the wind/tidal hybrid system suitable for meeting typical load demand of coastal area, consisting of 30,000 families, in Sandwip Island, Bangladesh. Mousavi and Gazafroodi [306] suggested wind/tidal hybrid system to optimally meet the typical variable AC load in Tehran, Iran. Presence of tides and wind flow availability are the key essentials for the system operation. Da and Khaligh [307] evaluated hybrid offshore wind and tidal turbine energy harvesting system with independently controlled rectifiers.

A hybrid wind-current power generation system was installed off the Japanese coast by Tokyo-based Mitsui Ocean Development & Engineering Company (MODEC), the device's developer [311, 312]. The 500-kW Savonius Keel & Wind Turbine Darrieus, or SKWID, was a hybrid system featuring an omnidirectional Darrieus wind turbine and a Savonius tidal turbine, which share a floating axis. A power generation assembly sat between them at deck level, anchored by a set of rubber mounts to a floating semi-submersible platform. The system was able to generate power from either or both sources. According to the company, the wind turbine had a hub height of 47 m (154 ft) and a rotor diameter of 15.2 m (49 ft).

Addressing the issue of stability, the company said the tidal turbine acted as ballast, making the assembly self-righting, and that the location of the generator at deck level created a low center of gravity as well as offering easy access for operations and maintenance. The Savonius tidal turbine was especially suited to harvesting energy from weak currents, MODEC said, and its rotation can be used to start the wind turbine spinning in low wind conditions.

According to Aziz et al. [121], wind-tidal system operates in the following modes:

1. When load demand is less than wind generation, wind generator supplies the load and surplus generation charges the batteries.
2. In case of wind fluctuations when load demand is greater than the unstable wind generation supply, tidal generator pays for the fluctuations to ensure the stable power supply.
3. In the absence of tidal generation and presence of fluctuations in the wind when mismatch between load demand and supply is large, batteries are used as backup to fulfill the load requirement [313–321].

Capacity factor of tidal generator is 30%–50% and that of wind generator is 25%–35%. Hybridization of these two renewable sources results in a system with overall good capacity factor (i.e., in the absence of wind, tidal generates the required power). As system is located in the ocean, there is no danger of pollution from this hybrid system. The lifetime of the system is expected to be 15–20 years, but it is still under the research phase. The cost of electricity can be in the range of $0.14–$0.46/kWh [121].

6.6.2.4 HRES Involving Biomass/Waste Energy

Besides the examples of biomass-solar, biomass-wind, and others cited above, there are other interesting examples of hybrid biomass energy reported in the literature [322–324]. NTE Energy has been awarded a patent [286] for its proprietary biomass technology that it says will allow the simultaneous operation of a biomass energy cycle and a traditional power plant. The technology integrates biomass and conventional steam electric generating technologies into a single hybrid renewable power generation facility, according to NTE. The addition of the hybrid technology to an existing generation facility provides substantial capital cost savings over standalone biomass facilities through the shared steam turbine and other common major equipment systems, according to NTE. "Hybrid technology allows utilities to provide renewable, baseload power from a native biomass fuel." The system can also be used to generate heat or power. Woody biomass will most likely be the most common feedstock for the hybrid system, although the company has looked into other biomass fuels for certain projects [322].

Although low-temperature fuel cells powered by methanol or hydrogen have been well studied, existing low-temperature fuel cell technologies cannot directly use biomass as a fuel because of the lack of an effective catalyst system for polymeric materials. Now, researchers at the Georgia Institute of Technology [323] have developed a new type of low-temperature fuel cell that directly converts biomass to electricity with assistance from a catalyst activated by solar or thermal energy. The hybrid fuel cell can use a wide variety of biomass sources, including starch, cellulose, lignin—and even switchgrass, powdered wood, algae, and waste from poultry processing. The device could be used in small-scale units to provide electricity for developing nations, as well as for larger facilities to provide power where significant quantities of biomass are available. The new solar-induced direct biomass-to-electricity hybrid fuel cell was described in the journal *Nature Communications*, on February 7, 2014.

The challenge for biomass fuel cells is that the carbon-carbon bonds of the biomass—a natural polymer—cannot be easily broken down by conventional catalysts, including expensive precious metals. To overcome that challenge, scientists have developed microbial fuel cells in which microbes or enzymes break down the biomass. But that process has many drawbacks: power output from such cells is limited, microbes or enzymes can only selectively break down certain types of biomass, and the microbial system can be deactivated by many factors. Deng and his research team at Georgia Tech got around those challenges by altering the chemistry to allow an outside energy source to activate the fuel cell's oxidation-reduction reaction.

In the new system, the biomass is ground up and mixed with a polyoxometalate (POM) catalyst in solution and then exposed to light from the sun—or heat. A photochemical and thermochemical catalyst, POM functions as both an oxidation agent and a charge carrier. POM oxidizes the biomass under photo or thermal irradiation and delivers the charges from the biomass to the fuel cell's anode. The electrons are then transported to the cathode, where they are finally oxidized by oxygen through an external circuit to produce electricity.

The system provides major advantages, including combining the photochemical and solar-thermal biomass degradation in a single chemical process, leading to high solar conversion and effective biomass degradation. It also does not use expensive noble metals as anode catalysts because the fuel oxidation reactions are catalyzed by the POM in solution. Finally, because the POM is chemically stable, the hybrid fuel cell can use unpurified polymeric biomass without concern for poisoning noble metal anodes.

The system can use soluble biomass or organic materials suspended in a liquid. In experiments, the fuel cell operated for as long as 20 hours, indicating that the POM catalyst can be re-used without further treatment. The researchers reported a maximum power density of 0.72 mW per sq. cm, which is nearly 100 times higher than cellulose-based microbial fuel cells and near that of the best microbial fuel cells. Deng believes the output can be increased five to ten times when the process is optimized. The researchers also need to compare operation of the system with solar energy and other forms of input energy, such as waste heat from other processes. Beyond the ability to directly use biomass as a fuel, the new cell also offers advantages in sustainability—and potentially lower cost compared to other fuel cell types [323].

Arabkoohsar and Sadi [324] examined a hybrid solar concentrating-waste incineration power plant for cost-effective and dispatchable renewable energy production. In this study, a steam power plant is proposed in which the heat demand is supplied by a hybrid system including a parabolic trough solar collector array and a waste incineration unit. The proposed hybrid power plant provides a uniform power production profile, making the produced intermittent solar energy well dispatchable, energy efficient, and cost-effective. The proposed plant is designed, sized, and analyzed over an entire year of operation in a case study in Denmark. The results of the simulations of the system are presented and discussed [324].

6.6.2.5 HRES Involving Microhydro Energy

Besides hybridization with solar and wind energy, microhydro can also be hybridized with other renewable sources as outlined below. Microhydro-biomass hybrid system consists of the combined operation of a MH generator and biomass backup system [325]. Microhydro generation depends on the availability of water flow. The more the water flow available on the site, the more power can be generated [121]. Sufficient water flow is present during the wet season to harness the required power from microhydro generator. But, during off/dry season, the water is insufficient to harness the required power from microhydro generator. Therefore, its hybridization is required with some other source like a biogas generator [203]. Biogas generation depends on the availability of biomass. Biomass (straw, wood, manure, wood waste, sugar cane, rice husk, corn cob, bamboo, etc.), sufficient water flow, and water head should be available. Water falls available especially in agricultural areas produce a handsome amount of biomass. Biomass produces biogas by digestion process which is then used to rev a generator. Biogas generator combined with batteries and diesel generator backup compensates the mismatch between load demand and generation if microhydro generation is not sufficient [326]. Different MH/biomass hybrid power generation systems have been introduced in literature for different purposes. Varshney et al. [327] reported that MH/biomass hybrid system was suitable for the electrification of the rural community, comprising of nine villages, in Narendra Nagar, India. Kanase-Patil et al. [328] suggested MH/biomass hybrid system for power generation to fulfill the electric load demand of distant area in the state of Uttarakhand, India. Gupta et al. [329] designed MH/biomass hybrid power generation system for meeting the variable load of typical area in Tehri-Garhwal district, India. Sanchez et al. [330] reported the implementation of MH/biomass hybrid system to meet the load demand of remote location, comprising of 120 households, in Amazon region, Brazil [121].

Bhatt et al. [331] examined feasibility and sensitivity analysis of an off-grid microhydro-photovoltaic-biomass and biogas-diesel-battery hybrid energy system for a remote area in Uttarakhand state, India. The aim of this paper was to study techno-economic feasibility of microhydro-photovoltaic-biomass and biogas-diesel-battery hybrid energy system (HES) in off-grid mode for a rural area in Uttarakhand state, India. The considered HES was designed for energy access in five unelectrified villages. Size optimization and sensitivity analysis of considered system was performed by using HOMER software in order to meet the electricity requirements of study area. The selection of optimum configuration is based on the lowest value of cost of energy (COE) and net present cost (NPC), along with the maximum value of renewable fraction (RF) and lowest harmful emissions (CO_2). By considering economy and environment as the main concerns, comparative analysis of four different types of models was presented. Based on sensitivity analysis, three configurations out of four different models were selected. By considering economy and environment as the main driving factors, one configuration out of three configurations was selected as the optimum configuration for the study area which resulted in total NPC $5,33,654, COE $0.197/kWh, RF 94%, and CO_2 emission 15,930 kg/year.

Biogas generation and water flow availability are the key engines for the MH-biomass system operation. In general, microhydro-biomass system operates in the following modes [63]:

1. When load demand is lower than the MH generation, only MH generator supplies the load with required power. Any surplus power charges the batteries.
2. When load demand is greater than the MH generation, power is supplied by the combined operation of MH generator and biomass backup system.
3. When mismatch between supply and load demand is so large that the combined operation of MH and biomass generation fails to fulfill the load demand, DG and battery bank are utilized as a backup [329, 332–335].

The attractive feature of this system is that it utilizes biomass which is of no use and has negative cost. Main advantages include the slightly increased output and less noisy operation. Microhydro generator has capacity factor of 51%–55%, while biomass has 80%–85% capacity factor. Hence, more reliable system results from the hybridization of these two sources. The lifetime of this system depends on the frequency of usage and is expected to be 15–30 years. The cost of electricity can be in the range of $0.09–$0.27kWh [121].

Microhydro-fuel cell hybrid system consists of the combined operation of a MH generator and fuel cell power generation bank system. In many cases, the standalone MH systems can be feasible for the electrification of far-flung and distant areas where the sufficient run of river flows or low heads are available and no grid supply and other sources of electricity generation are available. Availability of hydrogen supply (generation through electrolysis of water or reformation of natural gas), water flow, and water head are required. Water falls should be available in areas where hydrogen supply is easily available. The MH generation depends on the availability of water flows that keeps changing with the amount of rain fall. In case of escalated water flows, power has to be stored in some form to use it when water flows are less. Out of several, one viable choice is the production and storage of hydrogen when abundant water flows are available. Hydrogen, produced by the reformation of natural gas or electrolysis of water, can be stored in a hydrogen storage tank which then supplies the fuel cell for power generation [218]. Using fuel cell generation combined with batteries and a DG as backup makes the power supply a reliable process [129, 336–339]. Different MH/FCs hybrid power generation systems have been introduced in literature for different purposes. Soedibyo et al. [340] suggested MH/fuel cell hybrid system for meeting the typical variable load demand of a village area located in Leuwijawa, Indonesia. Tudu et al. [341] designed MH/fuel cell hybrid power generation system to fulfill the demand of typical variable load in Kolkata, India. Mostofi and Shayeghi [326] reported that MH/fuel cell hybrid power generation system was feasible for the electrification of rural community, consisting of 12 families, in Meshkinsha, Iran [121].

Barsoum and Petrus [342] examined the cost optimization of hybrid solar, micro-hydro, and hydrogen fuel cell using HOMER software. The main purpose of this study was to present the analysis of cost optimization of hybrid renewable energy system in remote area where grid connection is impossible. The analysis of hybrid system was modeled using HOMER software. HOMER software was utilized as an assessment with modeling tools because it simplified the task of evaluating design options for both off-grid and on-grid-connected systems and allowed load data and parameters to be inputted. In this study, HOMER was used to find the most cost-effective configuration among set of different simulated system to achieve the objective of least possible levelized cost of energy (LCOE) for a given system under a proposed load of 75 kW.

According to Aziz et al. [121], hydrogen generation and availability of water flows are the key requirements for the system operation. The system operates in the following modes [121]:

1. When load demand is less than MH generation, only MH generation is utilized by the load. Any excess/surplus power is utilized for hydrogen production and charging the batteries.
2. When load demand is greater than the MH generation or in case of unavailability of water flows, the stored hydrogen is used for power generation by the fuel cells.
3. When mismatch between load demand and supply is large, batteries and diesel generator are used as backup to fulfill the load requirement [121, 227, 326, 341].

Fuel cell is an emission-free power producing source and so is the microhydro generator. Fuel cell and microhydro have capacity factors of 90%–95% and 51%–55%, respectively. Hence, the resulted hybrid system is more robust, consistent, and environment friendly. System is noiseless and has smooth operation with reduced operation and maintenance costs. The lifetime of this system is expected to be 25–35 years. The cost of electricity is expected to be in the range of $0.13–$45 per kWh. More descriptions of all the hybrid systems described in above five sections can be found in references [2, 3, 78, 79, 117, 120–428].

6.6.3 Future Challenges in the Use of Renewable Hybrid Energy Resources

While HRES for off-grid energy has made significant progress, some challenges still remain [121, 148].

1. Renewable sources harness only limited amount of power. Therefore, a technological breakthrough is needed to harness a larger scale power from these sources. The efficiency of renewable energy sources like, PV, hydro,

wind, and tidal need to be improved to make these sources economically more attractive [402].

2. High capital cost of the renewable power generation sources makes their implementation difficult, particularly in developing countries. This also leads to an elevated payback period. A substantial reduction in production cost has to be realized in order to reduce the higher capital cost and to motivate the industry for the implementation of these sources [402].
3. More research to reduce power losses from power converters is needed. This requires more robust design of power electronic converters [403].
4. More improvement of life cycle of storage/backup devices, like batteries and supercapacitors, is needed. Also the use of hybrid energy storage devices need to be investigated [404].
5. More cost-effective hydrogen production, particularly by reforming, electrolysis, and thermal dissociation, is needed [405]. More effective production of hydrogen from other raw materials like natural gas and methanol is needed. The cost of electricity from fuel cell and its durability still needs to be further improved.
6. More research needs to be done for safety and protection issues of HRES systems. Significant variations in load for HRES system need to be protected. Safety concerns arise from the disposal of backup storage equipment like hydrogen storing tanks and batteries. Hydrogen storage itself is a major safety concern [404].
7. For hydro sources, transmission costs to the users need to be reduced [402].
8. More investigation of optimization of various HRES systems is needed. This subject is further treated in Chapter 7. More theoretical understanding of variable load profile for HRES systems is needed [406].Load fluctuations are less likely to be tackled by these stand-alone hybrid systems. The entire system can collapse when the variations in load are large [406].
9. More efforts are needed to improve the efficiency, cost, and durability of various storage devices in particular various types of batteries and supercapacitors.
10. More efforts are needed to harness dispatchable sources like hydro, geothermal, and biomass. These sources can provide more sustained power production.
11. More applications of renewable sources for various industries are needed to reduce carbon emission to the environment.

6.7 PERSPECTIVES, STRATEGIES, AND BRIEF GLOBAL OVERVIEW OF RURAL ELECTRIFICATION

One of the major uses of off-grid hybrid energy systems is the rural electrification. Rural electrification is the process of bringing electrical power to rural and remote areas. Rural communities are suffering from colossal market failures as the national grids fall short of their demand for electricity. As of 2017, over 1 billion people worldwide lack household electric power—14% of the global population. Figure 6.8

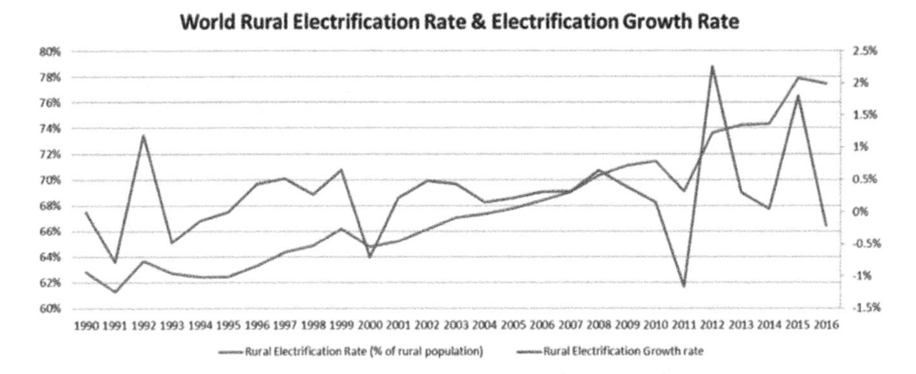

FIGURE 6.8 This graph shows the world rural electrification rate along with the electrification growth rate from 1990 to 2016 and synthesizes data from the World Bank [429].

shows the world rural electrification rate and the electrification growth rate between 1990 and 2016 [429].

Off-grid renewable energy technologies offer an economical and environmentally sustainable solution for bringing power and with it educational, job creation, and economic development, improved healthcare system, productivity and efficiency, and quality of life benefits to remote villages. All of this improves social structure and moral of villages. For example, in 2014, rural communities in India gained more than $21 million from increased economic activity driven by additions of electricity. There are also some additional benefits of rural electrification such as (a) reduced isolation and marginalization through telephone lines and television, (b) improved safety with the implementation of street lighting and lit road signs, and (c) reduced expenses on expensive fossil fuel lamps, i.e., kerosene [429].

Access to modern energy services and rural development are inextricably linked [16, 346]. Definitions of access vary [347]; Ranjit and O'Sullivan [350] define access as:

> *Access to modern energy can be defined as a household's ability to obtain an energy service, should it decide to do so. Access is a function of availability and affordability. For energy to be considered available to a household, the household must be within the economic connection and supply range of the energy network or supplier. Affordability refers to the ability of the household to pay the up-front connection cost (or first cost) and energy usage costs. A high up-front cost may discourage poor households from making a switch to a modern energy form.*

One can broaden this definition beyond households to include any potential consumer, from individuals to large organizations. Most rural societies experience limited access to modern energy services, due to problems of availability and/or affordability. Instead, they rely on traditional fuels—predominately animal dung, crop residues, and wood—for majority of their energy needs [351]. Such "energy poverty" has a serious impact on living standards and productivity. When burned, traditional fuels often produce hazardous chemicals with negative health impacts, especially when used indoors. For example, Ezzati and Kammen [352] provide strong evidence that exposure to indoor air pollution from the combustion of traditional fuels in Kenya enhances the risk of acute respiratory infection. They show that relatively affordable environmental interventions, such as use of an improved stove with traditional fuels, can reduce acute respiratory infection by 25% among infants and young children [345].

The fact that traditional fuels cannot produce a range of modern energy services such as mechanical power and electricity limits their ability to improve other aspects of life, including education and employment. Traditional fuels (generally biomass) also produce energy inefficiently. As a result, they require substantial time and effort to collect, and as local resource stocks decrease, they increasingly have to be sourced from further afield. This significantly reduces the time available for productive activities. If managed ineffectively, such resources use can also degrade the environment and create negative spillover effects in other sectors. Given the cultural practices in many rural areas, these impacts are often most felt by women and children [346, 351, 353–356, 429].

Although there are some methodological difficulties establishing a clear relationship between energy poverty and rural development [357], a common concept used is that of the "energy ladder" [346, 430]. Societies that depend on traditional energy activities are found at the bottom rung of the energy ladder. As they increasingly access modern energy services, they move up the energy ladder. For example, delivered energy (MJ/kg of fuel) for wood with stove is 5, charcoal wit stove is 10, kerosene is 12, biogas is 15 and LPG is 25–30 [346]. At the top of the ladder are societies that have full access to modern energy services and experience greater levels of economic development and higher income levels [346, 351, 430]. Table 6.4 shows the correlation between a country's dependence on biomass and its per capita gross national product (GNP).

Movement up the energy ladder can occur within various aspects of rural life: agriculture, household cooking, household lighting, and heating [346]. However, it is important to appreciate that Table 6.4 shows only a *correlation* between a dependency on biomass and per capita GNP—it does not necessarily indicate *causality* [346]. It

TABLE 6.4
Use of Biomass in Relation to GNP Per Person in 80 Developing Countries (after 156, 7) [431]

Avg. GNP Per Person	% Biomass Used
<200	95–100
200–400	90–95
400–500	75–80
500–1,000	Avg. 60 (40–70)
1,000–2,000	Avg. 40 (20–60)
2,000–3,000	Avg. 20 (0–40)

seems logical to assume that increased access to modern energy services (moving up the energy ladder) can catalyze rural development (measured in increased income). In fact, there is a co-dependent relationship: access to modern energy services can increase incomes (if used productively), and an increase in income can make modern energy services more affordable [344–358, 429–433].

For the past 10 years, it has been frequently estimated that around 2 billion people have no access to modern energy services and about 1.16 billion people live without access to electricity [5, 351]. Access to modern energy services and electricity is low in many developing countries, particularly in sub-Saharan Africa and parts of Asia (see Table 6.5 for figures on Africa). If the MDGs (millennium development goals) are to be achieved in these parts of the world, then significant efforts are needed to bring rural areas out of energy poverty [430]. This can be done in two ways: increasing access to energy for domestic use—essentially increasing access to technologies which use modern fuels or make use of traditional fuels in cleaner, safer, and more environmentally sound ways—and increasing access to electricity [344–358, 429–433].

6.7.1 Approaches to the Development of Rural Electrification

Recent international agreements support the access to affordable, sustainable, reliable, and modern energy for all [359, 360]. Poverty is the key factor limiting accessibility among urban households [142], while the remote communities beyond the extent of the grid have seen few successful electrification projects [143]. Since the 1980s, remote electrification projects have been completed through private and public sector investments and/or institutional loans [144]. Some of the electrification

TABLE 6.5
Levels of Electricity Access in Selected sub-Saharan African Countries [431]

	Population	Access to Electricity (% of Population)	
Country	People Living in Rural Areas (millions)	Urban	Rural
Benin	5.3	51	5.5
Cameroon	8.14	77	16.5
Ethiopia	65.89	86	2
Kenya	30.29	51.5	3.5
Malawi	11.36	34	2.5
Mali	8.34	41	2.5
Senegal	7.18	82	19
Uganda	26.94	8.5	2.5
Zambia	7.7	50	3.5

Source: World Bank [358]

projects worked well initially and some were plagued with different challenges which impacted the functionality of the systems and limited the intended returns [145, 146]. Access remains a problem as rural communities which received electricity services still have 20% to 25% of homes not connected [144].

Rural electrification faces many challenges. Some of these consist of but are not limited to inadequate maintenance, increase in demand, lack of commercial small-scale technologies, inadequate policy guidance, competing subsidies, mismatch of needs, insufficient finances, limited resources, and inadequate stakeholder management [147]. Meanwhile, practitioners and researchers have applied various techniques to mitigate these challenges using readily available technologies, but there is still a need for the design mechanism to adequately synchronize the technology, local resources, and the needs of the community [78]. The lack of community engagement matched with the unmanaged community expectations and an absence of extensive historical data for adequate system design negatively impedes rural electrification success [146].

Rural electrification has been conducted mainly through a top-down approach. This can be categorized into three distinct but overlapping areas: policy, economics, and engineering. Often, the community involvement in rural electrification is absent [146].

6.7.1.1 Top-Down Policy Approach

Policy implementations have tended to adopt a top-down approach. More than three decades ago, Derthick [2] argued that policy making is a combination of events among politicians debating with each other and the monotonous routines of limited consultations among the most affected [2]. Rydin [79] recognized that debate in the policy process relates to the power structure and level of interests of the few [79]. In addition, policy requires skills in separating desires from the language that shapes the policy [79]. Pierson [361] argued that policies demand an examination into the policy processes and the outcomes [361]. What has changed in the policy design and implementation? Dols et al. [3] argued that evidence-based-practices is a new process in organizational policy development and must be given more attention since it requires individuals who are committed and familiar with the issues and the meaning of the evidence [3]. The organizational level process allows for education, observation, and new learning [3]. Rural communities require procedures to address the collective action problem and the societal differences in generating consensus with limited professional capacities, in a manner that will provide dialogue and agreement.

6.7.1.2 Top-Down Economic Approach

In many developing nations, rural electrification has been financed by governments borrowing from the World Bank [117] and Inter-American Development Banks [362] and donations from nongovernmental organizations [363]. Rural electrification in America had its genesis in the rural electrification act of 1936, which allowed for the provision of loans from federal funds to cooperatives. These cooperatives were consumer owned and not for profit [364]. Compared to the American cooperatives, incurring external debt for rural electrification does not provide adequate institutional and financial support for all stakeholders in their various roles [364, 365]. Some communities received government subsidies of 50% or long-term loans with low or no interest rates [364, 365]. Zomers [366] noted that due to the relatively high costs of rural electrification, utilities find it unattractive to extend the service to these communities [366]. Why do utilities with all the expertise refuse to extend electricity service? The revenues from these services do not adequately compensate the investments due to expected low demand [367]. There is often a difficulty in extracting the rents for the assets, receiving payments for services, and costs of maintenance of the systems, as in the case of Tunisia [363]. Grid systems in the larger centers have connection fees and set access charges as well as cost per kWh usage charges [368]. The utility will have metering, billing, and collection operations [369]. If a household does not pay the power bill, then the utility can disconnect the power [369]. The utility also has maintenance operations essential for sustainable operation [369]. The difficulty in obtaining rents for services is not because consumers cannot afford to pay but because they do not feel any need to [363]. In summary, at a superficial level the economic issue for rural electrification might appear to be financial inability of users to pay, but at a deeper level there are issues around the motivation to pay. Hence it becomes difficult to recover capital costs and this becomes disincentive to central investment. The deeper issue remains: Why do people not want to pay? The answer to that appears to be rooted in the way the system is designed, either top-down or emergent, and the values that residents hold.

6.7.1.3 Top-Down Electricity Design Approach

Designing electrification networks have followed a top-down approach with utility engineers planning the implementation of electricity systems. While the main criteria for the design is the consumers expected/predicted demand; once the power system has been developed, the load quickly grows leaving the utility to expand the network in an unplanned and sometimes haphazard manner [370]. Besides, the unchecked demand surge, the nature of the fuel available is another limiting factor [370, 371]. Utilities are often left to plan for uncertainties that are related to prudent decision-making/forecasting of deterministic parameters (such as, load growth rates, fuel costs construction time, interest rates, financial constraints, economic growth, and environmental constraints), as well as probabilistic aspects of load and equipment analysis [372]. The typical electrification system (TES) design outlined by Cavallaro [371] supports the generation of economic returns based on an estimated demand and future increases in uses for electricity [371]. Traditionally, the planning process of electric utilities consisted of comparing the electric production capacity with the projected demand and building the additional production capacity needed to meet the expected demand in compliance with safety regulations and environmental standards [373]. Wilson and Biewald's [374] Integrated Resource Plan (IRP) system operates almost like the TES, with one notable difference of examining all possible sources and making a selection based on the most economically viable option [374]. The issue is that the utility is perceived as a project, and the economic expressed in economic utility.

6.7.1.4 Bottom-Up and Emergent Approach

Altogether, the literature points to a need for more active participation of residents in rural electrification design process. The challenge is motivating residents to engage, and this seems to require a re-think of the design process itself. The evidence-based policy practices identified by Dols et al. [3] provides for some form of autonomous application where individuals develop policies to align with their ways. This method was proposed for use in a professional organizational (company) setting, but has not been extended to the societal setting. Whether it be Dols et al.'s [3] organizational policy practices; Derthick [2], Rydin [79], and Richardson's [375] national policy methods; or Smit et al.'s [376] community-based system dynamics (CBSD), they are all limited by absence of a mechanism to achieve this. There is a need for a process methodology that can be applied to emergent settings and mechanisms to quantify outcomes.

Finally, the study of Blair et al. [359] recognizes the need for community engagement in rural electrification and presents a from-the-ground-up gamified community-based action research approach to rural electricity design entitled the Community Access Resource for Electricity Sustainability (CARES). CARES explores the community activity systems and values from the community's perspective to determine what residents need the most. Applying good conversational techniques and research methods in an unbiased yet structured way, CARES establishs essential electricity needs in remote communities. It explores a way for establishing transformation projects that will allow for affordability, sustainability, reliability, and adaptation of modern electricity services. The study applies this approach to a case study implemented in a small Amazonian village, Kabakaburi (Guyana).

6.7.2 Brief Global Overview of Off-grid Electrification

Off-grid electrification is nothing new. Diesel- or gasoline-fueled mini-grid and stand-alone off-grid generation systems are well established in many locations. Many small islands, for example, in Indonesia, the Philippines, and China, rely on diesel generation for electricity. Some 108 isolated diesel grids, from 46 operators, are in operation in the Philippines alone. Stand-alone solar home systems are providing a bottom rung that was previously not easily accessible on the energy access ladder. They are able to provide a very basic level of power that is affordable for many and enhances quality of life significantly in key respects. Before, the lowest rung was usually some form of mini-grid system, typically based on diesel generation. But up-front capital costs, running costs, and the need for project governance frameworks meant that this wasn't an option for many people. Now the mini-grid rung is also becoming potentially more accessible with lower cost renewable technologies, enabling the development of hybrid (diesel and renewables) or renewables-only mini-grids, both of which bring down running costs. Mini-grid development, however, remains difficult and hindered by policy and regulatory obstacles. But, if these can be overcome, then mini-grids have a logical role to play in providing needed electricity.

Renewable sources are being utilized more and more in mini-grid and stand-alone off-grid situations. A review of research by the International Renewable Energy

Agency (IRENA) suggests that small-scale hydropower, for example, is currently the largest generation source for mini-grids [1, 5, 7, 8, 343]. This is common in countries such as Nepal, where hydro availability provides a practical solution for remote rural communities. Mini-hydro is more prevalent in Asia than in Africa but has potential in hilly and mountainous location such as in Cameroon, the Democratic Republic of Congo, Ethiopia, Kenya, Rwanda, Tanzania, Uganda, and Zimbabwe. There are a limited number of operational sites in some of these countries. When resources are available, wind and biofuels are also used in the mini-grids. But the biggest technological change in recent years has been a rapid rise in the use of solar photovoltaic generation in both stand-alone home energy systems and mini-grids.

Falling solar costs have led to solar playing a center-stage role in providing many households with a first rung on the energy ladder with the spread of solar lanterns and small stand-alone solar home systems. New business models are emerging to support this development. For example, following the lead of mobile phone companies, solar developers are offering small basic home solar systems, sufficient for minimal LED lighting and device charging, on tariff terms that allow subscribers to keep the equipment once the contract term (typically twelve months) has expired.

Despite the rise of renewable sources, diesel remains an important fuel for off-grid generation. IRENA report that it is the second largest energy source for mini-grids [1, 5, 8, 9, 343]. Diesel has the advantage, subject to fuel being available, of being able to generate electricity at all times when renewable energy may not be on hand. For this reason, many installations are hybrid systems, combining renewable and diesel generation. A range of factors come into play in decisions about choice of technology—cost, availability and sustainability being foremost among them. Future developments in battery technology and pricing add another key variable in the "renewables versus diesel" balance.

But the lower running costs of renewable systems give them an advantage over diesel. The cost of diesel can vary substantially. In oil-producing countries such as Nigeria and Angola, it is as low as \$0.59–\$0.74 a liter, but in countries such as Zambia and Zimbabwe it is closer to \$1 a liter. These price variations shift the economics of different total system costs considerably, making the difference between diesel being more expensive or cheaper than renewables [1, 5, 7, 8, 343].

As we look to the future, the momentum is with renewable technologies. Already, it is often only fossil fuel subsidies that give diesel a cost advantage in some locations. In many isolated rural locations, the physical supply chain challenge of delivering diesel is a significant disadvantage. Unlike diesel, renewable generation involves no fuel cost to run. And with battery technologies rapidly evolving, renewable systems are beginning to overcome "always on" availability challenges.

Rates of electrification in sub-Saharan Africa have not kept pace with the population growth [14, 429, 434–454]. The region as a whole has now overtaken Asia as having the largest number of people lacking access to electricity. The countries with the largest populations currently without electricity are India, Nigeria, Ethiopia, Côte d'Ivoire, Democratic Republic of Congo, and Bangladesh. In the developing countries of Asia, there are an estimated 526 million people without electricity. But demographic pressures and different rates of electrification are ensuring the problem is becoming more concentrated in Africa and will be even more so in future decades. Population growth has led the International Energy Agency (IEA) [451, 454, 455] to increase its estimate for those without electricity in sub-Saharan Africa from 585 million back in 2009 to the latest estimate of 634 million. Also, it isn't until the mid-2020s that the IEA [451, 454] expects this trend of increasing numbers without energy access in sub-Saharan Africa to start to reverse.

With an estimated 80% of those without access to electricity living in rural areas, IEA asserts that off-grid energy systems provide the most viable means of access to electricity. IEA goes on to anticipate, in its "new policies" energy outlook, that 315 million people in rural areas will gain access to electricity by 2040, with most of this new electricity access coming from the development of mini-grids (140 million people) and other types of off-grid systems (80 million).While a detailed local energy-sector mapping is required to identify the most cost-effective route in particular locations, the importance of off-grid solutions is borne out by a study in Senegal that found that only 20% to at most 50% of the unconnected rural population could be most efficiently reached through grid extension investments. Off-grid electrification is the most realistic option for the remainder population [1, 5, 7, 8, 343, 456].

The important role the off-grid solutions can play is recognized by the power utilities industry in Africa. Majority of them believe that there is a medium to high probability that advances and cost reductions in green renewable off-grid technology will deliver an exponential increase in rural electrification levels by 2025. There are clear constraints to higher levels of rollout of mini-grids and off-grid systems. These include a shortage of proven business models, adequate and appropriate forms of financing, established supply chains, and implementation capacity.

Part of the success of stand-alone home systems in Africa has been due to the strongly customer-centric focus of pay-as-you-go (PAYG) SHS companies. Their business model, in many respects, is not about selling electricity, such as would be the case with a traditional utility or, indeed, a mini-grid operator. Instead, the focus is on the needs of the customer—lighting, access to information and communications, solar TV, etc.—and the pricing is pitched accordingly. The spread of customer-financing business models such as pay-as-you-go (PAYG) for solar home systems in Africa has particularly caught the imagination. Companies such as M-Kopa and Mobisol have been in the forefront, using payment systems such as M-PESA and Airtel MTN. This combination of solar and mobile technology is bringing affordable solar technologies to off-grid villages. Mobisol reports that it has installed over 30,000 solar home systems for households in Tanzania and Rwanda since its creation in 2010. Customers pay off the monthly installments using their mobile phones. After 1–3 years, they fully own their personal electricity source. Nairobi-based M-Kopa reported in January 2016 that it had reached the milestone of connecting over 300,000 homes to solar power [14, 429, 439–454].

A key factor behind the growth of PAYG home electrification systems has been business models that are designed to respond to the needs of rural households and their ability to pay. Pricing propositions take account of existing energy expenditure on items such as kerosene. Further, the development of flexible credit facilities tailored to what customers can afford and their income patterns enables customers to scale up their solar solutions, products, and services. Stand-alone home solar systems have also been able to develop in a relatively unregulated market context. Especially due to this unregulated market space, barriers to entering it are reduced and the business case for a profitable business is easier to establish. Access to customer mobile payment services is an important prerequisite for successful solutions, with countries offering the best developed payment infrastructure, and, ideally, interoperability between different mobile providers, being in the best position to increase the uptake of stand-alone electrification. Where this is the case, companies have not had problems financing growth. Good minimum safety and quality standards, clear access plans, and tax exemptions for clean technology products are enablers for the stand-alone home systems sector [429, 448–454].

If customers have higher energy requirements than can be met by stand-alone home systems, then a renewables-fed mini-grid can be an option. Mini-grids have a longer history than stand-alone home systems as a solution for electrifying rural or remote island communities out of reach of the main grid. With the exception of those located next to local sources of hydropower, such as in Nepal, they have typically been powered by diesel generators. Diesel relies on fuel transport for operation and entails a higher running cost compared to renewable generation. Partly because of the disadvantages of diesel, mini-grid development has been limited, but the advent of cheaper renewable power technology has seen the emergence of hybrid mini-grids (combining diesel and renewable generation) and renewables-only mini-grids. In contrast to the entrepreneurial impetus that is driving the growth of stand-alone home energy systems, mini-grids require more planning and institutional context. Because they serve a community of users, they rely on local governance frameworks or some other existing infrastructure that can provide a framework for their development. In short, they require some framework to ensure the agreement on planning, operating, pricing, and maintenance.

Different models for mini-grid deployment exist. They can be classified into four categories—utility operated, privately operated, community operated, or hybrids that combine a mix of the others. Due to the complex regulatory, process-related, technical, financial, and commercial challenges, there is no one-size-fits-all approach. The best structure depends on local circumstances. Often, much of the momentum for mini-grid development depends on government- or donor-led initiatives, and private sector involvement has been limited. Many mini-grids have been government- or donor-led and rely on some form of subsidy and the continuing commitment of the sponsoring agency.

Where mini-grids have been successful, their role is likely to have been clearly identified in national energy plans, and the projects themselves are rooted in funding and governance models that are sustainable and able to cover repair and maintenance costs and processes. Where these conditions exist, private sector interest in mini-grids is likely to grow. In Kenya, for example, the Kenya Energy Regulatory Commission has licensed Powerhive East Africa Ltd., making it the first private company in Kenya's history to receive a utility concession to generate, distribute, and sell electricity to the Kenyan public.

Historically, most mini-grids have been in the lower tiers, with a resultant high level of capital and overhead costs relative to power output. Mini-grids on this scale are unlikely to be as cost-effective for "entry level" power compared to stand-alone home systems. But stand-alone home systems are not adequate for households, and particularly businesses, that need more power. For higher consuming customers beyond the reach of the main grid, mini-grids provide a more feasible full electrification option. A key challenge for mini-grids is that, notwithstanding the possible presence of a main "anchor customer," they need to serve very low-income customers, and so the system design must center on affordability [14, 429, 434–452, 455, 457, 458].

Besides the use of stand-alone systems and mini-grids for customers who are not connected to the grid, the off-grid solutions are becoming a financially viable option for grid-connected customers, especially where electricity supply is not reliable or not sufficient, or where electricity tariffs are high. A broad transformation in the electricity sector in the coming years, both for grid-connected and "beyond the grid" customers, is expected. This will have a major impact on the future sustainability of incumbent generation, transmission, and distribution utilities, and these companies will need to adapt their business models accordingly.

The Global Market Outlook of the European Photovoltaic Industry Association (EPIA) for 2013–2017 estimates new PV off-grid installations in Africa to be between 70 and100 MW per annum [1, 5, 7, 8, 343, 456]. Countries in East Africa have emerged as leaders in the deployment of stand-alone solar solutions for providing electricity access for lighting and mobile charging as well as for other services such as radio, television, and productive uses (e.g., pumping). Global investments in the off-grid solar sector reached $284 million in 2017, with East Africa accounting for over half (57%) of all investments [343, 456]. An estimated 2,000 solar borehole pumps and 1,000 solar surface pumps (under 2.5 kW) are in operation in Kenya. Certain banks, microfinance institutions, and equipment suppliers offer credit lines for solar-powered irrigation systems [1, 5, 7, 8, 343, 456].

Off-grid systems have been used for decades to supply remote areas (e.g., rural villages, islands, and even small cities) that are not connected to any national electricity grid. For renewables, in particular solar PV, applications in off-grid systems were more common than grid-connected systems until 1995. Furthermore, mini-grids are used to supply power to remote industrial sites, telecommunication facilities, or public applications (e.g., lighthouses, national parks, schools, or mountain shelters) [1, 5, 7, 8, 343, 456]. According to IRENA, the usage market can be divided into six categories [14, 429, 434–455, 457, 458]:

a. Islands
b. Remote systems
c. Commercial/industrial—to ensure energy security (99.99% reliability) or provide cheap energy sources, especially if connected to heat production
d. Community/utility—often demonstration projects in the case of developed countries

e. Institutional/campuses—includes hospitals, government buildings, and other institutions with access to cheap capital and no short payback requirements
f. Military—US-specific market for 500 military facilities within and outside of the United States

Global analyses of off-grid renewable energy systems are scarce, but some data on global overview of mini- and nanogrids, stand-alone systems, and diesel generators do exist. Many of these overview reports are from market research companies like Navigant Consulting.

Islands and remote systems are an important market and opportunity for off-grid renewable energy systems. For islands or rural areas where the mini-grid infrastructure is already in place, there is the economic case to displace—or ensure more efficient use of—diesel generators [15]. The 37 Small Island Developing States (SIDS) had a total installed capacity of 28.4 GW at the end of 2012 [1, 5, 7, 8, 343]. This included 10.2 GW for Singapore, around 5 GW, each for Cuba and the Dominican Republic. This left around 8 GW for all other SIDS. But there are also other island power systems that are not part of SIDS. Worldwide, there are 2,056 islands with 1,000–100,000 inhabitants each, with an average electricity demand of 25.6 GWh. That equals around 12 GW of capacity [1, 5, 7, 8, 343]. Many of these island systems are hybrid mini-grid systems with some level of renewable energy.

Beyond islands, global statistics on rural electrification through off-grid systems (either based on diesel or renewables) are limited, although a number of countries are collecting or reporting national data. Recent estimates suggest a current market potential for upgrading existing diesel-based, off-grid systems with renewables in the range of 40–240 GW [1, 5, 7, 343, 456].

Off-grid renewable energy solutions have emerged as a mainstream solution to expand access to modern energy services in a timely and environmentally sustainable manner. Over the past decade, the deployment of stand-alone and mini-grid systems has witnessed tremendous progress as technology costs have plummeted, innovation in deployment and financing models has picked up, and a more diverse set of stakeholders, including local entrepreneurs, the international private sector, and financing institutions, have become engaged in the sector. Besides providing energy for lighting and cooking, off-grid solutions are being deployed to support the delivery of public services (e.g., education, water and primary health care), the development of livelihoods by powering productive end uses (e.g., agriculture), and other commercial and industrial needs [14, 429, 434–455, 457, 458].

IRENA estimates that some 133 million people accessed lighting and other electricity services using off-grid renewable energy solutions in 2016. This includes about 100 million using solar lights (<11 W), 24 million using solar home systems (>11 W), and at least 9 million connected to a mini-grid. A large part of the growth in off-grid renewables has occurred over the last 5 years, driven by the rapidly decreasing costs of solar lighting solutions and by the establishment of local supply chains making these solutions accessible. The adoption of innovative delivery and tailored financing models, such as pay-as-you-go (PAYG) and microfinancing, has driven the growth of solar home systems that offer a wider range of electricity services. The population served by hydropower-based mini-grids has more than doubled since 2007, reaching 6.4 million in 2016 due mainly to growth in Asia.

While solar lighting solutions and solar home systems serve a dominant share of the population served by off-grid renewable energy solutions, they account for a small fraction (4%) of the total installed capacity. High-capacity solar home systems and mini-grids based on solar, hydropower, and biomass have the potential to offer a wider range of electricity services, including for productive uses. Indeed, energy-efficient appliances enable off-grid systems to deliver a wider spectrum of electricity services at lower installed capacities.

Off-grid renewable energy capacity has witnessed a spectacular threefold increase from under 2 GW in 2008 to over 6.5 GW in 2017. While a proportion of the deployed capacity is to support household electrification, a majority (83%) is dedicated for industrial (e.g., cogeneration), commercial (e.g., powering telecommunication infrastructure), and public end uses (e.g., street lighting, water pumping). Countries in Africa and Asia account for most of the growth over the past years, with more than 53 million people in Africa and 76 million in Asia now using such power sources [14, 429, 434–455, 457, 458].

The African continent has seen rapid developments in off-grid renewable energy over the past 5 years. The population served by off-grid solutions in Africa grew rapidly from just over 2 million people in 2011 to over 53 million in 2016. Solar lights have driven this rapid growth, but about 10% of the population in Africa (5.4 million) obtained a higher level of electricity services from off-grid solar. Solar home systems reached over 4 million people in 2017, thanks in part to innovations in technology design and financing (e.g., PAYG) and to coupling with mobile payment platforms, mainly in East Africa [1, 5, 7, 8, 343, 456].

The cumulative off-grid renewable energy capacity has increased from 231 MW in 2008 to nearly 1.2 GW in 2017. The deployment of solar technologies has been a key driver of growth in off-grid capacity, with over 820 MW installed as solar lights, home systems, and mini-grids and for commercial/public services. The abundance of the resource, the distributed nature of technology, and decreasing costs are leading solar to become a default choice for meeting a wide range of electricity services in areas largely underserved by the national grid [1, 5, 7, 8, 343, 456].

The capacity of hydropower-based mini-grids has grown from 124 MW in 2008 to 162 MW in 2017, although its share in total off-grid capacity fell from 53% to under 15% during the same period. Besides electrification, off-grid capacity for self-consumption in industry has grown (e.g., bagasse-based cogeneration) from 29 MW in 2008 to nearly 200 MW in 2017.

Asia has dominated the deployment of off-grid renewables globally over the past decades. Total capacity reached nearly 4.3 GW in 2017, up from 1.3 GW in 2008. This growth has been largely a result of the increased use of industrial bioenergy, although the share of solar in total capacity has been significant, rising from 11% in 2008 to over 30% in 2017. Off-grid solar is being deployed to provide a wide range of services, including household electrification and industrial and commercial/public use (discussed in detail in the next section). Off-grid hydropower capacity has more than doubled between 2008 and 2017, reaching 127 MW.

In terms of the population served, off-grid renewables provided electricity services to under 10 million people in 2008, increasing almost eightfold to over 76 million in 2016. Solar lights have reached around 50 million people in the region, while solar home systems provided electricity services to over 20 million. Bangladesh is a prominent example of deploying solar home systems at scale, reaching over 18 million people through 4.1 million systems in 2017 [434–436, 455, 457, 458]. Mini-grid solutions can provide higher tiers of electricity services and have reached over 7 million people in 2016, although the majority of these systems are hydropower based. At least 3 million additional people have access to some form of electricity services through off-grid solutions in areas other than in Asia and Africa. Total installed capacity has risen from around 400 MW in 2011 to over 1.1 GW in 2017, with the South America sub-group accounting for the majority share of it [1, 5, 9, 343, 434–439, 455–458].

Electricity access rates in South America are among the highest in the developing world, and off-grid renewable energy solutions are considered key for last-mile electricity delivery and for industrial (e.g., mining) and commercial applications. Off-grid capacity on the continent rose from 256 MW in 2008 to 456 MW in 2017. The availability of hydropower resources has led to the development of related off-grid energy infrastructure, even as the use of bioenergy in industry for captive use has increased. The deployment of solar has grown remarkably since 2012, with capacity increasing sixfold to reach 88.5 MW in 2017.

The Oceania sub-group accounts for a marked share of total off-grid capacity as the comprising island nations transition away from fossil fuels toward a greater use of renewable energy. Capacity has grown from over 125 MW in 2010 to above 150 MW in 2017. During the same time frame, the share of solar in total off-grid renewables capacity has risen from 4.7% to over 21%.

Most of the members of the Caribbean Community (CARICOM) have universal or high rates of electricity access. Exceptions include Belize, Guyana, Haiti, and Suriname, which face enormous challenges related to rural electrification and/or energy poverty [1, 5, 7, 8, 343, 456]. Given the electrification and reliability issues in these four countries, self-generation is a common way for firms and larger consumers to ensure that they have reliable electricity access. In Guyana, several firms meet most or all their energy needs through self-generation. In Haiti, the unreliability of the existing grid system prompts even those consumers who are grid connected to rely entirely or partly on self-generation, primarily with inefficient diesel generators. As the economic case for renewables strengthens, islands in Oceania as well as in the Caribbean are expected to see a stronger transition toward a renewables-based power system. Off-grid renewable energy capacity has been deployed across a wide range of end-use sectors providing electricity services. Of the 6.6 GW of off-grid capacity in 2017, the industry sector dominates, followed by mixed-use and commercial/public services. Around 1.5 GW of off-grid capacity serves unknown sector(s) due to the lack of end-use disaggregated data. Bioenergy accounts for the majority share of industrial off-grid capacity with feedstock depending on local conditions, including agricultural and forestry residues. Solar photovoltaic (PV) accounts for the majority of use in the commercial and public sectors, as well as in residential and agriculture/forestry [14, 444–450].

Within commercial and public uses, most solar PV use is for powering telecommunication infrastructure, followed by schools, street lighting, health centers, and water pumping. The modular and distributed nature of solar PV enables it to be adapted to a wide range of off-grid applications, and several programs and initiatives have been launched to accelerate deployment. In India, where a strong policy focus has led to rapid development, especially in the use of solar for water pumping, solar pumps offer an attractive option to provide affordable and sustainable modern energy for meeting water pumping needs for irrigation and drinking water supply and are increasingly being deployed. Another area of growing interest is the use of solar PV for powering rural healthcare centers.

The new database of off-grid power plants included information about 38,600 small hydropower plants, 500 biogas generators, 8,100 solar mini-grids, 107,300 solar water pumps, and 168,300 other solar power plants. It also included 650 records of annual sales of small solar devices in different countries (amounting to about 51 million units in total), as well as information about the 42 million biogas digesters that have been built over the last 3 decades. IRENA has confirmed the existence of much of the off-grid solar PV currently estimated from import statistics and has identified the end uses of about 70% of off-grid power capacity. IRENA is currently in the process of integrating the new data into its existing off-grid power statistics and analyzing the data for further insights into the linkages between off-grid power supplies and the achievement of other sustainable development goals [14, 429, 434–455, 457, 458].

In *India,* which currently has about 650,000 telecom base stations (averaging 3–5 kWe each and requiring 1–3 GWe of capacity in total), some 40% of the power needed comes from the grid and 60% from diesel generators. The government mandated that 50% of rural sites be powered via renewables by 2015. In 2013, 9,000 towers were operating with renewable electricity. By 2020, 75% of rural and 33% of urban stations will need to run on alternative energy [1, 5, 7]. *Bangladesh* has 138 solar power telecom towers [446, 453].

In *Africa,* there are currently more than 240,000 telecom towers of which an estimated 145,000 are located in off-grid sites. Only 4,000 of these sites deploy renewable energy sources (e.g., solar, wind or hybrid combinations). In 2020, these numbers are expected to grow to 325,000 and 189,000, respectively [1, 5, 7, 8, 343]. *Africa* has had wind-driven water pumping systems for decades. Around 400,000 such systems are in operation across Africa [435]. Some systems use the wind force directly to drive a mechanical pump while others work with electricity as intermediate energy carrier. *India* installed 1,417 wind-driven water pumps in the fiscal year 2013/2014 [441]. *Bangladesh* had 150 solar PV-driven water pumps installed in July 2014 and targets 1,550 pumps by 2017 [14, 452]. However, the use of wind pumps has stalled with the emergence of low-cost solar PV pumping systems.

Lighting is another sector with high potential and already enjoys a rapidly growing deployment of renewables. The size of the overall commercial outdoor lighting market is estimated at $11 billion. Within this sector, there has been a pronounced shift to LED technologies. In 2012, for example, 54% of the 2 million luminaires installed along roadways and tunnels around the world used LED lights for outdoor area and street lighting [1, 5, 7, 343, 456]. Another forecast from Navigant Research predicts that shipments of smart, LED-based street lights will top 17 million by 2020 [14, 445, 446–449]. It should be noted that these figures are not consistent. A single US company had installed more than 60,000 systems in 60 different countries, representing more than 10 MW of solar capacity as of November 2013 [1, 5, 7, 343, 456]. In the health sector, 2 million domestic gas and kerosene refrigerators are operational worldwide, plus 100,000 vaccine coolers. The average consumption is 300 L of kerosene per year. PV systems have been designed to replace these systems [14, 444–449].

Growth in the use of solar PV for health care has increased fivefold since 2010, reaching over 10 MW by 2018. Access to electricity plays a critical role in the functionality of healthcare facilities and in the quality, accessibility, and reliability of health services delivered to rural communities. An estimated 1 billion people globally are served by health facilities without reliable electricity supply, 255 million of whom live in sub-Saharan Africa [1, 5, 7, 8].

In summary, there are three basic technical approaches to bring electricity to remote areas using solar, diesel generator, wind, hydropower, biopower, and thermoelectric sources [343]:

(1) *Energy home systems (EHS)*: These small power systems are designed to power individual households or small buildings and provide an easily accessible, relatively cheap and easy to maintain solution. They are often operated by nanogrids, pico grids (for equipment), or stand-alone systems. The dispersed character of rural settlements is an ideal setting for these solutions, in particular with renewable energy (RE) that is especially competitive in remote areas. Pico PV system (PPS), solar home systems (SHS), small hydro plants (SHP), or wind home systems (WHS) almost always offer a potential solution for providing electricity to isolated places. In these stand-alone systems, power generation is installed close to the load, and there are no transmission and distribution costs. Moreover, to keep prices affordable, components can be minimized and capacities maintained low, mainly serving small DC appliances for lighting and communication.

(2) *Mini-grids*: Mini-grids (sometimes referred to as isolated grid) provide centralized electricity generation at the local level, using village-wide distribution networks not connected to the main national grid. Mini-grids mostly use low AC current (230 V or 155 V), with a centralized production and storage with an installed capacity between 5 kW and 50 kW although larger systems exist. Mini-grids provide capacity for both domestic appliances and local businesses and have the potential to become one of the most powerful technological approaches for accelerated rural electrification. They can be powered by fossil fuel (mostly diesel), but they also lend themselves for utilizing local renewable energy (RE) resources. Diesel remains the most

used technology because it used to be the cheapest option, and it requires rather modest initial investments. However, RE present numerous competitive advantages including lower cost.

(3) *Hybrid power systems*: They use RE as a primary source and a genset (most of the time, diesel but potentially with gasoline and LPG) as a backup. This solution is especially interesting for isolated villages/small towns, away from the national electricity grid. Moreover, they provide enough power to satisfy modern domestic needs (lighting, communication, refrigeration, water supply) as well as public services (health centers, schools, etc.) and the development of a local economy (small industries, related services such as telecommunication towers, and water irrigation systems). Finally, the implementation of mini-grids has proved to have a positive social impact by fostering and improving local governance structures through the involvement of the community in the decision-making linked with the energy system.

A hybrid energy system can use several RE technologies and balance the specific advantages and shortcomings of each resource. Small hydropower continuously produces costs competitive electricity for villages close to water resources, but is very site specific and may be dependent on seasonal effects. As solar resources are abundant, PV can be used almost everywhere, especially in southern countries, but has rather high requirements for storage since there is no generation after nightfall. The generation of WHS follows the site-specific wind profile over the year. Diesel, gasoline, and LPG generators can be added as complementary sources to ensure the continuity of supply and maximize the lifetime of components by reducing the stress on the overall system thus reducing the overall costs [4].

Hybrid mini-grids are in many cases the most economical solution for village electrification. Numerous studies and simulations have shown that they are competitive in comparison with conventional energy supply systems based on fossil fuels. Renewable off-grid enterprises have emerged in many areas to meet the demand for electricity in rural communities. Due to their geographical location and relatively low-aggregate demand, expanding the nationwide grid to rural areas is expensive and challenging. Renewable energy-based mini-grids are less dependent on larger scale infrastructure and can be implemented faster and cheaper [14, 429, 434–455, 457, 458].

REFERENCES

1. *Off-grid renewable energy systems: Status and methodological issues.* Abu Dhabi: IRENA; 2015. IRENA report.
2. Derthick M. Policymaking for social security. *Journal of Politics.* 1980;42:1191–3.
3. Dols JD, Muñoz LR, Martinez SS, Mathers N, Miller PS, Pomerleau TA, Timmons A, White S. Developing policies and protocols in the age of evidence-based practice. *The Journal of Continuing Education in Nursing.* 2017;48:87–92.
4. Mini-grids. *Wikipedia, The Free Encyclopedia.* 2020 [last visited 28 February, 2020].
5. *Innovation Landscape Brief: Renewable Mini Grids.* Abu Dhabi: IRENA; 2019. ISBN 978-92-9260-146-1.

6. RMI (Rocky Mountain Institute). The economics of grid defection: When and where distributed solar generation plus storage competes with traditional utility service. 2014. Available from: http://www.rmi.org/electricity_grid_defection.
7. IRENA (International Renewable Energy Agency). International off-grid renewable energy conference. Key findings and recommendations. Abu Dhabi: IRENA. Available from: www.irena.org/DocumentDownloads/Publications/IOREC_Key%20Findings%20and%20Recommendations.pdf.
8. Renewable energy and jobs. Abu Dhabi; IRENA: 2013. IRENA report.
9. ICSHP/UNIDO (United Nations International Center on Small Hydro Power/United Nations Industrial Development Organization). World small hydropower report: Executive summary. Available from: www. smallhydroworld.org.
10. Szabo S, Bódis K, Huld T, Moner-Girona M. Energy solutions in rural Africa: Mapping electrification costs of distributed solar and diesel generation versus grid extension. *Environmental Research Letters*. 2011;6(3):034002. Available from: http://iopscience.iop.org/1748-9326/6/3/034002/pdf/1748-9326_6_3_034002.pdf.
11. Breyer C. Identifying off-grid diesel-grids on a global scale for economic advantageous upgrading with PV and wind power. IN: 5th ARE Workshop Academia meets Industry, 27th PVSEC, 25 September 2012, Frankfurt. Available from: www.ruralelec.org/fileadmin/DATA/Documents/07_Events/PVSEC/120925__ARE_Breyer2012_IdentifyingOff-gridDiesel-Grids_AREworkshop_27thPVSEC_Frankfurt_final_pdf.pdf.
12. Craine S, Mills E, Guay J. Clean energy services for all: Financing universal electrification. Sierra Club; 2014 [Accessed December 2014]. Available from: http://action.sierraclub.org/site/DocServer/0747_Clean_Energy_Services_Report_03_web.pdf?docID=15922.
13. Howe E, Blechinger P, Cader C, Breyer C. Analyzing drivers and barriers for renewable energy integration to small islands power generation – tapping a huge market potential for mini-grids. In: 2nd International Conference on Micro Perspectives for Decentralized Energy Supply, 28 February–1 March 2013, Berlin, Germany. Available from: www.reiner-lemoine-institut.de/sites/default/files/howe2013_paper_analyzing_drivers_and_barriers_0.pdf.
14. World Bank. Scaling up access to electricity: The case of Bangladesh. 2014. Available from: www.wdronline.worldbank.org/bitstream/handle/10986/18679/887020BRI0Live00Box385194B00PUBLIC0.pdf?sequence=1. A website report URI http://hdl.handle.net/10986/18679
15. ARE (Alliance for Rural Electrification). Best practices of the alliance for rural electrification. Alliance for Rural Electrification. 2013. Available from: www.ruralelec.org/fileadmin/DATA/Documents/06_ Publications/ARE_Best_Practises_2013_FINAL.pdf.
16. Rena R. Renewable energy for rural development—A namibian experience, rural development—Contemporary issues and practices. Dr. Adisa RS, editors. InTech. ISBN: 978-953-51- 0461-2. Available from: http://www.intechopen.com/books/rural-development-contemporary-issues-and-practices/renewable-energy-for-rural-development-a-namibian-experience.
17. Singh D, editor. *Renewable Energy for Village Electrification*. New Delhi: Goldline; 1997.
18. Khennas S, Barnett A. *Best Practices for Sustainable Development of Micro Hydro Power in Developing Countries*. Washington, DC: World Bank; 2000. xi p.
19. Lew D. Micro-hybrids in rural China: Rural electrification with wind/PV hybrids. *Renewable Energy-Focus*. 2001;2(3):30–33.
20. Davis M. *Institutional Frameworks for Electricity Supply to Rural Communities—A Literature Review*. Cape Town: Energy Development Research Center University of Capetown; 1995.

21. Wang S. PV experience in Tibet. In: Proceedings of the Village-Scale Hybrid System Design Integration Workshop, 29–31 August, Beijing, China; 2000.
22. Wu Y, Yu Q. Development and market prospect of wind/diesel hybrid power system in Chinese offshore islands. 2000.
23. Borbely AM, Kreider JF, editor. *Distributed Generation: The Power Paradigm for the New Millennium.* New York: CRC; 2001.
24. Cromwell G. What makes technology transfer? Small-scale hydropower in Nepal's public and private sectors. *World Development.* 1990;20(7):979–89.
25. Li J, editor. *Commercialization of Solar PV Systems in China.* Beijing, China: Environment Science Press; 2001.
26. Taghvaee MH, Radzi MAM, Moosavain SM, Hizam H, Hamiruce Marhaban M. A current and future study on non-isolated DC–DC converters for photo- voltaic applications. *Renewable and Sustainable Energy Reviews.* 2013;17:216–27. http://dx.doi.org /10.1016/j.rser.2012.09.023.
27. Tesfahunegn SG, Vie PJS, Ulleberg O, Undeland TM. A simplified battery charge controller for safety and increased utilization in standalone PV ap- plications. In: 2011 37th IEEE Photovoltaic Specialists Conference, Seattle, WA; 2011. pp. 002441–7. IEEE. http://dx.doi.org/10.1109/PVSC.2011.6186441.
28. Mamur H, Ahiska R. Application of a DC–DC boost converter with maximum power point tracking for low power thermoelectric generators. *Energy Conversion and Management.* 2015;97:265–72. http://dx.doi.org/10.1016/j.enconman.2015.03.068.
29. Van Campen B, Guidi D, Best G. *Solar Photovoltaics for Sustainable Agriculture and Rural Development.* Rome: Food Agricultural Organisation; 2000.
30. Nordman, B, Christensen K. Local power distribution with nanogrids. In: International Green Computing Conference Proceedings, IGCC 2013, Arlington, VA; 2013. pp. 1–8. 10.1109/IGCC.2013.6604464.
31. Nordman B. Nanogrids, power distribution, and building networks. Darnell Green Power Forum. January 2011. Available from: URL: http://nordman.lbl.gov/docs/nano4.pdf.
32. Nordman B. Nanogrids: Evolving our electricity systems from the bottom up, . May 2011. Available from: URL: BNordman@LBL.gov.
33. Milwaukee, WI, Nordman B, Christensen K, Meier A. Think globally, distribute power locally: The promise of nanogrids. *Computer* 2012; 45:89–91. DOI:10.1109/MC.2012.323. Corpus ID: 4021557
34. Silva WWAG, Donoso-Garcia PF, Seleme SI, Oliveira TR, Santos CHG, Bolzon AS. Study of the application of bidirectional dual active bridge converters in dc nanogrid energy storage systems. In: 2013 Brazilian Power Electronics Conference, Gramado, RS, Brazil; 2013. pp. 609–14. IEEE. http://dx.doi.org/10.1109/COBEP.2013.6785178.
35. NexTek Power Systems. Available from: http://www.nextekpower.com/.
36. Okabe Y. QoEn (quality of energy): Routing toward energy and demand service in the future internet. November 2009. ICE-IT, Academic Center for Computing and Media Studies, Kyoto University.
37. Bryan J, Duke R, Round S. Decentralized generator scheduling in a nanogrid using DC bus signaling. In: IEEE Power Engineering Society General Meeting. 2004. Vol. 2, pp. 977–82. IEEE. http://dx.doi.org/10.1109/PES.2004.1372983.
38. Nag SS, Adda R, Ray O, Mishra SK. Current-fed switched inverter based hybrid topology for DC nanogrid application. In: IECON 2013—39th Annual Conference of the IEEE Industrial Electronics Society, Vienna, Austria; 2013. pp. 7146–51. IEEE. http://dx.doi.org/10.1109/IECON.2013.6700320.
39. IEEE 802, LAN/MAN Standards Committee. IEEE Std 802.3at-2009, DTE Power Enhancements. Available from: http://www.ieee802.org/3/at/.

40. Emerge Alliance. Public overview of the emerge alliance standard. Available from: http://www.emergealliance.org/en/standard/.
41. Redwood Systems. 2013. Available from: http://www.redwoodsystems.com/.
42. Panepinto PA. Making power adapters smarter and greener. In: Greenplug, 2009 IEEE International Symposium on Sustainable Systems and Technology. May 2009.
43. Gershenfeld N, Krikorian R, Cohen D. The internet of things. *Scientific American.* 2004;291(4):76–81.
44. Internet Zero. Available from: http://cba.mit.edu/projects/I0/.
45. Elsayed AT, Mohamed AA, Mohammed OA. DC microgrids and distribution systems: An overview. *Electric Power Systems Research.* 2015;119:407–17. http://dx.doi.org/10.1016/j.jpgr.2014.10.017.
46. Adda R, Ray O, Mishra S, Joshi A. DSP based PWM control of switched boost inverter for DC nanogrid applications. In: IECON 2012—38th Annual Conference on IEEE Industrial Electronics Society. 2012. pp. 5285–90. IEEE. http://dx.doi.org/10.1109/IECON. 2012.6389539.
47. Lee FC, Boroyevich D, Mattavelli P. *Renewable Energy and Nanogrids (REN), a website report by* Renewable Energy and Nanogrids (REN) mini-consortium (2015).
48. Arun SU, Nag SS, Mishra SK. A multi-input single-control (MISC) battery charger for DC nanogrids. In: 2013 IEEE ECCE Asia Downunder, Melbourne, Australia; 2013. pp. 304–10. IEEE. http://dx.doi.org/10.1109/ECCE-Asia.2013.6579113.
49. Pelland S, Turcotte D, Colgate G, Swingler A. Nemiah valley photovoltaic- diesel minigrid: System performance and fuel saving based on one year of monitored data. *IEEE Transactions on Sustainable Energy.* 2011;3: 1–9.
50. Garimella N, Nair N-KC. Assessment of battery energy storage systems for small-scale renewable energy integration. TENCON 2009—2009 IEEE Region 10 Conference, Singapore; 2009. pp. 1–6. IEEE. http://dx.doi.org/10.1109/TENCON .2009. 5395831.
51. Burmester D. Nanogrid topology, control and interactions in a microgrid structure [PhD thesis]. Wellington, NZ: Victoria University of Wellington; 2018.
52. Burmester D, Rayudu R, Seah W, Akinyele D. A review of nanogrid topologies and technologies. *Renewable and Sustainable Energy Reviews.* 2017;67:760–75.
53. Dong D, Boroyevich D, Wang R, Cvetkovic I. A two-stage high power density single-phase AC-DC bi-directional PWM converter for renewable energy systems. In: 2010 IEEE Energy Conversion Congress and Exposition, Atlanta, Georgia; 2010. pp. 3862–9. IEEE. http://dx.doi.org/10.1109/ECCE.2010.5617767.
54. Goikoetxea, Ander, Canales JM, Sánchez Roberto, Zumeta Pablo. DC versus AC in residential buildings: Efficiency comparison. *IEEE EuroCon 2013.* 2013:1–5. 10.1109/EUROCON.2013.6625162.
55. Burmester D, Rayudu R, Seah W. A comparison between temperature and current sensing in photovoltaic maximum power point tracking. In: 2014 18th National Power Systems Conference (NPSC), Guwahati, India; 2014. pp. 1–6. IEEE. http://dx.doi.org /10.1109/NPSC.2014.7103779.
56. Kim H-S, Ryu M-H, Baek J-W, Jung J-H. High-efficiency isolated bidirectional AC–DC converter for a DC distribution system. *IEEE Transactions on Power Electronics.* 2013;28:1642–54. http://dx.doi.org/10.1109/TPEL.2012.2213347.
57. Liu J, Wong K, Allen S, Mookken J. Performance evaluations of hard-switching interleaved DC/DC boost converter with new generation silicon carbide MOSFETs. [Online]. Available from: http://www.cree.com/power/products/1200v-sic-mosfet-packaged/packaged/$/media/Files/Cree/Power/Articles%20and%20PapersFile:Power_Article_4.pdf.

58. Chuang YC, Ke YL. A novel high-efficiency battery charger with a buck zero- voltage-switching resonant converter. *IEEE Transactions on Energy Conversion*. 2007;22:848–54. http://dx.doi.org/10.1109/TEC.2006.882416.
59. Shwehdi MH, Mohamed SR. Proposed smart DC nano-grid for green buildings—A reflective view. In: 2014 International Conference on Renewable Energy Research and Applications. 2014. pp. 765–9. IEEE. http://dx.doi.org/10.1109/ICRERA.2014.7016488.
60. Cvetkovic I, Dong D, Zhang W, Jiang L, Boroyevich D, Lee FC, et al. A testbed for experimental validation of a low-voltage DC nanogrid for buildings. In: 2012 15th International Power Electronics and Motion Control Conference, Novi Sad, Serbia; 2012. pp. LS7c.5-1–LS7c.5-8. IEEE. http://dx.doi.org/10.1109/EPEPEMC.2012.6397514.
61. Garbesi K, Vossos V, Shen H. Catalog of DC appliances and power systems. U. S. Department of Energy, Office of Energy Efficiency and Renewable Energy. Berkeley,CA: Ernest Orlando Lawrence Berkeley National Laboratory; 2012. Report No.: LBNL-5364E.
62. Wu H, Wong S-C, Tse CK, Chen Q. Control and modulation of a family of bidirectional AC-DC converters with active power compensation. In: 2015 IEEE Energy Conversion Congress and Exposition. 2015. pp. 661–8. IEEE. http://dx.doi.org/10.1109/ECCE.2015.7309752.
63. Yi-Hung Liao A. Novel reduced switching loss bidirectional AC/DC converter PWM strategy with feed forward control for grid-tied microgrid systems. *IEEE Transactions on Power Electronics*. 2014;29:1500–13. http://dx.doi.org/10.1109/TPEL.2013.2260872.
64. Ganesan SI, Pattabiraman D, Govindarajan RK, Rajan M, Nagamani C. Control scheme for a bidirectional converter in a self-sustaining low-voltage DC nanogrid. *IEEE Transaction on Industrial Electronics*. 2015;62:6317–26. http://dx.doi.org/10.1109/TIE.2015.2424192.
65. Vossos V, Garbesi K, Shen H. Energy savings from direct-DC in U.S. residential buildings. *Energy and Buildings*. 2014;68:223–31. http://dx.doi.org/10.1016/j.enbuild.2013.09.009.
66. Keyhani H, Toliyat HA. Single-Stage Multistring PV Inverter with an isolated high-frequency link and soft-switching operation. *IEEE Transactions on Power Electronics*. 2014;29:3919–29. http://dx.doi.org/10.1109/TPEL.2013.2288361.
67. Horowitz N. Power supplies: A hidden opportunity for energy savings. Natural Resources Defense Council Technical Report 2002.
68. Chandrasena RPS, Shahnia F, Ghosh A, Rajakaruna S. Operation and control of a hybrid AC-DC nanogrid for future community houses. In: 2014 Australasian Universities Power Engineering Conference, Curtin University, Perth, Australia; 2014. pp. 1–6. IEEE. http://dx.doi.org/10.1109/AUPEC.2014.6966617)51.
69. Sajeeb MMH, Rahman A, Arif S. Feasibility analysis of solar DC nano grid for off grid rural Bangladesh. In: 2015 3rd International Conference on Green Energy Technology. 2015. pp. 1–5. IEEE. http://dx.doi.org/10.1109/ICGET.2015.7315109.
70. Cairoli P, Kondratiev I, Dougal RA. Controlled power sequencing for fault protection in DC nanogrids. In: 2011 International Conference on Clean Electrical Power. 2011. pp. 730–37. IEEE. http://dx.doi.org/10.1109/ICCEP.2011.6036384.
71. Pei X, Kang Y. Short-circuit fault protection strategy for high-power three- phase three-wire inverter. *IEEE Transactions on Industrial Informatics*. 2012;8:545–53. http://dx.doi.org/10.1109/TII.2012.2187913.
72. Bryan J, Duke R, Round S. *Decentralised Control of a Nanogrid*. Christchurch, Canterbury: Department of Electrical and Computer Engineering, University of Canterbury; 2002.

73. LBNL (Lawrence Berkely national Laboratory). The microgrid concept. Available from: http://der.lbl.gov/research/microgrid-concept.
74. Lasseter R, Akhil A, Marnay C, Stephens J, Dagle J, Guttromsom R, et al. *Integration of distributed energy resources: The CERTS micro grid concept.* Berkeley, CA: Lawrence Berkeley National Laboratory; April 2002. Report No.: LBNL-50829.
75. He MM, Reutzel EM, Jiang X, Katz RH, Sanders SR, Culler DE. An architecture for local energy generation, distribution, and sharing. In: IEEE Energy 2030 Conference Proceedings, Atlanta, Georgia; November 17–18, 2008.
76. USB Implementers Forum, Inc. Available from: http://usb.org.
77. Powered USB. Available from: http://poweredusb.org/.
78. Foley, G. Rural electrification in the developing world. *Energy Policy.* 1992;20:145–52.
79. Rydin, Y. Can we talk ourselves into sustainability? The role of discourse in the environmental policy process. *Environmental Values.* 1999;8:467–484.
80. Cimen H, Oguz E, Oguz Y, Oguz H. Power flow control of isolated wind-solar power generation system for educational purposes. In: 2012 22nd Australasian Universities Power Engineering Conference (AUPEC). 2012. pp. 1–5.
81. Grandjean A, Adnot J, Binet G. A review and an analysis of the residential electric load curve models. *Renewable and Sustainable Energy Reviews.* 2012;16:6539–65. http://dx.doi.org/10.1016/j.rser.2012.08.013.
82. Latha SH, Chandra Mohan S. Centralized power control strategy for 25 kw nano grid for rustic electrification. In: 2012 International Conference on Emerging Trends in Science, Engineering And Technology, Tiruchirap, India. 2012. pp. 456–61. IEEE. http://dx.doi.org/10.1109/INCOSET.2012.6513949.
83. Schonberger JK. Distributed control of a nanogrid using DC bus signalling [PhD dissertation]. Christchurch, NZ: Department of Electrical and Computer Engineering University of Canterbury; May 2005.
84. Sadabadi MS, Karimi A, Karimi H. Fixed-order decentralized/distributed control of islanded inverter-interfaced microgrids. *Control Engineering Practice.* 2015;45:174–93. http://dx.doi.org/10.1016/j.conengprac.2015.09.003.
85. Weaver WW, Robinett RD, Parker GG, Wilson DG. Energy storage requirements of DC microgrids with high penetration renewables under droop control. *International Journal of Electrical Power and Energy Systems.* 2015;68:203–9. http://dx.doi.org/10.1016/j.ijepes.2014.12.070.
86. Brocco A. Fully distributed power routing for an ad hoc nanogrid. In: 2013 IEEE International Workshop on Inteligent Energy Systems, Vienna, Austria; 2013. pp. 113–8. IEEE. http://dx.doi.org/10.1109/IWIES.2013.6698571.
87. Biabani M, Aliakbar Golkar M, Johar A, Johar M. Propose a home demand- side-management algorithm for smart nano-grid. In: Proceedings of the 4th Annual International Power Electronics, Drive Systems and Technologies Conference, Tehran, Iran; 2013. pp. 487–94. IEEE. http://dx.doi.org/10.1109/PEDSTC.2013.6506757.
88. Riffonneau Y, Bacha S, Barruel F, Ploix S. Optimal power flow management for grid connected PV systems with batteries. *IEEE Transactions on Sustainable Energy.* 2011;2:309–20. http://dx.doi.org/10.1109/TSTE.2011.2114901.
89. Mansour S, Joos G, Harrabi I, Maier M. Co-simulation of real-time decentralized vehicle/grid (RT-DVG) coordination scheme for e-mobility within nanogrids. In: 2013 IEEE Electrical Power Energy Conference, Halifax, Nova Scotia, Canada; 2013. pp. 1–6. IEEE. http://dx.doi.org/10.1109/EPEC.2013.6802916.
90. Logenthiran T, Srinivasan D, Shun TZ. Demand side management in smart grid using heuristic optimization. *IEEE Transactions on Smart Grid.* 2012;3:1244–52. http://dx.doi.org/10.1109/TSG.2012.2195686.

91. Barker S, Mishra A, Irwin D, Shenoy P, Albrecht J. SmartCap: Flattening peak electricity demand in smart homes. In: 2012 IEEE International Conference on Pervasive Computing and Communications, Lugano, Switzerland; 2012. pp. 67–75. IEEE. http://dx.doi.org/10.1109/PerCom.2012.6199851.
92. Wang H, Shu L. On a novel property of the earliest deadline first algorithm. In: 2011 8th International Conference on Fuzzy Systems and Knowledge Discovery. 2011. Vol. 1, pp. 197–201. IEEE. http://dx.doi.org/10.1109/FSKD.2011.6019496.
93. Fathima AH, Palanisamy K. Optimization in microgrids with hybrid energy systems—A review. *Renewable and Sustainable Energy Reviews*. 2015;45:431–46. http://dx.doi.org/10.1016/j.rser.2015.01.059.
94. Guerrero JM, Vasquez JC, Matas J, de Vicuna LG, Castilla M. Hierarchical control of droop-controlled AC and DC microgrids—A general approach toward standardization. *IEEE Transactions on Industrial Electronics*. 2011;58:158–72. http://dx.doi.org/10.1109/TIE.2010.2066534.
95. Bouzid AM, Guerrero JM, Cheriti A, Bouhamida M, Sicard P, Benghanem M. A survey on control of electric power distributed generation systems for microgrid applications. *Renewable and Sustainable Energy Reviews*. 2015;44:751–66. http://dx.doi.org/10.1016/j.rser.2015.01.016.
96. Palizban O, Kauhaniemi K. Hierarchical control structure in microgrids with distributed generation: Island and grid-connected mode. *Renewable and Sustainable Energy Reviews*. 2015;44:797–813. http://dx.doi.org/10.1016/j.rser.2015.01.008.
97. Planas E, Gil-de-Muro A, Andreu J, Kortabarria I, Martínez de Alegría I. General aspects, hierarchical controls and droop methods in microgrids: A review. *Renewable and Sustainable Energy Reviews*. 2013;17:147–59. http://dx.doi.org/10.1016/j.rser.2012.09.032.
98. Schonberger J, Duke R, Round SD. DC-bus signaling: A distributed control strategy for a hybrid renewable nanogrid. *IEEE Transactions on Industrial Electronics*. 2006;53:1453–60. http://dx.doi.org/10.1109/TIE.2006.882012.
99. Zhang W, Lee FC, Huang P-Y. Energy management system control and experiment for future home. In: 2014 IEEE Energy Converion Congress and Exposition, Pittsburgh, PA; 2014. pp. 3317–24. http://dx.doi.org/10.1109/ECCE.2014.6953851.
100. Qu D, Wang M, Sun Z, Chen G. An improved DC-bus signaling control method in a distributed nanogrid interfacing modular converters. In: 2015 IEEE 11th International Conference on Power Electronics, Drives and Energy Systems, Sydney, Australia; 2015. pp. 214–8. IEEE. http://dx.doi.org/10.1109/PEDS.2015.7203408.
101. Schonberger J, Round S, Duke R. Autonomous load shedding in a nanogrid using DC bus signalling. In: IECON 2006—32nd Annual Conference on IEEE Industrial Electronics, Paris, France; 2006. pp. 5155–5160. IEEE. http://dx.doi.org/10.1109/IECON.2006.347865.
102. Suzdalenko A, Vorobyov M, Galkin I. Development of distributed energy management system for intelligent household electricity distribution grid. In: Eurocon, Zagreb, Croatia; 2013. pp. 1474–8. IEEE. http://dx.doi.org/10.1109/EUROCON.2013.6625172.
103. Adda R, Ray O, Mishra SK, Joshi A. Synchronous-reference-frame-based control of switched boost inverter for standalone dc nanogrid applications. *IEEE Transactions on Power Electronics*. 2013;28:1219–33. http://dx.doi.org/10.1109/TPEL.2012.2211039.
104. Burmester D, Rayudu R, Exley T. Single ended primary inductor converter reliance of efficiency on switching frequency for use in MPPT application. In: 2013 IEEE PES Asia-Pacific Power and Energy Engineering Conference, Kowloon, Hong Kong; 2013. pp. 1–6. IEEE. http://dx.doi.org/10.1109/APPEEC.2013.6837187.

105. Kanakasabapathy P. Switched-capacitor/switched-inductor Ćuk-derived hy-brid converter for nanogrid applications. In: Proceedings of 2015 International Conference on Computation of Power, Energy, Information and Communication. 2015. pp. 0430–35. IEEE. http://dx.doi.org/10.1109/ICCPEIC.2015.7259497.
106. Nordman B, Christensen K. Local power distribution with nanogrids. In: 2013 International Green Computing Conference Proceedings. 2013. pp. 1–8. IEEE. http://dx.doi.org/10.1109/IGCC.2013.6604464.
107. Nordman B. Nanogrids: Evolving our electricity systems from the bottom up. IN: Darnell Green Power Forum. Berkeley, CA: Lawrence Berkeley National Laboratory; 2009.
108. Khodayar ME, Wu H. Demand forecasting in the smart grid paradigm: Features and challenges. *The Electricity Journal.* 2015;28:51–62. http://dx.doi.org/10.1016/j.tej.2015.06.001.
109. Boroyevich D, Cvetkovic I, Burgos R. Intergrid: A future electronic energy network? *IEEE Journal of Emerging and Selected Topics in Power Electronics.* 2013;1:127–38. http://dx.doi.org/10.1109/JESTPE.2013.2276937.
110. Werth A, Kitamura N, Matsumoto I, Tanaka K. Evaluation of centralized and distributed microgrid topologies and comparison to Open Energy Systems (OES). In: IEEE 15th International Conference on Environment and Electrical Engineering, Rome, Italy; 2015. pp. 492–97. IEEE. http://dx.doi.org/10.1109/EEE IC.2015.7165211.
111. Nordman B, Christensen K. The need for communications to enable DC power to be successful. In: IEEE 1st International Conference on DC Micro- Grids, Atlanta, GA; 2015. pp. 108–12. IEEE. http://dx.doi.org/10.1109/ICDCM.2015.7152019.
112. Werth A, Kitamura N, Tanaka K. Conceptual study for open energy systems: Distributed energy network using interconnected dc nanogrids. *IEEE Transactions on Smart Grid.* 2015;6:1621-1630. http://dx.doi.org/10.1109/TSG.2015.2408603.
113. Bani-Ahmed A, Weber L, Nasiri A, Hosseini H. Microgrid communications: State of the art and future trends. In: International Conference on Renewable Energy Research and Applications, Milwaukee, WI; 2014. pp. 780–85. IEEE. http://dx.doi.org/10.1109/ICRERA.2014.7016491.
114. Oliveira DQ, Zambroni de Souza AC, Almeida AB, Santos MV, Lopes BIL, Marujo D. Microgrid management in emergency scenarios for smart electrical energy usage. In: IEEE Eindhoven PowerTech, Eindhoven University, Eindhoven, Netherlands; 2015. pp. 1–6. IEEE. http://dx.doi.org/10.1109/PTC.2015.7232309.
115. File: Hybrid Power System.gif. *Wikipedia, The Free Encyclopedia.* 2020 [last visited May 17, 2017].
116. Stand-alone power system. *Wikipedia, The Free Encyclopedia.* 2020 [last visited October 3, 2019].
117. The World Bank. Honduras rural electrification. Washington, DC: World Bank; 2016. Technical Report.
118. Sovacool BK Deploying off-grid technology to eradicate energy poverty. *Science.* 2012;338(5). Available from: www.sciencemag.org/content/338/6103/47.summary.
119. Renewable hybrid power solution modular, containerized, digitally-connected power solution for off-grid electrification. New York: GE Power; April 2017. A website report, Report No.: GEA33009.
120. Ramakumar R. Electric power generation: Conventional methods, Chapter 2. In: Grigsby LL, editor. *The Electric Power Engineering Handbook.* Boca Raton, FL: CRC Press; 2001.
121. Aziz MS, Saleem U, Ali E, Siddiq K. A review on bi-source, off-grid hybrid power generation systems based on alternative energy sources. *Journal of Renewable and Sustainable Energy.* 2015;7:043142. https://doi.org/10.1063/1.4929703.

122. U.S. DOE Annual Energy Outlook 2014. Available from: http://www.eia.gov/.
123. OpenEI Database. Available from: http://en.openei.org/.
124. Energy Efficiency. Available from: http://www.mpoweruk.com/.
125. Levelized Cost of Energy—Version 8.0. Available from: http://www.lazard.com/PDF/Levelized%20Cost%20of%20Energy%20-%20Version%208.0.pdf.
126. Capacity Factors for Utility Scale Generators Primarily Using Fossil Fuels, January 2013–March 2015. Available from: http://www.eia.gov/electricity/monthly/epm_table_grapher.cfm?t1/4epmt_6_07_a.
127. Vosen SR, Keller JO. Hybrid energy storage systems for stand-alone electric power systems: Optimization of system performance and cost through control strategies. *International Journal of Hydrogen Energy*. 1999;24:1139–56. 043142-29.
128. Gomaa, S, Seoud AKA, Kheiralla HN. Design and analysis of photovoltaic and wind energy hybrid systems in Alexandria, Egypt. *Renewable Energy*. 1995;6(5–6):643–7.
129. Nayar CV, Phillips SJ, James WL, Pryor TL, Remmer D. Novel wind/diesel/battery hybrid energy system. *Solar Energy*. 1993;51(1):65–78.
130. Wichert B. PV-diesel hybrid energy systems for remote area power generation—A review of current and future developments. *Renewable and Sustainable Energy Reviews*. 1997;1(3):209–28.
131. Valentino T, McCluer P, Hughes E. Investigation of geothermal energy technologies and gas turbine hybrid system. *Geothermal Resources Council Transactions*. 1996;20:195–201.
132. Ramakumar R, Hughes WL. Renewable energy sources and rural development in developing countries. *IEEE Transactions on Education*. 1981;24(3):242–51.
133. Hozumi S. Development of hybrid solar systems using solar thermal, photovoltaics, and wind energy. *International Journal of Solar Energy*. 1986;4(5):257–80 [translated from J. Jpn. Sol. Energy Soc. 11(4), 15 (1986)].
134. Physik F, Oldenburg U. Wind-solar hybrid electrical supply systems. Results from a simulation model and optimization with respect to energy pay. *Solar & Wind Technology*. 1988;5(3):239–47.
135. Ramakumar R, Allison HJ, Hughes WL. Review paper prospects for tapping solar energy on a large scale. *Solar Energy*. 1974;16:107–15.
136. Ramakumar R, Abouzahr I, Krishnan K, Ashenayi K. Design scenarios for integrated renewable energy systems. *IEEE Transactions on Energy Conversion*. 1995;10(4):736–46.
137. Ramakumar R, Allison HJ, Hughes WL. Solar energy conversion and storage systems for the future. *IEEE Transactions on Power Apparatus and Systems*. 1975;94(6):1926–34.
138. Rahman S, Tam K. A feasibility study of photovoltaic-fuel cell hybrid energy system. *IEEE Transactions on Energy Conversion*. 1988;3(1):50–55.
139. Lahimer AA, Alghoul MA, Yousif F, Razykov TM, Amin N, Sopian K. Research and development aspects on decentralized electrification options for rural household. *Renewable and Sustainable Energy Reviews*. 2013;24:314–24.
140. Deshmukh MK, Deshmukh SS. Modeling of hybrid renewable energy systems. *Renewable and Sustainable Energy Reviews*. 2008;12(1):235–49.
141. Notton G, Muselli M, Louche A. Atonomous hybrid photovoltaic power plant using a back-up generator: A case study in a mediterranean island. *Renewable Energy*. 1996;7(4):371–391.
142. Moser CO. *Confronting Crisis: A Comparative Study of Household Responses to Poverty and Vulnerability in Four Poor Urban Communities; ESSD Environmentally & Socially Sustainable Development Work in Progress*. Environmentally Sustainable Development Studies and Monographs Series; No. 7. Washington, DC, USA: World Bank Group; 1996.

143. Ngubane IA, Nephawe K. Investigating factors affecting rural electrification project management within a South African municipality. *International Journal of Research Management and Business Studies*. 2017;4:35–40.
144. The World Bank. *The Welfare Impact of Rural Electrification: A Reassessment of the Costs and Benefits*. Washington, DC: The International Bank for Reconstruction and Development/The World Bank; 2008.
145. Sreekumar N, Dixit S. Challenges in rural electrification. *Economic And Political Weekly*. 2011;46:27–33.
146. Murni S, Whale J, Urmee T, Davis J, Harries D. The role of micro hydro power systems in remote rural electrification: A case study in the Bawan Valley, Borneo. *Procedia Engineering*. 2012;49:189–96.
147. Nieuwenhout F, Van Dijk A, Lasschuit P, Van Roekel G, Van Dijk V, Hirsch D, Arriaza H, Hankins M, Sharma B, Wade H. Experience with solar home systems in developing countries: A review. *Progress in Photovoltaics: Research and Applications*. 2001;9:455–74.
148. Ibrahim M, Khair A, Ansari S. A review of hybrid renewable energy systems for electric power generation. *International Journal of Engineering Research and Applications*. 2015;5(8): 42–8. www.ijera.com.
149. Reddy YJ, Kumar YP, Raju KP. Real time and high fidelity simulation of hybrid power system dynamics. In: Proceedings of the IEEE International Conference on Recent Advances in Intelligent Computational Systems, Trivandrum, India; Sep. 2011.
150. Zhang L, Barakat G, Yassine A. Design and optimal sizingof hybrid PV/wind/diesel system with battery storage by using DIRECT search algorithm. In: 15th International Power Electronics and Motion Control Conference, EPE-PEMC 2012 ECCE Europe, Novi Sad, Serbia; 2012.
151. Tudu B, Majumder S, Mandal KK, Chakraborty N. Optimal Unit Sizing of Stand-alone Renewable Hybrid Energy System Using Bees Algorithm, International Conference on Energy, Automation and Signal, Bhubaneswar, Odisha; 2011. pp.1–6. doi:10.1109/ICEAS.2011.6147175.
152. Kabouris J, Contaxis GC. Autonomous system expansion planning considering renewable energy sources—A computer package. *IEEE Transactions on Energy Conversion*. 1992;7(3):374–81.
153. Kabouris J, Contaxis GC. Optimum expansion planning of an unconventional generation system operating in parallel with alarge scale network. *IEEE Transactions on Energy Conversion*. 1991;6(3):394–400.
154. Kusakana K, Vermaak HJ, Numbi BP. Optimal sizing of a hybrid renewable energy plant using linear programming. In: Power and Energy Society Conference and Exposition in Africa: Intelligent Grid Integration of Renewable Energy Resources (PowerAfrica), July, 09–13, 2012, Johannesburg, South Africa; 2012.
155. Connoly D, Lund H, Mathiesen BV, Leahy M. A review of computer tools for analysing the integration of renewable energy into various energy systems. *Applied Energy*. 2011;87:1059–82.
156. NREL (National Renewable Energy Laboratory): HOMER. The Micropower Optimization Model [Online]. Available from: http://www.homerenergy.com.
157. Afzal A. Performance analysis of renewable energy system [PhD thesis].
158. Ruberti T. Off-grid hybrids: Fuel cell solar-PV hybrids. *Refocus*. 2003;4(5):54–57.
159. Choi BY, Noh YS, Ji YH, Lee BK, Won CY. Battery-integrated power optimizer for PV-battery. In: 2012 IEEE Vehicle Power and Propulsion Conference, October 9–12, 2012, Seoul, Korea; 2012.
160. Etxeberria A, Vechiu I, Camblong H, Vinassa JM. Hybrid energy storage systems for renewable energy sources integration in microgrids: A review.

161. Lee YJ, Han DH, Byen BJ, Seo HU, Choe GH, Kwon WS, Kim DJ. A new hybrid distribution system interconnected with PV array.
162. Goyal M, Gupta R. Operation and control of a distributed microgrid with hybrid system. 978-1-4673-0934-9/12.
163. Bashir M, Sadeh J. Size optimization of new hybrid stand-alone renewable energy system considering a reliability index.
164. Reliability Test System Task Force of the Application of Probability Methods Subcommittee. IEEE reliability test system. *IEEE Transactions on Power Apparatus and Systems*. 1979;PAS-98(6):273–82.
165. Reddy YJ, Kumar YP, Raju KP, Ramsesh A. Retrofitted hybrid power system design with renewable energy sources for buildings. *IEEE Transactions on Smart Grid*. 2012;3(4):2174–87.
166. Solar photovoltaics by Chetan Singh Solanki Second edition Prentice Hall of India (2011), ISBN-10:8120343867
167. Sao CK, Lehn PW. Control and power management of converter fed microgrids. *IEEE Transactions on Power Systems*. 2008;23:1088–98.
168. Maharjan L, Inoue S, Akagi H. A transformerless energy storage system based on a cascade multilevel pwm converter with star configuration. *IEEE Transactions on Industry Applications*. 2008;44(5):1621–30.
169. Nehrir MH, Wang C, Strunz K, Aki H, Ramakumar R, Bing Z, Miao Z, Salameh Z. A review of hybrid renewable/alternative energy systems for electric power generation: Configurations, control, and applications. *IEEE Transactions on Sustainable Energy*. 2011;2(4):392–403.
170. Sánchez V, Ramirez JM, Arriaga G. Optimal sizing of a hybrid renewable system. In: IEEE International Conference on Industrial Technology, Vine del Mar, Chile; 2010. pp. 949–954. doi:10.1109/CIT.2010.5472544.
171. Testa A, De Caro S, Scimone T. Optimal structure selection for small-size hybrid renewable energy plants. In: Proceedings of the 14th European Conference on Power Electronics and Applications, Birmingham, UK; 2011. pp.1–10.
172. bin Othman MM, Musirin I. Optimal sizing and operational strategy of hybrid renewable energy system using HOMER. In: The 4th International Power Engineering and Optimization Conference, Shah Alam, Malaysia.
173. Wind-battery systems incorporating resource uncertainty. *Applied Energy*. 2010;87(8):2712–27.
174. Castañeda M, Fernández LM, Sánchez H. Sizing methods for stand-alone hybrid systems based on renewable energies and hydrogen. In: 16th IEEE Mediterraanean Electrochemical Conference, Yasmine Hammamet , Tunisia; 2012. pp. 832–836. doi: 10.1109/MELCON.2012.6196558.
175. Cano A, Jurado F, Sánchez H. Sizing and energy management of a stand-alone PV/hydrogen/battery-based hybrid system. In: International Symposium on Power Electronics, Electrical Drives, Automation and Motion, Sorrento, Italy; 2012.
176. Huang R, Low SH, Topcu U, Chandy KM, Clarke CR. Optimal design of hybrid energy system with PV/ wind turbine/ storage: A case study. IN: IEEE International Conference on Smart Grid Communications (Smartgridcomm), Brussels, Belgium; 2011. pp. 511–516. doi: 10.1109/SmertGridComm.2011.6102376.
177. Kaundinya D, Balachandra P, Ravindranath N. Grid-connected versus stand-alone energy systems for decentralized power: A review of literature. *Renewable and Sustainable Energy Reviews*. 2009;13(8):2041–2050.
178. Ter-Gazarian AG, Kagan N. Design model for electrical distribution systems considering renewable, conventional and energy storage units. In: *IEEE Proceedings C*.

1992;139(6):499–504. Available from: http://digital-library.theiet.org/content/journals/10.1049/ip-c.1992.0069.
179. Razali NM, Hashim AH. Backward reduction application for minimizing wind power scenarios in stochastic programming. In: 4th International Power Engineering and Optimization Conference (PEOCO), 2010. 2010. pp. 430–4. IEEE.
180. Agbossou K, Kolhe M, Hamelin J, Bose TK. Performance of a stand-alone renewable energy system based on energy storage as hydrogen. *IEEE Transactions on Energy Conversion*. 2004;19(3):633–40.
181. Valente LCG, de Almeida SCA. Economic analysis of a diesel/photovoltaic hybrid system for decentralized power generation in northern Brazil. *Energy*. 1998;23(4):317–323.
182. Elhadidy MA, Shaahid SM. Parametric study of hybrid (wind+solar+diesel) power generating systems. *Renewable Energy*. 2000;21(2):129–139.
183. Chedid R, Akiki H, Rahman S. A decision support technique for the design of hybrid solar-wind power system. *IEEE Transactions on Energy Conversion*. 1998;13(1):76–83.
184. Deshmukh MK, Deshmukh SS. Modeling of hybrid renewable energy systems. *Renewable and Sustainable Energy Reviews*. 2008;12(1):235–49.
185. Celik AN. Optimisation and techno-economic analysis of autonomous photovoltaic–wind hybrid energy systems in comparison to single photovoltaic and wind systems. *Energy Conversion and Management*. 2002;43(18): 2453–68.
186. Ambia MN, Islam MK, Shoeb MA, Maruf MNI, Mohsin ASM. An analysis & design on micro generation of a domestic solar-wind hybrid energy system for rural & remote areas - perspective bangladesh solar power wind power. In: IEEE 2nd International Conference on Mechanical and Electronics Engineering (ICMEE 2010), Kyoto, Japan; 2010. Vol. 2, pp. 107–10.
187. Kasera J, Chaplot A, Maherchandani JK. Modeling and simulation of wind-pv hybrid power system using MATLAB/ Simulink. In: IEEE Students' Conference on Electrical, Electronics and Computer Science (SCEECS), Bhopal, India; 2012. pp. 1–4.
188. Valencia I, Pe~nalvo E, Carbini D, Navarro AP. Optimization of renewable sources in a hybrid systems operation. In: 13th Spanish Portuguese Conference on Electrical Engineering (13CHLIE), Valence, Chile. pp. 1–6.
189. Samarakou MT, Hennet JC. Simulation of a combined wind and solar power plant. *International Journal of Energy Research*. 1986;10:1–10.
190. Roy C, Majumder A, Chakraborty N. Optimization of a stand-alone solar PV-wind-DG hybrid system for distributed power generation at Sagar Island. *AIP Conference Proceedings*. 2010;1298(1):260–5.
191. Nandi SK, Ghosh HR. A wind–PV-battery hybrid power system at Sitakunda in Bangladesh. *Energy Policy*. 2009;37(9):3659–64.
192. Shiroudi AR, Rashidi R, Gharehpetian GB, Mousavifar SA, Akbari Foroud A. Case study: Simulation and optimization of photovoltaic-wind-battery hybrid energy system in Taleghan-Iran using homer software. *Journal of Renewable Sustainable Energy*.
193. Beyene G, Palm B. Feasibility study of solar-wind based standalone hybrid system for application in Ethiopia. In: World Renewable Energy Congress 2011, Linkoping, Sweden; 2011.
194. Bekele G, Boneya G. Design of a photovoltaic-wind hybrid power generation system for Ethiopian remote area. *Energy Procedia*. 2012:14:1760–65.
195. Sureshkumar U, Manoharan PS, Ramalakshmi APS. Economic cost analysis of hybrid renewable energy system using HOMER. In: 2012 International Conference on Advances in Engineering, Science and Management (ICAESM). 2012. pp. 94–9. IEEE.
196. Sood PK, Lipo TA, Hansen IG. A versatile power converter for high-frequency link systems. *IEEE Transactions on Power Electronics*. 1988;3(4):249–56.

197. Sawle Y, Gupta SC, Bohre AK. Review of hybrid renewable energy systems with comparative analysis of off-grid hybrid system. *Renewable and Sustainable Energy Reviews*. 2018;81(2):2217–35.
198. Sawle Y, Gupta SC, Bohre AK, Meng W (Reviewing Editor). PV-wind hybrid system: A review with case study. *Cogent Engineering*. 2016;3:1. DOI: 10.1080/23311916.2016.1189305.
199. Sawle Y, Gupta SC, Bohre AK. Optimal sizing of standalone PV/Wind/Biomass hybrid energy system using GA and PSO optimization technique. *Energy Procedia*. 2017;117:690–8.
200. Querikiol EM, Taboada EB. Performance evaluation of a micro off-grid solar energy generator for islandic agricultural farm operations using HOMER. *Journal of Renewable Energy*., 2018;Vol. 2018:article id 2828173 (9 pages). https://doi.org/10.1155/2018/2828173.
201. Babatunde OM, Adedoja OS, Babatunde DE, Denwigwe IH. Off-grid hybrid renewable energy system for rural healthcare centers: A case study in Nigeria. *Energy Science and Engineering*. 2019;7:676–93. DOI: 10.1002/ese3.314.
202. Esan AB, Agbetuyi AF, Oghorada O, Ogbeide K, Awelewa AA, Afolabi AE. Reliability assessments of an islanded hybrid PV-diesel-battery system for a typical rural community in Nigeria. *Heliyon*. 2019;5:e01632. DOI: 10.1016/j.heliyon.2019.e01632.
203. Kruangpradit P, Tayati W. Hybrid renewable energy system development in Thailand. *Renewable Energy*. 1996;8:514–7.
204. Muhida R, Mostavan A, Sujatmiko W, Park M, Matsuura K. The 10 years operation of a PV-micro-hydro hybrid system in Taratak, Indonesia. *Solar Energy Materials and Solar Cells*. 2001;67(1–4):621–27.
205. Hossan M, Shakawat M, Hossain MM. Optimization and modeling of a hybrid energy system for off-grid electrification. In: 2011 IEEE 10th International Conference on Environment and Electrical Engineering (EEEIC). 2011. pp. 1–4. IEEE.
206. Qais A, Othman MM, Khamis N, Musirin I. Optimal sizing and operational strategy of PV and micro-hydro. In: IEEE 7th International Power Engineering and Optimization Conference (PEOCO), Langkawi, Indonesia; 2013. pp. 714–7. IEEE.
207. Kenfack J, Neirac FP, Tatietse TT, Mayer D, Fogue M, Lejeune A. Microhydro-PV-hybrid system: Sizing a small hydro-PV-hybrid system for rural electrification in developing countries. *Renewable Energy*. 2009;34(10):2259–63.
208. Kusakana K, Munda JL, Jimoh AA. Feasibility study of a hybrid PV-micro hydro system for rural electrification. In: IEEEAFRICON, 2009, AFRICON'09, Nairobi, Kenya; 2009. pp. 1–5. IEEE.
209. Nfah EM, Ngundam JM. Feasibility of pico-hydro and photovoltaic hybrid power systems for remote villages in Cameroon. *Renewable Energy*. 2009;34(6):1445–50.
210. Ogueke NV, Ikpamezie II, Anyanwu E. The potential of a small hydro/photovoltaic hybrid system for electricity generation in FUTO, Nigeria. *International Journal of Ambient Energy*. 2014;37:1–10. DOI: 10.1080/01430750.2014.952841.
211. Syahputra R, Soesanti I. Planning of hybrid micro-hydro and solar photovoltaic systems for rural areas of central Java, Indonesia. *Journal of Electrical and Computer Engineering*. 2020; 1:1–16. DOI: 10.1155/2020/5972342.
212. Meshram S, Agnihotri G, Gupta S. Modeling of grid connected DC linked PV/hydro hybrid system. *Electrical and Electronics Engineering*. 2013;2(3):13–27.
213. Meshram S, Agnihotri G, Gupta S. Performance analysis of grid integrated hydro and solar based hybrid systems. *Advances in Power Electronics*. 2013;Vol. 2013:article id 697049, pp.1–7. https://doi.org/10.1155/2013/697049.
214. Ashok S. Optimised model for community-based hybrid energy system. *Renewable Energy*. 2007;32(7):1155–64.

215. Beluco A, Souza PK, Krenzinger A. PV hydro hybrid systems. *IEEE Latin America Transactions*. 2008;6(7):626–31.
216. Ehnberg SGJ, Bollen MHJ. Reliability of a small power system using solar power and hydro. *Electric Power Systems Research*. 2005;74(1):119–27.
217. Kauranen P, Lund P, Vanhanen J. Development of a self-sufficient solar-hydrogen energy system. *International Journal of Hydrogen Energy*. 1994;19(1):99–106.
218. Dienhart H, Siegel A. Hydrogen storage in isolated electrical energy systems with photovoltaic and wind energy. *Internaional Journal of Hydrogen Energy*. 1994;19(1):61–66.
219. Ro K, Rahman S. Two loop controller for maximizing performance of a grid-connected photovoltaic-fuel cell hybrid power plant. *IEEE Transactions on Energy Conversion*. 1998;13(3):276–81.
220. Chen HC, Chen PH, Chang LY, Bai WX. Stand-alone hybrid generation system based on renewable energy. *International Journal of Environmental Science and Development*. 2013;4(5):514–20.
221. Moziraji RK, Joneidi IA, Bagheri M. A dynamic model for photovoltaic (PV)/fuel cell (FC) hybrid energy system. *Journal of Basic and Applied Scientific Research*. 2013;3(2):811–9.
222. Turner J, Sverdrup G, Mann MK, Maness P-C, Kroposki B, Ghirardi M, Evans RJ, Blake D. Renewable hydrogen production. *International Journal of Energy Research*. 2008;32:379–407.
223. Yang WJ, Kuo CH, Aydin O. A hybrid power generation system: Solar-driven Rankine engine–hydrogen storage. *International Journal of Energy Research*. 2001;25: 1107–25.
224. Lagorse J, Simoes M, Miraoui A, Costerg P. Energy cost analysis of a solar-hydrogen hybrid energy system for stand-alone applications. *International Journal of Hydrogen Energy*. 2008;33(12):2871–79.
225. Krishna KM. Optimization analysis of microgrid using HOMER—A case study. In: 2011 Annual IEEE India Conference (INDICON), Hyderabad, India; 2011. pp. 1–5. IEEE.
226. Bruni G, Cordiner S, Galeotti M, Mulone V, Nobile M, Rocco V. Control strategy influence on the efficiency of a hybrid photovoltaic-battery-fuel cell system distributed generation system for domestic applications. *Energy Procedia*. 2014;45: 237–46.
227. Dufo-Lopez R, Bernal-Agustin JL, Contreras J. Optimization of control strategies for stand-alone renewable energy systems with hydrogen storage. Renewable Energy. 2007;32(7):1102–26.
228. Cetin E, Yilanci A, Ozturk HK, Colak M, Kasikci I, Iplikci S. A micro-DC power distribution system for a residential application energized by photovoltaic–wind/fuel cell hybrid energy systems. *Energy and Buildings*. 210;42(8),:1344–52.
229. El-Shatter TF, Eskandar MN, El-Hagry MT. Hybrid PV/Fuel cell system design and simulation. *Journal. of Renewable Energy*, 2002, 27, 479–485. DOI: 10.1016/S0960-1481(01)00062-3
230. Ghenai C, Bettayeb M. Grid-tied solar PV/fuel cell hybrid power system for university building. *Energy Procedia*. 2019;159:96–103. DOI: 10.1016/j.egypro.2018.12.025.
231. Scott JH. The development of fuel cell technology for electric power generation: From NASA's manned space program to the "hydrogen economy." *Proceedings of the IEEE*. 2006;94(10):1815–25.
232. Rajashekara K. Hybrid fuel-cell strategies for clean power generation. *IEEE Transactions on Industry Applications*. 2005;41(3):682–9.
233. El-Shatter TF, Eskandar MN, El-Hagry MT. Hybrid PV/fuel cell system design and simulation. *Renewable Energy*. 2002;27(3):479–85.

234. El-Shatter TF, Eskander MN, El-Hagry MT. Energy flow and management of a hybrid wind/PV/fuel cell gen- eration system. *Energy Conversion and Management.* 2006;47(9–10):1264–80.
235. Bayrak G, Cebeci M. Grid connected fuel cell and PV hybrid power generating system design with Matlab Simulink. *International Journal of Hydrogen Energy.* 2014;39(16):8803–12.
236. Tanrioven M. "Reliability and cost–benefits of adding alternate power sources to an independent micro-grid community." *Journal of Power Sources.* 2005;150(378):136–49.
237. Nelson DB, Nehrir MH, Wang C. Unit sizing and cost analysis of stand-alone hybrid wind/PV/fuel cell power generation systems. *Renewable Energy.* 2006;31(10):1641–56.
238. Ghosh P, Emonts B, Janßen H, Mergel J, Stolten D. Ten years of operational experience with a hydrogen-based renewable energy supply system. *Solar Energy.* 2003:75(6):469–78.
239. Obara S, Kawae O, Kawai M, Morizane Y. A study of the installed capacity and electricity quality of a fuel cell- independent microgrid that uses locally produced energy for local consumption. *International Journal of Energy Research.* 2013;37:1764–83.
240. Frebourg P, Ketjoy N, Nathakaranakul S, Pongtornkulpanich A, Rakwichian W, Laodee P. Feasibility study of a small-scale grid-connected solar parabolic biomass hybrid power plant in Thailand. Available from: e-nett.sut.ac.th/download/RE/RE17 .pdf.
241. Fiedler F, Nordlander S, Persson T, Bales C. Thermal performance of combined solar and pellet heating systems. *Renewable Energy.* 2006;31(1):73–88.
242. Kibaara S, Chowdhury S, Chowdhury SP.. A thermal analysis of parabolic trough CSP and biomass hybrid power system. In: 2012 IEEE PES Transmission and Distribution Conference and Exposition (T&D), Montevideo, Uruguay; 2012. pp. 1–6.
243. Chou TL, Shih Z, Hong H, Han C, Chiang K. Investigation of the thermal performance of high-concentration photovoltaic solar cell package. IEEE International Conference on Electron Materials Packaging. 2007:1–6.
244. Mahalakshmi M, Latha S. An economic and environmental analysis of biomass-solar hybrid system for the textile industry in India. *Turkish Journal of Electrical Engineering and Computer Sciences.*2015;23:1735–1747. Pdf available from: http://journals.tubitak .gov.tr/elektrik/inpress.htm.
245. Jradi M, Riffat S. Modelling and testing of a hybrid solar-biomass ORC-based micro-CHP system. *International Journal of Energy Research.* 2014;38:1039–52.
246. Pradhan SR, Bhuyan PP, Sahoo SK, Prasad GRKDS. Design of standalone hybrid biomass & PV system of an off-grid house in a remote area. *International Journal of Engineering Research and Applications.* 2013;3:433–7. Pdf available from: http://www .ijera.com/papers/Vol3_issue6/BU36433437.pdf.
247. Janardhan K, Srivastava T, Satpathy G. Hybrid solar PV and biomass system for rural electrification. *International Journal of ChemTech Research.* 2013;5:802–10.
248. Nixon JD, Dey PK, Davies PA. 2012. The feasibility of hybrid solar-biomass power plants in India. *Energy.* 2012;46:541–54. DOI: 10.1016/j.energy.2012.07.058.
249. Balamurugan P, Kumaravel S, Ashok S. Optimal operation of biomass gasifier based hybrid energy system. *ISRN Renewable Energy.* 2011;Vol. 2011:article id 395695. https://doi.org/10.5402/2011/395695.
250. Sou S, Siemers W, Exell RH. The design of a photovoltaic/biomass hybrid electrical energy system for a rural village in Cambodia. *International Journal of Ambient Energy.* 2010;31:3–12. DOI: 10.1080/01430750.2010.9675803.
251. Zhang C, Sun J, Ma J, Xu F, Qiu L. Environmental assessment of a hybrid solar-biomass energy supplying system: A case study. *International Journal of Environmental Research Public Health.* 2019;16(12):2222. DOI: 10.3390/ijerph16122222.

252. Ho WS, Hashim H, Lim JS. Integrated biomass and solar town concept for a smart eco-village in Iskandar Malaysia (IM). *Renewable Energy.* 2014;69:190–201.
253. Kaushika ND, Mishra A, Chakravarty MN. Thermal analysis of solar biomass hybrid co-generation plants. *International Journal of Sustainable Energy.* 2005;24(4):175–86.
254. Bansal M, Khatod DK, Saini RP. Modeling and optimization of integrated renewable energy system for a rural site. In: 2014 International Conference on Reliability Optimization and Information Technology, Faridabad, India; February 2014. pp. 25–8.
255. Neha Energie, Solar & Biomass Hybrid. Available from: http://www.nepad.org/system/files/Neha%20Energie.pdf.
256. Liu G, Rasul MG, Amanullah MT, Khan MM. Feasibility study of stand-alone PV-wind-biomass hybrid energy system in Australia. In: Asia-Pacific Power and Energy Engineering Conference, APPEEC, Wuhan, China; 2011. pp. 1–6. DOI: 10.1109/APPEEC.2011.574912575.
257. Srinivas T, Reddy BV. Hybrid solar–biomass power plant without energy storage. *Case Studies in Thermal Engineering.* 2014;2:75–81.
258. Prasartkaew B, Kumar S. Experimental study on the performance of a solar-biomass hybrid air-conditioning system. *Renewable Energy.* 2013;57:86–93.
259. cTigue JD, Zhu G, Turchi CS, Mungas G, Kramer N, King J et al. Hybridizing a geothermal plant with solar and thermal energy storage to enhance power generation. Golden, CO: NREL; June 2018. Technical Report No.: NREL/TP-5500-70862, Contract No. DE-AC36-08GO28308.
260. Garapati N, Adams BM, Bielicki JM, Schaedle P, Randolph JB, Kuehn TH. A hybrid geothermal energy conversion technology—A potential solution for production of electricity from shallow geothermal resources. *Energy Procedia.* 2017;114:7107–17.
261. Goel PK, Singh B, Murthy SS, Kishore N. Isolated wind–hydro hybrid system using cage generators and battery storage. *IEEE Transactions on Industrial Electronics.* 2011;58(4):1141–53.
262. Tripathy SC, Kalantar M, Balasubramanian M. Dynamics and stability of a hybrid wind-diesel power system. *Energy Conversion and Management.* 1992;33(12):1063–72.
263. Dhanalakshmi R, Palaniswami S. ANFIS based neuro-fuzzy controller in LFC of wind- micro hydro diesel hybrid power system. *International Journal of Computer Applications.* 2012;42(6):28–35.
264. Vieira F, Ramos MH. Optimization of operational planning for wind/hydro hybrid water supply systems. *Renewable Energy.* 2009;34(3):928–36.
265. Khan M, Jidin R, Pasupuleti J, Shaaya SA. Optimal combinations of PV, wind, micro-hydro, and diesel systems for a seasonal load demand. In: IEEE International Conference on Power and Energy (PECon), Kuching Sarawak, Malaysia; 2014. pp. 171–6. IEEE.
266. Sadiqi M, Pahwa A, Miller RD. Basic design and cost optimization of a hybrid power system for rural commun-ities in Afghanistan. In: North American Power Symposium (NAPS), Champaign, IL; 2012. 2012. pp. 1–6. IEEE.
267. Bekele G, Tadesse G. Feasibility study of small Hydro/PV/Wind hybrid system for off-grid rural electrification in Ethiopia. *Applied Energy.* 2012;97:5–15. https://doi.org/10.1016/j.apenergy.2011.11.059.
268. Atiqur Rahman MM, Al Awami AT, Rahim AHMA. Hydro-PV-wind-battery-diesel based stand-alone hybrid power system. In: International Conference on Electrical Engineering and Information & Communication Technology (ICEEICT), Dhaka, Bangladesh; 2014. pp. 1–6. IEEE.
269. Lal K, Dash BB, Akella AK. Optimization of PV/wind/micro-hydro/diesel hybrid power system in HOMER for the study area. *International Journal of Electrical Engineering and Informatics.* 2011;3(3):307–25.

270. Bhatti TS, Al-Ademi AAF, Bansal NK. Load-frequency control of isolated wind-diesel microhydro hybrid power systems (WDMHPS). *Energy.* 1997;22(5):461–70.
271. Kumar K, Ansari MA. Design and development of hybrid wind-hydro power generation system. In: IEEE International Conference on Energy Efficient Technologies for Sustainability (ICEETS), Nagercoil, India; April 2013. pp. 406–10.
272. Castronuovo ED, Lopes JAP. Bounding active power generation of a wind-hydro power plant. In: IEEE International Conference on Probabilistic Methods Applied to Power Systems, Ames Iowa; 2004. pp. 705–10.
273. Leach F, Munteanu RA, Vadan I, Capatana D. Didactic platform for the study of hybrid wind-hydro power plants. In: IEEE ELECTROMOTION 2009 – EPE Chapter "Electric Drives" Joint Symposium, Lille, France; July 2009, pp. 1–3.
274. Bakos GC. Feasibility study of a hybrid wind/hydro power-system for low-cost electricity production. *Applied Energy.* 2002;72(3–4):599–608.
275. Goel PK, Singh B, Murthy SS. Autonomous hybrid system using SCIG for hydro power generation and variable speed PMSG for wind power generation. In: IEEE International Conference on Power Electronics and Drive Systems, PEDS, Taipei, Taiwan; 2009. 2009. pp. 55–60.
276. Nasser M, Breban S, Courtecuisse V, Vergnol A, Robyns B, Radulescu MM. Experimental results of a hybrid wind / hydropower system connected to isolated loads. In: IEEE 13th Power Electronics and Motion Control Conference, EPE-PEMC 2008, Poznan, Poland; 2008. pp. 1896–1903.
277. Karki R, Member S, Hu P, Billinton R, Fellow L. Reliability evaluation considering wind and hydro power coordination. *IEEE Transactions on Power Systems.* 2010;25(2):685–93.
278. de la Nieta AAS, Contreras J, Mu~noz JI, Optimal coordinated wind-hydro bidding strategies in day- ahead markets. *IEEE Transactions on Power Systems.* 2013;28(2):798–809.
279. Wangdee W, Li W, Billinton R. Coordinating wind and hydro generation to increase the effective load carrying capability. IEEE 11th International Conference on Probabilistic Methods Applied to Power Systems, Sevastopol, Uktain; Jun 2010. pp. 337–42.
280. Sirasan K, Kamdi SY. Solar wind hydro hybrid energy system simulation. *International Journal of Soft Computing And Engineering.* 2013;2(6):500–3. Available from: http://www.ijsce.org/attachments/File/v2i6/F1250112612.pdf.
281. Islam SM. Increasing wind energy penetration level using pumped hydro storage in island micro-grid system. *International Journal of Energy and Environmental Engineering.* 2012;3(9):1–12 [043142-32 Aziz et al. *Journal of Renewable Sustainable Energy* 7, 043142 (2015)].
282. Perez-Navarro A, Alfonso D, Álvarez C, Ibáñez F, Sanchez C, Segura I. Hybrid biomass-wind power plant for reliable energy generation. *Renewable Energy.* 2010;35:1436–43. DOI: 10.1016/j.renene.2009.12.018.
283. Segura I, Perez-Navarro A, Sanchez C, Ibanez F, Paya J, Bernal E. Technical requirements for economical viability of electricity generation in stabilized wind parks. *Internartional Journal of Hydrogen Energy.* 2007;32(16):3811–9.
284. Raheem A, Hassan M, Shakoor R. Pecuniary optimization of biomass/wind hybrid renewable system. In: Proceedings of the 1st International e-Conference on Energies, Rome, Italy; March 14–31, 2014. Vol. 1, pp. 1–10.
285. Vani N, Khare V. Rural electrification system based on hybrid energy system model optimization using HOMER. *Canadian Journal of Basic and Applied Sciences.* 2013;1(1):19–25. Available from: http://www.cjbas.com/archive/CJBAS-13-01-01-03.pdf.

286. Kesraoui M. Designing a wind/solar/biomass electricity supply system for an Algerian isolated village. In: 13th European Conference on Power Electronics and Applications, 2009, EPE'09, Barcelona, Spain; 2009. pp. 1–6. IEEE.
287. Maherchandani JK, Agarwal C, Sahi M. Economic feasibility of hybrid biomass/PV/ wind system for remote village using HOMER. *International Journal of Advanced Research in Electrical, Electronics and Instrumentation Engineering*. 2012;1(2):49–53. Available from: http://www.ijareeie.com/upload/august/2_Economic%20Feasibility%20of%20Hybrid.pdf.
288. Balamurugan P, Ashok S, Jose TL. Optimal operation of biomass/wind/PV hybrid energy system for rural areas. *International Journal of Green Energy*. 2009;6(1):104–16.
289. Balamurugan P, Ashok S, Jose TL. An optimal hybrid wind-biomass gasifier system for rural areas. *Energy Sources, Part A: Recovery, Utilization, and Environmental Effects*. 2011;33(9):823–32. DOI: 10.1080/15567030903117646.
290. Hongkai L, Chenghong X, Jinghui S, Yuexi Y. Green power generation technology for distributed power supply. China International Conference on Electricity Distribution. 2008:1–4.doi:10.1109/CICED.2008.5211704
291. Onar OC, Uzunoglu M, Alam MS. Dynamic modeling, design and simulation of a wind/fuel cell/ultra-capacitor-based hybrid power generation system. *Journal of Power Sources*. 2006;161(1):707–22.
292. Muljadi E, Butterfield CP. Pitch-controlled variable-speed wind turbine generation. *IEEE Transactions on Industry Applications*. 2001;37(1):240–6.
293. Mohod SW, Aware MV. Micro wind power generator with battery energy storage for critical load. *IEEE Systems Journal*. 2012;6(1):118–25.
294. Lu MS, Chang CL, Lee WJ, Wang L. Combining the wind power generation system with energy storage equipments. Industry Applications Society Annual Meeting, 2008. IAS'08, San Diego, CA; 2008. pp. 1–6. IEEE.
295. Leva S, Zaninelli D. Hybrid renewable energy-fuel cell system: Design and performance evaluation. *Electric Power Systems Research*. 2009;79(2):316–24.
296. Alam MS, Gao DW. Modeling and analysis of a wind/PV/fuel cell hybrid power system in HOMER. In: 2nd IEEE Conference on Industrial Electronics and Applications, 2007, ICIEA, Harbin, China; 2007. 2007. pp. 1594–9. IEEE.
297. Panapakidis IP, Sarafianos DN, Alexiadis MC. Comparative analysis of different grid-independent hybrid power generation systems for a residential load. *Renewable and Sustainable Energy Reviews*. 2012;16(1):551–63.
298. Mills A, Al-Hallaj S. Simulation of hydrogen-based hybrid systems using Hybrid2. *International Journal of Hydrogen Energy*. 2004;29(10):991–9.
299. Khan MJ, Iqbal MT. Pre-feasibility study of stand-alone hybrid energy systems for applications in Newfoundland. *Renewable Energy*. 2005;30(6):835–54.
300. Sobotka K. A wind-power fuel cell hybrid system study: Model of energy conversion for wind energy system with hydrogen storage [Master's thesis]. Akureyri, Iceland: RES | the School for Renewable Energy Science in affiliation with University of Iceland & the University of Akureyri; February 2009.
301. Meisen P, Hammons T. Harnessing the untapped energy potential of the oceans: Tidal, wave, currents and otec. In: IEEE 2005 Power Engineering Society General Meeting, June 12–16, 2005, San Francisco; 2005. pp. 1–2.
302. Caraiman G, Nichita C, M^ınzu V, Dakyo B, Jo CH. Study of a real time emulator for marine current energy conversion. In: IEEE XIX International Conference on Electrical Machines—ICEM 2010, Rome; 2010.
303. Junginger M, Faaij A, Turkenburg WC. Cost reduction prospects for offshore wind farms. *Wind Engineering*. 2004;28(1):97–118.

304. Guo KW. Green nanotechnology of trends in future energy: A review. *International Journal of Energy Research.* 2012;36:1–17.
305. Caraiman G, Nichita C, M^ınzu V, Dakyio B, Jo CH. Concept study of offshore wind and tidal hybrid conversion based on real time simulation. In: EA4EPQ International Conference on Renewable Energies and Power Quality (ICREPQ'11), April 13–15, 2010, Las Palmas de Gran Canaria (Spain); 2010.
306. Mousavi Gazafroodi SM. An autonomous hybrid energy system of wind/tidal/micro-turbine/battery storage. *International Journal of Electrical Power and Energy Systems.* 2012;43(1):1144–54.
307. Da Y, Khaligh A. Hybrid offshore wind and tidal turbine energy harvesting system with independently controlled rectifiers. In: 35th Annual Conference IEEE Industrial Electronics, Porto, Portugal; November 2009. pp. 4577–82.
308. Lim JH. Optimal combination and sizing of a new and renewable hybrid generation system. *International Journal of Future Generation Communication and Networking.* 2012;5(2):43–59.
309. Sikder AK, Khan NA, Hoque A. Design and optimal cost analysis of hybrid power system for Kutubdia island of Bangladesh. In: 2014 International Conference on Electrical and Computer Engineering (ICECE), Dhaka, Bangladesh; 2014. pp.729–32. IEEE.
310. Khan NA, Sikder AK, Saha SS. Optimal planning of off-grid solar-wind-tidal hybrid energy system for sandwip island of Bangladesh. In: 2014 2nd International Conference on Green Energy and Technology (ICGET), Dhaka, Bangladesh; 2014. pp. 41–4. IEEE.
311. Off-shore wind power: Floating production solution. Japan: MODEC; 2013. A website report.
312. Hybrid wind-tidal turbine to be installed off Japanese coast. Renewable energy world; July 12, 2013. A website report.
313. Pierre S, Nichita C, Brossard J, Dakyo B. Overview and analysis of different offshore wind-tidal hybrid systems as starting points for a real time simulator development. In: XIII Spanish Portuguese Conference on Electrical Engineering (XIIICHLIE), July 3–5, 2013, Valencia, Spain; 2013 [Paper No. 202. 043142-35 Aziz et al. *Journal of Renewable Sustainable Energy* 7, 043142 (2015)].
314. Aboul-Seoud T, Sharaf AM. A dynamic voltage regulator compensation scheme for a grid connected village electricity hybrid wind/tidal energy conversion scheme. In: 2009 IEEE Electrical Power Energy Conference, Calgary, Canada; October 2009. pp. 1–6.
315. Rahman ML, Oka S, Shirai Y. Hybrid power generation system using offshore-wind turbine and tidal turbine for power fluctuation compensation (HOT-PC). *IEEE Transactions on Sustainable Energy.* 2010;1(2):92–98.
316. Aboul-Seoud T, Sharaf AM. Utilization of the modulated power filter compensator scheme for a grid connected rural hybrid wind/tidal energy conversion scheme. In: IEEE Electrical Power and Energy Conference, Calgary, Canada; August 2010. pp. 1–6.
317. Wang L, Li C-N, Chen Y-T, Kao Y-T, Wang S-W. Analysis of a hybrid offshore wind and tidal farm connected to a power grid using a flywheel energy storage system. In: 2011 *IEEE Power and Energy Society General Meetings*, Detroit, MI; July 2011. pp. 1–8.
318. Rahman ML, Shirai SOY. Hybrid offshore-wind and tidal turbine power system for complement-the- fluctuation(HOTCF). Available from: http://proceedings.ewea.org/ewec2010/allfiles2/113_EWEC2010presentation.pdf.
319. Rahman ML, Shirai Y. DC connected hybrid offshore-wind and tidal turbine generation system. In: *Zero-Carbon Energy Kyoto* 2009, *Green Energy and Technology.* Tokyo: Springer; 2010. pp. 141–50.

320. Rahman ML, Shirai Y. Hybrid offshore-wind and tidal turbine (hott) energy conversion II (6-pulse GTO rectifier DC connection and inverter). In: IEEE International Conference on Sustainable Energy Technologies (ICSET 2008), Singapore; November 24–27, 2008. 2008. pp. 650–5.
321. Rahman ML, Shirai Y. Design and analysis of a prototype HOTT generation system. *BUP Journal*. 2012;1(1):111–29. Available from: http://www.bup.edu.bd/journal/111-129.pdf.
322. NTE gets patent for biomass hybrid renewable energy technology. St. Augustine, FL: NTE Energy; 2012. A website report.
323. 482 Solar-Induced Hybrid Fuel Cell Produces Electricity Directly from Biomass. Renewable Energy World; February 25, 2014. Available from: www.renewableenergyworld.com.
324. Arabkoohsar A, Sadi M. A hybrid solar concentrating-waste incineration power plant for cost-effective and dispatchable renewable energy production. In: 2018 IEEE 7th International Conference on Power and Energy (PECon), Kuala Lumpur, Malaysia; 2018. pp. 1–6. DOI: 10.1109/PECON.2018.8684179.
325. Gupta A. Saini RP, Sharma MP. "Computerized modelling of hybrid energy system—Part I: Problem formulation and model development." In: IEEE 5th International Conference on Electrical and Computer Engineering (ICECE 2008), Xian, China; December 2008. pp. 7–12.
326. Mostofi F, Shayeghi H. Feasibility and optimal reliable design of renewable hybrid energy system for rural electrification in Iran. *International Journal of Renewable Energy Research*. 2012;2(4):574–82.
327. Varshney N, Sharma MP, Khatod DK. Sizing of hybrid energy system using HOMER. *International Journal of Emerging Technology and Advanced Engineering*. 2013;3(6):436–42.
328. Kanase-Patil AB, Saini RP, Sharma MP. Sizing of integrated renewable energy system based on load profiles and reliability index for the state of Uttarakhand in India. *Renewable Energy*. 2011;36(11):2809–21.
329. Gupta A, Saini RP, Sharma MP. Hybrid energy system sizing incorporating battery storage: An analysis via simulation calculation. In: IEEE Third International Conference on Power Systems, 2009 (ICPS '09). 2009. pp. 1–6. New Delhi, India
330. Sánchez AS, Torres EA, Kalid RD. Renewable energy generation for the rural electrification of isolated communities in the Amazon Region. *Renewable and Sustainable Energy Reviews*. 2015;49:278–90.
331. Bhatt A, Sharma MP, Saini RP. Feasibility and sensitivity analysis of an off-grid micro hydro–photovoltaic–biomass and biogas–diesel–battery hybrid energy system for a remote area in Uttarakhand state, India. *Renewable and Sustainable Energy Reviews*. 2016;61:53–69. DOI: 10.1016/j.rser.2016.03.030.
332. Hiremath RB, Shikha S, Ravindranath NH. Decentralized energy planning: Modeling and application—A review. *Renewable and Sustainable Energy Reviews*. 2007;11(5):729–52.
333. Barley CD, Winn CB. Optimal dispatch strategy in remote hybrid power systems. *Solar Energy*. 1996;58(4):165–79.
334. Arivalagan A, Raghavendra BG. Integrated energy optimization model for a cogeneration based energy supply system in the process industry. International Journal of Electrical Power and Energy Systems. 1995;17(4):227–33.
335. Ramakumar R, Abouzahr I, Ashenayi K. A knowledge-based approach to the design of integrated renewable energy systems. *IEEE Transactions on Energy Conversion*. 1992;7(4):648–59.

336. Little M, Thomson M, Infield D. Electrical integration of renewable energy into stand-alone power supplies incorporating hydrogen storage. *International Journal of Hydrogen Energy*. 2007;32(10–11):1582–8.
337. Onar OC, Uzunoglu M, Alam MS. Modeling, control and simulation of an autonomous wind turbine/photovoltaic/fuel cell/ultra-capacitor hybrid power system. *Journal of Power Sources*. 2008;185(2):1273–83.
338. Wang C, Nehrir MH. Power management of a stand-alone wind/photovoltaic/fuel cell energy system. *IEEE Transactions on Energy Conversion*. 2008;23(3):957–67.
339. Khan MJ, Iqbal MT. Analysis of a small wind-hydrogen stand-alone hybrid energy system. *Applied Energy*. 2009;86(11):2429–42.
340. Soedibyo S, Suryoatmojo H, Robandi I, Ashari M. Optimal design of fuel-cell, wind and micro-hydro hybrid system using genetic algorithm. *TELKOMNIKA*. 2012;10(4):695–702.
341. Tudu B, Mandal KK, Chakraborty N. Optimal design and performance evaluation of a grid independent hybrid micro hydro-solar-wind fuel cell energy system using meta-heuristic techniques. In: IEEE Proceedings of 2014 1st International Conference on Non-Conventional Energy (ICONCE 2014), Kalyani, India; 2014. pp. 89–93.
342. Barsoum N, Petrus PD. Cost optimization of hybrid solar, micro-hydro and hydrogen fuel cell using HOMER software. *Energy and Power Engineering*. 2015;7:337–47. Published Online August 2015 in SciRes. http://www.scirp.org/journal/epe, http://dx.doi.org/10.4236/epe.2015.78031.
343. Off-Grid Renewable Energy Solutions. Global and regional status and trends. Abu Dhabi: IRENA; 2018. IRENA report.
344. IRENA. *Renewable Energy and Jobs–Annual Review 2018*. Abu Dhabi: International Renewable Energy Agency; 2018.
345. Energy4Impact. SOGER (Scaling Up Off Grid Energy in Rwanda) mid term in-depth evaluation. 2018. Available from: www.energy4impact.org/file/2024/download?token=Qs5Bi2ge.
346. Barnes D, Floor W. Rural Energy in Developing Countries: A Challenge for Economic Development. Annual Review of Energy and Environment. Nov. 1996; 21:497–530. https://doi.org/10.1146/annurev.energy.21.1.497.
347. Brew-Hammond A. Challenges to increasing access to modern energy services in Africa. In: Background Paper, Conference on Energy Security and Sustainability, March 28–30, 2007, Maputo, Mozambique; 2007.
348. Brew-Hammond A, Crole-Rees A. Reducing Rural Poverty through Increased Access to Energy Services: A Review of the Multifunctional Platform Project in Mali. A website report by UNDP, www.ml.undp.org (2004).
349. Chaurey A, Ranganathana M, Mohanty P. Electricity access for geographically disadvantaged rural communities—Technology and policy insights. *Energy Policy*. 2004;32:1693–1705.
350. Ranjit L, O'Sullivan K. Energy, Chapter 21. In: Klugman J, editor. *A Sourcebook for Poverty Reduction Strategies*. Washington, DC: World Bank; 2002. Vol. 2.
351. World Bank. *Meeting the Challenge for Rural Energy and Development*. Washington, DC: World Bank; 1996.
352. Ezzati M, Kammen D. Evaluating the health benefits of transitions in household energy technologies in Kenya. *Energy Policy*. 2002;30(10):815–826.
353. Barnes D. *Transformative Power: Meeting the Challenge of Rural Electrification*. Washington, DC: Energy Sector Management Assistance Program (ESMAP), United National Development Programme/World Bank; 2005.

354. Cecelski E. *Enabling Equitable Access to Rural Electrification: Current Thinking and Major Activities in Energy, Poverty and Gender.* Briefing Paper. Washington, DC: World Bank; 2000.
355. Murphy JT. Making the energy transition in rural East Africa: Is leapfrogging an alternative? *Technological Forecasting and Social Change.* 2001:68:173–193.
356. Sagar A. Alleviating energy poverty for the world's poor. *Energy Policy.* 2005;33:1367–1372.
357. Cherni J, Hill Y. Energy and policy providing for sustainable rural livelihoods in remote locations—The case of Cuba. *Geoforum.* 2009; 40:645–54.
358. World Bank. *African Development Indicators.* Washington, DC: World Bank; 2006.
359. Blair N, Pons D, Krumdieck S. Electrification in remote communities: Assessing the value of electricity using a community action research approach in Kabakaburi, Guyana. *Sustainability.* 2019;11:2566. DOI: 10.3390/su11092566 www.mdpi.com/journal/sustainability.
360. The United Nations, The World Bank. *Global Tracking Framework—Progress towards Sustainable Energy 2017.* Washington, DC: International Bank for Reconstruction and Development/The World Bank and the International Energy Agency; 2017.
361. Pierson P. The study of policy development. *Journal of Policy History.* 2005;17:34–51.
362. Inter-American Development Bank. *IDB Approves $100 Million for Rural Electrification in Bolivia.* New York: Inter-American Development Bank; 2016.
363. Nieuwenhout F, Van Dijk A, Van Dijk V, Hirsch D, Lasschuit P, Van Roekel G, Arriaza H, Hankins M, Sharma B, Wade H. Monitoring and evaluation of solar home systems. Experiences with applications of solar PV for households in developing countries. Petten, The Netherlands: Netherlands Energy Research Foundation ECN; 2000. Technical Report.
364. Carmody JM. Rural electrification in the United States. *The Annals of the American Academy of Political and Social Science.* 1939;201:82–8.
365. Mala K, Schlä A, Pryor T. Better or worse? The role of solar photovoltaic (PV) systems in sustainable development: Case studies of remote atoll communities in Kiribati. *Renewable Energy.* 2008;34:358–61. DOI: 10.1016/j.renene.2008.05.013.
366. Zomers A. The challenge of rural electrification. *Energy for Sustainable Development.* 2003;7:69–76.
367. Bhattacharyya SC, Srivastava L. Emerging regulatory challenges facing the Indian rural electrification program. *Energy Policy.* 2009;37:68–79.
368. Belize Electricity Limited. *Rate Schedule.* Belize: Belize Electricity Limited; 2018.
369. Government of Belize. *Electricity Act.* Belize: Belize Electricity Limited; 2000. Chapter 221, Technical Report.
370. Weedy BM, Cory BJ, Jenkins N, Ekanayake JB, Strbac G. *Electric Power Systems.* New York: Wiley; 2012.
371. Cavallaro MB. *Power Distribution System Design.* New York: Current Solutions P.C.; 2002 [Acccessed April 25, 2019]. Available from: https://www.scribd.com/document/55218251/Power.
372. Gorenstin B, Campodonico N, Costa J, Pereira M. Power system expansion planning under uncertainty. *IEEE Transactions on Power Systems.* 1993;8:129–136.
373. Kreith F. Integrated resource planning. *Journal of Energy Resources Technology.* 1993;115:80–5.
374. Wilson R, Biewald B. *Best Practices in Electric Utility Integrated Resource Planning: Examples of State Regulations and Recent Utility Plans.* Montpelier, VT: Synapse Energy Economics and the Regulatory Assistance Project (RAP); 2013.

375. Richardson J. Government, interest groups and policy change. *Political Studies.* 2000;48:1006–25. DOI: 10.1111/1467-9248.00292.
376. Smit S, Musango JK, Brent AC. Understanding electricity legitimacy dynamics in an urban informal settlement in South Africa: A community based system dynamics approach. *Energy for Sustainable Development.* 2019;49:39–52.
377. Alazraque-Cherni J. Renewable energy for rural sustainability in developing countries. *Bulletin of Science, Technology & Society.* 2008;28(2):105–114.
378. Steger U. *Sustainable Development and Innovation in the Energy Sector.* Berlin: Springer; 2005.
379. World Bank. *Renewable Energy for Development: The Role of the World Bank Group.* Washington, DC: World Bank; 2004.
380. World Bank. *The World Bank Group's Energy Program: Poverty Reduction, Sustainability and Selectivity.* Washington, DC: World Bank; 2001.
381. Goldemberg J. Rural Energy in Developing Countries, Chapter 10. In: Goldemberg J, editor. *World Energy Assessment: Energy and the Challenges of Sustainability.* New York: United Nations Development Programme; 2000.
382. Havet I, Chowdhury S, Takada M, Cantano A. *Energy in National Decentralization Policies.* New York: United National Development Programme; 2009.
383. Alliance for Rural Electrification. *A Green Light for Renewable Energy in Developing Countries.* Brussels: Alliance for Rural Electrification; 2009.
384. Ockwell D, Watson J, MacKernon G, Pal P, Yamin F. *UK-India Collaboration to Identify the Barriers to the Transfer of Low Carbon Energy Technology.* London: Report by the Sussex Energy Group, University of Sussex, TERI and IDS for the United Kingdom Department for Environment, Food and Rural Affairs; 2006.
385. ESMAP. *Technical and Economic Assessment of Off-grid, Mini-grid and Grid Electrification Technologies.* Washington, DC: World Bank; 2007.
386. Byrne J, Shen B, Wallace. The economics of sustainable energy for rural development: A study of renewable energy in rural China. *Energy Policy.* 1998;26(1):45–54.
387. Ockwell D, Ely A, Mallett A, Johnson O, Watson J. *Low Carbon Development: The Role of Local Innovative Capabilities.* STEPS Working Paper 31. Brighton: STEPS Centre and Sussex Energy Group, SPRU, University of Sussex; 2009.
388. Martinot E, Reiche K. *Regulatory Approaches to Rural Electrification and Renewable Energy: Case Studies from Six Developing Countries.* Working Paper. Washington, DC: World Bank; 2000.
389. World Bank. *Project Paper on Proposed Additional Financing for Renewable Energy in the Rural Market Project.* Washington, DC: World Bank; 2008.
390. United Nations Framework Convention on Climate Change. CDM Interactive Map. 2009 [Accessed April 14, 2010]. Available from: http://cdm.unfccc.int/Projects/MapApp/index.html.
391. Elhadidy MA, Shaahid SM. Promoting applications of hybrid (wind+photovoltaic+diesel+battery) power systems in hot regions. *Renewable Energy.* 2004;29(4):517–28.
392. Marnay C,. Venkataramanan G, Stadler M, Siddiqui AS, Firestone R, Chandran B. Optimal technology selection and operation of commercial-building microgrids. *IEEE Transactions on Power Systems.* 2008;23(3):975–982.
393. Giraud F, Salameh ZM. Steady-state performance of a grid-connected rooftop hybrid wind—Photovoltaic power system with battery storage. *IEEE Transactions on Energy Conversion.* 2001;16(1):1–7.
394. Kellogg WD, Nehrir MH, Venkataramanan G, Gerez V. Generation unit sizing and cost analysis for stand- alone wind, photovoltaic, and hybrid wind/PV Systems. *IEEE Transactions on Energy Conversion.* 1998;13(1):70–75.

395. Borowy BS, Salameh ZM. Methodology for optimally sizing the combination of a battery bank and PV array in a wind/PV hybrid system. *IEEE Transactions on Energy Conversion*. 1996;11(2):367–75.
396. Notton G, Muselli M,Poggi P, Louche A. Decentralized wind energy systems providing small electrical loads in remote areas. *International Journal of Energy Research*. 2001;25:141–64.
397. Barnett A. The diffusion of energy technology in the rural areas of developing countries: A synthesis of recent experience. *World Development*. 1990;18(4):539–53.
398. Bhattacharyya S. Renewable energies and the poor: Niche or nexus? *Energy Policy*. 2006;34:659–63.
399. Sreekala CS, Mathew A. Voltage and frequency control of wind hydro hybrid system in isolated locations using cage generators. In: 2013 International Mutli-Conference on Automation, Computing, Communication, Control and Compressed Sensing (iMac4s), Kottayam, India; March 2013. pp. 132–7.
400. Goel PK, Singh B, Murthy SS, Kishore N. Autonomous hybrid system using PMSGs for hydro and wind power generation. In: 2009 35th Annual Conference on IEEE Industrial Electronics, Porto, Portugal; November 2009. pp. 255–60.
401. Nguyen Ngoc PD, Pham TTH, Bacha S, Roye D. Optimal operation for a wind-hydro power plant to participate to ancillary services. In: 2009 IEEE International Conference on Industrial Technology, Churchill, Victoria, Australia; February 2009. pp. 1–5.
402. Barriers to Renewable Energy Technologies. Available from: http://www.ucsusa.org/clean_energy/smart-energy-solutions/increase-renewables/barriers-to-renewable-energy.html#.VZ1mDbmqqkp.
403. Power and Energy Circuits. Available from: http://ieee-cas.org/community/technical-committees/pecas.
404. Patil A, Patil V, Shin DW, Choi J-W, Paik D-S, Yoon S-J. Issue and challenges facing rechargeable thin film lithium batteries. *Materials Research Bulletin*. 2008;43(8):1913–42.
405. Rowsell JLC, Yaghi OM. Strategies for hydrogen storage in metal–organic frameworks. *Angewandte Chemie International Edition*. 2005;44(30):4670–79.
406. Kennedy BW. Integrating wind power: Transmission and operational impacts. *Refocus*. 2004;5(1):36–7.
407. Flowers L, Green J, Bergey M, Lilley A, Mott L. Village power hybrid systems development in the United States. Prepared for European Wind Energy Conference, October 10–14, 1994, Thessaloniki, Greece; November 1994. Golden, CO: NREL. NREL/TP-442-7227 • UC Category: 1210 • DE95000247.
408. ZhongYing W, Hu G, Dadi Z. China's achievements in expanding electricity access for the poor. *Energy for Sustainable Development*. 2006;10(3):5–16.
409. Ashden Awards. Bringing affordable, high-quality solar lighting to rural China. In: *Case study for 2008 Ashden Awards for Sustainable Energy*. London: Ashden; 2008.
410. National Renewable Energy Laboratory. *Renewable Energy in China*: WB/GEF Renewable Energy Development Project. Golden, CO: National Renewable Energy; 2004.
411. Wohlgemuth N, Painuly JP. National and international renewable energy financing: What can we learn from experience in developing countries? *Energy Studies Review*. 2006;14(2):154.
412. World Bank. *Implementation Completion and Results Report for Renewable Energy Development Project in China*. Washington, DC: World Bank; 2009.
413. Fuentes M. Report on the situation in Argentina. Presented at the International Expert Workshop on Clean Development Mechanism (CDM) and Sustainable Energy Supply in Latin America, April 2005, Buenos Aires; 2005.

414. World Bank. *Project Paper on Proposed Additional Financing for Renewable Energy in the Rural Market Project.* Washington, DC: World Bank; 2008.
415. Alazraki R, Haselip J. Assessing the uptake of small-scale photovoltaic electricity production in Argentina: The PERMER project. *Journal of Cleaner Production.* 2007;15:131–42.
416. World Bank. *Project Appraisal Document (PAD) for Renewable Energy in the Rural Market Project.* Washington, DC: World Bank; 1999.
417. Reiche K, Covarrubias A, Martinot E. *Expanding Electricity Access to Remote Areas: Off-Grid Rural Electrification in Developing Countries. Fuel.* 2000; 1(1.1):1–4.
418. Smits M. Technography of pico-hydropower in the Lao People's Democratic Republic. Vientiane: Laos Institute of Renewable Energy (LIRE); 2008. Report No.: 003.
419. Smits M. Neglected decentralized rural electricity production in the Lao People's Democratic Republic: A technography of pico-hydropower analyzed in a political ecology framework [MSc thesis]. Wageningen, The Netherlands: Wageningen University; 2008.
420. Smits M, Bush S. A light left in the dark: The practice and politics of pico- hydropower in the Lao People's Democratic Republic. *Energy Policy.* 38, 2010, 116–127.
421. Hello Namibia. Namibia: Motorola and GSMA complete MTC trials. 2008 [January 4, 2008] [Accessed April 14, 2010]. Available from: http://www.hellonam.com/speeches-opinions/5073-namibia-motorola-gsma-complete-mtc-trials.html.
422. Motorola Inc. *White Paper on Alternatives for Powering Telecommunications Base Stations.* Schaumburg, IL: Motorola, Inc; 2007.
423. TMCNet, Azuri C. Green technology -Motorola successfully trials solar, wind power to run cellular base station in Africa. 2007 [November 28, 2007] [Accessed April 14, 2010]. Available from: http://green.tmcnet.com/topics/green/articles/15386-motorola-successfully-trials-solar-wind-power-run-cellular.htm.
424. Green Telecom, Chan T. Motorola eyes potential for alternative energy powered systems. 2008 [January 11, 2008] [Accessed April 14, 2010]. Available from: http://www.greentelecomlive.com/?p=32.
425. GSMA Development Fund. *Wind and Solar.* London: GSMA Development Fund; 2007 [Habtetsion S, Tsighe Z. The energy sector in Eritrea—institutional and policy options for improving rural energy services. *Energy Policy.* 2002;30].
426. LetsGoMobile. Motorola wind and solar system trial. 2007 [February 16, 2007] [Accessed April 14, 2010]. Available from: http://www.letsgomobile.org/en/0884/motorolatrial/.
427. Troxler G, Powell W, editor. North Carolina rural electrification authority. University of North Carolina; 2006. A website report.
428. Hunt J. Rural electrification, In: Powell W, editor. *Encyclopedia of North Carolina.* State of North Carolina, Raleigh, NC: University of North Carolina Press; 2006.
429. Rural electrification. *Wikipedia, The Free Encyclopedia.* 2020 [last visited April 28, 2020].
430. Modi V, McDade S, Lallement D and Saghir J. *Energy Services for the Millennium Development Goals.* Washington, DC and New York: The World Bank and United Nations Development Programme; 2005.
431. Renewable energy technologies for rural development., United Nations Conference On Trade and Development. UNCTAD Current Studies on Science, Technology and Innovation. No. 1. New York: United Nations; 2010. Report No.: UNCTAD/DTL/STICT/2009/4.
432. IMF and World Bank. An investment framework for clean energy and development: A progress report. Washington DC: World Bank; 2006. Development Committee Report No.: DC2006-0012.

433. IRENA. *Renewable Energy Jobs & Access.* Abu Dhabi: International Renewable Energy Agency; 2012. Available from: www.irena.org/DocumentDownloads/Publications/Renewable_Energy_Jobs_and_Access.pdf.
434. IDCOL (Infrastructure Development Company Limited). Project and Programs: Solar Home System Programme, Infrastructure Development Company Limited, Dhaka. 2018. Available from: www.idcol.org/home/solar.
435. Renewable power generation costs. Abu Dhabi: IRENA; 2014. IRENA report.
436. IRENA (International Renewable Energy Agency). *Renewable Energy Statistics 2017.* Abu Dhabi: IRENA; 2017.
437. IRENA. Global Energy Transition Prospects and the Role of Renewables, Chapter 3. In: *Perspectives for the Energy Transition—Investment Needs for a Low-Carbon Energy System.* Abu Dhabi: IRENA; 2017.
438. IRENA. *Renewable Energy Auctions: Analysing 2016.* Abu Dhabi: IRENA; 2017.
439. IFC (International Finance Corporation). *Utility-Scale Photovoltaic Power Plants.* Washington, DC: IFC; 2015.
440. MNRE (Ministry of New and Renewable Energy). *4-Year Achievement Booklet.* New Delhi:: Government of India, Ministry of New and Renewable Energy. 2018. Available from: https://mnre.gov.in/sites/default/files/uploads/MNRE-4-Year-Achievement-Booklet.pdf.
441. IRENA. *Unlocking Renewable Energy Investment: The Role of Risk Mitigation and Structured Finance.* Abu Dhabi: IRENA; 2016.
442. IMF (International Monetary Fund). People's Republic of China. Washington, DC: IMF; 2016. IMF Country Report No.: 16/271.
443. Off-grid renewable energy solutions to expand electricity access: An opportunity not to be missed. Abu Dhabi: IRENA; 2019. IRENA report. ISBN 978-92-9260-101-0.
444. EPIA (European Photovoltaic Industry Association). Global Market Outlook for Photovoltaics 2014–2018. 2014. Available from: www.epia.org/fileadmin/user_upload/Publications/EPIA_Global_Market_Outlook_for_Photovoltaics_2014-2018_-_Medium_Respdf, accessed December 2014. http://www.solarpowereurope.org/
445. UN DESA. A survey of international activities in rural energy access and electrification. United Nations Department of Economic and Social Affairs. March 2014. Available from: http://sustainabledevelopment.un.org/content/documents/1272A%20Survey%20of%20International%20Activities%20in%20Energy%20Access%20and%20Electrification.pdf. prepared by division of sustainable development, department of economics and social affairs, United Nations. New York
446. IRENA (International Renewable Energy Agency). *Renewable Capacity Statistics 2018.* Abu Dhabi: International Renewable Energy Agency; 2018. Available from: www.irena.org/publications/2018/Mar/Renewable-Capacity-Statistics-2018.
447. World Bank. *Nepal: Scaling Up Electricity Access through Mini and Micro Hydropower Applications.* Washington, DC: World Bank; 2015. Available from: http://documents.worldbank.org/curated/en/650931468288599171/pdf/96844-REVISED-v1-Micro-Hydro-Report-0625-2015-Final.pdf.
448. CIF (Climate Investment Funds). Renewable energy mini-grids and distributed power generation. CTF Project/Program Approval Request. 2014. www.climateinvestmentfunds.org/cif/sites/climateinvestmentfunds.org/files/140508%20ADB%20CTF%20Proposal%20-%20RE%20Mini-Grids%20Program%20PUBLIC_1.pdf, accessed December 2014.
449. Gerlach AK, Gaudchau E, Cader C, Wasgindt V, Breyer C. Comprehensive country ranking for renewable energy based mini-grids providing rural off-grid electrification. In: 28th European Photovoltaic Solar Energy Conference, 30 September – 4 October, Paris; 2013. Available from: www.reiner-lemoine-institut.de/sites/default/files/5do.1

3.2_gerlach2013_paper_compcountryrankingre-basedmini-gridsprovruraloff-gridelec _28theupvsec_5do.13.2_paris_preprint.pdf.
450. IED (Innovation Energie Développement). Low Carbon Mini Grids Identifying the gaps and building the evidence base on low carbon mini-grids–Final Report. London, UK: Department for International Development (DfID); 2013 [November 2013]. Available from: www.gov.uk/government/uploads/system/uploads/attachment_data/file/278021/IED-green-min-grids-support-study1.pdf.
451. IEA. World Energy Outlook 2017. Available from: https://www.iea.org/weo2017.
452. Bhattacharya A. One village solved Bangladesh's unreliable energy grid problem with "swarm electrification." Quartz. A website report, Nov. 30, 2016.
453. ASEAN Centre for Energy. ASEAN guideline on off-grid rural electrification approaches. 2013. Available from: http://aseanenergy.org/media/documents/2013/04/11/a/s/asean_guideline_on_off-grid_rural_electrification_ final_1.pdf.
454. Access to electricity-SDG7: Data and projections. Brussels, Belgium: IEA; 2019. A website report.
455. IEA (International Energy Agency) and World Bank. *Global Tracking Framework—Progress Toward Sustainable Energy.* Washington, DC: World Bank; 2017.
456. Global landscape of renewable energy finance. Abu Dhabi: IRENA; 2018. IRENA report.
457. WHO (World Health Organization). Trends in Maternal Mortality: 1990 to 2015. Estimates by WHO, UNICEF, UNFPA, World Bank Group and the United Nations Population Division, Geneva, Switzerland. 2015. Available from: http://apps.who.int/iris/bitstream/10665/194254/1/9789241565141_eng.pdf.
458. WHO. Health and sustainable development: Energy access and resilience. n.d. Available from: www.who.int/sustainable-development/health-sector/health-risks/energy-access/en/.

7 Simulation and Optimization of Hybrid Renewable Energy Systems

7.1 INTRODUCTION

The growth in renewable energy has largely occurred over the last two decades due to concerns over environment. The first village hybrid power systems consisting of PV and diesel generator was, however, installed on December 16, 1978, in Papago Indian Village, Schuchuli, Arizona, USA. The power produced by the system was used for providing electricity for community refrigerator, washing machine, sewing machine, water pumps, and lights until an electric grid was extended to the village in 1983 [1–8]. In recent years, more than one renewable form of energy is being used in hybrid renewable energy systems (HRES). Micro hydro power (MHP), PV, small Wind power sources and biomass (with rice husks in India) with or without energy storage devices are widely used for providing electric power to consumers in remote areas. Different alternative energy resources have different production characteristics such as water in river changes flow according to the seasons, the solar irradiation is greater in summer than winter and higher in day and not at night and similarly wind speed is greater in spring time and low in mid to late summer, biomass may have collection problems in certain seasons etc. This is why they are usually used in hybrid system configurations. The components of a carefully chosen hybrid system (like solar and wind) can complement each other. Storage helps to improve power quality. Two or more renewable energy sources can be integrated in one system, based on the local renewable energy potential. Properly designed and optimized hybrid power system can improve cost, efficiency, and reliability of power generation and distribution.

Renewable energy sources such as wind, solar, and biomass, which have received a growing attention from many firms and policy planners around the globe, are being considered for utilization and electrification in many geographical locations of the world. The hybrid energy systems which incorporate such sources are typically capable of electricity production for several applications such as commercial or office buildings, rural areas with difficult access to electricity, hospitals, telecommunication services, and many other facilities. In such systems, energy is produced using primary power system components, such as the WECSs, PVs, hydro power systems, and/or other conventional generators, which employ fossil fuel-based generators such as the diesel generator. Hybrid power systems, in terms of their energy production, are generally ranged from small to large scales. They are capable of generating electricity for small residences to commercial scale systems, which can electrify a whole village or an island. [1–23].

It is important to bear in mind that there are particular advantages contingent upon the configuration and proper design of the HRESs. A WT/PV system, for instance, can generate more electrical power output from the wind during the spring time, while further peak energy output can be yielded from solar arrays during the summer time. A WT/ Hydro system has the merit of generating electrical power by releasing it into a hydropower plant when it is required. This would particularly be useful for islands which are not connected to larger grids. Nevertheless, HRESs, despite their primary advantages, suffer from a few drawbacks with regard to their design and operation process. Some of the advantages and disadvantages of HRES can be summarized as [3]:

Advantages:

1. Utilization of the natural and renewable sources.
2. Low level of O&M costs.
3. No pollution or wastes produced by the natural sources.
4. Minimizing the intermittency and thereby increasing the reliability.
5. Lower atmosphere contamination.
6. Fuel Savings.

Disadvantages:

1. Dependency on the natural cycle
2. Higher initial costs of these systems than comparably sized traditional generators
3. Relatively high land requirements
4. The peak-loads cannot be managed without energy storage
5. Complexity of the design procedure
6. Monthly fee charge

In many situations, however, advantages outweigh disadvantages. While, the hybrid systems can overcome limitations of the individual generating technologies in terms of their fuel efficiency, economics, reliability and flexibility, main concerns are the stochastic nature of widely used photovoltaic (PV) and wind energy resources. Wind is often not correlated with load patterns and may be discarded sometimes when abundantly available. Also, solar energy is only available during the day time. Predictions of power productions from intermittent sources on a short or long term basis can be difficult. Weather conditions can play significant role in these predictions. Due to all these factors, a large number of random variables and parameters in a hybrid energy system requires a simulation tool and an optimization method that most efficiently sizes the hybrid system components to realize their economic, technical and design objectives. The simulation tool is also important to understand the behavior of HRES at different scales and under different set of operating conditions (grid-tied or stand-alone) [1–8].

HRES simulation and optimization is a research field with great areas to be explored, such as the creation of new methodologies that could help to inform decision makers in the design stages of projects. These methodologies can provide support tools for system sizing or evaluation of trade-offs among different alternatives due to their ability to tackle non-linear problems with relatively high calculation speeds. For example, system cost optimization of either grid-connected or stand-alone HRESs has been carried out either using HOMER optimization software which is based on the analytical simulation of several scenarios or by using heuristic methodologies, relying on evolutionary algorithms to optimize microgrids or HRESs size according to cost, environmental, or reliability parameters [1–8]. In order to study HRES in both grid connected or off-grid modes, significant number of software tools and optimization techniques with different sets of objectives have been recently developed [1–8].

Apart from various mathematical models used in analyzing the behaviors of different elements of hybrid systems, simulation and optimization software have also been found useful for varieties of applications. Areas of application include design, control strategy, economic optimization and multi-objective optimization. Many software packages have emerged but the most commonly used is Hybrid Optimization Model for Electric renewable (HOMER) [9–11]. HOMER has been used in many optimization analyses of HRES involving PV system, microhydro system, diesel generators, wind turbines, electrolyzer, fuel cell and hydrogen tank. This software is developed by the National Renewable Energy Laboratory (NREL) in the United States. Another useful software is improved Hybrid Optimization by Genetic Algorithms (iHOGA) which is a program developed for hybrid energy system simulation and optimization. iHOGA formerly known as HOGA [12] is employed mainly in systems of hybrid renewable electrical energy involving (DC and/or AC) and/ or Hydrogen. iHOGA has a very good reliability model for resources, economic and components. It can also resolve some complex issues relating to hybrid systems using Genetic Algorithms. It could also deal with multi-objective optimization approaches such as handling of emissions and minimization of present net cost of a hybrid system. HYBRID2 [13, 14] is another software package widely used for optimization as well as performing comprehensive economic analysis and long term performance of a hybrid system. Other software that can be used for the same or closely related purposes are Transient Energy System Simulation Program (TRNSYS) [15], hydrogen energy models (HYDROGEMS) [16, 17], hybrid system simulation models (HYBRIDS) [18], Solar simulator (SOLSIM) [19], INSEL [20], RAPSIM (Remote Area Power Supply Simulator) [21] and SOMES [22]. EMPS is (EFI's Multiarea Power market Simulator)is a computer tool developed for forecasting and energy planning in electricity markets [23]. Energy PLAN is a computer model designed for performing energy system analysis [23] .The uses of any of these software packages for simulation and optimization task are highly determined by the nature of input data characteristics and the task for which the software is designed to perform. In order to deal effectively with design problems related to hybrid power system, a proper load profile study of case study area is important. The load profile helps to reveal variations in power demand at various instances in days, months as well as seasons. A fluctuation in a typical load profile of any given community is usually in line with operation of domestic appliances and public facilities as well as business engagements. Designs of hybrid system for off-grid location require appropriate survey of load to balance between energy demand and supply including optimum cost control.

Optimization of hybrid renewable energy system is very important for energy security [24].

With reference to the capacity sizes of the HRESs, it is important to notice that they typically have a wide range of capacity sizes from kilowatt to hundreds of kilowatt based on the load, which can serve. Small-scale hybrid power systems which typically have a capacity lower than 5 kW can meet the load demand of a remote house, a telecommunication relay system, or any small-scale system. Medium-scale hybrid power systems with a capacity rated from five to hundred kW can be adequate alternatives to electrify areas with more considerable number of residences and families, or other community facilities. These hybrid systems sometimes can be connected to the grid, if they are close to a national utility grid. However, in most occasions, their operation is standalone. At last, the large-scale hybrid power systems which typically have a capacity of larger than 100 kW are linked to the electricity grid to permit transferring electricity from the grid to the hybrid power system and vice versa. This situation happens when the electricity surplus of the system is considerable. HRES can be classified based on their size. Small scale (less than 5kW) is suitable for remote homes and telecommunication systems. Medium scale (between 5kW and 100 kW) is used for the remotely located communities. Finally large scale (higher than 100kW) is used for regional loads [1–8, 23]. The design and optimization of HRES under all these varying conditions can be best facilitated by appropriate simulation and optimization models.

The purpose of this chapter is to briefly review available software and optimization tools for the modeling and simulation of HRES and outline various optimization techniques and objective functions that are explored in the literature. Research published using the available software tools for the hybrid energy systems are also reviewed. Finally, the effects of other system parameters within a hybrid energy system that can affect simulation, design, performance and optimization of the system are briefly reviewed. It should be pointed out that the literature has paid most attention to PV/Wind/battery hybrid system because it is the most widely preferred system.

7.2 SOFTWARE TOOLS FOR HYBRID SYSTEM ANALYSIS

Renewable energy systems are based on a single source or multiple sources of renewable generators. Storage of energy is a part of this mix. Single source-based renewable energy system incorporates only one electricity generation option based on wind/solar thermal/solar photovoltaic (PV)/hydro/biomass, etc., along with appropriate energy storage and electronic systems. In the presence of storage, this can be considered as a hybrid source. In most cases, a hybrid energy system incorporates two or more electricity generation options based either on renewable energy units or on fossil fuel-based units like diesel-electric generator or a small gas-turbine along with energy storage and electronic devices. As this chapter shows, several hybrid energy system configurations can be used for power generation like PV-wind-diesel systems, hydro-wind-PV-based systems, biomass-wind-PV installations, wind-PV-based installation, PV-wind-hydrogen/fuel cell hybrid energy systems, etc. A hybrid energy system has the following main advantages in comparison to a single source-based system [1–23]:

1. Higher reliability
2. Reduced need for energy storage capacity, especially where different sources have complementary behavior
3. Better efficiency
4. Minimum levelized life cycle electricity generation cost, when optimum design technique is used

But in most cases due to lack of optimum designing or proper sizing, a hybrid energy system is oversized or not properly planned or designed, which makes installation cost high. The technical and economic analyses of a hybrid system are essential for the efficient utilization of renewable energy resources. Due to multiple generation systems, hybrid system solution is complex and requires to be analyzed thoroughly. This requires software tools and models which can be used for the design, analysis, optimization, and economic planning. A number of software tools have been developed to assess the technical and economic potential of various hybrid renewable technologies to simplify the hybrid system design process and maximize the use of renewable resources. In this section, we briefly assess available software tools for simulation of HRES. The use of an appropriate software tool is important in order to get reliable simulation results, which can be used for design and optimization of the chosen HRES system. The applications and status of 22 software, namely, HOMER, HYBRID 2, RETScreen, iHOGA, INSEL, TRNSYS, iGRHYSO, HYBRIDS, RAPSIM, SOMES, SOLSTOR, HySim, HybSim, IPSYS, HySys, Dymola/Modelica, ARES, SOLSIM, HYDROGEM, Hybrid Designer, EMPS, and EnergyPlan, are presented in this section. A comparative analysis of these software tools along with the literature review of research carried out using these software worldwide is also presented. The limitations, availability, and areas of further research have also been identified. The analysis of a PV-battery and PV-wind-battery hybrid system is presented as case studies using HOMER and RETScreen.

A comprehensive understanding is essential about available hybrid system models and software tools, their features, shortcomings, user need, and choice for research studies. In this section, the main features of the 22 software developed for hybrid system design are discussed along with a comparative analysis. Turcotte et al. [25] classified the software tools related to hybrid systems in four categories: pre-feasibility, sizing, simulation, and open architecture research tools. The pre-feasibility tools are mainly used for rough sizing and a comprehensive financial analysis (e.g.,

RETScreen). The sizing tools are used for the determination of the optimal size of each component of the system and provide detailed information about energy flows among various components (e.g., HOMER). In a simulation tool, the user has to specify the details of each component in order to get the detailed behavior of the system (e.g., HYBRID 2). In open architecture research tool, the user is allowed to modify the algorithms and interactions of the individual components (e.g., TRNSYS). Klise and Stein [26] described various PV performance models, hybrid system performance models, and battery storage models in a Sandia National Laboratory report. Arribas et al. [27] carried out a survey of ten existing software tools based on the availability, features, and applications and presented guidelines and recommendations in an International Energy Agency (IEA) report. This report also categorized tools into four categories, namely, dimensioning, simulation, research, and mini-grid design tools. Connolly et al. [28] surveyed 37 computer tools for analyzing integration of energy systems. The survey also includes three hybrid simulation tools: HOMER, RETScreen, and TRNSYS. The survey was carried out in collaboration with tool developers, which included five components, namely, background information, users, tool properties, application, case studies, and further information. This study provides information for analyzing the integration of renewable energy into different objective-based energy systems.

Ibrahim et al. [29] briefly described the design and simulation models of the hybrid systems working with wind-diesel hybrid systems for remote area electrification. Bernal-Agustín and Dufo-López [4] have revised the simulation and optimization techniques, as well as the existing tools used for stand-alone hybrid system design. Zhou et al. [30] have described HOMER, HYBRID 2, HOGA, and HYBRIDS, which are used for evaluating performance of the hybrid solar–wind systems. Erdinc and Uzunoglu [31] discussed commercially available software tools for hybrid system sizing simulation. Brief description of each of the 20 software tools is given below. The description closely follows the excellent reviews by Sinha and Chandel [1] and Ganesan et al. [23].

7.2.1 HOMER

The Hybrid Optimization Model for Electric Renewables (HOMER) is most widely used, freely available, and user-friendly software. The software is suitable for carrying out quick pre-feasibility, optimization, and sensitivity analysis in several possible system configurations. The National Renewable Energy Laboratory (NREL), USA, has developed HOMER for both on-grid and off-grid systems in 1993, and from the date of release, HOMER has been downloaded by over 80,000 people in 193 countries [1, 23]. HOMER uses windows as computer platform with visual C++ as programming language. HDKR (Hay, Davies, Klucher, and Reindl) anisotropic model for the solar photovoltaic system is used by this software. It is able to optimize hybrid systems consisting of a photovoltaic generator, batteries, wind turbines, hydraulic turbines, AC generators, fuel cells, electrolyzers, hydrogen tanks, AC-DC bidirectional converters, and boilers. The loads can be AC, DC, and/or hydrogen loads, as well as thermal loads. The simulation is carried out in 1-hour intervals, during which all of the parameters (load, input and output power from the components, etc.) remain constant. The control strategies are developed by NREL. It can be downloaded and used free of charge. HOMER uses inputs like various technology options, component costs, resource availability, and manufacturers data to simulate different system configurations and generates results as a list of feasible configurations sorted by net present cost (NPC). This software can simulate a system for 8,760 hours in a year. HOMER also displays simulation results in a wide variety of tables and graphs which helps to compare configurations and evaluate them on their economic and technical merits. It can determine load serve policies with lowest cost source to meet the load. HOMER can suggest the design of various systems based on economic parameters. The tables and graphs made by HOMER simulation can also be exported. HOMER has been used extensively in literature for hybrid renewable energy system optimization and various case studies. The last updated version of HOMER is 2.81 (Nov 2010) [1, 23]. The major limitations of HOMER are as follows: HOMER allows only single objective function for minimizing the net present cost (NPC), as such the multi-objective problems cannot be formulated. After optimization process, HOMER makes chart for the optimized system configurations based on NPC and does not rank the hybrid systems as per levelized cost of energy. HOMER does not consider depth of discharge (DoD) of battery bank, which plays an important role in the optimization of hybrid system, as both life and size of battery bank decrease with the increase in DoD. Therefore, the DoD should either be optimized or be included in sensitivity inputs of the HOMER. HOMER does not consider intra-hour variability. HOMER does not consider variations in bus voltage. Including flexibility in selecting an optimization technique relevant to a particular study in HOMER will enhance its robustness and will facilitate the comparative study of results using different techniques [16]. The inputs and outputs for HOMER software are illustrated in Table 7.1 [1, 23].

7.2.2 HYBRID 2

HYBRID 2 was developed by Renewable Energy Research Laboratory (RERL) of the University of Massachusetts, USA, with the support from National Renewable Energy Laboratory [14]. After HYBRID 1 in 1994, HYBRID 2 was developed in 1996 and now the most recent version is 1.3b, which can be downloaded and installed with a password on Windows XP version. Some changes have been made in the latest version of HYBRID 2, and problems like curve fitting function on the insolation data

TABLE 7.1
Inputs and Outputs of HOMER Software

Inputs: Load demand, resources, components details with cost, constraints, system control, and emission data

Outputs: Optimal sizing, net present cost, cost of energy, capital cost, capacity shortage, excess energy generation, renewable energy fraction, and fuel consumption

graph and overflow error with low-load simulation have been fixed. This software is programmed in Microsoft Visual BASIC and uses a Microsoft Access Database. HYBRID 2 is a probabilistic/time series computer model and uses statistical methods to account for inter time step variations and can perform detailed long-term performance and economic analysis and predict the performance of various hybrid systems. HYBRID 2 has a provision of time series simulations for time steps typically between 10 minutes and 1 hour. HYBRID 2 allows systems based on three buses containing wind turbines, PV array, diesel, battery storage, power converters, and a dump load. HYBRID 2 mainly contains four parts, namely, the Graphical User Interface (GUI), the Simulation Module, the Economics Module, and the Graphical Results Interface (GRI). Using GUI, the user can construct projects easily and maintain an organized structure. The Simulation and Economics Modules allow the user to run simulations and input error checking. Users can view detailed graphical output data through GRI. This software tool has limited access to parameters and lack of flexibility but it has a library with various resource data files. A password is needed to install the demo version of HYBRID 2 [1, 14]. It is a hybrid system simulation software. Other components, such as, for example, fuel cells or electrolyzers, can be modeled in the software. The simulation is very precise, as it can define time intervals from 10 minutes to 1 hour. The possibilities with regard to control strategies are very high. NREL recommends optimizing the system with HOMER, and then, once the optimum system is obtained, improving the design using HYBRID 2. It can be downloaded and used free of charge. The inputs and outputs for HYBRID 2 are illustrated in Table 7.2 [1, 23].

7.2.3 RETScreen

RETScreen is a feasibility study tool and is freely downloadable software developed by Ministry of Natural Resources, Canada, for evaluating both financial and environmental costs and benefits of different renewable energy technologies for any location in the world. This software uses visual basic and C language as working platform. RETScreen was released in 1998 for on-grid applications. RETScreen PV model also covers off-grid PV applications and include stand-alone, hybrid, and water pumping systems also. It has a global climate data database of more than 6,000 ground stations (month-wise solar irradiation and temperature data for the year), energy resource maps (i.e., wind maps), hydrology data, product data like solar photovoltaic panel details, and wind turbine power curves. It also provides link to NASA's climate database. The program is accessible in more than 30 languages and has two separate versions: RETScreen 4 and RETScreen Plus. RETScreen 4, is a

TABLE 7.2
Inputs and Outputs for HYBRID 2 [1, 23]

Inputs: Load demand, resources, power system component details, and financial data
Outputs: Technical analysis, sizing optimization, and financial evaluation

Microsoft Excel-based energy project analysis software tool which can determine the technical and financial viability of renewable energy, energy efficiency, and cogeneration projects. There are a number of worksheets for performing detailed project analysis including energy modeling, cost analysis, emission analysis, financial analysis and sensitivity, and risk analyses sheets. RETScreen is used for the analysis of different types of energy efficient and renewable technologies (RETS), covering mainly energy production, life cycle costs, and greenhouse gas emission reduction [1, 23].

RETScreen Plus is a Windows-based energy management software tool to study energy performance. This program requires Microsoft Excel 2000, Microsoft Windows 2000, and Microsoft.NET Framework 2.0 or higher versions, and it is also possible to work on Apple Macintosh computers using Virtual Box for Mac. The main limitations of RETScreen are:

1. Does not take into account the effect of temperature for PV performance analysis
2. No option for time series data file import
3. Limited options for search, retrieval, and visualization features
4. Data sharing problem
5. Does not support more advanced calculations

The inputs and outputs for RETScreen are illustrated in Table 7.3 [1, 23].

7.2.4 iHOGA

Improved Hybrid Optimization by Genetic Algorithm (iHOGA), formerly known as HOGA (Hybrid Optimization by Genetic Algorithm) is a C++-based hybrid system optimization software tool developed by the University of Zaragoza, Spain. iHOGA is used for optimum sizing of hybrid energy system which may include photovoltaic system, wind turbines, hydroelectric turbines, fuel cells, H_2 tanks, electrolyzers, storage systems, and fossil fuel-based generating systems with multi- or mono-objective optimization using genetic algorithm and sensitivity analysis with a low computational time. iHOGA can optimize the slope of the PV panels, calculates life cycle emissions, and allows probability analysis, and has purchase and selling energy options to the electrical grid with net metering system. The new iHOGA version is upgraded and includes degradation effects, sensitivity analysis, new constraints, database of various components, and currency changing facility. It has two versions

TABLE 7.3
Inputs and Outputs for RETScreen [1, 23]

Inputs: Climate database, project database, product database, and hydrology database
Outputs: Technical, financial and environmental analysis, sensitivity and risk analysis, energy efficiency, and cogeneration

namely full professional version (PRO) and educational version (EDU). PRO+ is a priced version which can be used without any limitation with all features and full technical support, whereas EDU version, which is free, can be used for training or educational purposes only and is not permitted in projects, engineering work, installation work, and for any work involving economic transactions. The limitations of the EDU version are as follows [1, 23]:

1. It is only a representation of iHOGA.
2. It can only simulate within a total average daily load of 10 kWh.
3. Sensitivity analysis is not included.
4. Probability analysis is not included.
5. Net metering is not included.

The latest version of iHOGA is 2.2 (November 2013) and can run only in Windows XP, Vista, 7, or 8. All versions of iHOGA need an internet connection to get the validity of license; otherwise it will not run. The loads can be AC, DC, and/or hydrogen loads. The simulation is carried out using 1-hour intervals, during which all of the parameters remained constant. The control strategies are optimized using genetic algorithms. It can be downloaded and used free of charge. The inputs and outputs for iHOGA are illustrated in Table 7.4 [1, 23].

7.2.5 INSEL

A general purpose graphical modeling language INSEL (Integrated Simulation Environment Language) was developed by University of Oldenburg, Germany, which allows the users to make a structure with the help of its library with a specified execution time. This simulation software has the flexibility of creating system models and configurations for planning and monitoring of electrical and thermal energy systems. This software has its own database of meteorological parameters of almost 2,000 locations worldwide, photovoltaic systems, thermal systems, and other devices hourly irradiance, temperature, humidity, and wind speed data can be generated by using this software from monthly mean values for any given location and orientation. Solar thermal systems also can simulate using INSEL. It supports the users with datasets for PV modules, thermal collectors, and meteorological parameters, which is fully compatible with MATLAB and Simulink. The software is under continuous improvement during the past two decades. The user selects blocks from its library and connects them in order to define the structure of the system. The system

TABLE 7.4
Inputs and Outputs of iHOGA Software [1]

Inputs: Constraints, resource data, components data, and economic data
Outputs: Multi-objective optimization, life cycle emission, probability analysis, and buy-sell energy supply analysis

operation analysis can be carried out with a time frame specified by the user. The flexibility to create system models and configurations is a very interesting feature. It is a simulation, but not an optimization program. It is not free of charge [1, 23].

7.2.6 TRNSYS

In 1975, the University of Wisconsin and the University of Colorado (USA) jointly developed energy system simulation software named Transient Energy System Simulation Program (TRNSYS). TRNSYS was initially developed for thermal systems simulation, but with a span of more than 35 years, this software has upgraded and changed its features. It has now included photovoltaic, thermal solar, and other systems and has become a hybrid simulator. This simulation program is developed for modeling of thermal energy flows based on FORTRAN code. This is extremely flexible graphically based software used to simulate transient system behavior with two parts: one is kernel and another is library. Kernel processes the input file and solves the system with various techniques and determines convergence whereas, the second part, library, includes various models which can also be modified by the user. TRNSYS does not provide optimization facilities but it carries out simulation with great precision with graphics and other details. TRNSYS is used in solar systems (solar thermal and photovoltaic systems), low-energy buildings and HVAC systems, renewable energy systems, cogeneration, fuel cells, etc. It allows the user to program in FORTRAN code, which does not provide optimization of energy sources, but it can be used for carrying out the simulation part in designing the renewable energy systems TRNSYS 17.0 was released in 2010. The latest version of TRNSYS is 17.1 released in June 2012 and is priced. The standard TRNSYS library includes many of the components commonly found in thermal and electrical renewable energy systems. The simulation is carried out with great precision, allowing the viewing of graphics with great detail and precision. However, it does not allow the carrying out of optimizations. It is not free of charge. The inputs and outputs of this software are illustrated in Table 7.5 [1, 23].

7.2.7 iGRHYSO

iGRHYSO (improved Grid-connected Renewable HYbrid Systems Optimization) is the improved version of the GRHYSO, which is developed in Cþþ for grid-connected hybrid renewable energy systems optimization. This software is available only in Spanish. iGRHYSO simulates and optimizes various renewable energy systems like photovoltaic, wind, and small hydro, with storage batteries using different

TABLE 7.5
Inputs and Outputs of TRNSYS software [1, 23]

Inputs: Metrological data and models from own library
Outputs: Dynamic simulation and behavior of thermal and electrical system

technologies or hydrogen. It can simulate and perform the analysis to find the net present cost at low value. The NASA website [32] is connected with this software which is helpful for importing irradiation, wind, and temperature data. The effects of temperature on photovoltaic generation and production by wind turbines can also be studied by using this software. This tool also considers different types of sales/purchase of electricity from grid. The IRR (Internal Rate of Return) can be calculated. This software can export simulation data in excel spreadsheet format [1, 23].

7.2.8 HYBRIDS

HYBRIDS is a Microsoft Excel spreadsheet-based and commercially available renewable energy system assessment application and design tool produced by Solaris Homes. It requires daily average load and environmental data estimated for each month of the year. It can only simulate one configuration at a time and is not designed to provide an optimized configuration. The user can improve design skills about renewable energy system using HYBRIDS. HYBRIDS is a simulation and an economic evaluation program for PV-wind-diesel-battery systems. It uses 1-hour intervals in the simulation and calculates the NPC. This software is no longer available [1, 23].

7.2.9 RAPSIM

In 1996, the University Energy Research Institute (MUERI), Australia, developed Remote Area Power SIMulator, or RAPSIM, which is a Windows-based software package for a hybrid system model. The software can simulate the performance of a range of hybrid power systems comprising of PV arrays, wind turbines, and diesel generators with battery storage. Solar radiation, wind speed, ambient temperature, system load, etc., are the main inputs required by RAPSIM. In 1997, version 2 of this software was available but whether updates after 1997 have been made to the software is not clear. This software is used to select hybrid PV-wind-diesel-battery systems. The total costs throughout the lifespan are calculated. The user can modify the components in order to see the effect on the total cost. It is basically a simulation software (although the cost of the system throughout its lifespan is obtained). It is not free of charge [1, 23].

7.2.10 SOMES

Simulation and Optimization Model for Renewable Energy Systems (SOMES) was developed in 1987 at Utrecht University, Netherlands. This model can simulate on an hourly basis with average electricity production from the renewable energy generators. The model can perform optimization task for searching the lowest electricity costs within defined constraints. This model uses inputs like weather data and load demand to get technical and economic performance of a particular system configuration. The energy system can be composed of renewable energy sources (PV arrays and wind turbines), a motor generator, a grid, battery storage, and several types of converters. It is not free of charge [1, 23].

7.2.11 SOLSTOR

Sandia National Laboratory (SNL) developed a model in the late 1970s and the early 1980s to carry out economics and optimization analysis for various hybrid systems known as SOLSTOR [33]. This model includes renewable energy components like PV arrays and wind turbines [1, 23]; storage batteries and other power conditioning options and utility grid or fuel burning generator can also be used as a backup electricity provider. It can minimize the life cycle cost of energy by choosing the optimum number of solar panels, optimum tilt angles, and optimal wind energy system components. It can be suited for both on-grid and off-grid applications, but it is not widely used by the researchers for simulating the energy systems. SOLSTOR can also find rates for electricity purchased from the grid, time-of-day (TOD) energy charges, as well as time-of-day peak demand changes and sell back to the grid of excess collected energy. The model can be run with both on-grid and off-grid conditions. But now this simulation model is no longer used [1, 23].

7.2.12 HySim

HySim is a hybrid energy simulation model developed by Sandia National Laboratory [26] in 1987, for analyzing remote rural off-grid hybrid system with PV, diesel generators, and battery storage combination with good system reliability. HySim carries out financial analysis including life cycle, fuel, levelized cost of energy, operation and maintenance costs, and cost comparisons between different configurations. HySim has not been used after 1996 [1, 23].

7.2.13 HybSim

HybSim, developed by Sandia National Laboratory, is a hybrid energy software for cost benefit analysis of remotely a located hybrid system comprising of fossil-fueled electrical generation source with renewable source [34]. This tool requires detailed load profile, battery characteristics, economic details of the whole system, and weather characteristics. HybSim can use data measured at 15 minutes' intervals. It requires a detailed load demand profile along with weather characteristics, solar radiation, wind velocities, etc. Cost comparison and performance comparison of various system component combinations can be evaluated through this software. HybSim version 1 (2005) is available and is undergoing further development [1, 23].

7.2.14 IPSYS

Integrated Power System tool known as IPSYS is a hybrid simulation modeling tool for remote systems with a component library and is able to make simulation of electricity generation through PV arrays, wind turbines, diesel generators, energy storage batteries hydro-reservoirs, fuel cells, as well as biogas reservoirs, biomass plants, natural gas, etc. [1, 23]. Cþþ language is used for this model, and no current graphical user interface option is available, but some scripts can be used to analyze the graphical output.

7.2.15 HySys

The Hybrid Power System Balance Analyzer, also known as HySys, is a hybrid simulation tool developed by wind technology group, Centro de Investigaciones Energeticas, Medioambientalesy Technologicas (CIEMAT) Institute in Spain for sizing and long-term analysis of off-grid hybrid systems mainly comprising PV arrays, wind turbines, and diesel generators, and it can operate within MATLAB. In 2003, version 1.0 of this software was developed but now it is currently being used internally by CIEMAT only [1, 23].

7.2.16 Dymola/Modelica

Dymola/Modelica is used by the Fraunhofer Institute for Solar Energy (ISE) in Germany for modeling hybrid systems along with PV, wind turbines, generators, fuel cells, and storage batteries with weather and insolation data inputs. It can evaluate life cycle costs and calculate levelized cost of energy, but now update status of this software is unknown [1, 23].

7.2.17 ARES

Autonomous Renewable Energy Systems (ARES) is a program developed at the Cardiff School of Engineering, University of Wales, UK, for simulation of PV-wind-battery systems. This software is able to calculate the system loss of load probability and system autonomy through the prediction of the storage battery voltage if input load and basic weather profile is given. It can calculate the LPSP (loss of power supply probability) based on the input data provided by the user. It employs a separate subroutine program for each of the sources considered. The software has two versions: ARES-I, and the modified version of ARES-I by Morgan et al. [35] is known as ARES-II. ARES-I, consisted of subroutine program in the following order: (1) weather statistics, (2) photovoltaic generation, (3) wind generation, (4) load calculation, (5) combined source and load current, (6) battery voltage subroutine, (7) controller action, and (8) presentation of results. ARES-II required load and basic weather profile inputs and calculates the system loss of load probability and system autonomy using storage battery voltage prediction. This software is not available now. ARES very precisely simulates PV-wind-battery systems [1, 23].

7.2.18 SOLSIM

Fachhochschule Konstanz (Germany) developed a technically sophisticated and flexible tool named SOLSIM [5, 36], for hybrid renewable simulation using photovoltaic panels, wind turbines, diesel generators, batteries, and bioenergy systems for electricity and heat generation. This software can carry out an economic analysis with very limited control options (e.g., photovoltaic panel tilt angles). SOLSIM allows a large amount of very specific data to be entered for simulation adjustment. The large amount of data created from each simulation can be displayed in either hourly, daily, weekly, or monthly intervals including graphic user interface as a feature which makes the program easy to learn and to use [37]. There is the possibility of including biogas and biomass generators to generate electricity and heat. It simulates the operation of the system and carries out the economic analysis. The control options are very limited, optimizing only the panel inclination angles. It performs economic analysis with limited control options and uses a large amount of data to perform the simulation of Integrated Renewable Energy System (IRES). Nowadays, SOLSIM is not widely used to perform the energy generating options. This software is no longer available [1, 23].

7.2.19 Hybrid Designer

Hybrid Designer was developed by the Energy and Development Research Centre (EDRC) of the University of Cape Town in South Africa and was funded by the South African Department of Minerals & Energy. This tool is mainly used for off-grid applications in Africa's weather condition. It is user-friendly software based on genetic algorithm which can evaluate different configurations with minimum life cycle cost. Hybrid Designer can simulate different sources such as photovoltaics, wind generator, battery, and an engine generator and can produce a complete solution with technical aspects and life cycle costs. It is used for simulating the renewable energy models in off-grid mode employing genetic algorithm concepts for minimizing the net present cost of a system [1, 23].

7.2.20 HYDROGEMS

HYDROGEMS [16] is not a program, but a series of libraries developed at the Institute for Energy Technology (IFE, Norway). The libraries are used by TRNSYS and by Engineering Equation Solver (EES) software. The libraries developed by HYDROGEMS model the following components: photovoltaic generators, wind turbines, diesel generators, polymeric and alkaline fuel cells, electrolyzers, hydrogen tanks, lead-acid batteries, and DC/AC converters. It is possible to carry out economic optimization, if it is used with the GenOpt software, using the lineal simplex optimization method. These libraries are free for TRNSYS users [1, 23].

7.2.21 EMPS

EMPS is (EFI's Multiarea Power market Simulator) a computer tool developed for forecasting and energy planning in electricity markets. It has been actually developed for simulating and optimizing the hydrothermal energies with hydropower. It also considers the transmission constraints and hydrological differences between two areas. Its main objective is to minimize the total expected cost of the whole systems considering all the constraints like fuel cost, cost of energy, and emissions. EMPS software can also be used for analyzing the overflow losses, calculating energy balances, forecasting electricity prices, scheduling of power, etc. [23].

7.2.22 EnergyPLAN

EnergyPLAN is a computer model designed for performing energy system analysis. It is a deterministic model which can optimize the operation of a given energy system based on the inputs and outputs defined by the users. It was developed and maintained by Sustainable Energy Planning Research Group at Alaborg University, Denmark, in the year 2000. It simulates the operation of national energy systems on an hourly basis including all the energy sectors. The main advantage of EnergyPLAN tool is that it aids to design and develop the 100% renewable energy systems [23].

The main developers of 21 hybrid simulation software (except HYDROGEMS) are shown in Table 7.6. The comparative features of these software tools based on technical and economic analysis capabilities and renewable energy generator options are shown in Table 7.7. Finally, a comparative analysis of freely accessible and most used software for hybrid system research is given in Table 7.8. All of these data are derived from excellent review papers by Sinha and Chandel [1] and Ganesan et al. [23].

TABLE 7.6
Main Developers of Various Software Tools for Hybrid Systems [1, 23]

Software	Developed by
HOMER	NREL, USA (1993)
HYBRID 2	University of Massachusetts, USA and NREL (HYBRID 1 in 1994, HYBRID 2 in 1996)
RETScreen	Developed by Ministry of Natural Resources, Canada in 1998
iHOGA	University of Zaragoza, Spain
INSEL	German University of Oldenburg (1986–1991)
TRNSYS	University of Wisconsin and University of Colorado (1975)
iGRHYSO	University of Zaragoza, Spain
HYBRIDS	Solaris Homes
RAPSIM	University Energy Research Institute Australia(1996)
SOMES	Utrecht University, Netherlands (1987)
SOLSTOR	SNL (late 1970s and early 1980s)
HySim	SNL (late 1980s)
HybSim	SNL
IPSYS	Riso national lab., Denmark
HYSYS	Wind technology group (CIEMAT), Spain
Dymola/Modelica	Fraunhofer Institute for Solar Energy, Germany
ARES	Cardiff School of Engineering, University of Wales, UK
SOLSIM	Fachhochschule Konstanz (Germany)
Hybrid Designer	Energy and Development Research Centre (EDRC), University of Cape Town, SA
EnergyPlan	Alaborg University, Denmark

TABLE 7.7
Analysis Capabilities of Various Software* [1, 23]

Tools	Economical Analysis	PV System	Wind System
HOMER	X	X	X
HYBRID 2	–	X	X
iHOGA	X	X	X
RETScreen	X	X	X
HYBRIDS	–	X	–
SOMES	X	X	X
RAPSIM	–	X	X
SOLSIM	X	X	X
ARES-I &II	–	X	X
HYSYS	–	X	X
INSEL	–	X	X
SOLSIM	X	X	X
HybSim	X	X	–
Dymola/Modelica	X	X	X
SOLSTOR	X	X	X
HySim	X	X	–
IPSYS	–	X	X
Hybrid Designer	X	X	X
TRNSYS	X	X	X
iGRHYSO	X	X	X
EnergyPlan	–	X	X

* Only HOMER and SOLSIM carry out bioenergy
* Only HOMER, iHOGA, IPSYS, and iGRHYSO carry out hydro energy
* Only HYBRID3, INSEL, and TRNSYS carry out thermal system
* Except SOLSTOR all software simulate storage device
* Except RETScreen, HYBRIDS, SOMES, Dymola/Modelica, and iGRHYSO all software simulate generator set
* Except Dymola/Modelica and Hybrid manager all software simulate technical analysis

7.3 RESEARCH STUDIES USING HYBRID SYSTEM SOFTWARE TOOLS

Theoretical assessments of HRES for different renewable source combinations and under different environments (grid connected or stand-alone) have been going on for the past several decades. Nehrir et al. [38] reported the evaluation of general performance of stand-alone hybrid PV/wind system using computer-modeling approach (MATLAB/Simulink). Lim [39] presented a method to design the optimal combination and unit sizing for wind-PV and tide hybrid system. Notton et al. [40] presented a mathematical model for sizing hybrid PV system on the basis of LOLP. The authors have highlighted that the optimal solution can be obtained if PV contributed to 75%

TABLE 7.8
The Advantages and Disadvantages of Freely Accessible Software [1, 23]:

HOMER
Advantages
User friendly and easy to understand
Provides efficient graphical representation of results
Hourly data handling capacity
Disadvantages
"Black Box" code used
First degree linear equations-based models used
Time series data in a form of daily average cannot be imported
RETScreen
Advantages
Strong product database and meteorological database from NASA
Only financial analysis is the main strength
Easy to use as it is Excel-based software
Disadvantages
No time series data import
Optionless data input
Limited options for search, retrieval, and visualization features
HYBRID 2
Advantages
User friendly and multiple electrical load options
Detailed dispatching option
Disadvantages
Does not work on Windows platforms later than Windows XP, some simulation errors shown although project is written successfully
HOGA
Advantages
Uses multi- or mono-objective optimization using genetic algorithm and sensitivity analysis
Requires low computational time. Purchase and selling energy options to the electrical grid with net metering system available
Disadvantages
Free EDU version has some limitations in analysis and internet connection is required to activate license

of the energy requirements. Elhadidy and Shaahid [41] analyzed hybrid system consisting of PV, wind, and diesel generator with battery backup in hybrid energy system. They studied the impact of variation of PV array area, number of wind generator, and battery storage capacity of HRES. Chedid et al. [42] proposed a decision support technique for the policy maker about the factors influencing the design of grid linked hybrid PV-wind power system. They used analytic hierarchy process (AHP) to quantify various parameters that lead to the confusion in planning hybrid system. Their study was based on political, social, technical, and economical issues.

About 15 years ago, El-Hefnawi [43], with the help of the FORTRAN programming language, proposed a novel mathematical modeling technique to estimate the minimum number of storage days and PV array area by taking into account the pre-operating time of the DG (diesel generator), as one of the main design variables of the HRESs. In the same year, Shrestha and Goel [44] proposed a novel approach to determine the most efficient configuration of the PV array sizes and battery in order to satisfy the demand load of the system. In 2007, Ashok [45] developed a novel hybrid power system by using different components to find out an optimized configuration of each of the components by employing a nonlinear optimization algorithm, which had constraints and limitations. The HRES configuration proposed, which integrated the PV panels with a DG system, was fueled by animal manure. In a research conducted in 2013 [46], a multi-objective optimization model for a grid-connected hybrid power system with configuration of PV panels, DG, and the storage system with the aim of minimizing the life cycle costs and decreasing the carbon dioxide emissions of the hybrid power systems was introduced. In a research performed in 2016 [47], a novel HRES model to enhance and optimize the customized power systems, using a MATLAB-based tool, entitled "Matlab/Sim Power System™," was introduced and presented. In the same year, Shafiullah [48] designed an integrated HRES system to speed up and ease the transmission of considerable power production. Chauhan and Saini [49] conducted a comprehensive research study by employing a techno-economic optimization approach for energy management of a stand-alone integrated renewable energy system for utilization in remote areas of India. Bordin et al. [50] presented a linear programming approach to perform analysis of the battery degradation and its optimization for the off-grid power systems with integration of solar power. The main contribution of this study was to develop a method to include battery degradation processes inside the optimization models.

There have also been many other research studies which aim to examine the sensitivity analysis, as well as the optimization of the HRESs configuration through employing new advanced methods. In such research works, the utilization of the clean sources of power has been outlined using diverse ranges of mathematical and optimization methods [1–8, 23]. Furthermore, there are also many other research works demonstrating the importance of renewable sources of energy [1–8, 23]. Ramazankhani et al. [51] investigated utilization of the geothermal energy source for producing hydrogen energy, which was considered as a unique research in this regard. Zarezade and Mostafaeipour [52] performed a study for identifying factors, which affected implementing solar energy in Yazd, Iran. Goudarzi and Mostafaeipour [53] investigated a passive system as a renewable configuration for buildings which was economically feasible and could save a considerable amount of investments for the households.

It is evident from the literature that researchers had diverse viewpoints toward optimizing the net present cost of the HRESs. Most common financial indicators typically included in the analysis process are the LCOE, equivalent annualized costs (EAC), and the NPV, which are described later in Section 7.5. Furthermore, it is worth pointing out that of all the objective functions—selecting an efficient configuration of the HRESs, estimating the minimum number of storage days, minimizing the life cycle costs, and reducing the carbon dioxide emissions—have been cited as the most frequent optimization goals (see Section 7.5).

The hybrid system analysis software tools outlined above have been used by a number of researchers worldwide. HOMER is found to be the most widely used tool in the research studies followed by RETScreen, HOGA, HYBRID 2, TRNSYS, and ARES.

7.3.1 HOMER-Based Studies

In this section, some hybrid system analysis studies using HOMER, for various locations, are discussed to highlight the usefulness of HOMER.

Sopian et al. [54] investigated the optimization calculations by both HOMER and genetic algorithm and found almost the same results and time taken to execute the program. Al-Karaghouli and Kazmerski [55] used HOMER, to estimate the system size and its life cycle cost of a total daily 31.6-kW load of a health clinic in southern Iraq and suggested most economic systems consist of 6-kW PV modules, 80 batteries (225 Ah and 6 V), and a 3-kW inverter. Authors also suggested that this type of system is good for remote rural electrification, as electricity produced from diesel generator is four times costlier than PV electricity and also prevents the release of CO_2, CO, NO_x, hydrocarbons, SO_2, and suspended particles. Rehman and Al-Hadhrami [56] carried out the technical and economic analysis for a small village of Saudi Arabia with a PV-diesel-battery hybrid system which provides cost reduction, air pollutant minimization to the atmosphere, reduction of diesel consumption, and maintenance of a continuous supply of power as compared to only diesel system. Fulzele and Dutt [57] used HOMER for optimum planning of a proposed hybrid system based on mathematical modeling of each component for a site located in Dudhagaon village in Maharastra, India, and showed that solar PV generator with battery and inverter is the most economical solution over PV-wind with battery. This paper also concluded that, although different renewable energy systems are technically suitable and available in market, all are not financially viable.

Nema et al. [58] studied PV–solar and wind hybrid energy system for GSM/CDMA type mobile base station in Bhopal, Central India, which is suitably modeled using HOMER software. Sureshkumar et al. [59] presented a real-time optimal cost analysis of hybrid system based on the load profile, solar radiation, and wind speed for a location named Mandapam in Tamil Nadu, India, and optimized the system based upon the total net present cost. Akella et al. [60] developed an Integrated Renewable Energy System (IRES) model consisting of mini-hydropower, solar photovoltaic, wind energy system, and biomass energy system and optimized using LINDO (Linear, Interactive, and Discrete Optimizer) software 6.10 version. The results were verified using TORA software version 1.00 and also comparative results were also shown between LINDO and HOMER software. Kumaravel and

Ashok [61] carried out the size optimization of a hybrid energy system for remote area electrification in Kakkavayal, Kerala, India, and analyzed the economic feasibility using solar PV/biomass/pico-hydel hybrid energy system and found that a biomass gasifier hybrid energy system is more suitable for rural areas of Western Ghats region of India. Afzal et al. [24] carried out the sizing, sensitivity, optimization, and greenhouse gas emission analyses of hybrid renewable energy systems for two locations, namely, Lakshadweep and Uttarpradesh, with same load demand and identified different configurations. Sinha and Chandel [62] carried out pre-feasibility study to assess the potential for solar–wind hybrid systems for Hamirpur in the Western Himalayan state of Himachal Pradesh, India, using 1 year time series solar and wind data. The analysis shows that Hamirpur has an excellent solar resource but low wind potential, and as such, solar PV-micro-wind-battery storage system will be suitable for residential and institutional buildings to this location.

Prasad [63] used HOMER software to determine the optimum hybrid configuration and the levelized cost of energy for Vadravadra site in Fiji Island. A pre-feasibility study of hybrid energy systems with hydrogen storage was done by Khan and Iqbal [9] with various energy options like wind, solar, diesel generator, battery storage systems, and electrolyzer tank for applications in Newfoundland. The results showed the suitability of a wind-diesel-battery system for stand-alone applications, as wind resources in Newfoundland have excellent potential. Zoulias and Lymberopoulos [10] studied the feasibility of replacement of fossil fuel-based generator with hydrogen technologies and found it technically feasible but not economically viable due to higher cost of hydrogen technologies. Himri et al. [64] presented an economic feasibility study of an existing grid-connected diesel power plant in Algeria, by adding a wind turbine to reduce the diesel consumption and environmental pollution, and concluded that the wind-diesel hybrid system becomes feasible at a wind speed of 5.48 m/s or more and a fuel price of $0.162/L or more.

Liu et al. [65] carried out the modeling and sensitivity analysis of wind-solar hybrid system for Yantai city, China, and obtained optimal results which helped in future microgrid planning. The technical and financial viability of a large-scale grid-connected hotel (over 100 beds) using real load data has been presented by Dalton et al. [66], and the analysis and viability on the basis of net present cost, renewable fraction, and payback time were carried out. The results indicate that wind energy system is a more economically viable technology for large-scale grid-connected operations in this case. Dursun et al. [67] showed how a PV/wind/diesel-battery system consisting of 4 wind turbines, a 125-kW diesel generator, 96 batteries, a 100-kW converter, and a 120-kW PV array can minimize the dependence on fossil fuel source to meet 124-kW load, which leads to decrease in total net present cost and cost of energy and the amount of pollutants emitted, and create a pollution free environment.

Al-Badi et al. [68] designed a model to assess wind and solar power cost per kWh using various sizes of wind turbines and PV panels for two sites in Oman and calculated energy costs for the two sites which are almost the same for three different sizes of a wind turbine. Mohamed and Khatib [69] proposed an optimization method based on iterative simulation of a hybrid PV/wind/diesel energy and battery system for supplying a building load demand at minimum cost and maximum availability. The hybrid system was also simulated using HOMER, but the proposed optimization method provided more accurate results as compared to HOMER results. Kusakana et al. [70] showed that a hybrid PV-microhydro system can meet the energy demand of an isolated area in South Africa, successfully, and then compared with other supply options such as grid extension and diesel generation.

Kenfack et al. [71] discussed sizing a microhydro-PV-hybrid system for rural electrification in Cameroon with different combinations and discussed various cost effects of different systems. Razak et al. [72] carried out the optimization and sensitivity analysis of proposed hybrid renewable energy system for the Pulau Perhentian Kecil, Terengganu, Malaysia, based on sizing and operational strategy to obtain the optimal configuration of hybrid renewable energy based on different combinations of generating system. Güler et al. [73] proposed four different scenarios of a hybrid system for meeting Turkish hotels electrical energy demand, where for insufficient renewable energy resources, electrical energy is purchased from grid, and for extra generation, electricity is sold to the grid. Shaahid and El-Amin [74] used HOMER for the optimization of a PV-diesel-battery system to supply a remote village in Saudi Arabia. Cotrell and Pratt [75] analyzed systems involving PV generators, wind turbines, AC generators, electrolyzers, hydrogen tanks, fuel cells, and batteries; there are also some cases that utilized hydrogen internal combustion engines. The analysis is carried out using HOMER. In this paper, the authors carry out an exhaustive study of the purchase costs of the various components. Specifically, a system to supply a town in Alaska is considered, using the PV-wind-battery optimum design. The electricity supply to a telecommunications station is also optimized, obtaining the PV-wind-battery-hydrogen optimum design (taking into account better cost, efficiency, and lifespan values of the electrolyzer and of the fuel cell than the current ones).

Baniasad et al. [76] presented a techno-economic analysis for stand-alone applications between the hybrid PV/wind/diesel/battery systems and the hybrid PV/wind/fuel cell system with a yearly load of 24.4 MWh in Kerman having a total area of 500 m^2 at different fuel price scenarios. The comparison between hybrid PV/wind/fuel cell and hybrid PV/wind/diesel systems show that even at high fuel price, the hybrid PV/wind/diesel system is economically better than the hybrid PV/wind/fuel cell system.

A feasibility study of photovoltaic–wind, biomass, and battery storage-based hybrid renewable energy system for a residential area in Australia by Liu et al. [77] shows that net present cost (NPC), cost of energy (COE), and emissions of the system are lower than a diesel-based system. Ashourian et al. [78] proposed ecofriendly, economical, highly sustainable and reliable green energy systems for electricity generation of island resorts in Malaysia, with a combination of solar and wind energy as intermittent renewable energy sources and a fuel cell and a battery storage energy system as a backup, and compared the optimal configuration green energy system with a diesel-based energy system in terms of net present cost, sensitivity analysis, and pollutant gas emission.

Wies et al. [79] presented a simulation work, using Simulink, of a real hybrid PV-diesel-battery system located in Alaska, comparing it with a system with only a diesel generator and another diesel-battery system to supply energy for the same load. Contaminating emissions were evaluated (CO_2, NO_x, and particles) for various cases, comparing the results with those obtained by means of HOMER [11] software. Additionally, the global efficiency of the system and its costs were determined. The results obtained indicate that the system with only a diesel generator had a lower installation cost, but higher operation and maintenance costs; additionally, it was less efficient and released more contaminating emissions than the PV-diesel-battery system.

Zoulias and Lymberopoulos [10] used HOMER for the study of the possible replacement of diesel generators and batteries with hydrogen energy storage in stand-alone power systems. Their conclusion is that this is technically viable although, given the low energy efficiency of the electricity-hydrogen-electricity conversion, the renewable sources would be over dimensioned. Economically, a PV-hydrogen system would be better when compared with a PV-diesel system, as long as a 50% reduction in the cost of electrolyzers and a 40% reduction in the cost of hydrogen tanks are made, and the target of 300 s/kW for fuel cells in stationary applications is achieved. Beccali et al. [80] used HOMER in order to compare the possible options for the supply of the electric and thermal loads of a small residential district in Palermo (Italy). The cost of energy (COE, the average cost per kWh of useful electricity) in the wind-hydrogen system is four times higher than in the use of grid electricity. In the case of the wind-hydrogen system, the COE is nine times higher than in the use of grid electricity.

7.3.2 Other Software-Based Studies

Apart from a large number of HOMER-based research studies, only a few research studies have been reported in the literature using other simulation software. The studies carried out using HOGA, HYBRID 2, ARIES, TRNSYS, and RETScreen are presented in this section.

Castañeda et al. [81] presented a comparative study of sizing methods for a stand-alone PV-wind-storage hybrid system using Simulink design optimization, HOMER, and HOGA. Hybrid system designed by the first method gives the most economic hybrid system, whereas by the second one, the total system cost is highest with a minimum capacity hydrogen tank. The third method gave similar results as the first method, with slightly higher expense, whereas the fourth method presented the smallest hydrogen tank capacity. Rajkumar et al. [82] used Adaptive Neuro-Fuzzy Inference System (AN-FIS) to model a solar–wind-battery hybrid stand-alone system and compared the results with HOMER and HOGA using real meteorological data for two locations, Miri and Kuching, in East Malaysia. It is found that AN-FIS gives similar results as HOMER, whereas HOGA predicts lower battery and PV size with large wind turbine component than HOMER and ANFIS. López et al. [83] applied Strength Pareto Evolutionary Algorithm (SPEA) to the multi-objective optimization of a stand-alone PV-wind-diesel (or gasoline) hybrid system with battery storage to minimize the levelized cost of energy (LCOE) and the equivalent CO_2 life cycle emissions (LCE). The authors have developed HOGA as design tool and used it for the optimization of hybrid systems located in two different locations Zaragoza and Jaca in Spain with different load profiles.

Mills and Al-Hallaj [84] designed and simulated 6.5-kWp solar array,12-kW wind turbine, 2-kW fuel cell, 8-kW electrolyzer, and 3-kW hydrogen gas compressor hybrid system using HYBRID 2 software to meet the varying 1-kW load needs in Chicago, USA. The study signifies the use of fuel cell as storage device to reduce battery bank and reduction in the size of renewable energy generators. The system performed well, although there was an over-dimensioning of the renewable sources (PV and wind), which generated more energy than that demanded by the load, given the low efficiency of the electricity-hydrogen-electricity energy conversion. It is also verified that the current costs of these systems make them less than profitable when compared with the hybrid systems with renewable sources with batteries and diesel generators.

Kalogirou [85] has modeled a photovoltaic–thermal (PV/T) battery hybrid system for climatic conditions of Nicosia, Cyprus, using TRNSYS. The TRNSYS is used to study the hourly, daily, and monthly performance of the system and for optimization of the water flow rate. In the study, the electrical efficiency of system along with life cycle analysis has been carried out. The mean annual efficiency of a standard solar photovoltaic system is found to increase from 2.8% to 7.7% for a solar PV–thermal hybrid system. The solar contribution of the system is found to be 49% with respect to thermal energy. The payback of this system is found to be 4.6 years. Thus, TRNSYS can be used for the analysis of thermal-based hybrid system effectively. Rockendorf et al. [86] constructed thermoelectric generator–solar thermal collector system and photovoltaic–solar thermal collector hybrid systems. The simulation of both systems using TRNSYS 14.1 is carried out. The results show that PV–thermal hybrid system is a promising technology, as the electric output of PV-hybrid collector is found to be higher than the thermoelectric collector.

RETScreen is also used for hybrid energy systems, but this software is more widely used in single source-based renewable power system financial studies. Liqun and Chunxia [87] proposed a remote off-grid PV-wind-battery-hybrid system for Dongwangsha, Shanghai, and carried out feasibility analysis including GHG emission, financial viability, and risk analysis using RETScreen. Optimized system is expected to generate 82.1% electricity by wind turbines and 16.2% from solar PV. The research studies indicate that HOMER is a widely used tool for fast analysis in identifying the optimum combination of hybrid systems based on the renewable resource availability at a particular location [1].

McGowan and Manwell [88] describe the latest advances in PV-wind-diesel-batteries hybrid systems, using data from hybrid systems in various locations in the world. Additionally, the simulation tools applied to these systems are described (mainly focusing on HYBRID 2). In a later paper [89], the designs for PV-wind-diesel-battery systems for various applications in South America are described.

7.3.3 Case Study Comparing HOMER and RETScreen

Not all software gives the same predictions. Sinha and Chandel [1] examined a case study which used two different hybrid systems, using freely downloadable software—RETScreen and HOMER. The two hybrid systems examined were solar PV-battery system and solar PV-wind battery system. The purpose of the case study was to demonstrate the difference in the use and predictions of performance of the two hybrid systems with different software tools.

For study purpose, a location was identified in western Himalayan terrain, Shimla (latitude 31.11 N and longitude 77.21 E), at 2,650 m above sea level. Meteorological data base of RETScreen software was used for simulation in HOMER software also. On the basis of the project location and local meteorological data, the scaled annual average value of solar radiation is 5.28 kWh/m^2/day. Maximum and minimum monthly average solar radiation values are observed during May and December with 7.21 kWh/m^2/day and 3.520 kWh/m^2/day, respectively. The scaled annual average value of wind speed for the location is 5.05 m/s. The highest value of monthly average wind speed is observed during the month of December with a maximum of 5.7 m/s and the lowest value is observed during August, with 3.8 m/s monthly average wind speed. The scaled annual average value of ambient temperature is 8.68 1°C. The highest value of temperature is observed during the month of June, with a maximum of 18.8 1°C, and minimum in the month of December, with 0.2 1°C. A small residential load of an annual average of 3.4 kWh/day and a peak load of 0.489 kW are taken in both the studies. Same monthly average load demand is estimated for both studies. It is assumed that the residential load can be delivered by either the PV-battery system or the PV-wind-battery system [1].

For PV-battery system, HOMER showed a PV production of 3,504 kWh/year with an excess electricity generation of 1,819 kWh/year, whereas RETScreen showed 2,600 kWh/year electricity delivered to load. It has no option to calculate excess electricity generated. HOMER showed more PV energy production than RETScreen. HOMER and RETScreen have some similarities, like both take only global irradiation as input and synthesize the diffuse irradiation internally [1]. RETScreen uses Microsoft Excel to perform analysis based upon statistical monthly averages with lots of meteorological and geographical inbuilt information. RETScreen uses Evans electrical model with month-averaged ambient temperature and the PV panel material characteristics data to calculate power output, whereas HOMER uses basic relation model.

There are other differences between HOMER and RETScreen. HOMER has time series data import option while RETScreen has no option for time series data. HOMER calculates net present cost, cost of energy, operating cost, and initial cost while RETScreen carries out a detailed cost, financial risk, and emission analysis and has strong data base. This is the major strength of RETScreen. HOMER includes temperature effect on solar PV system, while RETScreen does not include temperature effect. HOMER includes maximum annual capacity shortage and has capability for hourly basis data, while RETScreen has no option to estimate capacity shortage and evaluates performance based upon monthly averages. HOMER has more flexibility for data input and graphical representation, while RETScreen has its own database with no flexibility for data input and less graphical representation option. HOMER shows the computational time taken to simulate a study while RETScreen does not show the computational time taken as it is Excel-based software. Overall HOMER is better suited for a more advanced user and can handle a much denser simulation, which makes it one of the most widely used hybrid system optimization tool [1, 23].

For the case study of PV-wind turbine-battery system, a 3.4 kWh/day residential load is simulated in HOMER. Polycrystalline PV system of 1 kW, 1-kW wind turbine, and ten 12 V batteries with 1-kW inverter are used in the simulation study. The same power curve of 1-kW wind turbine is assumed for both studies. Also 0%, 5%, and 10% capacity shortage is introduced as sensitivity parameters. The project lifetime is taken as 25 years, system fixed capital cost is $400, system fixed operation and maintenance cost is $5, and annual real interest rate is taken as 2.01% [1]. The first optimized result showed that a system comprised of 1 kW of PV panel, 1-kW wind turbine with 1 kW of converter and 10 batteries gives a minimum cost of energy $0.818. The second optimized result showed a system with only 2 kW of a PV system, and the third optimized result shows a system with 4-kW wind turbine and maximum cost of energy $0.951.Results showed that only PV system generates 3,504-kWh/year electricity with 1,881 kWh/year of excess electricity, whereas only wind-based system generates 3,477 kWh/year of electricity with 2,131 kWh/year excess production. The PV-wind hybrid system generates less amount of excess electricity of 1143 kWh/year and PV and wind production of 1,752 and 869 kWh/year, respectively. This study also showed that PV-wind hybrid system runs with minimum cost of energy than PV-only or wind-only systems.

The PV-wind-battery system analysis was repeated with the sensitivity analysis. The same load was also simulated using 0%, 5%, and 10% of capacity shortage. First and second optimized results showed the same configuration of 1 kW of PV panel, 1-kW wind turbine with 1 kW of inverter and 10 batteries with the cost of energy $0.818 for the hybrid system with no capacity shortage. Third optimized results showed a hybrid system with 10% capacity shortage comprising of 1 kW of PV system, 5-kW wind turbine, and 1-kW inverter with minimum cost of energy $0.336. In case of 0% and 5% capacity shortage, system annual PV and wind production was 1,752 kWh/year and 869 kWh/year, respectively, and also an excess electricity generation of 1,143 kWh/year. With a 10% capacity shortage, yearly PV and wind productions are 1,752 kWh and 4,347 kWh, respectively [1]. Without a sensitivity analysis, HOMER takes only 16 seconds to simulate the PV-wind battery system with all the necessary inputs. For simulation with 0%, 5%, and 10% capacity shortage, the sensitive input takes 49 seconds. So this variation in time shows that computational time may vary according to the size of inputs and sensitive parameters. More inputs and more complicated simulation need more time than a simple simulation study [1].

Based on the case study described above and other reported information [1–8], following conclusions can be drawn regarding the use of software tool for simulation and optimization of hybrid renewable energy systems [1, 23]:

1. Among the 22 software tools, HOMER is found to be most widely used tool as it has maximum combination of renewable energy systems and performs optimization and sensitivity analysis which makes it easier and faster to evaluate the many possible system configurations.
2. The status of software tools HySim, HySys, SOMES, SOLSTOR, HYBRIDS, RAPSIM, ARES, IPSYS, and INSEL is not reported; so their present status is unknown. These software tools could be more useful for hybrid energy system research, if updates with some modifications like more renewable energy generator options and user flexibility are available in the near future.
3. Some improvements like incorporation of other energy sources, allowing user to modify or change control techniques with full flexibility, being more user friendly need to be done for these simulation tools which will be helpful in further research and hybrid system applications.
4. The performance of software tools for hybrid energy system design can be improved through the implementation of various control methods, load demand management, economic planning, inclusion of various renewable and non-renewable sources of energy with various storage systems, etc. which promises to reduce the total cost of the system with optimized planning.
5. The continuous upgradation of models and more user flexibility in the tools developed will be helpful for further research and stimulation of hybrid system application activities. Further follow-up studies for demonstrating the capabilities of the simulation tools for hybrid system analysis can be taken up.

7.4 OPTIMIZATION METHODS AND THEIR FREQUENCY OF USE FOR HRES

A hybrid power generation system might consist of renewable energy conversion system like wind turbine, PV array, hydro turbine, fuel cells, and other conventional generators like diesel generator, microturbine, and storage devices like battery. A hybrid power system might consist all or part of it. To accurately size the individual components of the system, simulation of the system under real operating condition like appropriate weather, insolation, wind speed, and loads is necessary. The components and subsystems of an HRES are interconnected to optimize the whole system. The design of a hybrid system will depend on several requirements like location, stand-alone or grid-tied, and DC or AC load. Usually, most of distributed hybrid systems are designed to supply power to houses or small community for basic electrical use like elementary lighting, radio and televisions, small domestic electrical appliances, and street lighting. While HRES consists strictly of renewable energies only, the main objective in case of HRES is to utilize maximum proportion of renewable energy; other factors include the financial investment, reliability, and durability. The first step in optimization of hybrid system performance is the modeling of individual components. Modeling process enables one to recognize and improve understanding of a situation, identify the problem, and support the decision-making. The details of modeling are reflected by its correct prediction of performance; however, it is too complex or extremely time-consuming to design a perfect model. A sufficiently appropriate model should be trade-off between complexity and accuracy. Performance of individual component is either modeled by deterministic or probabilistic approaches. General methodology for modeling and optimization of HRES like PV, wind, diesel generator, and battery is described below [1–8].

Optimization algorithms are ways of computing maximum or minimum of mathematical functions. Different objectives can be considered when optimizing a system's design. Maximizing the efficiency of the system and minimizing the cost of its production are examples of such objectives. Optimization methods and techniques can help to solve complex problems. When designing an HRES, we have to consider its components' performances. The main goal is to have a better performance with reduced costs. These goals can be achieved through optimal modeling of the system [90]. The three commonly used modeling and optimization techniques for hybrid systems are classical algorithms, metaheuristic methods, and hybrid of two or more optimization techniques.

A. Classical techniques

Classical optimization algorithms use differential calculus to find optimum solutions for differentiable and continuous functions. The classical methods have limited capabilities for applications whose objective functions are not differentiable and/or continuous. Several conventional optimization methods have been used for hybrid energy systems. Linear programming model (LPM), dynamic programming (DP), and nonlinear programming (NLP) are examples of classical algorithms widely in use for optimizing HRESs. Linear programming model (LPM) studies the cases in which the objective function is linear and the design variable space is specified using only linear equalities and inequalities. This model has been used in several studies for HRES optimization [5, 90]. These studies take advantage of the LPM capabilities to stochastically perform reliability and economic analysis. However, the energy delivery capability of the overall system is adversely affected by failure of any of the renewables to function properly [90]. Nonlinear programming (NLP) model studies the general cases in which the objective functions or the constraints or both contain nonlinear parts. This model has been used in some studies [5, 90]. The model enables solving complex problems with simple operations. However, the high number of iterations for numerical methods such as NLP increases the computational burden of the problem [90]. Dynamic programming (DP) studies the cases in which the optimization strategy is based on, splitting the problem into smaller subproblems. This method helps solve sequential or multistage problems in which the stages are related together. One advantage of DP is the ability to optimize each stage. Therefore, it can address the complexity of larger systems. However, the high number of recursive functions for DP makes the coding and implementation complex and confusing [5, 90]. Several examples and case studies on the use of this technique are described by Siddaiah and Saini [90] and Ghofrani and Hosseini [5].

B. Heuristic or metaheuristic techniques

Metaheuristic search techniques have been extensively used for optimizing complex systems such as HRESs due to their capabilities to give efficient, accurate, and optimal solutions. These algorithms are nature inspired as their developments are based on the behavior of nature. Examples of metaheuristic optimization in use for HRESs include genetic algorithm (GA), particle swarm optimization (PSO), simulated annealing (SA), and ant colony algorithm (ACA).

Genetic algorithm (GA) is an evolutionary population-based algorithm that includes several operations such as initialization, mutation, crossover, and selection to ensure finding an optimal solution to a given problem. Several studies used GA to optimize the design and operation of HRESs [5]. GA may result in local optima if it is not initialized or designed properly. Particle swarm optimization (PSO) simulates the social behavior of how a swarm moves to find food in a specific area. It is an iterative algorithm with the goal of finding a solution for a given objective function within a given space. Its application for optimizing HRESs has been investigated in several studies [5]. PSO is efficient in solving the scattering and optimization problems. However, it requires several modifications due to its complex and conflicted nature [90].

Simulated annealing (SA) is based on the metal annealing processing. A metal gets melted at a very high temperature and then it gets cooled down and finally gets frozen into a crystalline state with the minimum amount of energy. As a result, the metal develops larger crystal sizes with a minimum amount of defects in its metallic structure. SA has been used for hybrid system sizing in several studies [5]. Ant colony (AC) algorithm is based on the behavior of ants to use a specific pheromone to mark the path for other ants. More pheromones are left on the path as more ants follow the same path. On the other hand, if a path is not used, then the smell of the last pheromone will disappear. Ants are more attracted to the paths with the most pheromone smells and it usually leads them to places with most foods. By following this method, ants mark the shortest path toward food. AC simulates this behavior to find the most optimal solution for a given objective function. This algorithm has been used for size optimization for hybrid systems [5]. AC algorithms have high convergence speed but require long-term memory space [90].

C. Hybrid techniques

Combination of two or more optimization techniques can overcome limitations of the individual techniques mentioned above to provide more effective and reliable solutions for HRESs. This combination is referred to as hybrid techniques. Examples of such techniques are SA-tabu search; Monte Carlo simulation (MCS)-PSO; hybrid iterative/GA; MODO (multi-objective design optimization)/GA; artificial neural fuzzy interface system (ANFIS); artificial neural network (ANN)/GA/MCS; PSO/DE (differential evolution); evolutionary algorithms, and simulation optimization-MCS, which have been used in several studies for optimizing HRESs [5, 90]. Although hybrid techniques enhance the overall performance of the optimization, they may suffer from some limitations. Examples of such limitations are the partial optimism of the hybrid MCS-PSO method, suboptimal solutions of the hybrid iterative/GA, cost-sizing compromise of the hybrid methods, design complexity of the hybrid ANN/GA/MCS method, random adjusting of the inertia weight of the evolutionary algorithm, and coding complexity of the optimization-MCS [5, 90].It should be noted that different researchers may breakdown these categories somewhat differently. More details on some of these optimization techniques mentioned above are discussed below. A well designed simulation program permits to determine the optimum size of battery bank, PV array, wind turbine, hydro generation capacity, and other generation systems for an autonomous or grid integrated HRES for a given load and a desired LPSP based on various criteria. Some of the criteria are minimum cost of the system, minimum capacity of system and storage devices, maximum power generation, and minimum LPSP and minimum LOLP. These and other criteria for optimization are described in more detail later in this section and in section 7.5. As mentioned above, various optimization techniques such as graphical construction, probabilistic approach, iterative technique, artificial intelligence (AI), dynamic programming, linear programming, multi-objective, and others were used by researchers to optimize hybrid PV/wind energy system. These are described in some more details below [1–8, 23].

7.4.1 Specific Optimization Methods

As mentioned above, under the general classification of optimization methods, there are several specific simulation algorithm and techniques. These are briefly described below:

A. Graphical construction

Problem with two design variables can be solved by observing graphically how they change with respect to one another. All constraint functions are plotted in the same chart. By visual inspection of the feasible region, the optimized point on the graph can be identified after objective function contours are drawn. Markvart et al. [91] used a long time series of solar radiation where the optimal sizing was determined by a superposition of contributions from climatic cycles of low daily solar radiation. Ai et al. [92] presented a method for determining the optimum size of the hybrid PV/wind energy system. Performance of hybrid PV/wind energy system was determined on an hourly basis, by fixing the wind generators capacity. Annual LOLP with different capacity of PV array and battery bank were calculated and optimum configuration (cost and LPSP) was found by drawing a tangent to the trade-off curve. In this method, only two decision variables were considered in the optimization, i.e., either photovoltaic and battery or photovoltaic and wind turbine. Some significant factors such as the numbers of photovoltaic modules, photovoltaic area, photovoltaic slope angle, windswept area, and the wind turbine altitude were totally discounted.

Borowy et al. [93] developed a methodology for determination of the optimum size of the PV array and a battery storage (BS) for a stand-alone hybrid RES application. Solar irradiance and wind speed registered each hour and each day for 30 years successively are used to determine the average power generated by a wind turbine and a PV module for every hour. The optimum size configuration of batteries and PV modules is calculated based on the minimum cost of the power system, by considering electrical profile load demand and loses of devices. Markvart [91] proposed a procedure capable of generating the sizes of the PV array and wind turbine in a hybrid system by exploiting the measured values of solar and wind energy at a given location. He called this method "graphical construction method." The method allowed for determining the optimum configuration of two RES, which satisfies the energy consumption of a user throughout the year.

B. Probabilistic approach

In probability approach, randomness is present depending on the collected data; thus variable states are not described by unique values, but rather using one of the statistical tools. The optimum size of a hybrid PV/wind energy system can be calculated on an hourly basis or daily average power per month, the day of minimum PV power per month, and the day of minimum wind power per month. Two advantages of this method are that the cost and time of environmental and load data collection are minimum.

Probabilistic approaches for sizing of hybrid system study the effect of solar radiation and fluctuation in wind speed for system design. In this approach, appropriate models for resource generation and/or demand are developed and finally a risk model is created by a combination of these models. However, this optimization technique cannot characterize the dynamic changing performance of the integrated/hybrid system. Lujano-Rojas et al. [94] developed a hybrid algorithm (Monte Carlo and ANN) for considering the uncertainty related to a different parameter such as solar radiation, wind speed, fuel prices, and battery lifetime in the photovoltaic-wind hybrid system (PWHS). Tina and Gagliano [95] evaluated the influence of a tracking system on the probability density function of PV output through the first four moments (mean, variance, skewness, and kurtosis) in PV–wind-based hybrid system. Tina et al. [96] presented a probabilistic approach based on the convolution technique to assess the long-term performance of a considered system and developed analytical expressions to find power generated by convolution of PV-wind output power. Yang et al. [97] found that BS with an energy storage capacity of 3 days was appropriate for guaranteeing the desired LPSP of 1% in solar–wind-based hybrid system, and an LPSP of 0% can be attained with a BS of 5 days storage capacity.

C. Hourly average generation capacity method

In this method, the hourly average wind, insolation, and power demand are used for optimization of the system sizing. This calculation of hourly average generation capacity is based on the average annual monthly data of sun and the wind. The size of the photovoltaic and wind components is given by the following equations. The objective function (F_c) which is to be minimized is given by Equation 7.1.

$$F_c = C_c + C_m \tag{7.1}$$

where C_c is the capital cost and C_m is the annual maintenance cost. In order to have a good balance over time, the difference (ΔP) between power generated (P_{gen}) and power demand (P_{dem}) should be minimum.

$$\Delta P = P_{gen} - P_{dem} \tag{7.2}$$

D. Most unfavorable month method

In this method, the size of the PV and wind generators is calculated in the most unfavorable month. The unfavorable irradiation month and unfavorable wind speed month are determined based on the available data.

E. Deterministic approach

In deterministic approach, every set of variable states is uniquely determined by parameters in the model and by sets of previous states of these variables, and thus there is always a unique solution for given parameters, unlike probabilistic approach. Bhandari and Stadler [98] calculated the system size and cost for PV system installed in Nepal using this approach.

F. Iterative approach

Iterative approach is a mathematical procedure generally performed using a computer generated sequence of improved approximate solution for the optimization problem until a termination criterion is reached. As the number of optimization variables rises, the computation time increases exponentially when using this approach. Li et al. [99] used this approach to optimize PV-wind-battery HRES based on the minimization of life cycle cost. Performance evaluation of PWHS in an iterative approach is achieved using a recursive program that finishes when the optimum system design is reached. Generally, the entire optimization model takes into consideration the LPSP model for power reliability, and in addition, the levelized cost of energy (LCOE) and/or net present value (NPV) model for system cost correspondingly. Most of the searchers considered parameters such as the capacity of PV panels and rated the power of a wind system and BS capacity. For the desired reliability level, the optimum configuration is one that has the lowest LCE/NPV from all the possible sets of configurations. In this method, system cost is minimized additionally by linearly varying the values of the parameters or by linear programming techniques. Furthermore, the iterative approach cannot provide optimization for the PV area. PV module slope angle, wind turbine swept area, and wind turbine installation height as these parameters enormously influenced the system costs (LCE and/or NPV).

Many works in the literature are focused on sizing a PWHS using the iterative approach. Sanajaoba and Fernandez [32] proposed a model of hybrid solar–wind system optimization sizing by using iterative optimization method. In evaluating PWHS cost and reliability in relation to the electrical power of parameters like wind rated power, PV array or BS, considerations should be given to levelized energy cost (LEC) and loss of power supply probability (LPSP). Zhang et al. [100] proposed an algorithm for component sizing in PV-diesel generator based on the optimization of the power switch simulations, they minimized the cost of energy counting capital depreciation, fuel, maintenance, and emissions impact. Li et al. [101] presented an algorithm to decide the minimal system configuration based on energy balance. Yang et al. [102] optimized the sizing of PWHS using LPSP for system reliability assessment and the LCE for cost analysis and selected optimal configuration based on the undermost LCE.

G. Artificial intelligence

Artificial intelligence (AI) is the branch of computer science that studies and develops intelligent machines and software. It is often defined as "the study and design of intelligent agents," where an intelligent agent takes actions that maximize the chance of success. AI consists of branches such as artificial neural networks (ANN), genetic algorithms (GA), fuzzy logic (FL), and hybrid systems combining two or more of the above branches. The appropriate use of intelligent technologies leads to useful systems with improved performance or other characteristics that cannot be achieved through traditional methods.

Artificial intelligence approaches need no availability of weather data for sizing of integrated energy systems in remote sites. Numerous approaches are reported in the literature such as genetic algorithms (GA), particle swarm optimization technique (PSO), harmony search algorithm (HSA), simulated annealing (SA), ant colony algorithms (ACA), bacterial foraging algorithm (BFO), artificial bee colony algorithm (ABC), cuckoo search (CS), or a hybrid of such techniques. These algorithms can handle the nonlinear variation of system components of RES or intermittent nature of solar and wind energy sources. Sinha and Chandel [62] presented the trend of new generation artificial intelligence algorithms, mostly used during the past decade as these require less computation time and have better accuracy and good convergence in comparison to traditional methods. They suggested mixing of two or more algorithms to overcome the limitations of a single algorithm. Additionally, some other techniques are identified for follow-up research in the design of PV-wind hybrid systems. Paliwal et al. [103] optimized photovoltaic–wind-diesel-based integrated system to the fulfill techno-socioeconomic criterion and then developed the relationship between the size of storage units, the number of cycles, and replacements over the project lifetime. Askarzadeh [104] developed a novel discrete chaotic harmony search-based SA algorithm for optimal sizing of

PV-wind-battery-based integrated system and compared the obtained annual cost with discrete harmony search.

Merei et al. [37] carried out an analysis for an optimal solution in PV-wind-diesel-based hybrid system with the combination of three battery technologies (lead-acid, lithium ion, and vanadium redox flow battery). Kumar et al. [105] proposed a biogeography-based optimization (BBO) algorithm for PV-wind-based system that converges to a global optimum solution with relative computational simplicity. They also compared the results with other optimization algorithms such as GA, PSO, and HOMER. Arabali et al. [106] minimized cost and increased efficiency of an integrated PV-wind system with BS using a genetic algorithm (GA) and two-point estimate method. They also calculated the maximum capacity of the storage system and excess energy for different load shifting percentages. Kaviani et al. [107] found that yearly simulation with 1 hour time step offered high accuracy with approximate evaluations of reliability indices in solar–wind-fuel cell-based system and discussed the impact of component outages on the reliability and system cost. Hakimi et al. [108] used the excess power of wind-fuel cell-based system in electrolyzer, and the deficit power was fulfilled by fuel cell when the generation was not able to meet the load demand. They used particle swarm optimization for simulation. Using a genetic algorithm, Yang et al. [109] proposed power reliability model based on LPSP and an economic model based on annualized system cost. They also suggested that the solar–wind-battery-based integrated system with 3–5 days' BS was found to be appropriate for the required LPSP of 1% and 2% for the studied case.

H. Particle swarm optimization algorithm

PSO is a multi-agent parallel search optimization technique, which was presented by Eberhart and Kennedy [110]. PSO is an evolutionary technique inspired by the social behavior of bird flocking, fish schooling, and swarm theory [111–113]. The PSO idea relies on imposing various particles for finding the optimum solution. Each particle in the PSO algorithm represents a potential solution; these solutions are assessed by the optimization objective function to determine their fitness. In the next iteration, the solutions number doubles until it gets the optimum one. Imposing more particles in each iteration encourage coming to the optimum solution, furthermore, decreases the number of optimization iterations. In order to move to the optimum solution, particles move around in a multidimensional search space. The best experience for each particle is stored in the particle memory ($pbest_i$) and the best global obtained among all particles is called as a global best particle (gbest). During flight, the current position (x_i) and velocity (v_i) of each particle (i) is adapted according to its own experience and the experience of neighboring particles as described by the following equations:

$$v_i^{(g+1)} = \omega v_i^{(g)} + c_1 a_1 \left(pbest_i - x_i^{(g)}\right) + c_2 a_2 \left(gbest - x_i^{(g)}\right) \tag{7.3}$$

$$x_i^{(g+1)} = x_i^{(g)} + v_i^{(g+1)} \tag{7.4}$$

where, g is the counter of generations, and ω is the inertia weight factor in a range of (0.5, 1) and almost 1 encourages the global search [111–113], c_1 and c_2 are positive acceleration constants in a range of (40, 4), designated as self-confidence factor and swarm confidence factor, respectively [111–113], a_1 and a_2 are uniform randomly generated numbers in a range of (10, 1). Swarm size, number of particles, ω, c_1, and c_2 are the main parameters of the PSO algorithm, which are initialized by the users, based on the problem being optimized.

To execute the proposed management and optimization procedures of the HRES, a new proposed program-based PSO (NPPBPSO) has been developed. NPPBPSO has been written using MATLAB software in a flexible fashion that is not available in the recent market software such as HySys, HOMER, iHOGA, iGRHYSO, HYBRIDS, RAPSIM, SOMES, HySim, IPSYS, ARES, and SOLSIM [10, 11, 74, 75]. To run NPPBPSO, the following information must be accessible:

1. Initial values of PSO parameters, swarm size, the number of particles, ω, c_1 and c_2
2. The optimum design values; $LOLP\_HP_{index}$, $P_{LL_{sum\ index}}$, $E_{dummy_{min}}$ and $E_{dummy_{max}}$. (lower P. in line with others)
3. The geographic data of the sites under study and meteorological data of wind speed, solar radiation, and temperature at these sites
4. Specification of WT, PV modules, inverter, batteries, and diesel generator
5. The load power data, high-priority load (HPL), and low-priority load (LPL)
6. Technical and economic data of system components: T, r, and i

Parallel implementation of particle swarm optimization (PIPSO) can automatically distribute the evaluation of the fitness function and constraints among the ready-made processors or cores. PIPSO is likely to be faster and more time saving than the serial implementation of particle swarm optimization (SIPSO), when the fitness function is time-consuming to be computed or when there are many particles. Otherwise, the overhead of distributing the evaluation can cause PIPSO to be slower than SIPSO [111–113]. To use PIPSO, a license for parallel computing toolbox software and a parallel worker pool (parpool) must be available.

Particle swarm optimization technique is an optimization searching algorithm due to its advantages over the other techniques for reducing the levelized cost of energy (LCOE) with an acceptable range of the production, taking in consideration the losses between production and demand sides. The problem is defined and the objective function is introduced, taking into consideration the fitness values sensitivity in the particle swarm process. The best input source for PSO is a metaheuristic simulating the collective motion of animals like flocks of birds or shoals of fish. PSO was developed by Kennedy [111–113] based on the research of bird and fish movement behavior.

PSO has many advantages such as a very fast researching speed and simple calculation compared to another method. The problems of non-coordinate system present limitations of this optimization algorithm because it cannot work out. The use of the method in the PWHS is just starting and many research works have used it. Lee and

Chen [114] used an evolutionary PSO algorithm to solve the wind and photovoltaic capacity with the aim of maximizing the benefit-cost ratio. Kashefi Kaviani et al. [107] optimized a hybrid wind-photovoltaic–fuel cell generation system over its 20 years of operation with PSO, in order to minimize the annual cost of the hybrid system subject to reliable supply to meet load demand. Bansal et al. [115] used meta particle swarm optimization (MPSO) to solve the PV-wind-battery hybrid system optimization problem, by using this improved PSO technique, it can be avoided thus proving it as an effective technique. Sharafi et al. [116] studied PSO simulation-based approach to tackle the multi-objective optimization problem for a hybrid system consisting of a wind turbine, photovoltaic panels, diesel generator, batteries, fuel cell, electrolyzer, and hydrogen tank. Borhanazad et al. [117] used multi-objective particle swarm optimization (MOPSO) in order to obtain the best configuration of the PV-wind-diesel-battery-based hybrid system for three stations in Iran.

I. Genetic algorithm

GA is a dynamic search technique used in computing to find true or approximate solutions for the optimization and search problems. The GA is categorized as global search heuristics. The GA is a particular class of evolutionary algorithms that use techniques inspired by evolutionary biology such as inheritance, mutation, selection, and recombination.

A typical GA requires two things to be defined:

a. A genetic representation of the solution domain
b. A fitness function to evaluate the solution domain

GA might be useful in problem domains that have a complex fitness landscape that a traditional hill climbing algorithm might fail. Xu et al. [118] used GA with elitist strategy for optimally sizing a stand-alone hybrid PV/wind power system for a year (8,760 hours). Their main objectives were to minimize the total capital cost of the system with constrained LPSP. Genetic algorithm is a search process that mimics the process of natural selection and was developed by John Holland in the 1960–1970 period [37, 109, 119–123]. The GA generates solutions to optimization problems using techniques inspired by natural evolution such as inheritance, mutation, selection, and crossover. GA has several advantages: it can solve problems with multiple solutions, easy to understand and can easily be transferred to existing simulation and models, etc. It has some limitations like a tendency to converge toward local optima or even arbitrary points rather than the global optimum of the problem and cannot assure constant optimization response time. Many researchers have used the application of GA for the optimal design and operation of PV-wind-based hybrid energy systems.

Optimal sizing of stand-alone PV-wind systems is proposed by Koutroulis et al. [122] using GA to select the optimal number of units with minimum cost, subject to load demand fulfillment. In another study, Bourouni et al. [119] presented a GA-based optimal sizing of desalination systems by PV-wind generators as a power supply unit. Yang et al. [109, 120] used GA to optimize the configurations of hybrid solar–wind-battery bank system, where the decision variables are the number of PV modules, wind turbines, and batteries, the PV slope angle, and wind turbine tower height. This method is proposed for a hybrid system, which supplies power to a telecommunication relay station. Ould Bilal et al. [121] proposed an optimized sizing of a hybrid solar–wind-battery system through multi-objective genetic algorithm, satisfying two principal aims of annualized cost minimization and minimization of the loss of power supply probability (LPSP). Nafeh et al. [124] used GA to yield optimum PV-wind and battery ratings with minimum cost and power reliability. Merei et al. [37] used GA to optimize PV-wind-diesel-battery hybrid system with three different battery technologies. A controlled elitist genetic algorithm has been applied to perform a multi-objective design of PV-wind-battery hybrid system in order to find the best compromise between three objectives: life cycle cost (LCC), the system embodied energy (EE), and loss of power supply probability (LPSP). Moreover, a multi-objective genetic algorithm is used to study the techno-economic performance of the PV-wind hybrid energy system and optimized three, objectives, e.g., total system cost, autonomy level, and wasted energy rate, with the PV array peak power, the wind generator rated power, and the rated capacitor of the battery as decisive variables.

J. Simulated annealing (SA)

Simulated annealing, which mimics material annealing processing, was developed by Kirkpatrick [125]. It is a trajectory-based random search technique for global optimization. The main advantage of simulated annealing is its ability to avoid being trapped in local minima. Simulated annealing is a robust and versatile technique that can deal with highly nonlinear models, chaotic and noisy data, and many constraints. The main weakness of SA is that the quality of the outcome may be poor. Until now, little literature has been reported using SA in this field. Ekren and Ekren [126] used simulated annealing (SA) algorithm for optimizing the size of a PV-wind-battery hybrid energy system to minimize the total cost. The decision variables were PV size, wind turbine rotor swept the area, and the battery capacity; the authors found that SA algorithm gives a better result than the response surface methodology (RSM). According to Sutthibun and Bhasaputra [127], simulated annealing (SA) is applied to solve the multi-objective optimal placement of distributed generation (DG), the model that was used to identify the optimal location and size of the DRG (distributed renewable generation) to minimize the real power loss (PL) and production (Epg).

K. Harmony search algorithm (HSA)

Harmony search is a derivative-free, real-parameter optimization technique algorithm employed for the optimization, with several evolutionary metaheuristic optimization techniques. HSA is one of the most recent population-based optimization technique that may be adopted in various fields of engineering applications [128].

Maleki et al. [129] used four heuristic algorithms, namely, particle swarm optimization (PSO), tabu search (TS), simulated annealing (SA), and harmony search (HS) for optimum sizing of a cost-effective PV-wind-fuel cell and PV-wind-battery-based hybrid systems. Shiva et al. [130] presented a novel quasi-oppositional harmony search (HAS) (QOHS) algorithm in the context of automatic generation control (AGC) of the power system.

L. Ant colony algorithms (ACA)

Ant colony algorithms are initially proposed by Marco Dorigo in 1992 in his PhD thesis [131]. The algorithm was aiming to search for an optimal path in a graph. Daming Xu et al. [118] proposed a specific graph-based ant system to minimize the total capital cost, subject to the constraint of the LPSP for sizing of stand-alone hybrid wind/PV power systems.

M. Bacterial foraging algorithm (BFO)

Bacterial foraging algorithm is an optimization algorithm inspired by the group foraging behavior of bacteria such as *Escherichia coli* and *Myxococcus xanthus* [132]. It is the chemotaxis behavior of bacteria that will perceive chemical gradients in the environment and move toward or away from specific signals.

N. Artificial bee colony algorithm (ABC)

Artificial bee colony algorithm (ABC) is an optimization algorithm based on the intelligent foraging behavior of honey bee swarm, proposed by Karaboga and Basturk [133]. In ABC, the position of a food source represents a possible solution to the optimization problem and the nectar amount of a food source corresponds to the quality (fitness) of the associated solution. Nasiraghdam and Jadid [134] presented an original multi-objective artificial bee colony algorithm to solve the distribution system reconfiguration and hybrid PV-wind-fuel cell energy system sizing. This article also found total power loss, the total electrical energy cost, and the total emissions produced by hybrid energy system and grid minimization and the voltage stability index of distribution system maximization. To optimally size a hybrid energy system based on PV-wind-fuel cell for Rafsanjan, Iran, an efficient artificial bee swarm optimization (ABSO) algorithm is proposed by Maleki and Askarzadeh [129].

O. Cuckoo search (CS)

Cuckoo search is a new metaheuristic algorithm that solves optimization problems, which is based on the obligate brood parasitic behavior of some cuckoo species in combination with the Lévy flight behavior of some birds and fruit flies. The sizing optimization can be figured as a multi-objective problem with economic, technical, and environmental constraints. Nadjemi et al [135] proposed an updated state-of-the-art optimization techniques used for the sizing and energy management of

PWHS which is based on a new sizing approach of cuckoo search algorithm for grid application systems.

P. Multi-objective approach

It is found that the multi-objective approach covers two mutual parts. The first one consists of merging all the individual objective functions into a single composite and in the second approach consists of the determination of Pareto optimal solution. Many studies have been carried out for optimization of the hybrid using multi-objective design. M. Fadaee et al. [136] presented state-of-the-art applied multi-objective methods that use the evolutionary algorithms; after analyzing, they showed that there are a few studies about optimization techniques using the multi-objective approach and the most current applied methods are genetic algorithm and particle swarm optimization. Owing to the natural fluctuation of weather, systems should be analyzed using probabilistic approaches rather than deterministic ones. Malekpour et al. [123] developed a new algorithm to solve the control problem in the distribution system with high wind power penetration which is based on a multi-objective probabilistic approach. Malekpour and Niknam [137] employed a probabilistic load flow approach using point estimate method (PESM) to model the uncertainty in load demands and electrical power generation. Maheri [138] proposed in their work two algorithms for multi-objective optimization in wind, PV, diesel hybrid system. In one scenario, the most reliable systems are created under a cost constraint, and in the second scenario, the most cost-effective system was obtained under reliability constraint. Tant et al. [139] proposed a method based on multi-objective optimization which was used to optimize a PV-battery hybrid system and then applied the proposed system in the grid distribution. Moura et al. [140] optimized the combination of the RES such as wind, solar, and hydro, in order to maximize its supply to the peak load and at the same time minimizing the cost of combined intermittence for the electric load demand and system management. Bernal-Agustín et al. [4] used a Pareto Evolutionary Algorithm to design a hybrid system based on PV, wind, and diesel; for electrical energy generation, the hybrid system is designed for different load profile.

Q. Analytical method

In this approach, computational models to find the feasibility of the system characterize components of integrated energy systems. Therefore, system's performance can be evaluated for a set of feasible system configurations for a specific size of components. The best configuration of an integrated energy system is evaluated by comparing single or multiple performance indexes of the different configurations. Khatod et al. [141] considered the intermittence correlated with the solar irradiance and wind speed, Beta, and Weibull distributions to model solar radiation and wind speed. They found that the proposed method was computationally very efficient and required less time and much less amount of meteorological data than Monte Carlo simulation method. Kaldellis et al. [142] developed an optimum sizing method based on the criterion of minimum embodied energy; it found that the share of the battery component exceeds 27% of the system life cycle in electrical demand, showing the difference between grid-connected and stand-alone for a system-based PV-battery.

R. Hybrid models

It is imperative to bear in mind that most of the optimization problems have a multidimensional nature. Therefore, in order to sufficiently address an optimization problem, a suitable method of optimizing is the one which is able to solve a multi-objective problem on the basis of the heuristic methods such as the GA, PSO, neural network, and the tabu search. Additionally, the hybrid methods have the advantage of combining two or even three optimization methods together to give the best and the most effective results [143].

S. Software-based approach

One of the popular commercial software for designing and analyzing hybrid power system is HOMER of National Renewable Energy Laboratory (NREL), USA. Solar insolation, electrical load, hybrid generator technical details, costs, constraints, controls, and type of dispatch strategy are used as the input to the HOMER software. Hrayshat [144] carried out a detailed techno-economic analysis using HOMER software to design an optimal hybrid PV-diesel-battery system for a remote house in Jordan. As mentioned before, HOMER software has also been used with an hourly time step and environmental data as input. It gave optimization based on net present cost taking into consideration the constraints and the senility variables of the RES. It does not allow the user to select an appropriate component for a system.

7.4.2 Use of Optimization Methods for HRES

The above breakdown of optimization methods are sometimes denoted by somewhat different terminologies in the literature [1–8]. In practice, capacity optimization techniques for hybrid renewable power system simultaneously use optimization models classified by storage forms such as battery storage, hybrid battery and supercapacitor storage, thermal energy storage, and pumped hydro storage, and the optimization methods described above such as GA, PSO, ACO (Ant colony optimization), SA, CSA (crowd search algorithm), and MLP (mixed integer linear programming), and application software such as CPLEX, HOMER, and HOGA. The optimization procedure establishes proper interrelationships among these three to achieve final capacity optimization.

As shown above, over the past decade, there has been a significant growth and interest in development of tools and methods which are used to perform the optimization of the performance and efficiency of the HRESs by employing the latest computational models (i.e., commercial software tools and/or numerical approximations of system components). Such tools which are developed to evaluate the performance of the HRESs aim to aid the designer to analyze the integration of multiple renewable sources. Optimization methods fall into three classifications, which are

TABLE 7.9
Frequency of the Various Utilized Algorithms [3]

Algorithm/Software	Frequency of Use
HOGA	2
HOMER	25
SA	4
Iterative optimization	4
Linear programming techniques	11
ANN-based methods	1
ACO	2
HSS	1
BA	1
GA	12
BBO	1
PSO	10

the heuristic-based methods, the simulation and sampling techniques, and other optimization tools and software. Faccio et al. [3] examined the frequency of use of various optimization methods with the help of 98 sample case studies. Table 7.9 demonstrates the frequency of the common optimization methods and algorithms which have been used for the design of the HRESs out of total 98 sample case studies. Evidently, the HOMER is the most common tool with its contribution being 25. Other commonly used methods during the design and optimization of the HRESs are GA, PSO, MILP, SA, and iterative optimization [3]. As mentioned above, heuristic optimization is one way for the optimization process, which is established by trial and error, and its main purpose is to generate adequate solutions and responses to a complex problem in a reasonably practical time. Such methods typically fall into two classifications, heuristic and metaheuristic methods, though their difference is insignificant. Generally, heuristic is defined as "to find" or "to explore by trial and error."

Faccio et al. [3] also concluded that of all studies performed, the PSO, the GA, and the linear programming have been the most frequent methods. GA, an optimization method inspired by natural selection and derived from the biological evolution, aims to solve both constrained and unconstrained optimization problems. The GA repeatedly modifies a population of individual solutions. This method has been used in 12 research studies out of 98 research cases. PSO, developed in 1995, has many similarities to the GA method. PSO is a population-based stochastic optimization method, and it is inspired by the social behavior of bird flocking or fish schooling. The PSO algorithm has been considered for utilization in 10 research studies out of 98 research cases. The linear programming (LP) is defined as a method for achieving the best outcome, such as maximum profit or the least cost value in a mathematical modeling process, whose requirements are represented with linear relationships. Many LP models can be solved through the simplex method, which is similar to the

Gauss-Jordan algorithm based on the linear algebra. The model can simply explore the global optimum solutions. Nevertheless, every function using this method should be linear, which can occasionally produce unrealistic solutions. The LP method is most commonly utilized to find the upper limit of solutions. The study by Faccio et al. [3] also indicates that 11 research projects (out of 98 papers) have employed the linear optimization techniques to design and optimize the HRESs. By way of an example, a research which was performed by Malheiro et al. [141] addressed the optimum sizing and scheduling of an isolated HRES by employing an advanced optimization framework which included a mixed integer linear programming model to analyze the performance of the system over a time horizon of 1 year, considering hourly variations in both the availability of the renewable resources and the energy demand as well. The purpose of the study was to design the configuration of the HRESs incorporating the WT, the PV arrays, the battery bank, and the DG as a backup system to provide electricity demand for the application considered. The optimal configuration considering the levelized cost of energy (LCOE) over a lifetime of 20 years was performed.

Often in literature, three assessment methods are identified under the group of sampling and simulation techniques, which are Hammersley sequence sampling (HSS), simulated annealing (SA), and the Monte Carlo simulation technique. According to Faccio et al. [3], out of 98 research studies investigating the configuration of the HRESs, 4 research papers have implemented the SA method. However, the HSS model as well as the Monte Carlo simulation has only once been proposed for the application of the HRESs in the literature. The SA method, proposed in 1983, is an intriguing technique for optimizing the functions of different design variables. This heuristic method which also operates as a sampling technique aims to optimize the NP-complete problems for which an exponential number of steps is required to generate a relatively accurate approach. As a matter of fact, the term "NP" stands for "nondeterministic polynomial time." The "NP-complete" which is defined as the class of decision problems, contains the hardest problems in NP. This decision problem belongs to both the NP as well as the NP-hard complexity classes. Additionally, the NP-hard is defined as the property of a class of problems, which are informally, "at least as hard as the hardest problems in the NP." An example of the NP-hard could be observed in decision-making problems. Another example can be observed when finding the least cost cyclic route through all nodes of a weighted graph, which is commonly known as the traveling salesman problem. However, the example of the NP-complete is observed in the isomorphism problems. Other types of the NPs which are utilized are NP-easy, NP-equivalent, and NP-intermediate.

It is worthy to mention that in a comprehensive research conducted by Ekren O and Ekren B Y [126], an SA model has been developed to perform optimal sizing of a PV/wind system, incorporating an energy storage system. In their study, the objective function was the minimization of the total cost of the HRESs. The decision variables were the PV array size, WT rotor swept area, and the battery's capacity. The optimum results which were achieved by the SA method were then compared with the results of another study which used the response surface methodology (RSM) in its analysis. Consequently, it was suggested that the SA approach employed provides more accurate solutions than the RSM mathematical model. Afterward, a case study was then taken into account by evaluating the effectiveness of the employed model for a campus area in Turkey.

In a recent research performed in 2017 [50], the Hammersley sequence sampling (HSS) method was employed as a multi-objective optimization framework, with the aim of decreasing the cost to the least value, enhancing the energy production, and then inventing an advanced reliability method, to adequately design the sketch of a PV/Wind/FC system, which also incorporates the DG as a backup source in its configuration. An evaluation of the energy supply reliability outlined a relation among the energy demand and the energy produced by the designed HRES. Furthermore, during connecting the HRES with the electricity grid, the model proposed estimated the dynamic response of each power unit, taking into account the WT, the PV array, and the FC in the HRES configuration. In addition, the fluctuation of the electricity produced by each power unit and the overall energy generated by the designed HRES was then analyzed by employing an evolutionary multi-objective optimization model. Another sampling method which has been identified in the current survey analysis is the Monte Carlo simulation technique, which relies on the repeated random sampling to estimate the numerical results. This method was used to investigate the optimum design and scheduling of an HRES under uncertainty conditions [145]. Furthermore, a method was then developed and implemented to investigate the decisions on the equipment installation of the PV/wind/DG/battery HRES. Specifically, the purpose of the model was the detection of the optimum size and configuration of the HRES proposed as well as the energy storage systems in each power station in order to achieve the minimum total cost expected, while satisfying the power demand of each area.

Al-Falahi et al. [146] studied three capacity optimization methods: classical algorithm, modern technology, and software tools. It is pointed out that in solving complex optimization problems, modern technology based on artificial intelligence algorithm (AI) is more flexible and practical than classical algorithms, and hybrid algorithms can get more ideal optimization results than single algorithms. Modern algorithms of capacity optimization of hybrid renewable power system can be summarized to artificial neural network algorithm (ANN), swarm intelligence algorithm, and so on. Swarm intelligence algorithm includes genetic algorithm (GA), particle swarm optimization (PSO), ant colony algorithm (ACA), simulated annealing algorithm (SA), crowd search algorithm (CSA) and many other algorithms. Among them, the most common optimization algorithms are GA and PSO. GA has a mature analysis method in convergence, and the convergence speed can be estimated. PSO is a fast algorithm, but its shortcoming is that it is easy to fall into local optimization and its search accuracy is not high. Liu et al [147] used the improved GA to solve the multi-objective optimization problem. By improving the fitness function of GA, the slow convergence speed of the conventional GA is solved. Ding et al. [148] used GA to solve the model and obtained the optimal capacity of each power supply. Chen et al. [149] proposed a multi-objective function considering the lowest cost of the system and LPSP, which was solved by an improved multi-objective composite differential evolution (DE) algorithm. This algorithm introduced the concepts of congestion degree and Pareto sorting into the DE algorithm, and has the capability of multi-objective optimization. In addition, the chaotic PSO algorithm is applied to the optimization problem to solve the maximum power point tracking problem of photovoltaic arrays under local shadows. Yang et al. [102, 109, 120] improved the PSO algorithm by optimizing the asymmetric acceleration factor. The simulation results show that the method accelerates the convergence speed and optimizes the working state. However, there are still some problems such as easy to fall into local optimum and low search accuracy. Mao et al. [150] proposed a MILP problem for self-generation and load scheduling models considering carbon emissions and time-of-use tariffs in order to obtain the best economic and environmental benefits. The carbon trading cost and electricity cost is set as the optimization goal of the model. OrEkran and Ekran [126] put forward a response surface methodology for capacity optimization of WT-PV combined power generation system.

The ANNs, which have a performance such as pattern classification methods, are computing systems, inspired by the biological neural networks, which constitute animal brains. Such systems learn to progressively enhance the performance of the model and to perform the tasks by considering examples, generally without task-specific programming. In the ANNs, there are various criteria which are proposed to select the number of hidden neurons. Afterward, the evolved criterion will lead to the design of an intelligent ensemble neural network structure which can then be proposed to predict climatic conditions or other necessary characteristics for any renewable energy applications including the HRESs. The ANNs are increasingly used in various areas thanks to their capability of handling complex systems specificities. Examples of ANNs applications can be observed during the design of the MPPT (maximum power point tracking) controllers of the photovoltaics, as well as wind energy systems [151].

Other primary and common methods which have been extensively used during the design of the HRESs are heuristic-based algorithms such as the GA, PSO, and SA, which have the great advantage of being strong and fast for solving a variety of mathematical problems, which typically have a complicated structure. Their application usually covers a wide range of issues including computer science, chemical science, automated design, control engineering, mechanical engineering, and even biology. On the downside, their coding procedure is typically a complicated and time-consuming process to implement and run.

Extensive mathematical modeling methods have been derived from different research papers to perform modeling of the WTs, solar cells, DGs, battery strings, and other primary components of the hybrid power system. The purpose of each model has been to look for an accurate, close mathematical model which represents a powerful methodology to estimate the power generation as well as viability of the hybrid system considered. When an approximate or near closer mathematical model is available, it would be easier to test the system for suitability without spending money prior to the fabrication. From a mathematical modeling viewpoint, the next stage would be the development of the computer simulation models to evaluate the performance of the system for stability and output results under various input conditions. These computer simulations can be used for further analysis or to change various components or to redesign the system for better results before fabricating the final prototype models. The results of the current review have shown that in eight research cases, the mathematical modeling techniques with no specific optimization model has been performed to analyze the efficiency and optimization of the HRESs.

For instance, in a research carried out in 2011, an economic approach, according to the concept of the levelized unit electricity cost (LUEC), was adopted to discover the best indicator of the economic profitability of the system. Afterward, a grid-independent hybrid PV/wind system was simulated by running the program developed to study the relationships between system power reliability and system configurations. The optimal configurations of the hybrid system were determined in terms of different desired system reliability requirements and the LUEC parameter.

With regard to mathematical modeling techniques, in another research conducted in 2017, Bordin et al. [50] investigated the application of the linear programming methodology to perform a comprehensive analysis and review concerning battery degradation for the optimization application of an off-grid power system with solar energy integration. They studied a mathematical formulation for an off-grid hybrid system to optimally manage the battery integration. During the optimization process, the constraints were considered to be the load demand, the DG demand, the minimum production of the system, the maximum DG capacity, the converter efficiency, the renewable source capacity, the initial values of battery variables, the minimum battery charge level, the charge and discharge processes management, and the battery maximum charge level. The final results of this study emphasized the importance of the economic operation of storage capacities, which took into account the relationship between the degradation of a battery and the operating pattern to meet the electricity requirements of an off-grid site.

Hybrid PV-wind-battery system is the most examined system in the literature. This is extensively reviewed b Sawle et al. [152], Moghaddam et al. [153], Ma et al. [154], and Arribas et al. [155]. This system is analyzed with the help of GA, PSO, HSA (harmony search algorithm), ant colony algorithm (ACA), bacterial forging algorithm (BFO), artificial bee colony algorithm (ABC), Cuckoo search (CS), fuzzy logic, and artificial neural network, using an evolutionary approach and a hybrid model for multi-objective design where all individual objective functions are merged into a single composite and entered as Pareto optimal solution set to be determined. The obtained solution is said to be Pareto optimal. Hybrid wind-PV system is also analyzed using an iterative method, hill climbing, dynamic programming, linear programming, and multi-objective to determine different conditions of the wind-solar combination using LPSP. For the same system, an analytical method has also been used to find the feasibility of the system in which components of integrated energy systems are characterized by computational models. The efficiency of the hybrid system has also been analyzed using a probabilistic method using the statistical data gathering and battery and PV array was optimized using graphical construction using two factors [152–155].

The combined utilization of wind and solar as renewable energy sources is, therefore, becoming increasingly attractive and are being widely used as an alternative to oil-produced energy. Economic aspects of these renewable energy technologies are sufficiently promising to include them for rising power generation capability in developing countries. Research and development in renewable energy technologies have an obligation to continue, for improving their performance, establishing techniques for accurately predicting their output, and reliably integrating them with other conventional generating sources. Nema et al. [58] has provided an extensive review of the current and the future state-of-the-art development of hybrid energy system using wind and PV solar. Siddaiah et al. [90] presented a review on planning, configurations, modeling, and optimization techniques of PWHS for off-grid applications.

7.4.3 Application of Software Tools

As for application software used in the solution of capacity optimization of a hybrid renewable power system, Mao et al. [150] proposed a MILP problem solved by CPLEX. The results show that the model is more economical and environmental friendly than the conventional model and can effectively reduce the total cost of the system and carbon emissions. Behera and Pati [156] reviewed grid-connected distributed wind-photovoltaic energy management. As pointed out earlier, in many wind-PV optimization study, the HOMER software developed by National Renewable Energy Laboratory (NREL), USA, was used. As shown by the case study described earlier, with the help of HOMER software, the monthly average wind speed and solar radiation data can be discretized into hourly average data as data sources of wind and solar resources. Hamanah et al. [157] evaluated optimum sizing of hybrid PV, wind battery, and diesel system using HOMER software and lightning search algorithm. The study presented a new methodology to optimize the configuration of the hybrid energy system with the wind farm, photovoltaic array, diesel generator, and battery bank. Minimizing the annual cost is considered as an objective function with different constraints considering energy not served and renewable energy fraction. The lightning search algorithm was employed to obtain the best cost value of hybrid power system construction. Annual load data, solar irradiation, and output power of wind turbines were used in the optimization analysis with a 10-minute resolution. The behavior of different components of the system has been investigated. Simulation results gave the optimal decision variables: number PV panels, wind turbines, battery banks, and the capacity of the diesel generator with the electrification costs (capital cost and fuel cost) over 1 year for ten scenarios of the hybrid energy system. The results confirmed the potential of the proposed approach for the cheapest renewable energy generation. Anurag Chauhan [49] introduced many applications for an integrated renewable energy system. Among them, Hybrid Optimization by Genetic Algorithm (HOGA) software is used to output multi-objective optimization, energy supply analysis according to constraints resource, and component data. HOGA was developed by Electric Engineering Department of the University of Zaragoza (Spain) [1–3].

According to the analysis performed for each of the classification drivers of the HRESs, it was concluded that the HOMER, the Lingo, and the MATLAB Simulink design optimization (MSDO) were the most frequent software employed in the optimization modeling of the HRESs. Other optimization tools, which have also proved to be useful are model predictive control (MPC), the branch and bound (B&B), the generalized reduced gradient (GRG) techniques, and multi-criteria decision-making. Furthermore, the smart grid theory, which enables interaction between the generation and load to optimally deliver energy based on the operating conditions, has also been considered in studies lately. Although smart grid solves many of the contemporary problems, it gives rise to new control and optimization problems, especially with the growing role of renewable power sources [1–3].

Some of the methods utilized also cover a wider variety of processes. For instance, in a study conducted in 2015 by Wang et al. [158], the receding horizon optimization (RHO), which is considered as a classification of MPC has been proposed to minimize a certain cost function over a moving time horizon to estimate the optimum energy generation of those of power-generating equipment, which are under the group of larger systems. Basically, the MPC method is widely utilized in engineering and design of process systems for industrial applications. In addition, the current analysis has demonstrated that out of 98 studies, 12 researches are performed with the support of the HOMER tool, two researches perform configuration of the hybrid power systems using the Lingo modeling software, one study deals with the application of the MSDO software, and another research performs an optimization analysis on the performance efficiency of an HRES by employing the GRG-B&B method.

As it is concluded by the literature review, the HOMER software has considerable merits. On the top priority, it is user-friendly. Second, it is extensively utilized for the techno-economic as well as the environmental emission analysis. The energy sources investigated in the HOMER could cover a wide variety of the natural sources, not just renewable ones, such as the PV, the WT, hydro, and biomass, but also the conventional sources of power such as diesel fuel. The software is capable of performing the thermal systems analysis as well [1–8].

Optimization and simulation of the HRESs using the iHOGA tool has also been performed in two research cases, one in 2011 [82] and the other in 2014 [159]. The iHOGA is a strong optimization tool, which can aid the achievement of the least cost design of a system throughout the whole lifetime operation of the HRES. Similar to the HOMER software, the NPC is one of the primary objective functions considered during the analysis. Minimizing carbon dioxide emissions and the unmet load, which refers to the energy, that is not served, should also be taken into account during the optimization process and analysis to provide better insights concerning the optimization of the HRESs. It is noted that both HOMER and iHOGA have been programmed using the C++ programming language. In some rare occasions, chemical methods such as the pinch analysis method, which are typically utilized to minimize the energy consumption in the chemical processes, and thermodynamic cycles have also been considered as an optimization method to reduce the net present cost of the hybrid power system to a minimum level.

Some primary software tools which have also been utilized recently to perform analysis on the hybrid renewable energy systems include RAPSIM, TRNSYS, RETScreen, INSEL, PV sol, iGRHYSO, SOMES, SOLSTOR, HySim, HybSim, IPSYS, HySys, Dymola/Modelica, ARES, SOLSIM, and the Hybrid Designer [1]. As pointed out in Section 7.2, there are particular advantages and disadvantages for each of the tools or methods used in the analysis of the HRESs. For instance, the HOMER tool, which is a user-friendly and freeware software, makes it a simple way to analyze the sizing procedure, with efficient output diagrams, which could then be used and is helpful for a better analysis of the optimization results and outcomes. On the downside, the tool lacks the use of coding analysis and development of different optimization models. Such an issue could possibly lead to inflexibilities in the development of the more comprehensive and advanced hybrid energy models, which could then aid experts and designers to supply more extensive outcomes and results. This situation is also correct for many other tools or software, which are exclusively made for the design of the HRESs, such as, HYBRID 2, and iHOGA.

Most used software tools in the literature are HOMER, HYBRID 2, HOGA, HYDROGEMS+TRANSYS, and HYBRIDS. Except for HYBRIDS, they all can be downloaded and used for free. All of them are used to simulate PV, diesel, battery, wind, mini-hydro, fuel cell, elctrolyzers, hydrogen tank, and hydrogen load. Except for HOGA, they are all used for economic optimization. Thermal load could not be analyzed by HOGA and HYDROGEMS+TRANSYS. Multi-objective optimization and genetic algorithm can only be carried out by HOMER and HYDROGEMS+TRANSYS [1–8].

The above literature review divulges that HRES do not have natural competitive cost advantage against conventional fossil fuel power systems. However, the need for cleaner power and improvements in alternative energy technologies presents a good potential for prevalent use of such systems. HRES have demonstrated the potential to make adequate preparation and support in some of the basic infrastructure needs in remote and urban areas for different application. Sizing and optimization techniques must perform an effective search for an optimum combination of the critical parameters like system cost, system reliability, PV array size, the tilt angle of PV panels BS size, and WT size; it is recognized that over-sizing causes an increase of system costs and under sizing causes insufficient power supply [1–8].

The literature also indicates that none of the individual methods could perform better than all the other methods on all kinds of environmental situations. However, optimization approaches based on artificial intelligence algorithms are found to be more acceptable than traditional approaches because of their ability to search local and global optima, their good calculation accuracy, and faster convergence speed. Nature-inspired optimization techniques and hybrid optimization techniques will be important for further exploration in future research to face complexity and challenges of PV-wind hybrid systems. Among the software tools, HOMER is found to be most largely used since it has a maximum combination of renewable energy systems and performs optimization and sensitivity analysis, which makes it easier and faster to evaluate many possible system configurations [1–8].

7.5 SIMULATION AND OPTIMIZATION OBJECTIVES FOR HRESS

Various criteria are also considered for optimal design and component sizing of HRESs. These criteria can be broadly categorized as economic and technical.

Economic criteria are used to minimize costs of HRESs. Technical criteria include reliability, efficiency, and environmental objectives to supply the load demand of HRESs at desired reliability levels with maximum efficiency and minimum greenhouse gas emissions.

7.5.1 Categorizing Optimization Goals

In this section, we present the classification and contribution of the most prevalent optimization goals examined in the literature as reported by Faccio et al. [3]. Once again, the study of Faccio et al. [3] was undertaken with 98 sample case study. As shown in Table 7.10, the study showed that frequency of optimization goals in the design process showed considerable variations in the contributions of the optimization functions. Evidently, the cost factor has been the most prevalent design variable with its allocation being 84. Sizing and configuration has also contributed to a majority allocation of 45 of the performed research works in the literature review. The other common optimization goals are load demand parameters, energy production, and environmental emissions, with contributions of 31, 32, and 17, respectively. The less commonly used optimization goals are independent generation points (IGP), grid generation points (GGP), BGED, voltage stability index (VSI), and charge scheduling. Some of the notable studies for economic optimization are capital cost, annualized cost of system, and operation cost [3]. Other notable studies are for the reliability of LPSP and LOLP [3]. The technical-economic analysis for the system cost was also carried out in [3, 82]. Finally, yearly, monthly, and hourly average method for hybrid system was analyzed in [3, 159].

TABLE 7.10
Frequency of Various Objective Functions [3]

Optimization Objective	Frequency of Use (98 studies)
Charge scheduling	2
VSI	2
BGED	2
Battery cycle	8
Grid parameters	3
Sizing and configuration	45
Diesel fuel consumption	4
Power loss	3
Autonomy	4
Environment emissions	17
Energy production	32
Load demand parameters	30
Cost	84

A. Cost optimization

HRESs often times include higher capital costs and lower operation and maintenance (O&M) costs which require an optimization to determine the compromise solution between the costs and benefits. Cost optimization of hybrid renewable energy systems includes minimizing energy cost, net present cost (NPC), and any other costs associated with such systems. To address the techno-economic viability of a renewable energy project, researchers have had different viewpoints on the analysis of the cost. Based on a review performed for the classification of sizing, seven cost parameters have been taken into account as the most prevalent ones.

1. **Energy cost minimization**
   Several studies have investigated minimizing levelized cost of energy (LCOE) for HRESs. LCOE is the ratio of the total cost of the hybrid system to the annual energy supplied by the system.
2. **Net present cost minimization**
   Net present cost (NPC) of an HRES is defined as the total present value of the system that includes the initial cost of the system components as well as the replacement and maintenance cost within the project lifetime. The objective here is to minimize the NPC of HRESs.
3. **EAC (Equivalent annualized costs)**
   The EAC of a hybrid power system can be obtained by multiplying the total NPV of the system by the "loan repayment factor," $A_{t,r}$, where NPV is the net present value of the system and $A_{t,r}$ is considered to be the loan repayment factor [151], t is the number of years and r is the annual interest rate [3].
4. **LCC (Life cycle cost)**
   The LCC is used for determining the most feasible alternatives and solutions among different available classifications of the HRESs. The calculation method usually considered to perform the estimation of the LCC, where the total LCC is present-value dollars of a given alternative.
5. **LCOE (Levelized cost of electricity)**
   During the financial analysis of the HRESs, the "LCOE" is a very common parameter used in the economic evaluation of these systems. It is defined as the sum of all the net present values of the unit-cost of electricity over the lifetime of a generating asset. It is often taken as a substitute for the average price producing the asset, which should obtain in a market to breakeven over its lifetime.
6. **Net present value (NPV)**
   In the field of finance, the NPV is a method for obtaining the profitability of a financial project, which is calculated by subtracting the present values of cash outflows (including initial cost) from the present values of cash inflows over a period of time [3].
7. **Other cost-related optimization**
   Other cost-related optimizations include minimizing life cycle cost (LCC), levelized unit electricity cost (LUEC), annualized cost of the system (ACS), and capital cost (CC) of the hybrid.

B. **Sizing and configuration methods**

The sizing and configuration of the HRESs have always been considered as the primary stage of the design and optimization process. A research conducted in 2012 by Luna-Rubio et al. [143] categorized the sizing and configuration methods into four main classifications: probabilistic, analytical, iterative, and the hybrid model. Faccio et al. [3] have reviewed the classification of the sizing methods from 2003 until 2010 in order to analyze the latest optimized indicators and the design constraints, which affect the sizing optimization of the designed HRESs. In this review, grid type as well as the investigated technology has also been taken into consideration.

Probabilistic methods are considered as the easiest ways of configuring the size of the HRESs. However, the results obtained by using this method are not always considered as the most appropriate ones to estimate the best solution or configuration of the HRESs. Generally, one or two system performance indicators are enhanced to achieve the optimal result in order to perform the sizing of each component. As an example, a study performed in 2003 employed a novel probabilistic approach according to a convolution technique by using the probability density function (PDF), with the aim of specifying the probability of a random variable falling within a particular range of values in order to evaluate the long-term performance of hybrid PV-wind systems and then to perform the sizing procedure [97].

Analytical methods utilize the computational models, which describe the size of the hybrid system as a function of its techno-economic viability. Using such methods, the performance of the system can be evaluated for each of the HRES components. The analytical method has the merit of letting the designer know how to establish simulations of the system's performance and efficiency of the several HRES configurations. The drawback of this method is that it requires a longer span of time for the weather data in order to perform the simulation and optimization process. On this account, recently, there have been many computer tools to carry out the performance of the HRESs, which aid the designer to perform analysis for the integration of different renewable sources. Example of these tools are HOMER, HYBRID 2, and HOGA [45].

Iterative methods employ the recursive process, with the aim of ceasing the optimization process in an occasion that the most efficient configuration is achieved on the basis of the design characteristics. For instance, in a study conducted in 2007 [160], the iterative method was used to achieve the optimal configuration of the HRESs for a rural village by decreasing the LCC to a minimum level and maximizing the reliability of the systems. For this purpose, a numerical algorithm based on the Quasi-Newton method (which are a series of methodologies employed to either find zeroes or local maximum and minimum of functions) was used to solve the optimization problem. Generally, the procedure for performing the sizing methods using the iterative optimization would fall into four main classifications: selecting the models and the numbers of each of the HRES components, which are commercially available; considering a constant number for each of the HRES components, and then increasing the number of the other components to achieve an optimal energy balance; performing the second stage again for the variety of WTs or PV arrays to reach all available configurations; and calculating the necessary storage capacity by summing up all energy variations between the HRES production and the storage [143].

Bajpal and Dash [160] have considered another viewpoint to classify the sizing and configuration. On this basis, two main categories are taken into account: conventional techniques, which require the weather data from the meteorological sites for a specific location, and artificial intelligence (AI) techniques, which are performed in an occasion when the weather data are not available. Examples of the AI models include artificial neural network (ANN), fuzzy logic, genetic algorithm, wavelet transforms, and hybrid models. Conventional techniques are based on the energy balance and the reliability of energy supply.

Another research conducted in 2015 by Wang et al. [158] has considered the energy flow optimization method in an occasion that weather data are achievable for a specific location. On this basis, the energy flows during a typical day with the averaged weather data are considered for the sizing methods. Using this method, the average weather data for a region are extracted by sampling the seasonal weather data. Afterward, such data are compared with the solar radiation, wind speed, and temperature values at the same hour of different days to estimate their mean values. The corresponding solar and wind energy production can be achieved as the upper limits when performing the power dispatch for a study.

HRESs require an optimal design for their component sizing to economically, efficiently, and reliably meet its objectives. Table 7.11 provides some examples of literature studies related to HRES optimal sizing along with details regarding the

TABLE 7.11
Some Reported Literature for HRES Optimization Sizing Results [5]

References	Components of the Hybrid System	Load Specifications	Sizing Results
[161]	Wind turbine (WT), photovoltaic (PV), and battery	225 kW peak, 25 kW base	195 kW WT, 85 kW PV, 230 kW microturbine, 2.14 kAh battery
[162]	WT, PV, microturbine, and battery	1.5 kW constant	6 kW WT, 12.8 kW PV, 6 kAh battery
[163]	WT, PV, diesel, and battery	26 kW peak, 5 kW base	15 kW WT, 24 kW PV, 50 kW diesel, 151 kWh battery
[120]	WT, PV, and battery	1,500 W	78 × 100 W PV, 2 × 6 kW WT, 5,000 Ah (24 V) battery
[164]	PV, diesel, and battery	3.5 kW peak, 0.25 kW base	2.8 kW DG, 4.2 m^2 PV, 2.75 kWh battery
[106]	Wind, PV, and energy storage	1 MW peak, 0.4 MW base	2.096 MW wind, 0 MW PV, 6.576 MWh energy storage
[165]	Wind, PV, and energy storage		2.42 MW wind, 0 MW PV, 6.7878 MWh energy storage

hybrid system components, their load characteristics, and sizing results. Numerous methods are used to optimize stochastic nature of HRES. In this section, we examine these methods in detail [3]:

1 **Sizing optimization technique using algorithm approach**

A current and future state-of-the-art development of hybrid energy system using the wind and PV solar is given by Nema et al. [58]. The review focuses on design, operation, and control requirement of the stand-alone PV solar–wind hybrid energy systems with conventional backup source, i.e., diesel or grid. In this section, an overview of various optimization techniques approach utilized in PWHS is provided. They can be applied to reach a techno-economically optimum PWHS

Solar and the wind energy have a common inconvenient; they are unpredictable in nature and depend on the weather and climatic changes, and the variations of solar and wind energy may not match with the time distribution of load demand; this unpredictable nature influences the PWHS performance and results in early batteries discharge. Huneke et al. [166] proposed an optimal configuration of the electrical power supply system, following characteristic restrictions as well as hourly weather and demand data. He proposed an optimal mix of solar- and wind-based power generators combined with storage devices and a diesel generator set. As a result, the operation of this model is tested in two real off-grid energy systems. Both optimization processes resulted in hybrid energy systems, utilizing photovoltaics (PV), lead-acid batteries, and a diesel generator as a load-balancing facility. Tina and Gagliano proposed [95] a procedure for the probabilistic treatment of solar irradiance, and wind speed data is reported as a method of evaluating, at a given site, the electric energy generated by both a photovoltaic system and a wind system [2].

Energy resources are very imperative from the economic and political perceptions for the whole world. This is why technological transformation in energy systems is exceedingly significant and an economical supply of the electric load [167] to consumers is essential. Researchers [62, 168, 169] have used genetic algorithms (GA) for evaluating the optimal configuration cost of a hybrid energy system. The result obtained from the study indicated a good system performance and effective cost scenario. An optimal system configuration is one of the several phenomenal concepts in hybrid power system. The optimization of wind-solar-battery hybrid system was established in the study conducted by [170]. Yang et al. [162] and Diaf et al. [171] underpinned iterative optimization technique (loss of power probability method) for optimal hybrid system configuration. For the same purpose of optimization, Mawardi and Ptchumani [172] proposed an integration of stochastic modeling with numerical optimization for realization of more robust hybrid energy scheme. This particular method helps in reducing the overall cost of hybrid system implementation and performances. Therefore, regardless of the methodology used, what is significant is that an applicable and robust hybrid system should be designed to perform reliably even under the influence of some uncertainties.

2 Sizing using deterministic and stochastic approach

In the literature, there are two methods for sizing a hybrid system: the first method is "yearly average monthly method," which consists in sizing PV panel and wind turbine by deriving from the yearly averaged monthly values; the load has represented by the yearly mean monthly value. In the second method, "worst months," it chooses the worst months for solar and wind energy system separately. This method is characterized by choice of the worst month as the one in which the largest total area of PV module and wind turbine occurs [2].

It is proved that renewable energy sources in combination with electrical BS help to ensure the power reliability of the system and to reduce the cost of energy compared to stand-alone with a diesel generator. The optimal solutions strongly depend on the actual load demand curve. As both PV and wind energy benefit from BS, the costs of the battery can be shared, and the two technologies complement each other. Protogeropoulos et al. [173] presented a PWHS performance simulation procedure, with an hourly time step, in order to obtain the autonomy levels of potentially optimum arrangements as the battery size is varied. A research work remains in the use of the time series simulation method for designing the hybrid system and the feasibility study; the time series uses data in the provenance of the metrological station. PWHS's behavior is calculated based on the time series meteorological input data, which usually have a resolution of 1-hour intervals. A methodology has been proposed among a list of commercially available system devices in order to determine the optimal number and type of units capable to guarantee that the 20-year-round total system cost is minimized, subject to the constraint that the load energy requirements are completely covered [2]. Their research work focused on the manner used to minimize the cost [2].

3 Software tools used in sizing optimization of PWHS

The optimized configuration and optimum control strategy of the hybrid system are interdependent. This complexity makes the hybrid systems more difficult to be sized, designed, and analyzed. The target in sizing a PWHS reminds the use of renewable energy resources in an efficient and economic way. It means that the PWHS can work at the optimum conditions in terms of investment and power reliability requirement. Providing an optimum sizing method can help to guarantee the best performance and the lowest investment with maximum use of the system component. The techno-economic analysis of the hybrid system is critical to the efficient utilization of renewable energy resources. Due to multiple generation systems, hybrid system analysis is quite complex and is required to be analyzed thoroughly. This requires software tools for the design, analysis, optimization, and economic viability of the systems. There are research works related to hybrid systems carried out using this software. Sinha et al. [1] provided the current status of this software to provide basic insight for a researcher to identify and utilize the suitable tool for research and development studies of hybrid systems [2].

Among various program tools, Hybrid Optimization Model for Electric Renewables (HOMER) is part of the most popular tools for sizing a hybrid system. HOMER allows a power system's physical behavior based on its life cycle cost, which is the sum of installation and maintenance cost of system components over the system's lifetime. HOMER ensures that the programmer compares many various design configurations based on their technical and economic merits. Then, HYBRID 2 is simulation software for a very high accuracy optimization, as it can define time intervals from 10 minute to 1 hour. HOGA is a single objective or multi-objective optimization problem and control strategies are solved by using genetic algorithms. All the parameters remained constant during simulation of 1 hour interval. Moreover, HYBRIDS is a Microsoft Excel spreadsheet-based application tool for a renewable energy assessment and requires daily average load and environmental data for each month of the year. It simulates one configuration at a time but wide ranging in terms of renewable energy system variables. It is not required for system optimization; however, it helps to improve the hybrid system design. In addition, TRNSYS software uses a programmer-defined time step that varies from 0.01 seconds to 1 hour for simulation. HYDROGEMS tool is utilized for the analysis of hydrogen energy systems in a time step of 1 minute [28]. Finally, RETScreen is a Microsoft Excel-based software tool that can evaluate renewable energy for its energy efficiency, and for its technical and financial viability for the cogeneration projects. This tool is used for the analysis of energy efficient integrated system covering mainly energy production, life cycle costs, and greenhouse gas emission reduction. Other computer tools are also available for designing of hybrid systems [4] such as the General Algebraic Modeling System (GAMS) [174], providing an optimal design for a hybrid solar–wind energy plant, where the variables that are optimized over include the number of photovoltaic modules, the wind turbine height, the number of wind turbines, and the turbine rotor diameter, and the goal is to minimize costs. Numerous studies on software optimization design are reported in the literature and summarized by Anoune et al. [2]. All these programs need an external weather data file (TMY2/3, EPW) to perform the calculation of the optimum sizing of the hybrid system.These programs consider many constraints with multi-objective optimization and they can work with multiple energy sources and multiple system storages [2].

C. **Energy production scenarios**

Parameters related to energy production have been considered in 32 research studies. As a matter of fact, different scenarios have been considered to evaluate energy production in different ways. Samples of such indexes which have been included in the current survey are energy index ratio (EIR) expected energy not supplied (EENS), wasted renewable energy (WRE), excess electricity fraction (EEF), and deficiency of the power supply probability (DPSP).

The EIR as well as the EENS is considered only in one of the studies out of the 98 researches conducted. The EIR has been used to evaluate the reliability of the HRESs, designed to aid the load demand for supplying the required electricity. It has been suggested that in order to have the best configuration of the HRESs, the system has to have an EIR of above 0.9, highlighting the fact that the system should be 90% reliable in its performance [175]. The WRE is considered as a design parameter in one of the studies performed in 2013. The EEF is defined in the following form and has been considered in the two research studies conducted lately (one study in 2011 [82] and the other in 2014 [176]). DPSP is a design variable for expressing the reliability of an HRES proposed and is defined as the probability that an insufficient power supply results when the other components of the HRES (for instance, PV array, WT, and battery) are not capable of satisfying the load demand. The DPSP is considered to be a technical criterion to perform sizing and evaluation of an HRES, which employs battery pack, and therefore, by using this parameter as the objective function in the optimization process, an HRES configuration with a high efficiency and reliability can be achieved. In the literature, the LA (level of autonomy) is considered four times as an objective function. This parameter is defined as one minus the ratio of the total number of hours, in which the loss of load occurs in a hybrid power system (D_{lol}) to the total hours of operation (D_{total}).

D. **Load demand parameters**

Among 98 studies performed in the survey by Faccio et al. [3] concerning the design and optimization of the HRESs, load demand was cited 30 times. The most important load factors which have been considered to affect the reliability of the system are the LLP (loss of load probability) and the LPSP which have a significant impact on the performance and reliability of the system. Loss of load is determined from energy deficit which is the amount of required energy provided by the load at a certain hour, which cannot be covered by various generation or storage sources.

The LPSP (loss of power supply probability) is defined as the ratio of all energy deficits to the total load demand during the period considered. An LPSP of zero is indicative of the fact that the load will always be satisfied, and the LPSP of one indicates that the load will never be satisfied. Rajkumar et al. [82] has defined the LPSP based on the interactions which arise from the relationship among the load demand parameter, the energy production generated by the renewable source and battery's state of charge. It is noted that the LPSP defined here does not include the energy deficit owing to the component breakdown or maintenance downtime. It is merely in relation to the size or capacity of the energy storage as well as the load demand [82].

In the survey analysis of Faccio et al. [3], the LST (load shifting technique) is considered one time as an optimization goal. As a matter of fact, the LST refers to a load parameter which falls into two classifications: the low-priority load (LPL) and the high-priority load (HPL). HPL is applied in an occasion where the energy production of the HRES is restricted to a constant level, therefore requiring any available sources (renewable sources, storage, or traditional sources of power) to supply the necessary electricity demand. Nevertheless, the LPL can be supplied when the production from renewable sources of power is available, and it can then be fed from the surplus generation time of the HRESs. On the occasion that the energy source produced by the renewable energy sources exceeds the power required for the HPL, the surplus energy will be stored in the batteries even up to its maximum level. The excess power will be considered for utilization to feed the LPL. The excess power above the LPL requirements will be considered for utilization to feed the dummy load, which is a device, for simulating the electrical load. The unmet LPL will be shifted to the time of surplus generation. Provided that the power required for the HPL is greater than the power generated by the renewable sources; an energy storage system could then be used to ensure meeting the demand for the HPL until the required energy decreases to its least level. The unmet LPL will be shifted to the time of surplus generation [177, 178].

E. **Reduction of environmental emissions**

Decreasing the concentration of the environmental pollutants produced by those of HRESs, which also employ the DG in their configuration, has been considered as one of the primary objective functions to adequately design environment friendly HRESs with the least release of the greenhouse gases to the atmosphere. The environmental emission factors which are considered in the design procedure in the literature review are as follows:

(1) **Global warming potential**

Global warming potential (GWP) is defined as the relative estimation methodology of how much heat a greenhouse gas is able to trap in the atmosphere. As a matter of fact, it performs estimations by comparing the amount of heat which is inside a certain mass of the gas in question and the amount of heat trapped by a similar mass of the greenhouse gas. The GWP is typically measured over a specific time interval, commonly 20, 100, or 500 years, outlining the emissions of the carbon dioxide, whose GWP is standardized to a unit value.

The GWP parameter is contingent upon the absorption of infrared radiation by a given species, the spectral location of its absorbing wavelengths, and the atmospheric lifetime of the species.

(2) **Equivalent CO_2 LCE**

As a matter of principle, it is important to bear in mind that analyzing the life cycle emissions takes into account estimating the global warming potential of the energy sources through calculating the life cycle of each energy source. Such estimations are usually presented in units of "global warming potential per unit of electrical energy generated by that source." This equivalent CO_2 LCE parameter employs the GWP unit, the carbon dioxide equivalent (CO_2 emissions), the unit of electrical energy, and the kilowatt-hour (kWh). The purpose of such evaluations is to cover the full life of the power source, from material and fuel mining through the construction to operation and waste management.

One aspect of the life cycle emissions analysis is the equivalent carbon dioxide life cycle emissions CO_2 LCE (gCO_2/kWh), which is defined as the equivalent carbon dioxide emissions, usually produced by the DG or any power-generating equipment, which releases carbon dioxide emissions into the atmosphere. This parameter is taken into account to manufacture, transport, and recycle the components of the system (i.e., PV panels, structures,

WT, diesel or gasoline generator, batteries, inverter, charge regulator, and rectifier). It also includes the emissions from the fuel combustion of the diesel or gasoline generator, fuel extraction and refining, and fuel transport.

In off-grid HRESs, the equivalent CO_2 LCE is considered to be emissions per kilowatt-hour utilized by the electrical load. In a research conducted by Dufo-López et al. [83] in 2011, optimization of this parameter was conducted by employing the HOGA optimization tool.

(3) **DG environmental pollutants**

The DGs generally produce gas pollutants into the atmosphere. The primary polluting gases released by the DG are CO_2, NO_x, SO_2, CO, and PM_{10}. Optimization of the environmental pollutants could then be performed specifically for every single contaminant such as the carbon dioxide, unburned hydrocarbons (HC) particulate matters, NO_x, SO_2, and other environmental emissions by taking into account an upper bound limit for each of the aforementioned pollutants. On this account, HOMER has a powerful option method which specifically performs emissions control and restrictions on each of the pollutants released by the HRESs, integrated with the DG by considering a maximum limit control on these emissions.

In a research conducted in 2012, Nasiraghdam and Jadid [134] employed a calculation methodology, for estimating the total annual emissions produced with the support of a grid-tied hybrid power system. Among all pollutions considered, carbon dioxide is always seen as the primary source of the greenhouse effect. In a latest research in 2015, conducted by Shi et al. [179], the impact of DG carbon dioxide emissions was also considered.

F. **Voltage stability index (VSI)**

VSI is a design parameter in most of the HRESs, which should be enhanced. For detecting a bus of radial distribution systems that is most sensitive to voltage collapse, the VSI has been introduced for each bus. VSI is a number between zero and one, which is expressed as unity for no load, and zero at the point of voltage collapse. It is, therefore, desirable that the VSI parameter be close to unity for each bus as much as possible. In detail, the VSI index is elaborated according to a formula outlined by Nasiraghdam and Jadid [134].

G. **Breakeven grid extension distance (BGED)**

The maximum distance from the grid which makes the net present cost of extending the grid equal to the net present cost of stand-alone system is called breakeven grid extension distance. Based on the current review, BGED is only considered one time as an optimization goal in a study performed by Hafez and Bhattacharya [180]. In this study, the effect of the distance from grid and the optimal breakeven distance was addressed.

H. **Analysis of power reliability**

The reliability of power depends on the power system to ensure the supply of electrical energy to the loads in an adequate and secure way. A fluctuation characterizes power reliability of PWHS due to its intermittent nature. Solar radiation and wind speed are dependent on locale climate condition, the shadow of the clouds, the inclination of the PV panel, as well as the ambient temperature, which influences the energy production of the photovoltaic system, adding also the fluctuation of wind speed along the day makes the energy production from hybrid system highly variable. It is for this reason that the use of power reliability analysis is considered as an important step in any system design process. Most of the probabilistic techniques available in the literature for reliability evaluation are in the scope of adequate assessment. Most of the adequacy indices employed in many other literatures have replaced reliability indices. Adequate indicators reflect various factors such as system component availability and capacity, load characteristics and uncertainty, and system configuration and operational conditions. Reliability indices are used to evaluate the reliability performance of a generation system as PWHS against some predetermined minimum requirements or reliability standards. It compares additional designs and allows identification of weak spots and determines ways for correction in the generation system and finally integrates cost and performance for the subjects for final decision [3].

7.6 DESIGN VARIABLES AFFECTING SIMULATION AND OPTIMIZATION OF HRES

While simulation, optimization methods and objectives and software tools are important, the components of the HRES system are also very important. Here we briefly outline some of the important ones.

7.6.1 Grid Classification

The concept of the grid covers a variety of applications such as the operational and energy dimensions including smart meters, smart appliances, renewable energy resources, and energy efficient resources. Electronic power conditioning as well as the management of the production and distribution of electricity are also considered as the important aspects of the smart grid. In addition to the aforementioned issues, grid extension and connection costs are considered as important factors to integrate the renewable power sources into an existing electricity network.

HRESs are either grid connected or off grid. The advantage of the on-grid HRES is that it provides the opportunity to sell the excess energy generated by the HRESs to the grid for recovering some of the costs for the energy purchased. However, the off-grid HRESs have also received a considerable amount of attention from around the globe, as they allow electricity access in remote rural communities at lower costs than on-grid systems. They are usually integrated with storage units, especially batteries. A key issue in the cost-effectiveness of such systems is the battery degradation during the time that the battery is charged and discharged.

The study by Faccio et al. [3] concluded that more than 70% of the reported papers are for off-grid applications, while remaining dealing with optimization of HRES tied to the grid. When taking into account the concept of the grid, the grid power price as well as the breakeven extension distance is typically considered during the grid analysis of the HRESs. In rare occasions, the authors have only taken into account both the off-grid and the on-grid configurations of the HRESs simultaneously, for investigating their performance and optimization.

Other parameters related to the system's grid during the optimization process are GGP (grid generation points) and IGP (independent generation points), which have been considered only in one study in 2013 during the present survey analysis. As a matter of fact, the GGP is considered as a situation where the energy is produced and distributed to other demand points through a distribution network (radial microgrid) while the IGP is defined as the points producing energy just for their own consumption and not connected to any microgrid [2, 3]. Finally, the location of the battery storage on the grid (on utility grid, distribution grid, or behind the meter) is also important.

7.6.2 Technology and Energy Resource

During the design of the HRES, one important criterion is the correct allocation of the energy sources which are considered in the integration of the HRES. As a matter of fact, an optimized allocation of energy sources of an HRES will lead to a higher efficiency rate and lower net present cost. One interesting conclusion derived from the evaluation of the classification drivers emphasized the existence of diverse configurations of the HRES, ranging from low numbers of the energy sources, such as PV/hydro and wind/hydro, to higher numbers, such as PV/wind/DG/battery and PV/wind/DG/hydrogen/battery (Table 7.12). In addition, one

TABLE 7.12
Frequency of Various HRES Configurations (total 80 samples) [3]

Configurations	Frequency
PV/wind/hydro/hydrogen/battery	2
PV/wind/battery	36
PV/wind/wood gas generator	2
PV/wind/DG/hydrogen/battery	2
Wind/hydro	2
PV/hydro	3
PV/wind	17
PV/wind/DG/battery	3
PV/wind/DG	13

special configuration observed was the incorporation of the wood gas generator with the intermittent energy sources (solar and wind). Such a combination was rarely investigated from previously addressed research papers. It is worth mentioning that two configurations—wind/PV and wind/PV/DG—were also the most frequent combinations observed from the literature review, with their frequencies being 17 and 13, respectively. As a matter of fact, the number of studies coping with the combination of wind and solar systems with integration of the battery storage was much more than the HRESs with other types of system configurations. In general, 36 research cases investigated the integration of wind and PV with battery simultaneously. Additionally, in 34 research studies, optimization analysis for the integration of the DG with the HRES was also performed. In future, more studies are needed where other renewable sources like hydro, biomass, waste, and geothermal energy are included. The inclusion of these sources can significantly change the final workable optimization. The study by Faccio et al. [3] concluded that out of the 66 research cases, 47 research studies have included the WECSs in their configuration of the HRESs. It is important to bear in mind that utilization of those of small WTs which have a lower cut-in wind speed will result in a higher efficiency in electrical production. This will enable the higher production of wind power in areas with even a lower level of wind speed values. To estimate the power output of the wind turbine, different methodologies exist. However, the most common model is the one which uses the integral form of the equation. More details on this is given in many studies [1–8].

7.6.3 Central Control Unit of the HRESs

The central control management unit plays a crucial role in the operation of the HRESs. Adequate design of the system will not just result in the enhancement of the energy production from the HRES, but its efficiency as well. Additionally, the lifetime of the battery could be increased significantly, and in the same time, the number of HRES deficit hours, the DG's operating hours and the amount of dumped power could be reduced. Some other important points, which should be considered in the management of the control system, are battery management, engine on/off cycling, maximum power point tracking of the available solar and wind energy, load management, quality of power during power generation, and the operation of different components of the HRESs [2, 3].

7.6.4 Photovoltaic System

In addition to the regional resource and the cost of the fuel (as well as the fuel escalation rate), many other parameters can influence the actual payback period of a hybridization investment. Solar PV arrays usually have a longer lifespan (more than 20 years). However, their energy yield gets slightly decreased by the passage of the time. The energy yield has to be estimated in the techno-economic analysis across the lifetime of the project. PV panel manufacturers usually guarantee a 90% of the initial performance after 10 years and 80% after 25 years. Furthermore, the possibility to resort a guarantee, if it is required after a few years, remains an open question in areas where distributors are not well organizing the manufacturing processes.

The duty of a PV system is to convert the energy coming from the sun's power directly into electricity. One important primary part of the PV array, which is the smallest compartment, is the solar cell. Such cells are typically configured and adjusted in the module, which is joined in series and or/or parallel fashion to form arrays. The DC electricity produced at the terminals of the arrays can be utilized in a variety of applications, such as the DC motors or lighting systems. The current-voltage (I-V) diagram is nonlinear. Different parts of the diagram for a typical solar PV array are categorized as series configuration, parallel configuration, and single configuration. In a single configuration, the MPP is at the highest current and voltage points. In a parallel configuration, currents are added at the same voltage and MPP is at the highest total current and voltage points. In series configuration, voltages are added at all currents, and the MPP is at the highest voltage and current levels. The power generated by the PV array can be expressed as a function of the solar radiation and the ambient temperature.

A related point to consider in the design of the solar cells and estimation of their performance is the determination of the parameter values of photovoltaic (PV) cell models, which plays a big part in the design process. The key parameters representing the performance of solar cells include generated photocurrent, saturation current, series resistance, shunt resistance, and ideality factor. Estimating these parameters accurately is an essential part of a precise modeling and performance evaluation of solar cells. So far, various computational intelligence methods, such as genetic algorithm, particle swarm optimization, simulated annealing, and harmony search, have been proposed for optimal estimation of solar cell parameters. Many studies have aimed to overcome the shortcomings of the conventional deterministic algorithms and to investigate the efficiency and applicability of the algorithms. Hybrid methods integrating two or more metaheuristic algorithms have also been applied lately to explore the capability of stochastic artificial intelligence algorithms for estimating the solar cell parameters. These algorithms could find relevant parameter values through minimizing the root mean square error (RMSE) as the objective function in the optimization process. Furthermore, the metaheuristic algorithms have also demonstrated a worthwhile level of applicability for estimating the solar cell parameters with good performance [181].

7.6.5 Solar Power Conditioning Unit (PCU)

A solar power conditioning unit (PCU) is a device which is intelligently designed to check and evaluate the output power of the PV panels. In the event that the energy output of the hybrid power system is sufficient to charge the battery, no power can be extracted from the primary units. However, if there is a deficit in the solar energy output afterward, the remaining power will be taken from these units. By way of an example, if solar panels are supplying 12 A, and a 150-Ah battery is connected to the system, it will extract the remaining 3 A from the grid.

A PCU typically consists of the following functional units: solar charger, inverter, grid (main utility charger), output selector mechanism, battery bank, control algorithm, and the solar charger. The solar PCUs are integrated systems, consisting of a solar charge controller, an inverter, and a grid charger. They supply the facility to charge the battery bank through either a solar or grid/DG set. The purpose of the PCU is to continuously monitor the state of battery voltage, solar power output, and the load, and it is integrated for utilization in the configuration of the HRESs [3].

One latest and innovative research work which has received the attention of designers to adjust the performance of the HRESs is the adjustment of the performance of the solar PCU with the purpose of enhancing the efficiency of the mixed grid-tied systems, as well as the off-grid systems. In such events, when the primary lines are off, the solar PCU will begin functioning to prevent power losses and supply backup. Obviously, the research in this aspect is still in progress to devise further innovative methods for effective utilization of the PCUs [3].

7.6.6 Maximum Power Point Tracking (MPPT) Methods

The electronic control of the PV modules can typically be managed through using the MPPT algorithms by the use of the sophisticated power electronic converters. A tracking method, which is implemented by the power electronic converters, would ramp up the operating point of the PV close to the MPP. An MPPT algorithm is commonly applied in the converters to maximize the power extracted from PV modules, under varying atmospheric conditions. The utilization of the MPPT will aid to achieve the highest energy output from the PV array, thus reducing the PV array cost through decreasing the number of solar modules required to obtain the same power output [182]. There are six different MPPT methods employed in the literature. Artificial intelligent method involves parameters that rely on the adopted method [182]. It is digital implementation and it depends on the PV module parameters. Short circuit current method involves current as a parameter, and it can be an analog or a digital implementation with PV module parameter dependency. Open circuit voltage method involves voltage as a parameter, and it is once again an analog or a digital implementation with PV module parameter dependency. The remaining three methods, increment conductance (INC), perturb and observe (P&O), and hill climbing, all require voltage and current as parameters with no dependency on PV module parameters [2, 3]. The last two are both analog and digital implementations while INC is the digital implementation.

7.6.7 Energy Storage

In the design and optimization of HRESs, an energy storage system, in most cases, a battery pack, is considered specifically when an HRES proposed is supposed to meet the electrical demand of a rural community, with no access to the electricity grid.

The lifespan of the battery bank is contingent upon many parameters, mostly those relating to the way they are operated and to external conditions, in particular the ambient temperature. For instance, the typical lead-acid batteries are designed for the applications of solar plants, which usually lead to the energy loss between 15% and 20% of their lifespan, which is directly related to the number of charge and discharge cycles they can perform; for each of them, it is considered to be five degrees above the standard temperature, which is 25°C. Furthermore, the deeper the battery is discharged at each cycle (depth of discharge), the shorter is its lifespan. This indicates that in order to reach an optimal battery lifespan, one has to install a large enough battery to achieve a suitable depth of discharge. Considering the battery cost (around 20% to 30% of total system cost), it is more desirable to design a battery bank whose operating conditions will last for 6 years minimum and ideally 8 to 10 years. Contingent upon the type, the configuration, the temperature, and other design variables of the battery, the maximum charge cycles can be between 1,300 and 1,500 cycles [183]. Looking more precisely through the design of the battery and its charging cycles, as a side note, in a latest research, a novel and comprehensive approach using the Harmony search (HS) optimization has been used for charge scheduling of an energy storage system (ESS) with renewable power generators to find the optimized ESS scheduling under time-of-use (TOU) pricing with demand charge. A comparison with the obtained results of the genetic algorithm (GA) has also been performed. The results of this research have suggested that the HS proposed is more efficient to save electricity cost compared to the GA method. For all the cases studied, the percentage of saving values of HS are also higher than those of the GA method [184].

During the design procedure of HRES with battery storage, two important parameters namely nominal capacity of the battery and state of charge (SOC) are invariably considered. It bears mentioning that in most of the designed energy storage devices which are used in the configuration of the HRESs often employed the lead-acid battery. The energy conversion during charging as well as discharging of the battery takes place with a reversible reaction. In this regard, the modeling of the lead-acid battery for the real-time analysis of the HRESs should account for the dependence of battery parameters on the state of charge, the battery storage capacity, the rate of charge/discharge, the ambient temperature, and life and other internal phenomenon, such as gassing, double layer effect, self-discharge, heating loss, and diffusion. Energy storage sizing is one of the most important stages in the design process of the HRESs. As a matter of fact, there are four key-element steps to be considered in battery sizing, which are battery bank voltage, days of the autonomy, depth of discharge, and temperature. In future hybrid energy storage devices (battery and supercapacitors) will also play an important role in HRES optimization [3].

7.6.8 Inverter

The inverter is a high-technology component and its replacement as a result of the failure is usually undertaken by a technician from the supplying company. It is worthy to note that the lifespan of an inverter could extend to more than 10 years. The specific complexity of the inverter often necessitates an adequate after-sales service plan (which is considered as an important source of revenue and profit for a proposed business), to be implemented to ensure the long-term sustainability of the system. Furthermore, risks, which are linked with the failure of an inverter, should, therefore, be considered, especially in remote locations or countries with very limited specialized suppliers.

To prolong the runtime of an inverter, batteries can be integrated with the inverter. When attempting to add more batteries to an inverter, there are two basic alternatives to be considered, which are series configuration and parallel configuration. As a matter of fact, when the objective of the system is to produce higher overall voltage values for the inverter, the series configuration is considered during the design process. In an occasion that the lifetime of a single battery ceases, the other configured batteries will not be able to power the required load. The third cell in the battery pack produces only 2.8 V instead of the full nominal of 3.6 V. With depressed operating voltage, this battery reaches the end-of-discharge point sooner than a normal pack [3].

On the other hand, with the parallel configuration, the capacity of the battery will be increased and more importantly, such a configuration will lead to prolonging the running time of the inverter. On such an occasion, the batteries are connected in a parallel way. This would aid ramping up the overall ampere-hour (Ah) rating of the battery set. On an occasion that the single battery is discharged, the other batteries will then discharge through it. This can eventually lead to a rapid discharge of the entire battery pack, or even an over-current and possible fire. To prevent such an event, relatively large paralleled batteries may be connected via diodes or the systems with intelligent monitoring equipped with automatic switching to isolate an under-voltage battery from the other. In a condition that one cell in the parallel configuration becomes weak, this will not necessarily have an effect on the voltage of the circuit. Nevertheless, it provides a low runtime operation owing to reduced capacity of battery [3].

7.6.9 Modeling of Hydrogen Tanks

In every important respect, it is worthwhile to mention that the hydrogen gas poses a great challenge not only to its extraction but to its storage as well. The low density of the hydrogen gas and low boiling point of the liquid hydrogen make it difficult to store hydrogen either in a gaseous or liquid form. Therefore, for all practical purposes, the hydrogen gas is stored either as a high-pressure gas or as a liquid cooled down to cryogenic temperatures, or as metal hydrides where hydrogen gas is bound to a certain metal. This is an important factor if fuel cell is to be used in HRES [3].

7.6.10 Diesel Generator (DG)

During the design process of the HRESs, when the electricity coming from HRESs proposed is not significant enough, the DG component can be considered in the design of the HRESs. On such an occasion, it is required to supply the additional electricity using the DG system [3].The downside, however, would be the fact that the DG occasionally contains environmental pollutants. Examples of such pollutants are CO_x, CO_2, PM, SO_2 and other environmental emissions [3].

A recent research study conducted in 2017 by Azaza and Wallin [185] expressed the fuel consumption of the DG as a function of the output power. To evaluate the contribution of the backup DG in the total production, the hours of generator operation can be estimated using the energy deficit per month, the rated power of the DG, and the generator load factor. It is assumed here that the generator operates at a constant load factor during hours of operation. To evaluate the contribution of the backup DG in the total production, the hours of generator operation can be estimated using the energy deficit per month, the rated power of the DG, and the generator load factor. It is assumed here that the generator operates at a constant load factor during hours of operation.

7.7 CLOSING PERSPECTIVES

Global warming, ozone layer depletion, and in general, environmental concerns have led the world to find alternate sources of energy other than fossil fuels. Renewable energy sources have recently been progressing steadily for their techno-economic and environmental impacts. Obviously, the emergence of novel and advanced big data tools has also revolutionized how renewable energy technologies are researched, developed, designed, demonstrated, and deployed. By way of an example, from computational chemistry and inverse material design to adoption, reliability, and correlation of insolation forecasts with load use patterns, data scientists have brought greater opportunities to dramatically impact the future scaling of renewable energy. Forecasting methods have also made it much simpler to streamline the design process and to meet consumer electricity demand for power as well as reliability. On top of this, the utilization of various probability density functions (PDFs) to represent the statistics of the resources has also been evaluated. This has led to more appropriate stochastic energy scheduling modeling and comprehensive power management strategies for load generation adequacy and security. Additionally, different optimization techniques have been analyzed lately, which has endorsed the study of optimal strategy of resource allocation to meet the load demand and ensure system security. Thereby, a robust stochastic approach with renewable energy resources and load demand has been developed, which has aided to enhance the security, reliability, and efficiency of power systems, and thereby, decreasing the dependency on fossil fuels [1–8, 23].

In this regard, the development of the HRESs has become noticeably considerable. It is, therefore, imperative to adopt innovative policies to aid the designers in developing and providing new solutions for environmentally friendly power systems. It is worthwhile to notice that recently several authors have addressed the optimization and design of the HRESs by employing different novel and advanced techniques. However, there is still a considerable gap with regard to this field of research. Furthermore, it is also observed that researchers have rarely made in-depth analyses of classification drivers of the HRESs during the design and optimization process [1–8, 23]. The literature has pointed out that a design of hybrid renewable energy system is site, objective function, and method specific. This point was made by Moghaddam et al. [152], among others, for most examined PV-wind-battery off-grid system. They designed a hybrid renewable energy PV/wind/battery system for improving the load supply reliability considering the net present cost (NPC) as the objective function to minimize. The NPC includes the costs related to the investment, replacement, operation, and maintenance of the hybrid system. The considered reliability index was the deficit power-hourly interruption probability of the load demand. The decision variables were the number of PV panels, wind turbines and batteries, capacity of transferred power by inverter, angle of PV panels, and wind tower height. To solve the optimization problem, a new algorithm named improved crow search algorithm (ICSA) was proposed. The design of the system was done for Zanjan city, Iran, based on real data of solar radiation and wind speed of this area. The performance of the proposed ICSA was compared with crow search algorithm (CSA) and particle swarm optimization methods in different combinations of system. This comparison showed that the proposed ICSA algorithm has better performance than other methods.

This chapter has given a current review of conducting a comprehensive evaluation of the common classification drivers, which affect the reliability as well as the performance of HRESs. At the first stage of this review, a detailed evaluation of the 22 software tools that were developed for simulation and optimization of HRES systems is described. The research carried out on these software tools and their applications to HRES are also delineated. In the second stage of the review, an evaluation of the common optimization goals and methods which were used in the design and optimization of the HRESs is performed. The number of the research cases examined by Faccio et al. [3] was 98 research papers, and the preliminary purpose was to identify the commonly used techniques and methods in the design and optimization process of the HRESs. Furthermore, in order to conduct an in-depth analysis over the sizing methods which are used for sizing configuration of each of the HRESs component including the PV, wind, battery, inverter, and other key-equipment, a review of the primary methods covering the optimized indicators, the design constraints such as the LPSP, LA, FLMPVS, WLR, GDB, grid type, as well as the investigated technology is performed [3]. This review emphasizes that there are four types of categories commonly considered during the sizing of the HRESs, which are, the probabilistic, the analytical, the hybrid, and the iterative [1–8, 23]. The review also considers other optimization objectives and techniques (including the use of software) used to pursue them and research behind them. Often optimization involves multiple objectives [1–8, 23].

In the last phase of this chapter, different compartments of the HRESs, such as the PV, WT, battery pack, inverter, and other related parts, which are typically considered as the primary objective in the design process are further discussed and elaborated in terms of their role in the optimization study. The current chapter concludes that the cost parameter has always been considered in the design process. However, the viewpoint toward using this parameter has always been different. For instance, in many research cases, the LCOE and the LCC have been the primary optimization objectives for the cost while in some other occasions, the NPV of the system is merely taken into account in the cost evaluation of the HRESs. It also becomes clear that among all optimization algorithms utilized, the heuristic-based design methods such as the GA, PSO, and linear-based algorithms are the most commonly used ones. Furthermore, an analysis of the software tools is also carried out, which emphasizes the fact that among these tools, HOMER, Lingo, HOGA and TRANSYS are the frequently used ones in the design process. Another result of this review highlights that the number of off-grid applications is much more than the on-grid cases. This issue has necessitated a requirement of using the component of the battery inside the configuration of the HRESs. It is also concluded that, of all the studies conducted, the PV-wind battery is the most frequent configuration of the HRESs [1–8, 23].

It should be noted that the simulation, design, and optimization of the HRESs has come a long way in terms of research and development. However, there are still considerable challenges with regard to the efficiency and the optimum utilization of these systems. Clearly, the future research has emphasized the importance of the primary design variables during the analysis and optimization of the configuration of the HRESs, and then outlining the key strategies, which could then be used for further development and optimization of the HRESs. Obviously, among all focal factors to be considered, optimization of the key variables such as the LPSP, DPSP, power loss, and specific grid parameters, such as IGP, and GGP, and the cycles of the battery are going to play a bigger part in the design and optimization of the HRESs. Specific sizing design constraints such as the state of charge of the battery, the level of autonomy, and the capacity the accumulator have also been considered as important objective functions during the design and optimization of the HRESs in future research. Moreover, the review is also indicative of the fact that metaheuristic methods and algorithms are more efficient than the conventional methods of optimization during the design of the key parameters which affect the reliability of the HRESs. In every important respect, they can automatically manage a wide range of complexities. In particular, multi-objective optimization (MOO) metaheuristics are the most appropriate ones for optimum design of HRESs, since the HRESs models involve multiple objectives at the same time such as cost, performance, supply/demand management, and grid limitations. In general, when it comes to the heuristic-based design methods, researchers have considered a considerably higher rate of convergence to reach to the optimal solutions [1–8, 23].

It is worthwhile to mention that future research directions concerning the design and optimization of the HRESs should also take into account the ongoing challenges which are encountered by decision makers. A few of these challenges include enhancing the efficiency of the solar PV arrays during the design and configuration of the HRESs; implementing the policies and the optimization models, with the aim of minimizing the cost of the each of the HRES components; applying innovative technologies and methods to enhance the life cycle of the storage devices utilized in the configuration of the HRESs; designing the adequate protection devices for installation in order to enhance the safety and the reliability of the HRESs; analysis and optimization of load fluctuations for designing the HRESs; devising innovative policies to improve the disposal of storage devices, such as batteries and hydrogen tanks; demonstrating and analyzing the correlation between the energy production and the design parameters; and designing and proposing automatic control systems with the aim of optimizing and enhancing the efficiencies of the storage systems of the HRESs [1–8, 23].

REFERENCES

1. Sinha S Chandel S. Review of software tools for hybrid renewable energy systems. *Renewable and Sustainable energy reviews.* 2014;32:192–205.
2. Anoune K, Bouya M, Astito A, Abdellah AB. Sizing methods and optimization techniques for PV-wind based hybrid renewable energy system: A review. *Renewable and Sustainable Energy Reviews.* 2018;93: 652–67.
3. Faccio M, Gamberi M, Bortolini M, Nedaei M. State-of-art review of the optimization methods to design the configuration of hybrid renewable energy systems (HRESs). *Frontiers in Energy.* 2018;12(4):591–622. https://doi.org/10.1007/s11708-018-0567-.
4. Bernal-Agustín JL, Dufo-Lopez R . Simulation and optimization of stand-alone hybrid renewable energy system. *Renewable and Sustainable Energy Reviews.* 2009;13:2111–8.
5. Ghofrani M, Hosseini NN. Optimizing hybrid renewable energy systems: A review. An open access intech paper Ch. 8. http://dx.doi.org/10.5772/65971.
6. Mohammed YS, Mustafa MW, Bashir N. Hybrid renewable energy systems for off-grid electric power: Review of substantial issues. *Renewable and Sustainable Energy Reviews.* 2014;35:527–39.
7. Aziz MS, Saleem U, Ali E, Siddiq K. A review on bi-source, off-grid hybrid power generation systems based on alternative energy sources. *Journal of Renewable Sustainable Energy.* 2015;7:043142. https://doi.org/10.1063/1.4929703.
8. Ibrahim M, Khair A, Ansari S. A review of hybrid renewable energy systems for electric power generation. *International Journal of Engineering Research and Applications,* www.ijera.com. 2015;5(8):42–8.
9. Khan MJ, Iqbal MT. Pre-feasibility study of stand-alone hybrid energy systems for applications in Newfoundland. *Renewable Energy.* 2005;30:835–54.
10. Zoulias EI, Lymberopoulos N. Techno-economic analysis of the integration of hydrogen energy technologies in renewable energy based stand-alone power systems. *Renewable Energy.* 2007;32:680–96.
11. HOMER (The Hybrid Optimization Model for Electric Renewables). Available from: http://www.nrel/gov/HOMER.
12. HOGA (Hybrid Optimization by Genetic Algorithms). Available from: http://www.unizar.es/rdufo/hoga-eng.htm; [accessed 21. 04. 2013].
13. HYBRID2. Available from: http://www.ceere.org/rerl/projects/software/hybrid2 (another version of HYBRID developed by University of Massachusetts with the help of NREL)/.
14. Green, HJ, Manwell, J. HYBRID2: A versatile model of the performance of hybrid power systems. WindPower'95, Washington, DC; 1995.
15. TRNSYS (defined in sections 7.1 and 7.2). Available from: http://sel.me.wisc.edu/trnsys/; [accessed 3. 01. 2013].
16. Ulleberg, O, Glöckner, R. HYDROGEMS. Hydrogen energy models. In: WHEC 2002–14th World Hydrogen Energy Conference; 2002.
17. HYDROGEMS (defined in sections 7.1 and 7.2). Available from: http://www.hydrogems.no/; [accessed 4. 02. 2013].
18. T. Berrill. *HYBRIDS.* 29 Burnett St., Wellington Point, Qld, Australia: Solaris solar powered home; 2005.

19. SOLSIM.(defined in sections 7.1 and 7.2) Available from: http://www.fh-konstanz.de; [accessed 7. 03. 2013].
20. INSEL (defined in sections 7.1 and 7.2). Available from: http://www.insel.eu/; [accessed 7. 03. 2013].
21. RAPSIM.(defined in sections 7.1 and 7.2) Available from: http://www.comm.murdoch .edu.au/synergy/9803/rapsim.html; [accessed 7. 03. 2013].
22. SOMES. (defined in sections 7.1 and 7.2) Available from http://www.web.co.bw/sib/ somes_3_2_description.pdf; [accessed 7. 03. 2013].
23. Ganesan G, Sagayaraj R, Nazar Ali A. Investigations on review of software tools for the integration of renewable energy systems for sustainable energy development. *International Journal for Research in Engineering Application & Management (IJREAM)*. 2018;4(4):705–10. DOI: 10.18231/2454-9150.2018.0568.
24. Afzal A, Mohibullah M, Sharma VK. Optimal hybrid renewable energy systems for energy security: A comparative study. *International Journal of Sustainable Energy*. 2009;29(1):48–58.
25. Turcotte D, Ross M, Sheriff F. Photovoltaic hybrid system sizing and simulation tools: Status and needs. In: PV Horizon: Workshop on Photovoltaic Hybrid Systems, Montreal; September 10, 2001. pp. 1–10.
26. Klise GT, Stein JS. Models used to assess the performance of photovoltaic systems. Sandia Report, Sand2009-8258; December 2009.
27. Arribas L, Bopp G, Vetter M, Lippkau A, Mauch K. World-wide overview of design and simulation tools for hybrid pv systems. International energy agency photovoltaic power systems program. IEA PVPS Task 11. Report IEAPVPS T11–01:2011; January 2011.
28. Connolly D, Lund H, Mathiesen BV, Leahy M. A review of computer tools for analysing the integration of renewable energy into various energy systems. *Applied Energy*. 2010;87:1059–82.
29. Ibrahim H, Lefebvre J, Methot JF, Deschenes JS. Numerical modeling wind–diesel hybrid system: Overview of the requirements, models and software tools. In: Proceedings of the IEEE Electrical Power and Energy Conference, Winnipeg, MB; October 3–5, 2011. p. 23–8.
30. Zhou W, Lou C, Li Z, Lu L, Yan H. Current status of research on optimum sizing of stand-alone hybrid solar–wind power generation systems. *Applied Energy*. 2010;87:380–9.
31. Erdinc O, Uzunoglu M. Optimum design of hybrid renewable energy systems: overview of different approaches. *Renewable and Sustainable Energy Reviews*. 2012;16:1412–25.
32. Sanajaoba S, Fernandez E. Maiden application of cuckoo search algorithm for optimal sizing of a remote hybrid renewable energy system. *Renewable Energy*. 2016;96:1–10.
33. Aronson EA, Caskey DL, Caskey BC. SOLSTOR description and user's guide. Sand 79-2330. Albuquerque, NM: Sandia National Laboratories; 1981.
34. Kendrick L, Pihl J, Weinstock I, Meiners D, Trujillo D. Hybrid generation model simulator (HybSim). In: Proceedings of the EESAT Conference, San Francisco, Sand2003–3790a; October 27–29, 2003.
35. Morgan TR, Marshall RH, Brinkworth BJ. ARES: A refined simulation program for the sizing an optimization of autonomous hybrid energy systems. *Solar Energy* 1997;59(4–6):205–15.
36. Schaffrin C, Knoblich I, Seeling-Hochmuth GC, Van Kuik E. Solsim and hybrid designer: Self optimizing software tools for simulation of solar hybrid applications. EuroSun 98; September 14–17, 1998. p. 1–7.

37. Merei G, Berger C, Sauer DU. Optimization of an off-grid hybrid PV-Wind-Diesel system with different battery technologies using genetic algorithm. *Solar Energy.* 2013;97:460–73.
38. Nehrir MH, Wang C, Strunz K, Aki H, Ramakumar R, Bing J, Miao Z, Salameh Z.A review of hybrid renewable/alternative energy systems for electric power generation: Configurations, control, and applications. *IEEE Transactions on Sustainable Energy.* 2011;2:392–403. DOI: 10.1109/TSTE.2011.2157540.
39. Lim JH, Optimal combination and sizing of a new and renewable hybrid generation system. *International Journal of Future Generation Communication and Networking.* 2012;5(2):43–59.
40. Notton G, Muselli M, Louche A. Atonomous hybrid photovoltaic power plant using a back-up generator: A case study in a mediterranean island. *Renewable Energy.* 1996;7(4):371–391.
41. Elhadidy MA, Shaahid SM. Promoting applications of hybrid (wind+photovoltaic+diesel+battery) power systems in hot regions. *Renewable Energy.* 2004;29(4):517–528.
42. Chedid R, Akiki H, Rahman S. A decision support technique for the design of hybrid solar-wind power system. *IEEE Transactions on Energy Conversion.* 1998;13(1):76–83.
43. El-Hefnawi SH. Photovoltaic diesel-generator hybrid powersystem sizing. *Renewable Energy.* 1998; 13(1):33–40.
44. Shrestha GB, Goel L. A study on optimal sizing of stand-alonephotovoltaic stations. *IEEE Transactions on Energy Conversion.* 1998; 13(4):373–78.
45. Ashok S. Optimized model for community-based hybrid energy system. *Renewable Energy.* 2007;32(7):1155–64.
46. Agarwal N, Kumar A, Varun B. Optimization of grid independenthybrid PV–diesel–battery system for power generation in remote villages of Uttar Pradesh, India. *Energy for Sustainable Development.* 2013;17(3):210–19.
47. Wang FC, Chen HC. The development and optimization of customized hybrid power systems. *International Journal of Hydrogen Energy.* 2016;41(28):12261–72.
48. Shafiullah GM. Hybrid renewable energy integration (HREI) system for subtropical climate in Central Queensland, Australia. *Renewable Energy.* 2016;96:1034–53.
49. Chauhan A, Saini RP. A review on integrated renewable energy system based power generation for stand-alone applications: Configurations, storage options, sizing methodologies and control. *Renewable and Sustainable Energy Reviews.* 2014;38:99–120.
50. Bordin C, Anuta HO, Crossland A, Gutierrez IL, Dent CJ, VigoD. A linear programming approach for battery degradation analysis and optimization in offgrid power systems with solar energy integration. *Renewable Energy.* 2017;101:417–30.
51. Ramazankhani ME, Mostafaeipour A, Hosseininasab H, Fakhrzad MB. Feasibility of geothermal power assisted hydrogen production in Iran. *International Journal of Hydrogen Energy.* 2016;41(41):18351–69.
52. Zarezade M, Mostafaeipour A. Identifying the effective factors onimplementing the solar dryers for Yazd province, Iran. *Renewable & Sustainable Energy Reviews.* 2016;57:765–75.
53. Goudarzi H, Mostafaeipour A. Energy saving evaluation of passive systems for residential buildings in hot and dry regions. *Renewable & Sustainable Energy Reviews.* 2017; 68:432–46.
54. Sopian K, Zaharim A, Ali Y, Nopiah ZM, Razak JA, Muhammad NS. Optimal operational strategy for hybrid renewable energy system using genetic algorithms. WSEAS Transactions on Mathematics 2008;7(4):130–40.
55. Al-Karaghouli A, Kazmerski LL. Optimization and life-cycle cost of health clinic PV system for a rural area in southern Iraq using HOMER software. *Solar Energy.* 2010;84:710–4.

56. Rehman S, Al-Hadhrami LM. Study of a solar PV–diesel–battery hybrid power system for a remotely located population near Rafha, Saudi Arabia. *Energy*. 2010;35:4986–95.
57. Fulzele JB, Dutt S. Optimum planning of hybrid renewable energy system using HOMER. *International Journal of Electrical and Computer Engineering*. 2012;2(1):68–74.
58. Nema P, Nema RK, Rangnekar S. A current and future state of art development of hybrid energy system using wind and PV-solar: A review. *Renewable and Sustainable Energy Reviews*. 2009;13(8):2096–103.
59. Sureshkumar U, Manoharan PS, Ramalakshmi APS. Economic cost analysis of hybrid renewable energy system using HOMER. In: Proceedings of the IEEE – International Conference on Advances in Engineering Science and Management; March 30–31, 2012. p. 94–9.
60. Akella AK, Sharma MP, Saini RP. Optimum utilization of renewable energy sources in a remote area. *Renewable and Sustainable Energy Reviews*. 2007;11:894–908.
61. Kumaravel S, Ashok S. An optimal stand-alone biomass/solar-PV/pico-hydel hybrid energy system for remote rural area electrification of isolated village in Western-Ghats region of India. *International Journal of Green Energy*. 2012;9:398–408.
62. Sinha S, Chandel SS. Review of recent trends in optimization techniques for solar photovoltaic–wind based hybrid energy systems. *Renewable and Sustainable Energy Reviews*. 2015;50:755–69.
63. Prasad RD. A case study for energy output using a single wind turbine and a hybrid system for Vadravadra site in Fiji Islands. *Online Journal on Power and Energy Engineering*. 2009;1(1):22–5.
64. Himri Y, Stambouli AB, Draoui B, Himri S. Techno-economical study of hybrid power system for a remote village in Algeria. *Energy*. 2008;33:1128–36.
65. Liu S, Wu Z, Dou X, Zhao B, Zhao S, Sun C. Optimal configuration of hybrid solar-wind distributed generation capacity in a grid-connected microgrid. In: 2013 IEEE PES Innovative Smart Grid Technologies (ISGT); February 24–27, 2013. pp. 1–6.
66. Dalton GJ, Lockington DA, Baldock TE. Feasibility analysis of renewable energy supply options for a grid-connected large hotel. *Renewable Energy*. 2009;34:955–64.
67. Dursun B, Gokcol C, Umut I, Ucar E, Kocabey S. Techno-economic evaluation of a hybrid PV–wind power generation system. *International Journal of Green Energy*. 2013;10:117–36.
68. Al-Badi AH, Albadi MH, Malik A, Al-Hilali M, Al-Busaidi A, Al-Omairi S. Levelized electricity cost for wind and PV–diesel hybrid system in Oman at selected sites. *International Journal of Sustainable Engineering*. 2014;7(2):92–102. http://dx.doi.org/19397038.2013.768714S.
69. Mohamed A, Khatib T. Optimal sizing of a PV/wind/diesel hybrid energy system for Malaysia. In: Proceedings of the IEEE International Conference on in Industrial Technology (ICIT), Cape Town, Western Cape, South Africa; February 25–28, 2013. pp. 752–7.
70. Kusakana K, Munda JL, Jimoh AA. Feasibility study of a hybrid PV-micro hydro system for rural electrification. In: Proceedings of the IEEE, Africon, Nairobi, Kenya; September 23–25, 2009. pp. 1–5.
71. Kenfack J, Neirac FP, Tatietse TT, Mayer D, Fogue MD, Lejeune A. Micro hydro–PV-hybrid system: Sizing a small hydro–PV-hybrid system for rural electrification in developing countries, technical note. *Renewable Energy*. 2009;34:2259–63.
72. Razak NABA, Othman MMB, Musirin I. Optimal sizing and operational strategy of hybrid renewable energy system using HOMER. In: Proceedings of the 4th International Power Engineering and Optimization Conference (PEOCO, Shah Alam, Selangor, Malaysia); June 23–24, 2010. pp. 495–501.

73. Güler Ö, Seyit AAG, Dincsoy ME. Feasibility analysis of medium-sized hotel's electrical energy consumption with hybrid systems. *Sustainable Cities and Society*. 2013;9:15–22.
74. Shaahid SM, El-Amin I. Techno-economic evaluation of off-grid hybrid photovoltaic–diesel–battery power systems for rural electrification in Saudi Arabia—A way forward for sustainable development. *Renewable and Sustainable Energy Reviews*. 2008;13(3):625–33. DOI: 10.1016/j.rser.2007.11.017.
75. Cotrell J, Pratt W. Modeling the feasibility of using fuel cells and hydrogen internal combustion engines in remote renewable energy systems. September 2003 • NREL/TP-500-34648, NREL Report, Contract No. DE-AC36-99-GO10337, Golden, CO.
76. Baniasad AI, Askari LB, Kaykhah MM, Askari HB. Optimization and technoeconomic feasibility analysis of hybrid (photovoltaic/wind/fuel cell) energy systems in Kerman, Iran; considering the effects of electrical load and energy storage technology. *International Journal of Sustainable Energy*. 2014;33(3):635–49. http://dx.doi.org/10.1080/ 14786451.2013.769991S
77. Liu G, Rasul MG, Amanullah MTO, Khan MMK. Feasibility study of stand-alone PV–wind–biomass hybrid energy system in Australia. In: Proceedings of the IEEE Power and Energy Engineering Conference (APPEEC), Wuhan, Asia Pacific; March 25–28, 2011. pp. 1–6.
78. Ashourian MH, Cherati SM, Mohd Zin AA, Niknam N, Mokhtar AS, Anwari M. Optimal green energy management for island resorts in Malaysia. *Renewable Energy*. 2013;51:36–45.
79. Wies RW, Johnson RA, Agrawal AN, Chubb TJ. Simulink model for economic analysis and environmental impacts of a PV with Diesel–Battery system for remote villages. *IEEE Transactions on Power Systems*. 2005;20(2):692–700.
80. Beccali M, Brunone S, Cellura M, Franzitta V. Energy, economic and environmental analysis on RET-hydrogen systems in residential buildings. *Renewable Energy*. 2008;33(3):366–82.
81. Castañeda M, Fernández LM, Sánchez HC. A sizing methods for stand-alone hybrid systems based on renewable energies and hydrogen. In: Proceedings of the Melecon 2012, Medina Yasmine, Hammamet, Tunisia; March 25–28, 2012. pp. 1–4.
82. Rajkumar RK, Ramachandaramurthy VK, Yong BL, Chia DB. Techno-economical optimization of hybrid PV/wind/battery system using Neuro-Fuzzy. *Energy*. 2011;36(8):5148–53. DOI: 10.1016/j.energy.2011.06.017.
83. Dufo-López R, Bernal-Agustín JL, Yusta-Loyo JM, DomínguezNavarro JA, Ramírez-Rosado IJ, Lujano J, Aso I. Multi-objective optimization minimizing cost and life cycle emissions of standalone PV-wind-diesel systems with batteries storage. *Applied Energy*. 2011;88(11):4033–41.
84. Mills A, Al-Hallaj S. Simulation of hydrogen-based hybrid systems using Hybrid2. *International Journal of Hydrogen Energy*. 2004;29:991–9.
85. Kalogirou SA. Use of TRNSYS for modelling and simulation of a hybrid–thermal solar system for Cyprus. *Renewable Energy*. 2001;23:247–60.
86. Rockendorf G, Sillmann R, Podlowski L, Litzenburger B. PV-hybrid and thermoelectric collectors. *Solar Energy*. 1999;67:227–37.
87. Liqun L, Chunxia L. Feasibility analyses of hybrid wind–PV-battery power system in Dongwangsha, Shanghai. *Przegląd Elektrotechniczny*. 2013; 1a:239–42.
88. McGowan JG, Manwell JF. Hybrid wind/PV/diesel system experiences. *Renewable Energy*. 1999;16(1–4):928–33.
89. McGowan JG, Manwell JF, Avelar C, Warner CL. Hybrid wind/PV/diesel hybrid power systems modelling and South American applications. *Renewable Energy*. 1996;9(1–4):836–47.

90. Siddaiah R, Saini R. A review on planning, configurations, modeling and optimization techniques of hybrid renewable energy systems for off grid applications. *Renewable and Sustainable Energy Reviews*. 2016;58:376–96.
91. Markvart T. Sizing of hybrid photovoltaic-wind energy systems. *Solar Energy*. 1996;57(4):277–81.
92. Ai B, Yang H, Shen H, and Liao X. Computer-aided design of PV/wind hybrid system. *Renewable Energy*. 2003;28(10):1491–1512.
93. Borowy BS, Salameh ZM. Methodology for optimally sizing the combination of a battery bank and PV array in a wind/PV hybrid system. *IEEE Transaction on Energy Conversion*. 1996;11(2):367–75.
94. Lujano-Rojas JM, Dufo-López R, Bernal-Agustín JL. Probabilistic modelling and analysis of stand-alone hybrid power systems. *Energy*. 2013;63:19–27.
95. Tina G, Gagliano S. Probabilistic analysis of weather data for a hybrid solar/wind energy system. *International Journal of Energy Research*. 2011;35(3):221–32.
96. Tina G, Gagliano S, Raiti S. Hybrid solar/wind power system probabilistic modelling for long-term performance assessment. *Solar Energy*. 2006;80(5):578–88.
97. Yang HX, Lu L, Burnett J. Weather data and probability analysis of hybrid photovoltaic-wind power generation systems in Hong Kong. *Renewable Energy*. 2003;28(11):1813–24.
98. Bhandari B, Lee KT, Lee GY, Cho YM, Ahn SH. Optimization of hybrid renewable energy power systems: A review. *International Journal of Precision Engineering and Manufacturing-Green Technology*. 2015;2(1):99–112. DOI: 10.1007/s40684-015-0013-z.
99. Li J, Wei W, Xiang J. A simple sizing algorithm for stand-alone PV/wind/battery hybrid microgrids. *Energies*. 2012;5:5307–23. DOI: 10.3390/en5125307.
100. Zhang XF, Luo JS, Li Y. Transient semianalytic analysis of a lossless multiconductor transmission line with nonlinear loads excited by a plane wave. *Journal of Engineering Mathematics*. 2013;80(1):165–72.
101. Li C-H, Zhu X-J, Cao G-Y, Sui S, Hu M-R. Dynamic modeling and sizing optimization of stand-alone photovoltaic power systems using hybrid energy storage technology. *Renewable Energy* 2009;34(3):815–26.
102. Yang H, Lu L, Zhou W. A novel optimization sizing model for hybrid solar-wind power generation system. *Solar Energy*. 2007;81(1):76–84.
103. Paliwal P, Patidar NP, Nema RK. Determination of reliability constrained optimal resource mix for an autonomous hybrid power system using particle swarm optimization. *Renewable Energy*. 2014;63:194–204.
104. Askarzadeh A. A discrete chaotic harmony search-based simulated annealing algorithm for optimum design of PV/wind hybrid system. *Solar Energy*. 2013;97:93–101.
105. Kumar R, Gupta RA, Bansal AK. Economic analysis and power management of a stand-alone wind/photovoltaic hybrid energy system using biogeography based optimization algorithm. *Swarm and Evolutionary Computation*. 2013;8:33–43.
106. Arabali A, Ghofrani M, Etezadi-Amoli M, Fadali MS, Baghzouz Y. Genetic algorithm-based optimization approach for energy management. *IEEE Transactions on Power Delivery* 2013;28(1):162–70.
107. Kashefi Kaviani A, Riahy GH, Kouhsari SM. Optimal design of a reliable hydrogen based stand-alone wind/PV generating system, considering component outages. *Renewable Energy*. 2009;34(11):2380–90.
108. Hakimi SM, Moghaddas-Tafreshi SM. Optimal sizing of a stand-alone hybrid power system via particle swarm optimization for Kahnouj area in south-east of Iran. *Renewable Energy*. 2009;34(7):1855–62.
109. Yang H, Zhou W, Lu L, Fang Z. Optimal sizing method for stand-alone hybrid solarwind system with LPSP technology by using genetic algorithm. *Solar Energy* 2008;82(4):354–67.

110. Eberhart R, Kennedy J. A new optimizer using particle swarm theory. In: MHS'95. Proceedings of the Sixth International Symposium on Micro Machine and Human Science, Nagoya, Japan; 1995. pp. 39–43. DOI: 10.1109/MHS.1995.494215.
111. Kennedy J. Particle swarm optimization. In: *Encyclopedia of Machine Learning*. New York: Springer; 2011.760–766.
112. Kennedy J, Eberhart R. Particle swarm optimization. In: Proceedings of the International Conference on Neural Networks, Vancouver, BC, Canada; 1995. Vol. 4, pp. 1942–48.
113. Poli R, Kennedy J, Blackwell T. Particle swarm optimization. *Swarm Intelligence*. 2007;1(1):33–57.
114. Lee T-Y, Chen C-L. Wind-photovoltaic capacity coordination for a time-of-use rate industrial user. *IET Renewable Power Generation*. 2009;3(2):152.
115. Bansal AK, Gupta RA, Kumar R. Optimization of hybrid PV/wind energy system using Meta Particle Swarm Optimization (MPSO). India International Conference on Power Electronics 2010 (IICPE2010); 2011. pp. 1–7.
116. Sharafi M, ELMekkawy TY. Multi-objective optimal design of hybrid renewable energy systems using PSO-simulation based approach. *Renewable Energy*. 2014;68:67–79.
117. Borhanazad H, Mekhilef S, Gounder Ganapathy V, Modiri-Delshad M, Mirtaheri A. Optimization of micro-grid system using MOPSO. *Renewable Energy*. 2014;71:295–306.
118. Xu D, Kang L, Chang L, Cao B. Optimal sizing of stand-alone hybrid wind/PV power systems using genetic algorithm. In: Proceedings of the Canadian Conference on Electrical and Computer Engineering, Saskatoon, SK, Canada; May 1–4, 2005. pp. 1722–25.
119. Bourouni K, Ben M'Barek T, Al Taee A. Design and optimization of desalination reverse osmosis plants driven by renewable energies using genetic algorithms. *Renewable Energy*. 2011;36(3):936–50.
120. Yang H, Wei Z, Chengzhi L. Optimal design and techno-economic analysis of a hybrid solar-wind power generation system. *Applied Energy*. 2009;86(2):163–9.
121. Ould Bilal B, Sambou V, Ndiaye PA, Kébé CMF, Ndongo M. Optimal design of a hybrid solar–wind-battery system using the minimization of the annualized cost system and the minimization of the loss of power supply probability (LPSP). *Renewable Energy*. 2010;35(10):2388–90.
122. Koutroulis E, Kolokotsa D, Potirakis A, Kalaitzakis K. Methodology for optimal sizing of stand-alone photovoltaic/wind-generator systems using genetic algorithms. *Solar Energy*. 2006;80(9):1072–88.
123. Malekpour AR, Tabatabaei S, Niknam T. Probabilistic approach to multi-objective Volt/Var control of distribution system considering hybrid fuel cell and wind energy sources using Improved Shuffled Frog Leaping Algorithm. *Renewable Energy*. 2012;39(1):228–40.
124. Nafeh AE-SA. Optimal Economical sizing of a PV-wind hybrid energy system using genetic algorithm. *International Journal of Green Energy*. 2011;8(1):25–43.
125. Kirkpatrick S. Optimization by simulated annealing: Quantitative studies. *Journal of Statistical Physics*. 1984;34(5–6):975–86.
126. Ekren O, Ekren BY. Size optimization of a PV/wind hybrid energy conversion system with battery storage using simulated annealing. *Applied Energy*. 2010;87(2): 592–8.
127. Sutthibun T, Bhasaputra P. Multi-objective optimal distributed generation placement using simulated annealing. In: ECTI-CON2010: The 2010 ECTI International Conference on Electrical Engineering/Electronics, Computer, Telecommunications and Information Technology; 2010. pp. 810–813.

128. Das S, Mukhopadhyay A, Roy A, Abraham A, Panigrahi BK. Exploratory power of the harmony search algorithm: Analysis and improvements for global numerical optimization. *IEEE Transactions on Systems, Man, and Cybernetics, Part B (Cybernetics).* 2011;41(1):89–106.
129. Maleki A, Askarzadeh A. Comparative study of artificial intelligence techniques for sizing of a hydrogen-based stand-alone photovoltaic/wind hybrid system. *International Journal of Hydrogen Energy.* 2014;39(19):9973–84.
130. Shiva CK, Mukherjee V. A novel quasi-oppositional harmony search algorithm for automatic generation control of power system. *Applied Soft Computing.* 2015;35:749–65.
131. Dorigo M, Gambardella LM. Ant colony system: A cooperative learning approach to the traveling salesman problem. *IEEE Transactions on Evolutionary Computation.* 1997;1(1):53–66.
132. Passino KM. Biomimicry of bacterial foraging for distributed optimization and control. *IEEE Control Systems Magazine.* 2002;22(3):52–67.
133. Karaboga D, Basturk B. A powerful and efficient algorithm for numerical function optimization: Artificial bee colony (ABC) algorithm. *Journal of Global Optimization.* 2007;39(3):459–71.
134. Nasiraghdam H, Jadid S. Optimal hybrid PV/WT/FC sizing and distribution system reconfiguration using multi-objective artificial bee colony (MOABC) algorithm. *Solar Energy.* 2012;86(10):3057–71.
135. Nadjemi O, Nacer T, Hamidat A, Salhi H. Optimal hybrid PV/wind energy system sizing: Application of cuckoo search algorithm for Algerian dairy farms. *Renewable and Sustainable Energy Reviews.* 2017;70:1352–65.
136. Fadaee M, Radzi MAM. Multi-objective optimization of a stand-alone hybrid renewable energy system by using evolutionary algorithms: A review. *Renewable and Sustainable Energy Reviews.* 2012;16(5):3364–9.
137. Malekpour AR, Niknam T. A probabilistic multi-objective daily Volt/Var control at distribution networks including renewable energy sources. *Energy.* 2011;36(5):3477–88.
138. Maheri A. Multi-objective design optimisation of standalone hybrid wind-PVdiesel systems under uncertainties. *Renewable Energy.* 2014;66:650–61.
139. Tant J, Geth F, Six D, Tant P, Driesen J. Multiobjective battery storage to improve PV integration in residential distribution grids. *IEEE Transactions on Sustainable Energy.* 2013;4(1):182–91.
140. Moura PS, de Almeida AT. Multi-objective optimization of a mixed renewable system with demand-side management. *Renewable and Sustainable Energy Reviews.* 2010;14(5):1461–8.
141. Khatod DK, Pant V, Sharma J. Analytical approach for well-being assessment of small autonomous power systems with solar and wind energy sources. *IEEE Transactions on Energy Conversion.* 2010;25(2):535–45.
142. Kaldellis JK, Zafirakis D, Kondili E. Optimum autonomous stand-alone photovoltaic system design on the basis of energy pay-back analysis. *Energy.* 2009;34(9):1187–98.
143. Luna-Rubio R, Trejo-Perea M, Vargas-Vázquez D, Ríos-Moreno GJ. Optimal sizing of renewable hybrids energy systems: A review of methodologies. *Solar Energy.* 2012;86(4): 1077–88.
144. Hrayshat ES. Techno-economic analysis of autonomous hybrid photovoltaic-diesel-battery system. *Energy for Sustainable Development.* 2009;13:143–50. DOI: 10.1016/j.esd.2009.07.003.
145. Chang KH, Lin G. Optimal design of hybrid renewable energysystems using simulation optimization. *Simulation Modelling Practice and Theory.* 2015;52:40–51.

146. Al-Falahi MDA, Jayasinghe SDG, Enshaei H. A review on recent size optimization methodologies for standalone solar and wind hybrid renewable energy system. *Energy Conversion and Management*. 2017;143:252–74.
147. Liu Y, Dong H, Lohse N, Petrovic S. A multi-objective genetic algorithm for optimization of energy consumption and shop floor production performance. *International Journal of Production Economics*. 2016; 179:259–72.
148. Ding F, Li P, Huang B, Gao F, Ding C, Wang C. Modeling and simulation of grid-connected hybrid photovoltaic/battery distributed generation system. In: China International Conference on Electricity Distribution (CICED), Nanjing, China; 2010. pp. 1–10. IEEE.
149. Chen B, Zeng W, Lin Y, Zhong Q. An enhanced differential evolution based algorithm with simulated annealing for solving multiobjective optimization problems. *Journal of Applied Mathematics*. 2014;931630, 13 pages. https://doi.org/10.1155/2014/931630.
150. Mao M, Ji M, Dong W, Chang L. Multi-objective economic dispatch model for a microgrid considering reliability. In: 2nd International Symposium on Power Electronics for Distributed Generation Systems, PEDG 2010; 2010. pp. 993–8.
151. Warwick K, Ekwue A, Aggarwal R. Artificial intelligence techniques in power systems. London: IET; 1997.
152. Sawle Y, Gupta SC, Bohre AK. PV-wind hybrid system: A review with case study. *Cogent Engineering*. 2016;3(1):1189305. DOI: 10.1080/23311916.2016.1189305.
153. Moghaddam S, Bigdeli M, Moradlou M, Siano P. Designing of stand-alone hybrid PV/wind/battery system using improved crow search algorithm considering reliability index. *International Journal of Energy and Environmental Engineering*. 2019;10:429–49. https://doi.org/10.1007/s40095-019-00319-y.
154. Ma T, Javed MS. Integrated sizing of hybrid PV-wind-battery system for remote island considering the saturation of each renewable energy resource. *Energy Conversion and Management*. 2019;182:178–90. DOI: 10.1016/j.enconman.2018.12.059.
155. Arribas L, Cano L, Cruz I, Mata M, Llobet E. PV–wind hybrid system performance: A new approach and a case study. *Renewable Energy*. 2010;35:128–37. DOI: 10.1016/j.renene.2009.07.002.
156. Behera S, Pati B. Grid-connected distributed wind-photovoltaic energy management: A review. An open access intech paper. (2019). DOI: 10,5772/intechopen.88923.
157. Hamanah WM, Abido MA, Alhems LM. Optimum sizing of hybrid PV, wind, battery and diesel system using lightning search algorithm. *Arabian Journal for Science and Engineering*. 2020;45: 1871–83. https://doi.org/10.1007/s13369-019-04292-w.
158. Wang X, Palazoglu A, El-Farra NH. Operational optimization and demand response of hybrid renewable energy systems. *Applied Energy*. 2015;143:324–35. DOI: 10.1016/j.apenergy.2015.01.004.
159. Belmili H, Haddadi M, Bacha S, Falmi M, Bendib B. Sizing stand-alone photovoltaic-wind hybrid system: Techno-economic analysis and optimization. *Renewable and Sustainable Energy Reviews*. 2014;30:821–32. DOI: 10.1016/j.rser.2013.11.011.
160. Bajpai P, Dash V. Hybrid renewable energy systems for power generation in stand-alone applications: A review. *Renewable and Sustainable Energy Reviews*. 2012;16(5):2926–39.
161. Kalantar M., Mousavi S.M.G. Dynamic behaviour of a stand-alone hybrid power generation system of wind turbine, microturbine, solar array and battery storage. Applied Energy 2010; 87: 3051–3064.
162. Yang HX, Zhou W, Lu L, Fang Z. Optimal sizing method for stand-alone hybrid solar-wind system with LPSP technology by using genetic algorithm. *Solar Energy*. 2008;82:354–67.

163. Boonbumroong U., Pratinthong N., Thepa S., Jivacate C., Pridasawas W. PSO for ac-coupling standalone hybrid power system. Solar Energy 2011; 85: 560–569.
164. Zhang X, Tan SC, Li G, Li J, Feng Z. Components sizing of hybrid energy systems via optimization of power dispatch simulations. *Energy.* 2013;52:165–72.
165. Arabali A., Ghofrani M., Etezadi-Amoli M., Fadali M.S. Stochastic performance assessment and sizing for a hybrid power system of solar/wind/energy storage. IEEE Transactions on Sustainable Energy 2014; 5(2): 363–371.
166. Huneke F, Henkel J, Benavides González JA, Erdmann G. Optimisation of hybrid off-grid energy systems by linear programming. *Energy, Sustainability and Society.* 2012;2(1):7.
167. Fung CC, Hoand SCY, Nayar CV. Optimisation of a hybrid energy system using simulated annealing technique. In: IEEE TENCON'93; 1993. Vol. 5, pp. 235–8.
168. Wu Y, Lee C, Liu L, Tsai S. Study of reconfiguration for the distribution system with distributed generators. *IEEE Transactions on Power Delivery.* 2010;25(3):1678–85.
169. Fetanat A, Ehsan K. Size optimization for hybrid photovoltaic-wind energy system using ant colony optimization for continuous domains based integer programming. *Applied Soft Computing.* 2015;31:196–209.
170. Rajendra AR, Natarajan E. Optimization of integrated photovoltaic–wind power generation systems with battery storage. *Energy.* 2006;31:1943–54.
171. Diaf S, Notton G, Belhamel M, Haddadic M, Louche A. Design and techno economical optimization for hybrid PV/wind system under various meteorological conditions. *Applied Energy.* 2008;85:968–87.
172. Mawardi A, Ptchumani R. Effects of parameter uncertainty on the performance variability of proton exchange membrane (PEM) fuel cells. *Journal of Power Sources.* 2006;160:232–45.
173. Protogeropoulos C, Brinkworth BJ, Marshall RH. Sizing and techno-economical optimization for hybrid solar photovoltaic/wind power systems with battery storage. *International Journal of Energy Research.* 1997;21(6):465–79.
174. Mousa K, AlZu'bi H, Diabat A. Design of a hybrid solar-wind power plant using optimization. In: Proceedings of the Second International Conference on Engineering System Management and Applications; 2010. pp. 1–6.
175. Kanase-Patil AB, Saini RP, Sharma MP. Integrated renewable energy systems for off grid rural electrification of remote area. *Renewable Energy.* 2010;35(6):1342–49.
176. Rohani G, Nour M. Techno-economic analysis of stand-alonehybrid renewable power system for RasMusherib in United Arab Emirates. *Energy.* 2014;64:828–41.
177. Eltamaly AM, Mohamed MA, Alolah AI. A novel smart gridtheory for optimal sizing of hybrid renewable energy systems. *Solar Energy*, 2016, 124: 26–38.
178. Aktas A, Erhan K, Ozdemir S, Ozdemir E. Experimental investigation of a new smart energy management algorithm for a hybrid energy storage system in smart grid applications. *Electric Power Systems Research.* 2017;144:185–196.
179. Shi Z, Wang R, Zhang T. Multi-objective optimal design of hybrid renewable energy systems using preference-inspired co-evolutionary approach. *Solar Energy.* 2015;118: 96–106.
180. Hafez O, Bhattacharya K. Optimal planning and design of a renewable energy based supply system for microgrids. *Renewable Energy.* 2012;5:7–15.
181. Yuvarajan S, Shoeb J. A fast and accurate maximum power pointtracker for PV systems. In: 23rd Annual IEEE Applied Power Electronics Conference and Exposition, Austin, TX, USA; 2008. pp. 67–72.
182. Yusof Y, Sayuti SH, Abdul Latif M, Wanik MZC. Modeling and simulation of maximum power point tracker for photovoltaic system. In: Proceedings of National Power and Energy Conference, Kuala Lumpur, Malaysia; 2004. pp. 8–93.

183. Geem ZW, Yoon Y. Harmony search optimization of renewable energy charging with energy storage system. *International Journal of Electrical Power & Energy Systems*. 2017;86:120–6.
184. Zhou T, Francois B. Modeling and control design of hydrogen production process for an active hydrogen/wind hybrid power system. *International Journal of Hydrogen Energy*. 2009;34(1):21–30.
185. Azaza M, Wallin F. Multi objective particle swarm optimization of hybrid micro-grid system: A case study in Sweden. *Energy*. 2017;123:108–18.

Index

A

B

C

D

E

F

M

N

O

P

Q

R

S

T

U

V

W

Z